Intermediate Algebra

Summary Notebook

Kirk Trigsted
University of Idaho

Randy Gallaher
Lewis & Clark Community College

Kevin Bodden
Lewis & Clark Community College

Editorial Director, Mathematics: Christine Hoag
Editor in Chief: Paul Murphy
Acquisitions Editor: Dawn Giovanniello
Sponsoring Editor: Dawn Murrin
Editorial Assistant: Kristin Rude
Design Manager: Andrea Nix
Art Director: Heather Scott
Production Manager: Ron Hampton
Senior Math Media Producer: Ceci Fleming
Project Manager, MathXL: Eileen Moore
Executive Marketing Manager: Michelle Renda
Marketing Assistant: Alica Frankel
Market Development Manager: Dona Kenly
Manufacturing Manager: Carol Melville

To my wife Wendy, and precious children,
Benjamin, Emily, Gabrielle, and baby to be.
— **Kirk Trigsted**

To my wife, Karen, and my children, Ethan,
Ben, and Annie, for their inspiration,
encouragement, and patience.
— **Randy Gallaher**

To my wife, Angie, and my children, Shawn,
Payton, and Logan, for their love and support;
and to my parents for their encouragement
and consultation.
— **Kevin Bodden**

Photograph Credits: Page 1, iStockphoto.com; Page 29, Shutterstock; Page 47, FreezeFrameStudio/ iStockphoto; Page 72, Ivan Cholakov/iStockphoto; Page 77, Toyota Motor Sales, USA, Inc.; Page 78, Chicagophoto/Dreamstime LLC-Royalty Free; Page 106, Evgeny Vasenev/Shutterstock; Page 156, European Space Agency; Page 197, European Space Agency; Page 240, Matthew Chambers/ iStockphoto.com; Page 300, Herbert Kratky/Shutterstock; Page 360, Shutterstock; Page 424, Shutterstock

ISBN-13: 978-0-321-65285-0
ISBN-10: 0-321-65285-1

Contents

Preface   vi

Chapter R Review   R-1

R.1 Sets of Numbers   R-1
R.2 Order of Operations and Properties of Real Numbers   R-12
R.3 Algebraic Expressions   R-27

Chapter 1 Equations and Inequalities in One Variable   1

1.1 Linear Equations in One Variable   1
1.2 Linear Inequalities in One Variable   12
1.3 Compound Inequalities; Absolute Value Equations
and Inequalities   25
1.4 Formulas and Problem Solving   37

Chapter 2 Graphs and Functions   47

2.1 The Rectangular Coordinate System and Graphing   47
2.2 Relations and Functions   57
2.3 Function Notation and Applications   66
2.4 Graphs of Linear Functions   78
2.5 Linear Equations in Two Variables   85
2.6 Linear Inequalities in Two Variables   101

Chapter 3 Systems of Linear Equations and Inequalities   106

3.1 Systems of Linear Equations in Two Variables   106
3.2 Systems of Linear Equations in Three Variables   123
3.3 More Problem Solving with Systems of Linear Equations   136
3.4 Systems of Linear Inequalities in Two Variables   146

Chapter 4 Polynomial Expressions and Functions   154

4.1 Rules for Exponents   154
4.2 Introduction to Polynomial Functions   168
4.3 Multiplying Polynomials   176
4.4 Dividing Polynomials   185

Chapter 5 Factoring   197

5.1 Greatest Common Factor and Factoring by Grouping   197
5.2 Factoring Trinomials   203
5.3 Special-Case Factoring; A General Factoring Strategy   217
5.4 Polynomial Equations and Models   226

Chapter 6 Rational Expressions, Equations, and Functions   240

6.1 Introduction to Rational Expressions and Functions   240
6.2 Multiplying and Dividing Rational Expressions   249
6.3 Adding and Subtracting Rational Expressions   256
6.4 Complex Rational Expressions   267
6.5 Rational Equations and Models   275
6.6 Variation   292

Chapter 7 Radicals and Rational Exponents   300

7.1 Radical Expressions   300
7.2 Radical Functions   307
7.3 Rational Exponents and Simplifying Radical Expressions   314
7.4 Operations with Radicals   328
7.5 Radical Equations and Models   338
7.6 Complex Numbers   349

Chapter 8 Quadratic Equations and Functions; Circles   360

8.1 Solving Quadratic Equations   360
8.2 Quadratic Functions and Their Graphs   372
8.3 Applications and Modeling of Quadratic Functions   387
8.4 Circles   403
8.5 Polynomial and Rational Inequalities   415

Chapter 9 Exponential and Logarithmic Functions and Equations   424

9.1 Transformations of Functions   424
9.2 Composite and Inverse Functions   441
9.3 Exponential Functions   455
9.4 The Natural Exponential Function   467
9.5 Logarithmic Functions   475
9.6 Properties of Logarithms   486
9.7 Exponential and Logarithmic Equations   495
9.8 Applications of Exponential and Logarithmic Functions   502

Appendix A Conic Sections   513

Introduction to Conic Sections   513
A.1 The Parabola   515
A.2 The Ellipse   525
A.3 The Hyperbola   538

Appendix B Sequences and Series   549

B.1 Introduction to Sequences and Series   549
B.2 Arithmetic Sequences and Series   558
B.3 Geometric Sequences and Series   566
B.4 The Binomial Theorem   579

Appendix C Basic Math Review—Fractions, Decimals, Proportions, Percents 586

C.1 Fractions   586
C.2 Decimals   597
C.3 Proportions   605
C.4 Percents   609

Glossary   G-1

Answers   A-1

Index   I-1

Preface

Introduction

This *Summary Notebook* contains the pages of Kirk Trigsted, Randy Gallaher, and Kevin Bodden's *Intermediate Algebra* eText in a portable, spiral-bound format. The structure of the *Summary Notebook* helps students organize their notes by providing them with space to summarize the videos and animations. Students can also use it to review the eText's material anytime, anywhere.

A Note to Students

This textbook was created for you! Unlike a traditional text, we wanted to create content that gives you, the reader, the ability to be an active participant in your learning. This eText was specifically designed to be read online. The eText pages have large, readable fonts and were designed without the need to scroll. Throughout the material, we have carefully placed thousands of hyperlinks to definitions, previous chapters, interactive videos, animations, and other important content. Many of the videos and animations allow you to actively participate and interact. Take some time to "click around" and get comfortable with the navigation of the eText and explore its many features.

Before you attempt each homework assignment, read the appropriate section(s) of the eText. At the beginning of each section (starting in Chapter R), you will encounter a feature called Things to Know. This feature includes all the prerequisite objectives from previous sections that you will need to successfully learn the material presented in the new section. If you do not have a basic understanding of these objectives, click on the desired hyperlinks and rework through the objectives, taking advantage of any videos or animations.

An additional feature of the eText is the inclusion of an audio icon ( ). By clicking on this icon, you can listen to the text as you read. While you read through the pages of the eText, use the margins of this *Summary Notebook* to take notes, summarize key points, and list helpful tips and reminders. An additional option, if made available by your instructor, is to use the *Guided Notebook* to guide your note taking as you work through the eText. The *Guided Notebook* provides more specific direction on how to proceed through the material while providing more space for note taking.

Try testing yourself by working through the corresponding You Try It exercises. Remember, you learn math by doing math! The more time you spend working through the videos, animations, and exercises, the more you will understand. If your instructor assigns homework in MyMathLab or MXL, rework the exercises until you get them right. Be sure to go back and read the eText at anytime while you are working on your homework. This text caters to your educational needs. We hope you enjoy the experience!

A Note to Instructors

Today's students have grown up in a technological world where everything is "clickable." We have taught with MyMathLab for many years, and have experienced first-hand how fewer and fewer students have been referring to their traditional textbooks, opting instead to use electronic resources. As the use of technology plays an ever increasing role in how we are teaching our students, it seems only natural to have a textbook that mirrors the way our students are learning. We are excited to have written a text that was conceived from the ground up to be used as an online interactive tool. Unlike traditional printed textbooks, this eText was specifically designed for students to read online while working in MyMathLab. Therefore, we wrote this text entirely from an online perspective, with MyMathLab and its existing functionality specifically in mind. Every hyperlink, video, and animation has been strategically integrated within the context of each page to engage the student and maximize his or her learning experience. All of the interactive media was designed so students can actively participate while they learn math. Many of the interactive videos and animations require student interaction, giving specific feedback for incorrect responses.

We are proponents of students learning terms and definitions. Therefore, we have created hyperlinks throughout the text to the definitions of important mathematical terms. We have also inserted a tremendous amount of "just-in-time review" throughout the text by creating links to prerequisite topics. Students have the ability to reference these review materials with just a click of the mouse. You will see that the exercise sets are concise and nonrepetitive. Since MyMathLab will be used as the main assessment engine, there is no need for a repetitive exercise set in the hardcopy version of the text. Every single exercise is available for you to assign within MyMathLab or MathXL. For the first time, instructors can assess reading assignments! We have created five conceptual reading assessment questions for every section of the eText, giving the student specific feedback for both correct and incorrect responses. Each feedback message directs them back to the appropriate location for review. Our hope is that every student that uses the eText will have a positive learning experience.

Acknowledgments

First of all, we want to thank our wives for their loving support, encouragement, and sacrifices. Thanks to our children for being patient when we needed to work and for reminding us when we needed to take breaks. We could not have completed this project without generous understanding and forgiveness from each of you.

Writing this textbook has been one of the most difficult and rewarding experiences of our lives. We truly could not have accomplished our goal without the support and contributions of so many wonderful, talented people. These extraordinary talents deserve every accolade we can give them. From Pearson, we would like to thank Paul Murphy, our editor and friend, for believing in us; Greg Tobin and Chris Hoag for their continued support; Dawn Murrin for tirelessly coordinating our efforts and keeping us on track; Heather Scott and Andrea Nix for their design brilliance; Dona Kenly, Michelle Renda, and Tracy Rabinowitz for their marketing expertise; Ceci Fleming and Ruth Berry for taking care of all the details; Eileen Moore, Rebecca Williams, and the rest of the MathXL development team for an amazing job; Tom Benfatti and the art team for their great eye for detail; and Ron Hampton, our production manager, for all his support.

Along the way, we had the help of so many people from around the world and from many different walks of life. The contributions of Susan Knights and Elaine Page cannot be measured. Their detailed reviews and numerous suggestions have improved this text considerably. Our "Friday conversations" were invaluable throughout the entire writing process. Thank you for everything! We appreciate everyone else who made this book a reality: Alice Champlin and Anthony T. J. Kilian at Magnitude Entertainment for creating quality videos and animations; Pamela Trim for reading the entire eText and catching the little things; Edumedia for creating the answers and accuracy checking the text; Donna Densmore for her work on the *Guided Notebook*.

The following list attempts to recognize all of the reviewers. Please accept our deepest apologies if we have inadvertently omitted anyone. We have benefited greatly from you honest feedback, constructive criticism, and thoughtful suggestions. Each of you has helped us create a resource we truly believe will be effective in helping students learn mathematics. We are deeply grateful and we genuinely thank you all from the bottom of our hearts.

Gregory Bloxom, Pensacola Junior College
Mary Ann Teel, University of North Texas
Mike May, SJ, Saint Louis University
Jeanette Shotwell, Central Texas College
Sharon Jackson, Brookhaven College
Tom Blackburn, Northeastern Illinois University
Disa Beaty, Rose State College
Charlotte Ellen Bell, Richland College
David Bell, Florida Community College at Jacksonville
Dr. Marsha Lake, Brevard Community College (FL)–Titusville Campus
Stacey Moore, Wallace State Community College–Hanceville, AL
Linda Parrish, Brevard Community College
Natalie Rivera, Estrella Mountain Community College
Denise Nunley, Glendale Community College
Arunas Dagys, Saint Xavier University
Karen Egedy, Baton Rouge Community College
Evelyn Porter, Utah Valley University
Mary Jo Anhalt, Bakersfield College, Delano Center
Debby Casson, University of New Mexico
Shawna Haider, Salt Lake Community College
Carla K. Kulinsky, Salt Lake Community College
Kenneth Takvorian, Mount Wachusett Community College
Phil Veer, Johnson County Community College
Amadou Hama, Kennedy-King College, Chicago
Al Hemenway, Los Angeles Mission College
Tom Pulver, Waubonsee Community College
Thomas Hartfield, Gainesville State College
Shirley Brown, Weatherford College
Rita LaChance, University of Maine–Augusta
Dr. Said Ngobi, Victor Valley College
Dr. Mary Wagner-Krankel, St. Mary's University
Gail Brooks, McLennan Community College
Daniela Kojouharov, Tarrant County College, SE
Amberlee Bosse, University of Phoenix
Mark Chapman, Baker College
Jeremy Coffelt, Blinn College
Donna Densmore, Bossier Parish Community College
Elizabeth Howell, North Central Texas College
Joyce Lindstrom, St. Charles Community College
Frank Marfai
Lisa Sheppard, Lorain County Community College

Review

CHAPTER R CONTENTS

R.1 Sets of Numbers

R.2 Order of Operations and Properties of Real Numbers

R.3 Algebraic Expressions

R.1 Sets of Numbers

OBJECTIVES

1 Identify Sets

2 Classify Real Numbers

3 Plot Real Numbers on a Number Line

4 Use Inequality Symbols to Order Real Numbers

5 Compute the Absolute Value of a Real Number

OBJECTIVE 1:   IDENTIFY SETS

A **set** is a collection of objects. Each object in a set is called an **element** or a **member** of the set. Typically, capital letters are used to name sets, and braces { } group the list of elements in a set. For example, the seven days of the week form a set. If we name this set D, we write

$$D = \{\text{Sunday, Monday, Tuesday, Wednesday, Thursday, Friday, Saturday}\}.$$

The **element symbol** ($\in$) indicates that an object is an element of a set. For example, Friday $\in D$ means "Friday is an element of the set D." A slash through an element symbol forms the **not-an-element symbol** ($\notin$), which indicates that an object is not an element of a set. For example, January $\notin D$ means "January is not an element of the set D."

A set that contains no elements is called an **empty set** or **null set** and is denoted by a pair of **empty braces** { } or the **null symbol** $\varnothing$. For example, suppose A represents the set of all days of the week that begin with the letter B. Since there are no such days, $A = \{\ \} = \varnothing$.

The symbol $\{\varnothing\}$ is not the empty set. It is the set containing the element $\varnothing$. Be sure to use the symbols { } or $\varnothing$ to represent the empty set, not $\{\varnothing\}$.

In algebra, the elements of a set are typically numbers. For example, if S represents the set of single-digit numbers, then

$$S = \{0, 1, 2, 3, 4, 5, 6, 7, 8, 9\}.$$

Both sets D and S are **finite**, meaning that each set has a fixed number of elements (7 days, 10 single-digit numbers). Both sets are written using the **roster method**, which displays the elements in a list. When a set is finite, it is possible to list all of its elements (though it may not be *practical* to list them all).

Some sets are **infinite**, meaning that they have an unlimited number of elements. For example, the set of **natural numbers**, or **counting numbers**, denoted by $\mathbb{N}$, is the set

$$\mathbb{N} = \{1, 2, 3, 4, 5, \dots\}.$$

The ellipsis ($\dots$) indicates that the list continues forever. Note that 0 is not included in the set of natural numbers.

We can also write sets using **set-builder notation**, which includes a general element of the set (usually a letter that represents the numbers in the set), a vertical bar (|) to indicate "such that," and any conditions that must be satisfied by each element of the set. For example, set S is written in set-builder notation as

$$S = \{x | x \text{ is a single-digit number}\}.$$

This statement reads as "S equals the set of all x such that x is a single-digit number."

$$S = \{\underbrace{x}_{\substack{\text{all } x}} \underbrace{|}_{\substack{\text{such that}}} \underbrace{x \text{ is a single-digit number}}\}$$
the set of   all x such that x is a single-digit number

Example 1 Using the Symbols $\in$ and $\notin$

Fill in the blank with the correct symbol, $\in$ or $\notin$.

a. 100 _____ $\mathbb{N}$ **b.** $\dfrac{2}{3}$ _____ $\mathbb{N}$ **c.** 0 _____ $\mathbb{N}$

d. 8 _____ $\{x | x \text{ is a number larger than } 5\}$ **e.** 11 _____ $\{x | x \text{ is an even number}\}$

Solution

a. $\mathbb{N}$ indicates the set of natural numbers. If we write out the list of natural numbers, we will eventually reach 100. So, $100 \in \mathbb{N}$.

b. No fractions are included within the set of natural numbers, so $\dfrac{2}{3} \notin \mathbb{N}$.

c. 0 is not a natural number, so $0 \notin \mathbb{N}$.

d. 8 is larger than 5, so $8 \in \{x | x \text{ is a number larger than } 5\}$.

e. 11 is an odd number, not even, so $11 \notin \{x | x \text{ is an even number}\}$.

You Try It   Work through this You Try It problem.

Work Exercises 1–6 in this textbook or in the ***MyMathLab***® Study Plan.

Example 2 Listing the Elements of a Set

Using the roster method, list the elements in the set described.

a. Months with exactly 30 days

b. Natural numbers between 4 and 8, inclusive

c. Natural numbers greater than 10

d. Months with 35 days

Solution

a. Table 1 shows the numbers of days in each month. We know that April, June, September, and November each have exactly 30 days, so we have the set $\{$April, June, September, November$\}$.

Month	Days	Month	Days
January	31	July	31
February	28 or 29	August	31
March	31	September	30
April	30	October	31
May	31	November	30
June	30	December	31

Table 1
Days per Month

b. The natural numbers *between* 4 and 8 are $5, 6,$ and 7. The word *inclusive* means that 4 and 8 are also included in the set. So, we have $\{4, 5, 6, 7, 8\}$.

c.–d. Try listing the elements of these sets on your own. Click here to check your answer, or watch this video for a complete solution.

You Try It   Work through this You Try It problem.

Work Exercises 7–10 in this textbook or in the ***MyMathLab***® Study Plan.

Two sets A and B are **equal sets**, denoted as $A = B$, if they contain the exact same elements. For example, if $A = \{1, 2, 3\}$ and $B = \{3, 2, 1\}$, then $A = B$ because both sets contain the same three elements, just listed in a different order.

If every element of set A is also an element of set B, then A is a **subset** of B, denoted as $A \subseteq B$, and B is a **superset** of A, denoted as $B \supseteq A$. If every element of A is an element of B, but A and B are not equal sets, then A is a **proper subset** or **strict subset** of B, denoted as $A \subset B$, and B is a **proper superset** or **strict superset** of A, denoted as $B \supset A$. For example, if $C = \{1, 2, 3, 4, 5, 6\}$ and $D = \{2, 4, 6\}$, then $D \subset C$ because every element in D is also in C. We can also write this as $C \supset D$.

If a **subset symbol** ($\subset$ or $\subseteq$) has a slash through it, the result is a **not-a-subset symbol** ($\not\subset$ or $\not\subseteq$), meaning that one set is not a subset of another. For example, since $A = \{1, 2, 3\}$ is not a subset of $D = \{2, 4, 6\}$, we write $A \not\subset D$. Likewise, if a **superset symbol** ($\supset$ or $\supseteq$) has a slash through it, the result is a **not-a-superset symbol** ($\not\supset$ or $\not\supseteq$).

Example 3 Using Subset and Superset Symbols

Let $A = \{0, 1, 2, 3, 4, 5\}, B = \{0, 2, 4\}, C = \{1, 3, 5, 7\}$, and $D = \{1, 3, 5\}$. Determine if each statement is true or false.

a. $B \subset A$ **b.** $C \supseteq D$ **c.** $C \subseteq A$ **d.** $D \not\subset A$

Solution

a. $B \subset A$ means "B is a strict subset of A." Every element in B is also in A, but A has elements not in B, so the statement $B \subset A$ is true.

b. $C \supseteq D$ says "C is a superset of D." Every element in D is also in C, so $C \supseteq D$ is true.

c. $C \subseteq A$ says "C is a subset of A." Since the element 7 is in C but not in A ($7 \in C, 7 \notin A$), the statement $C \subseteq A$ is false.

d. $D \not\subset A$ means "D is not a subset of A." Every element in D is also an element of A, so $D \not\subset A$ is false.

You Try It   **Work through this You Try It problem.**

Work Exercises 11–14 in this textbook or in the *MyMathLab*® Study Plan.

OBJECTIVE 2:   CLASSIFY REAL NUMBERS

Now let's formalize the definition of the set of *natural numbers*.

Definition   Natural Numbers

A **natural number**, or **counting number**, is an element of the set $\mathbb{N} = \{1, 2, 3, 4, 5, \dots\}$.

Adding the element 0 to the set of natural numbers gives the set of *whole numbers*, denoted by the symbol $\mathbb{W}$.

Definition   Whole Numbers

A **whole number** is an element of the set $\mathbb{W} = \{0, 1, 2, 3, 4, 5, \dots\}$.

Every natural number is also a whole number, so $\mathbb{N} \subset \mathbb{W}$.

Adding the negative numbers $-1, -2, -3, -4, \dots$ to the set of whole numbers gives the set of *integers*, denoted by $\mathbb{Z}$.

Definition   Integers

An **integer** is an element of the set $\mathbb{Z} = \{\dots, -4, -3, -2, -1, 0, 1, 2, 3, 4, \dots\}$.

Every whole number is also an integer, so $\mathbb{N} \subset \mathbb{W} \subset \mathbb{Z}$.

The integers are still "whole" in the sense that they are not fractions. If we include fractions, we get the set of *rational numbers*, denoted by $\mathbb{Q}$.

Definition   Rational Numbers

A **rational number** is an element of the set $\mathbb{Q} = \left\{ \dfrac{p}{q} \,\middle|\, p \text{ and } q \text{ are integers, } q \neq 0 \right\}$.

A rational number can be written as the quotient of two integers p and q as long as $q \neq 0$. $\dfrac{p}{q}$ is called the **fraction form** for the rational number. p is the **numerator,** and q is the **denominator**. Examples of rational numbers include $\dfrac{3}{4}, \dfrac{5}{6}, -\dfrac{7}{8},$ and $\dfrac{13}{11}$.

Every integer is also a rational number because we can write the integer with a denominator of 1. For example, $3 = \dfrac{3}{1}, 0 = \dfrac{0}{1},$ and $-8 = \dfrac{-8}{1}$. So, $\mathbb{N} \subset \mathbb{W} \subset \mathbb{Z} \subset \mathbb{Q}$.

Rational numbers are often written in **decimal form**. When a fraction is written as a decimal, the result is either a **terminating decimal**, such as $\dfrac{3}{4} = 0.75$ and $-\dfrac{7}{8} = -0.875$, or a **repeating decimal**, such as $\dfrac{5}{6} = 0.8333\ldots = 0.8\overline{3}$ and $\dfrac{13}{11} = 1.181818\ldots = 1.\overline{18}$. Every rational number can be expressed as a terminating or a repeating decimal. Conversely, every number that can be written as a terminating or a repeating decimal is a rational number.

Numbers that cannot be written as terminating or repeating decimals form the set of *irrational numbers*, denoted by $\mathbb{I}$.

Definition   Irrational Numbers

An **irrational number** is an element of the set $\mathbb{I} = \{x \mid x \text{ is a number that is not rational}\}$.

An irrational number cannot be written as the quotient of two integers, and its decimal form does not repeat or terminate. Square roots and other numbers expressed with radicals are often irrational numbers. For example, $\sqrt{3} = 1.7320508\ldots$ and $\sqrt[3]{5} = 1.7099759\ldots$ have decimal forms that continue indefinitely without a repeating pattern, so each is an irrational number. Another example of an irrational number is $\pi \approx 3.1415926\ldots$.

 Do not automatically assume that a number expressed with a radical is an irrational number. Some numbers expressed with radicals are rational. For example, $\sqrt{9} = 3, \sqrt{2.25} = 1.5,$ and $\sqrt[3]{8} = 2$ are all rational numbers. In fact, $\sqrt{9} = 3$ and $\sqrt[3]{8} = 2$ are also integers, whole numbers, and natural numbers.

If the set of rational numbers is combined with the set of irrational numbers, the result is the set of *real numbers*, denoted by $\mathbb{R}$.

Definition   Real Numbers

A **real number** is an element of the set $\mathbb{R} = \{x \mid x$ is a rational number or an irrational number$\}$.

We use real numbers in everyday activities such as computing salaries, making purchases, and measuring distances. As we progress through algebra, we will occasionally encounter numbers that are not real. For example, $\sqrt{-4}$ is not a real number because there is no real number whose square is -4. In Chapter 7, we will define such *non-real complex numbers*. Barring those, every number you can think of is a real number.

Figure 1 gives a visual of the set of real numbers that shows the following relationships:

- Every real number is either rational or irrational;
- All natural numbers are whole numbers;
- All whole numbers are integers; and
- All integers are rational numbers.

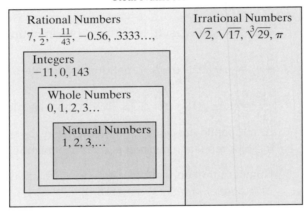

Figure 1
The real numbers

Example 4 Classifying Real Numbers

Each element of set A is a real number.

$$A = \left\{-7, -\frac{5}{9}, 0, 1.3, \frac{\pi}{2}, \sqrt{7}, 2.\overline{6}, 4\right\}$$

List the elements of A that are

a. Natural numbers   **b.** Whole numbers   **c.** Integers

d. Rational numbers   **e.** Irrational numbers

Solution

a. Comparing set A to $\mathbb{N} = \{1, 2, 3, 4, 5, \dots\}$, 4 is the only natural number in A.

b. If we compare set A to $\mathbb{W} = \{0, 1, 2, 3, 4, 5, \dots\}$, we see that 0 and 4 are the only whole numbers in A.

c. The integers in A, when compared to
$\mathbb{Z} = \{\dots, -4, -3, -2, -1, 0, 1, 2, 3, 4, \dots\}$, are $-7, 0,$ and 4.

d. To find the rational numbers, we look in A for numbers that can be written as the quotient of two integers (or as a terminating or repeating decimal). The rational numbers in A are $-7, -\dfrac{5}{9}, 0, 1.3, 2.\overline{6}$, and 4. Note that $\dfrac{\pi}{2}$ is in fraction form, but it cannot be written as the quotient of two integers (2 is an integer but π is not). So, $\dfrac{\pi}{2}$ is not rational.

e. There are two real numbers remaining in A that are not rational. So, the numbers $\dfrac{\pi}{2}$ and $\sqrt{7}$ are the irrational numbers.

Before classifying a real number, check to see if it can be simplified. For example, the rational number $\dfrac{6}{2}$ can be simplified as $\dfrac{6}{2} = 3$, so it is also an integer, a whole number, and a natural number. Likewise, the number $-\sqrt{16} = -4$ is an integer.

You Try It   **Work through this You Try It problem.**

Work Exercises 15–20 in this textbook or in the *MyMathLab*® Study Plan.

OBJECTIVE 3: PLOT REAL NUMBERS ON A NUMBER LINE

The **real number line**, or **real axis**, is a graph that represents the set of all real numbers. Every point on the number line corresponds to exactly one real number, and every real number corresponds to exactly one point on the number line. The point is called the **graph** of that real number, and the real number that corresponds to a point is called the **coordinate** of that point.

The point with coordinate 0 is called the **origin** of the number line. Numbers located to the left of the origin are **negative numbers**, whereas numbers located to the right of the origin are **positive numbers**. The number 0 is neither negative nor positive. The distance from 0 to 1 represents the **unit distance** (the length of one unit) for the number line. This distance establishes the **scale** for the number line. For example, Figure 2 shows us that any two consecutive integers are located at a scale of one unit apart.

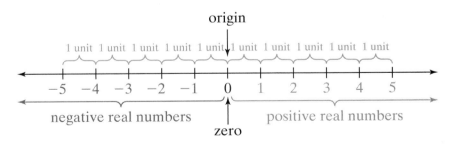

Figure 2 The real number line

We **plot**, or **graph**, a real number by placing a solid circle (•) at its location on the number line.

Example 5 Plotting Real Numbers on the Number Line

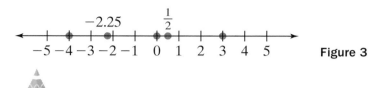 Plot the following set of numbers on the number line.

$$\left\{-4, 3, \frac{1}{2}, 0, -2.25\right\}$$

Solution These five real numbers are plotted on the number line in Figure 3. Watch this video for a complete solution.

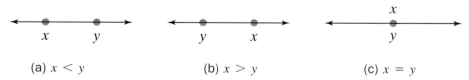

Figure 3

You Try It Work through this You Try It problem.

Work Exercises **21–22** in this textbook or in the *MyMathLab*® Study Plan.

My video summary

OBJECTIVE 4: USE INEQUALITY SYMBOLS TO ORDER REAL NUMBERS

To find the **order** of two real numbers x and y means to determine if the first number x is smaller than, larger than, or equal to the second number y. We can find the *order* of any two real numbers by comparing their locations on the real number line.

Order of Real Numbers

1. If x is located to the left of y on the real number line, shown in Figure 4(a), then x **is less than** y, denoted as $x < y$.

2. If x is located to the right of y on the real number line, shown in Figure 4(b), then x **is greater than** y, denoted as $x > y$.

3. If x and y are located at the same position on the real number line, shown in Figure 4(c), then x **is equal to** y, denoted as $x = y$.

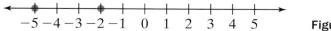

(a) $x < y$ (b) $x > y$ (c) $x = y$

Figure 4

The symbol for equality ($=$) is the **equal sign**. Symbols for less than ($<$) and greater than ($>$) are called **inequality symbols**. We use these symbols to state the order of real numbers. For example, since -5 is located to the left of -2 on the number line, as shown in Figure 5, we say -5 *is less than* -2 by writing $-5 < -2$. We say -2 *is greater than* -5 by writing $-2 > -5$.

Figure 5

Figure 6 shows that 0.5 and $\dfrac{1}{2}$ share the same position on the number line. So, we say 0.5 *equals* $\dfrac{1}{2}$ by writing $0.5 = \dfrac{1}{2}$.

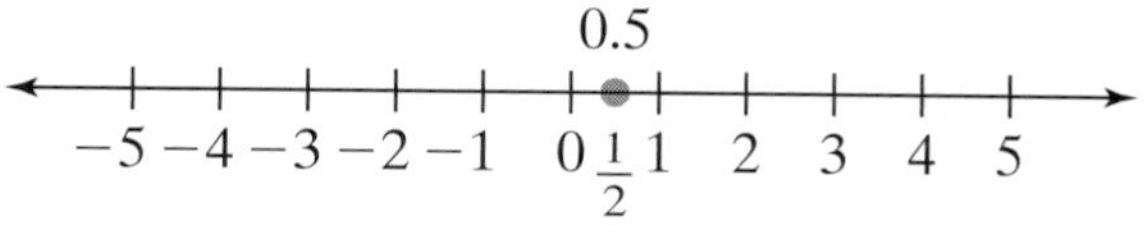

$$0.5$$

$$\overset{\longleftarrow}{\underset{-5\ -4\ -3\ -2\ -1\quad 0\ \frac{1}{2}\ 1\quad 2\quad 3\quad 4\quad 5}{\rule{6cm}{0.4pt}}}\longrightarrow$$

Figure 6

Example 6 Ordering Real Numbers Using Equality and Inequality Symbols

Fill in the blank with the correct symbol, $<, >$, or $=$, to make a true statement about the order of real numbers.

a. -1 ____ 0 **b.** -2.3 ____ -4.5 **c.** $-\dfrac{9}{4}$ ____ -2.25 **d.** $\dfrac{7}{8}$ ____ $\dfrac{5}{6}$

Solution

a. Click here to see -1 and 0 plotted on the real number line. Since -1 is to the left of 0 on the number line, -1 *is less than* 0. So, $-1 < 0$.

b. Click here to see -2.3 and -4.5 plotted on the real number line. Since -2.3 is to the right of -4.5 on the number line, -2.3 *is greater than* -4.5. So, $-2.3 > -4.5$.

c. Click here to see $-\dfrac{9}{4}$ and -2.25 plotted on the number line. Since the two numbers share the same position, they are equal. So, $-\dfrac{9}{4} = -2.25$.

d. To compare $\dfrac{7}{8}$ and $\dfrac{5}{6}$, first it is helpful to convert both fractions to decimals: $\dfrac{7}{8} = 0.875$ and $\dfrac{5}{6} = 0.8\overline{3}$. On the number line, 0.875 is to the right of $0.8\overline{3}$, so $\dfrac{7}{8}$ *is greater than* $\dfrac{5}{6}$. So, $\dfrac{7}{8} > \dfrac{5}{6}$. Click here to see the numbers plotted on the number line.

You Try It   Work through this **You Try It** problem.

Work Exercises 23–28 in this textbook or in the *MyMathLab*®️ Study Plan.

In addition to $<$ and $>$, there are three other inequality symbols that can be used to compare real numbers: *less than or equal to* ($\leq$), *greater than or equal to* ($\geq$), and *not equal to* ($\neq$). Notice that the $\leq$ and $\geq$ inequality symbols allow for equality as part of the comparison. For this reason, inequalities of the forms $x \leq y$ and $x \geq y$ are called **non-strict inequalities**. Similarly, inequalities of the forms $x < y, x > y$, and $x \neq y$ are called **strict inequalities** because they do not allow for equality as part of the comparison.

Example 7 Using Inequality Symbols

Determine if each statement is true or false.

a. $-9 \geq -10$ b. $\dfrac{3}{2} \neq 2.5$ c. $\dfrac{5}{3} < 1.\overline{6}$ d. $-\dfrac{3}{4} \leq -0.75$

Solution Try to answer these problems on your own. Click here to check your answers.

You Try It Work through this You Try It problem.

Work Exercises 29–32 in this textbook or in the **MyMathLab**® Study Plan.

OBJECTIVE 5: COMPUTE THE ABSOLUTE VALUE OF A REAL NUMBER

As defined earlier in this section, the distance from 0 to 1 on a real number line is the **unit distance** for the number line. See Figure 2. The concept of unit distance is related to the concept of *absolute value*.

> **Definition Absolute Value**
>
> The **absolute value** of a real number a, denoted as $|a|$, is the distance from 0 to a on the real number line.

The notation $|4|$ reads as "the absolute value of four," whereas $|-4|$ reads as "the absolute value of negative four." Since both 4 and -4 are 4 units from 0 on the number line, both absolute values equal 4. So, $|4| = 4$ and $|-4| = 4$. See Figure 7.

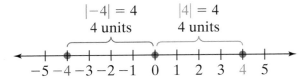

Figure 7

CAUTION
Because distance cannot be negative, the absolute value of a real number will always be non-negative.

Example 8 Computing Absolute Values

My video summary

Compute each absolute value.

a. $|2|$ b. $|-1.5|$ c. $\left|-\dfrac{7}{2}\right|$ d. $|0|$

Solution Try to find each absolute value on your own. Click here to check your answer, or watch this video for a complete solution.

You Try It Work through this You Try It problem.

Work Exercises 33–36 in this textbook or in the **MyMathLab**® Study Plan.

R.1 Exercises

In Exercises 1–6, fill in the blank with the correct symbol, $\in$ or $\notin$.

You Try It

1. e _____ $\{a, e, i, o, u\}$

2. 4.5 _____ $\mathbb{N}$

3. 25 _____ $\{3, 6, 9, \ldots, 45\}$

4. 75 _____ $\{5, 10, 15, 20, \ldots\}$

5. 13 _____ $\{x | x \text{ is an odd number}\}$

6. 6.4 _____ $\{x | x \text{ is a number smaller than } 7\}$

You Try It

In Exercises 7–10, use the roster method to list the elements in each set.

7. Natural numbers between 3 and 9, inclusive

8. Multiples of 4 that are greater than 1 and less than 30

9. Natural numbers greater than 20

10. Natural numbers less than 0.5

In Exercises 11–14, let $A = \{1, 2, 3, 4, 5, 6, 7, 8, 9\}$, $B = \{1, 3, 5, 7, 9\}$, $C = \{0, 2, 4\}$, and $D = \{9, 7, 5, 3, 1\}$. Determine if each statement is true or false.

You Try It

11. $C \subseteq A$

12. $D \supseteq B$

13. $A \supset B$

14. $B = D$

In Exercises 15–18, classify each number as a natural number, whole number, integer, rational number, irrational, and/or real number. Each number may belong to more than one set.

15. 259

16. $-11{,}401$

17. $-\dfrac{43}{7}$

18. $-\sqrt{7}$

In Exercises 19–20, given each set of real numbers, list the numbers that are (a) natural numbers, (b) whole numbers, (c) integers, (d) rational numbers, and/or (e) irrational numbers.

You Try It

19. $\left\{-17, -\sqrt{5}, -\dfrac{25}{29}, 0, 0.331, 1, \dfrac{\pi}{2}\right\}$

20. $\left\{-11, -3.\overline{2135}, -\dfrac{3}{9}, \dfrac{\sqrt{7}}{2}, 21.1\right\}$

You Try It

In Exercises 21–22, plot each set of numbers on the real number line.

21. $\left\{-1, 1.75, -\dfrac{9}{4}, 5\right\}$

22. $\left\{\dfrac{2}{3}, 3.8, 0, -\dfrac{5}{2}\right\}$

In Exercises 23–28, fill in the blank with the correct symbol $<, >$, or $=$.

You Try It

23. 0 _____ -7

24. -9 _____ -2

25. -2.6 _____ -3.2

26. $-\dfrac{7}{2}$ _____ -3.5

27. $\dfrac{9}{5}$ _____ $\dfrac{7}{4}$

28. $-\dfrac{1}{2}$ _____ $-\dfrac{2}{5}$

In Exercises 29–32, determine if each statement is true or false.

You Try It **29.** $-13 \le -12$ **30.** $\dfrac{15}{4} \ge 3.75$ **31.** $5 \neq -5$ **32.** $6.2 < 6.20$

In Exercises 33–36, compute the absolute value.
You Try It

 33. $|-18|$ **34.** $|34|$ **35.** $\left|\dfrac{3}{8}\right|$ **36.** $|-9.23|$

R.2 Order of Operations and Properties of Real Numbers

THINGS TO KNOW

Before working through this section, be sure you are familiar with the following concepts:

VIDEO   ANIMATION   INTERACTIVE

You Try It   **1.** Compute the Absolute Value of a Real Number (Section R.1, Objective 5)

OBJECTIVES

1 Perform Operations on Real Numbers

2 Simplify Numeric Expressions Containing Exponents and Radicals

3 Identify and Use the Properties of Real Numbers

4 Use Order of Operations to Evaluate Numeric Expressions

OBJECTIVE 1:   PERFORM OPERATIONS ON REAL NUMBERS

The four **basic operations** of arithmetic are *addition*, *subtraction*, *multiplication*, and *division*. Let's review the rules for each operation, beginning with **addition**.

Addition of Real Numbers

The result of adding two real numbers is called the **sum** of the numbers. The numbers being added are called **terms**, or **addends**. How we add terms depends on their signs (positive or negative).

Adding Two Real Numbers

1. If the signs of the two terms are the same (both positive or both negative), add their absolute values and use the common sign as the sign of the sum.

2. If the signs of the two terms are different (one positive and one negative), subtract the smaller absolute value from the larger absolute value and use the sign of the number with the larger absolute value as the sign of the sum.

Example 1 Adding Two Real Numbers

 Add.

a. $-9 + (-5)$ **b.** $2 + (-8)$ **c.** $-\dfrac{5}{6} + \dfrac{7}{8}$ **d.** $-3.4 + 2.8$

Solution For each sum, we apply the appropriate addition rule.

a. Both -9 and -5 are negative, so the sum is negative. We add the absolute values, $|-9| = 9$ and $|-5| = 5$, and keep the negative sign in the result. $9 + 5 = 14$, so $-9 + (-5) = -14$.

b. The signs are different: $|2| = 2$ and $|-8| = 8$. Since -8 has a larger absolute value than 2, the sum is negative. We subtract 2 from 8 and write a negative sign in the result. $8 - 2 = 6$, so $2 + (-8) = -6$.

c. The fractions have different signs: $\left|-\dfrac{5}{6}\right| = \dfrac{5}{6}$ and $\left|\dfrac{7}{8}\right| = \dfrac{7}{8}$. $\dfrac{7}{8}$ has a larger absolute value, so the sum is positive. Try to finish this problem by subtracting $\dfrac{5}{6}$ from $\dfrac{7}{8}$ and writing a positive sign in the result. Remember to find a common denominator. Click here to check your answer, or watch this video for the complete solution.

d. Try to work this problem on your own. Click here to check your answer, or watch this video for the complete solution.

You Try It   **Work through this You Try It problem.**

Work Exercises 1–4 in this textbook or in the *MyMathLab*® Study Plan.

Opposites or Additive Inverses

What if the two terms being added have the same absolute value but different signs? For example, consider $4 + (-4)$. Both 4 and -4 have the same absolute value, 4. Since they have different signs, we subtract their absolute values. Since $4 - 4 = 0$, the sum is $4 + (-4) = 0$, which is neither positive nor negative. Numbers such as 4 and -4 are called *opposites*, or *additive inverses*.

Definition   Opposites (or Additive Inverses)

Two numbers are **opposites**, or **additive inverses**, if their sum is zero.

When two numbers are opposites, each number is the opposite of the other. For example, -8 is the opposite of 8, and 8 is the opposite of -8 (because $8 + (-8) = 0$). We use a negative sign $(-)$ to denote "opposite of." So, -8 reads as "the opposite of eight," and $-(-8)$ reads as "the opposite of negative eight." So, $-(-8) = 8$. In general, if a is a number, then $-(-a) = a$.

So, the opposite of a positive number is the negative number with the same absolute value, and the opposite of a negative number is the positive number with the same absolute value. Zero is its own opposite.

Finding the Opposite of a Real Number

To find the opposite of a real number, change its sign.

Example 2 Finding Opposites of Real Numbers

Find the opposite of each number.

a. 17 **b.** -8.4 **c.** $\dfrac{2}{9}$ **d.** 0

Solution

a. The opposite of 17 is -17.

b. 8.4 is the opposite of -8.4 because $-(-8.4) = 8.4$.

c. The opposite of $\dfrac{2}{9}$ is $-\dfrac{2}{9}$.

d. 0 is the opposite of 0.

You Try It   Work through this You Try It problem.

Work Exercises 5–8 in this textbook or in the *MyMathLab*®️ Study Plan.

Subtraction of Real Numbers

The concept of opposites brings us to the next arithmetic operation: **subtraction**. The result of subtracting two real numbers is called the **difference** of the numbers. The number being subtracted is called the **subtrahend**, and the number being subtracted from is called the **minuend**.

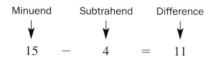

My video summary

Subtraction is defined in terms of addition. We subtract a number by adding its opposite.

Subtracting Two Real Numbers

If a and b represent two real numbers, then

$$a - b = a + (-b).$$

Example 3 Subtracting Two Real Numbers

Subtract.

a. $4 - 12$ **b.** $-2 - 7$ **c.** $\dfrac{1}{4} - \left(-\dfrac{2}{3}\right)$ **d.** $-2.1 - (-4.8)$

Solution For each subtraction, add the opposite.

a. Change to add the opposite of 12: $4 - 12 = 4 + (-12)$

We now use the appropriate addition rule. The signs are different. $|4| = 4$ and $|-12| = 12$. Since -12 has the larger absolute value, this sum is negative. We subtract 4 from 12 and write a negative sign in the result. $12 - 4 = 8$, so $4 + (-12) = -8$. The difference is $4 - 12 = -8$.

b. Change to add the opposite of 7: $-2 - 7 = -2 + (-7)$

Both -2 and -7 are negative, so the sum is negative also. $|-2| = 2$ and $|-7| = 7$. Since $2 + 7 = 9$, we get $-2 + (-7) = -9$. The difference is $-2 - 7 = -9$.

c. Change to add the opposite of $-\dfrac{2}{3}$: $\dfrac{1}{4} - \left(-\dfrac{2}{3}\right) = \dfrac{1}{4} + \dfrac{2}{3}$

Try to finish this problem on your own and remember to find a common denominator. Click here to check your answer, or watch this video for the complete solution.

d. Try to work this problem on your own. Click here to check your answer, or watch this video for the complete solution.

You Try It   Work through this You Try It problem.

Work Exercises 9–12 in this textbook or in the *MyMathLab*® Study Plan.

Multiplication of Real Numbers

The next arithmetic operation is **multiplication**. The result of multiplying two real numbers is called the **product** of the numbers. The numbers being multiplied are called **factors**. The sign of a product depends on the signs of the factors.

Multiplying Two Real Numbers

Multiply the absolute values of the two factors to get the absolute value of the product. Determine the sign of the product using the following rules:

1. If the signs of the two factors are the same (both positive or both negative), then the product is positive.

2. If the signs of the two factors are different (one positive and one negative), then the product is negative.

Example 4 Multiplying Two Real Numbers

 Multiply.

a. $(-9)(-3)$ **b.** $6(-7)$ **c.** $-\dfrac{2}{3}\left(\dfrac{5}{4}\right)$ **d.** $(-5.6)(-3.25)$

My video summary

Solution

a. $|-9| = 9$ and $|-3| = 3$. The absolute value of the product is $9 \cdot 3 = 27$. Since -9 and -3 are both negative, the product is positive. So, $(-9)(-3) = 27$.

b. One factor is positive and the other is negative, so the product is negative. Since $6 \cdot 7 = 42$, the product is $6(-7) = -42$.

c. The factors have different signs, so the product is negative. Try to finish this problem by multiplying $\dfrac{2}{3}$ and $\dfrac{5}{4}$, and writing a negative sign in the result.

Remember to write the fraction in lowest terms. Click here to check your answer, or watch this video for the complete solution.

d. Try to work this problem on your own. Click here to check your answer, or watch this video for the complete solution.

You Try It   Work through this You Try It problem.

Work Exercises 13–16 in this textbook or in the **MyMathLab**® Study Plan.

Reciprocals or Multiplicative Inverses

Consider the product $-5\left(-\dfrac{1}{5}\right) = 1$. Numbers such as -5 and $-\dfrac{1}{5}$ are called *reciprocals*, or *multiplicative inverses*.

Definition   Reciprocals (or Multiplicative Inverses)

Two numbers are **reciprocals**, or **multiplicative inverses**, if their product is 1.

Every nonzero real number b has a reciprocal, denoted as $\dfrac{1}{b}$. When two numbers are reciprocals, each is the reciprocal of the other. For example, $-\dfrac{1}{5}$ is the reciprocal of -5, and -5 is the reciprocal of $-\dfrac{1}{5}$.

 Zero does not have a reciprocal because there is no number that can be multiplied by 0 to result in 1.

Finding the Reciprocal of a Nonzero Real Number

Find the reciprocal of a nonzero real number as follows:

1. If the number is written as a fraction, interchange the numerator and denominator and simplify.
2. If the number is not written as a fraction, write it in the denominator of a fraction with 1 as the numerator and simplify.

Example 5 Finding Reciprocals

Find the reciprocal of each number.

a. -8 b. $\dfrac{5}{6}$ c. 0.4 d. $-\dfrac{1}{7}$ e. $\sqrt{2}$

Solution

a. Since -8 is not written as a fraction, we write it as the denominator of a fraction with a numerator of 1: $\dfrac{1}{-8}$ or $-\dfrac{1}{8}$. The reciprocal of -8 is $-\dfrac{1}{8}$.

b. Since $\dfrac{5}{6}$ is a fraction, we interchange the numerator 5 and the denominator 6. The reciprocal of $\dfrac{5}{6}$ is $\dfrac{6}{5}$.

Multiply by 1

c. The reciprocal of the decimal number 0.4 is $\dfrac{1}{0.4} = \dfrac{1}{0.4} \cdot \dfrac{10}{10} = \dfrac{10}{4} = \dfrac{5}{2}$ or 2.5.

d. The reciprocal of the fraction $-\dfrac{1}{7}$ is $-\dfrac{7}{1} = -7$.

e. The reciprocal of $\sqrt{2}$ is $\dfrac{1}{\sqrt{2}}$. We will learn to simplify numbers of this form further in Chapter 7.

The reciprocal of a number will have the same sign as the number. So, the reciprocal of a positive number is positive, and the reciprocal of a negative number is negative.

You Try It   Work through this You Try It problem.

Work Exercises 17–20 in this textbook or in the **MyMathLab**® Study Plan.

Division of Real Numbers

We use reciprocals to define the next arithmetic operation: division. The result of dividing two real numbers is called the **quotient** of the numbers. The number being divided is called the **dividend**, and the number we divide by is called the **divisor**.

$$\underset{\text{Dividend}}{20} \;\div\; \underset{\text{Divisor}}{4} \;=\; \underset{\text{Quotient}}{5}$$

$$\text{Dividend} \longrightarrow \dfrac{20}{4} = 5 \longleftarrow \text{Quotient} \qquad \text{Divisor} \longrightarrow$$

To divide by a number, we multiply by its reciprocal.

Dividing Two Real Numbers Using Reciprocals

If a and b represent two real numbers and $b \neq 0$, then

$$a \div b = \frac{a}{b} = a \cdot \frac{1}{b}.$$

Because division is defined in terms of multiplication, the rules for finding the sign of a quotient are the same as the rules for a product.

Dividing Two Real Numbers Using Absolute Value

Divide the absolute values of the numbers to obtain the absolute value of the quotient. Determine the sign of the quotient by the following rules:

1. If the signs of the two numbers are the same (both positive or both negative), then their quotient is positive.

2. If the signs of the two numbers are different (one positive and one negative), then their quotient is negative.

Example 6 Dividing Two Real Numbers

My video summary

Divide.

a. $-50 \div 10$ **b.** $\dfrac{-30}{-6}$ **c.** $\dfrac{1}{4} \div \left(-\dfrac{2}{3}\right)$ **d.** $-3 \div \left(-\dfrac{1}{5}\right)$

Solution

a. $|-50| = 50$ and $|10| = 10$. The absolute value of the quotient is $50 \div 10 = 5$. Since the signs of -50 and 10 are different, the quotient is negative. So, $-50 \div 10 = -5$.

b. Both numbers are negative, so the quotient is positive. Since $30 \div 6 = 5$, the quotient is $\dfrac{-30}{-6} = 5$.

c. To divide by a fraction, we multiply by its reciprocal. So, $\dfrac{1}{4} \div \left(-\dfrac{2}{3}\right) = \dfrac{1}{4} \cdot \left(-\dfrac{3}{2}\right)$. Try to finish this problem on your own. Click here to check your answer, or watch this video for the complete solution.

d. Try to work this problem on your own. Click here to check your answer, or watch this video for the complete solution.

You Try It   **Work through this You Try It problem.**

Work Exercises 21–24 in this textbook or in the *MyMathLab*® Study Plan.

OBJECTIVE 2:   **SIMPLIFY NUMERIC EXPRESSIONS CONTAINING EXPONENTS AND RADICALS**

Consider the product $2 \cdot 2 \cdot 2 \cdot 2 \cdot 2$. The factor 2 is repeated five times. We can write this product as the *exponential expression* 2^5. The superscript number 5 indicates that the factor 2 is repeated five times. The number 5 is called an *exponent*, or *power*, and the number 2 is called the *base*.

$$\underbrace{2 \cdot 2 \cdot 2 \cdot 2 \cdot 2}_{5 \text{ factors of } 2} = 2^5 \quad \longleftarrow \text{Exponent}$$
$$\uparrow$$
$$\text{Base}$$

We read 2^5 as "2 raised to the fifth power."

Definition   Exponential Expression

If a is a real number and n is a natural number, then the **exponential expression** a^n represents the product of n factors of a.

$$a^n = \underbrace{a \cdot a \cdot a \cdot \ldots \cdot a}_{n \text{ factors of } a}$$

a is the **base** of the expression, and n is the **exponent** or **power**.

We read a^n as "a raised to the nth power" or "a to the nth." An exponent of 2 is usually read as "squared," whereas an exponent of 3 is read as "cubed." For example, 5^2 is "five squared" and 5^3 is "five cubed."

An exponent of 1 is usually not written. For example, $2^1 = 2$.

We simplify numeric exponential expressions by multiplying the factors. For example, $2^5 = 2 \cdot 2 \cdot 2 \cdot 2 \cdot 2 = 32$. When a negative sign is involved, we must see if the negative sign is part of the base. For example, consider $(-5)^2$. The parentheses around -5 indicate that the negative sign is part of the base. So, $(-5)^2 = (-5)(-5) = 25$. In the expression -5^2, however, the negative sign is not part of the base. Instead, we need to find the opposite of 5^2. So, $-5^2 = -(5 \cdot 5) = -25$.

Example 7 Simplifying Numeric Expressions Containing Exponents

My video summary

 Simplify each exponential expression.

a. 5^3 b. $(-3)^4$ c. -3^4 d. $(-2)^5$ e. $\left(\dfrac{3}{4}\right)^2$

Solution

a. The exponential expression 5^3 means to multiply three factors of 5, so $5^3 = 5 \cdot 5 \cdot 5 = 125$.

b. The parentheses around the -3 indicate that the base is -3, and the negative sign is part of the base. So, $(-3)^4 = (-3)(-3)(-3)(-3) = 81$.

c. Since there are no parentheses around the -3, the negative sign is not part of the base. Instead we find the opposite of 3^4. So, $-3^4 = -(3 \cdot 3 \cdot 3 \cdot 3) = -81$.

d.–e. Try to work these problems on your own. Click here to check your answers, or watch this video for the complete solutions.

CAUTION A common error that occurs when simplifying exponential expressions is to multiply the base by the exponent. For example, $10^2 = 10 \cdot 10 = 100$, not $10 \cdot 2 = 20$.

You Try It Work through this You Try It problem.

Work Exercises 25–28 in this textbook or in the *MyMathLab* **Study Plan.**

A **radical expression** is an expression that contains a **radical sign** $\sqrt{\ }$. For example, $\sqrt{16}$, $\sqrt[3]{8}$, and $\sqrt[4]{81}$ are all radical expressions. The expression $\sqrt{16}$ represents the *principal*, or *positive*, *square root* of 16. The **principal square root** of a number a is a positive number b whose square is a. So, $\sqrt{16} = 4$ because 4 is positive and $4^2 = 16$.

The expression $\sqrt[3]{8}$ represents the *cube root* of 8. The **cube root** of a number a is a number b (positive or negative) whose cube is a. So, $\sqrt[3]{8} = 2$ because $2^3 = 8$.

The expression $\sqrt[4]{81}$ represents the *principal*, or *positive, fourth root* of 81. The **principal fourth root** of a number a is a positive number b whose fourth power is a. So, $\sqrt[4]{81} = 3$ because 3 is positive and $3^4 = 81$.

Example 8 Simplifying Numeric Expressions Containing Radicals

 Simplify each radical expression.

a. $\sqrt{25}$ b. $-\sqrt{100}$ c. $\sqrt{-4}$ d. $\sqrt[3]{64}$

e. $\sqrt[3]{-64}$ f. $-\sqrt[3]{-8}$ g. $\sqrt[4]{16}$ h. $\sqrt{\dfrac{1}{9}}$

My video summary

Solution

a. For $\sqrt{25}$ we must find a positive number whose square is 25. The number 5 is positive and $5^2 = 25$, so $\sqrt{25} = 5$.

b. The negative sign in front of the radical indicates that the square root is negative. $-\sqrt{100} = -10$ because $10^2 = 100$.

c. For $\sqrt{-4}$ we need to find a positive number whose square is -4. When a real number is squared, the result is always non-negative, so $\sqrt{-4}$ is not a real number.

d. For $\sqrt[3]{64}$ we find a number whose cube is 64. $4^3 = 64$, so $\sqrt[3]{64} = 4$.

e.–h. Try to simplify these radical expressions on your own. Click here to check your answers, or watch this video for the complete solutions.

As discussed in Section R.1, some radical expressions will not simplify to rational numbers. We study radical expressions in greater depth in Chapter 7.

You Try It   **Work through this You Try It problem.**

Work Exercises 29–34 in this textbook or in the *MyMathLab*® Study Plan.

OBJECTIVE 3:   IDENTIFY AND USE THE PROPERTIES OF REAL NUMBERS

In this section, we have learned to simplify numeric expressions involving addition, subtraction, multiplication, division, exponents, and radicals. Now we identify useful properties of real numbers to help with simplification. Numeric examples will be presented for each property to illustrate its truth, but the properties will also apply to algebraic expressions.

We begin with the *commutative properties*. The **commutative property of addition** states that the *order* of the terms in addition does not affect the sum. For example, $3 + 5 = 8$ and $5 + 3 = 8$, so $3 + 5 = 5 + 3$.

> **Commutative Property of Addition**
>
> If a and b are real numbers, then $a + b = b + a$.

The **commutative property of multiplication** states that the *order* of the factors in multiplication does not affect the product. For example, $(-3)(7) = -21$ and $7(-3) = -21$, so $(-3)(7) = 7(-3)$.

> **Commutative Property of Multiplication**
>
> If a and b are real numbers, then $a \cdot b = b \cdot a$.

CAUTION Subtraction and division do not have commutative properties: $a - b \neq b - a$; $a \div b \neq b \div a$. For example, $9 - 2 = 7$ but $2 - 9 = -7$, so $9 - 2 \neq 2 - 9$. Likewise, $6 \div 3 = 2$ but $3 \div 6 = 0.5$, so $6 \div 3 \neq 3 \div 6$.

Next, we look at the *associative properties*. The **associative property of addition** states that regrouping terms does not affect the sum. For example, $(3 + 5) + 4 = 8 + 4 = 12$ and $3 + (5 + 4) = 3 + 9 = 12$, so $(3 + 5) + 4 = 3 + (5 + 4)$.

> **Associative Property of Addition**
>
> If $a, b,$ and c are real numbers, then $(a + b) + c = a + (b + c)$.

The **associative property of multiplication** states that regrouping factors does not affect the product. For example, $(-2 \cdot 5) \cdot 3 = (-10) \cdot 3 = -30$ and $-2 \cdot (5 \cdot 3) = -2 \cdot (15) = -30$, so $(-2 \cdot 5) \cdot 3 = -2 \cdot (5 \cdot 3)$.

> **Associative Property of Multiplication**
>
> If $a, b,$ and c are real numbers, then $(a \cdot b) \cdot c = a \cdot (b \cdot c)$.

Subtraction and division do not have associative properties: $(a - b) - c \neq a - (b - c)$; $(a \div b) \div c \neq a \div (b \div c)$. For example, $(7 - 4) - 2 = 3 - 2 = 1$ but $7 - (4 - 2) = 7 - 2 = 5$, so $(7 - 4) - 2 \neq 7 - (4 - 2)$. Likewise, $(80 \div 8) \div 2 = 10 \div 2 = 5$ but $80 \div (8 \div 2) = 80 \div 4 = 20$, so $(80 \div 8) \div 2 \neq 80 \div (8 \div 2)$.

Example 9 Using the Commutative and Associative Properties

Use the given property to complete each statement.

a. Associative property of multiplication: $5(7y) = $ _____

b. Associative property of addition: $(4x + 2) + 7 = $ _____

c. Commutative property of addition: $8(3x + 2) = $ _____

d. Commutative property of multiplication: $x(y + 4) = $ _____

Solution

a. The associative property of multiplication states that we can regroup the factors. So $5(7y) = (5 \cdot 7)y$.

b. By the associative property of addition, $(4x + 2) + 7 = 4x + (2 + 7)$.

c. The commutative property of addition allows us to change the order of the terms being added. So, $8(3x + 2) = 8(2 + 3x)$.

d. By the commutative property of multiplication, $x(y + 4) = (y + 4)x$.

You Try It   Work through this You Try It problem.

Work Exercises 35–38 in this textbook or in the _MyMathLab_® Study Plan.

The **distributive property** states that multiplication distributes over addition (or subtraction). For example, compare $5(3 + 4) = 5(7) = 35$ with $(5 \cdot 3) + (5 \cdot 4) = 15 + 20 = 35$. Since both expressions simplify to 35, they are equivalent: $5(3 + 4) = (5 \cdot 3) + (5 \cdot 4)$.

> **Distributive Property**
>
>
>
> If $a, b,$ and c are real numbers, then $a(b + c) = ab + ac$.

Note: Because multiplication is commutative, we can also write the distributive property as

$$(b + c)a = ba + ca.$$

Example 10 Using the Distributive Property

 Use the distributive property to multiply.

a. $4(x + 9)$ **b.** $(x - 4) \cdot 2$ **c.** $-5(x + y - 4)$

Solution

a. Begin with the original expression: $4(x + 9)$

    Apply the distributive property: $= 4 \cdot x + 4 \cdot 9$

    Find each product: $= 4x + 36$

b.–c. Try to complete these problems on your own. Click here to check your answer, or watch this video for the complete solution.

You Try It **Work through this You Try It problem.**

Work Exercises 39–42 in this textbook or in the *MyMathLab*® Study Plan.

We complete our discussion of the properties of real numbers with *identity* and *inverse properties*.

If 0 is added to any real number, then the sum is the original number. For example, $5 + 0 = 5$ and $0 + 2 = 2$. So, the *identity* of a number is not changed by adding the number plus 0. Since 0 is the only real number for which this is true, 0 is called the **identity element for addition**, or the **additive identity**.

Identity Property of Addition

If a is a real number, then $a + 0 = a$ and $0 + a = a$.

Similarly, if 1 is multiplied by a real number, then the product is the original number. For example, $7 \cdot 1 = 7$ and $1 \cdot 3 = 3$. So, the *identity* of a number is not changed by multiplying the number and 1. Since 1 is the only real number for which this is true, 1 is called the **identity element for multiplication**, or the **multiplicative identity**.

Identity Property of Multiplication

If a is a real number, then $a \cdot 1 = a$ and $1 \cdot a = a$.

Earlier in this section, we defined opposites and reciprocals. For example, 3 and -3 are opposites because $3 + (-3) = 0$, and 3 and $\dfrac{1}{3}$ are reciprocals because $3 \cdot \dfrac{1}{3} = 1$. The opposite of a number is also called the **additive inverse** of the number. The reciprocal of a number is also called the **multiplicative inverse** of the number. The **inverse property of addition** states that every real number has a unique additive inverse.

My video summary

Inverse Property of Addition

If a is a real number, then there is a unique real number $-a$ such that

$$a + (-a) = -a + a = 0.$$

The **inverse property of multiplication** states that every nonzero real number has a unique multiplicative inverse.

Inverse Property of Multiplication

If a is a nonzero real number, then there is a unique real number $\dfrac{1}{a}$, such that

$$a \cdot \frac{1}{a} = \frac{1}{a} \cdot a = 1.$$

Example 11 Identifying Properties of Real Numbers

Identify the property of real numbers demonstrated in each statement.

a. $(2x + 0) - 4 = 2x - 4$ **b.** $\dfrac{1}{6}(6x) = \left(\dfrac{1}{6} \cdot 6\right)x$

c. $(-13 + 13) + 7 = 0 + 7$ **d.** $2x - 8 = 2(x - 4)$

Solution Try to identify each property on your own. Click here to check your answers.

You Try It   Work through this You Try It problem.

Work Exercises 43–46 in this textbook or in the *MyMathLab*® Study Plan.

OBJECTIVE 4:   USE ORDER OF OPERATIONS TO EVALUATE NUMERIC EXPRESSIONS

The numeric expression $2 + 3 \cdot 4$ contains two operations: addition and multiplication. Depending on which operation is performed first, a different answer will result.

Add first, then multiply: $2 + 3 \cdot 4 = \quad 5 \cdot 4 \quad = 20$

Multiply first, then add: $2 + 3 \cdot 4 = 2 + 12 = 14$

⟩ Different results

For this reason, mathematicians have agreed on a set **order of operations**, which determines the priority in which operations should be performed. For example, we perform multiplication before addition from left to right. So the correct evaluation for our numeric expression above is 14, not 20.

Order of Operations

1. **Parentheses (or other grouping symbols)**   Evaluate operations within parentheses (or other grouping symbols) first, starting with the innermost set and working out.

2. **Exponents and Radicals**   Work from left to right evaluating any exponential or radical expressions as they occur.

3. **Multiplication and Division**   Work from left to right and perform any multiplication or division operations as they occur.

4. **Addition and Subtraction**   Work from left to right and perform any addition or subtraction operations as they occur.

Example 12 Using Order of Operations to Evaluate Numeric Expressions

Simplify each expression.

a. $50 \div 5 \cdot 2$ **b.** $18 - 6(4) + \dfrac{12}{3}$ **c.** $7 - 4^2 + 6 \div 3(8 - 10)^2$

Solution For each expression, we follow the order of operations.

a. The two operations in this problem are division and multiplication, which have equal priority. So, we work from left to right and perform the division and multiplication in order.

$$\begin{aligned} \text{Begin with the original expression:}\quad & 50 \div 5 \cdot 2 \\ \text{Divide } 50 \div 5:\quad & = 10 \cdot 2 \\ \text{Multiply:}\quad & = 20 \end{aligned}$$

b. There are four operations in this problem: subtraction, multiplication, addition, and division. Working from left to right, we perform the multiplication and division first, in that order. Then we perform the addition and subtraction from left to right.

$$\begin{aligned} \text{Begin with the original expression:}\quad & 18 - 6(4) + \frac{12}{3} \\ \text{Multiply } 6(4);\ \text{divide } \frac{12}{3}:\quad & = 18 - 24 + 4 \\ \text{Subtract } 18 - 24:\quad & = -6 + 4 \\ \text{Add:}\quad & = -2 \end{aligned}$$

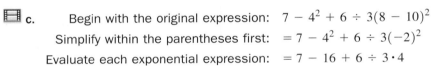

c.   Begin with the original expression: $7 - 4^2 + 6 \div 3(8 - 10)^2$
Simplify within the parentheses first: $= 7 - 4^2 + 6 \div 3(-2)^2$
Evaluate each exponential expression: $= 7 - 16 + 6 \div 3 \cdot 4$

There are four operations remaining in this expression. Try to finish simplifying this problem on your own. Click here to check your answer, or watch this video for a complete solution.

You Try It   **Work through this You Try It problem.**

Work Exercises 47–52 in this textbook or in the *MyMathLab*® Study Plan.

Recall that fraction bars −, absolute value symbols | |, and radicals √ can serve as grouping symbols in addition to being operations. We must simplify expressions separately in the numerator and denominator of a fraction before dividing. Likewise, we simplify expressions within absolute value symbols before evaluating the absolute value. Also, we simplify expressions beneath the radical sign before evaluating the radical expression.

Though not specified in the order of operations, absolute values should be evaluated in the same priority with exponents and radicals. Of course, any expression between the absolute value bars must be simplified first.

Example 13 Using Order of Operations with Special Grouping Symbols

Simplify each expression.

a. $\dfrac{8(-5) + 4^2}{-3(-2)}$

b. $48 \div \sqrt{6 \cdot 7 + 3(-2)} \div |11 - 3^2|$

Solution

a. The fraction bar serves as a grouping symbol, grouping expressions in the numerator and denominator. We begin by simplifying the numerator and denominator separately before dividing.

 *My video summary*

Begin with the original expression: $\dfrac{8(-5) + 4^2}{(-3)^2 - 5}$

Evaluate the exponents separately in the numerator and denominator: $= \dfrac{8(-5) + 16}{9 - 5}$

Multiply in the numerator: $= \dfrac{-40 + 16}{9 - 5}$

Add and subtract separately in the numerator and denominator: $= \dfrac{-24}{4}$

Divide: $= -6$

b. The radical and absolute value symbols act as grouping symbols. First, we simplify beneath the radical sign and within the absolute value symbols. Try to do this on your own. Click here to check your answer, or watch this video for a complete solution.

You Try It Work through this You Try It problem.

Work Exercises 53–54 in this textbook or in the *MyMathLab*® Study Plan.

Some numeric expressions contain **nested grouping symbols** (grouping symbols within grouping symbols). In such situations, we begin with the innermost set of grouping symbols and work our way outward.

Example 14 Using Order of Operations with Nested Grouping Symbols

My interactive video summary

 Simplify each expression.

a. $-2^3 + [3 - 5 \cdot (1 - 3)]$

b. $\dfrac{|2 - 3^3| + 5}{5^2 - 4^2}$

Solution Try to simplify each expression on your own. Click here to check your answers, or work through this interactive video for complete solutions.

You Try It Work through this You Try It problem.

Work Exercises 55–56 in this textbook or in the *MyMathLab*® Study Plan.

R.2 Exercises

You Try It

In Exercises 1–4, add.

1. $-7 + 10$ **2.** $-13 + (-9)$ **3.** $-\dfrac{8}{9} + \dfrac{5}{6}$ **4.** $5.1 + (-7.4)$

You Try It

In Exercises 5–8, find the opposite of each real number.

5. -13 **6.** 52 **7.** $-\dfrac{4}{15}$ **8.** -9.34

You Try It

In Exercises 9–12, subtract.

9. $-8 - 14$ **10.** $5 - (-6)$ **11.** $-\dfrac{5}{8} - \left(-\dfrac{3}{4}\right)$ **12.** $7.5 - 10.1$

You Try It

In Exercises 13–16, multiply.

13. $(-12)(-6)$ **14.** $8(-4)$ **15.** $\left(-\dfrac{5}{12}\right)\left(-\dfrac{3}{10}\right)$ **16.** $-7.2(6.75)$

You Try It

In Exercises 17–20, find the reciprocal of each real number.

17. 12 **18.** $-\dfrac{6}{11}$ **19.** $-\dfrac{1}{9}$ **20.** 0.8

You Try It

In Exercises 21–24, divide.

21. $200 \div (-8)$ **22.** $\dfrac{-75}{-5}$ **23.** $-\dfrac{9}{16} \div \dfrac{5}{12}$ **24.** $-3.9 \div (-0.2)$

You Try It

In Exercises 25–28, simplify each exponential expression.

25. 8^3 **26.** -5^4 **27.** $(-4)^3$ **28.** $\left(-\dfrac{5}{8}\right)^2$

You Try It

In Exercises 29–34, simplify each radical expression.

29. $\sqrt{169}$ **30.** $\sqrt{\dfrac{1}{25}}$ **31.** $\sqrt[3]{216}$

32. $\sqrt[3]{-125}$ **33.** $\sqrt[4]{625}$ **34.** $-\sqrt{121}$

In Exercises 35–38, use the given property to complete each statement.

35. Commutative property of addition: $ab + c = $ _____

You Try It **36.** Commutative property of multiplication: $6(v + 4) = $ _____

37. Associative property of addition: $4 + (a + 11) = $ _____

38. Associative property of multiplication: $45 \cdot (y \cdot z) = $ _____

In Exercises 39–42, use the distributive property to multiply.

39. $5(x + 8)$ **40.** $(x - 6) \cdot 3$ **41.** $-7(m - n + 5)$ **42.** $x(x + y - 2)$

In Exercises 43–46, identify the property of real numbers demonstrated in each statement.

  43. $(5 \cdot 1)x^2 = 5x^2$ **44.** $\left(\dfrac{3}{4} \cdot \dfrac{4}{3}\right)y = 1y$

45. $(5 + x) - 5 = (x + 5) - 5$ **46.** $(2x + 4) + (-4) = 2x + (4 + (-4))$

In Exercises 47–56, simplify each expression using the order of operations.

47. $4(2 - 5)^3$ **48.** $100 \div 25 \cdot 4$

49. $7 - (4 \cdot 6 - 26)$ **50.** $10^2 + 20^2 \div 5^2$

51. $3 \cdot (18 + 3)^2 - 4 \cdot (8 - 3)^2$ **52.** $-\dfrac{7}{6}\left(\dfrac{1}{2}\right) + \dfrac{5}{9} \div \dfrac{7}{6}$

 53. $18 \div \sqrt{4 \cdot 5 + 2^4} \div |-3^4 + 13 \cdot 6|$ **54.** $\dfrac{7 \cdot 6 + 2^2}{31 - 2^3}$

 55. $5^2 - 3 \cdot [2 + 3 \cdot (1 + 5)]$ **56.** $\dfrac{|-5^2 + (-3)^2| + 4 \cdot 5}{6 \div 2 - 3 \cdot 2^2}$

R.3 Algebraic Expressions

THINGS TO KNOW

Before working through this section, be sure you are familiar with the following concepts:

VIDEO   ANIMATION   INTERACTIVE

1. Perform Operations on Real Numbers (Section R.2, **Objective 1**)

2. Simplify Numeric Expressions Containing Exponents and Radicals (Section R.2, **Objective 2**)

3. Use Order of Operations to Evaluate Numeric Expressions (Section R.2, **Objective 4**)   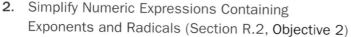

OBJECTIVES

1 Evaluate Algebraic Expressions

2 Simplify Algebraic Expressions

3 Write Verbal Descriptions as Algebraic Expressions

OBJECTIVE 1: EVALUATE ALGEBRAIC EXPRESSIONS

A **variable** is a symbol (usually a letter) that represents a changeable value. For example, the letter x might represent 5 in one situation and 5000 in another. A **constant** has a value that never changes. For example, 3 is a constant because the value of 3 is always the same.

An **algebraic expression** is a variable or a combination of variables, constants, and operations. Examples of algebraic expressions include

$$x, \quad y, \quad 5z, \quad x^2, \quad \sqrt{y}, \quad w + 7, \quad x + y, \quad a^2 + 2ab + b^2, \quad \frac{3m - 4n}{2m + n}.$$

Algebraic expressions can be used to describe quantities in a general way. For example, if the most common price for an iPhone app is $0.99 (*Source*: pinchmedia.com), then the algebraic expression $0.99x$ gives the cost for purchasing x apps at this price.

To find the cost for purchasing six apps, we substitute 6 in place of x and simplify the resulting numeric expression. This means that we **evaluate** the algebraic expression and find the **value** of the expression $0.99x$ when $x = 6$:

$$0.99x = 0.99(6) = 5.94$$

So, six apps cost $5.94.

The value of an algebraic expression changes for different values of the variables.

My video summary

> **Evaluate Algebraic Expressions**
>
> To **evaluate an algebraic expression**, substitute the given values for the variables and simplify the resulting numeric expression using the order of operations.

Example 1 Evaluating Algebraic Expressions

Evaluate each algebraic expression for the given values of the variables.

a. $2x^3 - 7$ for $x = 3$ **b.** $a^2 + 3ab - 5b^2$ for $a = -2$ and $b = 4$.

Solution

a. Begin with the original algebraic expression: $2x^3 - 7$

Substitute 3 for x: $= 2(3)^3 - 7$

Evaluate the exponent: $= 2(27) - 7$

Multiply: $= 54 - 7$

Subtract: $= 47$

b. Begin with the original algebraic expression: $a^2 + 3ab - 5b^2$

Substitute -2 for a and 4 for b: $= (-2)^2 + 3(-2)(4) - 5(4)^2$

Try to finish simplifying on your own by following the order of operations. Click here to check your answer, or watch this video for the complete solution.

You Try It   Work through this You Try It problem.

Work Exercises 1–12 in this textbook or in the *MyMathLab*® Study Plan.

OBJECTIVE 2:   SIMPLIFY ALGEBRAIC EXPRESSIONS

The **terms** of an algebraic expression are the quantities being added. For example, in the expression $3x^2 + 5xy + 7$, the terms are $3x^2, 5xy$, and 7. Because subtraction is defined by adding the opposite, subtracted quantities are "negative" terms. So, in the expression $9x - 4y$, the terms are $9x$ and $-4y$.

Two terms are **like terms** if they have variable factors that are exactly the same. Like terms can only differ by their numeric factors. For example, the terms $5x^2y$ and $4x^2y$ are like terms because they are alike except for the numeric factors 5 and 4. The numeric factor of a term is called the **coefficient** of the term.

When an algebraic expression contains like terms, it can be simplified by combining them. To combine like terms, we use the distributive property in reverse.

Example 2 Combining Like Terms

Simplify each algebraic expression by combining like terms.

a. $10x^2 + 2x^2$ **b.** $7xy + 8xy + xy$ **c.** $7y - 5y + 3y^2$

Solution

a. The terms $10x^2$ and $2x^2$ have the exact same variable factor x^2, so they are like terms.

$$\text{Begin with the original expression:} \quad 10x^2 + 2x^2$$
$$\text{Reverse the distributive property:} \quad = (10 + 2)x^2$$
$$\text{Add 10 and 2:} \quad = 12x^2$$

b. The three terms $7xy, 8xy$, and xy are like terms.

$$\text{Begin with the original expression:} \quad 7xy + 8xy + xy$$
$$\text{Reverse the distributive property:} \quad = (7 + 8 + 1)xy$$
$$\text{Add } 7 + 8 + 1: \quad = 16xy$$

c. The terms $7y$ and $-5y$ are like terms, but not $3y^2$ because y and y^2 are not exactly the same. We can only combine the two like terms.

$$\text{Begin with the original expression:} \quad 7y - 5y + 3y^2$$
$$\text{Reverse the distributive property:} \quad = (7 - 5)y + 3y^2$$
$$\text{Subtract } 7 - 5: \quad = 2y + 3y^2$$

You Try It   Work through this You Try It problem.

Work Exercises 13–16 in this textbook or in the *MyMathLab*® Study Plan.

Sometimes it is necessary to reorder the terms in an algebraic expression so that like terms are grouped together. Doing this is called **collecting like terms**. Collecting like terms is possible because of the commutative and associative properties of addition.

Example 3 Simplifying Algebraic Expressions

Simplify each algebraic expression.

a. $2x^2 + 9x + 4x^2 - 6x$ **b.** $8xy + 13 - 5xy + 3xy - 4$

Solution

a. The terms $2x^2$ and $4x^2$ are like terms, as are $9x$ and $-6x$.

$$\text{Begin with the original expression:}\quad 2x^2 + 9x + 4x^2 - 6x$$
$$\text{Rearrange the terms to collect like terms:}\quad = 2x^2 + 4x^2 + 9x - 6x$$
$$\text{Reverse the distributive property:}\quad = (2 + 4)x^2 + (9 - 6)x$$
$$\text{Add and subtract:}\quad = 6x^2 + 3x$$

 *My video summary*

 b. $8xy$, $-5xy$, and $3xy$ are like terms, as are 13 and -4.

$$\text{Begin with the original expression:}\quad 8xy + 13 - 5xy + 3xy - 4$$
$$\text{Rearrange the terms to collect like terms:}\quad = 8xy - 5xy + 3xy + 13 - 4$$

Try to finish this problem on your own. Click here to check your answer, or watch this video for a complete solution.

You Try It   **Work through this You Try It problem.**

Work Exercises 17–20 in this textbook or in the **MyMathLab** Study Plan.

Before we combine like terms, it is often necessary to remove grouping symbols. We can do this using the distributive property.

Example 4 Simplifying Algebraic Expressions

Simplify each algebraic expression.

a. $2(x - y) + 5x$ **b.** $7(2z + 4) - (5z - 7)$

 *My video summary*

Solution

a. We use the distributive property to remove the parentheses.

$$\text{Begin with the original expression:}\quad 2(x - y) + 5x$$
$$\text{Distribute the 2:}\quad = 2x - 2y + 5x$$
$$\text{Rearrange the terms to collect like terms:}\quad = 2x + 5x - 2y$$
$$\text{Reverse the distributive property:}\quad = (2 + 5)x - 2y$$
$$\text{Add:}\quad = 7x - 2y$$

b. We can think of $-(5z - 7)$ as $-1(5z - 7)$. So, we use the distributive property to remove both sets of parentheses.

$$\text{Begin with the original expression:}\quad 7(2z + 4) - (5z - 7)$$
$$\text{Rewrite } -(5z - 7) \text{ as } -1(5z - 7):\quad = 7(2z + 4) - 1(5z - 7)$$
$$\begin{array}{l}\text{Use the distributive property}\\ \text{to remove the parentheses:}\end{array}\quad = 14z + 28 - 5z + 7$$

Try to finish this problem on your own. Click here to check your answer, or watch this video for a complete solution.

You Try It   Work through this You Try It problem.

Work Exercises 21–24 in this textbook or in the **MyMathLab**® Study Plan.

OBJECTIVE 3:   WRITE VERBAL DESCRIPTIONS AS ALGEBRAIC EXPRESSIONS

In algebra, one key component of problem solving involves writing verbal descriptions as algebraic expressions. When translating the verbal descriptions, look for key word phrases that translate into arithmetic operations. Table 2 provides a list of phrases and their corresponding operations.

Addition	Subtraction	Multiplication	Division
add	subtract	multiply	divide
plus	minus	times	divided by
sum	difference	product	quotient
increased by	decreased by	of	per
more than	less	double	ratio
total	less than	triple	
		twice	

Table 2

Example 5 Writing Verbal Descriptions as Algebraic Expression

Write each word phrase as an algebraic expression.

a. Fourteen decreased by a number x

b. The sum of a number x and thirty

c. Eight more than twice a number x

d. Triple the sum of a number x and six

e. Three-fourths of a number x

f. The quotient of a number x and eight, increased by twice the number

Solution

a. Looking at Table 2, the phrase "decreased by" indicates subtraction. What is being subtracted? Fourteen is decreased by x, so x is being subtracted from 14. "Fourteen decreased by a number x" is written algebraically as $14 - x$.

b. The word *sum* means addition. What is to be added? x and 30. So, "the sum of a number x and thirty" is written algebraically as $x + 30$.

c. "More than" means addition. What is being added? 8 and twice x. That is, 8 and $2x$. So, written algebraically, "eight more than twice a number x" is $8 + 2x$.

d.–f. Try to finish these problems on your own. Click here to check your answers, or watch this video for a complete solution.

You Try It   Work through this You Try It problem.

Work Exercises 25–30 in this textbook or in the **MyMathLab**® Study Plan.

My video summary

R.3 Exercises

In Exercises 1–12, evaluate each algebraic expression for the given values of the variable.

1. $9w^2$ for $w = -5$

2. $\frac{2}{3}x - 5$ for $x = -6$

You Try It

3. $5x - 9y$ for $x = -3$ and $y = -7$

4. $36a + 60b$ for $a = -\frac{3}{4}$ and $b = \frac{2}{5}$

5. $m^2 + 6mn + 5n^2$ for $m = 3$ and $n = -2$

6. $a^3 - b^3$ for $a = 4$ and $b = -3$

7. $-z^2 - 6z$ for $z = -3$

8. $|8x^2 - 5y|$ for $x = -1$ and $y = 3$

9. $\frac{6b - 9b^2}{b^2 - 6}$ for $b = 3$

10. $5\sqrt{w} + 3z$ for $w = 100$ and $z = -2$

11. $b^2 - 4ac$ for $a = -4, b = 3$, and $c = -2$

12. $\frac{y_2 - y_1}{x_2 - x_1}$ for $x_1 = 4, x_2 = 6, y_1 = -2$, and $y_2 = 7$

In Exercises 13–24, simplify each algebraic expression.

You Try It **13.** $-8z + 3z$

14. $\frac{2}{3}z^2 - \frac{1}{4}z^2$

You Try It **15.** $10ab - ab + 4ab$

16. $17m - 5m + 13m^2$

17. $7a - 20 - 17a + 50$

18. $3k + 3k^2 + 7k + 7k^2$

You Try It **19.** $7x^2 - 8xy + y^2 - 3x^2 - 4y^2 + 6xy$

20. $\frac{3}{4}x - \frac{4}{3} + \frac{7}{2}x + \frac{5}{6}$

21. $5(3t - 5) - 6t$

22. $2.5(y^2 + 2y - 3) + 1.4y^2 + 3.6$

23. $2(x^2 + 4) - (10 - x^2)$

24. $-5(3w - 9) + 2(2w + 7)$

In Exercises 25–30, write each word phrase as an algebraic expression.

25. The sum of a number x and ten

You Try It **26.** A number x decreased by seventeen

27. The product of five and a number x, increased by nine

28. Twice the sum of a number x and four.

29. Twelve more than five-eighths of a number x

30. The quotient of a number x and two, minus the product of four and the number

CHAPTER ONE

Equations and Inequalities in One Variable

CHAPTER ONE CONTENTS

1.1   Linear Equations in One Variable
1.2   Linear Inequalities in One Variable
1.3   Compound Inequalities; Absolute Value Equations and Inequalities
1.4   Formulas and Problem Solving

1.1   Linear Equations in One Variable

THINGS TO KNOW

Before working through this section, be sure you are familiar with the following concepts:

| | | VIDEO | ANIMATION | INTERACTIVE |

You Try It   **1.** Evaluate Algebraic Expressions (Section R.3, Objective 1)

You Try It   **2.** Simplify Algebraic Expressions (Section R.3, Objective 2)

You Try It   **3.** Write Verbal Descriptions as Algebraic Expressions (Section R.3, Objective 3)

OBJECTIVES

1   Determine If a Given Value Is a Solution to an Equation

2   Solve Linear Equations in One Variable

3   Identify Equations That Are Contradictions and Those
    That Are Identities

4   Use Linear Equations to Solve Application Problems

OBJECTIVE 1:   DETERMINE IF A GIVEN VALUE IS A SOLUTION TO AN EQUATION

My interactive video summary

An **equation** is a statement that two quantities are equal. When an equation contains one or more variables, it is called an **algebraic equation**. An algebraic equation indicates that two algebraic expressions are equal. Algebraic equations contain an equal sign ($=$), while algebraic expressions do not. For example, $3x + 2 = x + 10$ is an algebraic *equation*. It is made up of the two algebraic *expressions* $3x + 2$ and $x + 10$. Work through this interactive video to distinguish between expressions and equations.

An equation is called an **equation in one variable** if it contains a single variable. Each of the following three equations is an example of an equation in one variable:

$$4a - 1 = 3, \quad x^2 + 4x - 5 = 0, \quad \text{and} \quad \frac{1}{3}y - \frac{3}{4} = \frac{1}{2}y$$

Even if the same variable appears multiple times, the equation is still considered to be "in one variable."

We will consider equations that contain more than one variable in Section 1.4, but for now we focus only on equations in one variable, specifically *linear* equations in one variable.

Definition   **Linear Equation in One Variable**

A **linear equation in one variable** is an equation that can be written in the form $ax + b = c$, where a, b, and c are real numbers and $a \neq 0$.

Linear equations are also **first-degree equations** because the exponent on the variable is understood to be 1. If an equation contains a variable raised to an exponent other than 1, then the equation is *nonlinear*. If an equation contains a variable under a radical sign or in the denominator of a fraction, then the equation is nonlinear also.

My interactive video summary

Watch this interactive video to learn how to distinguish between linear and nonlinear equations, or click here to see an example of how to identify linear and nonlinear equations.

When working with an equation in one variable, our usual goal is to solve it. We want to find all of its solutions or all values that, when substituted for the variable, make the equation true.

Example 1 Determining If a Given Value Is a Solution to an Equation

Determine if the given value is a solution to the equation.

a. $2x + 3 = 11$; $x = 4$ **b.** $3y + 8 = 5y - 4$; $y = 2$

c. $\dfrac{2}{3}w - \dfrac{1}{2} = \dfrac{1}{4}$; $w = \dfrac{3}{8}$

Solution We substitute the given value for the variable and simplify. If the resulting statement is true, then the value is a solution to the equation.

a. Begin with the original equation: $\quad 2x + 3 = 11$

Substitute 4 for x: $\quad 2(4) + 3 \stackrel{?}{=} 11$

Simplify: $\quad 8 + 3 \stackrel{?}{=} 11$

$11 = 11 \quad$ True

The final statement is true, so $x = 4$ is a solution to the equation.

b. Begin with the original equation: $\quad 3y + 8 = 5y - 4$

Substitute 2 for y: $\quad 3(2) + 8 \stackrel{?}{=} 5(2) - 4$

Simplify: $\quad 6 + 8 \stackrel{?}{=} 10 - 4$

$14 = 6 \qquad$ False

The final statement is false, so $y = 2$ is not a solution to the equation.

 c. Begin with the original equation: $\quad \dfrac{2}{3}w - \dfrac{1}{2} = \dfrac{1}{4}$

Substitute $\dfrac{3}{8}$ for w: $\quad \dfrac{2}{3}\left(\dfrac{3}{8}\right) - \dfrac{1}{2} \stackrel{?}{=} \dfrac{1}{4}$

Finish the simplification. Then click here or watch this video to check your work.

You Try It Work through this You Try It problem.

Work Exercises 1–6 in this textbook or in the *MyMathLab*® Study Plan.

OBJECTIVE 2: SOLVE LINEAR EQUATIONS IN ONE VARIABLE

We now introduce some common terms to help us solve linear equations in one variable.

Once all solutions to an equation are found, we list them in a solution set. Two or more equations with the same solution set are called **equivalent equations**. When solving a linear equation in one variable, we look for simpler equivalent equations until we find one that ends with an isolated variable of the form

$$variable = value.$$

To find simpler equivalent equations, we use the properties of equality to add, subtract, multiply, and/or divide both sides of an equation by the same quantity without changing its solution set.

 CAUTION Multiplying or dividing both sides of an equation by zero is not allowed. Do you see why?

My interactive video summary

For a full explanation of the properties of equality, watch this interactive video.

Example 2 Solving Linear Equations in One Variable

Use the properties of equality to solve each equation.

a. $3x - 1 = 5$ b. $8 = \dfrac{1}{2}n + 3$

Solution

a. Begin with the original equation: $3x - 1 = 5$

   Add 1 to both sides: $3x - 1 + 1 = 5 + 1$

   Simplify: $3x = 6$

   Divide both sides by 3: $\dfrac{3x}{3} = \dfrac{6}{3}$

   Simplify: $x = 2$

The solution set is $\{2\}$ as long as we have made no errors. We check this result to make sure it satisfies the original equation.

Check   Begin with the original equation: $3x - 1 = 5$

   Substitute 2 for x: $3(2) - 1 \overset{?}{=} 5$

   Simplify: $6 - 1 \overset{?}{=} 5$

           $5 = 5$ True

Since $x = 2$ satisfies the original equation, the solution set is $\{2\}$.

b. Begin with the original equation: $8 = \dfrac{1}{2}n + 3$

   Subtract 3 from both sides: $8 - 3 = \dfrac{1}{2}n + 3 - 3$

   Simplify: $5 = \dfrac{1}{2}n$

   Multiply both sides by 2: $2(5) = 2\left(\dfrac{1}{2}n\right)$

   Simplify: $10 = n$

Check   Confirm that the solution set is $\{10\}$ by substituting $n = 10$ into the original equation. When finished, click here to check your work.

You Try It   **Work through this You Try It problem.**

Work Exercises 7–12 in this textbook or in the **MyMathLab**® Study Plan.

Some equations will have algebraic expressions on both sides of the equal sign. When this occurs, we use the properties of equality to move all variables to one side.

Example 3 Solving Linear Equations with Variables on Both Sides

Solve: $6x - 5 = 2x - 3$

Solution

Begin with the original equation:	$6x - 5 = 2x - 3$
Subtract $2x$ from both sides:	$6x - 5 - 2x = 2x - 3 - 2x$
Simplify:	$4x - 5 = -3$
Add 5 to both sides:	$4x - 5 + 5 = -3 + 5$
Simplify:	$4x = 2$
Divide both sides by 4:	$\dfrac{4x}{4} = \dfrac{2}{4}$
Simplify:	$x = \dfrac{1}{2}$

Check Confirm that the solution set is $\left\{\dfrac{1}{2}\right\}$ by substituting $x = \dfrac{1}{2}$ into the original equation. When finished, click here to check your work.

You Try It **Work through this You Try It problem.**

Work Exercises 13–16 in this textbook or in the *MyMathLab*® Study Plan.

Sometimes equations contain non-simplified expressions. When solving such equations, it is usually best to simplify each side first before using the properties of equality to isolate the variable.

Example 4 Solving Equations Containing Non-Simplified Expressions

Solve: $5(x - 6) - 2x = 3 - (x + 1)$

Solution

Begin with the original equation:	$5(x - 6) - 2x = 3 - (x + 1)$
Use the distributive property:	$5x - 30 - 2x = 3 - x - 1$
Combine like terms:	$3x - 30 = 2 - x$

Now the equation is simplified and looks like the equation in Example 3. Finish solving this equation. Click here to check your answer, or watch this video for a fully worked solution.

Note: Though it is usually best to simplify each side of an equation before using the properties of equality to isolate the variable, it is not always necessary. Watch this video for an explanation.

You Try It Work through this You Try It problem.

Work Exercises 17–20 in this textbook or in the _MyMathLab_® Study Plan.

When an equation contains fractions, it is usually best to remove the fractions first. To do this, we multiply both sides of the equation by an appropriate common multiple of all the denominators, usually the least common denominator (LCD) of all the fractions.

Example 5 Solving Linear Equations Containing Fractions

My video summary

Solve: $\dfrac{x}{3} - \dfrac{5}{12} = \dfrac{5}{6}x - \dfrac{11}{12}$

Solution This equation contains fractions with the denominators 3, 6, and 12, so the LCD is 12. We multiply both sides of the equation by 12 to clear the fractions.

Begin with the original equation:
$$\frac{x}{3} - \frac{5}{12} = \frac{5}{6}x - \frac{11}{12}$$

Multiply both sides by 12:
$$12\left(\frac{x}{3} - \frac{5}{12}\right) = 12\left(\frac{5}{6}x - \frac{11}{12}\right)$$

Use the distributive property:
$$12\left(\frac{x}{3}\right) - 12\left(\frac{5}{12}\right) = 12\left(\frac{5}{6}x\right) - 12\left(\frac{11}{12}\right)$$

Clear the fractions:
$$4x - 5 = 10x - 11$$

Now the equation looks like the equation in Example 3. Finish solving this equation. Click here to check your answer, or watch this video for a fully worked solution.

Example 6 Solving Linear Equations Containing Fractions

My video summary

Solve: $\dfrac{1}{3}(1 - x) - \dfrac{x + 1}{2} = -2$

Solution Try solving the equation on your own. Click here to check your answer, or watch this video for a fully worked solution.

Note: When solving an equation that contains fractions, it is usually best to clear the fractions first. However, clearing the fractions is not always necessary. Click here for an explanation.

You Try It Work through this You Try It problem.

Work Exercises 25–28 in this textbook or in the _MyMathLab_® Study Plan.

When an equation contains decimals, we remove the decimals by multiplying both sides of the equation by an appropriate power of 10, such as $10^1 = 10$, $10^2 = 100$, and $10^3 = 1000$. We can usually determine the appropriate power of 10 by looking at the constants in the equation and choosing the constant with the greatest number of decimal places. We count those decimal places and then raise 10 to that power. Multiplying the equation by that power of 10 will usually clear all the decimals.

Example 7 Solving Linear Equations Containing Decimals

 Solve: $0.5n - 0.25 + 0.075n = 0.5 - 0.025n$

Solution The coefficients 0.075 and -0.025 each have three decimal places. This is the greatest number of decimal places. We can clear the decimals by multiplying both sides of the equation by $10^3 = 1000$.

Begin with the original equation: $0.5n - 0.25 + 0.075n = 0.5 - 0.025n$

Multiply both sides by 1000: $1000(0.5n - 0.25 + 0.075n) = 1000(0.5 - 0.025n)$

Distribute to clear the decimals: $500n - 250 + 75n = 500 - 25n$

Combine like terms: $575n - 250 = 500 - 25n$

Finish solving this equation. Click here to check your answer, or watch this video for a fully worked solution.

Example 8 Solving Linear Equations Containing Decimals

Solve: $0.1(y - 2) + 0.03(y - 4) = 0.02(10)$

Solution Try solving the equation on your own. Click here to check your answer, or watch this video for a fully worked solution.

You Try It Work through this You Try It problem.

Work Exercises 29–32 in this textbook or in the **MyMathLab**® Study Plan.

In the previous examples, we used the following guidelines for solving linear equations in one variable. If a particular step does not apply to a given equation, simply skip it.

Guidelines for Solving Linear Equations in One Variable

Step 1. Clear all fractions from the equation by multiplying both sides by the LCD. Clear all decimals by multiplying both sides by the appropriate power of 10.

Step 2. Remove grouping symbols using the distributive property.

Step 3. Simplify each side of the equation by combining like terms.

Step 4. Use the Addition Property of Equality to move all variable terms to one side of the equation and all constant terms to the other side.

Step 5. Use the Multiplication Property of Equality to isolate the variable.

Step 6. Confirm that the result satisfies the original equation.

If an equation contains fractions or decimals within grouping symbols, multiplying by the LCD or power of 10 may not clear the fractions or decimals within the grouping symbols. In such a situation, remove all grouping symbols and multiply both sides by the LCD for all remaining fractions or the appropriate power of 10 for all remaining decimals.

OBJECTIVE 3:   IDENTIFY EQUATIONS THAT ARE CONTRADICTIONS AND THOSE THAT ARE IDENTITIES

A **conditional equation** is an equation that is true for some values of the variable but not for others. So far, the equations we have solved have been conditional equations with one solution. However, not every linear equation in one variable has a single solution. There are two other possible cases: no solution and the set of all real numbers.

Consider the equation $x = x + 1$. No matter what value is substituted for x, the resulting value on the right side will always be one greater than the value on the left side. Therefore, the equation can never be true. We call such an equation a **contradiction**. It has *no solution*. Its solution set is the *empty* or *null set*, denoted by $\{\ \}$ or $\varnothing$, respectively.

Now consider the equation $x + x = 2x$. The expression on the left side of the equation simplifies to the expression on the right side. No matter what value we substitute for x, the resulting values on both the left and right sides will always be the same. Therefore, the equation is always true. We call such an equation an **identity**. Its solution set is the set of all real numbers, denoted by $\mathbb{R}$ or $\{x \mid x \text{ is a real number}\}$.

Example 9 shows how to identify contradictions and identities when solving equations.

Example 9 Identifying Contradictions and Identities

Use the properties of equality to determine if the equation is a contradiction or an identity.

a. $5(x + 4) = 3(x + 7) + 2x$ **b.** $3(x - 4) = x + 2(x - 6)$

Solution

a. Begin with the original equation: $5(x + 4) = 3(x + 7) + 2x$

Use the distributive property: $5x + 20 = 3x + 21 + 2x$

Combine like terms: $5x + 20 = 5x + 21$

Subtract $5x$ from both sides: $5x + 20 - 5x = 5x + 21 - 5x$

Simplify: $20 = 21$ False

All the variable terms subtract out, leaving a false statement. Therefore, the equation is a contradiction and has no solution. Its solution set is $\{\ \}$ or $\varnothing$.

b. Begin with the original equation: $3(x - 4) = x + 2(x - 6)$

Use the distributive property: $3x - 12 = x + 2x - 12$

Combine like terms: $3x - 12 = 3x - 12$

The left and right sides of the equation are identical. Since any value of x will make the equation true, the equation is an identity. Its solution set is $\mathbb{R}$ or $\{x \mid x \text{ is a real number}\}$.

If we did not notice that the left and right sides of the equation were equivalent and continued "solving," we would obtain the following:

Subtract $3x$ from both sides: $\quad 3x - 12 - 3x = 3x - 12 - 3x$

Simplify: $\qquad\qquad\qquad -12 = -12 \quad$ True

All the variable terms subtract out, leaving a true statement.

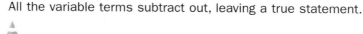

You Try It **Work through this You Try It problem.**

Work Exercises 21–24 in this textbook or in the *MyMathLab*® Study Plan.

OBJECTIVE 4: USE LINEAR EQUATIONS TO SOLVE APPLICATION PROBLEMS

When solving application problems, we use mathematical models. The following strategy can be used for solving application problems involving linear equations.

Problem-Solving Strategy for Applications of Linear Equations

Step 1. Define the Problem. Read the problem carefully, or multiple times if necessary. Identify what you are trying to find and determine what information is available to help you find it.

Step 2. Assign Variables. Choose a variable to assign to an unknown quantity in the problem. For example, use p for price. If other unknown quantities exist, express them in terms of the selected variable.

Step 3. Translate into an Equation. Use the relationships among the known and unknown quantities to form an equation.

Step 4. Solve the Equation. Determine the value of the variable and use the result to find any other unknown quantities in the problem.

Step 5. Check the Reasonableness of Your Answer. Check to see if your answer makes sense within the context of the problem. If not, check your work for errors and try again.

Step 6. Answer the Question. Write a clear statement that answers the question(s) posed.

Example 10 NFL Touchdown Passes

Roger Staubach and Terry Bradshaw were both NFL quarterbacks. In 1973, Staubach threw three touchdown passes more than twice the number of touchdown passes thrown by Bradshaw. If the total number of touchdown passes thrown between Staubach and Bradshaw was 33, then how many touchdown passes did each player throw?

Solution We follow the problem-solving strategy for applications of linear equations.

Step 1. Define the Problem. We need to find the number of touchdown passes thrown by each quarterback. Staubach threw three more than twice the number of touchdown passes thrown by Bradshaw. Together, the two quarterbacks threw a total of 33 touchdown passes.

Step 2. Assign Variables.   We choose the variable B to represent the unknown number of touchdown passes thrown by Bradshaw. The number of touchdown passes thrown by Staubach is $2B + 3$ (three more than twice the number thrown by Bradshaw).

Step 3. Translate into an Equation.   The total number of touchdown passes thrown is 33. So, we have the following equation:

$$\underbrace{B}_{\text{Bradshaw's TD passes}} + \underbrace{(2B + 3)}_{\text{Staubach's TD passes}} = \underbrace{33}_{\text{Total}}$$

Step 4. Solve the Equation.

$$B + (2B + 3) = 33$$

Combine like terms: $3B + 3 = 33$

Subtract 3 from both sides: $3B + 3 - 3 = 33 - 3$

Simplify: $3B = 30$

Divide both sides by 3: $\dfrac{3B}{3} = \dfrac{30}{3}$

Simplify: $B = 10$

If $B = 10$, then $2B + 3 = 2(10) + 3 = 20 + 3 = 23$.

Step 5. Check the Reasonableness of Your Answer.   Bradshaw threw 10 touchdown passes and Staubach threw 23 touchdown passes. Since 23 is three more than twice 10, and the sum of 10 and 23 is 33, these results make sense within the context of the problem.

Step 6. Answer the Question.   In 1973, Bradshaw threw 10 touchdown passes and Staubach threw 23 touchdown passes.

You Try It **Work through this You Try It problem.**

Work Exercises 33–34 in this textbook or in the *MyMathLab*® Study Plan.

Example 11 Fuel Stop

My video summary

  Before filling up her car, Kenya sees the price of gasoline as \$3.26 per gallon. When she finished filling the tank, Kenya owed \$40.75. How many gallons of gas did she purchase?

Solution

Step 1. Define the Problem.   We need to find the amount of gas purchased. We know the price per gallon (\$3.26) and the total amount spent (\$40.75).

Step 2. Assign Variables.   We choose the variable g to represent the number of gallons of gas. This is the only unknown for the problem.

Step 3. Translate into an Equation.   We obtain the total cost by multiplying the price per gallon times the number of gallons. So, we have the following equation:

$$\underbrace{3.26}_{\text{price per gallon}} \cdot \underbrace{g}_{\text{number of gallons}} = \underbrace{40.75}_{\text{total cost}}$$

Step 4. **Solve the Equation.** Try solving the equation and finishing the problem on your own. Click here to check your answer, or watch this video for a fully worked solution.

You Try It Work through this You Try It problem.

Work Exercises 35–36 in this textbook or in the *MyMathLab*® Study Plan.

Example 12 Original Price

 A laptop computer is on sale for 15% off the original price. If the sale price is $552.33, find the original price of the laptop.

Solution Work through the problem-solving strategy to solve this problem on your own. Click here to check your answer, or watch this video for a fully worked solution.

You Try It Work through this You Try It problem.

Work Exercises 37–38 in this textbook or in the *MyMathLab*® Study Plan.

1.1 Exercises

In Exercises 1–6, determine if the given value is a solution to the equation.

You Try It

1. $3x - 8 = -23;\quad x = -5$

2. $8y - 7 = 3(y - 1);\quad y = -2$

3. $z^2 - 7 = z + 5;\quad z = 4$

4. $5(t + 1) - t = 6t - 7;\quad t = 6$

5. $\dfrac{3}{5}n + \dfrac{17}{10} = \dfrac{3}{10} - \dfrac{1}{3}n;\quad n = -\dfrac{3}{2}$

6. $0.12m + 0.25 = 0.07m;\quad m = 5$

In Exercises 7–32, solve each equation and check your answer. Identify all identities and contradictions.

You Try It

7. $3x + 32 = 71$

8. $-17 - 5w = -52$

9. $25 = 4a - 7$

10. $6y + 5 = 7$

11. $15 = 12 - \dfrac{1}{3}m$

12. $\dfrac{1}{4}z + 5 = 2$

13. $9x + 7 = 3x - 11$

14. $5t - 3 = 7t + 11$

15. $11 - t = t - 9$

16. $3 - 8y = 12 - 2y$

17. $9 + 2x + 8 = 2x + 12 - 5x$

18. $6(a + 5) = 5(a + 6) - 4$

19. $3x - 2(x - 1) = 5x + 20 - 3x$

20. $4(x + 1) - 7x = -(x - 7) + (2x + 3)$

21. $2(3x - 5) = 6(x - 2)$

22. $5m + 2 - 7m = -2(m - 1)$

23. $8(m - 6) - 3m = 5(m - 9) - 3$

24. $7n + 1 - 9n = 3(n + 2) - 5n$

25. $\dfrac{n}{5} = \dfrac{n}{2} + \dfrac{3}{10}$

26. $\dfrac{x}{3} - \dfrac{5}{4} = \dfrac{x}{6} - \dfrac{1}{12}$

27. $\frac{1}{2}y - \frac{1}{3}(y - 1) = 5y$

28. $\frac{w - 3}{4} + \frac{w + 5}{7} = \frac{5}{14}$

 29. $0.3z + 5.4 = 0.1z + 7$

30. $0.025p - 1 = 0.15(p - 5)$

 31. $0.12x + 0.3(x - 4) = 0.01(2x - 3)$

32. $0.001(1 - k) + 0.005(k - 3) = 0.5$

In Exercises 33–38, solve each application problem and check your answer.

You Try It
33. Making Investments Sabrina invested $20,000 in stocks and bonds. Her investment in bonds is $2000 more than half her investment in stocks. How much did Sabrina invest in stocks? How much did she invest in bonds?

34. Grand Slam Titles During her tennis career, Martina Navratilova won a total of 59 Grand Slam titles in three categories: women's singles, women's doubles, and mixed doubles. Out of this total, her number of wins in women's singles is eight more than her number of wins in mixed doubles. Her number of wins in women's doubles is one more than three times her number of wins in mixed doubles. How many Grand Slam titles did Martina win in each category?

You Try It
35. Parking Fees It costs $8.00 to park in a parking garage for the first two hours. Then it costs $3.50 per hour for every hour afterwards. If Anna's parking fee is $22.00, how long has her car been parked?

 36. Using a Coupon Judy purchased apples from a grocery store for $1.75 per pound. After using a coupon for $0.35 off the total price, her cost was $4.55. How many pounds of apples did Judy buy?

You Try It
37. Buying a Jacket Latoya purchased a leather jacket on sale for 20% off the original price. If the sale price of the jacket was $127.96, what was its original price?

 38. Bookstore Markup A college bookstore marks up the price of textbooks by 35%. This results in the store selling textbooks for 35% more than the original cost to the store. If the bookstore charges $102.33 for your intermediate algebra textbook, what is the book's original cost to the store?

1.2 Linear Inequalities in One Variable

THINGS TO KNOW

Before working through this section, be sure you are familiar with the following concepts:

	VIDEO	ANIMATION	INTERACTIVE

You Try It
1. Identify Sets (Section R.1, Objective 1)

You Try It
2. Use Order of Operations to Evaluate Numeric Expressions (Section R.2, Objective 4)

You Try It
3. Evaluate Algebraic Expressions (Section R.3, Objective 1)

You Try It
4. Simplify Algebraic Expressions (Section R.3, Objective 2)

OBJECTIVES

1 Determine If a Given Value Is a Solution to an Inequality

2 Graph the Solution Set of an Inequality on a Number Line

3 Use Interval Notation to Express the Solution Set of an Inequality

4 Solve Linear Inequalities in One Variable

5 Use Linear Inequalities to Solve Application Problems

OBJECTIVE 1:   DETERMINE IF A GIVEN VALUE IS A SOLUTION TO AN INEQUALITY

While the equal sign is used in equations to indicate when two quantities are equal, inequality symbols are used in **inequalities** to indicate when two quantities are unequal. We saw examples of numeric inequalities in Section R.1 when we ordered real numbers. Examples of inequalities involving algebraic expressions are $x < 5$, $2x - 5 \geq 8$, $x^2 \leq 3x + 5$, and $|x + 6| - 9 > 3$. The only difference between equations and inequalities is that the equal sign in equations is replaced by an inequality symbol in inequalities.

My interactive video summary

Watch this interactive video to work on learning how to distinguish between equations and inequalities.

As with equations, we can determine if a given value of a variable is a **solution** to an inequality by substituting the value for the variable and checking the resulting statement. If a false statement results, then the value is not a solution to the inequality. If a true statement results, then the value is a solution to the inequality.

Example 1 Determining If a Given Value Is a Solution to an Inequality

My interactive video summary

Determine if the given value is a solution to the inequality.

a. $3x + 4 < 8$; $x = 2$ **b.** $n^2 + 5n \geq 4$; $n = -6$

Solution

a. This inequality is called a strict inequality because the left side is "strictly" less than the right side. We substitute 2 for x in the inequality. Since the resulting statement, $10 < 8$, is false, $x = 2$ is not a solution to the inequality.

b. This inequality is called a non-strict inequality because the left side is greater than *or* equal to the right side. There is a possibility of equality for some values of n. We substitute -6 for n in the inequality. Since the resulting statement, $6 \geq 4$, is true, $n = -6$ is a solution to the inequality.

To see all the steps for these solutions, watch this interactive video.

You Try It   Work through this You Try It problem.

Work Exercises 1–6 in this textbook or in the *MyMathLab*® Study Plan.

OBJECTIVE 2: GRAPH THE SOLUTION SET OF AN INEQUALITY ON A NUMBER LINE

Although most equations have a finite number of solutions, most *inequalities* have an infinite number of solutions. For example, the inequality $x < 4$ has infinitely many values of x that are less than 4. The collection of all solutions to an inequality forms the **solution set** of the inequality. Click here for an explanation of the difference between "finite" and "infinite."

When a solution set contains an infinite number of values, we cannot list each solution, but we can use set-builder notation to express the solution set. Also, we can graph the solution set on a number line. When graphing a solution set, we use an open circle ($\circ$) to indicate that a value is not included in the solution set and a closed circle ($\bullet$) to indicate that a value is included in the solution set. For example, the graph of $\{x \mid x < 4\}$ is shown as follows:

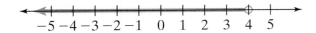

Example 2 Graphing the Solution Set of an Inequality on a Number Line

Graph each solution set on a number line.

a. $\{x \mid x \geq 0\}$ b. $\{x \mid 1 < x \leq 7\}$ c. $\{x \mid x < 3\}$

d. $\{x \mid 0 < x < 4\}$ e. $\{x \mid x \neq -2\}$ f. $\{x \mid -1 \leq x \leq 5\}$

g. $\{x \mid -3 \leq x < 2\}$ h. $\{x \mid x \text{ is any real number}\}$

Solution

a. The solution set $\{x \mid x \geq 0\}$ reads as "the set of all values for x such that x is greater than or equal to 0." Because the inequality is non-strict, we place a closed circle at 0 to show that 0 is a solution. Then we shade the number line to the right, the direction that indicates "greater than," to show that all values greater than 0 are also solutions.

b. The solution set $\{x \mid 1 < x \leq 7\}$ reads as "the set of all values for x such that x is greater than 1 and less than or equal to 7." The inequality on the left is strict, so we place an open circle at 1 to show that 1 is not a solution. Because the inequality on the right is non-strict, we place a closed circle at 7 to show that 7 is a solution and then shade the number line between the two circles to indicate that all values between 1 and 7 are also solutions.

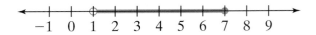

Click here to see the remaining solution sets and their corresponding graphs.

You Try It   **Work through this You Try It problem.**

Work Exercises 7–10 in this textbook or in the *MyMathLab*® Study Plan.

OBJECTIVE 3: USE INTERVAL NOTATION TO EXPRESS THE SOLUTION SET OF AN INEQUALITY

We can express the solution set of an inequality using **interval notation**. First, each solution set has a **lower bound** and an **upper bound**, separated by a comma, which make up the **endpoints** of the interval. An interval will have the form

$$\text{lower bound, upper bound.}$$

Then we indicate whether or not these endpoints are included in the interval. A parenthesis—"(" for the lower bound or ")" for the upper bound—shows that the endpoint is not included in the solution set. This is similar to using an open circle when graphing on a number line. A square bracket—"[" for the lower bound or "]" for the upper bound—shows that the endpoint is included in the solution set. This is similar to using a closed circle when graphing on a number line.

For example, the solution set in Example 2(b), $\{x\,|\,1 < x \le 7\}$, is written as $(1, 7]$ in interval notation. We use a parenthesis on 1 because it is not included in the solution set, and we use a square bracket on 7 because it is included in the solution set.

Table 1 summarizes the three ways of expressing intervals used in this text: set-builder notation, number line graph, and interval notation. Typically we use interval notation and graphs when expressing solution sets for inequalities.

CAUTION

A parenthesis is always used for $-\infty$ and ∞ because these symbols are not numbers. Instead, they indicate that the interval is unbounded in a particular direction.

My interactive video summary

 Watch this interactive video to help you determine the lower and upper bounds of an interval or solution set.

Table 1

Graph	Interval Notation	Set-Builder Notation
open circle at a, open circle at b	(a, b)	$\{x\,|\,a < x < b\}$
closed circle at a, closed circle at b	$[a, b]$	$\{x\,|\,a \le x \le b\}$
open circle at a, closed circle at b	$(a, b]$	$\{x\,|\,a < x \le b\}$
closed circle at a, open circle at b	$[a, b)$	$\{x\,|\,a \le x < b\}$
open circle at a, arrow right	(a, ∞)	$\{x\,|\,x > a\}$
open circle at b, arrow left	$(-\infty, b)$	$\{x\,|\,x < b\}$
closed circle at a, arrow right	$[a, \infty)$	$\{x\,|\,x \ge a\}$
closed circle at b, arrow left	$(-\infty, b]$	$\{x\,|\,x \le b\}$

Example 3 Using Interval Notation to Express the Solution Set of an Inequality

Write each solution set using interval notation.

a. $\{x \mid x < 5\}$　　　**b.** $\{x \mid 2 \leq x < 10\}$　**c.** $\{x \mid x \geq -3\}$

d. $\{x \mid -6 < x < 0\}$　**e.** $\{x \mid -1 \leq x \leq 5\}$　**f.** $\{x \mid x \text{ is any real number}\}$

Solution

a. The solution set $\{x \mid x < 5\}$ has no lower bound and an upper bound of 5. The interval notation is $(-\infty, 5)$. We use a parenthesis on $-\infty$ to show that there is no lower bound. We use a parenthesis on 5, the upper bound, to show that 5 is not included in the solution set.

b. The solution set $\{x \mid 2 \leq x < 10\}$ has a lower bound of 2 and an upper bound of 10. The interval notation is $[2, 10)$. We use a square bracket on 2, the lower bound, to show that 2 is included in the solution set and a parenthesis on the upper bound, 10, to show that 10 is not included in the solution set.

Click here to see the remaining solution sets and their corresponding interval notation.

You Try It　**Work through this You Try It problem.**

Work Exercises 11–14 in this textbook or in the *MyMathLab*® Study Plan.

OBJECTIVE 4:　SOLVE LINEAR INEQUALITIES IN ONE VARIABLE

The inequality $4x - 7 \geq 3$ is called a **linear inequality in one variable**.

Definition

A **linear inequality in one variable** is an inequality that can be written in the form $ax + b < c$, where a, b, and c are real numbers and $a \neq 0$.

Note: The inequality symbol $<$ can be replaced with $>$, $\leq$, $\geq$, or $\neq$.

Why do we require $a \neq 0$ in our definition? Click here for an explanation.

My interactive video summary

Watch this interactive video to learn how to identify linear and nonlinear inequalities, or click here to see an example of how to recognize linear and nonlinear inequalities.

Solving linear inequalities is similar to solving linear equations. We use properties of inequalities to isolate the variable on either side of the inequality symbol.

Properties of Inequalities:

Let a, b, and c be real numbers.

1. Addition Property of Inequality: Adding or subtracting the same quantity from both sides of an inequality results in an equivalent inequality.

$$\text{If } a < b, \text{ then } a + c < b + c \text{ and } a - c < b - c.$$

2. Multiplication Property of Inequality: Multiplying or dividing both sides of an inequality by a positive number results in an equivalent inequality.

$$\text{If } a < b \text{ and } c > 0, \text{ then } ac < bc \text{ and } \frac{a}{c} < \frac{b}{c}.$$

Multiplying or dividing both sides of an inequality by a negative number, and switching the direction of the inequality, results in an equivalent inequality.

$$\text{If } a < b \text{ and } c < 0, \text{ then } ac > bc \text{ and } \frac{a}{c} > \frac{b}{c}.$$

Note: The inequality symbol $<$ can be replaced with $>$, $\leq$, $\geq$, or $\neq$.

Click here to view some examples on how to use the inequality properties.

We handle inequalities involving non-simplified expressions, fractions, or decimals in the same way as equations. We just need to be careful when multiplying or dividing both sides of an inequality by a negative number. You may want to review the guidelines for solving linear equations in one variable before using a similar process when solving linear inequalities in one variable.

Example 4 Solving a Linear Inequality in One Variable

Solve the inequality $4x - 8 \geq 3(x + 2) + 3x$. Graph the solution set on a number line and write the solution set in interval notation.

Solution Our approach is similar to the one used when solving linear equations. We want to isolate the variable on one side of the inequality symbol.

Begin with the original inequality:	$4x - 8 \geq 3(x + 2) + 3x$
Use the distributive property:	$4x - 8 \geq 3x + 6 + 3x$
Simplify:	$4x - 8 \geq 6x + 6$
Subtract $6x$ from both sides:	$4x - 6x - 8 \geq 6x - 6x + 6$
Simplify:	$-2x - 8 \geq 6$
Add 8 to both sides:	$-2x - 8 + 8 \geq 6 + 8$
Simplify:	$-2x \geq 14$
Divide both sides by -2, and switch the direction of the inequality:	$\dfrac{-2x}{-2} \leq \dfrac{14}{-2}$
Simplify:	$x \leq -7$

The graph of the solution set appears on the following number line:

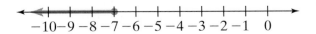

In interval notation, the solution set is $(-\infty, -7]$. Remember to switch the direction of the inequality if you multiply or divide both sides of an inequality by a negative number. Click here to see why this is important.

Check Since this solution set contains an infinite number of values, we cannot check every solution. Therefore, we pick one test value from the proposed solution set as a "check." For example, let's choose $x = -10$ as our test value and substitute it for x in the original inequality to see if a true statement results. Since the resulting statement, $-48 \geq -54$, is true, $x = -10$ is a solution to the inequality. Click here for the steps in this process.

 Checking one solution, or even several solutions, does not guarantee that the solution set of the inequality is correct. It only determines whether or not the tested values are solutions. However, it is still a good idea to check one or two test values so that we are comfortable with the solution set.

You Try It Work through this You Try It problem.

Work Exercises 15–18 in this textbook or in the *MyMathLab*® Study Plan.

Example 5 Solving a Linear Inequality

 Solve the inequality $2 - 5(x - 2) < 4(3 - 2x) + 7$. Write the solution set in set-builder notation.

Solution Click here to check your answer, or watch this video to see the full solution worked out.

You Try It Work through this You Try It problem.

Work Exercises 19–24 in this textbook or in the *MyMathLab*® Study Plan.

Example 6 Solving a Linear Inequality Containing Fractions

 Solve the inequality $\dfrac{m}{2} - 5 + 2m > -\dfrac{m}{4} + \dfrac{1}{2}$ and write the solution set in interval notation.

Solution Click here to check your answer, or watch this video to see the detailed solution.

You Try It Work through this You Try It problem.

Work Exercises 25–26 in this textbook or in the *MyMathLab*® Study Plan.

When solving inequalities involving fractions or decimals, some students like to clear these right away. This is not required. The same solution set will result whether or not fractions or decimals are cleared first.

As with equations, a linear inequality can have no solution or all real numbers as its solution. Remember that a contradiction has no solution, and an identity has the set of all real numbers as its solution.

Example 7 Solving Special Cases of Linear Inequalities

Solve the following inequalities. Write each solution set in interval notation.

a. $3 + 4(x - 5) \leq 7x - 3(x + 8)$

b. $2(3 - x) - 7 > 4(x - 1) - 6x$

Solution

a. We use the properties of inequalities to isolate the variable on one side of the inequality.

Begin with the original inequality:	$3 + 4(x - 5) \leq 7x - 3(x + 8)$
Use the distributive property:	$3 + 4x - 20 \leq 7x - 3x - 24$
Simplify:	$4x - 17 \leq 4x - 24$
Subtract $4x$ from both sides:	$4x - 4x - 17 \leq 4x - 4x - 24$
Simplify:	$-17 \leq -24$ False

Since the final statement is a contradiction, the inequality has no solution. The solution set is the empty set { } or null set, $\varnothing$.

My video summary

b. Watch the video to see the full solution process and confirm that the result is an identity.

You Try It   **Work through this You Try It problem.**

Work Exercises 27–28 in this textbook or in the *MyMathLab*® Study Plan.

An inequality statement can involve more than two parts. For example, a Google product search in February 2009 showed the range in price of a Nintendo Wii Super Accessory Bundle Pack from $193 to $465. Letting W be the price of a Wii Bundle Pack, we can use the three-part inequality $193 \leq W \leq 465$ to show this relationship.

To solve a three-part inequality, we use the properties of inequalities to isolate the variable between the two inequality symbols. Remember that *what we do to one part, we must do to all three parts* in order to obtain an equivalent inequality.

Example 8 Solving a Three-Part Inequality

Solve the inequality $-2 < \dfrac{3x - 5}{4} \leq 3$. Graph the solution set on a number line; write this solution set in interval notation.

Solution We use the properties of inequalities to isolate the variable between the inequality symbols.

Begin with the original inequality:	$-2 < \dfrac{3x - 5}{4} \leq 3$
Multiply all parts by 4 to clear the fraction:	$4(-2) < 4 \cdot \dfrac{3x - 5}{4} \leq 4(3)$

$$\text{Simplify:} \qquad -8 < 3x - 5 \le 12$$

$$\text{Add 5 to all parts:} \quad -8 + 5 < 3x - 5 + 5 \le 12 + 5$$

$$\text{Simplify:} \qquad -3 < 3x \le 17$$

$$\text{Divide all parts by 3 to isolate the variable:} \qquad \frac{-3}{3} < \frac{3x}{3} \le \frac{17}{3}$$

$$\text{Simplify:} \qquad -1 < x \le \frac{17}{3}$$

The graph of this solution set is shown as follows:

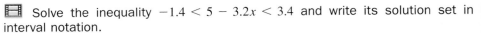

The solution set in interval notation is $\left(-1, \frac{17}{3}\right]$.

CAUTION When solving three-part inequalities, remember that what you do to one part of the inequality must be done to **all three parts** in order to obtain an equivalent inequality.

Example 9 Solving a Three-Part Inequality

 Solve the inequality $-1.4 < 5 - 3.2x < 3.4$ and write its solution set in interval notation.

Solution First, we multiply all parts by 10 to clear the decimals and obtain the equivalent inequality $-14 < 50 - 32x < 34$. Solving this inequality results in the solution set $\left(\frac{1}{2}, 2\right)$ in interval notation. Watch this video to see the solution worked out in detail.

You Try It **Work through this You Try It problem.**

Work Exercises 29–30 in this textbook or in the *MyMathLab*® Study Plan.

**CAUTION** When writing a solution set in set-builder notation or in interval notation, we want the values to increase as we read from left to right.

It is important to distinguish the different ways in which solution sets can be expressed, particularly the difference between set-builder notation and interval notation. Click here for an explanation on how to recognize these.

At this point, you may wish to review the strategy for solving application problems with linear equations. We present a variation of this strategy here using mathematical models for linear inequalities.

OBJECTIVE 5: USE LINEAR INEQUALITIES TO SOLVE APPLICATION PROBLEMS

Strategy for Solving Application Problems Involving Linear Inequalities

Step 1. Define the Problem. Read the problem carefully, or multiple times if necessary. Identify what you are trying to find and determine what information is available to help you find it.

Step 2. Assign Variables. Choose a variable to assign to an unknown quantity in the problem. For example, use p for price. If other unknown quantities exist, express them in terms of the selected variable.

Step 3. Translate into an Inequality. Use the relationships among the known and unknown quantities to form an inequality. It is helpful to consider different key words or phrases for each inequality symbol:

$<$ means "less than," "not as much as," "fewer than," …
$>$ means "greater than," "more than," "exceeds," …
$\geq$ means "at least," "greater than or equal to," …
$\leq$ means "less than or equal to," "no more than," "at most," …
$\neq$ means "not equal to"

Step 4. Solve the Inequality. Determine the solution set of the inequality.

Step 5. Check the Reasonableness of Your Answer. Check to see if your results make sense within the context of the problem. If not, check your work for errors and try again.

Step 6. Answer the Question. Write a clear statement that answers the question(s) posed.

Example 10 Wireless Plan Anytime Minutes

Suppose AT&T Wireless offers a Nation 900 monthly plan that includes 900 anytime minutes and unlimited nights and weekends for $60. Each additional anytime minute (or fraction of a minute) costs the user $0.40. If Antoine subscribes to this plan, how many anytime minutes can he use each month while keeping his total monthly cost to no more than $75 (before taxes)? (Source: www.att.com)

Solution We can follow the problem-solving strategy for applications of linear inequalities.

Step 1. Define the Problem. We need to find the number of anytime minutes that Antoine can use each month. We know that his monthly charge is $60, which includes 900 anytime minutes. Also, each additional anytime minute (or fraction of a minute) costs $0.40. Antoine wants his total monthly cost to be no more than $75 before taxes.

Step 2. Assign Variables. We choose the variable m to represent the unknown number of anytime minutes that Antoine can use each month. The charge for any additional minutes used will be $0.4(m - 900)$. (Click here to see why.)

Step 3. Translate into an Inequality. The sum of the monthly fee and the charge for additional minutes must not exceed $75. So, we have the following inequality:

Monthly fee   Charge for additional minutes   Total cost (before taxes)

$$60 \quad + \quad 0.4(m - 900) \quad \leq \quad 75$$

Step 4. Solve the Inequality.

$$60 + 0.4(m - 900) \leq 75$$

Multiply through by 10: $\quad 10 \cdot 60 + 10(0.4)(m - 900) \leq 10 \cdot 75$

Simplify: $\quad 600 + 4(m - 900) \leq 750$

Distribute: $\quad 600 + 4m - 3600 \leq 750$

Simplify: $\quad 4m - 3000 \leq 750$

Add 3000 to both sides: $\quad 4m - 3000 + 3000 \leq 750 + 3000$

Simplify: $\quad 4m \leq 3750$

Divide both sides by 4: $\quad \dfrac{4m}{4} \leq \dfrac{3750}{4}$

Simplify: $\quad m \leq 937.5$

Step 5. Check the Reasonableness of Your Answer. Because any fraction of a minute is billed as a whole minute, we must round our answer down to 937. Otherwise, the total cost would be slightly higher than $75. Click here for an explanation.

Step 6. Answer the Question. Antoine can use no more than 937 anytime minutes per month if he wants to keep his monthly cost to no more than $75. Click here to see the check for this example.

You Try It **Work through this You Try It problem.**

Work Exercises 31–32 in this textbook or in the *MyMathLab*® Study Plan.

How you define your variable(s) will affect the inequality model. However, the actual answer to the problem question will remain the same. For example, examine an alternate solution to Example 10 by clicking here.

Example 11 Making a Profit

My video summary

An online retailer sells plush toys. She purchases the toys at the wholesale price of $2.75 each and sells them online for $7.25. If her fixed costs are $900, how many plush toys must she sell in order to make a profit? Solve the inequality $R > C$ with R as her revenue and C as her cost.

Solution

Step 1. Define the Problem. We need to find the number of plush toys that must be sold in order to make a profit. We know that the price online is $7.25 each, and the wholesale cost to the retailer is $2.75 each. The fixed costs are $900.

Step 2. Assign Variables. We let the variable x represent the number of plush toys that are sold. The retailer's revenue, R, is $7.25x$ because the number sold times the price online equals the revenue. The retailer's cost, C, is the sum of the fixed costs and the wholesale cost per toy. Therefore, C is $2.75x + 900$.

Step 3. Translate into an Inequality. To create a profit, the retailer's revenue must be greater than her cost. So, we have the following inequality:

$$\overbrace{7.25x}^{\text{Revenue, }R} \quad > \quad \overbrace{2.75x + 900}^{\text{Cost, }C}$$

Step 4. Solve the Inequality. The solution set is $\{x \mid x > 200\}$, or $(200, \infty)$, in interval notation. Watch this video to see the rest of the solution completed in detail. The online retailer needs to sell more than 200 plush toys to realize a profit.

You Try It   Work through this You Try It problem.

Work Exercise 33–34 in this textbook or in the ***MyMathLab***® Study Plan.

Example 12 Taxable Income

In 2008, a single person in the 25% tax bracket was required to pay a tax of $0.25x - 3656.25$, where x is the amount of taxable income. For this tax bracket, the tax due was more than \$4481.25 but not more than \$16,056.25. Determine the range of taxable income for this tax bracket.

Solution We use the problem-solving strategy for applications of linear inequalities. We know that x is the amount of taxable income and that the tax in this bracket can be expressed by $0.25x - 3656.25$. The tax is more than \$4481.25 but not more than \$16,056.25. So, we get the following three-part inequality:

$$4481.25 < 0.25x - 3656.25 \leq 16056.25$$

Begin with the original inequality: $4481.25 < 0.25x - 3656.25 \leq 16056.25$

Add 3656.25 to all 3 parts: $4481.25 < 0.25x - 3656.25 \leq 16056.25$
$$+3656.25 \qquad\qquad + 3656.25 \quad + 3656.25$$

Simplify: $8137.50 < 0.25x \leq 19712.50$

Divide through by 0.25: $\dfrac{8137.50}{0.25} < \dfrac{0.25x}{0.25} \leq \dfrac{19712.50}{0.25}$

Simplify: $32{,}550 < x \leq 78{,}850$

Therefore, the 25% tax bracket covers taxable income that is more than \$32,550 but not more than \$78,850.

You Try It   Work through this You Try It problem.

Work Exercises 35–36 in this textbook or in the ***MyMathLab***® Study Plan.

1.2 Exercises

In Exercises 1–6, determine if the given value is a solution to the inequality.

You Try It

1. $4x - 7 \leq 10$; $x = 3$

2. $10 - (3 - 5m) > 3m + 2$; $m = -2$

3. $3(5 - n) + 7 < 4 - 2(n + 3)$; $n = 4$

4. $y^2 + 4y \geq y + 10$; $y = -5$

5. $0.2(x + 0.5) + 3.25 \leq 0.4x + 3.1$; $x = 0.5$

6. $\dfrac{2}{3}v + \dfrac{5}{4} > \dfrac{3}{4} - \dfrac{1}{6}v$; $v = \dfrac{3}{5}$

In Exercises 7–10, graph the solution set on a number line.

You Try It

7. $\{x | x < 3\}$

8. $\{x | x \geq -4\}$

9. $\{x | 2 < x \leq 9\}$

10. $\{x | 0 < x < 5\}$

In Exercises 11–14, write the solution set in interval notation.

You Try It

11. $\{x | x \leq 1\}$

12. $\{x | -6 \leq x \leq 11\}$

13. $\left\{ x \,\middle|\, \dfrac{3}{2} \leq x < \dfrac{7}{3} \right\}$

14. $\{x | x > -5.4\}$

In Exercises 15–18, solve the inequality. Graph the solution set and write it in interval notation.

You Try It

15. $4x + 15 > 9$

16. $0.2x + 1.2 \leq 0.4 - 0.2x$

17. $\dfrac{7}{5}(b - 20) \geq \dfrac{3}{2}(4b - 9)$

18. $7(x - 3) - 4(x - 1) < 3(x - 5)$

In Exercises 19–30, solve the inequality. Write the solution set in interval notation.

You Try It

19. $12 + 3x \geq 2x - 3$

20. $3(b + 1) - 7b > 5b + 3$

21. $0.75 - (0.25m + 1) \leq 1.5 - 0.5m$

22. $11 + 6(z + 1) \geq (5z - 6) - 8$

23. $3(2 - x) + 8x < 12 + 2(x - 3)$

24. $2.9x + 1.2(2x + 3) \leq 3.9x - 4.1$

You Try It

25. $\dfrac{2 - 3w}{5} - \dfrac{w}{2} < -4$

26. $\dfrac{a}{8} - \dfrac{7}{4} \leq \dfrac{a}{4} - 1$

You Try It

27. $9(2x + 3) - 5x > 10x + 3(x + 5)$

28. $5(n - 3) > 5n + 2$

29. $-10 \leq \dfrac{3p + 1}{2} < 5$

30. $-6 < \dfrac{7 - 5x}{3} < -1$

You Try It

In Exercises 31–36, solve each application problem.

You Try It

31. **Wireless Plan** Suppose AT&T Wireless offers a Nation 1350 monthly plan that includes 1350 anytime minutes and unlimited nights and weekends for $80. Each additional anytime minute (or fraction of a minute) costs the user $0.35. If Logan subscribes to this plan, how many anytime minutes can she use each month while keeping her total cost to no more than $125 (before taxes)? (Source: www.att.com)

32. **Checked Baggage** As of June 15, 2008, all checked baggage on American Airlines must not exceed 62 inches or an additional fee is charged. The size of baggage is determined by adding *length + width + height*. If a bag has a height of 14 inches and a width that is 12 inches less than the length, what lengths are acceptable for the bag to be checked free of charge? (Source: www.aa.com)

You Try It

33. **Retail Sales** An online retailer sells plush toys. He purchases the toys at the wholesale price of $3.15 each and sells them online for $14.75. If his fixed costs are $1060, how many plush toys must he sell in order to make a profit? That is, solve the inequality $R > C$, with R as his revenue and C as his cost.

34. **Truck Rental** Glider Truck Rentals will rent a 24-foot truck for a daily rate of $85 plus $0.10 per mile with 200 free miles. Y'All Rentals will rent a 24-foot truck for a daily rate of $40 plus $0.20 per mile. For what number of miles driven (in one day) will Glider Truck Rentals be cheaper?

You Try It

35. **2008 Taxes** In 2008, a married couple filing jointly in the 28% tax bracket had to pay a tax of $0.28x - 11,256$, where x is the amount of taxable income. For this tax bracket, the tax due was more than $25,550 but not more than $44,828. Determine the range of taxable income for this tax bracket. (Source: www.irs.gov)

36. **Repair Costs** A mechanic's estimate for replacing a part on a car ranges from $1300 to $2260. If the part costs $400 and the mechanic charges $60 per hour for labor, what is the range of the mechanic's estimates for the time it will take to replace the part?

1.3 Compound Inequalities; Absolute Value Equations and Inequalities

THINGS TO KNOW

Before working through this section, be sure you are familiar with the following concepts:

	VIDEO	ANIMATION	INTERACTIVE

You Try It

1. Compute the Absolute Value of a Real Number (Section R.1, **Objective 5**)

You Try It

2. Solve Linear Equations in One Variable (Section 1.1, **Objective 2**)

You Try It

3. Solve Linear Inequalities in One Variable (Section 1.2, **Objective 4**)

OBJECTIVES

1 Find the Union and Intersection of Two Sets

2 Solve Compound Linear Inequalities in One Variable

3 Solve Absolute Value Equations

4 Solve Absolute Value Inequalities

OBJECTIVE 1:   FIND THE UNION AND INTERSECTION OF TWO SETS

The words *and* and *or* are sometimes used when working with sets of numbers.

Intersection

For any two sets A and B, the **intersection** of A and B is given by $A \cap B$ and represents the elements that are in set A **and** in set B.

$$A \cap B = \{x | x \text{ is an element of } A \textbf{ and } \text{an element of } B\}$$

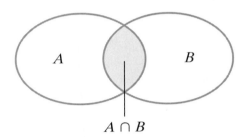

$$A \cap B$$

From the diagram, we see that the intersection of two sets is the overlap of the sets. Much like the intersection of two roads is the region common to both roads, the intersection of two sets is the set of elements that are common to both sets. If there is no overlap, the intersection is the empty set, and we write $A \cap B = \{ \ \}$ or $A \cap B = \varnothing$.

Example 1 Finding the Intersection of Two Sets

Let $A = \{1, 3, 4, 5, 7, 10, 12\}$ and $B = \{2, 4, 6, 8, 10, 12\}$. Find $A \cap B$, the intersection of the two sets.

Solution The set $A \cap B$ is the set of elements that are in both A and B. Both sets contain the numbers 4, 10, and 12. So,

$$A \cap B = \{1, 3, 4, 5, 7, 10, 12\} \cap \{2, 4, 6, 8, 10, 12\} = \{4, 10, 12\}.$$

You Try It   Work through this You Try It problem.

Work Exercises 1–2 in this textbook or in the *MyMathLab*®® Study Plan.

Example 2 Finding the Intersection of Two Sets

Let $A = \{x | x > -2\}$ and $B = \{x | x \le 5\}$. Find $A \cap B$, the intersection of the two sets.

Solution Click here to check your answer, or watch the video for the solution.

You Try It   Work through this You Try It problem.

Work Exercise 5 in this textbook or in the *MyMathLab*®® Study Plan.

My video summary

Union

For any two sets A and B, the **union** of A and B is given by $A \cup B$ and represents the elements that are in set A **or** in set B.

$$A \cup B = \{x \mid x \text{ is an element of } A \textbf{ or } \text{an element of } B\}$$

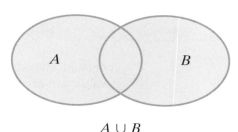

$A \cup B$

From the diagram, we see that the union of two sets is the combination of the sets. The union of two sets contains elements that are just in set A, just in set B, or in both A and B. Elements that appear in both sets A and B are only listed once when writing the union of the sets.

Example 3 Finding the Union of Two Sets

Let $A = \{1, 3, 4, 5, 7, 10, 12\}$ and $B = \{2, 4, 6, 8, 10, 12\}$. Find $A \cup B$, the union of the two sets.

Solution The union is the set of all unique values that are in either set A or in set B.

$$A \cup B = \{1, 2, 3, 4, 5, 6, 7, 8, 10, 12\}$$

Notice that 4, 10, and 12 occurred in both sets, but were only listed once in the union.

You Try It   Work through this You Try It problem.

Work Exercises 3–4 in this textbook or in the *MyMathLab*® Study Plan.

Example 4 Finding the Union of Two Sets

My video summary

 Let $A = \{x \mid x < -2\}$ and $B = \{x \mid x \geq 5\}$. Find $A \cup B$, the union of the two sets.

Solution Click here to check your answer, or watch the video for the solution.

You Try It   Work through this You Try It problem.

Work Exercise 6 in this textbook or in the *MyMathLab*® Study Plan.

Example 5 Finding the Intersection of Intervals

My video summary

 Find the intersection of the following intervals and graph the set on a number line.

a. $[0, \infty) \cap (-\infty, 5]$ **b.** $\big((-\infty, -2) \cup (-2, \infty) \big) \cap [-4, \infty)$

Solution Click here to check your answer, or watch the video for the solution.

You Try It Work through this You Try It problem.

Work Exercises 7–8 in this textbook or in the *MyMathLab*® Study Plan.

OBJECTIVE 2: SOLVE COMPOUND LINEAR INEQUALITIES IN ONE VARIABLE

A **compound inequality** consists of two inequalities that are joined together using the words *and* or *or*. Figure 1 shows two examples of compound inequalities.

> **a.** $x + 2 < 5$ and $3x \geq -6$
>
> **b.** $x + 3 \leq 1$ or $2x - 5 > 7$

Figure 1 Examples of compound inequalities

The word *and* indicates intersection. A number is a solution to a compound inequality using *and* if it is a solution to *both* inequalities. For example, $x = 2$ is a solution to the compound inequality in Figure 1(a) because it is a solution to both $x + 2 < 5$ and $3x \geq -6$. Click here to check the solution process.

The word *or* indicates union. A number is a solution to a compound inequality using *or* if it is a solution to *either* inequality. For example, $x = 8$ is a solution to the compound inequality in Figure 1(b) because it is a solution to at least one of the inequalities. Click here to check the solution process.

My interactive video summary

For more practice on checking solutions to compound inequalities, work through this interactive video.

We now present a general strategy for solving compound inequalities.

Guidelines for Solving Compound Linear Inequalities

Step 1. Solve each inequality separately.

Step 2. Graph each solution set on a number line.

Step 3. For compound inequalities using *and*, the solution set is the intersection of the individual solution sets.

For compound inequalities using *or*, the solution set is the union of the individual solution sets.

Example 6 Solving a Compound Linear Inequality Using *and*

My video summary

Solve $3x - 5 < -2$ and $4x + 11 \geq 3$. Graph the solution set and then write it in interval notation.

Solution Let's follow the guidelines for solving compound linear inequalities.

Step 1. Solve each inequality separately.

$$\begin{aligned} \text{Original inequality:} \qquad & 3x - 5 < -2 \\ \text{Add 5 to both sides:} \quad & 3x - 5 + 5 < -2 + 5 \\ \text{Simplify:} \qquad\qquad & 3x < 3 \end{aligned}$$

$$\text{Divide both sides by } 3: \qquad \frac{3x}{3} < \frac{3}{3}$$

$$\text{Simplify:} \qquad x < 1$$

and

$$\text{Original inequality:} \qquad 4x + 11 \geq 3$$

$$\text{Subtract } 11 \text{ from both sides:} \quad 4x + 11 - 11 \geq 3 - 11$$

$$\text{Simplify:} \qquad 4x \geq -8$$

$$\text{Divide both sides by } 4: \qquad \frac{4x}{4} \geq \frac{-8}{4}$$

$$\text{Simplify:} \qquad x \geq -2$$

Step 2. Graph each solution set on a number line.

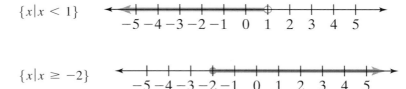

$\{x \mid x < 1\}$

$\{x \mid x \geq -2\}$

Continue to follow the solution process as shown here and/or watch this video to see the solution worked out in detail.

Step 3. Since the compound inequality uses *and*, the solution set is the intersection of the two graphs. We look for all *x*-values that are common to both solution sets. The first solution set includes all values of *x* that are less than 1. The second solution set includes all values of *x* that are greater than or equal to -2. The *x*-values common to both solution sets include all *x*-values that are greater than or equal to -2 and less than 1, or $\{x \mid -2 \leq x < 1\}$. The following graph shows this solution set:

$\{x \mid -2 \leq x < 1\}$

The solution set of the compound inequality, written in interval notation, is $[-2, 1)$.

You Try It Work through this You Try It problem.

Work Exercises 9–10 in this textbook or in the *MyMathLab*® Study Plan.

Example 7 Solving a Compound Linear Inequality Using *or*

Solve $9 - 4x < -7$ or $5x + 6 < 3(x + 2)$. Graph the solution set and then write it in interval notation.

Solution

Step 1. Solve each inequality separately.

My video summary

$$9 - 4x < -7 \quad \text{or} \quad 5x + 6 < 3(x + 2)$$

$$-4x < -16 \qquad\qquad 5x + 6 < 3x + 6$$

$$x > 4 \qquad\qquad\qquad 5x < 3x$$

$$5x - 3x < 3x - 3x$$

$$2x < 0$$

$$x < 0$$

The inequality switched directions because we divided both sides by a negative number.

Steps 2–3. Graph each solution set on a number line. Since the compound inequality uses *or*, the solution set is the union of the two graphs. We look for all x values that appear in either of the two solution sets. Complete the solution; then click here to check your answer. Or, watch the video for a detailed solution.

You Try It   Work through this You Try It problem.

Work Exercises 11–12 in this textbook or in the *MyMathLab*® Study Plan.

Like other inequalities, a compound inequality can have no solution or the set of all real numbers as its solution set. For example, the solution set of the compound inequality

$$2x - 3 \le -1 \quad \text{and} \quad x - 7 \ge -3$$

is the null set, $\varnothing$ (click here to see why), whereas the solution set of the compound inequality

$$10x + 7 > 2 \quad \text{or} \quad 3x - 6 \le 9$$

is the set of all real numbers, $\mathbb{R}$ or $(-\infty, \infty)$ in interval notation (click here to see why).

You Try It   Work through this You Try It problem.

Work Exercises 13–16 in this textbook or in the *MyMathLab*® Study Plan.

OBJECTIVE 3:   SOLVE ABSOLUTE VALUE EQUATIONS

Before reading further, you may want to review absolute value in Section R.1. Consider how to solve the equation $|x| = 5$. First, we look for values of x that are 5 units away from zero on a number line. The two numbers that are 5 units away from zero on a number line are $x = -5$ and $x = 5$, as shown in Figure 2.

five units from zero five units from zero

```
        ↓              ↓
◄───────●───────┼───────●───────►
       -5       0       5
```

If $|x| = 5$, then $x = -5$ or $x = 5$.
The solution set is $\{-5, 5\}$.

Figure 2 Solution to $|x| = 5$

How do we solve the equation $|x + 2| = 5$? First, we consider the expression inside the absolute value bars as a single quantity. To solve the equation, we look for values of x that make that quantity equal to -5 or 5. So, we write the following:

$$x + 2 = 5 \quad \text{or} \quad x + 2 = -5$$

This idea can be applied to all algebraic expressions, as shown in the following box.

Absolute Value Equation Property

If u is an algebraic expression and c is a real number such that $c > 0$ then $|u| = c$ is equivalent to $u = -c$ or $u = c$.

Some special situations arise if $c = 0$ or $c < 0$. If $c = 0$, then $|u| = 0$ and $u = 0$. If $c < 0$, then $|u| = c$ has no solution. Click here for an explanation of these results.

Example 8 Solving an Absolute Value Equation

Solve: $|m + 4| = 8$

Solution Using the form $|u| = c$, where $u = m + 4$ and $c = 8$, we look for values of m such that $m + 4$ is 8 units away from zero on a number line. The absolute value equation property shows that $|m + 4| = 8$ is equivalent to

$$m + 4 = -8 \quad \text{or} \quad m + 4 = 8$$

Now we can solve the two equations separately and combine the two solution sets.

$$
\begin{aligned}
m + 4 &= -8 \quad &\text{or} \quad m + 4 &= 8 \\
m &= -12 \quad & m &= 4
\end{aligned}
$$

The solution set is $\{-12, 4\}$. Click here to check this answer.

Example 9 Solving an Absolute Value Equation

 Solve: $|1 - 3x| = 4$

Solution Click here to check your answer, or watch this video for a detailed solution.

You Try It   Work through this You Try It problem.

Work Exercises 17–20 in this textbook or in the *MyMathLab*® Study Plan.

Example 10 Solving an Absolute Value Equation

Solve: $|2x - 5| = 0$

Solution Click here to check your answer, or watch this video for a detailed solution.

Example 11 Solving an Absolute Value Equation

Solve: $|3x + 7| = -4$

Solution Using the form $|u| = c$, where $u = 3x + 7$ and $c = -4$, we see that the absolute value expression is equal to a negative number. This means that there is no solution, and the solution set is the empty set, $\{\ \}$.

You Try It Work through this You Try It problem.

Work Exercises 21–22 in this textbook or in the *MyMathLab*® Study Plan.

Care must be taken to not just blindly apply the absolute value equation property. Click here to see what would happen in Example 11 if we mistakenly applied the absolute value equation property using $c < 0$.

When applying the absolute value equation property, remember to write the absolute value equation in the form $|u| = c$. If the equation is not in this form, then we must isolate the absolute value expression first before applying the property.

Now we present a general strategy for solving absolute value equations.

Strategy for Solving Absolute Value Equations

Step 1. Isolate the absolute value expression on one side of the equation to obtain the form $|u| = c$.

Step 2. Apply the absolute value equation property.

If $c > 0$, write $|u| = c$ as $u = -c$ or $u = c$.

If $c = 0$, write $|u| = c$ as $u = c$.

If $c < 0$, the equation has no solution.

Step 3. Solve any equations from step 2 to find the solution set.

Step 4. Check your answer within the original absolute value equation to confirm the solution set.

Example 12 Solving an Absolute Value Equation

Solve: $2|w - 1| + 3 = 11$

Solution Following the strategy for solving absolute value equations, first we isolate the absolute value expression on one side of the equation.

Start with the original equation:	$2|w - 1| + 3 = 11$
Subtract 3 from both sides:	$2|w - 1| + 3 - 3 = 11 - 3$
Simplify:	$2|w - 1| = 8$
Divide both sides by 2:	$\dfrac{2|w - 1|}{2} = \dfrac{8}{2}$
Simplify:	$|w - 1| = 4$

Finish solving the equation. Click here to check your answer, or watch the video to see the full solution.

Example 13 Solving an Absolute Value Equation

 Solve: $-3|2 - m| + 8 = 2$

Solution Click here to check your answer, or watch the video to see the full solution.

You Try It Work through this You Try It problem.

Work Exercises 23–24 in this textbook or in the _MyMathLab_® Study Plan.

OBJECTIVE 4: **SOLVE ABSOLUTE VALUE INEQUALITIES**

If $|u|$ and c are real numbers, then the trichotomy property tells us that exactly one of the following is true:

$$|u| = c, \quad |u| < c, \quad \text{or} \quad |u| > c$$

Now that we know how to solve absolute value equations of the form $|u| = c$, let's turn our attention to solving absolute value inequalities of the forms $|u| < c$ or $|u| > c$.

When solving $|x| = 5$, we looked for values of x that were 5 units away from zero on a number line. To solve the inequality $|x| < 5$, we look for values of x that are *less than* 5 units away from zero on a number line. Similarly, to solve the inequality $|x| > 5$, we look for values of x that are *greater than* 5 units away from zero on a number line. The solutions to these two inequalities are illustrated in Figures 3 and 4, respectively.

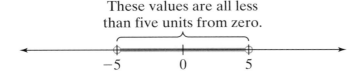

These values are all less
than five units from zero.

If $|x| < 5$, then $-5 < x < 5$.
The solution set is $\{x | -5 < x < 5\}$ in set-builder notation or $(-5, 5)$ in interval notation.

Figure 3 Solution to $|x| < 5$

These values are more
than five units from zero.   These values are more
than five units from zero.

If $|x| > 5$, then $x < -5$ or $x > 5$.
The solution set is $\{x | x < -5 \text{ or } x > 5\}$ in set-builder notation or $(-\infty, -5) \cup (5, \infty)$ in interval notation.

Figure 4 Solution to $|x| > 5$

Figures 3 and 4 help us see that absolute value inequalities can be converted into alternate forms that do not include absolute value expressions. We present these alternate forms in the following summary.

Absolute Value Inequality Property

If u is an algebraic expression and c is a real number such that $c > 0$, then

   1. $|u| < c$ is equivalent to $-c < u < c$

     and

   2. $|u| > c$ is equivalent to $u < -c$ or $u > c$

Similar forms exist for the non-strict cases $|u| \leq c$ and $|u| \geq c$.

As with absolute value equations, special cases arise when working with absolute value inequalities. Two special cases are when $|u| < 0$ and $|u| \geq 0$. Click here to find the solution sets of these two cases.

Example 14 Solving an Absolute Value Inequality

  Solve: $|2m - 1| \leq 5$

Solution Using the form $|u| \leq c$, where $u = 2m - 1$ and $c = 5$, we look for values of m such that $2m - 1$ is less than or equal to 5 units away from zero on a number line. Next we use the absolute value inequality property to write $|2m - 1| \leq 5$ is equivalent to $-5 \leq 2m - 1 \leq 5$.

Now solve the resulting three-part inequality. Click here to compare your answer and see a check, or watch this video to confirm your results.

You Try It   **Work through this You Try It problem.**

Work Exercises 25–30 in this textbook or in the *MyMathLab*®️ Study Plan.

Example 15 Solving an Absolute Value Inequality

  Solve: $|5x + 1| > 3$

Solution Click here to compare your answer and see a check, or watch this video to confirm your results.

In Example 15, $|5x + 1| > 3$ is *not* equivalent to $-3 > 5x + 1 > 3$. A three-part inequality must be true from far left to far right: $-3 > 3$ is a false statement. Another common error in this type of problem is to write $5x + 1 > -3$ for the first inequality, instead of $5x + 1 < -3$. Think carefully about the meaning of the inequality before writing it.

You Try It   **Work through this You Try It problem.**

Work Exercises 31–36 in this textbook or in the *MyMathLab*®️ Study Plan.

As with absolute value equations, not every absolute value inequality will be written in a form that will allow us to use one of the absolute value inequality properties immediately. The following general strategy for solving absolute value inequalities can be used as a guide.

Strategy for Solving Absolute Value Inequalities

Step 1. Isolate the absolute value expression to obtain one of the following forms: $|u| < c$, $|u| > c$, $|u| \le c$, or $|u| \ge c$.

Step 2. Apply the absolute value inequality property.

Step 3. Solve the resulting compound inequality.

Example 16 Solving an Absolute Value Inequality

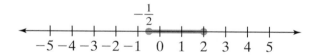 Solve: $|4x - 3| + 2 \le 7$

Solution Following the strategy for solving absolute value inequalities, first we isolate the absolute value expression.

$$\text{Write the original inequality:} \quad |4x - 3| + 2 \le 7$$
$$\text{Subtract 2 from both sides:} \quad |4x - 3| \le 5$$

Now we have the form $|u| \le c$, where $u = 4x - 3$ and $c = 5$. We look for values of x such that $4x - 3$ is less than or equal to 5 units away from zero on a number line. Using the absolute value inequality property, $|4x - 3| \le 5$ is equivalent to $-5 \le 4x - 3 \le 5$.

We find the solution set by solving the three-part inequality:

$$\text{Write the three-part inequality:} \quad -5 \le 4x - 3 \le 5$$
$$\text{Add 3 to all three parts:} \quad -2 \le 4x \le 8$$
$$\text{Divide all three parts by 4 and simplify:} \quad -\frac{1}{2} \le x \le 2$$

The solution set is $\left\{ x \,\middle|\, -\dfrac{1}{2} \le x \le 2 \right\}$ in set-builder notation or $\left[-\dfrac{1}{2}, 2 \right]$

in interval notation. Click here to see a check for this solution set. The following shows the graph of the solution set on a number line:

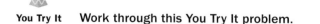

Example 17 Solving an Absolute Value Inequality

Solve: $5|1 - 2x| - 3 > 12$

Solution Follow the strategy for solving absolute value inequalities; then click here to check your solution set, or work through this interactive video.

You Try It **Work through this You Try It problem.**

Work Exercises 37–42 in this textbook or in the *MyMathLab*® Study Plan.

My video summary

My interactive video summary

1.3 Exercises

In Exercises 1–4, find the indicated set given $A = \{-2, 0, 1, 5, 6, 9, 15\}$, $B = \{-6, -4, -2, 0, 2, 4, 6\}$, and $C = \{-1, 3, 7, 11, 20\}$.

You Try It **1.** $A \cap B$ **2.** $A \cap C$

You Try It **3.** $A \cup B$ **4.** $A \cup B \cup C$

You Try It **5.** Find $A \cap B$ if $A = \{x | x < 9\}$ and $B = \{x | x \geq 2\}$.

6. Find $A \cup B$ if $A = \{x | x \leq -3\}$ and $B = \{x | x \geq 5\}$.

You Try It **7.** Find the intersection: $(-8, 5] \cap (-12, 3)$

8. Find the intersection: $((-\infty, 0) \cup (0, \infty)) \cap [-3, \infty)$

You Try It

In Exercises 9–16, solve the compound inequality. Write your answer in interval notation or state that there is no solution.

9. $2x - 3 \leq 5$ and $5x + 1 > 6$ **10.** $\dfrac{n-1}{3} \geq 1$ and $\dfrac{4n-2}{2} \leq 9$

You Try It **11.** $4k + 2 \leq -10$ or $3k - 4 > 8$ **12.** $2x + 1 > 5$ or $5 - 3x < 11$

13. $2y - 1 \leq 7$ or $3y - 5 > 1$ **14.** $3b - 5 \leq 1$ and $1 - b < -3$

You Try It **15.** $x + 1 < -5$ and $2x - 3 < -11$ **16.** $0.3x + 3 \leq x + 0.2$ or $4(2x - 5) > 3x + 5$

In Exercises 17–24, solve each absolute value equation.

17. $|x| = 3$ **18.** $|x - 1| = 5$

You Try It

19. $|6x + 7| = 3$ **20.** $\left| \dfrac{3x+1}{5} \right| = \dfrac{2}{3}$

You Try It **21.** $|7x + 3| = 0$ **22.** $|x - 2| = -4$

23. $3|4x - 3| + 1 = 10$ **24.** $-2|3x + 2| + 7 = 5$

You Try It

In Exercises 25–30, solve each absolute value inequality. Write your answer in interval notation.

25. $|9x| \leq 3$ **26.** $|x - 2| < 3$ **27.** $|3 - 4x| < 11$

You Try It **28.** $\left| \dfrac{2x-5}{3} \right| \leq \dfrac{1}{5}$ **29.** $|3x - 7| \leq 0$ **30.** $|x - 1| < -2$

In Exercises 31–36, solve each absolute value inequality. Write your answer in interval notation.

31. $|2x| > 6$ **32.** $|x + 4| > 5$ **33.** $\left| \dfrac{2x-6}{7} \right| \geq \dfrac{3}{14}$

You Try It **34.** $|2 - 5x| \geq 3$ **35.** $|9x - 4| > 0$ **36.** $|3x + 1| \geq -5$

In Exercises 37–42, solve each absolute value inequality. Write your answer in interval notation.

37. $|4x - 3| + 2 \leq 9$

38. $|1 - 2x| - 4 > 3$

39. $3|x + 5| - 1 \geq 8$

You Try It

40. $-|3x + 2| + 5 > -6$

41. $-4|5x + 2| - 3 < 5$

42. $2|3 - 4x| + 1 < 4$

1.4 Formulas and Problem Solving

THINGS TO KNOW

Before working through this section, be sure you are familiar with the following concepts:

| | | VIDEO | ANIMATION | INTERACTIVE |

You Try It

1. Write Verbal Descriptions as Algebraic Expressions (Section R.3, Objective 3)

You Try It

2. Solve Linear Equations in One Variable (Section 1.1, Objective 2)

You Try It

3. Use Linear Equations to Solve Application Problems (Section 1.1, Objective 4)

You Try It

4. Solve Linear Inequalities in One Variable (Section 1.2, Objective 4)

You Try It

5. Use Linear Inequalities to Solve Application Problems (Section 1.2, Objective 5)

OBJECTIVES

1 Solve a Formula for a Given Variable

2 Use Formulas to Solve Application Problems

OBJECTIVE 1: SOLVE A FORMULA FOR A GIVEN VARIABLE

A **formula** is an equation that describes the relationship between two or more variables. Typically formulas apply to physical or financial situations that relate quantities such as length, area, volume, time, speed, money, interest rates, and so on. For example, $P = 2l + 2w$ is the formula for finding the perimeter of (or distance around) a rectangle, and it relates three variables: perimeter P, length l, and width w. If we know the values of any two of these variables, then we can use the formula to find the third value.

My interactive video summary

Watch this interactive video to learn how to use a formula to find the value of a variable.

To **solve a formula for a given variable** means to isolate the variable on one side of the equation. For example the formula $P = 2l + 2w$ is solved for P.

When working with application problems, it may help to solve a given formula for a different variable. For example if we need to find the length l of a rectangle when the perimeter P and width w are known, then we solve the formula $P = 2l + 2w$ for the variable l. To solve a formula for a given variable, we follow the steps for solving an equation.

Example 1 Solving a Formula for a Given Variable

Solve each formula for the given variable.

a. Area of a triangle: $A = \dfrac{1}{2}bh$ for b

b. Perimeter of a rectangle: $P = 2l + 2w$ for l

Solution

a.

Begin with original formula:	$A = \dfrac{1}{2}bh$
Multiply both sides by 2 to clear the fraction:	$2(A) = 2\left(\dfrac{1}{2}bh\right)$
Simplify:	$2A = bh$
Divide both sides by h:	$\dfrac{2A}{h} = \dfrac{bh}{h}$
Simplify:	$\dfrac{2A}{h} = b$

The area of a triangle formula, $b = \dfrac{2A}{h}$, is solved for the variable b.

My video summary

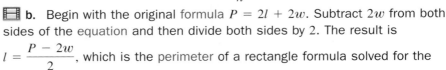 **b.** Begin with the original formula $P = 2l + 2w$. Subtract $2w$ from both sides of the equation and then divide both sides by 2. The result is $l = \dfrac{P - 2w}{2}$, which is the perimeter of a rectangle formula solved for the variable l. Watch this video to see a fully worked solution.

CAUTION

To solve a formula for a given variable, the variable must be isolated on one side of the equation and must be the only variable of its type left in the equation.

Note: The relationship among the variables of a formula does not change when the formula is solved for a different variable. For example, the relationship among the perimeter, length, and width of a rectangle remains the same whether the formula is solved for l, w, or P. If we know any two of these variables, we can always find the third. Click here to see an example of why this is true.

You Try It   Work through this You Try It problem.

Work Exercises 1–10 in this textbook or in the **MyMathLab**® Study Plan.

(EBook Screens 1.4-37–1.4-46)

OBJECTIVE 2:   USE FORMULAS TO SOLVE APPLICATION PROBLEMS

For the rest of this section, we will use various types of formulas to solve application problems. The formulas in Examples 1 through 4 are from geometry. Click here to review common formulas for area and perimeter or here to review common formulas for volume and surface area.

When using formulas to solve application problems, we can follow the problem-solving strategy for applications of linear equations first introduced in Section 1.1.

Example 2 Basketball Court Dimensions

The length of a college basketball court (rectangle) is 6 feet less than twice its width. If the perimeter is 288 feet, then what are the dimensions of the court?

Solution

Step 1. Define the Problem.   We need to find both the length and the width of a college basketball court (rectangle). We know that the perimeter is 288 feet, and the length is 6 feet less than twice the width. We will use the formula for the perimeter of a rectangle to solve this problem.

Step 2. Assign Variables.   If we let w represent the width of the court, then the length is $2w - 6$ (6 feet less than twice the width).

Step 3. Translate into an Equation.   The perimeter is 288 feet, so we have the equation:

$$2l + 2w = P$$
$$2(2w - 6) + 2w = 288$$

Step 4. Solve the Equation.   Finish solving the equation to find the width of the court, then use the width to find the length. When finished, click here to check your answer, or watch this video for a fully worked solution.

You Try It   **Work through this You Try It problem.**

Work Exercises 11–12 in this textbook or in the *MyMathLab*® Study Plan.

Example 3 Coffee Can Height

A 13-ounce Maxwell House coffee can has a surface area of 186π cm². Find the height of the can if its radius is 5.0 cm.

Solution

Step 1. Define the Problem.   We need to find the height of the can. We know that the can is a circular cylinder, the radius is 5.0 cm, and the surface area is 186π cm². We refer to the common formulas for volume and surface area to find the formula for the surface area of a cylinder: $SA = 2\pi r^2 + 2\pi rh$.

1.4 Formulas and Problem Solving   39

Step 2. Assign Variables. We know values for all of the variables expect h. Let h represent the height of the can.

Step 3. Translate into an Equation. Substituting for the variables, we have the equation:

$$2\pi r^2 + 2\pi rh = SA$$
$$2\pi(5.0)^2 + 2\pi(5.0)h = 186\pi$$

Step 4. Solve the Equation. Finish solving the equation to find the height of the can. Check the result for reasonableness and write a clear answer. When finished, click here to check your answer, or watch this video for a fully worked solution.

You Try It **Work through this You Try It problem.**

Work Exercises 13–14 in this textbook or in the *MyMathLab*® Study Plan.

A well-known property of triangles is that the sum of the measures of the three angles equals 180°. We can use this fact as a formula to solve problems involving triangles.

Example 4 A Triangular Field

Ken owns a triangular-shaped plot of land. The measure of the largest angle on the edge of the land is 15° more than twice the measure of the middle-sized angle. The measure of the smallest angle is two-thirds the measure of the middle-sized angle. Find the three angles on the edge of the land.

Solution Work through the problem-solving strategy to solve this problem on your own. Click here to check your answer, or watch this video for a fully worked solution.

You Try It **Work through this You Try It problem.**

Work Exercises 15–16 in this textbook or in the *MyMathLab*® Study Plan.

We have used only geometry formulas so far. However, there are many other types of formulas that are used when solving applications. In the remaining examples, we look at other formulas. For example, some applications involve finding the cost of a collection of items. Suppose you purchase 3 candy bars that cost $0.75 each. Your total cost for the 3 candy bars will be 3($0.75), or $2.25.

The following formula helps us to solve this problem:

$$\text{Total value} = (\text{value per item})(\text{number of items})$$

Note: We have used the term *total value* in the above formula instead of *total cost* because in an application, the "value" could be cost, profit, revenue, or earnings.

Example 5 Candle Sale

For a fundraiser, Anna's softball team sold candles in two sizes: large and small. The team earned $6.50 in profit for each large candle sold and $3.75 in profit for each small candle sold. If the team sold a total of 90 candles and earned $480.50 in profit, how many candles of each size were sold?

Solution

Step 1. Define the Problem. We need to find the number of candles sold for each size. We know that a total of 90 candles were sold and that the team earned $480.50. We also know that the team earned a profit of $6.50 per large candle and $3.75 per small candle.

Step 2. Assign Variables. Let x be the number of large candles sold. Then $90 - x$ is the number of small candles sold.

Step 3. Translate into an Equation. The profit from selling x large candles is $6.50x$. The profit from selling $90 - x$ small candles is $3.75(90 - x)$. Adding these two amounts results in the total profit of $480.50. So, we have the equation:

$$\underbrace{6.50x}_{\substack{\text{Profit from} \\ \text{large candles}}} + \underbrace{3.75(90 - x)}_{\substack{\text{Profit from} \\ \text{small candles}}} = \underbrace{480.50}_{\substack{\text{Total} \\ \text{profit}}}$$

Finish solving the equation to find the number of large candles sold, then use the result to find the number of small candles sold. Click here to check your answer, or watch this video to see the rest of the solution.

You Try It Work through this You Try It problem.

Work Exercises 17–18 in this textbook or in the _MyMathLab_® Study Plan.

A common banking formula, $I = Prt$, is used for computing simple interest. In this formula, the variables can be defined as follows: I is the *simple interest* earned on an investment or paid for a loan; P is the *principal,* or the amount that is invested or borrowed; r is the *interest rate* in decimal form; and t is the *time* that the money is invested or borrowed.

Example 6 Investing Money

Elizabeth inherited $25,000 from her grandparents with the promise that she invest the money and use it to attend college. She invests some of the money in a certificate of deposit (CD) that pays 5% simple interest and the rest in a money market account that pays 3.5% simple interest. If the total interest earned was $1130 after one year, how much did she invest in each account?

Solution Watch this video for a fully worked solution. Click here to view the answer only.

You Try It Work through this You Try It problem.

Work Exercises 19–20 in this textbook or in the _MyMathLab_® Study Plan.

Example 7 illustrates a uniform motion problem. To solve it, we use the distance formula $d = rt$ or distance = rate $\times$ time. When using the distance formula with applications, it is often helpful to create a table to organize the information.

When working with the distance formula, the units for rate and time must be consistent. For example, if the unit for rate is miles per hour, then the unit for time must be hours.

(EBook Screens 1.4-37–1.4-46)

My video summary

Example 7 Fishing Trip

 At 9:00 A.M., Rick left his house to go fishing. He rode his scooter at an average speed of 10 mph. At 9:15 A.M., his girlfriend, Deb, went to Rick's house and then followed his path on her bike at a rate of 15 mph. If Deb caught up with Rick at the same time that they both reached the fishing hole, how far is it from Rick's house to the fishing hole? At what time did Rick and Deb arrive at the fishing hole?

Solution Watch this video for a fully worked solution. Click here to view the answer only.

You Try It Work through this You Try It problem.

Work Exercises 21–24 in this textbook or in the *MyMathLab*® Study Plan.

Now we will look at mixture problems. Suppose you purchase a 50-pound bag of lawn fertilizer with a 12% nitrogen concentration. The description "12% nitrogen concentration" means that out of all the components in the fertilizer, 12% is pure nitrogen. The 50-pound bag contains a total of $0.12(50) = 6$ pounds of pure nitrogen. So, if all the components of the fertilizer were separated, there would be 6 pounds of pure nitrogen and 44 pounds of other components.

We can find the amount of a particular component in a mixture by multiplying its concentration (in decimal form) by the amount of mixture:

$$\text{amount of component} = (\text{concentration})(\text{amount of mixture})$$

We can also write this in the following equivalent form:

$$\text{concentration} = \frac{\text{amount of component}}{\text{amount of mixture}}$$

The key to solving most mixture problems is recognizing that the total *amount* of a component does not change when two or more substances are mixed together. However, the *concentration* of that component might change. Example 8 helps to illustrate this concept.

Example 8 Bleach Concentration

 Suppose 2 gallons of a 10% bleach solution is mixed with 3 gallons of a 25% bleach solution. What is the concentration of bleach in the new 5-gallon mixture?

Solution Watch this animation to see a detailed solution to this problem. Click here for the answer only.

CAUTION The concentration of a mixture must always be in between the concentrations of the two mixed solutions. Do you see why?

Example 9 Alcohol Concentration

My video summary

How many milliliters of a 70% alcohol solution must be mixed with 30 mL of a 40% alcohol solution to result in a mixture that is 50% alcohol?

Solution

Step 1. **Define the Problem.** We mix two solutions together to obtain a third solution. We know that 30 mL of a 40% alcohol solution will be mixed with an unknown amount of a 70% alcohol solution to result in a 50% alcohol solution.

Step 2. Assign Variables. Let x be the unknown amount of the 70% alcohol solution (in mL). Since we have to mix this amount with 30 mL of a 40% alcohol solution, the resulting amount of 50% alcohol solution is $30 + x$ (in mL). This is illustrated in Figure 5.

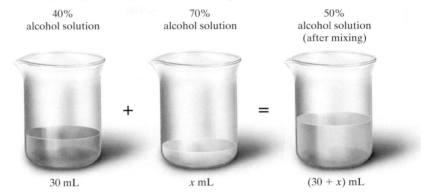

40% alcohol solution   70% alcohol solution   50% alcohol solution (after mixing)

30 mL x mL $(30 + x)$ mL

Figure 5

Step 3. Translate into an Equation. To set up the equation, we write the number of milliliters of pure alcohol in the 40% solution plus the number of milliliters of pure alcohol in the 70% solution to equal the number of milliliters of pure alcohol in the 50% solution. The equation is as follows:

Amount of pure alcohol in the 40% solution		Amount of pure alcohol in the 70% solution		Amount of pure alcohol in the 50% solution
$0.40(30)$	$+$	$0.70x$	$=$	$0.50(30 + x)$

Step 4. Solve the Equation. Finish solving the equation and determine the solution to the problem. Click here to check your answer or watch this video to see a detailed solution.

You Try It   Work through this **You Try It** problem.

Work Exercises 25–28 in this textbook or in the _MyMathLab_® Study Plan.

All of the examples in this section have involved equations, but sometimes we may have to use formulas when solving applications of inequalities. Consider Example 10.

Example 10 Perimeter of a Fence

The perimeter of a rectangular fence must be at least 80 feet and no more than 140 feet. If the width of the fence is 8 feet, what is the range of values allowed for the length of the fence?

Solution Let l be the length of the rectangle, as shown in Figure 6. The perimeter P of the rectangle is $P = 2(8) + 2l = 16 + 2l$.

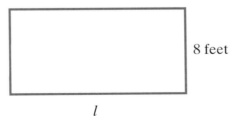

8 feet

l

Figure 6

The perimeter must be greater than or equal to 80 and less than or equal to 140. This can be written as $80 \le P \le 140$. Because the perimeter is $16 + 2l$, we substitute for P to get the following three-part inequality $80 \le 16 + 2l \le 140$.

Now we solve this three-part inequality:

Rewrite the three-part inequality:

Subtract 16: $\quad 80 - 16 \le 16 - 16 + 2l \le 140 - 16$

Simplify: $\quad 64 \le 2l \le 124$

Divide by 2: $\quad \dfrac{64}{2} \le \dfrac{2l}{2} \le \dfrac{124}{2}$

Simplify: $\quad 32 \le l \le 62$

The length of the fence must be between 32 feet and 62 feet, inclusive. Note that the word *inclusive* means that the fence could be 32 feet or 62 feet as well as any value in between. $\qquad 80 \le 16 + 2l \le 140$

You Try It   Work through this You Try It problem.

Work Exercises 29–30 in this textbook or in the *MyMathLab*® Study Plan.

1.4 Exercises

In Exercises 1–10, solve the formula for the given variable.

You Try It

1. $A = lw$ for w

2. $I = Prt$ for P

3. $V = \dfrac{1}{3}\pi r^2 h$ for h

4. $Ax + By = C$ for y

5. $E = I(r + R)$ for r

6. $A = \dfrac{1}{2}h(a + b)$ for a

7. $A = 2\pi r^2 + 2\pi rh$ for h

8. $C = \dfrac{5}{9}(F - 32)$ for F

9. $z = \dfrac{x - \mu}{\sigma}$ for x

10. $A = 2lw + 2lh + 2wh$ for l

In Exercises 11–30, use the problem-solving strategy for applications of linear equations to solve the application problem.

You Try It

11. Volleyball Court Dimensions   A volleyball court is twice as long as it is wide. If the perimeter of the rectangular court is 54 meters, determine the dimensions of the court.

12. Sail Height   The area of a triangular sail for a boat is 90 square feet. If the base of the sail is 10 feet long, find its height.

13. **Volume of a Cone**   The volume of a cone is 84π cubic inches. If the radius of the cone is 6 inches, determine its height.

14. **Surface Area of a Box**   A rectangular box is 36 cm wide and 15 cm high. If the surface area of the box is 3630 square centimeters, find the length of the box.

15. **Angles in a Triangle**   The measure of the largest angle of a triangle is twice the measure of the smallest angle. The measure of the middle-sized angle is 20° more than the smallest angle. Find the measures of the three angles.

16. **Isosceles Triangles**   An isosceles triangle has two equal angles. If the measure of the third angle is 20° less than the sum of the two equal angles, find the measures of the three angles.

17. **Coin Collection**   Jose has been collecting quarters and dimes. He empties his wallet to find that he has 33 coins (only quarters and dimes) worth $5.40. How many quarters and how many dimes does Jose have?

18. **Concession Stand**   A basketball concession stand sells a soda for $1.25, a bag of popcorn for $0.75, and a candy bar for $1.00. At one game, 14 fewer candy bars were sold than bags of popcorn. That same night, the number of sodas sold was 10 more than the number of candy bars and bags of popcorn sold altogether. If the concession stand made $117 that night, how many of each item was sold?

19. **Investing Money**   Juanita has $28,000 to invest in two accounts that pay simple interest on an annual basis. One account pays 4% in simple interest, and the other pays 5% simple interest. How much would she have to invest in each account to earn a total interest of $1320?

20. **Borrowing Money**   Jamal borrowed $4000 to buy a used car. He borrowed some of the money from a bank that charged 8.5% simple interest and the rest from a friend who charged 10% simple interest. If the total interest for a one-year loan was $362.50, how much did Jamal borrow at each rate?

21. **Catching Up**   A carload of fraternity brothers leaves their house for a long weekend road trip. Two hours later a second carload of brothers leaves the same house and travel the same path. If the second car drives an average of 20 mph faster than the first, what is the average speed of each car if it takes the second car three hours to catch up to the first?

22. **Going to Mexico**   Manuel traveled 31 hours nonstop to Mexico, a total of 2300 miles. He took a train part of the way, which averaged 80 mph, and then took a bus the remaining distance, which averaged 60 mph. How long was Manuel on the train?

23. **Hiking**   Abbas and Muhammad are at opposite ends of a 16-mile hiking trail. If Abbas' average hiking speed on the trail is 1.5 mph whereas Muhammad's average hiking speed is 2.5 mph, how long will it be before they meet?

24. **River Transport**   A grain barge travels on a river from point A to point B loading and unloading grain. The barge travels at a rate of 5 mph relative to the water. The river flows downstream at a rate of 1 mph. If the trip upstream takes two hours longer than the trip downstream, how far is it from point A to point B?

You Try It 25. **Mixing Alcohol**   Suppose 8 pints of a 12% alcohol solution is mixed with 2 pints of a 60% alcohol solution. What is the concentration of alcohol in the new 10-pint mixture?

26. **Mixing Orange Juice**   How much of an 80% orange juice drink must be mixed with 20 gallons of a 20% orange juice drink to obtain a mixture that is 50% orange juice?

27. **Mixing Acid**   How many liters (L) of pure water should be mixed with a 5-L solution of 80% acid to produce a mixture that is 70% water?

28. **Blending Coffee**   A coffee shop owner blends a gourmet brand of coffee with a cheaper brand. The gourmet coffee usually sells for $9.00 per pound. The cheaper brand sells for $5.00 per pound. How much of each type should be mixed in order to have 30 pounds of coffee that is worth $6.50 per pound?

You Try It

29. **Perimeter**   The perimeter of a rectangular fence must be at least 100 feet and no more than 180 feet. If the width of the fence is 22 feet, what is the range of values allowed for the length of the fence?

30. **Perimeter**   The perimeter of a rectangular fence must be at least 120 feet and no more than 180 feet. If the length of the fence must be twice the width, what is the range of values allowed for the width of the fence?

Graphs and Functions

CHAPTER TWO CONTENTS

2.1 **The Rectangular Coordinate System and Graphing**

2.2 **Relations and Functions**

2.3 **Function Notation and Applications**

2.4 **Graphs of Linear Functions**

2.5 **Linear Equations in Two Variables**

2.6 **Linear Inequalities in Two Variables**

2.1 The Rectangular Coordinate System and Graphing

THINGS TO KNOW

Before working through this section, be sure you are familiar with the following concepts:

| | VIDEO | ANIMATION | INTERACTIVE |

You Try It

1. Plot Real Numbers on a Number Line (Section R.1, Objective 3)

You Try It

2. Determine If a Given Value Is a Solution to an Equation (Section 1.1, Objective 1)

You Try It

3. Solve Linear Equations in One Variable (Section 1.1, Objective 2)

OBJECTIVES

1 Plot Ordered Pairs in the Rectangular Coordinate System

2 Determine If an Ordered Pair Is a Solution to an Equation

3 Find Unknown Coordinates

4 Graph Equations by Plotting Points

5 Find x- and y-Intercepts

OBJECTIVE 1:   PLOT ORDERED PAIRS IN THE RECTANGULAR COORDINATE SYSTEM

Graphs can be used to show relationships between two variables. Figure 1 shows the selling price of Google stock over time since it went public in August 2004. By looking at the graph, we can see how the value of the stock has changed over time. This information might be useful to someone who wants to invest in Google.

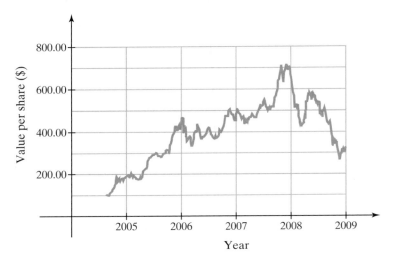

Figure 1 Google Stock Value

In algebra, we use graphs to show relationships defined by *equations in two variables*. An **equation in two variables** contains two distinct variables. Three examples of such equations follow:

$$p + q = 1, \quad y = x^2 + 3x - 4, \quad \text{and} \quad rt = 12$$

My interactive video summary

📹 Watch this interactive video to practice identifying equation in two variables.

To graph such equations, we use the **rectangular coordinate system**, also known as the *Cartesian coordinate system* in honor of its inventor René Descartes. Two perpendicular real number lines form the **coordinate axes**. The horizontal axis is called the ***x*-axis**, and the vertical axis is called the ***y*-axis**. These two axes intersect at a point called the **origin**.

The entire plane represented by this system is called the **coordinate plane**. It is also known as the **Cartesian plane** or the ***xy*-plane**. The axes divide the plane into four regions called **quadrants**. We call the upper-right region Quadrant I, the upper-left region Quadrant II, the lower-left region Quadrant III, and the lower-right region Quadrant IV. Figure 2 provides a visual of the rectangular coordinate system.

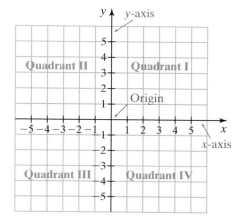

Figure 2

The Rectangular Coordinate System

We use the rectangular coordinate system to identify points on the coordinate plane. Each point on the plane represents an **ordered pair** of real numbers in the form (x, y).

The first number, x, is called the **x-coordinate** or **abscissa** and represents the number of units that the point lies to the left or right of the origin. The second number, y, is called the **y-coordinate** or **ordinate** and represents the number of units that the point lies above or below the origin.

 Watch this video for a brief overview of the rectangular coordinate system.

Figure 3 shows two ordered pairs located on the coordinate plane. Point A is 3 units to the left and 4 units above the origin, so A is identified by the ordered pair $(-3, 4)$. The x-coordinate for A is negative because the point lies to the left of the origin. The y-coordinate for A is positive because the point lies above the origin. Point B is 4 units to the right and 3 units below the origin, so B is identified by the ordered pair $(4, -3)$. The x-coordinate for B is positive because the point lies to the right of the origin. The y-coordinate for B is negative because the point lies below the origin.

My video summary

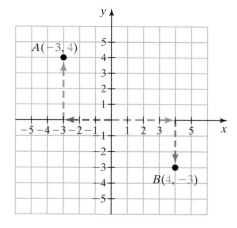

Figure 3

Compare the locations of points A and B in Figure 3. Both points include -3 and 4 as part of the ordered pair, but the numbers, or **coordinates**, appear in reverse order. Also, point A is located in Quadrant II, whereas point B is located in Quadrant IV. So the *order* of the *pair* of coordinates is just as important as the coordinates themselves and helps to explain the meaning of the term *ordered pair*.

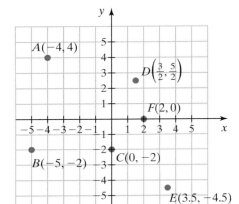

When a point lies in Quadrant II, its x-coordinate is negative, whereas its y-coordinate is positive $(x < 0, y > 0)$. When a point lies in Quadrant IV, its x-coordinate is positive, whereas its y-coordinate is negative $(x > 0, y < 0)$. Watch this video on how to locate a specific ordered pair when you know the signs of the coordinates.

When we graph an ordered pair in the coordinate plane, we are **plotting** that point.

Example 1 Plotting Ordered Pairs

Plot each ordered pair in the coordinate plane. In which quadrant or on which axis does each point lie?

a. $A(-4, 4)$ **b.** $B(-5, -2)$ **c.** $C(0, -2)$

d. $D\left(\dfrac{3}{2}, \dfrac{5}{2}\right)$ **e.** $E(3.5, -4.5)$ **f.** $F(2, 0)$

My video summary

Solution The points are plotted in Figure 4. Point A lies in Quadrant II. Point B lies in Quadrant III. Point C lies on the y-axis. Point D lies in Quadrant I. Point E lies in Quadrant IV. Point F lies on the x-axis. Watch this video for a detailed solution.

Figure 4

You Try It Work through this You Try It problem.

Work Exercise 1 in this textbook or in the *MyMathLab*® Study Plan.

OBJECTIVE 2: DETERMINE IF AN ORDERED PAIR IS A SOLUTION TO AN EQUATION

In Section 1.1 we learned that a solution to an equation in one variable is a value that, when substituted for the variable, makes the equation true. A **solution to an equation in two variables** is an ordered pair of values that, when substituted for the variables, makes the equation true.

Example 2 Determining If an Ordered Pair Is a Solution to an Equation

Determine if each ordered pair is a solution to the equation $2x - y = 7$.

a. $(2, -3)$ **b.** $(3, 1)$ **c.** $\left(-\dfrac{3}{2}, -10\right)$

My interactive video summary

Solution We substitute the value of the x-coordinate for x and the value of the y-coordinate for y and then simplify. If the resulting statement is true, then the ordered pair is a solution to the equation.

a. Begin with the original equation: $2x - y = 7$
   Substitute 2 for x and -3 for y: $2(2) - (-3) \overset{?}{=} 7$
                        Simplify: $4 + 3 \overset{?}{=} 7$
                                   $7 = 7$ True

The final statement is true, so $(2, -3)$ is a solution to the equation.

b. Substitute 3 for x and 1 for y. Work through the simplification to determine if the ordered pair is a solution. Click here or watch this interactive video to see the detailed solution.

c. Substitute the values for the variables and work through the simplification to determine if the ordered pair is a solution. Click here or watch this interactive video to see the detailed solution.

You Try It   Work through this You Try It problem.

Work Exercises 2–4 in this textbook or in the *MyMathLab*® Study Plan.

OBJECTIVE 3:   FIND UNKNOWN COORDINATES

The **graph of an equation in two variables** includes all points whose coordinates are solutions to the equation. To make such a graph, we can plot several points that satisfy the equation. Then we connect the points with a line or smooth curve.

In Example 2, we determined if a given ordered pair was a solution to an equation in two variables. But how do we find such ordered pair solutions if we only know one coordinate? We choose a convenient value for one variable, substitute it into the equation, and solve for the other variable. The resulting values for both variables together form an **ordered pair solution** to the equation.

Example 3 Finding Unknown Coordinates

My interactive video summary

Find the unknown coordinate so that each ordered pair satisfies $3x + 4y = 20$.

a. $(8, ?)$ **b.** $(?, 2)$ **c.** $\left(-\dfrac{2}{3}, ?\right)$

Solution

a. To find the y-coordinate when the x-coordinate is 8, substitute 8 for x and solve for y.

            Begin with the original equation: $3x + 4y = 20$
                      Substitute 8 for x: $3(8) + 4y = 20$
                                Simplify: $24 + 4y = 20$
              Subtract 24 from both sides: $4y = -4$
                  Divide both sides by 4: $y = -1$

The ordered pair $(8, -1)$ is a solution to the equation $3x + 4y = 20$.

b. Begin with the original equation: $3x + 4y = 20$

Substitute 2 for y: $3x + 4(2) = 20$

Simplify: $3x + 8 = 20$

Finish solving for x to find the x-coordinate when the y-coordinate is 2. Once completed, click here or watch this interactive video to check your work.

c. Substitute the x-coordinate into the equation and solve for y to determine the y-coordinate. Once finished, click here or watch this interactive video to check your work.

You Try It Work through this You Try It problem.

Work Exercises 5–7 in this textbook or in the MyMathLab® Study Plan.

OBJECTIVE 4:   GRAPH EQUATIONS BY PLOTTING POINTS

Along with finding solutions to equations in two variables, we can identify other key points to help us graph equations. For example, an **endpoint** represents the end of the graph in one direction. Other key points include **maximum points** (high points on a graph) or **minimum points** (low points on a graph). Let's use the following strategy to graph equations by plotting points:

Strategy for Graphing Equations by Plotting Points

Step 1. Find several points that satisfy the equation. The exact number of points to find depends on the equation being graphed. Find enough to form a pattern. Try to locate key points such as endpoints and maximum or minimum points.

Step 2. Plot the points found in step 1.

Step 3. Connect the points with a straight line or smooth curve, depending on the pattern formed.

Note: We will discuss specific types of graphs as we move through this text. For example, we discuss lines in Section 2.4 and parabolas in Section 8.2. In these sections, we provide specialized techniques for finding key points that can be used for their graphs.

Example 4 Graphing Equations by Plotting Points

My interactive video summary

 Graph each equation by plotting points.

a. $2x + y = 1$ **b.** $y = x^2 - 4$ **c.** $y = |x|$

Solution

a. Step 1. To find points that satisfy $2x + y = 1$, let $x = -2, -1, 0, 1$, and 2. Substitute each x-coordinate into the original equation and solve for y to find the y-coordinate. The results appear in Table 1. Watch this interactive video to see a detailed solution.

Table 1

x	Original Equation:	$2x + y = 1$	y	Point
-2	Substitute -2 for x: Solve for y:	$2(-2) + y = 1$ $y = 5$	5	$(-2, 5)$
-1	Substitute -1 for x: Solve for y:	$2(-1) + y = 1$ $y = 3$	3	$(-1, 3)$
0	Substitute 0 for x: Solve for y:	$2(0) + y = 1$ $y = 1$	1	$(0, 1)$
1	Substitute 1 for x: Solve for y:	$2(1) + y = 1$ $y = -1$	-1	$(1, -1)$
2	Substitute 2 for x: Solve for y:	$2(2) + y = 1$ $y = -3$	-3	$(2, -3)$

Step 2. Plot the points found in step 1. See Figure 5(a).

Step 3. The points form a straight pattern, so we connect them with a straight line. We draw arrows on each end to show that the line continues forever. See Figure 5(b).

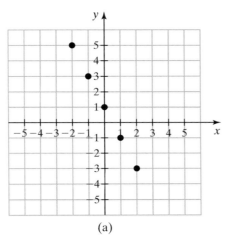

(a)

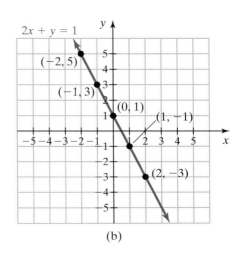

(b)

Figure 5

b. **Step 1.** To find points that satisfy $y = x^2 - 4$,
let $x = -3, -2, -1, 0, 1, 2,$ and 3. The results appear in Table 2.

x	$y = x^2 - 4$	Point
-3	$y = (-3)^2 - 4 = 5$	$(-3, 5)$
-2	$y = (-2)^2 - 4 = 0$	$(-2, 0)$
-1	$y = (-1)^2 - 4 = -3$	$(-1, -3)$
0	$y = (0)^2 - 4 = -4$	$(0, -4)$
1	$y = (1)^2 - 4 = -3$	$(1, -3)$
2	$y = (-2)^2 - 4 = 0$	$(2, 0)$
3	$y = (3)^2 - 4 = 5$	$(3, 5)$

Table 2

Step 2. Plot the points found in step 1. Click here to check your work.

Step 3. Do you see a pattern? Connect the points appropriately with a straight line or smooth curve. Click here to check your graph, or watch this interactive video for a detailed solution.

c. The equation $y = |x|$ is called the absolute value equation. Try graphing the equation yourself by plotting points. Substitute some negative values for x as well as positive values. When done, check your graph by clicking here or watching this interactive video.

You Try It   Work through this You Try It problem.

Work Exercises 8–15 in this textbook or in the _MyMathLab_® Study Plan.

OBJECTIVE 5:   FIND x- AND y-INTERCEPTS

Look at Figure 5(b). Notice that the graph of $2x + y = 1$ crosses the y-axis at the point $(0, 1)$ and also crosses the x-axis at the point $\left(\frac{1}{2}, 0\right)$. The points where a graph crosses or touches the axes are called **intercepts**. The graph of $2x + y = 1$ has two intercepts: $(0, 1)$ and $\left(\frac{1}{2}, 0\right)$.

A **y-intercept** is the y-coordinate of a point where a graph crosses or touches the y-axis. An **x-intercept** is the x-coordinate of a point where a graph crosses or touches the x-axis. So for the graph of $2x + y = 1$, the y-intercept is 1 and the x-intercept is $\frac{1}{2}$. Figure 6 illustrates examples of x- and y-intercepts.

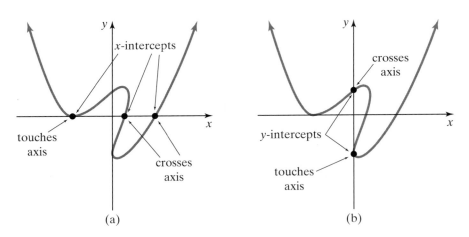

Figure 6 x- and y-Intercepts

Example 5 Finding x- and y-Intercepts

Find the intercepts of the graph shown in Figure 7. What are the x-intercepts? What are the y-intercepts?

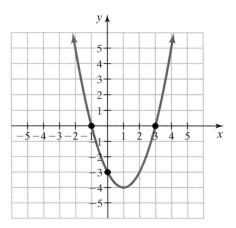

Figure 7

Solution The graph crosses the y-axis at $(0, -3)$ and crosses the x-axis at $(-1, 0)$ and $(3, 0)$, so the intercepts are $(0, -3), (-1, 0)$, and $(3, 0)$. The x-intercepts are -1 and 3. The y-intercept is -3.

Note: When the type of intercept is not specified, we must list the intercept as an ordered pair. If the type of intercept is specified, then we list only the coordinate as the intercept. For example, in Figure 7, an intercept is $(3, 0)$. However, the x-intercepts are -1 and 3, and the y-intercept is -3.

You Try It Work through this You Try It problem.

Work Exercises 16–18 in this textbook or in the **MyMathLab**® Study Plan.

2.1 Exercises

1. Plot each ordered pair in the coordinate plane. In which quadrant or on which axis does each point lie?

You Try It

a. $A(-1, 5)$ **b.** $B(0, 4)$ **c.** $C(1, 3)$

d. $D\left(\dfrac{7}{2}, -\dfrac{5}{2}\right)$ **e.** $E(-3, 0)$ **f.** $F(-5, -2)$

In Exercises 2–4, determine if the given ordered pairs are solutions to the equations.

2. $3x - 2y = 8$

You Try It

a. $(2, -1)$

b. $(0, 4)$

c. $\left(3, \dfrac{1}{2}\right)$

d. $(-4, -10)$

3. $y = 2x^2 - x + 5$

a. $(-1, 4)$

b. $(-2, 11)$

c. $(0, 5)$

d. $\left(\dfrac{1}{2}, 5\right)$

4. $y = |x - 3|$

a. $(2, 5)$

b. $(-1, 4)$

c. $(5, 2)$

d. $(1, 2)$

In Exercises 5–7, find the unknown coordinate so that each ordered pair is a solution to the given equation.

You Try It

5. $4x - 3y = 6$

   a. $(-3, ?)$

   b. $\left(?, -\dfrac{2}{3}\right)$

6. $y = 2x - 5$

   a. $\left(-\dfrac{1}{2}, ?\right)$

   b. $(?, 7)$

7. $y = \dfrac{2}{3}x + 2$

   a. $(-9, ?)$

   b. $(?, 5)$

In Exercises 8–15, graph each equation by plotting points.

8. $y = 2x + 1$

You Try It **10.** $3x - 2y = 8$

12. $y = 3x^2$

14. $y = |x + 3|$

9. $y = \dfrac{1}{4}x - 2$

11. $3x + y = 3$

13. $y = x^2 - 2x - 3$

15. $y = |x| - 2$

In Exercises 16–18, find the x- and y-intercepts of each graph.

16.

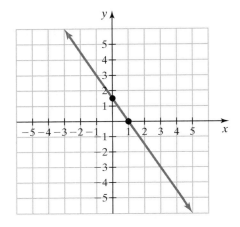

17.

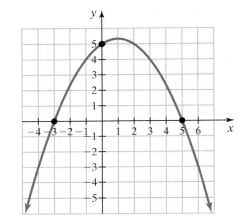

18.

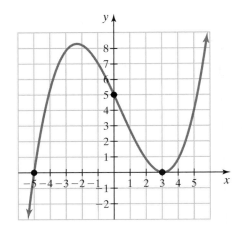

2.2 Relations and Functions

THINGS TO KNOW

Before working through this section, be sure you are familiar with the following concepts:

VIDEO   ANIMATION   INTERACTIVE

You Try It

1. Identify Sets (Section R.1, Objective 1)

You Try It

2. Use Interval Notation to Express the Solution Set of an Inequality (Section 1.2, Objective 3)

You Try It

3. Solve a Formula for a Given Variable (Section 1.4, Objective 1)

OBJECTIVES

1 Identify Independent and Dependent Variables

2 Find the Domain and Range of a Relation

3 Determine If Relations Are Functions

4 Determine If Graphs Are Functions

5 Solve Application Problems Involving Relations and Functions

---

OBJECTIVE 1:   IDENTIFY INDEPENDENT AND DEPENDENT VARIABLES

When applying math to everyday life, often we encounter situations where one quantity is related to another. The cost to fill a gas tank is related to the number of gallons purchased. The amount of simple interest paid on a loan is related to the amount owed. The total cost to manufacture LCD televisions is related to the number of televisions produced, and so on.

In Section 1.4 we solved formulas for a given variable. In doing so, we had to express a relationship between the given variable and any remaining variables. When we solve for a variable, that variable is called the **dependent variable** because its value *depends on* the value(s) of the remaining variable(s). Any remaining variables are called **independent variables** because we are free to select their values.

If the average price per gallon of regular unleaded gas is $1.94 on a given day, then two gallons would cost $1.94 \cdot (2) = \$3.88$. Similarly, three gallons would cost $1.94 \cdot (3) = \$5.82$. Therefore, the cost to fill a tank is related to the number of gallons of gas purchased. We can model the situation with the equation

$$y = 1.94x,$$

where $y =$ cost in dollars and $x =$ number of gallons of regular unleaded gas purchased.

Since the equation is solved for y, we identify y (cost) as the dependent variable and x (number of gallons) as the independent variable. When an equation involving x and y is not solved for either variable, we will call x the independent variable and y the dependent variable.

Example 1 Identifying Independent and Dependent Variables

For each of the following equations, identify the dependent variable and the independent variable(s).

a. $y = 3x + 5$ **b.** $w = ab + 3c^2$ **c.** $3x^2 + 9y = 12$

Solution

a. Since the equation is solved for y, we identify y as the dependent variable. The remaining variable, x, is the independent variable.

b. Since the equation is solved for w, we identify w as the dependent variable. The remaining variables, a, b, and c, are independent variables.

c. Since the equation is not solved for either variable, we identify x as the independent variable and y as the dependent variable.

You Try It   **Work through this You Try It problem.**

Work Exercises 1–4 in this textbook or in the *MyMathLab*® Study Plan.

In Example 1 we see that it is possible to have more than one independent variable in an equation. However, we will limit our discussion mainly to situations involving one dependent and one independent variable. Our gasoline cost model, $y = 1.94x$, is an example of such a situation.

OBJECTIVE 2:   FIND THE DOMAIN AND RANGE OF A RELATION

A **relation** is a correspondence between two sets of numbers that can be represented by a set of ordered pairs.

Definition

A **relation** is a set of ordered pairs.

In Section 2.1 we learned that equations in two variables define a set of ordered pair solutions. When writing ordered pair solutions, we write the value of the independent variable first, followed by the value of the dependent variable. For example, if 1 gallon of regular unleaded gas is purchased, the total cost is $1.94(1) = \$1.94$, which gives the ordered pair $(1, 1.94)$. If 2 gallons are purchased, the total cost is $1.94(2) = \$3.88$, which gives the ordered pair $(2, 3.88)$. We can create more ordered pair solutions by following this process. Click here to see a table of values.

Since a graph is a visual representation of the ordered pair solutions to an equation, we consider equations and graphs to be relations because they define a set of ordered pairs. A relation shows the connection between the set of values for the independent variable, called the **domain** (or *input values*), and the set of values for the dependent variable, called the **range** (or *output values*).

At this point, you may wish to review sets in Section R.1 and interval notation in Section 1.2.

Definitions

The **domain** is the set of all values for the independent variable. These are the first coordinates in the set of ordered pairs and are also known as *input values*.

The **range** is the set of all values for the dependent variable. These are the second coordinates in the set of ordered pairs and are also known as *output values*.

Example 2 Finding the Domain and Range of a Relation

Find the domain and range of each relation.

My interactive video summary

 a. $\{(-5, 7), (3, 5), (6, 7), (12, -4)\}$

b.

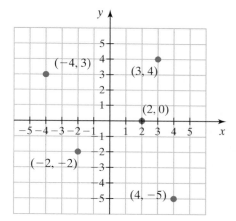

Solution For parts a and b, identify the first coordinates of each ordered pair to find the domain and the second coordinates of each ordered pair to find the range. Click here to check your answers, or watch this interactive video for detailed solutions.

You Try It Work through this You Try It problem.

Work Exercises 5–6 in this textbook or in the *MyMathLab*® Study Plan.

Example 3 Finding the Domain and Range of a Relation

Find the domain and range of each relation.

My interactive video summary **a.**

b.

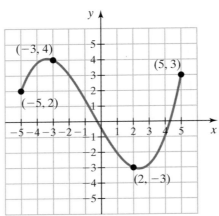

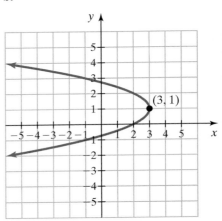

c. $y = |x - 1|$

Solution Click here to check your answers, or watch this interactive video for detailed solutions.

You Try It   **Work through this You Try It problem.**

Work Exercises 7–8 in this textbook or in the *MyMathLab*® Study Plan.

When working with application problems, we often need to restrict the domain to use only those values that make sense within the context of the situation. This restricted domain is called the **feasible domain**. The feasible domain is the set of values for the independent variable that make sense, or are *feasible*, in the context of the application. For example, in our gasoline cost model, $y = 1.94x$, the domain of the equation is all real numbers. However, it doesn't make sense to use negative numbers in the domain since x represents the number of gallons of gas purchased. Therefore, the feasible domain would be all real numbers greater than or equal to 0, written as $\{x \mid x \geq 0\}$ or $[0, \infty)$.

OBJECTIVE 3: DETERMINE IF RELATIONS ARE FUNCTIONS

A relation *relates* one set of numbers, the domain, to another, the range. When each value in the domain corresponds to (is paired with) exactly one value in the range, we have a special type of relation called a **function**.

Definition
A **function** is a special type of relation in which each value in the domain corresponds to exactly one value in the range.

Given a set of ordered pairs, we can determine if the relation is a function by looking at the x-coordinate. If no x-coordinate is repeated, then the relation is a function because each input value corresponds to exactly one output value. If the same x-coordinate corresponds to two or more different y-coordinates, then the relation is not a function.

Given an equation, we can test input values to see if, when substituted into the equation, there is more than one output value. In order for an equation to be a function, each input value must correspond to one and only one output value.

Example 4 Determining If Relations Are Functions

Determine if each of the following relations is a function.

My interactive video summary

a. $\{(-3, 6), (2, 5), (0, 6), (17, -9)\}$

b. $\{(4, 5), (7, -3), (4, 10), (-6, 1)\}$

c. $\{(-2, 3), (0, 3), (4, 3), (6, 3), (8, 3)\}$

d. $|y - 5| = x + 3$

e. $y = x^2 - 3x + 2$

f. $4x - 8y = 24$

Solution Click here to see which relations are functions, or watch this interactive video for detailed explanations.

You Try It Work through this You Try It problem.

Work Exercises 9–16 in this textbook or in the **MyMathLab**® Study Plan.

OBJECTIVE 4: DETERMINE IF GRAPHS ARE FUNCTIONS

If a relation appears as a graph, we can determine if the relation is a function by using the **vertical line test**.

> **Vertical Line Test**
>
> If a vertical line intersects (crosses or touches) the graph of a relation at more than one point, then the relation is not a function. If every vertical line intersects the graph of a relation at no more than one point, then the relation is a function.

 Why does the vertical line test work? Watch this animation to find out.

 Example 5 Determining If Graphs Are Functions

Use the vertical line test to determine if each graph is a function.

a.
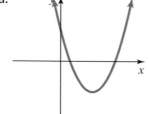

b.
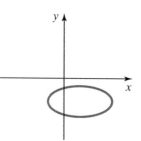

c.
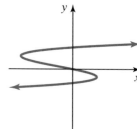

d.
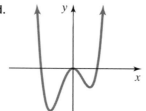

e.
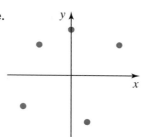

f.
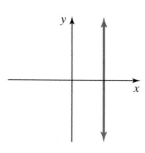

Solution Apply the vertical line test to each graph. Click here to check your answers or work through this animation.

You Try It Work through this You Try It problem.

Work Exercises 17–24 in this textbook or in the *MyMathLab*® Study Plan.

How many x- and y-intercepts can the graph of a function have? Click here to find out.

OBJECTIVE 5: SOLVE APPLICATION PROBLEMS INVOLVING RELATIONS AND FUNCTIONS

Real-world situations can be described by relations and functions. We use mathematical models to describe these situations.

Example 6 Video Entertainment and Sleep

The data in the following table represent the average daily hours of sleep and average daily hours of video entertainment for six students at a local college.

Video Entertainment	Sleep	Video Entertainment	Sleep
8	4	5	7
7	5	4	8
2	9	7	6

a. If a researcher believes the number of hours of video entertainment affects the number of hours of sleep, identify the independent variable and the dependent variable.

b. What are the ordered pairs for this data?

c. What are the domain and range?

d. Is this relation a function? Explain.

Solution

a. Since the researcher believes hours of sleep is affected by, or depends on, the hours of video entertainment, the independent variable would be number of hours of video entertainment, and the dependent variable would be number of hours of sleep.

b. The independent variable is hours of video entertainment and the dependent variable is hours of sleep, so the corresponding ordered pairs are $\{(8, 4), (7, 5), (2, 9), (5, 7), (4, 8), (7, 6)\}$.

c. The domain is the set of first coordinates (x-coordinates) from the ordered pairs in part (b), and the range is the set of second coordinates (y-coordinates). Therefore, the domain is $\{2, 4, 5, 7, 8\}$ and the range is $\{4, 5, 6, 7, 8, 9\}$.

d. This relation is not a function because one value from the domain, 7, corresponds to more than one value from the range, 5 and 6. This is shown in the ordered pairs $(7, 5)$ and $(7, 6)$.

You Try It Work through this You Try It problem.

Work Exercise 25 in this textbook or in the *MyMathLab*® Study Plan.

Example 7 High-Speed Internet Access

My video summary

The percent of households, y, with high-speed Internet access in 2007 can be modeled by the equation $y = 0.70x + 20.03$, where x is the annual household income (in $1000s). (*Source:* U.S. Department of Commerce)

a. Identify the independent and dependent variables.

b. Use the model equation to estimate the percent of households in 2007 with high-speed Internet access (to the nearest whole percent) if the annual household income was $50,000. What point would this correspond to on the graph of the equation?

c. Is the relation a function? Explain.

d. Determine the feasible domain.

Solution

a. Since the equation is solved for y in terms of x, the independent variable is the annual household income (in $1000s) and the dependent variable is the percent of households in 2007 with high-speed Internet access.

Watch this video to see the rest of the solution.

You Try It Work through this You Try It problem.

Work Exercises 26–28 in this textbook or in the *MyMathLab*® Study Plan.

2.2 Exercises

In Exercises 1–4, identify the independent and dependent variables.

You Try It

1. $y = 3x^2 - 7x + 5$ **2.** $k = 12v^2$ **3.** $4x + 5y = 18$ **4.** $e = \dfrac{\sqrt{a^2 - b^2}}{a}$

In Exercises 5–8, find the domain and range of each relation.

u Try It

5. $\{(4, -2), (1, -1), (0, 0), (1, 1), (4, 2)\}$ **6.**

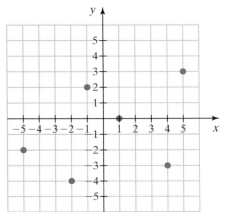

You Try It

7.

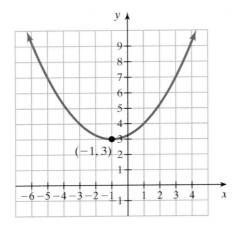

8.

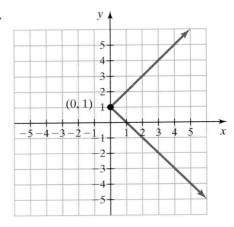

In Exercises 9–16, determine if the relation is a function. Assume x is the independent variable.

You Try It

9. $\{(2, 9), (3, 2), (7, 4), (8, 1), (10, -8)\}$

10. $\{(3, 4), (4, 5), (6, 3), (4, 1), (9, 12)\}$

11. $\{(3, 7), (5, 7), (8, 10), (11, 10), (15, 6)\}$

12. $\{(4, -5), (4, 0), (4, 3), (4, 7), (4, 11)\}$

13. $3x - y = 5$ **14.** $y = |x + 3| - 1$ **15.** $|y - 1| + x = 2$ **16.** $x^2 + y = 1$

In Exercises 17–24, use the vertical line test to determine if each graph is a function.

You Try It

17.

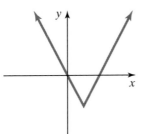

18.

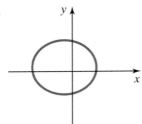

19.

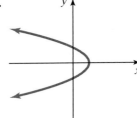

20.

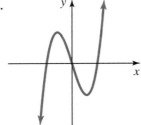

21.

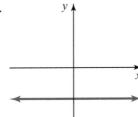

22.

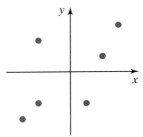

23.

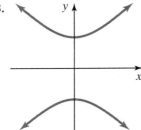

24.

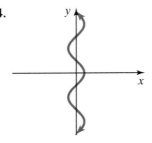

In Exercises 25–28, solve each application problem involving relations and functions.

You Try It

25. Population Data   The data in the following table represent the percent of the population with access to clean water and the percent undernourished for several South American countries in 2006.

% population with access to clean water	% undernourished
96	3
91	7
77	15
84	12
92	8

a. If a researcher believes the percent of the population with access to clean water affects the percent undernourished, identify the independent and dependent variables.

b. What are the corresponding ordered pairs for the data?

c. What are the domain and range?

d. Is this relation a function? Explain.

ou Try It

26. Medical Expenses   The equation $y = 4.09x - 321.96$ relates the amount of out-of-pocket medical expenses (in billions of dollars), x, to the amount paid by insurance (in billions of dollars), y. (*Source: U.S. Centers for Medicare & Medicaid Services, 2007*)

a. Identify the independent and dependent variables.

b. What is the feasible domain?

c. Is the relation a function? Explain.

d. Use the model to estimate the amount paid by insurance if out-of-pocket expenses were $270 billion. What point would this correspond to on the graph of the equation?

e. Solve the equation for x.

f. Use your model to estimate the amount of out-of-pocket expenses if total insurance payments were $1 trillion.

27. Gas Prices   The following graph shows the average price per gallon of regular unleaded gasoline in the United States for the years 2002–2007. (*Source: U.S. Energy Information Administration, 2008*)

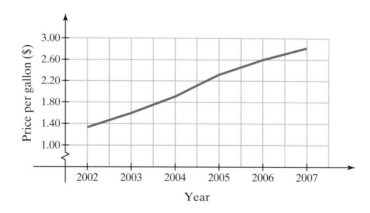

a. Is this graph a function? Explain.

b. Identify the independent and dependent variables.

c. Use the graph to estimate the average price per gallon in 2006.

d. If the data can be modeled by the equation $y = 0.30x + 0.72$, where x is the number of years since 2000 and y is the price per gallon (in dollars), in what year does the model predict that the price per gallon will reach $5.00?

28. **College Instructors**   The percentage of full-time instructors at colleges and universities can be modeled by the equation $y = -0.036x^2 - 0.066x + 61.002$, where x is the number of years since 1990. (*Source:* U.S. National Center for Educational Statistics)

   a. Sketch a graph of the equation for the domain $0 \leq x \leq 25$ by plotting points.

   b. What is the y-intercept? What does it mean in the context of the problem?

   c. Use the model to estimate the percentage of full-time instructors in the year 2030.

   d. How would you interpret an x-intercept in this case? Is it reasonable within the context of the problem?

2.3 Function Notation and Applications

THINGS TO KNOW

Before working through this section, be sure you are familiar with the following concepts:

VIDEO   ANIMATION   INTERACTIVE

You Try It   1. Use Order of Operations to Evaluate Numeric Expressions (Section R.2, **Objective 4**)

You Try It   2. Evaluate Algebraic Expressions (Section R.3, **Objective 1**)

You Try It   3. Solve a Formula for a Given Variable (Section 1.4, **Objective 1**)

OBJECTIVES

1 Express Equations of Functions Using Function Notation

2 Evaluate Functions

3 Graph Simple Functions by Plotting Points

4 Interpret Graphs of Functions

5 Solve Application Problems Involving Functions

OBJECTIVE 1:   EXPRESS EQUATIONS OF FUNCTIONS USING FUNCTION NOTATION

Functions expressed as equations are often named using letters such as f, g, and h. The symbol $f(x)$ is read as "f of x" and is an example of **function notation**. We can use function notation in place of the dependent variable in the equation of a function. For example, the function $y = 2x + 3$ may be written as $f(x) = 2x + 3$.

The symbol $f(x)$ represents the value of the dependent variable (output) for a given value of the independent variable (input). For $y = f(x)$, we can interpret $f(x)$ as follows: f is the name of the function that relates the independent variable x to the dependent variable y.

Do not confuse function notation with multiplication. $f(x)$ does not mean $f \cdot x$.

When using function notation, any symbol can be used to name the function, and any other symbol can be used to represent the independent variable. For example, consider the function $y = 2x + 3$, which tells us that the value of the dependent variable is obtained by multiplying the value of the independent variable by 2 and then adding 3. We can express this function in many ways:

$$f(x) = 2x + 3 \quad \text{function name: } f; \quad \text{independent variable: } x$$
$$H(t) = 2t + 3 \quad \text{function name: } H; \quad \text{independent variable: } t$$
$$P(r) = 2r + 3 \quad \text{function name: } P; \quad \text{independent variable: } r$$
$$\Phi(n) = 2n + 3 \quad \text{function name: } \Phi; \quad \text{independent variable: } n$$

All four of these functions are equivalent even though they have different function names and different letters representing the independent variable. **Equivalent functions** represent the same set of ordered pairs.

When possible, we choose letters for the variables to provide meaning. If writing a function that represents the perimeter of a square, for example, we might use P to represent "perimeter" and s to represent "side length." The function notation $P(s)$ represents the perimeter of a square with side length s. For this function, s is the independent variable, P is the dependent variable, and $P(s) = 4s$ is the function that tells us how to find the perimeter P from the side length s.

One benefit of using function notation is that it clearly shows the relationship between the independent and dependent variables of an equation. Any equation of a function can be written in function notation using the following procedure:

Expressing Equations of Functions Using Function Notation

Step 1. Choose an appropriate name for the function.

Step 2. Solve the equation for the dependent variable.

Step 3. Replace the dependent variable with equivalent function notation.

When letters are used to name functions, the case of the letters matters (upper or lowercase). For example, f and F are different symbols, so $f(x)$ and $F(x)$ represent different functions. Lowercase and uppercase letters should not be switched within a problem.

Example 1 Expressing Equations of Functions Using Function Notation

Write each function using function notation. Let x be the independent variable and y be the dependent variable.

a. $y = 2x^2 - 4$ **b.** $y - \sqrt{x} = 0$ **c.** $3x + 2y = 6$

Solution

a. We name the function f. The equation is already solved for y, so we simply replace y with $f(x)$:

Begin with the original formula: $y = 2x^2 - 4$

Replace y with $f(x)$: $f(x) = 2x^2 - 4$

b. We name the function g. We solve for y and then replace y with $g(x)$:

Begin with the original equation: $y - \sqrt{x} = 0$

Solve for y by adding $\sqrt{x}$ to both sides: $y = \sqrt{x}$

Replace y with $g(x)$: $g(x) = \sqrt{x}$

c. Try working through this process yourself. Name the function h. Click here to check your answer, or watch this video to see a fully worked solution.

You Try It   **Work through this You Try It problem.**

Work Exercises 1–10 in this textbook or in the *MyMathLab*® Study Plan.

OBJECTIVE 2:   EVALUATE FUNCTIONS

For $y = f(x)$, the symbol $f(x)$ represents the value of the dependent variable y for a given value of the independent variable x. For this reason, we call $f(x)$ the **value of the function**. For example, $f(2)$ represents the value of the function f when $x = 2$. When we determine such a function value, we *evaluate the function*.

To **evaluate a function**, we substitute the given value for the independent variable and simplify.

Example 2 Evaluating Functions

My interactive video summary

 If $f(x) = 4x - 5$, $g(t) = 3t^2 - 2t + 1$, and $h(r) = \sqrt{r} - 9$, evaluate each of the following.

a. $f(3)$ **b.** $g(-1)$ **c.** $h(16)$ **d.** $f\left(\dfrac{1}{2}\right)$

Solution

a. The notation $f(3)$ represents the value of the function f when x is 3. We substitute 3 for x in the function f and simplify:

Substitute 3 for x in the function f: $f(3) = 4(3) - 5$

Simplify: $= 12 - 5$

$= 7$

So, $f(3) = 7$, meaning that the value of f is 7 when x is 3.

b. The notation $g(-1)$ represents the value of the function g when t is -1. We substitute -1 for t in the function g and simplify.

Substitute -1 for t in the function g: $g(-1) = 3(-1)^2 - 2(-1) + 1$

Finish the problem by simplifying the right side. Click here to check your answer or watch this interactive video to see the fully worked solution.

c. The notation $h(16)$ represents the value of the function h when r is 16. We substitute 16 for r in function h and simplify. Try working through this solution by yourself. Click here to check your answer or watch this interactive video to see the fully worked solution.

d. The notation $f\left(\dfrac{1}{2}\right)$ represents the value of the function f when x is $\dfrac{1}{2}$.

We substitute $\dfrac{1}{2}$ for x in the function f and simplify:

Substitute $\dfrac{1}{2}$ for x in the function f: $f\left(\dfrac{1}{2}\right) = 4\left(\dfrac{1}{2}\right) - 5$

Finish the problem by simplifying the right side. Click here to check your answer or watch the interactive video to see the fully worked solution.

You Try It   **Work through this You Try It problem.**

Work Exercises 11–18 in this textbook or in the *MyMathLab*® Study Plan.

OBJECTIVE 3:   GRAPH SIMPLE FUNCTIONS BY PLOTTING POINTS

In Example 2, part a, we see that $f(3) = 7$. For the function f, this means that an input of 3 into the function f results in an output of 7, which corresponds to the ordered pair $(3, 7)$. We say that $(3, 7)$ **belongs** to the function f. The **graph of a function** is the graph of all ordered pairs that belong to the function. Since $f(3) = 7$, the point $(3, 7)$ lies on the graph of f. This leads us to the following theorem.

Theorem

The point (a, b) lies on the graph of a function f if and only if $f(a) = b$.

In Section 2.1 we learned to graph equations by plotting points. We now use a similar strategy to graph simple functions expressed as equations.

Strategy for Graphing Simple Functions by Plotting Points

Step 1. Find several points that belong to the function. The exact number of points to find depends on the function being graphed. Find enough to form a pattern. Try to locate key points such as endpoints and maximum and minimum points.

Step 2. Plot the points found in step 1.

Step 3. Connect the points with a straight line or smooth curve, depending on the pattern formed.

Note: We will discuss different types of functions as we move through this text. For example, we discuss linear functions in Section 2.4 and quadratic functions in Section 8.2. In these sections, we provide specific techniques that can be used to graph those functions.

Example 3 Graphing Simple Functions by Plotting Points

Graph each function by plotting points.

a. $f(x) = 2x - 1$ **b.** $g(x) = x^2 + 2x - 3$ **c.** $h(x) = 2|x| - 1$

Solution

a. Step 1. Note that the domain of f is the set of all real numbers. We can evaluate $f(x)$ for $x = -2, -1, 0, 1,$ and 2. We organize the work in Table 3.

x	$y = f(x)$	(x, y)
-2	$f(-2) = 2(-2) - 1 = -5$	$(-2, -5)$
-1	$f(-1) = 2(-1) - 1 = -3$	$(-1, -3)$
0	$f(0) = 2(0) - 1 = -1$	$(0, -1)$
1	$f(1) = 2(1) - 1 = 1$	$(1, 1)$
2	$f(2) = 2(2) - 1 = 3$	$(2, 3)$

Table 3

Step 2. Plot the points found in Table 3.

Step 3. Do you see a pattern? Connect the points appropriately with a straight line or smooth curve. Click here to check your graph or work through this animation to see the complete solution.

b. **Step 1.** The domain of g is the set of all real numbers. Evaluate $g(x)$ for $x = -4, -3, -2, -1, 0, 1,$ and 2 to create a table of ordered pairs that belong to the function. Click here to check your table.

Step 2. Plot the points found in step 1.

Step 3. Connect them with a line or smooth curve as appropriate. Click here to check your answer, or work through this animation to see the fully worked solution.

c. Create a table of ordered pairs that belong to the function. Be sure to evaluate the function for negative values of x as well as positive values of x. Plot the ordered pairs and connect them with a line or smooth curve as appropriate. Click here to check your answer, or work through this animation to see the fully worked solution.

You Try It **Work through this You Try It problem.**

Work Exercises 19–24 in this textbook or in the *MyMathLab*® Study Plan.

OBJECTIVE 4: INTERPRET GRAPHS OF FUNCTIONS

Graphs of functions can be used to show a variety of everyday situations visually. In the next three examples, we use graphs of functions to model common situations.

Example 4 Straight-Line Depreciation

The word **depreciation** means a loss in value. **Straight-line depreciation** is one of several accounting methods allowed by the Internal Revenue Service for deducting losses in the value of equipment as it ages. Claiming a loss in **book value** allows the owner to avoid paying taxes on the amount of the loss. The graph in Figure 8 uses straight-line depreciation to depict the book value of a machine as it ages. Use the graph to answer the questions.

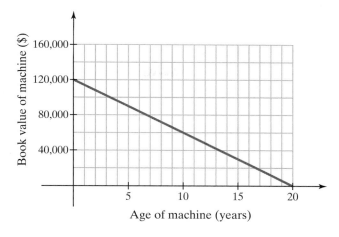

Figure 8
Straight-Line Depreciation

a. What was the book value of the machine when it was new?

b. How old will the machine be when it no longer has a book value?

c. What will the book value of the machine be when it is 5 years old?

d. At what age will the machine have lost half of its original book value?

e. By how much does the book value of the machine decrease each year?

Solution

a. The graph contains the point $(0, 120000)$, which represents the book value of the machine when it was new. This book value was $120,000.

b. The graph contains the point $(20, 0)$, which shows that the machine will have a book value of $0 (no book value) after 20 years.

c. The graph contains the point $(5, 90000)$, which represents the book value of the machine after 5 years. This book value will be $90,000.

d. From part (a), we know the original book value of the machine was $120,000. Half of this is $60,000. The graph contains the point $(10, 60000)$. So, the machine will have lost half of its original book value when it is 10 years old.

e. Since the machine's original value of $120,000 is lost evenly over a 20-year period, the book value decreases by $\dfrac{\$120,000}{20 \text{ years}} = \$6000$ each year.

Example 5 Spring Temperatures

My video summary

The graph of the function in Figure 9 gives the outside temperatures over one 24-hour period in spring. Use the graph to answer the questions.

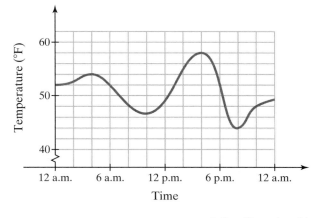

Figure 9
Spring Temperatures

a. Over what time periods was the temperature rising?

b. Over what time periods was the temperature falling?

c. What was the highest temperature for the day? At what time was it reached?

d. What was the lowest temperature for the day shown? At what time was it reached?

e. Over what time period did the temperature decrease most rapidly?

Solution Work through the questions and then watch this video to check your answers.

You Try It   **Work through this You Try It problem.**

Work Exercises 25–28 in this textbook or in the *MyMathLab*®️ Study Plan.

Example 6 Flight Altitude

A Boeing 757 jet took off and climbed steadily for 20 minutes until it reached an altitude of 18,000 feet. The jet maintained that altitude for 30 minutes. Then it climbed steadily for 10 minutes until it reached an altitude of 26,000 feet. The jet remained at 26,000 feet for 40 minutes. Then it descended steadily for 20 minutes until it reached an altitude of 20,000 feet, where it remained for 30 more minutes. During the final 20 minutes of the flight, the jet descended steadily until it landed at its destination airport. Draw a graph of the 757's altitude as a function of time.

Solution The 757's altitude is a function of time, so the independent variable *time* (in minutes) is represented by the horizontal axis and the dependent variable *altitude* is represented by the vertical axis. To draw the graph, identify key points. Plot these points and connect them. For example, the ordered pair $(0, 0)$ represents an altitude of 0 feet (ground) at time 0 minutes (before takeoff). The ordered pair $(20, 18000)$ represents an altitude of 18,000 feet at 20 minutes after takeoff. To show the initial climb of the airplane, we connect the points $(0, 0)$ and $(20, 18000)$ with a straight line segment. Continue this process. Click here to check your answer, or watch this video to see the fully worked solution.

You Try It   **Work through this You Try It problem.**

Work Exercises 29–32 in this textbook or in the *MyMathLab*®️ Study Plan.

OBJECTIVE 5:   SOLVE APPLICATION PROBLEMS INVOLVING FUNCTIONS

Functions are used to model a variety of real-world applications in fields of study such as physics, biology, business, and economics. As we move through this text, we will develop methods for creating such models ourselves. However, for the next few examples, the models used to solve applications are given.

Example 7 Falling Rock

A rock is dropped from the top of a cliff. Its height, above the ground, in feet, at t seconds is given by the function $h(t) = -16t^2 + 900$. Use the model to answer the following questions.

a. Evaluate $h(0)$. What does this value represent?

b. Evaluate $h(2)$. What does this value represent? How far has the rock fallen at this time?

c. Evaluate $h(10)$. Is this possible? Explain.

d. Evaluate $h(7.5)$. Interpret this result.

e. Determine the feasible domain and the range that makes sense (or feasible range) within the context of the problem.

f. Graph the function.

My interactive video summary

Solution

a. $h(0) = -16(0)^2 + 900 = 900$

The height of the rock at $t = 0$ seconds (before the rock is dropped) is 900 feet above the foot of the cliff. Therefore, the height of the cliff is 900 feet.

b. $h(2) = -16(2)^2 + 900 = 836$

Two seconds after being dropped, the rock is 836 feet above the ground. Therefore, the rock has fallen $900 - 836 = 64$ feet.

c. $h(10) = -16(10)^2 + 900 = -700$

A negative height means that the rock has fallen below ground level. Our result shows that the rock would be 700 feet below the ground at the base of the cliff 10 seconds after being dropped. This result is not possible (assuming there is not a 700-foot-deep hole at the base of the cliff). Therefore, $t = 10$ seconds is outside the feasible domain of the function.

d. $h(7.5) = -16(7.5)^2 + 900 = 0$

After 7.5 seconds, the rock is 0 feet above the ground, or literally on the ground. Therefore, the rock hits the ground at the foot of the cliff 7.5 seconds after being dropped.

e. The feasible domain includes only those domain values that make sense within the context of the application problem. Since t represents time, its values are nonnegative. From part (d) we know that the rock hits the ground at $t = 7.5$ seconds. This means that function values for times after 7.5 seconds will not make sense (as we saw in part (c)). The feasible domain is $\{t \mid 0 \leq t \leq 7.5\}$ or $[0, 7.5]$.

Likewise, the feasible range includes only those range values that make sense within the context of the application problem. For example, height is non-negative. From part (a) we know that $h(0) = 900$, which is the largest possible value for h. Therefore, the feasible range is $\{h \mid 0 \leq h \leq 900\}$ or $[0, 900]$.

f. Graph the function by plotting several points and connecting them with a smooth curve. Click here to check your graph, or watch the interactive video for the fully worked solution.

Example 8 Rent in Queens, New York

The average monthly rent, R, for apartments in Queens, New York, is modeled by the function $R(a) = 2.2a$, where a is the floor area of the apartment in square feet. Use the model to answer the following questions. (*Source:* apartments.com, 2008)

a. What is the average monthly rent for apartments in Queens, New York, with a floor area of 800 square feet?

b. What is the floor area of an apartment if its rent is $1430 per month?

c. Determine the feasible domain and the feasible range of the function.

d. Graph the function.

My interactive video summary

Solution

a. $R(800) = 2.2(800) = 1760$
The average monthly rent for an 800-square-foot apartment in Queens is \$1760.

b. In this case, we solve for a when $R(a) = 1430$.

Rewrite the original function: $R(a) = 2.2a$

Substitute 1430 for $R(a)$: $1430 = 2.2a$

Divide both sides by 2.2: $\dfrac{1430}{2.2} = \dfrac{2.2a}{2.2}$

Simplify: $650 = a$

So, the floor area of an apartment with a monthly rent of \$1430 is 650 square feet.

c. For the application situation to make sense, the floor area of an apartment must not be negative. In a practical sense, it is not likely to find apartments with floor areas of 1 square foot or 100,000 square feet. However, there are no definite limits. Therefore, the feasible domain is $\{a|a \geq 0\}$ or $[0, \infty)$.

Also, the problem does not make sense if the monthly rent is negative. As with floor area, you are not likely to find apartments with monthly rents of more than \$1,000,000. However, there is no definite maximum limit. Therefore, the feasible range is $\{R|R \geq 0\}$ or $[0, \infty)$.

d. Graph the function by plotting points. Click here to check your answer or watch this interactive video for the fully worked solution.

You Try It   Work through this You Try It problem.

Work Exercises 33–36 in this textbook or in the *MyMathLab*® Study Plan.

2.3 Exercises

In Exercises 1–10, write each function using function notation. Let x be the independent variable and y be the dependent variable. Use the letter f to name each function.

 You Try It

1. $y = |2x - 5|$ **2.** $y = 3x^2 + 2x - 5$ **3.** $y + \sqrt{x} = 1$ **4.** $2x + y = 3$

 5. $3x + 4y = 12$ **6.** $-6x + 18y = 12$ **7.** $4x^2 - 2y = 10$

8. $3y + 6\sqrt{x - 5} = 0$ **9.** $\dfrac{3y - 7}{2} = 3x^2 + 1$ **10.** $\dfrac{5y - 8}{3} = \dfrac{10x^2 + 4}{6}$

 You Try It   In Exercises 11–18, evaluate each function.

11. $f(x) = 3x - 5;$ $f(6)$ **12.** $h(x) = 2x^2 + 5x - 17;$ $h(-4)$

13. $F(z) = 2|z - 3| - 5;$ $F(0)$ **14.** $T(t) = \dfrac{5}{6}t + \dfrac{1}{3};$ $T(8)$

15. $r(x) = 3 + \sqrt{x - 5};$ $r(9)$ **16.** $c(x) = \sqrt{25 - x^2};$ $c(3)$

17. $\Phi(p) = (p - 1)p^3;$ $\Phi(3)$ **18.** $R(x) = 8x^2 - 2x + 1;$ $R\left(-\dfrac{1}{2}\right)$

In Exercises 19–24, graph each function by plotting points.

You Try It

19. $f(x) = 2x - 3$

20. $g(x) = -\dfrac{1}{2}x + 4$

21. $h(x) = x^2 - 4x + 3$

22. $F(x) = |x + 1| - 3$

23. $G(x) = -2|x| + 3$

24. $H(x) = \sqrt{x + 4}$

In Exercises 25–28, solve each application problem.

You Try It

25. Text Messaging The graph of the function shown illustrates a text messaging plan offered by AT&T Wireless. Use the graph to answer the following questions.

   a. Under this plan, what is the monthly cost for the first 200 text messages?

   b. After the first 200 text messages, what is the cost per text message under this plan?

   c. What is the monthly cost for 600 text messages?

   d. What are the feasible domain and the feasible range of this function?

   e. A second AT&T Wireless plan offers unlimited texting for $20 per month. Describe the number of text messages that must be used per month in order to make this second plan a better deal.

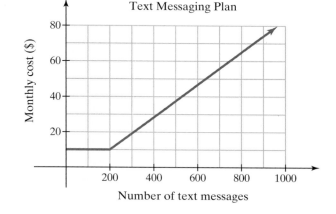

26. Straight-Line Depreciation A rancher bought a new truck and will use the truck's depreciation in book value as a tax write-off. The graph of the function shown illustrates the rancher's plan. Use the graph to answer the following questions.

   a. What did the rancher pay for the truck?

   b. How old will the truck be when its book value is $0?

   c. What are the feasible domain and the feasible range of this function?

   d. What will the book value be when the truck is 4 years old?

   e. How much money (in book value depreciation) can the rancher write off on taxes each year?

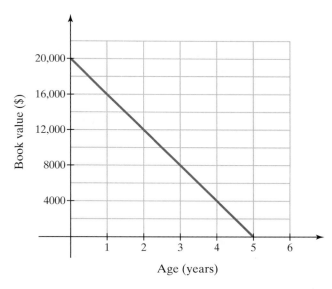

27. Dow Jones Industrial Average The graph of the function shown illustrates the 2008 Dow Jones Industrial Average (DJIA). Each tick mark on the horizontal axis represents the first opening day of the market for that given month. Use the graph to answer the following questions.

a. Over which month(s) did the DJIA increase?

b. Over which month(s) did the DJIA stay relatively the same?

c. Over which month(s) did the DJIA decrease?

d. Over which month did the DJIA decrease most rapidly?

e. Approximate the DJIA's value at the beginning of 2008.

f. Approximate its value at the end of 2008.

g. Approximately how much of the DJIA's value was lost over 2008?

h. Approximate the average loss per month.

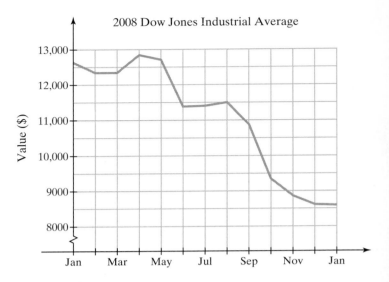

2008 Dow Jones Industrial Average

28. Fishing Rod Production   A fishing rod manufacturer developed a model for the profit expected when a given number of fishing rods is produced. The graph of the profit function is shown. Use this graph to answer the following questions.

a. If the manufacturer produces 7000 fishing rods per year, what is the expected profit?

b. For what interval(s) of fishing rods produced will the expected profit increase?

c. For what interval(s) of fishing rods produced will the expected profit decrease?

d. What is the maximum profit that can be expected in a year? How many fishing rods should it produce to achieve this profit?

e. If the company achieves the maximum profit found in part (d), what would be the average profit per fishing rod produced?

f. If the manufacturer produces 4000 fishing rods per year, what would be the average profit per fishing rod produced?

g. If you were the company's chief executive officer (CEO), how many fishing rods would you have produced? Explain why you gave this answer.

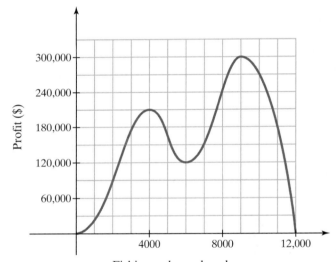

Fishing rods produced per year

In Exercises 29–30, choose the graph that best describes the situation.

29. Ferris Wheel   Jasmine rides on a Ferris wheel. Which graph best represents Jasmine's height above ground (while on the ride) as a function of time?

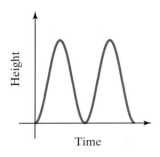
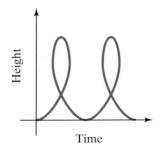
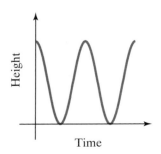

30. Falling Ball While playing in his tree house, Tyrone drops a ball out of its window. Which graph best represents the height of the ball above ground as a function of time?

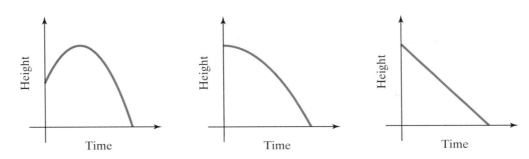

In Exercises 31–32, draw a graph of a function that represents each application situation.

You Try It

31. Exercise One morning, Karen took a walk. She left her home and walked 4 blocks in 4 minutes at a steady speed. Then she realized that she did not have her cell phone. Being on call for work, she jogged home in 2 minutes to get it. Just as she arrived home, her cell phone rang, so Karen remained at home for 3 minutes during the call. Now pressed for time, she decided to jog instead of walk. She jogged 10 blocks in 5 minutes when her cell phone rang again. She then stopped for 4 minutes to take the call. Now being even more pressed for time, Karen ran 9 blocks back towards her home in 3 minutes. To cool down, she walked the last block home in 1 minute. Draw a graph of Karen's distance from home (in blocks) as a function of time (in minutes).

32. Stock Price Bill bought stock in a company for $10 per share. After his purchase, the price remained constant for two days. At that time, a negative report was released about the company, so the price declined steadily for three days until it had lost a total of $6 per share. The price remained steady for four days. Then, a positive report was released about the company, so the price rose sharply for two days, gaining $8 per share. The price continued gaining gradually for four more days, ultimately rising $1 more per share. At that point, the price began to decline slowly for six days, losing $2 per share. Then, Bill sold the stock. Draw a graph of the stock's value (in dollars per share) as a function of the time (in days).

In Exercises 33–36, solve each application problem.

You Try It

33. Car Payments Car companies occasionally offer qualified buyers an incentive of 0% financing and no money down. Ethan just purchased a new automobile under such an agreement. The amount of money, A, that he will still owe the company after n monthly payments have been made is given by the function, $A(n) = 34{,}515 - 575.25n$. Use the model to answer the following questions:

   a. Evaluate $A(0)$. What does this value represent?

   b. Evaluate $A(24)$. What does this value represent?

   c. Evaluate $A(72)$. Is this possible? Explain.

   d. After 3 years of making payments, how much will Ethan still owe on the car?

   e. How long will it take Ethan to pay off the car?

   f. Determine the feasible domain and the feasible range of the function in the context of the problem.

   g. Graph the function by plotting points.

34. **Gas Mileage** The amount of gas, G (in gallons), that remains in the tank of a 2009 Toyota Prius after it has been driven m miles, starting with a full tank, is given by the function, $G(m) = \dfrac{571.2 - m}{48}$. (*Source:* cars.com)

  a. Evaluate $G(100)$. What does this value represent?
  b. Evaluate $G(200)$. What does this value represent?
  c. How much gas does a full tank hold?
  d. How much gas will remain in the tank after it has been driven 400 miles, starting with a full tank?
  e. After filling up, how far can the Prius travel before it runs out of gas?
  f. Determine the feasible domain and the feasible range of the function in the context of this problem.
  g. Graph the function by plotting points.

35. **Willis Tower** If an object is dropped from the roof of the Willis Tower (formerly known as the Sears Tower), its height, h, in meters, at t seconds would be given approximately by the function $h(t) = -4.9t^2 + 442.225$. Use the function to answer the following questions. (*Source:* www.willistower.com).

  a. Evaluate $h(0)$. What does this value represent?
  b. Evaluate $h(5)$. What does this value represent?
  c. Evaluate $h(10)$. Is this possible? Explain.
  d. Evaluate $h(9.5)$. What does this value represent?
  e. Determine the feasible domain and the feasible range of the function in the context of this problem.
  f. Graph the function by plotting points.

36. **Area of a Circle** The area A of a circle with circumference C is given by the function $A(C) = \dfrac{C^2}{4\pi}$. Use this function to answer the following questions:

  a. Evaluate $A(4\pi)$. What does this value represent?
  b. Evaluate $A(12)$. What does this value represent?
  c. Determine the area of a circle with a circumference of 16π (in inches).
  d. Determine the feasible domain and the feasible range of the function in the context of this problem.
  e. Graph the function by plotting points. (Hint: To make graphing easier, substitute multiples of π for C.)

2.4 Graphs of Linear Functions

THINGS TO KNOW

Before working through this section, be sure you are familiar with the following concepts:

| | VIDEO | ANIMATION | INTERACTIVE |

You Try It

1. Graph Equations by Plotting Points
   (Section 2.1, **Objective** 4)

You Try It

2. Find x- and y-intercepts
   (Section 2.1, **Objective** 5)

You Try It

3. Graph Simple Functions by Plotting Points
   (Section 2.3, **Objective** 3)

OBJECTIVES

1 Graph Linear Functions by Plotting Points

2 Graph Linear Functions by Using Intercepts

3 Graph Vertical and Horizontal Lines

OBJECTIVE 1:   GRAPH LINEAR FUNCTIONS BY PLOTTING POINTS

We looked at graphing equations in two variables by plotting points in Section 2.1. In Example 4 of that section, we graphed the equation $2x + y = 1$ (click here to see the graph). This is an example of a **linear equation in two variables.**

> **Definition**   **Linear Equation in Two Variables (Standard Form)**
>
> A **linear equation in two variables** is an equation that can be written in the standard form $Ax + By = C$, where A, B, and C are real numbers, and A and B are not both equal to 0.

Looking at the graph of $2x + y = 1$, we see that the graph is a function because it passes the vertical line test. Solving for y, we get $y = -2x + 1$. Replacing y with $f(x)$ allows us to write the equation in function notation: $f(x) = -2x + 1$. This is an example of a **linear function.**

> **Definition**   **Linear Function**
>
> A **linear function** is a function of the form $f(x) = ax + b$, where a and b are real numbers.

Note: There is no restriction on the constants a and b in our definition of a linear function unlike those we have seen in other definitions. Why is this the case? Click here to find out.

If we evaluate a linear function for any real number x, the result is a real number. This means that the domain of every linear function is the set of all real numbers, $\mathbb{R}$ or $(-\infty, \infty)$ in interval notation. To graph linear functions by plotting points, we can follow the procedure for graphing simple functions by plotting points from Section 2.3. For a linear function, we only need to plot two points to completely determine the graph. In practice, we will include a third point as a check.

Example 1 Graphing Linear Functions by Plotting Points

Graph $4x - 3y = 3$ by plotting points.

Solution First, we solve the equation for y.

$$\text{Original Equation:}\quad 4x - 3y = 3$$

$$\text{Subtract } 4x \text{ from both sides:}\quad -3y = -4x + 3$$

$$\text{Divide both sides by } -3:\quad y = \frac{4}{3}x - 1$$

Now we create a table of values using three different values for x: -3, 0, and 3. The resulting ordered pairs are $(-3, -5)$, $(0, -1)$ and $(3, 3)$. Click here to see the complete table of values.

The three ordered pairs are plotted in Figure 10(a). Notice that the points line up, so we connect them with a straight line, as shown in Figure 10(b).

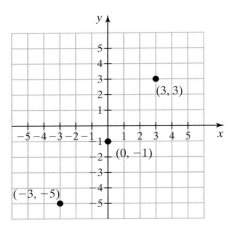

Figure 10(a)   **Figure 10(b)**

Since the coefficient of x is a fraction with a denominator of 3, we selected values for x that are multiples of 3 to simplify computations. However, any three values for x would work since the domain of the linear function is all real numbers.

You Try It   **Work through this You Try It problem.**

Work Exercises 1–3 in this textbook or in the *MyMathLab*® Study Plan.

Example 2 Graphing Linear Functions by Plotting Points

Graph $f(x) = -\dfrac{3}{5}x + 2$ by plotting points.

Solution Create a table of values by selecting three distinct values for x and evaluating the function for each value. Using multiples of 5 can simplify the computations since the coefficient of x has a denominator of 5. Plot the resulting points and connect them with a straight line. Click here to compare your graph or watch this video to see a detailed solution.

You Try It   **Work through this You Try It problem.**

Work Exercises 4–8 in this textbook or in the *MyMathLab*® Study Plan.

OBJECTIVE 2:   GRAPH LINEAR FUNCTIONS BY USING INTERCEPTS

In Section 2.1 we used a graph to identify intercepts. (You may wish to review this topic before reading further.) Now we consider how to find intercepts algebraically so that we can use them to graph linear functions.

My video summary

To find x-intercepts of an equation, we let $y = 0$ and solve for x because every point on the x-axis has a y-coordinate of 0. To find y-intercepts, we let $x = 0$ and solve for y because every point on the y-axis has an x-coordinate of 0.

Finding x- and y-Intercepts of a Graph Given an Equation

- To find an x-intercept, let $y = 0$ (or $f(x) = 0$) and solve for x.
- To find a y-intercept, let $x = 0$ and solve for y (or evaluate $f(0)$).

Intercepts are often easy to find and plot because one of the coordinates is 0. This makes them useful points to find when graphing equations or functions. Since the graph of a linear function can be drawn using only two points, we can use the x- and y-intercepts along with a third point to check. When both the x- and y-intercept are 0, the same point results, $(0, 0)$. In these cases, the second point needed to graph the line would not be an intercept. This concept will be explored later in Example 5.

My video summary

Example 3 Graphing Linear Functions by Using Intercepts

Graph $2x - 5y = 8$ by using intercepts.

Solution Watch the video, or continue reading to see the solution.

To find the x-intercept, we let $y = 0$ and solve for x. To find the y-intercept, we let $x = 0$ and solve for y. As a check, we let $x = -1$ and find the corresponding y-value.

x-intercept:

$$\text{Let } y = 0: \quad 2x - 5(0) = 8$$
$$\text{Simplify:} \quad 2x = 8$$
$$\text{Divide both sides by 2:} \quad x = 4$$

The x-intercept is 4, so the corresponding point is $(4, 0)$.

y-intercept:

$$\text{Let } x = 0: \quad 2(0) - 5y = 8$$
$$\text{Simplify:} \quad -5y = 8$$
$$\text{Divide both sides by } -5: \quad y = -\frac{8}{5}$$

The y-intercept is $-\frac{8}{5}$, so the corresponding point is $\left(0, -\frac{8}{5}\right)$.

Check point:

$$\text{Let } x = -1: \quad 2(-1) - 5y = 8$$
$$\text{Simplify:} \quad -2 - 5y = 8$$
$$\text{Add 2 to both sides:} \quad -5y = 10$$
$$\text{Divide both sides by } -5: \quad y = -2$$

The corresponding check point is $(-1, -2)$.

We can plot the three points and connect the points with a straight line. The resulting graph is shown in Figure 11.

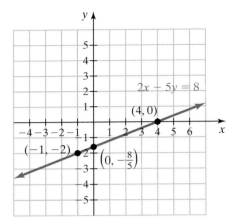

$2x - 5y = 8$

$(4, 0)$

$(-1, -2)$

$\left(0, -\frac{8}{5}\right)$

Figure 11

You Try It   Work through this You Try It problem.

Work Exercises 9–13 in this textbook or in the **MyMathLab**® Study Plan.

Example 4   Graphing Linear Functions by Using Intercepts

Graph $f(x) = -\dfrac{2}{3}x + 2$ by using intercepts.

Solution Find the x-intercept by letting $f(x) = 0$ and solving for x. Then find the y-intercept by letting $x = 0$ and evaluating $f(0)$. As a check, evaluate the function for a third value, such as $x = -3$, and find the corresponding point. Plot the intercepts and the check point. Then connect the points with a straight line. Click here to check your graph, or watch this video for a detailed solution.

You Try It   Work through this You Try It problem.

Work Exercises 14–17 in this textbook or in the **MyMathLab**® Study Plan.

The domain of a linear function is all real numbers, so every linear function has at least one intercept, the y-intercept, because $f(0)$ exists. When the graph of a linear function has only one intercept, we need to find at least one more point in order to graph the function.

Example 5   Graphing Linear Functions by Using Intercepts

Graph $6x = 4y$ by using intercepts.

Solution Find the x-intercept by letting $y = 0$ and solving for x. Find the y-intercept by letting $x = 0$ and solving for y.

Let $y = 0$: $6x = 4(0)$ Let $x = 0$: $6(0) = 4y$

Simplify: $6x = 0$ Simplify: $0 = 4y$

Divide by 6: $x = 0$ Divide by 4: $0 = y$

The x-intercept is 0.   The y-intercept is 0.

My video summary

My video summary

The graph has only one intercept: $(0, 0)$. Select two more values for x and find the corresponding y-values to plot two additional points on the graph. Use the points to graph the function. Click here to check your graph, or watch this video for a detailed solution.

You Try It Work through this You Try It problem.

Work Exercises 18–20 in this textbook or in the *MyMathLab*® Study Plan.

Compare the equations in Examples 3 and 5. How can we tell when the graph of a linear function will go through the origin? Click here to find out.

OBJECTIVE 3: GRAPH VERTICAL AND HORIZONTAL LINES

In the definition of a linear equation in two variables, there is a restriction that A and B cannot both be equal to 0. But what happens if only one is equal to 0? In either case, we still have a linear equation in two variables because the equation can be written in standard form. $Ax = C$ can be written as $Ax + 0 \cdot y = C$, and $By = C$ can be written as $0 \cdot x + By = C$.

The equation $Ax = C$ can be solved for x giving $x = \dfrac{C}{A} = a$, where a is the x-intercept. The equation $By = C$ can be solved for y giving $y = \dfrac{C}{B} = b$, where b is the y-intercept. When graphed, the special linear equations $x = a$ and $y = b$ are vertical lines and horizontal lines, respectively.

Example 6 Graphing Horizontal Lines

Graph $y = 2$.

Solution The equation $y = 2$ can be written as $0x + y = 2$ to show that it is a linear equation in two variables. For any real number x in this equation, y will always equal 2. Therefore, every point on the graph of the equation will have a y-coordinate of 2. If we choose x-values of -3, 0, and 4, the corresponding ordered pairs are $(-3, 2)$, $(0, 2)$, and $(4, 2)$. The resulting graph, shown in Figure 12, is a horizontal line with y-intercept 2 and no x-intercept.

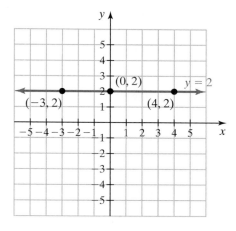

Figure 12

In all of our examples so far, the graphs of linear functions have had one or two intercepts. When does the graph have an infinite number of intercepts? Find out by clicking here.

Example 7 Graphing Vertical Lines

 Graph $x = -4$.

Solution The equation $x = -4$ can be written as $x + 0y = -4$ to show that it is a linear equation in two variables. For any real number y in this equation, x will always equal -4. Therefore, every point on the graph of the equation will have an x-coordinate of -4. Choose three values for y and find the corresponding points. Plot the points and connect them with a line. Click here to check your graph, or watch this video to see a detailed solution.

You Try It   Work through this **You Try It** problem.

Work Exercises 21–23 in this textbook or in the *MyMathLab* ® Study Plan.

Are horizontal lines functions? What about vertical lines? Sketch a graph for each type of line and apply the vertical line test to find out. Click here to check your answer.

We now summarize what we know about the graphs of vertical and horizontal lines.

> **Graphs of Vertical and Horizontal Lines**
>
> The graph of $y = b$ is a **horizontal line** through the point $(0, b)$. The constant b is the y-intercept of the graph. Horizontal lines are functions and can be written in function notation as $f(x) = b$.
>
> The graph of $x = a$ is a **vertical line** through the point $(a, 0)$. The constant a is the x-intercept of the graph. Vertical lines are not functions.

My video summary

2.4 Exercises

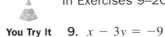
You Try It   In Exercises 1–8, graph each function by plotting points.

1. $y = 2x - 4$

2. $3x - 2y = -16$

3. $0.2x + 0.1y = 2$

4. $f(x) = -4x + 1$

5. $f(x) = 0.25x + 2$

6. $g(x) = \dfrac{3x - 1}{4}$

You Try It

7. $g(x) = -\dfrac{2}{5}x - 3$

8. $4x = 12 - 3y$

In Exercises 9–20, graph each function by using intercepts.

You Try It   **9.** $x - 3y = -9$

10. $3x + y = 6$

11. $y = 3.2x + 1.5$

12. $-x + 2y = 5$

13. $1.2x - 2.4y = 4.8$

14. $f(x) = x - 5$

You Try It **15.** $g(x) = -4x - 3$

16. $f(x) = -\dfrac{2}{5}x + 1$

17. $g(x) = \dfrac{1}{3}x + \dfrac{1}{4}$

18. $y = 3x$

19. $5y = -3x$

20. $f(x) = \dfrac{1}{2}x$

You Try It

In Exercises 21–23, graph each equation by plotting points.

21. $y = 5$

22. $f(x) = 3$

23. $x = 3$

You Try It

2.5 Linear Equations in Two Variables

THINGS TO KNOW

Before working through this section, be sure you are familiar with the following concepts:

VIDEO   ANIMATION   INTERACTIVE

You Try It

1. Solve a Formula for a Given Variable
(Section 1.4, **Objective 1**)

You Try It

2. Find x- and y-Intercepts
(Section 2.1, **Objective 5**)

You Try It

3. Graph Linear Functions by Plotting Points
(Section 2.4, **Objective 1**)

OBJECTIVES

1 Find the Slope of a Line

2 Graph a Line Using the Slope and a Point

3 Determine the Relationship between Two Lines

4 Write the Equation of a Line from Given Information

5 Write Equations of Parallel and Perpendicular Lines

6 Use Linear Models to Solve Application Problems; Direct Variation

---

OBJECTIVE 1:   FIND THE SLOPE OF A LINE

In Section 2.4 we graphed linear functions by plotting points. Although the graph of every linear function is a straight line, there are differences between the graphs, as illustrated in Figure 13.

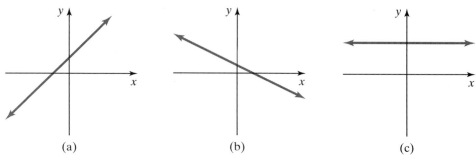

Figure 13

The graph of a linear function may rise (Figure 13(a)), fall (Figure 13(b)), or remain the same (Figure 13(c)) as we read the graph from left to right. How quickly the graph rises or falls is determined by the *steepness* of the line, known as the **slope**. If the graph is rising or increasing, then the slope is positive. If the graph

My video summary

is falling or decreasing, then the slope is negative. If the graph is staying the same, then it is constant, and the slope is zero.

Between any two fixed points on the line, the slope is the same and represents the **rate of change** of y with respect to x. To compute the slope, we form the ratio between the vertical change in y (called the **rise**) and the horizontal change in x (called the **run**) as we move from one of the fixed points to the other. The slope tells us how much the value of y changes for a given change in the value of x. Watch this video for a brief overview of the concept of slope.

Definition Slope

Given two points, (x_1, y_1) and (x_2, y_2), on the graph of a line, the **slope** m of the line containing the two points is given by the formula

$$m = \frac{\text{rise}}{\text{run}} = \frac{\text{change in } y}{\text{change in } x} = \frac{y_2 - y_1}{x_2 - x_1},$$

where $x_1 \neq x_2$.

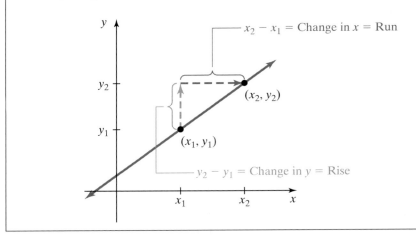

Why do we use the letter m to denote slope? Click here for an explanation.

Example 1 Finding the Slope of a Line Given Two Points

Find the slope of the line containing the points $(-2, 1)$ and $(3, 5)$.

Solution Let $(x_1, y_1) = (-2, 1)$ and $(x_2, y_2) = (3, 5)$. Now substitute the coordinates into the slope formula:

$$m = \frac{\text{rise}}{\text{run}} = \frac{\text{change in } y}{\text{change in } x} = \frac{y_2 - y_1}{x_2 - x_1}$$

$$= \frac{5 - 1}{3 - (-2)} = \frac{5 - 1}{3 + 2} = \frac{4}{5}$$

The slope of the line is $\frac{4}{5}$. This means that for every horizontal change, or run, of 5 units, there is a corresponding vertical change, or rise, of 4. Since the slope is positive, the graph of the line will increase as we read from left to right. The slope of this line is shown in Figure 14.

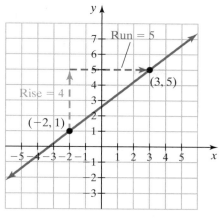

Figure 14

You Try It Work through this You Try It problem.

Work Exercises 1–2 in this textbook or in the *MyMathLab*® Study Plan.

When the slope is a fraction, we consider the fraction as rise over run. How do we interpret the slope $m = 3$? Click here to find out.

Example 2 Finding the Slope of a Line Given Two Points

Find the slope of the line containing the points $(-4, 5)$ and $(3, -4)$.

Solution Substitute the values for the coordinates into the slope formula to find the slope. Click here to check your answer or watch this video to see a detailed solution.

You Try It Work through this You Try It problem.

Work Exercises 3–4 in this textbook or in the *MyMathLab*® Study Plan.

Would our result from Example 2 be different if we switched the order of the two points? Click here to find out.

Example 3 Finding the Slope of a Line Given Two Points

Find the slope of the line containing the points $(-3, 1)$ and $(5, 1)$.

Solution Let $(x_1, y_1) = (-3, 1)$ and $(x_2, y_2) = (5, 1)$. Now substitute the coordinates into the slope formula:

$$m = \frac{\text{rise}}{\text{run}} = \frac{\text{change in } y}{\text{change in } x} = \frac{y_2 - y_1}{x_2 - x_1} = \frac{1 - 1}{5 - (-3)} = \frac{1 - 1}{5 + 3} = \frac{0}{8} = 0$$

The slope of the line is 0, so the graph neither rises nor falls as we read from left to right. The line is horizontal. The y-coordinates of the points on a horizontal line are always the same, so there is no vertical change. This makes the numerator in the slope formula 0, and the slope is equal to zero.

You Try It Work through this You Try It problem.

Work Exercises 5–6 in this textbook or in the *MyMathLab*® Study Plan.

My video summary

Based on the results of Example 3, what can you say about the slope of a vertical line? Click here to check.

 CAUTION Avoid saying that a line has "no slope" since it is not always clear whether this means a slope of zero or an undefined slope. These are distinct concepts and should be clearly stated when they occur.

In Example 4 of Section 2.4 we used intercepts to plot the graph of the linear function $f(x) = -\frac{2}{3}x + 2$ or $y = -\frac{2}{3}x + 2$. The x-intercept is 3, and the y-intercept is 2, so the points $(3, 0)$ and $(0, 2)$ are located on the graph of the line. Using these two points, we can find the slope of the line, $m = \frac{2 - 0}{0 - 3} = \frac{2}{-3} = -\frac{2}{3}$. Notice the slope of this linear function is equal to the coefficient of x in the function and the y-intercept is the constant term. This result can be generalized by the following:

Slope-Intercept Form

A linear equation in two variables of the form

$$y = mx + b \text{ or } f(x) = mx + b$$

is written in **slope-intercept form**, where m is the slope of the line and b is the y-intercept.

Watch this video to see how the slope-intercept form is derived from the slope formula.

CAUTION Be careful not to assume that the coefficient of x is always the slope. This is true when the equation is in slope-intercept form, but is not true in general.

Example 4 Finding the Slope and *y*-Intercept of a Line from an Equation

Find the slope and y-intercept of the line $6x + 5y = 10$.

Solution To find the slope and y-intercept of the given line, we solve the equation for y and write it in slope-intercept form.

$$\begin{aligned}
\text{Begin with the original equation:} \quad & 6x + 5y = 10 \\
\text{Subtract } 6x \text{ from both sides:} \quad & 5y = -6x + 10 \\
\text{Divide both sides by 5:} \quad & y = \frac{-6x + 10}{5} \\
\text{Distribute:} \quad & y = \frac{-6}{5}x + \frac{10}{5} \\
\text{Simplyfy:} \quad & y = -\frac{6}{5}x + 2
\end{aligned}$$

With the equation now written in slope-intercept form, the slope is the coefficient of x, $m = -\frac{6}{5}$, and the constant term, $b = 2$, is the y-intercept.

Example 5 Finding the Slope and *y*-Intercept of a Line from an Equation

 Find the slope and *y*-intercept of the line $4x - 2y = -14$.

Solution Click here to check your answer or watch this video for a detailed solution.

You Try It   Work through this You Try It problem.

Work Exercises 7–8 in this textbook or in the *MyMathLab*® Study Plan.

Not every linear equation in two variables can be written in slope-intercept form. Think about what lines cannot be written in this form. Then click here for an example of such a line.

We summarize the relationship between the slope and the graph of a linear equation in Figure 15.

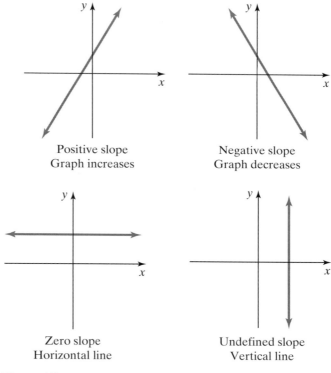

Positive slope
Graph increases

Negative slope
Graph decreases

Zero slope
Horizontal line

Undefined slope
Vertical line

Figure 15 Types of Lines

OBJECTIVE 2:   GRAPH A LINE USING THE SLOPE AND A POINT

In Section 2.4 when graphing linear functions by plotting points, we found that only two points are required to find the graph of a line. If we only have one point and the slope, we can use the slope and the given point to determine a second point on the graph.

Recall that slope $= \dfrac{\text{change in } y}{\text{change in } x}$. From the given point, we can apply the change in *x* to the *x*-coordinate and the change in *y* to the *y*-coordinate. This will allow us to obtain a second point for the graph. Watch this animation, which illustrates the concept, and then work through Example 6.

Example 6 Graphing a Line Using the Slope and a Point

Graph the line with slope $m = -\dfrac{1}{3}$ that passes through the point $(-1, 2)$.

Solution We are given one point on the graph, $(-1, 2)$, but we need a second point in order to sketch the graph. We can use the slope to find a second point. The slope $m = -\dfrac{1}{3}$ can be written as $m = \dfrac{-1}{3}$. This means that a vertical change of -1 unit corresponds to a horizontal change of 3 units.

Starting at the point $(-1, 2)$, we move down 1 unit and right 3 units to arrive at the point $(2, 1)$. Then we connect the points with a line, as shown in Figure 16(a).

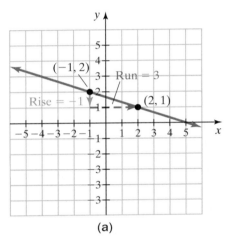
(a)

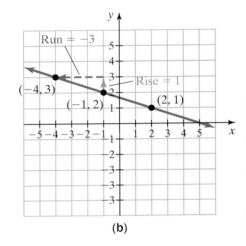
(b)

Figure 16

Notice that we could also write the slope as $m = \dfrac{1}{-3}$. In this form, the slope tells us that a vertical change of 1 unit corresponds to a horizontal change of -3 units. Starting at the point $(-1, 2)$, we move up 1 unit and left 3 units to arrive at the point $(-4, 3)$, which also lies on the graph, as shown in Figure 16(b).

Example 7 Graphing a Line Using the Slope and a Point

My video summary

📽 Graph the line with slope $m = \dfrac{2}{3}$ that passes through the point $(-1, -4)$.

Also, find three more points located on the line.

Solution Click here to see the graph, or watch this video for a complete solution.

🔺

You Try It **Work through this You Try It problem.**

Work Exercises 11–12 in this textbook or in the *MyMathLab*® **Study Plan.**

Example 8 Graphing a Line Using the Slope and the *y*-Intercept

My interactive video summary

 Graph $f(x) = -\dfrac{3}{5}x - 2$ using the slope and the *y*-intercept.

Solution The equation is in slope-intercept form, which means that the slope is $m = -\dfrac{3}{5}$ and the *y*-intercept is -2. The *y*-intercept gives us the point $(0, -2)$ on the graph. Use this point and the given slope to sketch the graph of the linear equation. Click here to check your answer or work through this interactive video.

You Try It Work through this You Try It problem.

Work Exercises 9–10 in this textbook or in the *MyMathLab*® Study Plan.

OBJECTIVE 3: DETERMINE THE RELATIONSHIP BETWEEN TWO LINES

Slopes and *y*-intercepts can be used to determine the relationship between two lines. **Parallel lines** have the same slope but different *y*-intercepts. Parallel lines never cross because the rate of change, or steepness, is the same for each line, but each line crosses the *y*-axis at a different point. **Coinciding lines** have the same slope *and* the same *y*-intercept. Coinciding lines appear to be the same line because their graphs lie on top of each other. **Perpendicular lines** intersect each other at right angles $(90°)$. If two lines are perpendicular, the product of their slopes is -1. This will occur if the slopes are opposite reciprocals. To obtain the opposite reciprocal of a number, we change the sign (opposite) and flip it over (reciprocal). **Intersecting lines** have different slopes.

CAUTION Perpendicular lines do intersect. We consider them separately from only intersecting lines because they intersect in a unique way (right angles).

What does it mean if perpendicular or intersecting lines have the same *y*-intercept? Click here to find out. In Table 1 we summarize how the slope and *y*-intercept allow us to determine the possible relationships between two lines.

	Parallel Lines	Coinciding Lines	Intersecting and Perpendicular Lines	Only Intersecting Lines (not Perpendicular)
Slopes are	*Same*	*Same*	opposite reciprocals $\left(m_1 \cdot m_2 = -1 \text{ or } m_1 = -\dfrac{1}{m_2}\right)$	*Different (but* $m_1 \cdot m_2 \neq -1$*)*
y-Intercepts are	*Different*	*Same*	Same or Different	*Same or Different*

Table 1 Relationship between Two Lines

Example 9 Determining the Relationship between Two Lines

My interactive video summary

 For each pair of lines, determine if the lines are parallel, perpendicular, coinciding, or only intersecting.

a. $3y = -4x + 10$
$16x + 12y = 27$

b. $y = \dfrac{3}{2}x - 7$
$6x - 4y = 28$

c. $2x - y = 3$
$3x + 6y = 5$

d. $5x + 2y = 8$
$x - 3y = 7$

Solution

a. Begin by writing each equation in slope-intercept form.

Original equation: $3y = -4x + 10$

Divide by 3: $y = -\dfrac{4}{3}x + \dfrac{10}{3}$

Original equation: $16x + 12y = 27$

Subtract $16x$: $12y = -16x + 27$

Divide by 12: $y = -\dfrac{4}{3}x + \dfrac{9}{4}$

In slope-intercept form, we see that the two lines have the same slope, $-\dfrac{4}{3}$, so the lines are either parallel or coinciding. The lines have different y-intercepts, $\dfrac{10}{3}$ and $\dfrac{9}{4}$. Therefore, these lines are parallel because they have the same slope but different y-intercepts.

b.–d. Click here to check your answers or watch this interactive video for a more detailed solution.

You Try It Work through this You Try It problem.

Work Exercises 13–18 in this textbook or in the *MyMathLab*® Study Plan.

OBJECTIVE 4: WRITE THE EQUATION OF A LINE FROM GIVEN INFORMATION

My video summary

How we write the equation of a line depends on what information is given. For example, if we are given the slope of a line and a point on the line, we can find the equation by using the **point-slope form**. (Watch this video to see how the point-slope form is derived.)

Point-Slope Form

Given the slope m of a line and a point (x_1, y_1) on the line, the **point-slope form** of the equation of the line is given by

$$y - y_1 = m(x - x_1).$$

Does it matter which point we choose when we use the point-slope form to write the equation of a line? Click here for the answer.

To write the equation of a line, we generally require the slope and a point on the line. Once we have this information, we use the point-slope form to write the equation of the line. Typically, we will have at least one point on the line, but obtaining the slope may require the use of the slope formula and a second point, or other information such as that the line is parallel to some given line.

Example 10 Writing the Equation of a Line Given the Slope and the *y*-Intercept

Write the equation of the line with slope $m = 6$ and passing through the point $(0, 4)$.

Solution We are given the slope $m = 6$ and a point $(0, 4)$. Notice that the *x*-coordinate of the point is 0 so the *y*-coordinate, 4, is the *y*-intercept, b. Using the slope-intercept form, $y = mx + b$, we substitute in the values of the slope and *y*-intercept to obtain the equation $y = 6x + 4$.

You Try It Work through this You Try It problem.

Work Exercises 19–22 in this textbook or in the *MyMathLab*® Study Plan.

Example 11 Writing the Equation of a Line Given the Slope and a Point

Write the equation of the line with slope $m = \dfrac{2}{5}$ and passing through the point $(5, 8)$.

Solution Given the slope $m = \dfrac{2}{5}$ and a point on the graph, $(x_1, y_1) = (5, 8)$, we can use the point-slope form, $y - y_1 = m(x - x_1)$. We substitute in the value of the slope and the coordinates of the point yielding the equation $y - 8 = \dfrac{2}{5}(x - 5)$.

$$\text{Point-slope form:} \quad y - 8 = \frac{2}{5}(x - 5)$$

$$\text{Distribute:} \quad y - 8 = \frac{2}{5}x - 2$$

$$\text{Add 8 to both sides:} \quad y = \frac{2}{5}x + 6$$

The equation in slope-intercept form is $y = \dfrac{2}{5}x + 6$.

You Try It Work through this You Try It problem.

Work Exercises 23–26 in this textbook or in the *MyMathLab*® Study Plan.

My video summary

Given the slope m and a point (x_1, y_1), could we use the slope-intercept form instead of the point-slope form to write the equation of the line? Watch this video to find out.

Click here to see Example 11 reworked using the slope-intercept form.

(EBook Screens 2.5-1–2.5-36)

Example 12 Writing the Equation of a Line Given Two Points

My interactive video summary

Write the equation of the line passing through the points $(3, -8)$ and $(5, -2)$.

Solution We are given two points, but we don't know the slope. We can compute the slope by using the slope formula: $m = \dfrac{-2 - (-8)}{5 - 3} = \dfrac{-2 + 8}{5 - 3} = \dfrac{6}{2} = 3$. Now that we know the slope, we can use it together with *either* of the two given points to write the equation of the line. Choose one of the points and find the equation of the line. Click here to check your answer or watch this interactive video to see a detailed solution.

You Try It **Work through this You Try It problem.**

Work Exercises 27–30 in this textbook or in the *MyMathLab*® Study Plan.

Example 13 Writing the Equation of a Vertical or Horizontal Line

Write the equation of a vertical line passing through the point $(4, 10)$ and then write the equation of a horizontal line passing through the same point.

Solution Vertical lines have the form $x = a$, which means that every point on the line has the same x-coordinate, a. The line passes through $(4, 10)$, which has x-coordinate 4. Therefore, $a = 4$. The equation of the vertical line through $(4, 10)$ is $x = 4$. Horizontal lines have the form $y = b$, which means that every point on the line has the same y-coordinate, b. The line passes through $(4, 10)$, which has y-coordinate 10. Therefore, $b = 10$. The equation of the horizontal line through $(4, 10)$ is $y = 10$.

You Try It **Work through this You Try It problem.**

Work Exercises 31–32 in this textbook or in the *MyMathLab*® Study Plan.

OBJECTIVE 5: WRITE EQUATIONS OF PARALLEL AND PERPENDICULAR LINES

At this point, you may want to review the discussion of parallel lines and perpendicular lines earlier in this section. When writing the equation of a parallel line or perpendicular line, the key ideas to remember are that parallel lines have the same slope and perpendicular lines have opposite-reciprocal slopes.

Example 14 Writing the Equation of a Perpendicular Line

My interactive video summary

Write the equation of the line that passes through the point $(-3, 1)$ and is perpendicular to $7x - 3y = 2$.

Solution We are given a point on the graph, but we do not know the slope. However, we know that the line is perpendicular to $7x - 3y = 2$, which gives us information about the slope. Perpendicular lines have opposite-reciprocal slopes, so we need to determine the slope of the given line first.

Original equation:	$7x - 3y = 2$
Subtract $7x$ from both sides:	$-3y = -7x + 2$
Divide both sides by -3:	$y = \dfrac{7}{3}x - \dfrac{2}{3}$

The slope of the given line is $\dfrac{7}{3}$. The opposite reciprocal is $-\dfrac{3}{7}$, so the slope of our line is $-\dfrac{3}{7}$. Now that we know the slope and a point on the graph, we can follow the process for writing the equation of a line as in the previous examples. Click here to check your answer, or watch this interactive video for a detailed solution.

You Try It   **Work through this You Try It problem.**

Work Exercises 35–36 in this textbook or in the *MyMathLab*® Study Plan.

Example 15 Writing the Equation of a Parallel Line

Write the equation of the line that passes through the point $(3, -2)$ and is parallel to $y = -3x + 5$.

Solution Click here to check your answer or watch this interactive video for the detailed solution.

You Try It   **Work through this You Try It problem.**

Work Exercises 33–34 in this textbook or in the *MyMathLab*® Study Plan.

We summarize the different forms for equations of lines in Table 2

Table 2 Equations of Lines

$m = \dfrac{y_2 - y_1}{x_2 - x_1}$	**Slope** Rate of change
$y - y_1 = m(x - x_1)$	**Point-Slope Form** Slope is m, and (x_1, y_1) is a point on the line.
$y = mx + b$	**Slope-Intercept Form** Slope is m, and y-intercept is b.
$Ax + By = C$	**Standard Form** A, B, and C are real numbers, with A and B not both zero and $A \geq 0$.
$y = b$	**Horizontal Line** Slope is zero, and y-intercept is b.
$x = a$	**Vertical Line** Slope is undefined, and x-intercept is a.
Perpendicular Lines	Opposite-Reciprocal Slopes $(m_1 \cdot m_2 = -1)$
Parallel Lines	Same slope, different y-intercept

OBJECTIVE 6:   USE LINEAR MODELS TO SOLVE APPLICATION PROBLEMS; DIRECT VARIATION

Graphs that display data as a set of points are called **scatter plots**. For example, Figure 17 shows a scatter plot of the percent of Internet users in the years 2005–2009 who use Microsoft Internet Explorer 6 (IE6) as their primary Web browser (*Source:* w3schools.com).

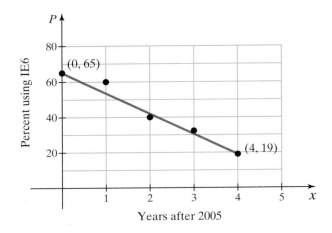

Years after 2005

Figure 17

Scatter plots can be used to show overall trends in data. When a trend is identified, analysts summarize the data by fitting an appropriate model to the data. The simplest model that can be used is a linear model such as $y = mx + b$. This model can be used to summarize data that fall on or near a line such as the data in Figure 17. Once the model is fit, we can make statements about data trends and use the model to make predictions.

Example 16 Primary Web Browser IE6

a. Use the scatter plot in Figure 17 to write a function of the form
   $P(x) = mx + b$ that models the percent, P, of Internet users who use IE6 as their primary Web browser x years after 2005. (*Source:* w3schools.com)
b. Interpret the slope and the y-intercept in the context of the problem.
c. Use the model to predict the percent of Internet users using IE6 as their primary browser in 2010.

Solution

a. Looking at the scatter plot, we see that the line passes through the points $(0, 65)$ and $(4, 19)$. Using these points, we can find the slope:

$$m = \frac{19 - 65}{4 - 0} = \frac{-46}{4} = -\frac{23}{2} \text{ or } -11.5$$

With this slope and the point $(4, 19)$ we can use the point-slope form of the equation of a line to obtain

Point-slope form:	$y - y_1 = m(x - x_1)$
Substitude in the slope and point:	$y - 19 = -11.5(x - 4)$
Distribute:	$y - 19 = -11.5x + 46$
Add 19 to both sides:	$y = -11.5x + 65$
Write in function form:	$P(x) = -11.5x + 65$

Notice that the y-intercept is 65, which we could have recognized from the given point $(0, 65)$.

b. The slope is $m = -11.5$, which can be written as

$$m = \frac{-11.5}{1} = \frac{\text{change in } y}{\text{change in } x} = \frac{\text{change in percent}}{\text{change in years}}.$$

This slope means that the percent of Internet users who use IE6 as their primary browser is decreasing by 11.5 percentage points each year.

Since x represents the number of years after 2005, $x = 0$ represents 2005. The y-intercept, 65, occurs when $x = 0$. This means that IE6 was the primary browser for 65% of Internet users in 2005.

c. The year 2010 is 5 years after 2005 so we have $x = 5$. Substitute 5 for x in $P(x) = -11.5x + 65$:

$$P(5) = -11.5(5) + 65 = -57.5 + 65 = 7.5$$

The model predicts that IE6 will be the primary browser for 7.5% of Internet users in 2010.

You Try It   **Work through this You Try It problem.**

Work Exercises 37–38 in this textbook or in the *MyMathLab*® Study Plan.

Another application of linear models is called **direct variation**. Direct variation means that one variable is a constant multiple of another.

Direct Variation

For a linear model of the form

$$y = kx,$$

we say that y **varies directly** with x, or y is *proportional to* x. The constant k is called the **constant of variation** or the *proportionality constant*.

We can use the strategy for solving application problems involving linear equations from Chapter 1 in the following examples.

Example 17 Water Pressure on a Scuba Diver

During scuba diving, water pressure varies directly with the depth of the diver. A diver 40 feet below the surface will face 17.8 pounds per square inch (psi) of pressure from the water. How much water pressure will a diver face if she is 100 feet below the surface?

Solution

Step 1. **Define the Problem.** We want to find the water pressure on the diver at 100 feet below the surface if the pressure is 17.8 psi at 40 feet below the surface. We know that pressure varies directly with the depth of the diver.

Step 2. **Assign Variables.** We let p = water pressure (psi) and d = depth (feet).

Step 3. **Translate to an Equation.** Since water pressure p varies directly with depth d, we can write the relation as

$$p = kd,$$

where k is the constant of variation. To determine k, we use the fact that the water pressure is 17.8 psi at a depth of 40 feet.

$$p = kd$$
$$17.8 = k(40)$$
$$\frac{17.8}{40} = k$$
$$0.445 = k$$

So, the constant of variation is 0.445. The equation is $p = 0.445d$.

Step 4. Solve the Equation. To determine the water pressure at a depth of 100 feet, we substitute 100 for d and solve for p.

$$p = 0.445d$$
$$p = 0.445(100)$$
$$p = 44.5$$

Step 5. Check the Reasonableness of the Answer. The constant of variation (0.445) is slightly less than $\frac{1}{2}$, so we expect the numeric value for water pressure (44.5) to be slightly less than half the numeric value for depth (100). Since 44.5 is slightly less than half of 100, the result seems reasonable.

Step 6. State the Answer. The diver will experience 44.5 psi of water pressure if she is 100 feet below the surface.

Example 18 Simple Interest Earned

 Simple interest after 1 year varies directly with the amount of principal. If $3,200 earns $240 in interest, then how much interest will $4,800 earn after one year? (**Note:** In this problem, the constant of variation represents the annual interest rate in decimal form.)

My video summary

Solution Click here to check your answer or watch this video for a detailed solution.

You Try It **Work through this You Try It problem.**

Work Exercise 39–40 in this textbook or in the *MyMathLab* Study Plan.

How is direct variation related to the slope-intercept form of a line? Click here for the details.

2.5 Exercises

In Exercises 1–6, find the slope of the line containing the given points.

You Try It

1. $(-5, 2)$ and $(2, -6)$

2. $(0, -7)$ and $(-4, -9)$

3.

You Try It

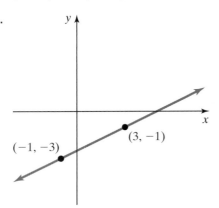

4.

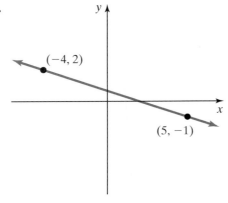

5. $\left(-\dfrac{3}{2}, -\dfrac{1}{4}\right)$ and $\left(\dfrac{5}{3}, -\dfrac{1}{4}\right)$

6. $(4, 9)$ and $(4, 5)$

In Exercises 7–8, find the slope and y-intercept for each equation.

7. $4x - y = 12$

8. $5x + 7y = 12$

In Exercises 9–10, find the slope and y-intercept and use them to graph the linear function.

9. $f(x) = \dfrac{1}{3}x + 3$

10. $f(x) = -\dfrac{3}{5}x + 4$

In Exercises 11–12, graph the line given the slope and a point on the line.

11. Slope $= \dfrac{1}{2}$; $(-2, -2)$

12. Slope $= -\dfrac{4}{3}$; $(3, 8)$

In Exercises 13–18, determine if the two lines are parallel, perpendicular, coinciding, or only intersecting.

13. $-4x - 2y = -9$
$\quad\;\; 5x - 10y = 7$

14. $y = 3x - 2$
$\quad\;\; 9x - 3y = 1$

15. $y = -2$
$\quad\;\; 3y = 12$

16. $x = -1$
$\quad\;\; y = -1$

17. $4x - 3y = 7$
$\quad\;\; 3x - 4y = -6$

18. $2x + 3y = 3$
$\quad\;\; y = -\dfrac{2}{3}x + 1$

In Exercises 19–30, write the equation of the line from the given information. Write the line in slope-intercept form and standard form.

19. Slope $= 1$; passes through $(0, -2)$

20. Slope $= -\dfrac{1}{6}$; y-intercept $= \dfrac{1}{2}$

21. Slope $= \dfrac{7}{9}$; passes through $\left(0, -\dfrac{8}{7}\right)$

22. Slope $= 0$; y-intercept $= 5$

23. Slope $= 3$; passes through $(-5, 3)$

24. Slope $= -3$; passes through $(-2, 3)$

25. Slope $= -\dfrac{4}{3}$; passes through $(3, 8)$

26. Slope $= -\dfrac{2}{3}$; passes through the origin

27. Passing through $(1, 2)$ and $(-2, 5)$

28. Passing through $(-5, 7)$ and $(3, -5)$

29. Passing through $\left(-\dfrac{1}{2}, 1\right)$ and $\left(-3, \dfrac{2}{3}\right)$

30. Passing through $(-3, 4)$ and $(2, 4)$

31. Write the equation of the horizontal line passing through the point $(5, -2)$.

32. Write the equation of the vertical line passing through the point $(5, -2)$.

33. Write the equation of the line parallel to $y = \dfrac{1}{4}x - 2$ that passes through the point $(-2, 3)$.

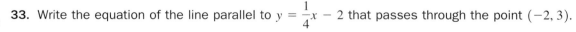

34. Write the equation of the line parallel to $3x - 5y = 1$ that passes through the point $(1, -4)$.

You Try It

35. Write the equation of the line perpendicular to $y = \dfrac{1}{4}x - 2$ that passes through the point $(-2, 3)$.

36. Write the equation of the line perpendicular to $3x - 5y = 1$ that passes through the point $(1, -4)$.

In Exercises 37–40, use linear models to solve application problems.

You Try It

37. College Tuition and Fees The following scatter plot shows the average annual tuition and fees for public 4-year colleges and universities in the years 2004–2008. (*Source:* College Board)

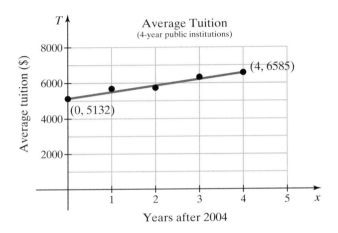

a. Use the two points labeled in the scatter plot to write a function of the form $T(x) = mx + b$ that models the average annual tuition and fees T at 4-year public colleges and universities x years after 2004.

b. Interpret the slope and the y-intercept in the context of the problem.

c. Use the model to predict the average annual tuition and fees at 4-year colleges and universities in 2015.

38. Daily Newspapers The following scatter plot shows the number of daily U.S. newspapers in the years 1980–2005. (*Source: Statistical Abstract*, 2009)

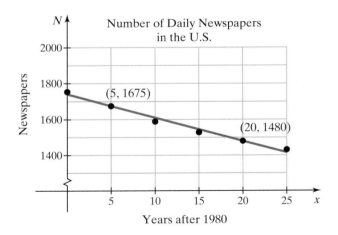

a. Use the two points labeled in the scatter plot to write a function of the form $N(x) = mx + b$ that models the number N of daily U.S. newspapers x years after 1980.

b. Interpret the slope and the y-intercept in the context of the problem.

c. Use the model to predict the number of daily U.S. newspapers in 2016.

39. **Hooke's Law**   The amount that a spring is stretched when a load is applied varies directly with the amount of the load. If a load of 50 pounds stretches a spring by 2 inches, how much will the spring stretch if a 180-pound load is applied?

ou Try It

40. **Calories from Fat**   The number of calories from fat in peanut butter varies directly with the number of ounces. If a 3-ounce serving of peanut butter contains 435 calories from fat, how many calories from fat are there in a 7-ounce serving?

2.6 Linear Inequalities in Two Variables

THINGS TO KNOW

Before working through this section, be sure you are familiar with the following concepts:

VIDEO   ANIMATION   INTERACTIVE

You Try It

1. Determine If a Given Value Is a Solution to an Inequality (Section 1.2, **Objective** 1)

You Try It

2. Solve Linear Inequalities in One Variable (Section 1.2, **Objective** 4)

You Try It

3. Graph a Line Using the Slope and a Point (Section 2.5, **Objective** 2)

OBJECTIVES

1 Determine If an Ordered Pair Is a Solution to a Linear Inequality in Two Variables

2 Graph a Linear Inequality in Two Variables

OBJECTIVE 1:   DETERMINE IF AN ORDERED PAIR IS A SOLUTION TO A LINEAR INEQUALITY IN TWO VARIABLES

In Section 1.2, we solved linear inequalities in one variable. The solution set for such an inequality is the set of all values that make the inequality true. Typically, we graph the solution set of a linear inequality in one variable on a number line.

In this section, we solve **linear inequalities in two variables**. A linear inequality in two variables looks much like a linear equation in two variables except that an inequality symbol replaces the equal sign.

Definition   Linear Inequality in Two Variables

A **linear inequality in two variables** is an inequality that can be written in the form $Ax + By < C$, where A, B, and C are real numbers, and A and B are not both equal to zero.

Note: The inequality symbol "$<$" can be replaced with $>$, $\leq$, or $\geq$.

An ordered pair is a **solution to a linear inequality in two variables** if, when substituted for the variables, it makes the inequality true.

Example 1 Determining If an Ordered Pair Is a Solution to a Linear Inequality in Two Variables

My interactive video summary

Determine if the given ordered pair is a solution to the inequality $3x + y > 2$.

a. $(1, 2)$ **b.** $(-2, -1)$ **c.** $(1, -1)$

Solution   We substitute the x- and y-coordinates for the variables and simplify. If the resulting statement is true, then the ordered pair is a solution to the inequality.

a. Begin with the original inequality: $3x + y > 2$

Substitute 1 for x and 2 for y: $3(1) + 2 \overset{?}{>} 2$

Simplify: $5 \overset{?}{>} 2$ True

The final statement is true, so $(1, 2)$ is a solution to the inequality.

Work through parts (b) and (c) by yourself. Click here to check your answers, or watch this interactive video for detailed solutions.

You Try It   **Work through this You Try It problem.**

Work Exercises 1–4 in this textbook or in the *MyMathLab*® Study Plan.

OBJECTIVE 2:   GRAPH A LINEAR INEQUALITY IN TWO VARIABLES

Now let's look for all solutions to a linear inequality in two variables. First, we focus on the related linear equation. For example, to find solutions to $x + y > 3$ or $x + y < 3$, we look at the equation $x + y = 3$. For an ordered pair to satisfy this equation, the sum of its coordinates must be 3. So $(-2, 5)$, $(1, 2)$, $(3, 0)$, and $(5, -2)$ are all solutions to the equation. (Click here if you are unsure why.) Plotting and connecting these points gives the line shown in Figure 18.

Notice that the line in Figure 18 divides the coordinate plane into two **half-planes**, an *upper half-plane* shaded blue and a *lower half-plane* shaded red. Let's examine some points in each half-plane to see which ones satisfy the inequalities $x + y > 3$ or $x + y < 3$.

Choose any point that lies in the upper half-plane, say $(4, 2)$. The sum of its coordinates is $4 + 2 = 6$, which is larger than 3. Choose any other point in this region, such as $(0, 4)$. The sum of its coordinates is $0 + 4 = 4$, which is also larger than 3.

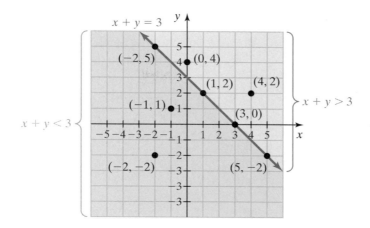

$x + y = 3$

$(-2, 5)$ $(0, 4)$

$(1, 2)$ $(4, 2)$

$(-1, 1)$ $(3, 0)$

$x + y > 3$

$x + y < 3$

$(-2, -2)$ $(5, -2)$

Figure 18

If we choose any point in the upper half-plane, the sum of its coordinates will be larger than 3. This means that any ordered pair from the upper half-plane is a solution to the inequality $x + y > 3$. The area shaded blue in Figure 18 represents the set of all ordered pair solutions to the inequality $x + y > 3$.

Repeat this process for the lower half-plane to see that the area shaded red in Figure 18 represents the set of all ordered pair solutions to the inequality $x + y < 3$. For example, test the points $(-1, 1)$ and $(-2, -2)$ from the lower half-plane. Click here to check.

The equation $x + y = 3$ acts as a **boundary line** that separates the solutions of the two inequalities $x + y > 3$ and $x + y < 3$.

Based on this information, we can define a set of **steps for graphing linear inequalities in two variables**.

Steps for Graphing Linear Inequalities in Two Variables

Step 1. Find the boundary line for the inequality by replacing the inequality symbol with an equal sign and graphing the resulting equation. If the inequality is strict, graph the boundary using a dashed line. If the inequality is non-strict, graph the boundary using a solid line.

Step 2. Choose a **test point** that does not belong to the boundary line and determine if it is a solution to the inequality.

Step 3. If the test point is a solution to the inequality, then shade the half-plane that contains the test point. If the test point is not a solution to the inequality, then shade the half-plane that does not contain the test point. The shaded area represents the set of all ordered pair solutions to the inequality.

Note: When graphing a linear inequality in two variables, using a dashed line is similar to using an open circle when graphing a linear inequality in one variable on a number line. Likewise, using a solid line is similar to using a solid circle. Do you see why?

Example 2 Graphing a Linear Inequality in Two Variables

 Graph each inequality.

My interactive video summary

a. $x - 2y \geq 4$ **b.** $3y < 2x$ **c.** $x < -2$

Solution

a. We follow the three-step process.

Step 1. The boundary line is $x - 2y = 4$. We graph the line as a solid line because the inequality is non-strict. See Figure 19(a).

Step 2. We choose the test point $(0, 0)$ and check to see if it satisfies the inequality.

Begin with the original inequality: $\quad x - 2y \geq 4$

Substitute 0 for x and for y: $\quad 0 - 2(0) \overset{?}{\geq} 4$

Simplify: $\quad 0 \overset{?}{\geq} 4 \quad$ False

The point $(0, 0)$ is not a solution to the inequality.

Step 3. Since the test point is not a solution to the inequality, we shade the half-plane that does not contain $(0, 0)$. See Figure 19(b). The shaded region, including the boundary line, represents all ordered pair solutions to the inequality $x - 2y \geq 4$.

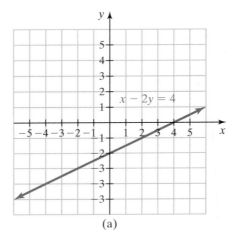

(a)

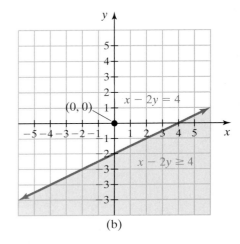

(b)

Figure 19

b. The boundary line is $3y = 2x$. Because the inequality is strict, graph the boundary line using a dashed line. Choose a test point and complete the graph. Click here or watch this interactive video to check your answer.

c. For $x < -2$, the boundary line is the vertical line $x = -2$. Because the inequality is strict, graph the boundary line using a dashed line. Choose a test point and complete the graph. Click here or watch this interactive video to check your answer.

CAUTION It is important to choose a test point that does not belong to the boundary line. Do you see why?

You Try It   **Work through this You Try It problem.**

Work Exercises 5–14 in this textbook or in the *MyMathLab*®️ Study Plan.

2.6 Exercises

In Exercises 1–4, determine if each ordered pair is a solution to the given inequality.

You Try It

1. $x - 4y < 8$

   a. $(2, -3)$

   b. $(4, -1)$

   c. $(-6, -2)$

2. $5x + 2y \geq 20$

   a. $(5, -1)$

   b. $(3, 2)$

   c. $(2, 5)$

3. $2x + 5y > 10$

   a. $(-5, 4)$

   b. $(2, 3)$

   c. $\left(\dfrac{5}{2}, \dfrac{3}{5}\right)$

4. $6x - 5y \leq 30$

   a. $(0, 0)$

   b. $(2, -4)$

   c. $(2.5, -1.5)$

In Exercises 5–14, graph each inequality.

You Try It

5. $2x - y \geq -3$

6. $3x + 2y > 6$

7. $-x + 3y < -9$

8. $x + y \leq 0$

9. $y \leq \dfrac{5}{2}x - 1$

10. $4y < -3x$

11. $\dfrac{1}{2}x + \dfrac{2}{3}y > \dfrac{5}{6}$

12. $0.6x - 1.8y \leq 2.4$

13. $x \geq -1$

14. $y < 3$

CHAPTER THREE

Systems of Linear Equations and Inequalities

CHAPTER THREE CONTENTS

3.1 Systems of Linear Equations in Two Variables

3.2 Systems of Linear Equations in Three Variables

3.3 More Problem Solving with Systems of Linear Equations

3.4 Systems of Linear Inequalities in Two Variables

3.1 Systems of Linear Equations in Two Variables

THINGS TO KNOW

Before working through this section, be sure you are familiar with the following concepts:

	VIDEO	ANIMATION	INTERACTIVE

You Try It
1. Determine If an Ordered Pair Is a Solution to an Equation (Section 2.1, Objective 2) —

You Try It
2. Graph a Line Using the Slope and a Point (Section 2.5, Objective 2) —

You Try It
3. Determine the Relationship between Two Lines (Section 2.5, Objective 3)

OBJECTIVES

1 Determine If an Ordered Pair Is a Solution to a System of Linear Equations in Two Variables

2 Solve Systems of Linear Equations in Two Variables by Graphing

3 Solve Systems of Linear Equations in Two Variables by Substitution

4 Solve Systems of Linear Equations in Two Variables by Elimination

5 Solve Inconsistent and Dependent Systems

6 Use Systems of Linear Equations in Two Variables to Solve Application Problems

OBJECTIVE 1:   DETERMINE IF AN ORDERED PAIR IS A SOLUTION TO A SYSTEM OF LINEAR EQUATIONS IN TWO VARIABLES

Figure 1 shows that the average annual expenditures on cellular phones have increased steadily since 2001, whereas expenditures on residential phones have decreased. The x-coordinate of the point where the two lines intersect gives the year when expenditures were the same for both phone types. The y-coordinate of the intersection point gives the average annual expenditure when the value was the same for both phone types.

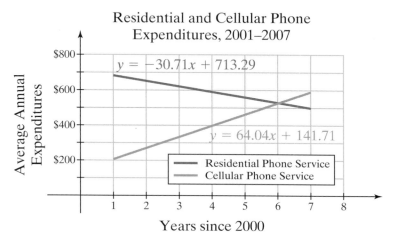

Figure 1
Source: Bureau of Labor Statistics Consumer Expenditure Survey, 2009.

Figure 1 illustrates the graph of a **system of linear equations in two variables**.

Definition   System of Linear Equations in Two Variables

A **system of linear equations in two variables** is a collection of two or more linear equations in two variables considered together. The **solution to a system** of linear equations in two variables is the set of all ordered pairs for which both equations are true.

The following three examples all represent systems of linear equations in two variables:

$$\begin{cases} 2x + 3y = 12 \\ x + 2y = 7 \end{cases} \qquad \begin{cases} 2a - 5b = 7 \\ \dfrac{1}{2}a + \dfrac{2}{3}b = -\dfrac{1}{6} \end{cases} \qquad \begin{cases} 4p - q = 4 \\ q = 2 \end{cases}$$

Within a given system of linear equations in two variables, the same two variables will be used. For example, the variables x and y are used in both equations of the first system. The solution to this system will be an ordered pair (x, y). To determine if a given ordered pair is a solution to a system, we check to see if the ordered pair makes both equations true when substituted for the variables.

Example 1 Determining If an Ordered Pair Is a Solution to a System of Linear Equations in Two Variables

 *My interactive video summary*

 Determine if each ordered pair is a solution to the following system:

$$\begin{cases} 2x + 3y = 12 \\ x + 2y = 7 \end{cases}$$

a. $(-3, 6)$　　　　　　　　b. $(3, 2)$

Solution Check each ordered pair to see if it makes both equations true. If both equations are true, then the ordered pair is a solution to the system. If either equation is false, then the ordered pair is not a solution to the system.

a.

	First Equation	Second Equation
Begin with the original equations:	$2x + 3y = 12$	$x + 2y = 7$
Substitute -3 for x and 6 for y in each equation:	$2(-3) + 3(6) \overset{?}{=} 12$	$(-3) + 2(6) \overset{?}{=} 7$

Finish simplifying both equations to see if $(-3, 6)$ is a solution to the system. Click here to check your answer, or watch this interactive video for the full solution.

b. Substitute 3 for x and 2 for y in each equation. Simplify to see if the resulting equations are true. Click here to check your answer, or watch this interactive video for a full solution.

You Try It　**Work through this You Try It problem.**

Work Exercises 1–4 in this textbook or in the *MyMathLab*® Study Plan.

OBJECTIVE 2:　SOLVE SYSTEMS OF LINEAR EQUATIONS IN TWO VARIABLES BY GRAPHING

In this section, we look at three methods for solving systems of linear equations in two variables: *graphing, substitution*, and *elimination*.

To **solve a system of linear equations in two variables by graphing**, we graph each line and find the intersection point, if any. When two linear equations are graphed, there are three possible outcomes:

1. The two lines intersect at one point. See Figure 2(a). The system has one solution.

2. The two lines are parallel and do not intersect at all. See Figure 2(b). The system has no solution.

3. The two lines coincide and have an infinite number of intersection points. See Figure 2(c). The system has an infinite number of solutions.

A system with at least one solution is **consistent**. A system without a solution is **inconsistent**.

When the equations in a system are coinciding lines, the system is **dependent**. When the equations in a system are different lines, the system is **independent**.

For now, we will focus on consistent systems with only one solution. In Objective 5 we will examine inconsistent and dependent systems.

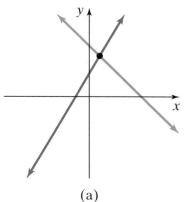

(a)
Intersecting Lines
One solution
Consistent, Independent
Two lines are different, having one common point. The lines have different slopes.

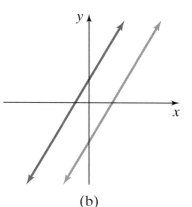

(b)
Parallel Lines
No solution
Inconsistent, Independent
Two lines are different, having no common points. The lines have the same slope but different y-intercepts.

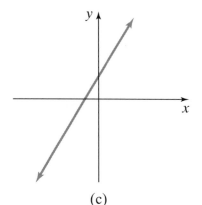

(c)
Coinciding Lines
Infinitely many solutions
Consistent, Dependent
Two lines are the same, having infinitely many common points. The lines have the same slope and the same *y*-intercept.

Figure 2

To solve systems of linear equations by graphing, we use a three-step process.

Solving Systems of Linear Equations in Two Variables by Graphing

Step 1. Graph the two equations on the same set of axes.

Step 2. If the lines intersect, then find the coordinates of the intersection point. The ordered pair is the solution to the system.

Step 3. Check the ordered-pair solution in both of the original equations.

Note: If the lines are parallel, then the system has no solution. If the lines coincide, then the system has infinitely many solutions.

 It can be difficult to identify the exact intersection point by graphing, so it is essential to check the solution.

Example 2 Solving Systems of Linear Equations in Two Variables by Graphing

Solve the following system by graphing:

$$\begin{cases} 2x + y = -4 \\ x + 3y = 3 \end{cases}$$

Solution We follow the three-step process.

Step 1. Graph each line, as shown in Figure 3.

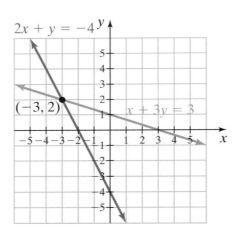

Figure 3

Graph of System $\begin{cases} 2x + y = -4 \\ x + 3y = 3 \end{cases}$

Step 2. The intersection point is $(-3, 2)$.

Step 3. Check $(-3, 2)$ in both equations to see if it is the solution to the system.

Watch this video to see the fully worked solution.

You Try It   Work through this You Try It problem.

Work Exercises 5–7 in this textbook or in the *MyMathLab*® Study Plan.

OBJECTIVE 3: SOLVE SYSTEMS OF LINEAR EQUATIONS IN TWO VARIABLES
BY SUBSTITUTION

The **substitution method** involves solving one of the equations for one variable, substituting the resulting expression into the other equation, and then solving for the remaining variable. The substitution method can be summarized in four steps.

Solving Systems of Linear Equations in Two Variables by Substitution

Step 1. Choose an equation and solve for one variable in terms of the other variable.

Step 2. Substitute the expression from step 1 into the other equation.

Step 3. Solve the equation in one variable from step 2.

Step 4. Substitute the solution from step 3 into one of the original equations to find the value of the other variable.

 It is good practice to check the ordered-pair solution by substituting it into the original equation that was not used in step 4.

Example 3 Solving Systems of Linear Equations in Two Variables by Substitution

▣ Use the substitution method to solve system: $\begin{cases} 5x + 2y = 8 \\ x + 6y = 10 \end{cases}$

Solution

Step 1. Choose an equation and solve for one of the variables. If possible, choose a variable that has a coefficient of 1. In this example, it is easiest to solve the second equation for x:

$$\text{Begin with the second equation:} \quad x + 6y = 10$$
$$\text{Subtract } 6y \text{ from both sides:} \quad x = -6y + 10$$

Step 2. Substitute $-6y + 10$ for x in the first equation.

$$\text{Rewrite the first equation:} \quad 5x + 2y = 8$$
$$\text{Substitute } -6y + 10 \text{ for } x: \quad 5(-6y + 10) + 2y = 8$$

Step 3. Solve for y.

$$\text{Equation from step 2:} \quad 5(-6y + 10) + 2y = 8$$
$$\text{Distribute:} \quad -30y + 50 + 2y = 8$$
$$\text{Simplify:} \quad -28y + 50 = 8$$
$$\text{Subtract 50 from both sides:} \quad -28y = -42$$
$$\text{Divide both sides by } -28 \text{ and simplify:} \quad y = \frac{-42}{-28} = \frac{3}{2}$$

Step 4. Find x by substituting $y = \frac{3}{2}$ into one of the original equations.

$$\text{Begin with the original second equation:} \quad x + 6y = 10$$
$$\text{Substitute } \frac{3}{2} \text{ for } y: \quad x + 6\left(\frac{3}{2}\right) = 10$$
$$\text{Simplify:} \quad x + 9 = 10$$
$$\text{Subtract 9 from both sides:} \quad x = 1$$

The solution to this system is the ordered pair $\left(1, \frac{3}{2}\right)$.

Even though we used the second original equation in step 4, we use the first original equation to check the final answer. Click here to see the check, or watch this video for a fully worked solution.

Figure 4 shows the graph of the system in Example 3. We see that the solution $\left(1, \dfrac{3}{2}\right)$ is the intersection point of the two lines in the system.

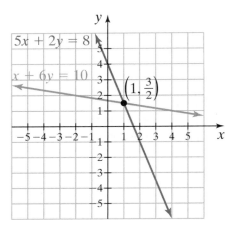

Figure 4

Graph of System $\begin{cases} 5x + 2y = 8 \\ x + 6y = 10 \end{cases}$

Example 4 Solving Systems of Linear Equations in Two Variables by Substitution

 Use the substitution method to solve the following system:

$$\begin{cases} 2x - 3y = -5 \\ x + y = 5 \end{cases}$$

Solution Try solving this system on your own. Click here to check your answer, or watch this video for a fully worked solution.

You Try It   Work through this You Try It problem.

Work Exercises 8–12 in this textbook or in the *MyMathLab*® Study Plan.

OBJECTIVE 4:   SOLVE SYSTEMS OF LINEAR EQUATIONS IN TWO VARIABLES BY ELIMINATION

The **elimination method** involves adding the two equations together in a way that will *eliminate* one of the variables. This method is based on the following logic.

Logic for the Elimination Method

If $A = B$ and $C = D$, then $A + C = B + D$.

This logic means that if two true equations are added, then the result will be a third true equation. Because equations are added, the elimination method is also known as the **addition method**.

Example 5 Solving Systems of Linear Equations in Two Variables by Elimination

Solve the following system:

$$\begin{cases} x - y = 6 \\ x + y = 12 \end{cases}$$

Solution Based on the logic for the elimination method, we can add two true equations together to result in a third true equation.

$$\begin{array}{r} x - y = 6 \\ \underline{x + y = 12} \\ 2x + 0 = 18 \end{array}$$

Notice that y is eliminated. Solving the resulting equation for x gives the x-coordinate of the ordered-pair solution.

$$\begin{array}{lr} \text{Rewrite the equation:} & 2x + 0 = 18 \\ \text{Simplify:} & 2x = 18 \\ \text{Divide both sides by 2:} & x = 9 \end{array}$$

The value of x is 9. By substituting 9 for x in one of the original equations, we can find the value of y.

$$\begin{array}{lr} \text{Begin with the second original equation:} & x + y = 12 \\ \text{Substitute 9 for } x\text{:} & 9 + y = 12 \\ \text{Subtract 9 from both sides:} & y = 3 \end{array}$$

The ordered-pair solution is $(9, 3)$. We can check $(9, 3)$ in both of the original equations.

	First Equation	Second Equation
Begin with the original equations:	$x - y = 6$	$x + y = 12$
Substitute 9 for x and 3 for y in each equation:	$9 - 3 \overset{?}{=} 6$	$9 + 3 \overset{?}{=} 12$
Simplify:	$6 = 6$ True	$12 = 12$ True

Since both equations are true, $(9, 3)$ is the solution to the system.

In Example 5, the y-variable was eliminated easily when we added the two equations together. Unfortunately, a variable will not always be eliminated so conveniently. Consider the system from Example 3. Adding the two original equations does not eliminate a variable.

$$\begin{array}{r} 5x + 2y = 8 \\ \underline{x + 6y = 10} \\ 6x + 8y = 18 \end{array}$$

In this case, we must write equivalent equations in a system before adding them so that a variable can be eliminated.

To eliminate a variable, the coefficients of the variable must differ only in sign. We can make this happen by multiplying one or both of the equations by a suitable nonzero constant.

Example 6 Solving Systems of Linear Equations in Two Variables by Elimination

Use elimination to solve the following system from Example 3:

$$\begin{cases} 5x + 2y = 8 \\ x + 6y = 10 \end{cases}$$

Solution If we multiply both sides of the second equation by -5, this will change the coefficient of the x-variable in the second equation to the opposite of the coefficient of the x-variable in the first equation. When the result is added to the first equation, the x-variable will be eliminated.

$$5x + 2y = 8 \xrightarrow{\text{No change needed}} 5x + 2y = 8 \xrightarrow{\text{No change needed}} 5x + 2y = 8$$

$$x + 6y = 10 \xrightarrow{\text{Multiply by } -5} -5(x + 6y) = -5(10) \xrightarrow{\text{Simplify}} -5x - 30y = -50$$

Add the new equations: $-28y = -42$

Solve for y: $y = \dfrac{-42}{-28} = \dfrac{3}{2}$

The value of y is $\dfrac{3}{2}$. We find the value of x as we did in step 4 of Example 3. So, the value of x is 1, and the solution to the system is the ordered pair $\left(1, \dfrac{3}{2}\right)$.

You Try It   **Work through this You Try It problem.**

Work Exercises 13–14 in this textbook or in the *MyMathLab*® Study Plan.

The elimination method can be summarized in five steps.

Solving Systems of Linear Equations in Two Variables by Elimination

Step 1. Choose a variable to eliminate.

Step 2. Multiply one or both equations by an appropriate nonzero constant so that the sum of the coefficients of one of the variables is zero.

Step 3. Add the two equations from step 2 together to obtain an equation in one variable.

Step 4. Solve the equation in one variable from step 3.

Step 5. Substitute the value found in step 4 into one of the original equations to solve for the other variable.

It is good practice to check the ordered-pair solution by substituting it into the original equation that was not used in step 5.

Example 7 Solving Systems of Linear Equations in Two Variables by Elimination

My video summary

📽 Use the elimination method to solve the following system:

$$\begin{cases} 5x - 2y = -12 \\ 3x + 4y = -2 \end{cases}$$

Solution

Step 1. Choose which variable to eliminate. We eliminate the y-variable because the coefficients of the y-variables will be opposites if we multiply the first equation by 2.

Step 2. Multiply both sides of the first equation by 2.

$$5x - 2y = -12 \xrightarrow{\text{Multiply by 2}} 2(5x - 2y) = 2(-12) \xrightarrow{\text{Simplify}} 10x - 4y = -24$$

$$3x + 4y = -2 \xrightarrow{\text{No change}} 3x + 4y = -2 \xrightarrow{\text{No change}} 3x + 4y = -2$$

Step 3. Add the two new equations together.

$$\begin{array}{r} 10x - 4y = -24 \\ 3x + 4y = -2 \\ \hline 13x = -26 \end{array}$$

Step 4. Solve the resulting equation for x.

Rewrite the resulting equation from step 3: $13x = -26$

Divide both sides by 13: $x = -2$

Step 5. Find y by substituting $x = -2$ into one of the original equations.

Begin with the original first equation: $5x - 2y = -12$

Substitute -2 for x: $5(-2) - 2y = -12$

Simplify: $-10 - 2y = -12$

Add 10 to both sides: $-2y = -2$

Divide both sides by -2: $y = 1$

The solution to this system is the ordered pair $(-2, 1)$.

Since we used the original first equation in step 5, we use the original second equation to check the final answer. Click here to see the check, or watch this video for a fully worked solution.

Example 8 Solving Systems of Linear Equations in Two Variables by Elimination

My video summary

📽 Use the elimination method to solve the following system:

$$\begin{cases} -2x + 5y = 29 \\ 3x + 2y = 4 \end{cases}$$

Solution Notice that, if we multiply the first equation by 3 and the second equation by 2, we can eliminate the x-variable. Try this and finish solving this system on your own. Click here to check your answer, or watch this video for a fully worked solution.

You Try It   Work through this You Try It problem.

Work Exercises 15–16 in this textbook or in the *MyMathLab*®️ Study Plan.

If the equations in a system have fraction coefficients, then it may be helpful to clear the fractions before proceeding. Likewise, it may be helpful to clear decimal coefficients.

Example 9   Solving Systems of Linear Equations in Two Variables Involving Fractions

Use the elimination method to solve the following system:

$$\begin{cases} \dfrac{2}{9}x - \dfrac{5}{6}y = -\dfrac{8}{3} \\ \dfrac{3}{4}x - \dfrac{1}{3}y = \dfrac{11}{12} \end{cases}$$

Solution   Try solving this system on your own. Click here to check your answer, or watch this video for a fully worked solution.

You Try It   Work through this You Try It problem.

Work Exercises 17–18 in this textbook or in the *MyMathLab*®️ Study Plan.

OBJECTIVE 5:   SOLVE INCONSISTENT AND DEPENDENT SYSTEMS

Each system in Examples 1 through 9 had only one solution. Now we look at systems without a solution and systems with an infinite number of solutions. Recall from Objective 1 that a system without a solution is inconsistent, and a system with an infinite number of solutions is dependent.

Example 10   Solving an Inconsistent System

Solve the following system:

$$\begin{cases} x - 2y = 11 \\ -2x + 4y = 8 \end{cases}$$

Solution   We can solve this system by graphing, substitution, or elimination. In this case, we use the elimination method.

Step 1. We choose to eliminate the x-variable because it has a coefficient of 1.

Step 2. Multiply both sides of the first equation by 2. This makes the coefficients of the x-terms in the two equations 2 and -2, respectively.

$$x - 2y = 11 \xrightarrow{\text{Multiply by 2}} 2x - 4y = 22$$

$$-2x + 4y = 8 \xrightarrow{\text{No change}} -2x + 4y = 8$$

Step 3. Add the two equations from step 2 together.

$$2x - 4y = 22$$
$$\underline{-2x + 4y = 8}$$
$$0 = 30 \quad \text{False}$$

Since both variables are eliminated, we are left with the contradiction $0 = 30$. When this happens, the system has no solution and is an inconsistent system. Figure 5 shows that the graphs of the two linear equations in this system are parallel. Recall from Objective 1 and Figure 2(b) that systems of linear equations in two variables involving parallel lines have no solution. We use the null symbol $\varnothing$ or empty set { } to show that a system has no solution.

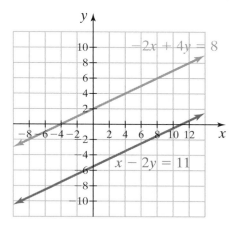

Figure 5

Graph of System $\begin{cases} x - 2y = 11 \\ -2x + 4y = 8 \end{cases}$

Note: If we had used the substitution method to solve this system, a similar contradiction would have occurred. Click here to see a solution using the substitution method.

You Try It   **Work through this You Try It problem.**

Work Exercises 21–26 in this textbook or in the *MyMathLab*® Study Plan.

Example 11 Solving a Dependent System

Solve the following system:

$$\begin{cases} -3x + 6y = 9 \\ x - 2y = -3 \end{cases}$$

Solution We can solve this system by graphing, substitution, or elimination. In this case, we use the substitution method.

Step 1. Solve the second equation for x.

Begin with the second equation: $x - 2y = -3$

Add $2y$ to both sides: $x = 2y - 3$

Step 2. Substitute $2y - 3$ for x in the first equation.

Rewrite the first equation: $-3x + 6y = 9$

Substitute $2y - 3$ for x: $-3(2y - 3) + 6y = 9$

Step 3. Solve for y.

$$\text{Distribute:} \quad -6y + 9 + 6y = 9$$

$$\text{Combine like terms:} \quad\quad\quad 9 = 9 \quad \text{True}$$

The equation simplifies to the identity $9 = 9$, which is true for any value of y. The system has an infinite number of solutions and is a dependent system. Any ordered pair that is a solution to one of the equations is a solution to both equations.

Figure 6 shows that the graphs of the two equations in the system are coinciding lines. The solution to the system is the set of all ordered pairs that lie on the graph of either equation.

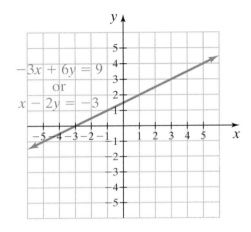

Figure 6

Graph of System $\begin{cases} -3x + 6y = 9 \\ x - 2y = -3 \end{cases}$

My interactive video summary

There are two common ways to write the solution to a dependent system: set-builder notation and **ordered-pair notation**. Watch this interactive video for an explanation.

With *set-builder notation*, we indicate the solutions to a dependent system by using one of the equations from the system (or any equivalent equation). The solution to the system in Example 11 can be written as $\{(x, y)\,|\,x - 2y = -3\}$. We read this as "the set of all ordered pairs (x, y), such that $x - 2y = -3$." Also, we can write the solution set equivalently as $\{(x, y)\,|\,-3x + 6y = 9\}$ or $\left\{(x, y)\,\middle|\,y = \frac{1}{2}x + \frac{3}{2}\right\}$. Do you see why?

With *ordered-pair notation*, we indicate the solutions to a dependent system by expressing one coordinate in terms of the other in an ordered pair. In step 1 of Example 11, we wrote x in terms of y as $x = 2y - 3$. So, the solution to the system is the set of all ordered pairs of the form $(2y - 3, y)$, where y is any real number.

If $y = -2$, then $x = 2(-2) - 3 = -7$ and $(-7, -2)$ is a solution to the system. If $y = 0$, then $x = 2(0) - 3 = -3$ and $(-3, 0)$ is a solution to the system.

Also, we can write the solution as $\left(x, \frac{1}{2}x + \frac{3}{2}\right)$. Do you see why?

Note: If we had used the elimination method to solve this system, a similar identity would have occurred. Click here to see a solution using the elimination method.

It is incorrect to say that the solution to a dependent system is "the set of all real numbers" or "the set of all ordered pairs." Although dependent systems have an infinite number of ordered-pair solutions, these solutions only consist of those ordered pairs whose points lie on the coinciding lines.

You Try It   Work through this You Try It problem.

Work Exercises 19–26 in this textbook or in the *MyMathLab*® Study Plan.

OBJECTIVE 6:   USE SYSTEMS OF LINEAR EQUATIONS IN TWO VARIABLES TO SOLVE APPLICATION PROBLEMS

Sometimes we can use a single equation in one variable to solve application problems, such as those in Section 1.1. However, it is often easier to use two variables and create a system of linear equations. The following steps are based on the problem-solving strategy for applications of linear equations from Section 1.1.

Problem-Solving Strategy for Applications Using Systems of Linear Equations

Step 1. Define the Problem.   Read the problem carefully; multiple times if necessary. Identify what you need to find and determine what information is available to help you find it.

Step 2. Assign Variables.   Choose variables that describe each unknown quantity.

Step 3. Translate into a System of Equations.   Use the relationships among the known and unknown quantities to form a system of equations.

Step 4. Solve the System.   Use graphing, substitution, or elimination to solve the system.

Step 5. Check the Reasonableness of Your Answers.   Check to see if your answers make sense within the context of the problem. If not, check your work for errors and try again.

Step 6. Answer the Question.   Write a clear statement that answers the question(s) posed.

Let's look back at the solution of Example 10 in Section 1.1. Now compare that solution process with the one shown in Example 12 on this page. Which approach do you prefer?

Example 12 NFL Touchdown Passes

Roger Staubach and Terry Bradshaw were both NFL quarterbacks. In 1973, Staubach threw three touchdown passes more than twice the number of touchdown passes thrown by Bradshaw. If the total number of touchdown passes thrown between Staubach and Bradshaw was 33, then how many touchdown passes did each player throw?

Solution We follow the problem-solving strategy using systems of equations.

Step 1. Define the Problem.   We need to find the number of touchdown passes thrown by each quarterback. Staubach threw three more than twice the number of touchdown passes thrown by Bradshaw. Together the two quarterbacks threw a total of 33 touchdown passes.

Step 2. **Assign Variables.**

Let S = the number of touchdowns thrown by Staubach.

Let B = the number of touchdowns thrown by Bradshaw.

Step 3. **Translate into a System of Equations.**

The combined number of touchdown passes is 33, so the first equation is

Staubach's touchdown passes	+	Bradshaw's touchdown passes	=	Total touchdown passes
S	+	B	=	33

To find the second equation, we use the fact that Staubach threw three more than twice the number of touchdowns thrown by Bradshaw. Therefore, the second equation is $S = 2B + 3$. This equation can be rewritten as $S - 2B = 3$, which fits the format of the first equation. So, we have the following system of equations:

$$\begin{cases} S + B = 33 \\ S - 2B = 3 \end{cases}$$

Step 4. **Solve the System.**

Using the elimination method and multiplying the first equation by 2, we obtain the system:

$$\begin{cases} 2S + 2B = 66 \\ S - 2B = 3 \end{cases}$$

Add the two equations together:

$$\begin{array}{r} 2S + 2B = 66 \\ \underline{S - 2B = 3} \\ 3S \quad\quad = 69 \end{array}$$

Dividing both sides of this equation by 3 gives $S = 23$. To find B, replace S with 23 in the original first equation.

Rewrite the original first equation:	$S + B = 33$
Substitute 23 for S:	$23 + B = 33$
Subtract 23 from both sides:	$B = 10$

Step 5. **Check the Reasonableness of Your Answers.** Bradshaw threw 10 touchdown passes and Staubach threw 23 touchdown passes. Since 23 is three more than twice 10, and the sum of 10 and 23 is 33, these results make sense within the context of the problem.

Step 6. **Answer the Question.** In 1973, Bradshaw threw 10 touchdown passes and Staubach threw 23 touchdown passes.

When using systems of linear equations to solve real-world problems, the solutions do not always work out to nice, exact answers. Sometimes we need to approximate or round the answer. It is best to wait to round your answer until the end of the problem-solving process, if possible. If a rounded answer is used in later calculations, then those calculations will have errors due to rounding. We face this issue in Example 13.

My interactive video summary

Example 13 Phone Expenditures

At the start of this section, we saw that Figure 1 showed the average annual expenditures on cellular phones and residential phones since 2001. Residential phone expenditures are modeled by $y = -30.71x + 713.29$, and cellular phone expenditures are modeled by $y = 64.04x + 141.71$, where x represents the number of years since 2000 and y represents the average annual expenditures. Use these models to find when the average annual expenditures for the two phone types were equal and what the expenditure was at that time.

Solution The two models together form the following system:

$$\begin{cases} y = -30.71x + 713.29 \\ y = 64.04x + 141.71 \end{cases}$$

We follow the problem-solving strategy using systems. Solve this system by substitution or elimination. Remember not to round until the end of the problem, if possible. Click here to see the answer, or watch this interactive video for a fully worked solution.

Example 14 Counting Calories

My interactive video summary

Malcolm and Jamal went to Burger King for lunch. Malcolm ate two cheeseburgers and a medium order of fries for a total of 1160 calories. Jamal ate one cheeseburger and two medium orders of fries for a total of 1330 calories. How many calories are in a Burger King cheeseburger? How many calories are in a medium order of fries? (*Source:* Burger King U.S. Nutrition Information, January 2008)

Solution We follow the problem-solving strategy using systems of equations.

Step 1. Define the Problem. We need to find the number of calories in a Burger King cheeseburger and the number of calories in a medium order of fries. From Malcolm's lunch, we know that two cheeseburgers and one medium order of fries have 1160 calories. From Jamal's lunch, we also know that one cheeseburger and two medium orders of fries have 1330 calories.

Step 2. Assign Variables.

Let C = the number of calories in a Burger King cheeseburger.

Let F = the number of calories in a medium order of fries.

Step 3. Translate into a System of Equations.

For Malcom's lunch, we write the first equation as $2C + F = 1160$. For Jamal's lunch, we write the second equation as $C + 2F = 1330$. So, we have the following system of equations:

$$\begin{cases} 2C + F = 1160 \\ C + 2F = 1330 \end{cases}$$

Step 4. Solve the System of Equations. Finish solving the system of equations on your own. Use the results to answer the question. Click here to see this answer, or watch this interactive video for a fully worked solution.

You Try It Work through this **You Try It** problem.

Work Exercises 27–32 in this textbook or in the *MyMathLab*®️ Study Plan.

3.1 Exercises

In Exercises 1–4, determine if each ordered pair is a solution to the given system.

You Try It

1. $\begin{cases} 3x - 2y = -2 \\ 4x + 3y = 37 \end{cases}$ **a.** $(4, 7)$ **b.** $(-4, -5)$

2. $\begin{cases} 3x + 18y = 15 \\ -5x - 30y = -25 \end{cases}$ **a.** $(12.8, -1.3)$ **b.** $\left(\dfrac{1}{2}, \dfrac{3}{4}\right)$

3. $\begin{cases} 0.3x + 0.4y = 2.3 \\ 0.5x - 0.2y = -0.5 \end{cases}$ **a.** $(-3, 8)$ **b.** $(1, 5)$

4. $\begin{cases} \dfrac{1}{9}x - \dfrac{1}{4}y = -\dfrac{2}{3} \\ \dfrac{1}{3}x - \dfrac{1}{2}y = -1 \end{cases}$ **a.** $(3, 4)$ **b.** $(12, 8)$

In Exercises 5–7, solve each system by graphing.

You Try It

5. $\begin{cases} y = 2x \\ y = -3x - 5 \end{cases}$ **6.** $\begin{cases} y = \dfrac{3}{2}x - 4 \\ y = 2x - 6 \end{cases}$ **7.** $\begin{cases} 3x - y = -5 \\ 4x + y = -2 \end{cases}$

In Exercises 8–12, solve each system by substitution.

8. $\begin{cases} x = 4 - y \\ 3x - 2y = -3 \end{cases}$ **9.** $\begin{cases} x + 4y = 5 \\ 2x - y = -8 \end{cases}$ **10.** $\begin{cases} 2x - y = 3 \\ -8x + 3y = -8 \end{cases}$

You Try It 11. $\begin{cases} 5x + 4y = -6 \\ 8x - 4y = -72 \end{cases}$ **12.** $\begin{cases} -\dfrac{2}{3}x + \dfrac{1}{2}y = -\dfrac{1}{3} \\ -\dfrac{1}{2}x + \dfrac{3}{4}y = -\dfrac{1}{2} \end{cases}$

In Exercises 13–18, solve each system by elimination.

You Try It 13. $\begin{cases} x + 2y = 10 \\ x - 2y = -6 \end{cases}$ **14.** $\begin{cases} 5x - y = 7 \\ 3x + 4y = 18 \end{cases}$

15. $\begin{cases} 8x - 6y = -6 \\ 5x - 7y = -33 \end{cases}$ **16.** $\begin{cases} 8x - 9y = 12 \\ 4x + 6y = -1 \end{cases}$

You Try It

17. $\begin{cases} \dfrac{3}{2}x + \dfrac{5}{6}y = \dfrac{2}{3} \\ \dfrac{5}{8}x - \dfrac{1}{6}y = \dfrac{19}{24} \end{cases}$ **18.** $\begin{cases} 0.6x + 0.4y = 21 \\ 0.3x - 0.6y = -1.5 \end{cases}$

You Try It

In Exercises 19–20, write the solution of each dependent system using (a) set-builder notation and (b) ordered-pair notation.

You Try It 19. $\begin{cases} 2x - y = -6 \\ 8x - 4y = -24 \end{cases}$ **20.** $\begin{cases} \dfrac{3}{4}x + \dfrac{1}{2}y = \dfrac{5}{8} \\ 0.75x + 0.5y = 0.625 \end{cases}$

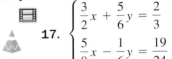

In Exercises 21–26, use the substitution or elimination method to solve each system. If the system is dependent, write the solution in ordered-pair notation.

You Try It

21. $\begin{cases} x + y = 3 \\ -2x - 2y = 1 \end{cases}$

22. $\begin{cases} 3x - 12y = 6 \\ -2x + 8y = -4 \end{cases}$

23. $\begin{cases} 8x - y = -13 \\ y = -8x \end{cases}$

24. $\begin{cases} -\frac{1}{2}x + \frac{2}{3}y = -\frac{1}{3} \\ \frac{2}{3}x - 2y = 1 \end{cases}$

25. $\begin{cases} 8x + 4y = 5 \\ 2x + y = -4 \end{cases}$

26. $\begin{cases} y = \frac{5}{7}x - 9 \\ -15x + 21y = -189 \end{cases}$

In Exercises 27–32, use a system of linear equations in two variables to solve each application problem.

27. Basketball Score   During a game of one-on-one, Rodney and Justin scored 47 points altogether. If Justin lost the game by 5 points, how many points did each player score?

You Try It

28. Cashing a Check   Cedric cashed a $250 check at the bank. The teller gave Cedric his money in $10 and $20 bills only. If the teller gave Cedric twice as many $20 bills as $10 bills, how many of each type did Cedric receive?

29. Counting Calories   Lefkios went to McDonald's for breakfast. He ate an Egg McMuffin® and two hash browns for a total of 600 calories. If an Egg McMuffin has twice as many calories as one hash brown, how many calories are in each? (*Source:* McDonald's USA Nutrition Facts, March 2009)

30. Going Swimming   One hot summer day, 492 people used the public swimming pool. The daily prices were $1.50 for children and $2.50 for adults. The admission receipts totaled $929. How many children and how many adults swam at the public pool that day?

31. Dimensions of a Rectangle   The perimeter of a rectangle is 300 inches. If the length is 10 inches less than three times the width, find the dimensions of the rectangle.

32. Ladies' Night   Thursday is ladies' night at the Slurp and Burp Bar and Grill. All beverages are $2.50 for men and $1.50 for women. A total of 956 beverages were sold last Thursday night. If the Slurp and Burp sold a total of $1996 in beverages last Thursday night, how many beverages were sold to women?

3.2 Systems of Linear Equations in Three Variables

THINGS TO KNOW

Before working through this section, be sure you are familiar with the following concepts:

VIDEO   ANIMATION   INTERACTIVE

You Try It

1. Determine If an Ordered Pair Is a Solution To a System of Linear Equations in Two Variables (Section 3.1, Objective 1)

You Try It

2. Solve Systems of Linear Equations in Two Variables by Elimination (Section 3.1, Objective 4)

You Try It

3. Use Systems of Linear Equations in Two Variables to Solve Application Problems (Section 3.1, Objective 6)

OBJECTIVES

1 Determine If an Ordered Triple Is a Solution to a System of Linear Equations in Three Variables

2 Solve Systems of Linear Equations in Three Variables

3 Use Systems of Linear Equations in Three Variables to Solve Application Problems

---

OBJECTIVE 1:   DETERMINE IF AN ORDERED TRIPLE IS A SOLUTION TO A SYSTEM OF LINEAR EQUATIONS IN THREE VARIABLES

Now that we know how to solve systems of linear equations in two variables, we can apply this knowledge to **systems of linear equations in three variables**. An equation such as $4x + 2y - 7z = 12$ is called a **linear equation in three variables** because there are three variables and each variable term is linear. Comparing this equation to our definitions for linear equations in one variable and linear equations in two variables leads us to the following definition.

> **Definition   Linear Equation in Three Variables**
>
> A **linear equation in three variables** is an equation that can be written in the form $Ax + By + Cz = D$, where A, B, C, and D are real numbers, and A, B, and C are not all equal to 0.

A solution to a linear equation in three variables is an **ordered triple** (x, y, z) that makes the equation true. As with the two-variable case, there are infinitely many solutions to these equations. The graph of the solution set to a linear equation in two variables is a line in two-dimensional space. With three variables, the graph of the solution set becomes a plane in three-dimensional space.

In Section 3.1, we saw that an ordered pair was an ordered-pair solution to a system of linear equations in two variables if it was a solution to both equations in the system. Similarly, an ordered triple is an **ordered-triple solution** to a **system of linear equations in three variables** if it is a solution to all three equations in the system.

> **Definition   System of Linear Equations in Three Variables**
>
> A **system of linear equations in three variables** is a collection of linear equations in three variables considered together. A **solution to a system** of linear equations in three variables is an ordered triple that satisfies all equations in the system.

Example 1 Determining If an Ordered Triple Is a Solution to a System

Determine if each ordered triple is a solution to the given system:

$$\begin{cases} 3x + y - 2z = 4 \\ 2x - 2y + 3z = 9 \\ x + y - z = 5 \end{cases}$$

a. $(3, 9, 7)$ **b.** $(2, -4, -1)$

My interactive video summary

Solution

a. To determine if the ordered triple $(3, 9, 7)$ is a solution to the system, substitute the values for the variables into each equation to see if a true statement results. To be a solution to the system, the ordered triple must be a solution to each equation in the system.

First Equation

Original equation: $\qquad 3x + y - 2z = 4$

Subsitute: $\quad 3(3) + (9) - 2(7) \overset{?}{=} 4$

Simplify: $\qquad\quad 9 + 9 - 14 \overset{?}{=} 4$

$\qquad\qquad\qquad\qquad 4 = 4 \quad$ True

Second Equation

Original equation: $\qquad 2x - 2y + 3z = 9$

Subsitute: $\quad 2(3) - 2(9) + 3(7) \overset{?}{=} 9$

Simplify: $\qquad\quad 6 - 18 + 21 \overset{?}{=} 9$

$\qquad\qquad\qquad\qquad 9 = 9 \quad$ True

Third Equation

Original equation: $\qquad x + y - z = 5$

Subsitute: $\quad (3) + (9) - (7) \overset{?}{=} 5$

Simplify: $\qquad\quad 3 + 9 - 7 \overset{?}{=} 5$

$\qquad\qquad\qquad\qquad 5 = 5 \quad$ True

Since the ordered triple makes each of the three equations true, $(3, 9, 7)$ is a solution to the system.

b. Check the ordered triple $(2, -4, -1)$ on your own. Click here to check your answer, or watch this interactive video for a detailed solution.

You Try It   **Work through this You Try It problem.**

Work Exercises 1–2 in this textbook or in the *MyMathLab*® Study Plan.

OBJECTIVE 2:   SOLVE SYSTEMS OF LINEAR EQUATIONS IN THREE VARIABLES

Graphically, we can visualize solutions to systems of linear equations in two variables as intersection points of lines. These systems could have one solution (intersecting lines), no solution (parallel lines), or infinitely many solutions (coinciding lines). These same situations apply when solving systems of linear equations in three variables, but graphically, we view the solutions as intersection points of planes. Figure 7 illustrates the possibilities for systems of linear equations in three variables.

To solve systems of linear equations in three variables by graphing would require us to graph planes in three dimensions, which is not a practical task. Instead, we will focus on solving systems of linear equations in three variables by applying the elimination method used for solving systems of linear equations in two variables.

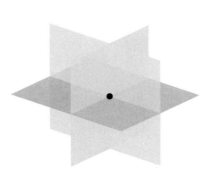

Consistent System
One unique solution

Consistent System
Infinitely many solutions
(Planes intersect at a line)

Consistent System
Infinitely many solutions
(Three equations describing
the same plane)

Inconsistent System
No solution
(Planes intersect two at a time)

Inconsistent System
No solution
(Three planes are parallel)

Inconsistent System
No solution
(Two planes are parallel)

Figure 7

When solving systems of linear equations in two variables by elimination, the goal is to reduce the system of two equations in two variables to a single linear equation in one variable. We then solve this equation for the remaining variable, substitute this value into either of the original equations, and solve for the variable that had been eliminated. See Figure 8. This process of substituting back into an original equation to find values of remaining variables is called **back substitution**.

Back-substitute to find the
value of the other variable

| System of linear equations in two variables | Linear equation in one variable | Solve for a variable |

Substitution
or elimination

Figure 8 Solving a system of linear equations in two variables

A similar process is used to solve a system of linear equations in three variables. The goal is to reduce the system of three equations in three variables down to a system of two equations in two variables. At that point, we can reduce the two-equation system to a single equation in one variable and easily solve that equation. Using back substitution, we can find the values of the other two variables. See Figure 9.

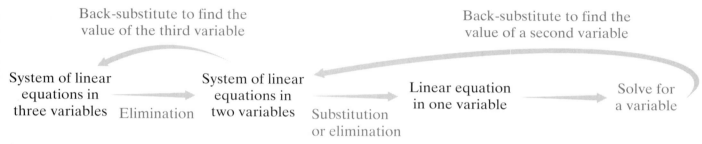

Back-substitute to find the
value of the third variable

Back-substitute to find the
value of a second variable

System of linear equations in three variables — Elimination — System of linear equations in two variables — Substitution or elimination — Linear equation in one variable — Solve for a variable

Figure 9 Solving a system of linear equations in three variables

Before looking at an example, we present some guidelines for solving systems of linear equations in three variables by elimination.

> **Guidelines for Solving a System of Linear Equations in Three Variables by Elimination**
>
> **Step 1. Write each equation in standard form.** Write each equation in the form $Ax + By + Cz = D$, lining up the variable terms. Number the equations to keep track of them.
>
> **Step 2. Eliminate a variable from one pair of equations.** Use the elimination method to eliminate a variable from any two of the original three equations, leaving one equation in two variables.
>
> **Step 3. Eliminate the same variable again.** Use a different pair of the original equations and eliminate the same variable again, leaving one equation in two variables.
>
> **Step 4. Solve the system of linear equations in two variables.** Use the resulting equations from steps 2 and 3 to create and solve the corresponding system of linear equations in two variables by substitution or elimination.
>
> **Step 5. Use back substitution to find the value of the third variable.** Substitute the results from step 4 into any of the original equations to find the value of the remaining variable.
>
> **Step 6. Check the solution.** Check the proposed solution in each equation of the system and write the solution set.

Example 2 Solving Systems of Linear Equations in Three Variables

My video summary

📽 Solve the following system:

$$\begin{cases} 2x + 3y + 4z = 12 \\ x - 2y + 3z = 0 \\ -x + y - 2z = -1 \end{cases}$$

Solution We follow our guidelines for solving systems of linear equations in three variables. Follow the steps below, or watch this video for a detailed solution.

Step 1. The equations are already in standard form and all the variables are lined up. We rewrite the system and number each equation.

$$\begin{cases} 2x + 3y + 4z = 12 & (1) \\ x - 2y + 3z = 0 & (2) \\ -x + y - 2z = -1 & (3) \end{cases}$$

Step 2. We can eliminate any of the variables. For convenience, we will eliminate the variable x from equations (1) and (2). We can do this by multiplying equation (2) by -2 and adding the equations together.

$$
\begin{array}{rl}
(1): & 2x + 3y + 4z = 12 \\
\text{Multiply } (2) \text{ by } -2: & \underline{-2x + 4y - 6z = 0} \\
\text{Add:} & 7y - 2z = 12 \quad (4)
\end{array}
$$

How do you know which variable to eliminate? Click here to find out.

Step 3. We need to eliminate the same variable, x, from a different pair of equations. We will use equations (2) and (3) for the second pair. Since the coefficients of x in these equations have the same absolute value, but opposite signs, we can eliminate the variable by simply adding the equations.

$$
\begin{array}{rl}
(2): & x - 2y + 3z = 0 \\
(3): & \underline{-x + y - 2z = -1} \\
\text{Add:} & -y + z = -1 \quad (5)
\end{array}
$$

Step 4. Combining equations (4) and (5), we form a system of linear equations in two variables.

$$\begin{cases} 7y - 2z = 12 & (4) \\ -y + z = -1 & (5) \end{cases}$$

To solve this system, we use the elimination method by multiplying equation (5) by 7 and adding the result to equation (4) to eliminate the variable y.

$$
\begin{array}{rl}
(4): & 7y - 2z = 12 \\
\text{Multiply } (5) \text{ by } 7: & \underline{-7y + 7z = -7} \\
\text{Add:} & 5z = 5 \\
\text{Divide by 5:} & z = 1
\end{array}
$$

Since $z = 1$, we back-substitute this into (5) to solve for y.

$$
\begin{aligned}
(5): && -y + z &= -1 \\
\text{Substitute 1 for } z: && -y + (1) &= -1 \\
\text{Simplify:} && -y + 1 &= -1 \\
\text{Subtract 1:} && -y &= -2 \\
\text{Divide by } -1: && y &= 2
\end{aligned}
$$

When back-substituting, we can use either of the equations, but often one equation is preferred over another. Can you see why we chose to use equation (5) instead of equation (4)? Click here for an explanation.

Step 5. Substitute 2 for y and 1 for z in any of the original equations, (1), (2), or (3), and solve for x. We will back-substitute using equation (2).

$$
\begin{aligned}
(2): && x - 2y + 3z &= 0 \\
\text{Substitute 2 for } y \text{ and 1 for } z: && x - 2(2) + 3(1) &= 0 \\
\text{Multiply:} && x - 4 + 3 &= 0 \\
\text{Simplify:} && x - 1 &= 0 \\
\text{Add 1:} && x &= 1
\end{aligned}
$$

The solution to the system is the ordered triple $(1, 2, 1)$.

Step 6. Click here to check your answer.

You Try It   Work through this You Try It problem.

Work Exercises 3–5 in this textbook or in the *MyMathLab* Study Plan.

Example 3 Solving Systems of Linear Equations in Three Variables with Missing Terms

My video summary

Solve the following system:

$$
\begin{cases}
2x + \ y = 13 \\
3x - 2y + \ z = \ 8 \\
x + 2y - 3z = \ 5
\end{cases}
$$

Solution We follow our guidelines for solving a system of linear equations in three variables.

Step 1. First, we rewrite and number each equation, lining up the variables. Notice that the first equation has a variable term missing, so we put a gap in its place.

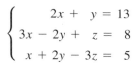

Step 2. We can eliminate any of the variables, but we notice that one equation already has z eliminated. By selecting z as the variable to eliminate, we can move directly to step 3.

Step 3. Looking at equations (2) and (3), we might be tempted to add these equations to eliminate y. However, in step 2, we selected z as the variable to eliminate, so we need to eliminate z again in this step.

Continue working through the solution. Click here to check your answer, or watch this video for a detailed solution.

You Try It   Work through this You Try It problem

Work Exercises 9–11 in this textbook or in the *MyMathLab* Study Plan.

Example 4 Solving Systems of Linear Equations in Three Variables Involving Fractions

 Solve the following system:

$$\begin{cases} \dfrac{1}{2}x + \quad y + \dfrac{2}{3}z = 2 \\[2mm] \dfrac{3}{4}x + \dfrac{5}{2}y - \ 2z = -7 \\[2mm] \quad x + 4y + 2z = 4 \end{cases}$$

Solution If an equation in the system contains fractions, then it is often helpful to clear the fractions first. After doing this, follow the guidelines for solving a system of linear equations in three variables. Click here to check your answer, or watch this video for a detailed solution.

You Try It   Work through this You Try It problem.

Work Exercises 6–8 in this textbook or in the *MyMathLab* Study Plan.

Systems of linear equations in three variables may be inconsistent or may include dependent equations. If, in our solution process, we find a contradiction, then the system is inconsistent and has no solution. In the two-variable case, this occurred if we had parallel lines because the lines had no points in common.

For the three-variable case, the system will be inconsistent if all three planes have no points in common. This would happen if all the planes were parallel, but would also happen if two or none of the planes are parallel, as shown in Figure 10.

Inconsistent System
No solution
(Planes intersect two at a time)

Inconsistent System
No solution
(Three planes are parallel)

Inconsistent System
No solution
(Two planes are parallel)

Figure 10

When solving systems of linear equations in two variables, encountering an identity meant the system was dependent and had an infinite number of ordered-pair solutions. This happened when the lines were coinciding.

However, unlike the two-variable case, identifying dependent systems in three variables takes a bit more work because we must consider all possible pairings of the equations in the system. Obtaining an identity with one pairing is

My video summary

not sufficient to say the system is dependent. This idea is illustrated in the following example.

Example 5 Solving Systems of Linear Equations in Three Variables with No Solution

Solve the following system:

$$\begin{cases} x - y + 2z = 5 \\ 3x - 3y + 6z = 15 \\ -2x + 2y - 4z = 7 \end{cases}$$

Solution We follow our guidelines for solving a system of linear equations in three variables.

Step 1. The equations are already in standard form with all the variables lined up. We rewrite the system and number each equation.

$$\begin{cases} x - y + 2z = 5 & (1) \\ 3x - 3y + 6z = 15 & (2) \\ -2x + 2y - 4z = 7 & (3) \end{cases}$$

Step 2. For convenience, we will eliminate the variable x from equations (1) and (2). To do this, we multiply equation (1) by -3 and add the equations together.

$$\begin{array}{ll} \text{Multiply (1) by } -3: & -3x + 3y - 6z = -15 \\ (2): & \underline{3x - 3y + 6z = 15} \\ \text{Add:} & 0 = 0 \quad \text{True} \end{array}$$

The last line is an identity. In the two-variable case, we would stop and say that the system had an infinite number of solutions. However, in systems of three variables, this is not necessarily the case.

Step 3. Let's continue our process and eliminate the variable x from the pairing of equations (1) and (3). To do this, we multiply equation (1) by 2 and add the equations.

$$\begin{array}{ll} \text{Multiply (1) by 2:} & 2x - 2y + 4z = 10 \\ (3): & \underline{-2x + 2y - 4z = 7} \\ \text{Add:} & 0 = 17 \quad \text{False} \end{array}$$

The last line is a contradiction, so the system is inconsistent and has no solution. Click here to find out why this occurs.

You Try It Work through this You Try It problem.

Work Exercises 14–15 in this textbook or in the **MyMathLab**® Study Plan.

My video summary

As we saw in Example 5, finding an identity in the three-variable case does not necessarily mean the system is dependent and has an infinite number of solutions. Watch this video for a more detailed explanation.

Example 6 Solving Systems of Dependent Linear Equations in Three Variables

Solve the following system.

$$\begin{cases} x - y + 2z = 5 \\ 3x - 3y + 6z = 15 \\ -2x + 2y - 4z = -10 \end{cases}$$

Solution

$$\begin{cases} x - y + 2z = 5 & (1) \\ 3x - 3y + 6z = 15 & (2) \\ -2x + 2y - 4z = -10 & (3) \end{cases}$$

For this system, notice that multiplying equation (1) by 3 gives us equation (2) and multiplying equation (1) by -2 gives us equation (3). Therefore, all three equations describe the same plane and the system is dependent.

 *My interactive video summary*

 Recall from the previous section that we can write the solutions to a dependent system in two ways: set-builder notation and ordered-pair notation (or in the three-variable case, ordered-triple notation). Watch this interactive video to review. The solution set for this system can be written as $\{(x, y, z)\mid x - y + 2z = 5\}$ in set-builder notation or $(5 + y - 2z, y, z)$ in ordered-triple notation. Click here to see some variations.

You Try It   Work through this You Try It problem.

Work Exercises 12–13 in this textbook or in the **MyMathLab** Study Plan.

Which notation should be used to express the solution of a dependent system? Click here for a suggestion.

Example 7 Solving Systems of Dependent Linear Equations in Three Variables

My video summary

Solve the following system.

$$\begin{cases} 4x + y - 5z = -1 & (1) \\ -2x - y + z = 3 & (2) \\ x + y + z = -4 & (3) \end{cases}$$

Solution Eliminating y using equations (1) and (2) results in the equation $2x - 4z = 2$. Eliminating y again using equations (1) and (3) results in the equation $3x - 6z = 3$. Combining these two equations gives a two-variable system:

$$\begin{cases} 2x - 4z = 2 & (4) \\ 3x - 6z = 3 & (5) \end{cases}$$

If we divide equation (4) by 2 and divide equation (5) by 3, then we see that the two equations in the system are equivalent. Therefore, this two-variable system has an infinite number of solutions. Solving either equation (4) or (5) for x gives the relationship $x = 2z + 1$. When we substitute this result into any of the three original equations and solve for y, we have $y = -3z - 5$. Since we have written two of the variables in terms of the third, the ordered-triple solution is $(2z + 1, -3z - 5, z)$. Watch this video to see the complete solution.

You Try It   Work through this You Try It problem.

Work Exercises 16–17 in this textbook or in the *MyMathLab*®️ Study Plan.

Having an infinite number of ordered-triple solutions does not mean that all ordered triples are solutions to a system of linear equations in three variables. Click here for an explanation.

OBJECTIVE 3:   USE SYSTEMS OF LINEAR EQUATIONS IN THREE VARIABLES TO SOLVE APPLICATION PROBLEMS

Building on the problem-solving strategy using systems of equations from Section 3.1, we now look at applications involving systems of linear equations in three variables.

Example 8 Real-Time Strategy Game

While playing a real-time strategy game, Joel created military units to defend his town: warriors, skirmishers, and archers. Warriors require 20 units of food and 50 units of gold. Skirmishers require 25 units of food and 35 units of wood. Archers require 32 units of wood and 32 units of gold. If Joel used 506 units of gold, 606 units of wood, and 350 units of food to create the units, how many of each type of military unit did he create?

Solution

Step 1. Define the Problem.   We want to find the number of each type of military unit created. There are three types of units: warriors, skirmishers, and archers. Each unit requires a certain amount of gold, wood, and food. We know how much of each resource is needed for each unit, and we know the total amount of each resource that is used. We summarize this information in the following table.

	Each Warrior	Each Skirmisher	Each Archer	Total
Units of Gold	50	0	32	506
Units of Wood	0	35	32	606
Units of Food	20	25	0	350

Step 2. Assign Variables.   Let W, S, and A represent the number of warrior, skirmisher, and archer units, respectively.

Step 3. Translate into a System of Equations.   We need to translate the given information into three equations to form a system. When writing a system of equations, totals are often a good place to start. We know the total units of gold, wood, and food used, and we know how much of each resource is required by the individual units. Therefore, we can write one equation based on total gold, a second equation based on total wood, and a third equation based on total food. A total of 506 units of gold were used. Since warriors require 50 units of gold and archers require 32 units of gold, we write the following equation.

$$\underbrace{50W}_{\text{Gold for warriors}} + \underbrace{32A}_{\text{Gold for archers}} = \underbrace{506}_{\text{Total gold}}$$

Note that there is no variable term involving S because skirmishers do not require gold.

A total of 606 units of wood were used. Since skirmishers require 35 units of wood and archers require 32 units of wood, we write a second equation:

$$\underbrace{35S}_{\text{Wood for skirmishers}} + \underbrace{32A}_{\text{Wood for archers}} = \underbrace{606}_{\text{Total wood}}$$

Again notice that there is a variable term missing because warriors do not require wood.

A total of 350 units of food were used. Since warriors require 20 units of food and skirmishers require 25 units of food, we write a third equation:

$$\underbrace{20W}_{\text{Food for warriors}} + \underbrace{25S}_{\text{Food for skirmishers}} = \underbrace{350}_{\text{Total food}}$$

Step 4. Solve the System.   Using the three equations, we form the system

$$\begin{cases} 50W & + 32A = 506 \\ 35S + 32A = 606. \\ 20W + 25S & = 350 \end{cases}$$

Now solve this system using the elimination method. Remember to check the reasonableness of your answer. Click here to check your answer, or watch this video for a detailed solution.

You Try It   Work through this You Try It problem.

Work Exercises 18–19 in this textbook or in the *MyMathLab*®️ Study Plan.

Example 9 Buy Clothes Online

My video summary

Wendy ordered 30 T-shirts online for her three children. Small T-shirts cost $4 each, medium T-shirts cost $5 each, and large T-shirts are $6 each. She spent $40 more for the large T-shirts than for the small T-shirts. Wendy's total bill was $154. How many T-shirts of each size did she buy?

Solution Following the problem-solving strategy, we begin by defining some variables. Let S, M, and L represent the number of small, medium, and large T-shirts, respectively. With the given information, we can write and solve a system of three linear equations. One equation uses the total number of shirts purchased; a second equation uses the total amount spent, and a third equation uses the fact that Wendy spent $40 more on large T-shirts than on small T-shirts. Finish the problem on your own. Click here to check your answer, or watch this video for a detailed solution.

You Try It   Work through this You Try It problem.

Work Exercises 20–23 in this textbook or in the *MyMathLab*®️ Study Plan.

3.2 Exercises

In Exercises 1 and 2, two ordered triples are given. Determine if each ordered triple is a solution to the given system.

You Try It

1. $(-1, 1, -2), (1, -1, 2)$
$$\begin{cases} x + y + z = -2 \\ -x - 3y - 2z = 2 \\ 2x - 2y + 5z = -14 \end{cases}$$

2. $(2, -1, 4), (-2, 1, -4)$
$$\begin{cases} \frac{1}{2}x + 3y - z = 6 \\ -x + y + \frac{1}{4}z = 2 \\ x - 4y + z = -10 \end{cases}$$

In Exercises 3–11, solve each system of linear equations.

3. $\begin{cases} x + y + z = 4 \\ 2x - y - 2z = -10 \\ -x - y + 3z = 8 \end{cases}$

4. $\begin{cases} x - 2y + z = 6 \\ 2x + y - 3z = -3 \\ x - 3y + 3z = 10 \end{cases}$

5. $\begin{cases} x - 2y + 2z = 2 \\ 3x + 2y - 2z = -1 \\ x - y - 2z = 0 \end{cases}$

You Try It

You Try It

6. $\begin{cases} x - \frac{1}{2}y + \frac{1}{2}z = -3 \\ x + y - z = 0 \\ -3x - 3y + 4z = 1 \end{cases}$

7. $\begin{cases} \frac{1}{3}x - \frac{2}{3}y + z = 0 \\ \frac{1}{2}x - \frac{3}{4}y + z = -\frac{1}{2} \\ -2x - y + z = 7 \end{cases}$

8. $\begin{cases} x + y + 10z = 3 \\ \frac{1}{2}x - y + z = -\frac{5}{6} \\ -2x + 3y - 5z = \frac{7}{3} \end{cases}$

9. $\begin{cases} -4x + 5y + 9z = -9 \\ x - 2y + z = 0 \\ 2y - 8z = 8 \end{cases}$

10. $\begin{cases} 2x + 2y + z = 9 \\ x + z = 4 \\ 4y - 3z = 17 \end{cases}$

11. $\begin{cases} x - y = 7 \\ y - z = 2 \\ x + z = 1 \end{cases}$

You Try It

 In Exercises 12–17, determine if the system has no solution or infinitely many solutions. If the system has infinitely many solutions, describe the solution with the equation of a plane or an ordered triple in terms of one variable.

12. $\begin{cases} 2x + 6y - 4z = 8 \\ -x - 3y + 2z = -4 \\ x + 3y - 2z = 4 \end{cases}$

13. $\begin{cases} 2x - y + z = -6 \\ x - \frac{1}{2}y + \frac{1}{2}z = -3 \\ 4x - 2y + 2z = -12 \end{cases}$

You Try It

14. $\begin{cases} x - 4y + 2z = 7 \\ \frac{1}{2}x - 2y + z = 1 \\ -3x + y - 4z = 9 \end{cases}$

15. $\begin{cases} 4x - y + z = 8 \\ x + y + 3z = 2 \\ 3x - 2y - 2z = 5 \end{cases}$

You Try It

16. $\begin{cases} x + 2y - z = 11 \\ x + 3y - 2z = 14 \\ 3x + 7y - 4z = 36 \end{cases}$

17. $\begin{cases} x - y + z = 5 \\ 2x + 3y - 3z = -5 \\ 3x + 2y - 2z = 0 \end{cases}$

You Try It

 In Exercises 18–23, solve using a system of linear equations in three variables.

You Try It

18. Real-Time Strategy Game While playing a real-time strategy game, Arvin created military units for a battle: long swordsmen, spearmen, and crossbowmen. Long swordsmen require 60 units of food and 20 units of gold. Spearmen require 35 units of food and 25 units of wood. Crossbowmen require 25 units of wood and 45 units of gold. If Arvin used 1975 units of gold, 1375 units of wood, and 1900 units of food to create the units, how many of each type of military unit did he create?

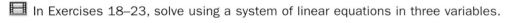

19. Concession Stand The concession stand at a school basketball tournament sells hot dogs, hamburgers, and chicken sandwiches. During one game, the stand sold 16 hot dogs, 14 hamburgers, and 8 chicken sandwiches for a total of $89.00. During a second game, the stand sold 10 hot dogs, 13 hamburgers, and 5 chicken sandwiches for a total of $66.25. During a third game, the stand sold 4 hot dogs, 7 hamburgers, and 7 chicken sandwiches for a total of $49.75. Determine the price of each product.

You Try It

20. Ordering Pizza Ben ordered 35 pizzas for an office party. He ordered three types: cheese, supreme, and pepperoni. Cheese pizza costs $9 each, pepperoni pizza costs $12 each, and supreme pizza costs $15 each. He spent exactly twice as much on the pepperoni pizzas as he did on the cheese pizzas. If Ben spent $420, how many pizzas of each type did he buy?

21. Theater Tickets On opening night of the play *The Music Man*, 1010 tickets were sold for a total of $10,300. Adult tickets cost $12 each, children's tickets cost $10 each, and senior citizen tickets cost $7 each. If the total number of adult and children tickets sold exceeded twice the number of senior citizen tickets sold by 170 tickets, then how many tickets of each type were sold?

22. NCAA Basketball Tyler Hansbrough was the leading scorer of the 2009 NCAA Basketball champions, the North Carolina Tar Heels. Hansbrough scored a total of 722 points during the 2009 season. He made 26 more one-point free throws than two-point field goals, and his number of two-point field goals was two less than 25 times his number of three-point field goals. How many free throws, two-point field goals, and three-point field goals did Tyler Hansbrough make during the 2009 season? (*Source:* espn.com)

23. Facebook Users The number of new Facebook users, y (in millions), between September 2008 and March 2009 can be modeled by the equation $y = ax^2 + bx + c$, where x represents the age of the user. Using the ordered-pair solutions $(15, 1)$, $(35, 7)$, and $(55, 3)$, create a system of linear equations in three variables for a, b, and c. Do this by substituting each ordered-pair solution into the model, creating an equation in three variables. Solve the resulting system to find the coefficients of the model. Then use the model to predict the number of new Facebook users who were 25 years old. (*Source:* www.facebook.com)

3.3 More Problem Solving with Systems of Linear Equations

THINGS TO KNOW

Before working through this section, be sure you are familiar with the following concepts:

	VIDEO	ANIMATION	INTERACTIVE

You Try It

1. Solve a Formula for a Given Variable
   (Section 1.4, Objective 1)

You Try It

2. Use Formulas to Solve Application Problems
   (Section 1.4, Objective 2)

You Try It

3. Use Systems of Linear Equations in Two
   Variables to Solve Application Problems
   (Section 3.1, Objective 6)

You Try It

4. Use Systems of Linear Equations in Three
   Variables to Solve Application Problems
   (Section 3.2, Objective 3)

OBJECTIVES

1 Use Systems of Linear Equations to Solve Uniform Motion Problems

2 Use Systems of Linear Equations to Solve Geometry Problems

3 Use Systems of Linear Equations to Solve Mixture Problems

OBJECTIVE 1: **USE SYSTEMS OF LINEAR EQUATIONS TO SOLVE UNIFORM MOTION PROBLEMS**

When the problem-solving strategy for applications of linear equations was first introduced in Chapter 1, we saw situations with more than one unknown quantity. At that time, we defined one variable and expressed all other unknowns in terms of that one variable. In this chapter, we have used more than one variable when solving applications involving systems of linear equations. Now, let's apply the problem-solving strategy for systems of linear equations to earlier application problems where we used a single equation and one variable. We start with uniform motion problems.

Example 1 Geocaching

▣ *Geocaching* is a worldwide hide-and-seek treasure hunt in which players (called *geocachers*) hide a cache (collection of items or trinkets such as coins or buttons) and share its GPS location online. Other players use a GPS device to help them locate the cache. Andrea receives an e-mail about a new geocache with a Travel Bug (a trackable item) that is within walking distance. She leaves to find the cache at the same time that another geocacher, Colin, leaves to locate the cache. They both arrive at the cache 45 minutes later. If the combined distance traveled is 4.5 miles and Andrea walks 0.4 mph faster than Colin, find their average speeds.

Solution

We apply our problem-solving strategy using systems of linear equations.

Step 1. Define the Problem. We want to find the average speed of both Andrea and Colin. We know the total distance they walk is 4.5 miles and that Andrea walks 0.4 mph faster than Colin. We also know that they each walk for 45 minutes (0.75 hours).

Step 2. Assign Variables. We know the time traveled and the distance traveled but not the average speeds. We can define two variables to represent the average speeds.

$$x = \text{Andrea's average speed}$$
$$y = \text{Colin's average speed}$$

Step 3. Translate into a System of Equations. This is a uniform motion problem, so we will need the distance formula $d = r \cdot t$. The distance traveled by each person is given by the product of the person's speed and time traveled. We know the total distance is 4.5 miles, so we write the following equation:

$$\underbrace{0.75x}_{\substack{\text{Distance Andrea}\\\text{traveled}}} + \underbrace{0.75y}_{\substack{\text{Distance Colin}\\\text{traveled}}} = 4.5$$

We also know that Andrea walks 0.4 mph faster than Colin. This gives us the following second equation:

$$\underset{\substack{\text{Andrea's}\\\text{speed}}}{x} = \underset{\substack{\text{Colin's}\\\text{speed}}}{y} + 0.4$$

Writing the two equations together gives the system:

$$\begin{cases} 0.75x + 0.75y = 4.5 \\ x = y + 0.4 \end{cases}$$

Note that in the first equation we used 0.75 (hours) instead of 45 (minutes). Click here to see why.

Step 4. Solve the System. Solve the system by graphing, substitution, or elimination. The second equation is already solved for a variable, so substitution may be preferable in this situation. Complete steps 5 and 6 on your own. Click here to check your answer, or watch this video for a detailed solution.

You Try It **Work through this You Try It problem.**

Work Exercises 1–2 in this textbook or in the *MyMathLab*® Study Plan.

Example 2 High Adventure Trek

Dimitri and Ricardo joined a high adventure trek in the Joshua Tree backcountry. One day the participants decided to hike to a site where they would go rock climbing. Ricardo's group left for the site at 8:30 A.M. and hiked at a rate of 3 mph. Dimitri's group had to finish packing gear and did not leave for the site until 25 minutes later, following the same path. If Dimitri's group, hiking at a rate of 4 mph, reached the rock climbing site at the same time as Ricardo's group, how far was the site from their base camp? At what time did the two groups arrive at the rock climbing site?

Solution We apply our problem-solving strategy using systems of linear equations.

Step 1. Define the Problem. We must find the distance between the camp and the rock climbing site, and the time when the groups arrived at the site. We know that Ricardo's group left at 8:30 A.M. traveling at 3 mph, while Dimitri's group left 25 minutes later traveling at 4 mph. Since they traveled the same path, the distance is the same for both groups. (Note that 1 hour = 60 minutes.)

Step 2. Assign Variables. We know the rate of travel, so we can find the distance traveled if we know the travel time. Let's define two variables to represent the travel times for Ricardo's group and Dimitri's group.

$$x = \text{hiking time for Ricardo's group}$$
$$y = \text{hiking time for Dimitri's group}$$

My video summary

Step 3. **Translate into a System of Equations.** This is a uniform motion problem, so we use the distance formula $d = r \cdot t$. The distance traveled by each group was the same, so we get the following equation:

$$\underset{\substack{\text{Distance for} \\ \text{Ricardo's group}}}{\underline{3x}} = \underset{\substack{\text{Distance for} \\ \text{Dimitri's group}}}{\underline{4y}}$$

We know that Ricardo's group hiked 25 minutes longer than Dimitri's group. Since 25 minutes $= \frac{25}{60} = \frac{5}{12}$ hours, we get the following second equation:

$$\underset{\substack{\text{Time for} \\ \text{Ricardo's} \\ \text{group}}}{\underline{x}} = \underset{\substack{\text{Time for} \\ \text{Dimitri's} \\ \text{group}}}{\underline{y}} + \frac{5}{12}$$

Writing the two equations together gives the following system:

$$\begin{cases} 3x = 4y \\ x = y + \frac{5}{12} \end{cases}$$

Step 4. **Solve the System.** Solve the system by graphing, substitution, or elimination. Since the variables are travel times, multiply one of the times by the corresponding rate to find the distance traveled. To get the arrival time, add the travel time to the appropriate starting time. Complete steps 5 and 6 on your own. Click here to see the answer, or watch this video for a detailed solution.

Compare this approach to the one-variable approach by reviewing Example 7 from Section 1.4.

You Try It   **Work through this You Try It problem.**

Work Exercises 3–4 in this textbook or in the *MyMathLab*® Study Plan.

Some uniform motion problems involve one motion that works with or against another. Walking up the down escalator or paddling upstream are examples of two motions working against each other. Kicking a football with the wind or walking in the direction of motion on a moving sidewalk are examples of two motions working together. For these problems, it is important to remember that when motions work together their rates are added, but when they work against each other, their rates are subtracted.

Example 3 Airplane Speed

My video summary

A small airplane flies from Seattle, Washington, to Portland, Oregon—a distance of 150 miles. Because the pilot faced a strong headwind, the trip took 1 hour and 15 minutes. On the return flight, the wind was still blowing at the same speed and had not changed direction. If the return trip took 45 minutes, what was the average speed of the airplane in still air? What was the speed of the wind?

Solution To solve this problem we need the distance traveled formula, $d = r \cdot t$. We are given the distance between the cities, which is the same on the return flight as it is on the flight from Seattle to Portland. We are also given the time spent on each flight, so we only need to express the rate of travel for the flight and return flight.

When the plane is flying into the wind, we must subtract the speed of the wind from the speed of the plane. However, when the plane is flying with the wind, we add the two speeds. If we let r = speed of the plane in still air and w = speed of the wind, then the rate of travel from Seattle to Portland is given by $r - w$ (against the wind). The rate of travel on the return flight will be $r + w$ (with the wind). Finish solving the problem, then click here to see the answer, or watch this video for a detailed solution.

You Try It   **Work through this You Try It problem.**

Work Exercises 5–6 in this textbook or in the *MyMathLab*® Study Plan.

OBJECTIVE 2:   USE SYSTEMS OF LINEAR EQUATIONS TO SOLVE GEOMETRY PROBLEMS

Geometry formulas, such as those for area and perimeter, often involve more than one variable. In Chapter 1, we expressed all unknown quantities in terms of one variable for such problems. Now we can use systems of linear equations in more than one variable to solve these same types of problems.

Example 4 Soccer Field Dimensions

My video summary

 Although most sports have set field dimensions, women's professional soccer in the United States does not. Dimensions of a rectangular soccer field can vary. For international play, a soccer field must have a width between 64 m and 75 m and a length between 100 m and 110 m. The Los Angeles Sol play on a field whose perimeter is 354 m. If the length is 27 m less than twice the width, what are the dimensions of the field? (*Source:* Federation Internationale de Football Association)

Solution We apply our problem-solving strategy using systems of linear equations.

Step 1. **Define the Problem.**   We must find the dimensions (length and width) of the soccer field. We know that the rectangular field has a perimeter of 354 m. We also know that the length is 27 m less than twice the width.

Step 2. **Assign Variables.**   To find the length and width, we can define our variables as follows:

$$L = \text{length of the field}$$
$$W = \text{width of the field}$$

Step 3. **Translate into a System of Equations.**   Since we know the perimeter of the rectangular field is 354 m, we can write one equation using the formula for the perimeter of a rectangle:

$$2L + 2W = 354$$

We also know that the length is 27 m less than twice the width, so our second equation is

$$L = \underbrace{2W}_{\text{Twice the width}} - \underbrace{27.}_{\text{27 meters less}}$$

Writing the two equations together gives the following system:

$$\begin{cases} 2L + 2W = 354 \\ L = 2W - 27 \end{cases}$$

Step 4. Solve the System. Solve the system by graphing, substitution, or elimination. Complete steps 5 and 6 on your own. Watch this video for a detailed solution, or click here to see the answer.

You Try It Work through this You Try It problem.

Work Exercises 7–8 in this textbook or in the _MyMathLab_® Study Plan.

Geometry problems sometimes involve *complementary* and *supplementary* angles. The measures of two **complementary angles** add to 90°. The measures of two **supplementary angles** add to 180°. Complementary angles form right angles and supplementary angles form a straight line.

Example 5 Complementary Angles

Find the measures of two complementary angles if the measure of the smaller angle is 15 degrees more than half the measure of the larger angle.

Solution Click here to see the answer, or watch this interactive video for a detailed solution.

You Try It Work through this You Try It problem.

Work Exercises 9–10 in this textbook or in the _MyMathLab_® Study Plan.

Example 6 Interior Angle Measures

Find the measures of the three interior angles, marked by x, y, and z in the following triangle. Remember that two angles forming a straight line are supplementary angles, and the measures of the three interior angles of a triangle must add to 180°.

Solution

Step 1. Define the Problem. We want to find the measures of the three interior angles in the given triangle. We know that the measures of the three interior angles add to 180° and that two angles forming a straight line are supplementary angles, which means that the sum of their measures is also 180°.

Step 2. Assign Variables. We want to find the measures of the three interior angles. In the given triangle, these angles are already defined as the variables x, y, and z.

Step 3. Translate into a System of Equations. Since there are three variables, we need a system of three linear equations. The sum of the measures of the three interior angles is 180°, so our first equation is

$$x + y + z = 180.$$

From the given triangle, we know that there are two sets of supplementary angles, which we can use to write the remaining two equations. Since supplementary angles have measures that add to 180°, we get the following two equations:

$$x + (2y - 5) = 180$$
$$z + (3y + 25) = 180$$

Now we have three equations to form the following system:

$$\begin{cases} x + y + z = 180 \\ x + (2y - 5) = 180 \\ z + (3y + 25) = 180 \end{cases}$$

Writing each equation in standard form, $Ax + By + Cz = D$, gives

$$\begin{cases} x + y + z = 180 \\ x + 2y = 185. \\ 3y + z = 155 \end{cases}$$

Step 4. Solve the System. Solve the system by elimination. Complete steps 5 and 6 on your own. Click here to see the answer, or watch this video for a detailed solution.

You Try It **Work through this You Try It problem.**

Work Exercises 16–17 in this textbook or in the *MyMathLab*® Study Plan.

OBJECTIVE 3:   USE SYSTEMS OF LINEAR EQUATIONS TO SOLVE MIXTURE PROBLEMS

Example 7 Fragrant Oil Mixture

My video summary

Nina owns a fragrant oil shop and needs to make 6 ounces of a 30% sandalwood essential oil mixture for a customer. She has a 10% sandalwood mixture and a 40% sandalwood mixture available. How much of each mixture should she combine to create the customer's desired product?

Solution

Step 1. Define the Problem. We want to find out how many ounces of each sandalwood mixture should be combined to obtain 6 ounces of a 30% sandalwood mixture.

Step 2. Assign Variables. We are combining two different mixtures, so we can define two variables.

$$x = \text{ounces of 10\% sandalwood mixture}$$
$$y = \text{ounces of 40\% sandalwood mixture}$$

Step 3. Translate into a System of Equations. The mixture will contain 6 ounces, so our first equation is based on this total amount:

$$x + y = 6$$

The amount of sandalwood oil in the final mixture must equal the total amount of sandalwood oil from the two original mixtures. We can write the second equation as

$$\underbrace{0.1x}_{\substack{\text{Sandalwood oil} \\ \text{in 10\% mixture}}} + \underbrace{0.4y}_{\substack{\text{Sandalwood oil} \\ \text{in 40\% mixture}}} = \underbrace{0.3(6).}_{\substack{\text{Sandalwood oil} \\ \text{in 30\% mixture}}}$$

Writing the two equations together gives the following system:

$$\begin{cases} x + y = 6 \\ 0.1x + 0.4y = 1.8 \end{cases}$$

Step 4. Solve the System. Solve the system by graphing, substitution, or elimination. Complete steps 5 and 6 on your own. Click here to see the answer, or watch this video for a detailed solution.

You Try It Work through this You Try It problem.

Work Exercises 11–12 in this textbook or in the *MyMathLab*® Study Plan.

Example 8 Bean Mixture

Twin City Foods, Inc., has a 10-lb bean mixture (lima beans and green beans) that sells for $5.75. If lima beans sell for $0.70 per pound and green beans sell for $0.50 per pound, how many pounds of each bean went into the mixture?

Solution Click here to see the answer, or watch this video for a detailed solution.

You Try It Work through this You Try It problem.

Work Exercises 13–15 in this textbook or in the *MyMathLab*® Study Plan.

Example 9 Investing Money

Tremayne had $40,000 to invest and decided to split up the money into three different accounts that pay 3%, 5%, and 6% in annual simple interest. The combined amount invested in the 5% account and the 6% account was three times the amount invested in the 3% account. If the total interest for one year was $1980, how much did Tremayne invest in each account?

Solution

Step 1. Define the Problem. We want to find the amount invested in each of the three accounts. The accounts pay 3%, 5%, and 6% in annual simple interest. Tremayne had $40,000 to invest, and total interest for one year was $1980. The combined amount in the 5% and 6% accounts was three times the amount in the 3% account.

Step 2. Assign Variables. There are three unknowns: the amount invested in each account. Let's define these three variables:

$$x = \text{amount invested at } 3\%$$
$$y = \text{amount invested at } 5\%$$
$$z = \text{amount invested at } 6\%$$

Step 3. Translate into a System of Equations. There are three variables, so we need three equations in our system. As with other systems, totals are a good place to start. We know the total amount invested. Also, we know that the sum of the three amounts invested must equal the total amount invested, so our first equation is

$$\underbrace{x}_{\substack{\text{Amt invested} \\ \text{at } 3\%}} + \underbrace{y}_{\substack{\text{Amt invested} \\ \text{at } 5\%}} + \underbrace{z}_{\substack{\text{Amt invested} \\ \text{at } 6\%}} = \underbrace{40{,}000.}_{\substack{\text{Total amt} \\ \text{invested}}}$$

We also know the total interest after one year. The total interest must equal the sum of the interest from each investment. Our second equation is

$$\underbrace{0.03x}_{\substack{\text{Interest from} \\ 3\% \text{ investment}}} + \underbrace{0.05y}_{\substack{\text{Interest from} \\ 5\% \text{ investment}}} + \underbrace{0.06z}_{\substack{\text{Interest from} \\ 6\% \text{ investment}}} = \underbrace{1980.}_{\substack{\text{Total} \\ \text{interest}}}$$

To find the third equation, we recall that the combined amounts in the 5% and 6% accounts was three times the amount in the 3% account. Our third equation is

$$\underbrace{y}_{\substack{\text{Amt invested} \\ \text{at } 5\%}} + \underbrace{z}_{\substack{\text{Amt invested} \\ \text{at } 6\%}} = \underbrace{3x.}_{\substack{\text{Three times the amt} \\ \text{invested at } 3\%}}$$

Writing the three equations together gives the following system:

$$\begin{cases} x + y + z = 40{,}000 \\ 0.03x + 0.05y + 0.06z = 1980. \\ y + z = 3x \end{cases}$$

Writing each equation in standard form, $Ax + By + Cz = D$, gives

$$\begin{cases} x + y + z = 40{,}000 \\ 0.03x + 0.05y + 0.06z = 1980. \\ 3x - y - z = 0 \end{cases}$$

Step 4. Solve the System. Solve the system by elimination. Complete steps 5 and 6 on your own. Click here to see the answer, or watch this video for a detailed solution.

You Try It   Work through this You Try It problem.

Work Exercise 18 in this textbook or in the **MyMathLab**® Study Plan.

3.3 Exercises

In Exercises 1–18, solve each application problem using the problem-solving strategy for systems of linear equations.

You Try It

1. **Traveling Speed**   Yessica leaves her house on her bicycle to locate a geocache at the same time that Zoe leaves her house heading for the same cache. Zoe lives further away than Yessica so she takes her car and travels 16 mph faster than Zoe. Their combined distance traveled is 17 miles. If both women arrive at the geocache site after 30 minutes, find their average speeds.

2. **Jogging Speed**   Starting at the same point, Angie and Payton jog in opposite directions. Angie can jog 2.5 miles per hour faster than Payton. After 1.5 hours, they are 16 miles apart. How fast can each person jog?

You Try It

3. **Amazing Chase**   Rachel and Lindsay are in third place in the Amazing Race. Due to a time penalty, they must start the next leg 90 minutes after the second-place team. If Rachel and Lindsay travel at an average rate of 13.2 mph and the second-place team travels at an average rate of 11 mph, how long will it take the girls to catch up to the second-place team?

4. **Aerobic Exercise**   As part of her training routine for basketball, Shaylle alternates between cycling and running for exercise. She cycles at a rate of 16 mph and runs at a rate of 6 mph. If she spends 8.5 hours exercising and covers a total of 96 miles, how much time did she spend on each exercise?

You Try It

5. **Airplane Speed**   An airplane faced a headwind during a flight between Joppetown and Jawsburgh that took 3 hours and 36 minutes. The return flight took 3 hours. If the distance from Joppetown to Jawsburgh is 1800 miles, find the air speed of the plane (the speed of the plane in still air) and the speed of the wind, assuming both remain constant.

6. **Boat Speed**   While preparing for the 2009 Purgatory Challenge kayak race, Marta practiced on the Willamette River in northwestern Oregon. On one trip she traveled 2.5 miles upstream in 40 minutes with the return trip taking 22.9 minutes. What was her speed in still water and the speed of the river current?

You Try It

7. **Soccer Field Dimensions**   The St. Louis Athletica women's soccer team began their inaugural season at Korte Stadium in Edwardsville, IL. The perimeter of the soccer field is 390 yards. If the length is 30 yards less than twice the width, what are the dimensions of the field? (*Source:* SIU-Edwardsville)

8. **Isosceles Triangle**   An isosceles triangle has an angle that measures 20 degrees more than twice the two common angles. What are the three angle measures?

You Try It

9. **Complementary Angles**   The measure of the larger of two complementary angles is 2° more than three times the measure of the smaller angle. What are the angle measures?

10. **Supplementary Angles**   A 2009 intersection study for the city of Marysville, ohio, found that Milford Avenue and Maple Street intersected at an angle much smaller than design standards. Find the measures of the angle of intersection and its supplement if the supplement measures 30 degrees more than nine times the angle of intersection.

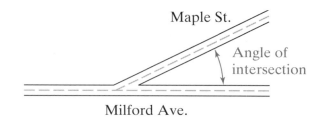

You Try It

11. Mixing Massage Oils   Kristen wants to make 4.5 ounces of a 25% jasmine essential oil. She only has a 15% jasmine mixture and a 30% jasmine mixture available. How much of each mixture should she combine to create the desired oil?

12. Blending Wines   A winery blends two wines to make a new 12% alcohol product. The base wine is 9% alcohol and the fortifier wine is 14% alcohol. If the winery wants to make 250 gallons of the new blend, how much of the base and fortifier wines should be mixed?

You Try It

13. Mixing Tea   A tea shop mixes Irish Breakfast tea that sells for $2.50 per ounce with buttered rum tea that sells for $1.75 per ounce to make 100 ounces of a new tea blend that sells for $2.10 per ounce with no loss of revenue. How many ounces of each type should be mixed to create the new tea blend?

14. Trail Mix   A bulk foods store makes a trail mix by combining two ingredients: dried tropical fruit that sells for $5 per pound and raw natural nuts that sell for $8.75 per pound. How much of each ingredient should be combined to get 60 pounds of a trail mix that sells for $6.50 per pound?

15. Investments   Scott invested a total of $3000 at two separate banks. One bank pays simple interest at 3% per year, while the other bank pays simple interest at 4% per year. If Scott earned $110.50 in interest during a single year, how much did he deposit at each bank?

You Try It

16. Interior Angles   Find the measures of the three interior angles in the following triangle.

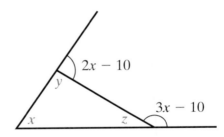

17. Triangular Yard   The backyard of a house is in the shape of a triangle. The longest side is 10 feet shorter than twice the shortest side. The third side is 10 feet less than the longest side. If the perimeter of the yard is 167 feet, what are the lengths of the sides?

You Try It

18. Investing Money   Amanda invested a total of $2200 into three separate accounts that pay 6%, 8%, and 9% annual simple interest. Amanda has three times as much money invested in the 9% account as she does in the 6% account. If the total interest for the year is $178, how much money did Amanda invest in each account?

3.4 Systems of Linear Inequalities in Two Variables

THINGS TO KNOW

Before working through this section, be sure you are familiar with the following concepts:

VIDEO   ANIMATION   INTERACTIVE

You Try It

1. Determine If an Ordered Pair Is a Solution to a Linear Inequality in Two Variables (Section 2.6, Objective 1)

You Try It

2. Graph a Linear Inequality in Two Variables (Section 2.6, Objective 2)

OBJECTIVES

1 Determine If an Ordered Pair Is a Solution to a System of Linear Inequalities in Two Variables

2 Graph Systems of Linear Inequalities in Two Variables

OBJECTIVE 1: DETERMINE IF AN ORDERED PAIR IS A SOLUTION TO A SYSTEM OF LINEAR INEQUALITIES IN TWO VARIABLES

In Section 2.6 we graphed linear inequalities in two variables. In this section, we learn how to graph **systems of linear inequalities in two variables**.

> **Definition** System of Linear Inequalities in Two Variables
>
> A **system of linear inequalities in two variables** is a collection of two or more linear inequalities in two variables considered together. A **solution to a system of linear inequalities in two variables** is an ordered pair that satisfies all inequalities in the system.

Example 1 Determining If an Ordered Pair Is a Solution to a System of Linear Inequalities in Two Variables

My interactive video summary

 Determine if the given ordered pair is a solution to the given system of inequalities.

$$\begin{cases} 2x - 3y \le 9 \\ 2x - y \ge -1 \end{cases}$$

a. $(1, -2)$ **b.** $(-1, 2)$ **c.** $(3, -1)$

Solution We substitute the x- and y-coordinates for the variables in each inequality and simplify. If the resulting statements are both true, the ordered pair is a solution to the system.

a.

	First Inequality	Second Inequality
Begin with the original inequalities:	$2x - 3y \le 9$	$2x - y \ge -1$
Substitute 1 for x and -2 for y:	$2(1) - 3(-2) \overset{?}{\le} 9$	$2(1) - (-2) \overset{?}{\ge} -1$
Simplify:	$8 \le 9$ True	$4 \ge -1$ True

The ordered pair makes both inequalities true, so $(1, -2)$ is a solution to the system of inequalities.

Work through parts (b) and (c) yourself. Click here to check your answers, or watch this interactive video for detailed solutions.

You Try It Work through this You Try It problem.

Work Exercises 1–6 in this textbook or in the **MyMathLab**® Study Plan.

OBJECTIVE 2:   GRAPH SYSTEMS OF LINEAR INEQUALITIES IN TWO VARIABLES

The **graph of a system of linear inequalities in two variables** is the intersection of the graphs of each inequality in the system. This graph represents the set of all solutions to the system of inequalities. Watch this animation to see how to graph the following system of linear inequalities from Example 1.

$$\begin{cases} 2x - 3y \le 9 \\ 2x - y \ge -1 \end{cases}$$

We obtain the graph of a system of linear inequalities by graphing each inequality in the system and finding the region they all have in common, if any.

Steps for Graphing Systems of Linear Inequalities in Two Variables

Step 1. Use the Steps for Graphing Linear Inequalities in Two Variables from Section 2.6 to graph each inequality in the system on the same coordinate plane.

Step 2. Determine the region where the shaded areas overlap, if any. This region represents the set of all solutions to the system of inequalities.

Example 2 Graphing Systems of Linear Inequalities in Two Variables

Graph the system of linear inequalities in two variables.

$$\begin{cases} x + y > 2 \\ 2x - y \le 6 \end{cases}$$

Solution   We follow the Steps for Graphing Systems of Linear Inequalities in Two Variables.

Step 1. **Graph $x + y > 2$.** The inequality $x + y > 2$ is strict, so we graph the boundary line $x + y = 2$ using a dashed line. The test point $(0, 0)$ does not belong to the inequality since $0 + 0 > 2$ is false, so we shade the half-plane that does not contain $(0, 0)$. See Figure 11(a).

Graph $2x - y \le 6$. The inequality $2x - y \le 6$ is non-strict, so we graph the boundary line $2x - y = 6$ using a solid line. The test point $(0, 0)$ belongs to the inequality ($2 \cdot 0 - 0 \le 6$ is true), so we shade the half-plane that contains $(0, 0)$. See Figure 11(b).

Step 2. We determine the region shared by both inequalities in the system. This is the darkest-shaded region in Figure 11(c). Any point in this region is a solution to the system of inequalities.

Watch this video for a detailed solution.

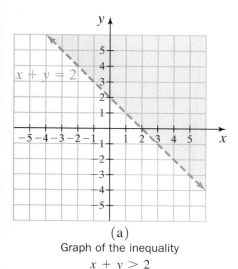

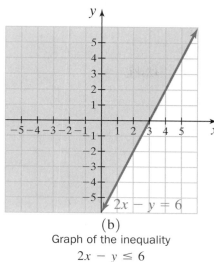

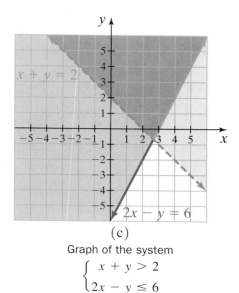

(a)	(b)	(c)
Graph of the inequality	Graph of the inequality	Graph of the system
$x + y > 2$	$2x - y \leq 6$	$\begin{cases} x + y > 2 \\ 2x - y \leq 6 \end{cases}$

Figure 11

 CAUTION Only the ordered pairs in the darkest-shaded region in Figure 11(c), including the darkest-shaded portion of the solid line $2x - y = 6$, are solutions to the system. Note that none of the points on the boundary line $x + y = 2$ are solutions to the system.

You Try It Work through this You Try It problem.

Work Exercises 7–12 in this textbook or in the *MyMathLab*® Study Plan.

Example 3 Graphing Systems with More Than Two Linear Inequalities

My video summary

Graph the system of linear inequalities in two variables.

$$\begin{cases} 4x > y \\ x - 3y < 9 \\ x + y < 4 \end{cases}$$

Solution Each inequality in the system is strict, so all of the boundary lines are dashed lines. Graph all three inequalities on the same coordinate plane and find the region shared by all three inequalities, if any. Click here to see the answer, or watch this video to see a detailed solution.

You Try It Work through this You Try It problem.

Work Exercises 19–21 in this textbook or in the *MyMathLab*® Study Plan.

Example 4 Identifying a System of Linear Inequalities without a Solution

My video summary

Graph the system of linear inequalities in two variables.

$$\begin{cases} x - 3y > 6 \\ 2x - 6y < -9 \end{cases}$$

Solution We follow the Steps for Graphing Systems of Linear Inequalities in Two Variables.

Step 1. **Graph $x - 3y > 6$.** The inequality $x - 3y > 6$ is strict, so we graph the boundary line $x - 3y = 6$ using a dashed line. The test point $(0, 0)$ does not belong to the inequality, so we shade the half-plane that does not contain $(0, 0)$. See Figure 12(a).

Graph $2x - 6y < -9$. The inequality $2x - 6y < -9$ is strict, so we graph the boundary line $2x - 6y = -9$ using a dashed line. The test point $(0, 0)$ does not belong to the inequality, so we shade the half-plane that does not contain $(0, 0)$. See Figure 12(b).

Step 2. The boundary lines $x - 3y = 6$ and $2x - 6y = -9$ are parallel, and the shaded regions are on opposite sides of these parallel lines. Therefore, no region is shared by both inequalities in the system. See Figure 12(c). There are no ordered pairs that satisfy both inequalities, so there is no solution to the system. We indicate this with the null symbol $\varnothing$ or empty set $\{\ \}$.

Watch this video for a more detailed solution.

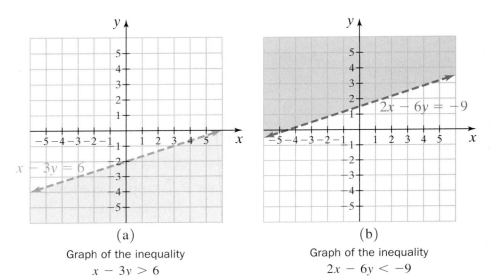

(a)

Graph of the inequality
$x - 3y > 6$

(b)

Graph of the inequality
$2x - 6y < -9$

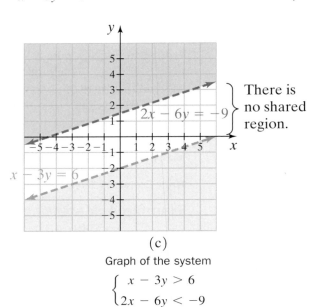

(c)

Graph of the system
$$\begin{cases} x - 3y > 6 \\ 2x - 6y < -9 \end{cases}$$

Figure 12

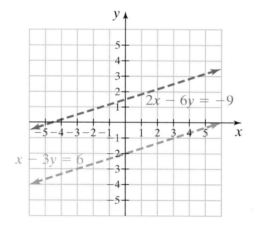

You Try It   Work through this You Try It problem.

Work Exercises 13–14 in this textbook or in the *MyMathLab*® Study Plan.

In Example 4, the system contained two linear inequalities whose boundary lines were parallel. Since the shaded regions for the two inequalities did not overlap, the system had no solution. In Example 5, we consider three systems of linear inequalities in two variables with parallel boundary lines similar to those in Example 4. However, the results are very different.

Example 5 Graphing Systems of Linear Inequalities with Parallel Boundary Lines

My interactive video summary

Graph each system of linear inequalities in two variables.

a. $\begin{cases} x - 3y < 6 \\ 2x - 6y < -9 \end{cases}$ b. $\begin{cases} x - 3y < 6 \\ 2x - 6y > -9 \end{cases}$ c. $\begin{cases} x - 3y > 6 \\ 2x - 6y > -9 \end{cases}$

Solution The three systems in this example are very similar to each other and to the system in Example 4. The graphs of all three systems have the boundary lines $x - 3y = 6$ and $2x - 6y = -9$. Since all of the inequalities are strict, the boundary lines will be dashed lines. See Figure 13.

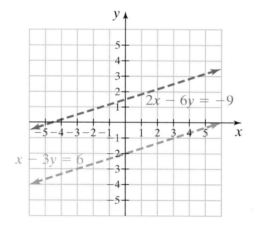

Figure 13

a. **Step 1.** **Graph $x - 3y < 6$.** The test point $(0, 0)$ belongs to the inequality, so we shade the half-plane that contains $(0, 0)$. See Figure 14(a).

   Graph $2x - 6y < -9$. The test point $(0, 0)$ does not belong to the inequality, so we shade the half-plane that does not contain $(0, 0)$. See Figure 14(b).

   Step 2. We determine the region shared by both inequalities in the system. This is the darkest-shaded region in Figure 14(c). Any ordered pair that is a solution to the inequality $2x - 6y < -9$ is a solution to the system.

   Watch part (a) of this interactive video for a more detailed solution.

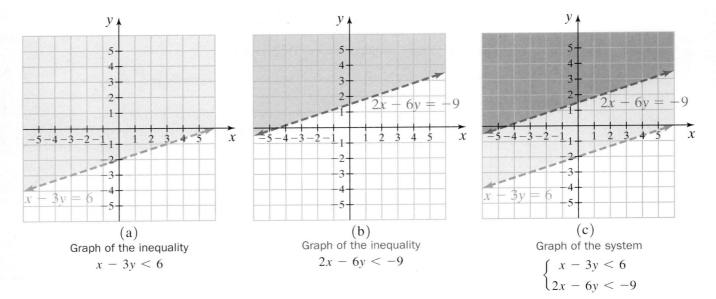

(a)
Graph of the inequality
$x - 3y < 6$

(b)
Graph of the inequality
$2x - 6y < -9$

(c)
Graph of the system
$$\begin{cases} x - 3y < 6 \\ 2x - 6y < -9 \end{cases}$$

Figure 14

b. Try graphing this system of inequalities on your own. Click here to see the answer, or watch part (b) of this interactive video for a detailed solution.

c. Graph this system of inequalities on your own. Click here to see the answer, or watch part (c) of this interactive video for a detailed solution.

You Try It   Work through this You Try It problem.

Work Exercises 15–18 in this textbook or in the *MyMathLab* ® Study Plan.

3.4 Exercises

In Exercises 1–6, determine if each ordered pair is a solution to the given system of linear inequalities in two variables.

You Try It

1. $\begin{cases} x - 2y < 6 \\ 3x + y > 2 \end{cases}$

   a. $(1, 0)$

   b. $(4, -1)$

   c. $(3, -2)$

2. $\begin{cases} 3x - 2y \geq 6 \\ 4x + 3y \leq 9 \end{cases}$

   a. $(5, -3)$

   b. $(0, 0)$

   c. $(2, -2)$

3. $\begin{cases} x - y > -8 \\ 5x - 2y \leq 10 \\ 4x + 3y > 12 \end{cases}$

   a. $(2, 4)$

   b. $(1, 1)$

   c. $(-1, 6)$

4. $\begin{cases} 2x + 3y < 6 \\ 3x - 4y \leq -4 \\ 3x - y \geq -6 \end{cases}$

   a. $(1, 3)$

   b. $(-2, 0)$

   c. $(3, 2)$

5. $\begin{cases} x + 2y \leq 6 \\ 4x + y \leq 10 \\ x \geq 0 \\ y \geq 0 \end{cases}$

   a. $(1, 6)$

   b. $(3, 4)$

   c. $(2, 2)$

6. $\begin{cases} x + y < 8 \\ x + y < 3 \\ x < 5 \\ y < 7 \end{cases}$

   a. $(5, -3)$

   b. $(0, 0)$

   c. $(2, -2)$

In Exercises 7–21, graph each system of linear inequalities in two variables.

You Try It

7. $\begin{cases} x + y \ge 4 \\ 2x - y \le 5 \end{cases}$

8. $\begin{cases} y < -2x \\ y > -3x \end{cases}$

9. $\begin{cases} 2x - y < 6 \\ y \ge \dfrac{3}{2}x + 3 \end{cases}$

You Try It

10. $\begin{cases} x - 3y \le 6 \\ x < 3 \end{cases}$

11. $\begin{cases} \dfrac{1}{3}x - \dfrac{1}{2}y \le 1 \\ y \ge -2 \end{cases}$

12. $\begin{cases} x > -4 \\ y \ge 1 \end{cases}$

You Try It

13. $\begin{cases} 3x - 5y \ge 15 \\ -3x + 5y \ge 15 \end{cases}$

14. $\begin{cases} x > 3 \\ x < -2 \end{cases}$

15. $\begin{cases} 6x - 8y < 4 \\ -3x + 4y \le 12 \end{cases}$

16. $\begin{cases} y \ge 2 \\ y \le 5 \end{cases}$

17. $\begin{cases} 2x - y > 7 \\ y < 2x + 5 \end{cases}$

18. $\begin{cases} 2x - 3y \ge 12 \\ -x + 1.5y < 6 \end{cases}$

You Try It

19. $\begin{cases} 3x + 2y \le 4 \\ 3x - 2y \ge -16 \\ x - 2y \le 4 \end{cases}$

20. $\begin{cases} 2x - y > -3 \\ 5x + y > 0 \\ x > 1 \end{cases}$

21. $\begin{cases} 2x + y \le 4 \\ x + 3y \le 6 \\ x \ge 0 \\ y \ge 0 \end{cases}$

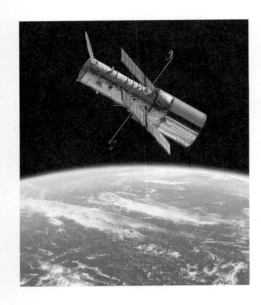

CHAPTER FOUR

Polynomial Expressions and Functions

CHAPTER FOUR CONTENTS

4.1 **Rules for Exponents**

4.2 **Introduction to Polynomial Functions**

4.3 **Multiplying Polynomials**

4.4 **Dividing Polynomials**

4.1 Rules for Exponents

THINGS TO KNOW

Before working through this section, be sure you are familiar with the following concepts:

VIDEO   ANIMATION   INTERACTIVE

You Try It

1. Perform Operations on Real Numbers (Section R.2, Objective 1)

You Try It

2. Simplify Numeric Expressions Containing Exponents and Radicals (Section R.2, Objective 2)

OBJECTIVES

1 Simplify Exponential Expressions Using the Product Rule

2 Simplify Exponential Expressions Using the Quotient Rule

3 Use the Zero-Power Rule

4 Use the Negative-Power Rule

5 Use the Power-to-Power Rule

6 Use the Product-to-Power and Quotient-to-Power Rules

7 Simplify Exponential Expressions Using a Combination of Rules

8 Use Rules for Exponents with Scientific Notation

OBJECTIVE 1: SIMPLIFY EXPONENTIAL EXPRESSIONS USING THE PRODUCT RULE

An **exponential expression** is a constant or algebraic expression that is raised to a power. The constant or algebraic expression makes up the **base**, and the power is the **exponent** on the base. From Section R.2, we know that exponents can be used to show repeated multiplication. For example, the expression 5^4 can be written as follows:

$$5^4 = 5 \cdot 5 \cdot 5 \cdot 5$$

Exponent

Base

The exponent 4 indicates to multiply the base 5 four times.

The same is true when the base is a variable or algebraic expression.

Exponent

$$a \cdot a \cdot a \cdot a = a^4$$

Base

Exponent

$$(2x) \cdot (2x) \cdot (2x) \cdot (2x) = (2x)^4$$

Base

The exponent 4 indicates to multiply the base, either a or $2x$, four times. We use parentheses (or other grouping symbols) when working with an algebraic expression such as $2x$. This reminds us to raise the entire expression, $2x$, to the fourth power, not just the variable x.

$$2 \cdot x \cdot x \cdot x \cdot x = 2x^4 \neq (2x)^4 = 2x \cdot 2x \cdot 2x \cdot 2x$$

If we multiply exponential expressions with the same base, such as $a^4 \cdot a^5$, we write:

$$a^4 \cdot a^5 = (a \cdot a \cdot a \cdot a)(a \cdot a \cdot a \cdot a \cdot a)$$
$$= a \cdot a \cdot a \cdot a \cdot a \cdot a \cdot a \cdot a \cdot a$$
$$= a^9.$$

We multiply four factors of a by five factors of a, so we are really multiplying the factor a nine times, which is the sum of the two individual exponents, 4 and 5. Therefore,

$$a^4 \cdot a^5 = a^{4+5} = a^9$$

This result is true in general and demonstrates the **Product Rule for Exponents**.

The Product Rule for Exponents

When multiplying exponential expressions with the same base, we add the exponents and keep the common base.

$$a^m \cdot a^n = a^{m+n}$$

If the exponential expressions have more than one base, we use the commutative and associative properties for multiplication to group like bases and then apply the product rule. For example, $(3x^2y)(2xy^3) = (3 \cdot 2)(x^2 \cdot x)(y \cdot y^3) = 6x^3y^4$. Click here to see how this process works.

CAUTION When no exponent is written, we assume that the exponent is 1. For example, $y = y^1$.

Example 1 Simplifying Exponential Expressions Using the Product Rule

My interactive video summary

 Simplify using the product rule.

a. $b^6 \cdot b^{11}$ **b.** $5m^7 \cdot 3m^2n^3$ **c.** $(-2)^3(-2)^5$

d. $(-x^2)(x^4)$ **e.** $(4x^5y^2)(3xy^4)$ **f.** $(-3b^4c^5)(-2b^5c^3)$

Solution

a. Both exponential expressions have the same base, so we apply the product rule. We add the exponents, 6 and 11, and keep the common base, b.

$$b^6 \cdot b^{11} = b^{6+11} = b^{17}$$

b. Since we have more than one base, m and n, we use the commutative property for multiplication to group like bases and then apply the product rule.

Begin with original expression: $5m^7 \cdot 3m^2n^3 = 5 \cdot m^7 \cdot 3 \cdot m^2 \cdot n^3$

Use commutative property to group like bases: $= 5 \cdot 3 \cdot m^7 \cdot m^2 \cdot n^3$

Multiply the constants: $= 15 \cdot m^7 \cdot m^2 \cdot n^3$

Apply the product rule: $= 15 \cdot m^{7+2} \cdot n^3$

Simplify: $= 15m^9n^3$

c. Both exponential expressions have the same base, so we apply the product rule. We add the exponents, 3 and 5, and keep the common base, -2.

$$(-2)^3(-2)^5 = (-2)^{3+5} = (-2)^8 = 256$$

d.–f. Click here to check your answers, or watch this interactive video for detailed solutions.

CAUTION When we multiply exponential expressions, we do not multiply the bases. For example, $2^2 \cdot 2^3 = 2^5$ not $(2 \cdot 2)^5$. Click here to see why.

You Try It Work through this You Try It problem.

Work Exercises 1–4 in this textbook or in the *MyMathLab* Study Plan.

OBJECTIVE 2: SIMPLIFY EXPONENTIAL EXPRESSIONS USING THE QUOTIENT RULE

When we divide exponential expressions with the same base, we expand each expression and simplify by cancelling, or dividing out, common factors. For example,

$$\text{Expand:} \quad \frac{x^7}{x^2} = \frac{x \cdot x \cdot x \cdot x \cdot x \cdot x \cdot x}{x \cdot x}$$

$$\text{Cancel common factors:} \quad = \frac{\cancel{x}^1 \cdot \cancel{x}^1 \cdot x \cdot x \cdot x \cdot x \cdot x}{{}_1\cancel{x} \cdot \cancel{x}_1}$$

$$\text{Simplify:} \quad = x \cdot x \cdot x \cdot x \cdot x$$

$$= x^5$$

In this example, each factor in the denominator cancels, or divides into, one of the factors in the numerator. We can find the same result by subtracting the two exponents.

$$\frac{x^7}{x^2} = x^{7-2} = x^5$$

As long as the denominator does not equal 0, this result is true in general and demonstrates the **Quotient Rule for Exponents**.

The Quotient Rule for Exponents

When dividing exponential expressions with the same base, we subtract the denominator exponent from the numerator exponent and keep the common base, as long as $a \neq 0$.

$$\frac{a^m}{a^n} = a^{m-n} \quad (a \neq 0)$$

If the exponential expressions have more than one base, we group like bases by forming quotients and then apply the quotient rule to each quotient. For example,

$\frac{6x^3y^5z^2}{2xy^2} = \frac{6}{2} \cdot \frac{x^3}{x^1} \cdot \frac{y^5}{y^2} \cdot \frac{z^2}{1} = 3x^2y^3z^2$. Click here to see how this process works.

Example 2 Simplifying Exponential Expressions Using the Quotient Rule

My interactive video summary

Simplify using the quotient rule.

a. $\dfrac{3^8}{3^6}$ b. $\dfrac{n^5}{n^2}$ c. $\dfrac{4x^8y^3}{2x^5y}$

d. $\dfrac{-15x^5z^4}{3x^2z^3}$ e. $\dfrac{24x^3y^2z^7}{9x^2z^4}$ f. $\dfrac{8a^{12}b^8c^3}{36a^7b^5c}$

Solution

a. Both exponential expressions have the same base, so we apply the quotient rule. We subtract the exponents, 8 and 6, and keep the common base, 3.

$$\frac{3^8}{3^6} = 3^{8-6} = 3^2 = 9$$

b. Since both exponential expressions have the same base, n, we apply the quotient rule.

$$\frac{n^5}{n^2} = n^{5-2} = n^3$$

c. These exponential expressions have more than one base, so we group like bases by forming quotients and then apply the quotient rule to each quotient.

Begin with original expression: $\dfrac{4x^8y^3}{2x^5y}$

Group like bases and form quotients: $= \dfrac{4}{2} \cdot \dfrac{x^8}{x^5} \cdot \dfrac{y^3}{y^1}$

Apply the quotient rule to each quotient: $= 2 \cdot x^{8-5} \cdot y^{3-1}$

Simplify: $= 2x^3y^2$

d.–f. Try the remaining problems on your own. Click here to check your answers, or watch this interactive video for detailed solutions.

⟨CAUTION⟩ Remember that when we divide exponential expressions, we subtract the exponent of the expression in the denominator *from* the exponent of the expression in the numerator.

You Try It   **Work through this You Try It problem.**

Work Exercises 5–8 in this textbook or in the *MyMathLab*® Study Plan.

OBJECTIVE 3:   USE THE ZERO-POWER RULE

If we multiply 5^3 by 5^0, we can use the product rule and write:

$$5^3 \cdot 5^0 = 5^{3+0} = 5^3.$$

From the multiplicative identity property we know that $5^3 \cdot 1 = 5^3$. Therefore, it makes sense that $5^0 = 1$ because $5^3 \cdot 5^0 = 5^3 = 5^3 \cdot 1$. This rule can be applied to all bases except 0 and is called the **Zero-Power Rule**.

The Zero-Power Rule

Any real number raised to the 0 power equals 1, except for 0^0.

$$a^0 = 1 \quad (a \neq 0)$$

Click here to see how the zero-power rule can also be derived from the quotient rule.

Example 3 Using the Zero-Power Rule

My video summary

▣ Evaluate. Assume that all variable bases are non-zero.

a. 10^0 b. -10^0 c. $5x^0$ d. $(5x)^0$

e. $(x+3)^0$ f. $a^0 + b^0$ g. $(x^2y^5z)^0$

Solution

a. The base 10 is non-zero, so $10^0 = 1$.

b. The base is 10, not -10, so we multiply the exponential expression 10^0 by -1. Using the order of operations, we get

$$-10^0 = -1(10)^0 = -1 \cdot 1 = -1.$$

c. We multiply the exponential expression x^0 by 5 and assume that the variable base x is non-zero. Applying the zero-power rule,

$$5x^0 = 5 \cdot x^0 = 5 \cdot 1 = 5.$$

d.–g. Try the remaining problems on your own. Click here to check your answers, or watch this video for detailed solutions.

You Try It   **Work through this You Try It problem.**

Work Exercises 9–12 in this textbook or in the *MyMathLab*® Study Plan.

 Remember that a^0 is an exponential expression. It does not mean $a \cdot 0$.

OBJECTIVE 4:   USE THE NEGATIVE-POWER RULE

When we use the quotient rule to simplify an expression such as $\dfrac{x^3}{x^5}$, the result involves a **negative exponent**.

$$\frac{x^3}{x^5} = x^{3-5} = x^{-2}$$

The exponent -2 is a negative number. To understand the meaning of a negative exponent, simplify $\dfrac{x^3}{x^5}$ by expanding the numerator and denominator and cancelling common factors.

$$\frac{x^3}{x^5} = \frac{x \cdot x \cdot x}{x \cdot x \cdot x \cdot x \cdot x} = \frac{\overset{1}{\cancel{x \cdot x \cdot x}}}{\underset{1}{\cancel{x \cdot x \cdot x}} \cdot x \cdot x} = \frac{1}{x \cdot x} = \frac{1}{x^2}$$

We can rewrite this result as $\dfrac{1}{x^2} = x^{-2}$, which means that x^{-2} is the reciprocal of x^2.

This rule applies to all bases except 0 and is called the **Negative-Power Rule**.

The Negative-Power Rule

When the exponent of a base is negative, take the reciprocal of the base by switching its location (numerator or denominator), then make the exponent positive.

$$a^{-n} = \frac{1}{a^n} \quad \text{and} \quad \frac{1}{a^{-n}} = a^n \quad (a \neq 0)$$

Click here to see how the negative-power rule can also be derived from the product rule.

 A negative exponent does not mean the result will be negative. For example, $2^{-3} = \dfrac{1}{2^3} = \dfrac{1}{8}$, which is positive.

Example 4 Using the Negative-Power Rule

My interactive video summary

Simplify. Use only positive exponents in your answers.

a. $(3)^{-2}$ **b.** m^{-5} **c.** $\dfrac{1}{5x^{-4}}$ **d.** $(7n)^{-1}$ **e.** $8a^{-1}$

f. -2^{-4} **g.** $4a^{-3}b^{2}$ **h.** $\dfrac{x^{-2}y}{z^{-5}}$ **i.** $2^{-1} - 3^{-1}$

Solution

a. The base is 3, and the exponent is -2. Since the exponent is negative, we take the reciprocal of the base and make the exponent positive. The reciprocal of 3 is $\dfrac{1}{3}$, so $(3)^{-2} = \dfrac{1}{3^{2}} = \dfrac{1}{9}$.

b. The base is m, and the exponent is -5. Since the exponent is negative, we take the reciprocal of the base and make the exponent positive. The reciprocal of m is $\dfrac{1}{m}$, so $m^{-5} = \dfrac{1}{m^{5}}$.

c. The negative exponent in the denominator is only on the variable x. Rewriting, we get $\dfrac{1}{5x^{-4}} = \dfrac{1}{5} \cdot \dfrac{1}{x^{-4}}$. In the fraction on the right, the base is x, and the exponent is -4. Since the exponent is negative, we take the reciprocal of the base by moving it to the numerator, then we make the exponent positive.

$$\dfrac{1}{x^{-4}} = x^{4}$$

Exponent is now positive

Base is now in the numerator

Simplifying gives

$$\dfrac{1}{5x^{-4}} = \dfrac{1}{5} \cdot \dfrac{1}{x^{-4}} = \dfrac{1}{5} \cdot x^{4} = \dfrac{x^{4}}{5}.$$

d.–i. Try the remaining parts on your own. Click here to check your answers, or watch this interactive video for detailed solutions.

You Try It   **Work through this You Try It problem.**

Work Exercises 13–16 in this textbook or in the *MyMathLab*® Study Plan.

OBJECTIVE 5:   USE THE POWER-TO-POWER RULE

Now we examine exponential expressions that are raised to a power. We can simplify an expression such as $\left(h^{5}\right)^{3}$ by expanding and then using the product rule.

$$\left(h^{5}\right)^{3} = h^{5} \cdot h^{5} \cdot h^{5} = h^{5+5} \cdot h^{5} = h^{5+5+5} = h^{15}$$

We can also find the same result by multiplying the exponents since $5 \cdot 3 = 15$. This rule is true in general and is called the **Power-to-Power Rule**.

The Power-to-Power Rule

When an exponential expression is raised to a power, we multiply the exponents.

$$\left(a^{m}\right)^{n} = a^{m \cdot n}$$

Example 5 Using the Power-to-Power Rule

My interactive video summary

 Simplify. Use only positive exponents in your answers.

a. $\left(a^3\right)^7$ **b.** $\left(x^{-2}\right)^4$ **c.** $\left(m^{-3}\right)^{-2}$

Solution

a. Since we are raising the exponential expression, a^3, to the seventh power, we apply the power-to-power rule and multiply the exponents.
$$\left(a^3\right)^7 = a^{3\cdot7} = a^{21}$$

b. Applying the power-to-power rule gives x^{-8}. The resulting exponent is negative, so we can apply the negative-power rule to obtain $\frac{1}{x^8}$.

c. Click here to check your answer, or watch this interactive video for detailed solutions.

You Try It Work through this You Try It problem.

Work Exercises 17–19 in this textbook or in the *MyMathLab*® Study Plan.

OBJECTIVE 6: USE THE PRODUCT-TO-POWER AND QUOTIENT-TO-POWER RULES

Product-to-Power Rule

To simplify a product raised to a power, such as $(2d)^4$, we expand the expression and combine like factors using the commutative property for multiplication.

$$\begin{aligned}(2d)^4 &= 2d \cdot 2d \cdot 2d \cdot 2d \\ &= 2 \cdot d \cdot 2 \cdot d \cdot 2 \cdot d \cdot 2 \cdot d \\ &= 2 \cdot 2 \cdot 2 \cdot 2 \cdot d \cdot d \cdot d \cdot d \\ &= 2^4 d^4\end{aligned}$$

We can get the same result by raising each factor of the base $2d$ to the common exponent, 4. For example, $(2d)^4 = 2^4 d^4$. This rule is called the **Product-to-Power Rule**.

The Product-to-Power Rule

When raising a product to a power, we raise each factor of the base to the common exponent.

$$(ab)^n = a^n b^n$$

Example 6 Using the Product-to-Power Rule

My interactive video summary

 Simplify. Use only positive exponents in your answers.

a. $(2x)^3$ **b.** $\left(c^2 d\right)^5$ **c.** $\left(-2a^3 b^4\right)^4$ **d.** $\left(x^{-3} y^2\right)^{-2}$

Solution

a. The base is $2x$. We apply the product-to-power rule and simplify.

$$\underbrace{(2x)^3}_{\text{Product-to-power}} = \underbrace{2^3 x^3}_{\text{Simplify}} = 8x^3$$

b. The base is $\left(c^2 d\right)$. We apply the product-to-power rule, raising each of the factors, c^2 and d, to the common exponent 5. To simplify, we apply the power-to-power rule.

$$\underbrace{\left(c^2 d\right)^5}_{\text{Product-to-power}} = \underbrace{\left(c^2\right)^5 d^5}_{\text{Power-to-power}} = c^{10} d^5$$

c.–d. Click here to check your answers, or watch this interactive video for detailed solutions.

You Try It   Work through this You Try It problem.

Work Exercises 20–23 in this textbook or in the **MyMathLab**® Study Plan.

Quotient-to-Power Rule

To simplify a quotient raised to a power, such as $\left(\dfrac{x}{y}\right)^4$, we expand the expression and multiply the resulting fractions.

$$\left(\frac{x}{y}\right)^4 = \frac{x}{y} \cdot \frac{x}{y} \cdot \frac{x}{y} \cdot \frac{x}{y}$$

$$= \frac{x \cdot x \cdot x \cdot x}{y \cdot y \cdot y \cdot y}$$

$$= \frac{x^4}{y^4}$$

We can get the same result by raising both the numerator and denominator of the base to the common exponent 4. For example, $\left(\dfrac{x}{y}\right)^4 = \dfrac{x^4}{y^4}$. This result is true in general and is called the **Quotient-to-Power Rule**.

The Quotient-to-Power Rule

When raising a quotient to a power, we raise both the numerator and denominator to the common exponent.

$$\left(\frac{a}{b}\right)^n = \frac{a^n}{b^n} \quad (b \neq 0)$$

For a quotient raised to a negative power, we have a variation of the quotient-to-power rule:

$$\left(\frac{a}{b}\right)^{-n} = \frac{b^n}{a^n}$$

Click here to see an explanation.

My interactive video summary

Example 7 Using the Quotient-to-Power Rule

 Simplify. Use only positive exponents in your answers.

a. $\left(\dfrac{y^3}{z^2}\right)^5$ b. $\left(\dfrac{2x^2y}{3z}\right)^3$ c. $\left(\dfrac{-mn^4}{p^2}\right)^3$ d. $\left(\dfrac{x^4}{y^3}\right)^{-2}$ e. $\left(\dfrac{5a^3}{b^{-4}}\right)^3$

Solution

a. We apply the quotient-to-power rule first and then apply additional exponent rules as needed.

$$\text{Quotient-to-power:} \quad \left(\dfrac{y^3}{z^2}\right)^5 = \dfrac{\left(y^3\right)^5}{\left(z^2\right)^5}$$

$$\text{Power-to-power:} \qquad\qquad = \dfrac{y^{15}}{z^{10}}$$

b. We apply the quotient-to-power rule first and then apply additional exponent rules as needed.

$$\text{Quotient-to-power:} \quad \left(\dfrac{2x^2y}{3z}\right)^3 = \dfrac{\left(2x^2y\right)^3}{\left(3z\right)^3}$$

$$\text{Product-to-power:} \qquad\qquad = \dfrac{2^3\left(x^2\right)^3y^3}{3^3z^3}$$

$$\text{Power-to-power:} \qquad\qquad = \dfrac{2^3x^6y^3}{3^3z^3}$$

$$\text{Simplify:} \qquad\qquad\qquad = \dfrac{8x^6y^3}{27z^3}$$

c.–e. Click here to check your answers, or watch this interactive video for detailed solutions.

You Try It Work through this You Try It problem.

Work Exercises 24–28 in this textbook or in the *MyMathLab* Study Plan.

OBJECTIVE 7: SIMPLIFY EXPONENTIAL EXPRESSIONS USING A COMBINATION OF RULES

In several examples we have seen how more than one rule for exponents was needed to simplify the exponential expressions. In general, we will consider an exponential expression to be **simplified** when the following conditions are met:

- No parentheses or grouping symbols are present.
- No zero or negative exponents are present.
- No powers are raised to powers.
- Each base occurs only once.

Now let's summarize the rules for exponents and then work through some additional examples of how to simplify exponential expressions.

Rules for Exponents

Product Rule	$a^m \cdot a^n = a^{m+n}$
Quotient Rule	$\dfrac{a^m}{a^n} = a^{m-n}$ $\qquad (a \neq 0)$
Zero-Power Rule	$a^0 = 1$ $\qquad (a \neq 0)$
Negative-Power Rule	$a^{-n} = \dfrac{1}{a^n}$ or $\dfrac{1}{a^{-n}} = a^n$ $\quad (a \neq 0)$
Power-to-Power Rule	$\left(a^m\right)^n = a^{m \cdot n}$
Product-to-Power Rule	$(ab)^n = a^n b^n$
Quotient-to-Power Rule	$\left(\dfrac{a}{b}\right)^n = \dfrac{a^n}{b^n}$ $\qquad (b \neq 0)$

CAUTION Don't forget to use the order of operations when simplifying exponential expressions.

Example 8 Simplifying Exponential Expressions Using a Combination of Rules

My interactive video summary

 Simplify using the rules for exponents. Use only positive exponents in your answers.

a. $\dfrac{x^2 y^0 z^{-3}}{x^4 z^{-2}}$

b. $\dfrac{24x^5 y^{-4}}{-3x^{-3} y^6}$

c. $\left(\dfrac{a^3 b^{-4}}{2a^4 b^3}\right)^3$

d. $\left(\dfrac{8m^6 n^3}{4m^{-3} n^7}\right)^{-4}$

e. $\left(-3m^3 n^{-8}\right)\left(-2m^{-5} n^3\right)^4$

f. $\left(\dfrac{3a^2 b}{c^{-3}}\right)^2 \left(\dfrac{-4a^{-5} b^3}{c^5}\right)^3$

Solution

a.–f. Click here to check your answers, or watch this interactive video for detailed solutions.

You Try It   Work through this You Try It problem.

Work Exercises 29–34 in this textbook or in the *MyMathLab*® Study Plan.

Note: There is often more than one way to simplify an exponential expression. Do not feel that you need to always work problems using the exact same steps as in the examples. Just make sure that the conditions for being simplified are met.

OBJECTIVE 8:   USE RULES FOR EXPONENTS WITH SCIENTIFIC NOTATION

An ArF excimer laser is used in refractive surgery to correct various eye disorders. This laser has a wavelength of 0.000000193 meters. A space explosion detected by NASA's Swift satellite on April 23, 2009, was the farthest object viewed in the known universe. This explosion was about 78,000,000,000,000,000,000,000 miles away. To help make such small and large numbers more manageable, we use **scientific notation**.

Scientific Notation

A number is written in **scientific notation** if it has the form

$$a \times 10^n,$$

where a is a real number, such that $1 \le |a| < 10$, and n is an integer.

Note: When writing scientific notation, typically we use a times sign ($\times$) instead of a dot ($\cdot$) to indicate multiplication.

In scientific notation, we write the wavelength of the ArF laser as 1.93×10^{-7} meters and the distance of the space explosion as 7.8×10^{22} miles. Notice that for very small numbers, the exponent on 10 is negative, while for very large numbers, the exponent on 10 is positive. The following procedures are used to convert between standard form and scientific notation.

Converting between Standard Form and Scientific Notation

To change a number from standard form to scientific notation,

1. Move the decimal point so that $|a|$ is greater than or equal to 1 but less than 10. To do this, place the decimal point to the right of the first non-zero digit.

2. Multiply the number by 10^n, where $|n|$ is the number of places that the decimal point moves. If the decimal point moves to the left, then $n > 0$ (n is positive). If the decimal point moves to the right, then $n < 0$ (n is negative). Remove any zeros lying to the right of the last non-zero digit or to the left of the first non-zero digit.

To change a number from scientific notation to standard form,

Move the decimal point $|n|$ places, inserting zero placeholders as needed. If $n > 0$, move the decimal point to the right. If $n < 0$, move the decimal point to the left.

Example 9 Converting between Standard Form and Scientific Notation

 Write each number in scientific notation.

a. 630,000

b. 0.000000329

c. 0.00000407

d. 70,920,000

Solution

a. Move the decimal point to the right of the first non-zero digit.

$$630{,}000.$$

We move the decimal point five places to the left, which means that the exponent on 10 will be 5. All remaining zeros after 3 are deleted.

$$630{,}000 = 6.3 \times 10^5$$

b.–d. Click here to check your answers, or watch this video for detailed solutions.

You Try It   **Work through this You Try It problem.**

Work Exercises 35–38 in this textbook or in the *MyMathLab*® Study Plan.

My video summary

Example 10 Converting between Standard Form and Scientific Notation

 Write each number in standard form.

a. 2.81×10^{-4} 　　　　　　　　b. 3.7×10^{6}

c. 7.02×10^{5} 　　　　　　　　d. 6.045×10^{-7}

Solution

a. Since the exponent on 10 is -4, we move the decimal point four places to the left, filling in zero placeholders as needed.

$$0002.81$$

So,

$$2.81 \times 10^{-4} = 0.000281.$$

b.–d. Click here to check your answers, or watch this video for detailed solutions.

You Try It   **Work through this You Try It problem.**

Work Exercises 39–42 in this textbook or in the *MyMathLab*® Study Plan.

Rules for exponents can be used when multiplying or dividing numbers written in scientific notation.

Example 11 Using Rules for Exponents with Scientific Notation

My interactive video summary

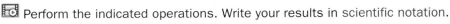 Perform the indicated operations. Write your results in scientific notation.

a. $\left(2 \times 10^{6}\right)\left(4 \times 10^{-3}\right)$ 　　b. $\left(7.2 \times 10^{-4}\right)\left(9.5 \times 10^{9}\right)$ 　　c. $\dfrac{8.4 \times 10^{7}}{2.1 \times 10^{3}}$

d. $\dfrac{1.8 \times 10^{5}}{4.5 \times 10^{-3}}$ 　　　　e. $(520{,}000{,}000)(0.00004)$ 　　f. $\dfrac{0.000000056}{0.0000000004}$

Hint: In parts (e) and (f), write each number in scientific notation before completing the operation.

Solution

a. Begin with original expression: $\left(2 \times 10^{6}\right)\left(4 \times 10^{-3}\right)$

　　　　　　　Regroup factors: $(2 \times 4) \times \left(10^{6} \times 10^{-3}\right)$

　　　Multiply constant factors: $8 \times \left(10^{6} \times 10^{-3}\right)$

　　　Apply the product rule: $8 \times \left(10^{6+(-3)}\right)$

　　　　　　　　　　Simplify: 8×10^{3}

b.

Begin with original expression:	$\left(7.2 \times 10^{-4}\right)\left(9.5 \times 10^{9}\right)$
Regroup factors:	$\left(7.2 \times 9.5\right) \times \left(10^{-4} \times 10^{9}\right)$
Multiply constant factors:	$68.4 \times \left(10^{-4} \times 10^{9}\right)$
Write 68.4 in scientific notation:	$6.84 \times 10^{1} \times \left(10^{-4} \times 10^{9}\right)$
Apply the product rule:	$6.84 \times 10^{1+(-4)+9}$
Simplify:	6.84×10^{6}

c.

Begin with original expression:	$\dfrac{8.4 \times 10^{7}}{2.1 \times 10^{3}}$
Regroup factors:	$\left(\dfrac{8.4}{2.1}\right) \times \left(\dfrac{10^{7}}{10^{3}}\right)$
Divide constant factors:	$4 \times \left(\dfrac{10^{7}}{10^{3}}\right)$
Apply the quotient rule:	$4 \times 10^{7-3}$
Simplify:	4×10^{4}

d.–f. Click here to check your answers, or watch this interactive video for detailed solutions.

You Try It Work through this You Try It problem.

Work Exercises 43–48 in this textbook or in the *MyMathLab*® Study Plan.

4.1 Exercises

Before working on the exercises, you may wish to review the conditions of simplified exponential expressions.

You Try It

In Exercises 1–4, simplify using the product rule.

1. $(-3)^{3} \cdot (-3)^{2}$ **2.** $x^{3} \cdot x^{8}$ **3.** $\left(4a^{3}b^{2}\right)\left(5a^{4}b^{7}\right)$ **4.** $\left(-5x^{4}y\right)\left(3x^{2}z^{3}\right)$

You Try It

In Exercises 5–8, simplify using the quotient rule.

5. $\dfrac{5^{7}}{5^{3}}$ **6.** $\dfrac{c^{9}}{c^{4}}$ **7.** $\dfrac{2x^{5}y^{6}}{-8x^{3}y^{2}}$ **8.** $\dfrac{15m^{8}n^{3}}{5m^{7}}$

You Try It

In Exercises 9–12, simplify using the zero-power rule.

9. 8^{0} **10.** -5^{0} **11.** $(12x)^{0}$ **12.** $3x^{0} + 2y^{0}$

You Try It

In Exercises 13–16, write each expression using only positive exponents. Simplify if possible.

13. 7^{-2} **14.** -4^{-2} **15.** $\dfrac{m^{-4}}{n^{-7}}$ **16.** $\dfrac{x^{-4}y^{3}}{z^{-5}}$

You Try It

In Exercises 17–19, simplify using the power-to-power rule.

17. $\left(h^{4}\right)^{8}$ **18.** $\left(q^{-5}\right)^{3}$ **19.** $\left(y^{-7}\right)^{-2}$

You Try It

In Exercises 20–23, simplify using the product-to-power rule.

 20. $(3x)^5$ **21.** $(-4m^3)^4$ **22.** $(4a^2b^4)^3$ **23.** $(2x^{-3}y^4)^{-3}$

You Try It

In Exercises 24–28, simplify using the quotient-to-power rule.

24. $\left(\dfrac{m^4}{n^7}\right)^5$ **25.** $\left(\dfrac{2a^2}{6b^5}\right)^3$ **26.** $\left(\dfrac{-4x^2}{y^5}\right)^3$ **27.** $\left(\dfrac{p^3}{q^9}\right)^{-4}$ **28.** $\left(\dfrac{2a^3}{b^{-2}}\right)^5$

You Try It

In Exercises 29–34, simplify using the rules for exponents. Use only positive exponents in your answers.

29. $\dfrac{a^3b^{-4}c^0}{a^5c^3}$ **30.** $\dfrac{-15x^{-6}y^7}{5x^{-2}y^3}$ **31.** $\left(\dfrac{4m^{-2}n^{-1}}{m^6n^5}\right)^3$

32. $\left(\dfrac{6x^{-4}y^6}{42x^2y^{-11}}\right)^{-2}$ **33.** $(7a^4b^{-6})(-4m^7n^{-5})^3$ **34.** $\left(\dfrac{-2a^4b^{-3}}{c^4}\right)^2\left(\dfrac{9a^{-1}b^8}{c^{-2}}\right)^{-2}$

You Try It

In Exercises 35–38, write each number in scientific notation.

 35. 6,300,000   **36.** 0.00000042   **37.** 0.000000000507   **38.** 300,190,000

You Try It

In Exercises 39–42, write each number in standard form.

39. 3.15×10^{-7} **40.** 7.38×10^5 **41.** 4.09×10^9 **42.** 8.0031×10^{-4}

You Try It

In Exercises 43–48, perform the indicated operations. Write your results in scientific notation.

 43. $(3.2 \times 10^4)(2.5 \times 10^7)$ **44.** $(8.3 \times 10^{-4})(5.6 \times 10^8)$ **45.** $\dfrac{7.5 \times 10^{11}}{1.5 \times 10^6}$

46. $\dfrac{2.7 \times 10^{10}}{8.1 \times 10^{-4}}$ **47.** $(71,000,000)(0.000006)$ **48.** $\dfrac{0.00000189}{0.000009}$

4.2 Introduction to Polynomial Functions

THINGS TO KNOW

Before working through this section, be sure you are familiar with the following concepts.

VIDEO   ANIMATION   INTERACTIVE

You Try It

1. Simplify Algebraic Expressions
    (Section R.3, Objective 2)

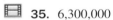

You Try It

2. Evaluate Functions
    (Section 2.3, Objective 2)

OBJECTIVES

1 Find the Coefficient and Degree of a Monomial

2 Find the Leading Coefficient and Degree of a Polynomial

3 Evaluate a Polynomial Function for a Given Value

4 Add Polynomials

5 Subtract Polynomials

6 Add and Subtract Polynomial Functions

OBJECTIVE 1: FIND THE COEFFICIENT AND DEGREE OF A MONOMIAL

A **term** is a constant, a variable, or the product of a constant and one or more variables raised to powers. If a term contains a single constant factor (possibly a simplified fraction) and if none of the variable factors can be combined using the rules for exponents, then it is called a **simplified term**. Examples of simplified terms include

$$4x^3, -\frac{2}{5}x^2y^3, x^2, 35x^{-3}, -3yz^4, 9y^{1/2}, \frac{8}{x}, \text{ and } 10.$$

The terms $7x^3x^2$ and $\frac{4x^4}{6}$ are not simplified. Do you see why?

Some simplified terms are also *monomials*.

> **Definition Monomial**
>
> A **monomial** is a simplified term in which all variables are raised to non-negative integer powers and no variables appear in any denominator.

My interactive video summary

In the preceding list of terms, $4x^3$, $-\frac{2}{5}x^2y^3$, x^2, $-3yz^4$, and 10 are all monomials.

The terms $35x^{-3}$, $9y^{1/2}$, and $\frac{8}{x}$ are not monomials. Do you see why? Watch this interactive video to determine if a term is a monomial.

Every monomial has both a *coefficient* and a *degree*.

> **Definition Coefficient of a Monomial**
>
> The **coefficient of a monomial** is the constant factor. It is sometimes called the **numerical coefficient**.

For example, the coefficient of $4x^3$ is 4, and the coefficient $-\frac{2}{5}x^2y^3$ is $-\frac{2}{5}$.

> **Definition Degree of a Monomial**
>
> The **degree of a monomial** is the sum of the exponents on the variables.

For example, the degree of $4x^3$ is 3, and the degree of $-\frac{2}{5}x^2y^3$ is $2 + 3 = 5$.

Example 1 Finding the Coefficient and Degree of a Monomial

Find the coefficient and degree of each monomial.

a. $4x^5y^2$ **b.** x^3 **c.** $\frac{3}{5}x$

d. $-x^4y^2z^5$ **e.** 8.5 **f.** $3.5xyz$

Solution

a. The coefficient of $4x^5y^2$ is the constant factor 4. The degree is the sum of the exponents on the variables: $5 + 2 = 7$.

b. Since $x^3 = 1x^3$, the coefficient is 1. Since the exponent is 3, the degree is 3.

c. The coefficient of $\frac{3}{5}x$ is $\frac{3}{5}$. The exponent of x is understood to be 1, so the degree is 1.

d.–f. Try answering these questions on your own, then click here to check.

 The answer to Example 1e demonstrates that the degree of any nonzero constant is 0.

You Try It   Work through this You Try It problem.

Work Exercises 1–6 in this textbook or in the *MyMathLab*® Study Plan.

OBJECTIVE 2:   FIND THE LEADING COEFFICIENT AND DEGREE OF A POLYNOMIAL

In Section R.3, we studied algebraic expressions. We now look at special kinds of algebraic expressions called *polynomials*.

> **Definition Polynomial**
>
> A **polynomial** is a monomial or a finite sum of monomials.

My interactive video summary

 Watch this interactive video to determine if an algebraic expression is a polynomial.

The monomials that make up a polynomial are called the **terms of the polynomial.** A polynomial is a **simplified polynomial** if all of its terms are simplified and none of its terms are like terms.

> **Definition Degree of a Polynomial**
>
> The **degree of a polynomial** is the largest degree of its terms.

Example 2 Finding the Degree of a Polynomial

Find the degree of each polynomial.

a. $7x - 5x^2y + 9x^3y^2$ **b.** $16x^4 - 8x^3 + 4x^2 - 2x + 1$

Solution

a. The degree of $7x$ is 1, the degree of $-5x^2y$ is $2 + 1 = 3$, and the degree of $9x^3y^2$ is $3 + 2 = 5$. The largest degree is 5, so the degree of the polynomial is 5.

b. Try finding the degree on your own, then click here to check your answer.

You Try It   Work through this You Try It problem.

Work Exercises 7–10 in this textbook or in the *MyMathLab*® Study Plan.

The polynomial $16x^4 - 8x^3 + 4x^2 - 2x + 1$ is an example of a **polynomial in one variable**. It is written in **descending order** by the degree of its terms. This is called the **standard form** for polynomials in one variable. When a polynomial is in standard form we can find its *leading coefficient*.

Definition   Leading Coefficient of a Polynomial in One Variable

When a polynomial in one variable is written in standard form, the coefficient of the first term (the term with the highest degree) is called the **leading coefficient**.

Example 3 Finding the Leading Coefficient and Degree of a Polynomial in One Variable

Write each polynomial in standard form. Then find its degree and leading coefficient.

a. $13 - 8x^2 + 9x^4 + 7x^3 - \dfrac{1}{2}x$ b. $\dfrac{1}{2}p + \dfrac{2}{3}p^2 - \dfrac{5}{8}p^3$

Solution

a. To put the polynomial in standard form, we write the terms in order of degree from the largest degree, 4, to the smallest degree, 0 (the constant term): $9x^4 + 7x^3 - 8x^2 - \dfrac{1}{2}x + 13$. The degree of the polynomial is 4, the largest degree. The leading coefficient is 9, the coefficient of the first term.

b. Try working this problem on your own. Click here to check your answers, or watch this interactive video for detailed solutions.

You Try It   Work through this You Try It problem.

Work Exercises 11–14 in this textbook or in the *MyMathLab*® Study Plan.

OBJECTIVE 3:   EVALUATE A POLYNOMIAL FUNCTION FOR A GIVEN VALUE

We now combine what we know about polynomials in one variable with what we learned in Section 2.2 about functions to introduce *polynomial functions*.

Definition   Polynomial Function

A function is a **polynomial function** if it is defined by a polynomial.

The function $P(x) = 2x^2 - 5x + 3$ is a polynomial function because it is defined by the polynomial $2x^2 - 5x + 3$. The **degree of the polynomial function** is the degree of the polynomial (the largest exponent).

We evaluate polynomial functions in the same way that we evaluated functions in Section 2.3. We substitute the given value for the variable and simplify.

Example 4 Evaluating Polynomial Functions

My interactive video summary

 If $P(x) = 4x^3 - 2x^2 + 8x + 7$, evaluate each of the following.

a. $P(4)$ b. $P(-2)$ c. $P\left(-\dfrac{1}{2}\right)$

Solution

a. Substitute 4 for x in the function P: $P(4) = 4(4)^3 - 2(4)^2 + 8(4) + 7$

   Simplify the exponents: $= 4(64) - 2(16) + 8(4) + 7$

   Simplify the multiplication: $= 256 - 32 + 32 + 7$

   Add and subtract: $= 263$

   The value of P is 263 when x is 4.

b. Substitute -2 for x
   in the function P: $P(-2) = 4(-2)^3 - 2(-2)^2 + 8(-2) + 7$

   Finish the problem on your own. Click here to check your answer, or watch this interactive video to see the complete solution.

c. Try to evaluate the function on your own. Click here to check your answer, or watch this interactive video to see the complete solution.

You Try It   Work through this You Try It problem.

Work Exercises 15–18 in this textbook or in the **MyMathLab**® Study Plan.

OBJECTIVE 4:   ADD POLYNOMIALS

In Section R.3, we learned how to simplify algebraic expressions. To do this, we often must combine like terms.

To add polynomials, we remove all grouping symbols, rearrange the terms so that like terms are grouped together, and combine all like terms.

Adding Polynomials

To add polynomials, remove all grouping symbols and combine like terms.

Example 5 Adding Polynomials

Add.

a. $\left(4x^2 - 3x + 6\right) + \left(9x^2 + 3x + 1\right)$

b. $\left(2y^3 + y^2 - 7\right) + \left(4y^3 - 3y^2 + 2y\right) + \left(6y^2 + 2\right)$

c. $\left(5p^2q + 7pq + 4pq^2\right) + \left(p^2q - 8pq - 3pq^2\right)$

My interactive video summary

Solution We follow the procedure for adding polynomials.

a. Write the original expression: $\left(4x^2 - 3x + 6\right) + \left(9x^2 + 3x + 1\right)$

Remove grouping symbols: $= 4x^2 - 3x + 6 + 9x^2 + 3x + 1$

Rearrange terms: $= 4x^2 + 9x^2 - 3x + 3x + 6 + 1$

Combine like terms: $= 13x^2 + 7$

b. Write the original expression: $\left(2y^3 + y^2 - 7\right) + \left(4y^3 - 3y^2 + 2y\right) + \left(6y^2 + 2\right)$

Remove grouping symbols: $= 2y^3 + y^2 - 7 + 4y^3 - 3y^2 + 2y + 6y^2 + 2$

Rearrange terms: $= 2y^3 + 4y^3 + y^2 - 3y^2 + 6y^2 + 2y - 7 + 2$

Finish this problem on your own. Click here to check your answer, or watch this interactive video to see the complete solution.

c. Try adding these polynomials on your own. Click here to check your answer, or watch this interactive video to see the complete solution.

Sometimes it is helpful to add polynomials vertically. To do this, line up like terms in columns and then combine like terms. We repeat Example 5a to show this method:

$$\begin{array}{r} 4x^2 - 3x + 6 \\ + \ 9x^2 + 3x + 1 \\ \hline 13x^2 \qquad\quad + 7 \end{array}$$

Repeat Examples 5b and c by adding vertically. Click here to see their solutions.

You Try It   **Work through this You Try It problem.**

Work Exercises 19–23 in this textbook or in the *MyMathLab*® Study Plan.

OBJECTIVE 5:   **SUBTRACT POLYNOMIALS**

Recall from Section R.2 that two real numbers are opposites if their sum is zero. In the same way, two polynomials are **opposite polynomials** if they add to zero. For example, the polynomials $2x^2 - 8x - 7$ and $-2x^2 + 8x + 7$ are opposites because

$$\left(2x^2 - 8x - 7\right) + \left(-2x^2 + 8x + 7\right) = 2x^2 - 2x^2 - 8x + 8x - 7 + 7 = 0.$$

A negative sign can also be used to indicate opposite polynomials. Just as -5 represents the opposite of 5, the notation $-\left(2x^2 - 8x - 7\right)$ represents the opposite of $2x^2 - 8x - 7$. So, we can write

$$-\left(2x^2 - 8x - 7\right) = -2x^2 + 8x + 7.$$

Notice that $2x^2 - 8x - 7$ and $-2x^2 + 8x + 7$ look the same except each pair of like terms have opposite signs. Therefore, we find the opposite of a polynomial by changing the sign of each term.

As we did with real numbers in Section R.2, we can subtract a polynomial by adding its opposite.

Subtracting Polynomials

To subtract a polynomial, add its opposite.

Example 6 Subtracting Polynomials

My interactive video summary

 Subtract.

a. $\left(10x^2 - 7x + 9\right) - \left(8x^2 - 4x + 1\right)$

b. $\left(2a^2 + 5ab - 9b^2\right) - \left(-4a^2 + 8ab - 5b^2\right)$

Solution We follow the procedure for subtracting polynomials.

a. Begin with the original expression: $\left(10x^2 - 7x + 9\right) - \left(8x^2 - 4x + 1\right)$

Change to add the opposite
polynomial: $= \left(10x^2 - 7x + 9\right) + \left(-8x^2 + 4x - 1\right)$

Remove grouping symbols: $= 10x^2 - 7x + 9 - 8x^2 + 4x - 1$

Rearrange terms: $= 10x^2 - 8x^2 - 7x + 4x + 9 - 1$

Combine like terms: $= 2x^2 - 3x + 8$

b. Begin with the original
expression: $\left(2a^2 + 5ab - 9b^2\right) - \left(-4a^2 + 8ab - 5b^2\right)$

Change to add the
opposite polynomial: $= \left(2a^2 + 5ab - 9b^2\right) + \left(+4a^2 - 8ab + 5b^2\right)$

Try finishing this problem on your own. Click here to check your answer, or watch this interactive video to see the detailed solution.

As with addition, polynomials may be subtracted vertically. We repeat Example 6a to show this method:

$$
\begin{array}{r}
10x^2 - 7x + 9 \\
- (8x^2 - 4x + 1) \\
\hline
\end{array}
\quad
\xrightarrow[\text{polynomial}]{\substack{\text{Add the} \\ \text{opposite}}}
\quad
\begin{array}{r}
10x^2 - 7x + 9 \\
+ \ -8x^2 + 4x - 1 \\
\hline
2x^2 - 3x + 8
\end{array}
$$

Repeat Example 6b by subtracting vertically. Click here to see the solution.

You Try It Work through this You Try It problem.

Work Exercises 24–28 in this textbook or in the *MyMathLab*®️ Study Plan.

OBJECTIVE 6: ADD AND SUBTRACT POLYNOMIAL FUNCTIONS

Now that we can add and subtract polynomials, we can learn how to add and subtract polynomial functions.

Definition **The Sum and Difference of Functions**

If we let f and g represent two functions, then

the **sum of f and g** is $(f + g)(x) = f(x) + g(x),$ and

the **difference of f and g** is $(f - g)(x) = f(x) - g(x).$

Example 7 Adding and Subtracting Polynomial Functions

My interactive video summary For $P(x) = x^4 - 9x^2 + 7$ and $Q(x) = 3x^4 - 4x^2 + 2x - 10$, find each of the following.

a. $(P + Q)(x)$ **b.** $(P - Q)(x)$

Solution

a.   Rewrite as a sum of 2 functions: $(P + Q)(x) = P(x) + Q(x)$

Substitute for the function notation: $= \left(x^4 - 9x^2 + 7\right) + \left(3x^4 - 4x^2 + 2x - 10\right)$

Remove grouping symbols: $= x^4 - 9x^2 + 7 + 3x^4 - 4x^2 + 2x - 10$

Rearrange terms: $= x^4 + 3x^4 - 9x^2 - 4x^2 + 2x + 7 - 10$

Combine like terms: $= 4x^4 - 13x^2 + 2x - 3$

b. Rewrite as a difference of 2 functions: $(P - Q)(x) = P(x) - Q(x)$

Substitute for the function notation: $= \left(x^4 - 9x^2 + 7\right) - \left(3x^4 - 4x^2 + 2x - 10\right)$

Add the opposite polynomial: $= \left(x^4 - 9x^2 + 7\right) + \left(-3x^4 + 4x^2 - 2x + 10\right)$

Try finishing this problem on your own. Click here to check your answer, or watch this interactive video to see the detailed solution.

 You Try It   Work through this You Try It problem.

Work Exercises 29–34 in this textbook or in the *MyMathLab*® Study Plan.

4.2 Exercises

In Exercises 1–6, find the coefficient and degree of each monomial.

You Try It

1. $6x^4$ **2.** $25a^5b^3$ **3.** $\dfrac{3x}{8}$ **4.** -13 **5.** $-xy$ **6.** $-2.8x^3y^6z^7$

In Exercises 7–10, find the degree of each polynomial.

You Try It

7. $15x^3y^4 + 13x^2y^5 - 11x^7y$ **8.** $27p^3 - 54p^2q + 36pq^2 - 8q^3$

9. $\dfrac{9}{10}x^2y^2z + \dfrac{8}{11}x^3y^3 - \dfrac{7}{12}xy^2z$ **10.** $17 - 10x$

In Exercises 11–14, write each polynomial in standard form. Then find its degree and leading coefficient.

You Try It

 11. $8 - 5m + m^5 - 9m^3 - \dfrac{2}{3}m^7$ **12.** $5 + 2y - 9y^2 + 6y^3$

13. $5x^2 - 9x + 4$ **14.** $4w - w^5 + 0.7w^2 - 10.4$

You Try It

In Exercises 15–18, evaluate each polynomial function for the given value.

 15. $P(x) = 18x^2 - 6x + 5;\ \ P\left(\dfrac{1}{3}\right)$

16. $Q(t) = t^5 + 3t^4 - 5t^3 - 8t^2 + t - 15;\ \ Q(-2)$

17. $R(p) = 0.4p^3 - 2.6p^2 - 6.8p + 52;\ \ R(10)$

18. $S(x) = 16.4x^4 - 10.6x^2 + 20.1;\ \ S(2.5)$

You Try It

In Exercises 19–23, add.

 19. $\left(14x^3 - 5x^2 - 9x - 7\right) + \left(-6x^3 + 2x^2 + 4x - 8\right)$
20. $\left(9x^2y - 6xy\right) + \left(8x^2y - xy\right)$

21. $\left(4w^2 - 7w + 5\right) + \left(3w^3 + 5w - 8\right)$
22. $\left(0.4m^3n^2 - 6.8mn\right) + \left(1.2m^2n^3 + 2.4mn\right)$

23. $\left(\dfrac{2}{5}x^2 + \dfrac{5}{6}xy - \dfrac{3}{8}y^2\right) + \left(-\dfrac{4}{5}x^2 + \dfrac{1}{9}xy - \dfrac{1}{4}y^2\right)$

You Try It

In Exercises 24–28, subtract.

 24. $\left(10y^3 - 3y + 8\right) - \left(-2y^3 - 5y^2 + 5y - 13\right)$
25. $\left(17p^2q + 21pq^2\right) - \left(10pq^2 - 13p^2q\right)$

26. $\left(m^3 - 2m + 8\right) - \left(-3 + m^2 - 6m^3\right)$
27. $\left(7.1x^2 - 3.2x + 5.8\right) - \left(4.8x^2 + 2.9\right)$

28. $\left(\dfrac{9}{10}a - \dfrac{1}{4}b\right) - \left(\dfrac{3}{10}a - \dfrac{5}{8}b\right)$

You Try It

In Exercises 29–34, let $P(x) = x^3 - 5x^2 + 2x - 1$, $Q(x) = x^3 - 9x^2 - 12$, and $R(x) = 9x^2 + 4x - 8$. Add or subtract the polynomial functions, as indicated.

 29. $(P + Q)(x)$

30. $(Q - P)(x)$

31. $(Q + R)(x)$

32. $(P - R)(x)$

33. $(Q - R)(x)$

34. $(P + R - Q)(x)$

4.3 Multiplying Polynomials

THINGS TO KNOW

Before working through this section, be sure you are familiar with the following concepts:

	VIDEO	ANIMATION	INTERACTIVE

You Try It

1. Simplify Algebraic Expressions
    (Section R.3, Objective 2)

You Try It

2. Simplify Exponential Expressions Using
    a Combination of Rules
    (Section 4.1, Objective 7)

OBJECTIVES

1 Multiply Monomials

2 Multiply a Polynomial by a Monomial

3 Multiply Two Binomials

4 Multiply Two Binomials Using Special Product Rules

5 Multiply Two or More Polynomials

6 Multiply Polynomial Functions

OBJECTIVE 1:   MULTIPLY MONOMIALS

Multiplying monomials is like simplifying exponential expressions using a combination of the rules for exponents that we saw in Section 4.1. We use the commutative and associative properties to group factors with like bases. We then apply the product rule for exponents.

Multiplying Monomials

Rearrange the factors to group the coefficients and to group like bases. Then, multiply the coefficients and apply the product rule for exponents.

Example 1 Multiplying Monomials

Multiply.

a. $\left(5x^4\right)\left(9x^3\right)$ **b.** $\left(-8p^3q^2r^4\right)\left(-\dfrac{1}{2}pq^6\right)$

Solution

a.   Begin with the original expression: $\left(5x^4\right)\left(9x^3\right)$

Rearrange the factors to group coefficients
and like bases: $= (5 \cdot 9)\left(x^4x^3\right)$

Multiply coefficients; apply the product rule: $= 45x^{4+3}$

Simplify: $= 45x^7$

b. Try to work this problem on your own. Click here for the complete solution.

You Try It   **Work through this You Try It problem.**

Work Exercises 1–3 in this textbook or in the *MyMathLab*®️ Study Plan.

OBJECTIVE 2:   MULTIPLY A POLYNOMIAL BY A MONOMIAL

To multiply a polynomial with more than one term by a monomial, we use the distributive property.

Multiplying Polynomials by Monomials

If m is a monomial and $p_1, p_2, p_3, \ldots, p_n$ are the terms of a polynomial, then

$$m(p_1 + p_2 + p_3 + \cdots + p_n) = mp_1 + mp_2 + mp_3 + \cdots + mp_n$$

Simplify each product $mp_1, mp_2, mp_3, \ldots, mp_n$ using the method for multiplying monomials.

Example 2 Multiplying a Polynomial by a Monomial

Multiply.

a. $2x^3\left(3x^2 - 7x + 9\right)$ **b.** $-0.7xy^2\left(6x^3y - 5x^2y^2\right)$

Solution

a.

Begin with the original expression:	$2x^3\left(3x^2 - 7x + 9\right)$
Distribute the monomial $2x^3$:	$= 2x^3 \cdot 3x^2 + 2x^3 \cdot (-7x) + 2x^3 \cdot 9$
Rearrange the factors:	$= 2 \cdot 3 \cdot x^3 \cdot x^2 + 2 \cdot (-7) \cdot x^3 \cdot x + 2 \cdot 9 \cdot x^3$
Multiply coefficients; apply the product rule:	$= 6x^{3+2} - 14x^{3+1} + 18x^3$
Add the exponents:	$= 6x^5 - 14x^4 + 18x^3$

b.

Begin with the original expression:	$-0.7xy^2\left(6x^3y - 5x^2y^2\right)$
Distribute the monomial $-0.7xy^2$:	$= \left(-0.7xy^2\right) \cdot 6x^3y + \left(-0.7xy^2\right) \cdot \left(-5x^2y^2\right)$

Try to finish this problem on your own. Click here to check your answer, or watch this video to see the complete solution.

You Try It   Work through this You Try It problem.

Work Exercises 4–6 in this textbook or in the **MyMathLab**® Study Plan.

OBJECTIVE 3:   MULTIPLY TWO BINOMIALS

A simplified polynomial with two terms is called a **binomial**. To multiply two binomials, we can use the distributive property twice. We can distribute the first binomial to each term in the second binomial. We can then distribute each term from the second binomial through the first binomial (from the back).

$$(a + b)(c + d) = (a + b)c + (a + b)d = ac + bc + ad + bd$$

The end result is that each term in the first binomial gets multiplied by each term in the second binomial.

$$(a + b)(c + d) = ac + ad + bc + bd$$

Multiplying Two Binomials

To multiply two binomials, multiply each term of the first binomial by each term of the second binomial. Combine like terms.

Example 3 Multiplying Two Binomials

Multiply $(2x + 5)(3x + 4)$.

Solution We multiply each term in the first binomial by each term in the second binomial.

Begin with the original expression: $(2x + 5)(3x + 4)$

Multiply $2x$ by $3x$ and 4; multiply 5
by $3x$ and 4: $= 2x \cdot 3x + 2x \cdot 4 + 5 \cdot 3x + 5 \cdot 4$

Simplify: $= 6x^2 + 8x + 15x + 20$

Combine like terms: $= 6x^2 + 23x + 20$

As with addition and subtraction of polynomials, we can organize the multiplication of binomials vertically. Click here to see Example 3 worked in a vertical format.

The acronym FOIL summarizes the process used in Example 3. Watch this animation about using the **FOIL method** to multiply two binomials.

The FOIL Method

FOIL reminds us to multiply the two **F**irst terms, the two **O**utside terms, the two **I**nside terms, and the two **L**ast terms.

Example 4 Multiplying Two Binomials Using the FOIL Method

Multiply using FOIL.

a. $(6x + 7y)(5x - 3y)$ **b.** $(4m^2 - 5n)(3m^2 - 9n)$

My video summary

Solution

a. The two first terms are $6x$ and $5x$. The two outside terms are $6x$ and $-3y$. The two inside terms are $7y$ and $5x$. The two last terms are $7y$ and $-3y$.

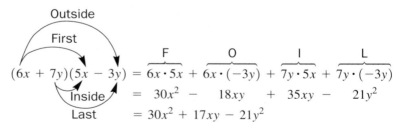

$$(6x + 7y)(5x - 3y) = \overset{F}{\overbrace{6x \cdot 5x}} + \overset{O}{\overbrace{6x \cdot (-3y)}} + \overset{I}{\overbrace{7y \cdot 5x}} + \overset{L}{\overbrace{7y \cdot (-3y)}}$$
$$= 30x^2 - 18xy + 35xy - 21y^2$$
$$= 30x^2 + 17xy - 21y^2$$

b. Try working this problem on your own. Click here to check your answer, or watch this video to see the complete solution.

CAUTION The FOIL method can be used only when multiplying two binomials.

You Try It   Work through this You Try It problem.

Work Exercises 7–13 in this textbook or in the **MyMathLab**® Study Plan.

OBJECTIVE 4:   MULTIPLY TWO BINOMIALS USING SPECIAL PRODUCT RULES

A **binomial sum** is a binomial in which the two terms are added. A **binomial difference** is a binomial in which one term is subtracted from the other. When a binomial sum and a binomial difference are made from the same two terms, they are called **conjugates** of each other. For example, the binomial sum $5x + 3y$ and the binomial difference $5x - 3y$ are conjugates of each other.

In the following example, we derive rules for three **special products**. In part a, we find the product of conjugates. In part b, we find the square of a binomial sum. In part c, we find the square of a binomial difference.

Example 5 Deriving the Special Product Rules for Binomials

Multiply.

a. $(A + B)(A - B)$ b. $(A + B)^2$ c. $(A - B)^2$

Solution   Use the FOIL method in each case.

$$
\begin{aligned}
\text{a.} \quad (A + B)(A - B) &= \overset{F}{\overbrace{A \cdot A}} + \overset{O}{\overbrace{A \cdot (-B)}} + \overset{I}{\overbrace{B \cdot A}} + \overset{L}{\overbrace{B \cdot (-B)}} \\
&= A^2 - AB + AB - B^2 \\
&= A^2 - B^2
\end{aligned}
$$

b. Write $(A + B)^2$ as $(A + B)(A + B)$. Finish multiplying. Click here for the complete solution.

c. Write $(A - B)^2$ as $(A - B)(A - B)$. Finish multiplying. Click here for the complete solution.

The result, $A^2 - B^2$, from multiplying conjugates is a called a **difference of two squares**, and the results $A^2 + 2AB + B^2$ and $A^2 - 2AB + B^2$ from squaring binomials are called **perfect square trinomials**. In general, a **trinomial** is a simplified polynomial with three terms.

Summarizing the results from Example 5 gives us the following **special product rules for binomials**.

Special Product Rules for Binomials

The product of
conjugates $(A + B)(A - B) = A^2 - B^2$ Difference of two squares

The square of
a binomial sum $(A + B)^2 = A^2 + 2AB + B^2$ Perfect square trinomial

The square of
a binomial difference $(A - B)^2 = A^2 - 2AB + B^2$ Perfect square trinomial

---

Recognizing these special cases will allow us to find their products more quickly by using the rules.

Example 6 Multiplying Binomials Using the Product of Conjugates

Multiply.

a. $(2x + 7)(2x - 7)$ **b.** $(10a^2 + b^3)(10a^2 - b^3)$

Solution

a. $(2x + 7)(2x - 7)$ is a product of conjugates. We use the rule for $(A + B)(A - B)$ with $A = 2x$ and $B = 7$.

Write the product of conjugates rule: $(A + B)(A - B) = A^2 - B^2$

Substitute $2x$ for A and 7 for B: $(2x + 7)(2x - 7) = (2x)^2 - (7)^2$

Simplify: $= 4x^2 - 49$

My video summary   **b.** Try working this problem on your own. Click here to check your answer, or watch this video to see the complete solution.

You Try It   Work through this You Try It problem.

Work Exercises 14–16 in this textbook or in the *MyMathLab*® Study Plan.

Example 7 Multiplying Binomials Using the Square of a Binomial Sum and the Square of a Binomial Difference

Multiply.

a. $(3x + 8y)^2$ **b.** $(9p^3 - 11q)^2$

Solution

a. $(3x + 8y)^2$ is the square of a binomial sum. We use the rule for $(A + B)^2$ with $A = 3x$ and $B = 8y$.

Write the square of a binomial
sum rule: $(A + B)^2 = A^2 + 2AB + B^2$

Substitute $3x$ for A and $8y$ for B: $(3x + 8y)^2 = (3x)^2 + 2(3x)(8y) + (8y)^2$

Simplify: $= 9x^2 + 48xy + 64y^2$

My video summary   **b.** $(9p^3 - 11q)^2$ is the square of a binomial difference. We use the rule for $(A - B)^2$ with $A = 9p^3$ and $B = 11q$. Try to finish this problem on your own. Click here to check your answer, or watch this video to see the complete solution.

 Be careful to avoid these two common errors:
$(A + B)^2 = A^2 + 2AB + B^2$, **not** $A^2 + B^2$
$(A - B)^2 = A^2 - 2AB + B^2$, **not** $A^2 - B^2$

You Try It   Work through this You Try It problem.

Work Exercises 17–22 in this textbook or in the *MyMathLab*® Study Plan.

OBJECTIVE 5: MULTIPLY TWO OR MORE POLYNOMIALS

We can apply what we have learned in Objective 3 to multiply any two polynomials. Watch this animation for an explanation.

Multiplying Two or More Polynomials

To multiply two polynomials, multiply each term of the first polynomial by each term of the second polynomial. Combine like terms. If there are more than two polynomials, multiply two polynomials at a time following this procedure.

Example 8 Multiplying Two Polynomials

Multiply.

a. $(3x + 5)(2x^2 + 7x + 9)$ **b.** $(y^2 + 2y - 9)(2y^2 - 4y + 7)$

Solution

a. We multiply each term in the binomial by each term in the trinomial.

$$\text{Begin with the original expression:} \quad (3x + 5)(2x^2 + 7x + 9)$$

Multiply $3x$ by $2x^2$, $7x$, and 9; multiply 5 by $2x^2, 7x$, and 9:

$$= 3x \cdot 2x^2 + 3x \cdot 7x + 3x \cdot 9 + 5 \cdot 2x^2 + 5 \cdot 7x + 5 \cdot 9$$

$$\text{Simplify:} \quad = 6x^3 + 21x^2 + 27x + 10x^2 + 35x + 45$$

$$\text{Combine like terms:} \quad = 6x^3 + 31x^2 + 62x + 45$$

Click here to see this problem solved by multiplying vertically.

b. Try to solve this problem on your own. Click here to check your answer, or watch this video to see detailed solutions.

You Try It Work through this You Try It problem.

Work Exercises 23–26 in this textbook or in the *MyMathLab*® Study Plan.

To multiply three or more polynomials, we multiply two polynomials at a time.

Example 9 Multiplying Three or More Polynomials

Multiply.

a. $2x^2(5x + 7)(x - 8)$ **b.** $(x + 5)^3$

Solution

a. Because multiplication is associative, we can group the multiplication in the order we wish. For this problem, we multiply the two binomials and then distribute the monomial through the resulting product.

$$\text{Begin with the original expression:} \quad 2x^2(5x + 7)(x - 8)$$

$$\text{Multiply } 5x + 7 \text{ and } x - 8: \quad = 2x^2(5x^2 - 33x - 56)$$

$$\text{Distribute } 2x^2: \quad = 10x^4 - 66x^3 - 112x^2$$

Watch this interactive video to see the complete solution process.

b. $(x + 5)^3$ is an exponential expression with the base $x + 5$ and the exponent 3, so we multiply $x + 5$ three times.

$$\begin{aligned} \text{Begin with the original expression:} \quad & (x + 5)^3 \\ \text{Group two factors to multiply:} \quad &= (x + 5)^2(x + 5) \\ \text{Multiply out } (x + 5)^2: \quad &= (x^2 + 10x + 25)(x + 5) \end{aligned}$$

Finish the problem by multiplying each term of the trinomial by each term of the binomial and then combining like terms. Click here to check your answer, or watch this interactive video for the complete solution.

You Try It   Work through this You Try It problem.

Work Exercises 27–32 in this textbook or in the **MyMathLab**® Study Plan.

In some cases, special product rules for binomials can be used to multiply two trinomials.

Example 10   Using Special Product Rules to Multiply Trinomials

 Use special product rules to multiply.

a. $(x + 8 + 3y)(x + 8 - 3y)$ **b.** $(2x + 3y + 5)^2$

Solution

a. Grouping the first two terms within each pair of parentheses, we can use the rule for a product of conjugates.

$$(x + 8 + 3y)(x + 8 - 3y) = [(x + 8) + 3y][(x + 8) - 3y]$$

We write the problem in the form $(A + B)(A - B)$ with $A = x + 8$ and $B = 3y$.

$$\begin{aligned} \text{Write the special} \\ \text{product rule:} \quad & (A + B)(A - B) = A^2 - B^2 \\ \text{Substitute } (x + 8) \\ \text{for } A \text{ and } 3y \text{ for } B: \quad & [(x + 8) + 3y][(x + 8) - 3y] = (x + 8)^2 - (3y)^2 \end{aligned}$$

Simplify the resulting expression to finish the problem. Click here to check your answer, or watch this interactive video for the complete solution.

b. Grouping the first two terms, we can use the rule for the square of a binomial sum.

$$(2x + 3y + 5)^2 = [(2x + 3y) + 5]^2$$

We write the problem in the form $(A + B)^2$ with $A = 2x + 3y$ and $B = 5$. Try to finish this problem on your own. Click here to check your answer, or watch this interactive video to see the complete solution.

You Try It   Work through this You Try It problem.

Work Exercises 33–34 in this textbook or in the **MyMathLab**® Study Plan.

OBJECTIVE 6:   MULTIPLY POLYNOMIAL FUNCTIONS

Now that we know how to multiply polynomials, we can also multiply polynomial functions.

Definition   The Product of Functions

If f and g represent functions, then the **product of f and g** is
$(f \cdot g)(x) = f(x) \cdot g(x)$.

Example 11 Multiplying Polynomial Functions

For $P(x) = 5x - 2$ and $Q(x) = 4x - 9$, find $(P \cdot Q)(x)$.

Solution

Define the original problem: $(P \cdot Q)(x) = P(x) \cdot Q(x)$

Substitute for the function
notation: $= (5x - 2)(4x - 9)$

$$\overset{F}{\overbrace{}} \quad \overset{O}{\overbrace{}} \quad \overset{I}{\overbrace{}} \quad \overset{L}{\overbrace{}}$$

Use the FOIL method: $= 5x \cdot 4x + 5x \cdot (-9) + (-2) \cdot 4x + (-2) \cdot (-9)$

Simplify the multiplication: $= 20x^2 - 45x - 8x + 18$

Combine like terms: $= 20x^2 - 53x + 18$

You Try It   Work through this You Try It problem.

Work Exercises 35–36 in this textbook or in the *MyMathLab*® Study Plan.

4.3 Exercises

In Exercises 1–3, multiply the monomials.

You Try It   **1.** $(5x^3)(6x^4)$

2. $\left(-\dfrac{3}{4}a^5b^3\right)\left(\dfrac{2}{5}a^2b^6\right)$

3. $(8x^3yz^5)(3xy^4z^2)$

You Try It   In Exercises 4–6, multiply the monomial and the polynomial.

4. $7x^3(5x^2 - 9)$

5. $3.4a^3b(2a^2b^3 + 7ab)$

6. $-2x^3y^2z(6xy - 12x^2yz^3 + 5y^2z^4)$

You Try It   In Exercises 7–13, multiply the binomials.

7. $(x + 10)(x + 3)$

8. $(y + 9)(y - 4)$

9. $(3x + 2)(2x + 5)$

10. $(6x - 5y)(2x + 7y)$

11. $(3m^2 - 2n^2)(m^2 - 4n^2)$

12. $(a - 4)(a^2 - 5)$

13. $\left(x + \dfrac{1}{3}\right)\left(x - \dfrac{1}{2}\right)$

You Try It In Exercises 14–22, multiply using the special product rules for binomials.

14. $(x - 10)(x + 10)$ 　　　　**15.** $(7x + 9)(7x - 9)$ 　　　　**16.** $(3m^2 + 2n)(3m^2 - 2n)$

17. $(x - 8)^2$ 　　　　**18.** $(11y - 4)^2$ 　　　　**19.** $(7p - 2q)^2$

20. $(y + 5)^2$ 　　　　**21.** $(4z + 5)^2$ 　　　　**22.** $(10x^2 + 11y)^2$

You Try It In Exercises 23–34, multiply the polynomials.

23. $(x + 4)(x^2 - 3x + 5)$ 　　　　**24.** $(3m + 2)(9m^2 - 11m + 1)$

25. $(4a + b)(3a^2 - 5ab + 2b^2)$ 　　　　**26.** $(x^2 - 3x + 7)(x^2 + x - 5)$

You Try It
27. $2x^2(5x + 7)(x - 8)$ 　　　　**28.** $(2y + 3)(2y - 3)(4y^2 + 9)$

29. $(x + 1)(x + 2)(x + 3)$ 　　　　**30.** $(x - 3)(x^2 + 3x + 9)$

31. $(x + 4)^3$ 　　　　**32.** $(x - 3)^4$

33. $(3x + 2y - 5)(3x + 2y + 5)$ 　　　　**34.** $(x + y + 7)^2$

In Exercises 35–36, find $(P \cdot Q)(x)$.

You Try It **35.** $P(x) = 2x - 5;$ 　$Q(x) = x + 8$ 　　　　**36.** $P(x) = x - 4;$ 　$Q(x) = x^2 + 3x - 5$

4.4 Dividing Polynomials

THINGS TO KNOW

Before working through this section, be sure you are familiar with the following concepts:

VIDEO　　　ANIMATION　　　INTERACTIVE

You Try It 1. Perform Operations on Real Numbers
(Section R.2, **Objective 1**)　　　

You Try It 2. Multiply Two or More Polynomials
(Section 4.3, **Objective 5**)　　　　　　

OBJECTIVES

1 Divide a Polynomial by a Monomial

2 Divide Polynomials Using Long Division

3 Divide Polynomials Using Synthetic Division

4 Divide Polynomial Functions

5 Use the Remainder and Factor Theorems

OBJECTIVE 1:   DIVIDE A POLYNOMIAL BY A MONOMIAL

In Section 4.1, we used the quotient rule to simplify exponential expressions such as $\frac{15x^4y^2z^5}{5x^3yz^3} = 3xyz^2$. To do this, we divide the coefficients and apply the quotient rule for like bases. In Section 4.2 we said a monomial is a simplified term in which all variables are raised to non-negative integer powers. Therefore, we have already seen how to divide one monomial by another. Now let's apply this knowledge and learn how to divide a polynomial by a monomial.

Recall that when we add or subtract fractions with a common denominator, we add or subtract the numerators and keep the common denominator.

$$\frac{A}{C} + \frac{B}{C} = \frac{A + B}{C} \quad \text{or} \quad \frac{A}{C} - \frac{B}{C} = \frac{A - B}{C}$$

Since a polynomial is a finite sum of monomials, working in the other direction (right to left), allows us to see the method by which we can divide a polynomial by a monomial.

Dividing a Polynomial by a Monomial

To divide a polynomial by a monomial, divide each term of the polynomial by the monomial and simplify the quotients using the rules for exponents.

$$\frac{A + B}{C} = \frac{A}{C} + \frac{B}{C} \quad \text{or} \quad \frac{A - B}{C} = \frac{A}{C} - \frac{B}{C}$$

Example 1 Dividing a Polynomial by a Monomial

Divide $15x^7 + 20x^6 - 5x^4 + 35x^3$ by $5x^3$.

Solution

To divide a polynomial by a monomial, divide each term of the polynomial by the monomial and simplify the quotients using the rules for exponents.

Begin with the original expression: $\dfrac{15x^7 + 20x^6 - 5x^4 + 35x^3}{5x^3}$

Divide each term of the polynomial by the monomial: $= \dfrac{15x^7}{5x^3} + \dfrac{20x^6}{5x^3} - \dfrac{5x^4}{5x^3} + \dfrac{35x^3}{5x^3}$

Simplify quotients using the rules for exponents: $= 3x^4 + 4x^3 - x + 7$

You Try It   Work through this You Try It problem.

Work Exercises 1–2 in this textbook or in the ***MyMathLab***® Study Plan.

Example 2 Dividing a Polynomial by a Monomial

 Divide $\dfrac{9a^3b^4 - 12a^2b^5 + 6ab^7}{6a^2b^3}$.

My video summary

Solution

Begin with the original expression: $\dfrac{9a^3b^4 - 12a^2b^5 + 6ab^7}{6a^2b^3}$

Divide each term of the polynomial by the monomial: $= \dfrac{9a^3b^4}{6a^2b^3} - \dfrac{12a^2b^5}{6a^2b^3} + \dfrac{6ab^7}{6a^2b^3}$

Try to finish this problem on your own. Click here to check your answer, or watch this video for the complete solution.

You Try It Work through this You Try It problem.

Work Exercises 3–4 in this textbook or in the **MyMathLab**® Study Plan.

OBJECTIVE 2: DIVIDE POLYNOMIALS USING LONG DIVISION

When dividing a polynomial by a polynomial, we use long division. **Polynomial long division** follows the same approach as long division for real numbers. Click here to review an example of how to divide two real numbers using long division.

We can think of long division as a four-step process: **divide, multiply, subtract,** and **drop**. The process is repeated until the remainder can no longer be divided, which occurs when the degree of the remainder is less than the degree of the divisor. The result of this process can be written in the form

$$\frac{\text{dividend}}{\text{divisor}} = \text{quotient} + \frac{\text{remainder}}{\text{divisor}}.$$

We can check the result of long division by noting that

$$\text{dividend} = (\text{divisor})(\text{quotient}) + \text{remainder}.$$

CAUTION

Before performing polynomial long division, write the dividend and divisor in descending order. If any powers are missing, then insert them with a coefficient of 0 as a placeholder.

We demonstrate long division involving polynomials in the following example.

Example 3 Dividing Polynomials Using Long Division

Divide $\left(8x^2 - 6x - 38\right) \div (2x - 5)$.

Solution

First, we note that the dividend is $8x^2 - 6x - 38$ and the divisor is $2x - 5$. Next, we apply the four-step process of divide, multiply, subtract, and drop until the remainder can no longer be divided.

Divide the first term of the dividend by the first term of the divisor. So, we divide $8x^2$ by $2x$. Since $\dfrac{8x^2}{2x} = 4x$, $4x$ is the first term in the quotient.

$$\begin{array}{r} 4x \\ 2x - 5 \overline{)8x^2 - 6x - 38} \end{array}$$

Multiply the result of the division, $4x$, by the divisor, $2x - 5$, to get $4x(2x - 5) = 8x^2 - 20x$. Write this result under the dividend, making sure to line up like terms.

$$
\begin{array}{r}
4x \\
2x - 5 \overline{)8x^2 - 6x - 38} \\
8x^2 - 20x
\end{array}
$$

Subtract the result of the multiplication from the first two terms of the dividend. So, we subtract $8x^2 - 20x$ from $8x^2 - 6x$. To do the subtraction, we can change the sign of each term in $\left(8x^2 - 20x\right)$ and add.

$$
\begin{array}{r}
4x \\
2x - 5 \overline{)8x^2 - 6x - 38} \\
-\left(8x^2 - 20x\right)
\end{array}
\quad \rightarrow \quad
\begin{array}{r}
4x \\
2x - 5 \overline{)8x^2 - 6x - 38} \\
\text{add} \rightarrow \quad -8x^2 + 20x \\
\hline
14x
\end{array}
$$

Drop down the next term, -38, and repeat the process.

$$
\begin{array}{r}
4x \\
2x - 5 \overline{)8x^2 - 6x - 38} \\
-\left(8x^2 - 20x\right) \\
\hline
14x - 38
\end{array}
$$

Divide $14x$ by $2x$. Since $\dfrac{14x}{2x} = 7$, the next term in the quotient is 7.

$$
\begin{array}{r}
4x + 7 \\
2x - 5 \overline{)8x^2 - 6x - 38} \\
-\left(8x^2 - 20x\right) \\
\hline
14x - 38
\end{array}
$$

Multiply 7 by $2x - 5$ to get $7(2x - 5) = 14x - 35$.

$$
\begin{array}{r}
4x + 7 \\
2x - 5 \overline{)8x^2 - 6x - 38} \\
-\left(8x^2 - 20x\right) \\
\hline
14x - 38 \\
14x - 35
\end{array}
$$

Subtract $14x - 35$ from $14x - 38$. We can do this by changing the sign of each term in $(14x - 35)$ and adding.

$$
\begin{array}{r}
4x + 7 \\
2x - 5 \overline{)8x^2 - 6x - 38} \\
-\left(8x^2 - 20x\right) \\
\hline
14x - 38 \\
-(14x - 35)
\end{array}
\quad \rightarrow \quad
\begin{array}{r}
4x + 7 \\
2x - 5 \overline{)8x^2 - 6x - 38} \\
-\left(8x^2 - 20x\right) \\
\hline
14x - 38 \\
\text{add} \rightarrow \quad -14x + 35 \\
\hline
-3
\end{array}
$$

Since the degree of the remainder is less than the degree of the divisor, our division is complete. The quotient is $4x + 7$, and the remainder is -3.

$$
\frac{\text{dividend}}{\text{divisor}} = \text{quotient} + \frac{\text{remainder}}{\text{divisor}} \quad \longrightarrow \quad \frac{8x^2 - 6x - 38}{2x - 5} = 4x + 7 - \frac{3}{2x - 5}
$$

Check: We can check our result by showing that
dividend $=$ (divisor)(quotient) $+$ remainder.

$$\underbrace{(2x - 5)}_{\text{Divisor}}\underbrace{(4x + 7)}_{\text{Quotient}} + \underbrace{(-3)}_{\text{Remainder}} = 8x^2 + 14x - 20x - 35 - 3 = \underbrace{8x^2 - 6x - 38}_{\text{Dividend}}$$

Our result checks, so the division is correct.

You Try It Work through this You Try It problem.

Work Exercises 5–8 in this textbook or in the **MyMathLab**® Study Plan.

Example 4 Dividing Polynomials Using Long Division

Divide $6x^2 + 8x^3 - 149x + 105$ by $x + 5$.

Solution First, we rewrite the dividend as $8x^3 + 6x^2 - 149x + 105$ so that the powers are in descending order. The divisor is $x + 5$. Next, we apply the four-step process of divide, multiply, subtract, and drop until the remainder can no longer be divided.

Divide the first term of the dividend by the first term of the divisor. So, we divide $8x^3$ by x. Since $\dfrac{8x^3}{x} = 8x^2$, $8x^2$ is the first term in the quotient.

$$\begin{array}{r} 8x^2 \\ x + 5 \overline{)8x^3 + 6x^2 - 149x + 105} \end{array}$$

Multiply the result of the division, $8x^2$, by the divisor, $x + 5$, to get $8x^2(x + 5) = 8x^3 + 40x^2$. Write this result under the dividend and line up like terms.

$$\begin{array}{r} 8x^2 \\ x + 5 \overline{)8x^3 + 6x^2 - 149x + 105} \\ 8x^3 + 40x^2 \end{array}$$

Subtract the result of the multiplication from the first two terms of the dividend. So, we subtract $8x^3 + 40x^2$ from $8x^3 + 6x^2$. We can do this by changing the sign of each term in $\left(8x^3 + 40x^2\right)$ and adding.

$$\begin{array}{r} 8x^2 \\ x + 5 \overline{)8x^3 + 6x^2 - 149x + 105} \\ -\left(8x^3 + 40x^2\right) \\ \hline -34x^2 \end{array}$$

Drop down the next term, $-149x$, and repeat the process.

$$\begin{array}{r} 8x^2 \\ x + 5 \overline{)8x^3 + 6x^2 - 149x + 105} \\ -\left(8x^3 + 40x^2\right) \\ \hline -34x^2 - 149x \end{array}$$

Try finishing the problem on your own. Click here to check your answer, or watch this video to see the complete solution.

Check: We can check our result by showing that
dividend $=$ (divisor)(quotient) $+$ remainder. Click here to see the check.

You Try It   **Work through this You Try It problem.**

Work Exercises 9–10 in this textbook or in the *MyMathLab*® Study Plan.

Example 5 Dividing Polynomials Using Long Division

 Divide $3x^4 + 5x^3 - 6x - 15$ by $x^2 - 2$.

My video summary

Solution The dividend is $3x^4 + 5x^3 - 6x - 15$ and the divisor is $x^2 - 2$.
Before we apply our four-step process, we write both polynomials in standard
form, inserting coefficients of 0 for any missing terms.

$$x^2 + 0x - 2 \overline{)3x^4 + 5x^3 + 0x^2 - 6x - 15}$$

Try finishing this problem on your own. Click here to check your answer, or watch
this video for the complete solution.

You Try It   **Work through this You Try It problem.**

Work Exercises 11–14 in this textbook or in the *MyMathLab*® Study Plan.

OBJECTIVE 3:   DIVIDE POLYNOMIALS USING SYNTHETIC DIVISION

If we divide a polynomial by a binomial of the form $x - c$, then a shortcut method
called **synthetic division** can be used instead of long division. Synthetic division
removes the repetition involved with long division. If we use zeros as placeholders for missing powers and line up like terms, we only need to work with the
coefficients because the position of the coefficient will determine the power of
the term.

Watch this animation to see how synthetic division works as compared to
polynomial long division.

We illustrate the synthetic division process in the following example.

Example 6 Dividing Polynomials Using Synthetic Division

Divide $2x^4 + 9x^3 - 12x + 1$ by $x + 5$ using synthetic division.

Solution

Step 1. Write both polynomials in standard form. Both polynomials are already
in standard form, so we move on to step 2.

Step 2. Rewrite the divisor as a binomial of the form $x - c$.
Since $x + 5 = x - (-5), c = -5$.

Step 3. Write down c and the coefficients of the dividend. If there are powers
"missing," insert a 0 coefficient as a placeholder for the missing term(s).

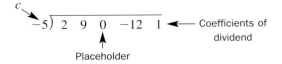

Step 4. Leave some space under the coefficients, draw a horizontal line, and drop down the leading coefficient under the line.

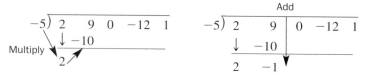

$$
\begin{array}{r|rrrrr}
-5) & 2 & 9 & 0 & -12 & 1 \\
& \downarrow \\
\hline
& 2
\end{array}
\quad
\begin{array}{l}
\text{Row 1} \\
\text{Row 2} \\
\text{Row 3}
\end{array}
$$

Step 5. Multiply c by the entry in Row 3 and put the result in the next position to the right in Row 2. Add the values from Rows 1 and 2 to get the next entry in Row 3.

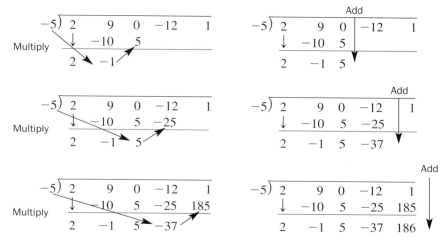

Step 6. Continue with step 5 until Row 3 is complete.

The quotient and remainder are determined by looking at the completed Row 3 of the synthetic division. Because the divisor is a polynomial of degree 1, the quotient will have a degree that is one less than the dividend. Since the dividend has a degree of 4, the quotient will have a degree of 3. The last entry in the last row is the remainder. The remaining entries in that row are the coefficients of the quotient when written in standard form.

$$
\begin{array}{r|rrrrr}
-5) & 2 & 9 & 0 & -12 & 1 \\
& \downarrow & -10 & 5 & -25 & 185 \\
\hline
& 2 & -1 & 5 & -37 & 186
\end{array}
$$

$$\underbrace{2 \quad -1 \quad 5 \quad -37}_{\text{Coefficients of quotient}} \quad \underbrace{186}_{\text{Remainder}}$$

Here, the quotient is $2x^3 - 1x^2 + 5x - 37$, and the remainder is 186.

$$\frac{2x^4 + 9x^3 - 12x + 1}{x + 5} = 2x^3 - x^2 + 5x - 37 + \frac{186}{x + 5}$$

Click here to see how to solve this same problem using long division.

Example 7 Dividing Polynomials Using Synthetic Division

 Divide $2x^4 - 3x^2 + 5x - 30$ by $x - 2$ using synthetic division.

Solution The divisor is a binomial of the form $x - c$ where $c = 2$. We start the synthetic division process as follows:

$$2\overline{)\,2 \quad 0 \quad -3 \quad 5 \quad -30}$$

⟵ Products

$$2$$

⟵ Sums

Complete this process on your own. Click here to check your answer, or watch this video for a detailed solution.

You Try It Work through this You Try It problem.

Work Exercises 15–22 in this textbook or in the *MyMathLab* ® Study Plan.

Example 8 Dividing Polynomials Using Synthetic Division

Divide $4x^3 - 8x^2 + 7x - 4$ by $x - \dfrac{1}{2}$ using synthetic division.

Solution The divisor is a binomial of the form $x - c$ with $c = \dfrac{1}{2}$. Complete the synthetic division on your own. Click here to check your answer, or watch this video for the full solution.

You Try It Work through this You Try It problem.

Work Exercises 23–24 in this textbook or in the *MyMathLab* ® Study Plan.

 Synthetic division can only be used when dividing a polynomial by a binomial of the form $x - c$.

OBJECTIVE 4: DIVIDE POLYNOMIAL FUNCTIONS

Now that we know how to divide polynomials, we can also divide polynomial functions.

> **Definition   The Quotient of Functions**
>
> Let f and g represent two functions such that the **quotient of f and g** is
> $\left(\dfrac{f}{g}\right)(x) = \dfrac{f(x)}{g(x)}$, as long as $g(x) \neq 0$.

Example 9 Dividing Polynomial Functions

For $P(x) = 15x^3 + 41x^2 + 4x + 3$ and $Q(x) = 5x + 2$, find $\left(\dfrac{P}{Q}\right)(x)$. State any values that cannot be included in the domain of $\left(\dfrac{P}{Q}\right)(x)$. (Note that $Q(x)$ cannot be 0.)

Solution

Begin with the original problem: $\left(\dfrac{P}{Q}\right)(x) = \dfrac{P(x)}{Q(x)}$

Substitute for the function notation: $= \dfrac{15x^3 + 41x^2 + 4x + 3}{5x + 2}$

Complete this problem using polynomial long division. To find the values that cannot be included in the domain, set $Q(x)$ equal to 0 and solve. Click here to check your answer, or watch this video for a detailed solution.

You Try It   Work through this You Try It problem.

Work Exercises 25–26 in this textbook or in the **MyMathLab**® Study Plan.

OBJECTIVE 5:   USE THE REMAINDER AND FACTOR THEOREMS

The Remainder Theorem

In Example 6, we divided $2x^4 + 9x^3 - 12x + 1$ by $x + 5$ to find a remainder of 186. Using $f(x) = 2x^4 + 9x^3 - 12x + 1$ and $c = -5$, we find $f(c) = f(-5) = 2(-5)^4 + 9(-5)^3 - 12(-5) + 1 = 186$, which is the remainder. Similarly, in Example 7, we divided $2x^4 - 3x^2 + 5x - 30$ by $x - 2$ to find a remainder of 0. Using $f(x) = 2x^4 - 3x^2 + 5x - 30$ and $c = 2$, we see that $f(c) = f(2) = 2(2)^4 - 3(2)^2 + 5(2) - 30 = 0$, which is the remainder.

These results suggest that if we divide a polynomial function by the binomial $x - c$, then the remainder will be $f(c)$. This result is true in general and is known as the **Remainder Theorem**.

> **The Remainder Theorem**
>
> If a polynomial $f(x)$ is divided by $x - c$, then the remainder is $f(c)$.

Click here to see a justification of the theorem.

Example 10 Using the Remainder Theorem

Use the remainder theorem to find the remainder when $f(x)$ is divided by $x - c$,

a. $f(x) = 5x^4 - 8x^2 + 3x - 1$; $x - 2$
b. $f(x) = 3x^3 + 5x^2 - 5x - 6$; $x + 2$

Solution

a. We could use polynomial long division or synthetic division to find the remainder, but the remainder theorem says that the remainder must be $f(2)$. Since $f(2) = 53$, the remainder is 53. Click here to see the work.

b. To find the remainder when $f(x) = 3x^3 + 5x^2 - 5x - 6$ is divided by $x + 2 = x - (-2)$, we evaluate $f(-2)$. Since $f(-2) = 0$, the remainder is 0. Click here to see the work.

You Try It Work through this You Try It problem.

Work Exercises 27–30 in this textbook or in the **MyMathLab**® Study Plan.

The Factor Theorem

In Example 10, we saw that when $f(x) = 3x^3 + 5x^2 - 5x - 6$ is divided by $x + 2$, the remainder is 0. The remainder of 0 means that $x + 2$ divides $f(x)$ evenly, and the binomial $x + 2$ is a factor of the polynomial function $f(x)$. From the remainder theorem, we can develop another important concept known as the **Factor Theorem**.

The Factor Theorem

For a polynomial function $f(x)$,

1. If $f(c) = 0$, then $x - c$ is a factor of $f(x)$.

2. If $x - c$ is a factor of $f(x)$, then $f(c) = 0$.

Example 11 Using the Factor Theorem

My video summary

Use the factor theorem to determine if $x - c$ is a factor of $f(x)$. If so, use synthetic division to write $f(x)$ in the form $f(x) = (x - c) \cdot q(x)$.

a. $f(x) = 2x^3 - 5x^2 + 7x - 4;\quad x - 1$

b. $f(x) = x^4 - 3x^3 + 15x + 45;\quad x + 3$

Solution

a. The divisor is $x - 1$, so we have $c = 1$. Evaluate $f(1)$.

$$f(1) = 2(1)^3 - 5(1)^2 + 7(1) - 4 = 2 - 5 + 7 - 4 = 0$$

Since $f(1) = 0$, the factor theorem states that $x - 1$ is a factor of $f(x)$. Use synthetic division to find the quotient when $f(x)$ is divided by $x - 1$, and then write $f(x)$ in the form $f(x) = (x - 1) \cdot q(x)$.

Click here to check your answer, or watch this video for a complete solution.

b. The divisor is $x + 3 = x - (-3)$, so we have $c = -3$. Evaluate $f(-3)$.

$$f(-3) = (-3)^4 - 3(-3)^3 + 15(-3) + 45 = 81 + 81 - 45 + 45 = 162$$

Since $f(-3) \neq 0$, $x + 3$ is not a factor of $f(x)$.

You Try It Work through this You Try It problem.

Work Exercises 31–34 in this textbook or in the **MyMathLab**® Study Plan.

4.4 Exercises

You Try It

In Exercises 1–4, divide.

You Try It

1. $\dfrac{20x^7 + 12x^5 - 8x^4 + 32x^2}{4x^2}$

2. $\dfrac{8y^5 - 3y^4 + 2y^2}{-4y^3}$

You Try It

3. $\dfrac{8x^4y^2 + x^3y^3 - 2x^2y^4}{2x^2y^2}$

4. $\dfrac{a^5b^3 - 2a^4b + 3a^2b^2}{3a^3b^2}$

In Exercises 5–14, divide using long division.

You Try It

5. $\dfrac{x^2 + 20x + 91}{x + 7}$

6. $\dfrac{10x^2 + 27x - 22}{2x + 7}$

You Try It

7. $\dfrac{2x^3 - 5x^2 + 3x + 6}{2x - 3}$

8. $\dfrac{6a^3 + 7a^2 + 12a + 3}{3a - 1}$

You Try It

9. $\dfrac{2x^2 + x^3 - 9x - 4}{x + 5}$

10. $\dfrac{23x^2 - 13x^3 + 6x^4 - 57x + 30}{3x - 5}$

You Try It

11. $\dfrac{3x^4 - 11x^3 + 9x^2 + 22x - 33}{x^2 - 2}$

12. $\dfrac{2x^5 + 3x^4 + 5x^3 + 6x^2 + 2x}{x^2 + 2}$

13. $\dfrac{9x^4 - 2x^2 + 2x + 9}{x^2 + 5}$

14. $\dfrac{6x^6 - x^4 + 6x^3 - x^2 + 2x}{2x^3 - x + 2}$

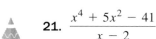
You Try It

In Exercises 15–24, divide using synthetic division.

15. $x^2 - x - 17$ divided by $x - 5$

16. $x^2 + \dfrac{16}{3}x - 4$ divided by $x + 6$

17. $\left(4x^3 + 5x - 7\right) \div (x + 2)$

18. $\left(3x^3 + 11x^2 + 2x + 31\right) \div (x + 4)$

19. $\dfrac{x^3 - 12x + 9}{x - 3}$

20. $\dfrac{7x^4 - 12x^3 - 8x^2 + 7x + 2}{x - 2}$

You Try It

21. $\dfrac{x^4 + 5x^2 - 41}{x - 2}$

22. $\dfrac{x^4 + 15x^2 - 16}{x + 1}$

You Try It

23. $\dfrac{15x^3 + 14x^2 - 43x + 14}{x - \dfrac{2}{5}}$

24. $\dfrac{3x^3 - 20x^2 - x - 7}{x + \dfrac{1}{3}}$

You Try It

In Exercises 25–26, find $\left(\dfrac{P}{Q}\right)(x)$ and state any values that cannot be included in the domain of $\left(\dfrac{P}{Q}\right)(x)$.

25. $P(x) = 3x^2 - 7x + 5;\quad Q(x) = x - 2$

26. $P(x) = 6x^3 + 11x^2 - 9x - 2;\quad Q(x) = 3x + 1$

In Exercises 27–30, use the remainder theorem to find the remainder if $f(x)$ is divided by $x - c$.

27. $f(x) = 3x^2 - 7x + 12; \quad x - 1$

28. $f(x) = 3x^2 - x - 14; \quad x + 2$

You Try It **29.** $f(x) = x^4 + 5x^3 - 5x + 29; \quad x + 3$

30. $f(x) = 20x^3 - 13x^2 + 14x - 3; \quad x - \dfrac{1}{4}$

In Exercises 31–34, use the factor theorem to determine if $x - c$ is a factor of $f(x)$. If so, use synthetic division to write $f(x)$ in the form $f(x) = (x - c) \cdot q(x)$.

You Try It

31. $f(x) = x^2 - 11x + 24; \quad x - 8$

32. $f(x) = x^3 + 8; \quad x + 2$

33. $f(x) = 2x^3 - 5x^2 + 7x - 4; \quad x + 1$

34. $f(x) = 4x^3 - 5x^2 - 8x + 4; \quad x - 2$

Factoring

CHAPTER FIVE CONTENTS

5.1 **Greatest Common Factor and Factoring by Grouping**

5.2 **Factoring Trinomials**

5.3 **Special-Case Factoring; A General Factoring Strategy**

5.4 **Polynomial Equations and Models**

5.1 Greatest Common Factor and Factoring by Grouping

THINGS TO KNOW

Before working through this section, be sure you are familiar with the following concepts:

		VIDEO	ANIMATION	INTERACTIVE

You Try It

1. Identify and Use the Properties of Real Numbers (Section R.2, **Objective 3**)

You Try It

2. Multiply a Polynomial by a Monomial (Section 4.3, **Objective 2**)

You Try It

3. Divide a Polynomial by a Monomial (Section 4.4, **Objective 1**)

OBJECTIVES

1 Factor Out the Greatest Common Factor from a Polynomial

2 Factor by Grouping

OBJECTIVE 1: FACTOR OUT THE GREATEST COMMON FACTOR FROM A POLYNOMIAL

Factoring a polynomial is the reverse process of multiplying polynomials. When a polynomial is written as an equivalent expression that is a product of polynomials, we say that the polynomial has been **factored** or written in **factored form**. In Example 9 of Section 4.3, we multiplied $2x^2, 5x + 7$, and $x - 8$ to result in the trinomial $10x^4 - 66x^3 - 112x^2$. Reversing the process, we factor $10x^4 - 66x^3 - 112x^2$ into the product $2x^2(5x + 7)(x - 8)$.

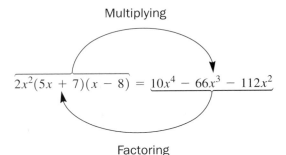

Multiplying

$$2x^2(5x + 7)(x - 8) = 10x^4 - 66x^3 - 112x^2$$

Factoring

Therefore, $2x^2, 5x + 7$, and $x - 8$ are **factors** of $10x^4 - 66x^3 - 112x^2$.

The word *factor* can be used as a noun or verb. As a noun, a *factor* is one of the numbers or expressions being multiplied to form a product. As a verb, *factor* means to rewrite the polynomial as a product.

In this section, we present procedures for factoring polynomials. We will always **factor over the integers**, which means that all coefficients in both the original polynomial and its factors will be integers.

When factoring a polynomial, we first find the *greatest common factor* of its terms.

Definition   Greatest Common Factor

The **greatest common factor (GCF)** of a polynomial is the expression with the largest coefficient and highest degree that divides each term of the polynomial evenly.

Consider the binomial $10x^3 + 15x^2$. Looking at the coefficients, 10 and 15, 5 is the largest integer that divides both numbers evenly. So, 5 is the coefficient part of the GCF. Looking at the exponential expressions x^3 and x^2, x^2 is the largest power of x that divides both expressions evenly. So, x^2 is the variable part of the GCF. Putting the two parts together, $5x^2$ is the GCF of the binomial $10x^3 + 15x^2$.

Note that if a variable appears in all terms of a polynomial, then the lowest power of that variable in the polynomial will be a factor of the GCF. Looking at our binomial $10x^3 + 15x^2$, the variable x appears in both terms, and the lowest power of x is x^2. So, x^2 is a factor of the GCF. A variable that does not appear in all terms of the polynomial will not be in the GCF.

Now that we know the GCF, we can factor it out of the binomial. We write each term of the binomial as a product that includes the factor $5x^2$.

$$10x^3 + 15x^2 = 5x^2 \cdot 2x + 5x^2 \cdot 3$$

Reversing the distributive property, we factor $5x^2$ from each term.

$$5x^2 \cdot 2x + 5x^2 \cdot 3 = 5x^2(2x + 3)$$

So, $10x^3 + 15x^2 = 5x^2(2x + 3)$.

Notice that the final result is the GCF times a binomial. This is expected since the original expression was a binomial. When the GCF is factored from a polynomial, the polynomial factor of the product will have the same number of terms as the original polynomial. Do you see why?

Let's generalize this example into the following steps for factoring out the GCF from a polynomial.

Steps for Factoring Out the Greatest Common Factor from a Polynomial

Step 1. Find the GCF of all terms in the polynomial.

Step 2. Write each term as a product that includes the GCF.

Step 3. Use the distributive property in reverse to factor out the GCF.

Step 4. Check the answer. The terms of the polynomial factor should have no more common factors. Multiplying the answer back out should give the original polynomial.

Example 1 Factoring Out the Greatest Common Factor from a Polynomial

Factor out the GCF.

a. $24m^5 + 32m^3 + 8m^2$ **b.** $18x^3y^5 - 36x^2y^7 + 27xy^9$

Solution We follow the four-step process.

a. **Step 1.** We have three terms: $24m^5$, $32m^3$, and $8m^2$. The largest integer that divides evenly into the three coefficients 24, 32, and 8 is 8. The variable m appears in all three terms and its smallest power is m^2. So, the GCF is $8m^2$.

Step 2. We write each term as the product of $8m^2$ and another factor. Since the third term $8m^2$ is the GCF, we write it multiplied by 1.

$$24m^5 + 32m^3 + 8m^2 = 8m^2 \cdot 3m^3 + 8m^2 \cdot 4m + 8m^2 \cdot 1$$

Step 3. Factor out $8m^2$ using the distributive property in reverse.

$$8m^2 \cdot 3m^3 + 8m^2 \cdot 4m + 8m^2 \cdot 1 = 8m^2(3m^3 + 4m + 1)$$

Step 4. The terms of the polynomial factor $3m^3 + 4m + 1$ have no common factors, so our GCF should be correct. Multiply the final answer back out.

$$8m^2(3m^3 + 4m + 1) = 8m^2 \cdot 3m^3 + 8m^2 \cdot 4m + 8m^2 \cdot 1$$
$$= 24m^5 + 32m^3 + 8m^2$$

This is the original polynomial, so our answer checks.
So, $24m^5 + 32m^3 + 8m^2 = 8m^2(3m^3 + 4m + 1)$.

My video summary

 b. Try factoring out the GCF on your own. Click here to check your answer, or watch this video for a complete solution.

⬦**CAUTION** When checking answers after factoring out the GCF, multiplying is not enough. We must also make sure that the terms of the polynomial factor have no more common factors. For example, consider $6x + 12 = 2(3x + 6)$. The GCF has not been factored out of the binomial correctly because the terms $3x$ and 6 still have a common factor of 3. However, multiplying the right side results in the left side, so the problem appears to check. The correct answer is $6x + 12 = 6(x + 2)$.

You Try It   **Work through this You Try It problem.**

Work Exercises 1–8 in this textbook or in the *MyMathLab*® Study Plan.

When the leading coefficient of a polynomial is negative, we may wish to factor out the negative sign with the GCF.

Example 2 Factoring Out a Negative Sign with the GCF

Factor out the negative sign with the GCF. $-25x^4 + 15x$

Solution   We again follow the four-step process.

Step 1. The largest integer that divides into -25 and 15 is 5. The variable x appears in both terms with its smallest power as x. Since the leading coefficient is negative, we include a negative sign as part of the GCF. So, $-5x$ is the GCF.

Step 2. We write each term as the product of $-5x$ and another factor. Since the second term, $15x$, has a positive coefficient, both factors must be negative.

$$-25x^4 + 15x = (-5x) \cdot 5x^3 + (-5x) \cdot (-3)$$

Step 3. Factor out $-5x$ using the distributive property in reverse.

$$(-5x) \cdot 5x^3 + (-5x) \cdot (-3) = -5x(5x^3 - 3)$$

So, $-25x^4 + 15x = -5x(5x^3 - 3)$. Click here to see the check for this answer.

You Try It   **Work through this You Try It problem.**

Work Exercises 9–12 in this textbook or in the *MyMathLab*® Study Plan.

In each example so far, the GCF has been a monomial. In the next example, we factor out a binomial as the GCF.

Example 3 Factoring Out a Binomial as the GCF

Factor out the common binomial factor as the GCF.

a. $5x(2x + 3) + 6y(2x + 3)$ **b.** $7x(x + y) - (x + y)$

Solution

a. Treating $5x(2x + 3)$ and $6y(2x + 3)$ as terms, the only common factor is the binomial factor $2x + 3$. So, the GCF is $2x + 3$. Each term is already written as a product of $2x + 3$, so we factor it out as follows:

$$5x(2x + 3) + 6y(2x + 3) = (2x + 3)(5x + 6y)$$

So, $5x(2x + 3) + 6y(2x + 3) = (2x + 3)(5x + 6y)$.

My video summary

b. Try factoring out the common binomial factor as the GCF on your own. Click here to check your answer, or watch this video for a complete solution.

You Try It Work through this You Try It problem.

Work Exercises 13–16 in this textbook or in the *MyMathLab*® Study Plan.

OBJECTIVE 2: FACTOR BY GROUPING

Suppose the polynomial $5x(2x + 3) + 6y(2x + 3)$ from Example 3a had been simplified so that its terms did not contain the common binomial factor $2x + 3$. Can we factor the simplified polynomial $10x^2 + 15x + 12xy + 18y$? The answer is yes. To do this, we use a method called **factoring by grouping**. Let's examine this method using the polynomial

$$10x^2 + 15x + 12xy + 18y.$$

Looking at all four terms in the polynomial, there are no common factors other than 1. However, if we group the first two terms together and consider them by themselves, they have a common factor of $5x$. If we factor out the $5x$ from those two terms, the result is

$$10x^2 + 15x = 5x(2x + 3).$$

Similarly, if we group the last two terms, they have a common factor of $6y$. Factoring out the $6y$ gives us

$$12xy + 18y = 6y(2x + 3).$$

Using the grouped pairs, we arrive at the original polynomial in Example 3a. We can then factor out the common binomial factor to complete the process:

$$10x^2 + 15x + 12xy + 18y = (10x^2 + 15x) + (12xy + 18y)$$
$$= 5x(2x + 3) + 6y(2x + 3)$$
$$= (2x + 3)(5x + 6y)$$

We summarize how to *factor by grouping* with the following steps:

Steps to Factor a Polynomial by Grouping

Step 1. Group terms with a common factor. It may be necessary to rearrange the terms.

Step 2. For each group, factor out the greatest common factor.

Step 3. Factor out the common polynomial factor, if there is one.

Step 4. Check your answer by multiplying out the factors.

Example 4 Factoring by Grouping

Factor by grouping.

a. $x^3 - 2x^2 + 7x - 14$ **b.** $4x^2 + 20wx - xy - 5wy$

Solution

a. There is no common factor for all four terms, but the first two terms have a common factor of x^2, and the last two terms have a common factor of 7. We follow our four-step process.

Begin with the original polynomial expression: $x^3 - 2x^2 + 7x - 14$

Group the first two terms and last two terms: $= (x^3 - 2x^2) + (7x - 14)$

Factor out the GCF from each group: $= x^2(x - 2) + 7(x - 2)$

Factor out the common binomial factor: $= (x - 2)(x^2 + 7)$

The answer is $x^3 - 2x^2 + 7x - 14 = (x - 2)(x^2 + 7)$. Click here to see the check.

My video summary

b. Try factoring this polynomial on your own. Click here to check your answer, or watch this video for a complete solution.

 If minus signs are involved when factoring by grouping, pay close attention to the placement of grouping symbols. For example, consider $wx - wy - xz + yz$. It is incorrect to write the grouping as $(wx - wy) - (xz + yz)$ because $-(xz + yz) = -xz - yz$, not $-xz + yz$. It is correct to write the grouping as $(wx - wy) + (-xz + yz)$ or $(wx - wy) - (xz - yz)$.

You Try It   Work through this You Try It problem.

Work Exercises 17–26 in this textbook or in the **MyMathLab**® Study Plan.

5.1 Exercises

You Try It   In Exercises 1–8, factor out the GCF from each polynomial.

1. $20x + 8$

2. $5y^6 + 10x^2y^8$

3. $12x^2y + 4xy$

4. $28m^6n^2 + 32m^4n^5$

5. $4x^4 - 20x^3 + 12x^2$

6. $9a^3b^5 - 6a^2b^4 + 3a^2b^2$

7. $16x^2y^3z^6 - 12x^4y^4z^4 + 20x^3y^2z^5$

8. $24w^6y^4 - 40w^2y^2 + 16w^2y - 48w^3y^4$

In Exercises 9–12, factor out the negative sign with the GCF.

9. $-8x^3 + 48x^2$

10. $-14x^5y^6 - 21x^2y^8$

You Try It **11.** $-4x^2 - 16x + 18$

12. $-2x^3 + 10x^2 - 14x$

You Try It   In Exercises 13–16, factor out the common binomial factor as the GCF.

13. $11(x + 7) + 3a(x + 7)$

14. $7x(z + 5) + (z + 5)$

15. $9y(x^2 + 11) - 7(x^2 + 11)$

16. $(x + 5y)(x + 7) + (3x - 2y)(x + 7)$

You Try It In Exercises 17–26, factor by grouping.

17. $xy + 5x + 3y + 15$

18. $ab + 7a - 2b - 14$

19. $9xy + 12x + 6y + 8$

20. $3xy - 2x - 12y + 8$

21. $5x^2 + 4xy + 15x + 12y$

22. $3x^2 + 3xy - 2x - 2y$

23. $x^3 + 9x^2 + 4x + 36$

24. $x^3 - x^2 - 7x + 7$

25. $x^3 + 8x^2 + x + 8$

26. $a^2 + 4b + ab + 4a$

5.2 Factoring Trinomials

THINGS TO KNOW

Before working through this section, be sure you are familiar with the following concepts:

		VIDEO	ANIMATION	INTERACTIVE

You Try It
1. Find the Leading Coefficient and Degree of a Polynomial (Section 4.2, Objective 2) — INTERACTIVE

You Try It
2. Multiply Two Binomials (Section 4.3, Objective 3) — VIDEO, ANIMATION

You Try It
3. Factor Out the Greatest Common Factor from a Polynomial (Section 5.1, Objective 1) — VIDEO

You Try It
4. Factor by Grouping (Section 5.1, Objective 2) — VIDEO

OBJECTIVES

1 Factor Trinomials of the Form $x^2 + bx + c$

2 Factor Trinomials of the Form $ax^2 + bx + c$ Using Trial and Error

3 Factor Trinomials of the Form $ax^2 + bx + c$ Using the ac Method

4 Factor Trinomials Using Substitution

OBJECTIVE 1:   FACTOR TRINOMIALS OF THE FORM $x^2 + bx + c$

In this section, we continue learning how to factor polynomials by focusing on trinomials. From Section 4.3, we know the product of two binomials is often a trinomial. Since factoring is the reverse of multiplication, trinomials will often factor into the product of two binomials.

Let's look at trinomials with a leading coefficient of 1 and a degree of 2. These trinomials have the form $x^2 + bx + c$. Begin by watching this animation.

Using FOIL, pay close attention to the multiplication process for two binomials. For example, look at the product $(x + 4)(x + 5)$.

$$(x + 4)(x + 5) = \overset{F}{\overbrace{x^2}} + \overset{O}{\overbrace{5x}} + \overset{I}{\overbrace{4x}} + \overset{L}{\overbrace{20}} = x^2 + 9x + 20$$

The product is a trinomial of the form $x^2 + bx + c$. Study the terms of the trinomial and the terms of each binomial factor, and then answer the following questions:

- Where does the term x^2 in the trinomial come from? Click here to check.
- Where does the term 20 in the trinomial come from? Click here to check.
- Where does the term $9x$ in the trinomial come from? Click here to check.

Reversing the multiplication, the trinomial $x^2 + 9x + 20$ factors into the product $(x + 4)(x + 5)$. The numbers 4 and 5 are the "last" terms in the binomial factors because 4 and 5 multiply to 20 and add to 9.

$$x^2 \quad + \quad 9x \quad + \quad 20 = (x + 4)(x + 5)$$
$$\uparrow \qquad\qquad \uparrow \qquad\quad \uparrow$$
$$x \cdot x \qquad 4x + 5x \quad 4 \cdot 5$$

The relationship between the terms of the trinomial and the terms of the binomial factors leads us to the following steps for *factoring trinomials of the form* $x^2 + bx + c$. Remember that we are factoring over integers.

Factoring Trinomials of the Form $x^2 + bx + c$

Step 1. Find two integers, n_1 and n_2, whose product is the constant term c and whose sum is the coefficient b. So, $n_1 \cdot n_2 = c$ and $n_1 + n_2 = b$.

Step 2. Write the trinomial in the factored form $(x + n_1)(x + n_2)$.

Step 3. Check the answer by multiplying out the factored form.

Example 1 Factoring Trinomials of the Form $x^2 + bx + c$

Factor.

a. $x^2 + 16x + 28$ **b.** $x^2 + 11x + 28$ **c.** $x^2 + 29x + 28$

My interactive video summary

Solution Follow the three-step process.

a. Step 1. We need to find two integers whose product is 28 and whose sum is 16. Since 28 and 16 are both positive, we begin by listing the pairs of positive factors for 28:

Factors of 28	Sum of Factors
$1 \cdot 28 = 28$	$1 + 28 = 29$
$2 \cdot 14 = 28$	$2 + 14 = 16$
$4 \cdot 7 = 28$	$4 + 7 = 11$

Step 2. Since 2 and 14 multiply to 28 and add to 16, we have

$$x^2 + 16x + 28 = (x + 2)(x + 14).$$

Step 3. We check our answer by multiplying.

$$\text{Check: } (x + 2)(x + 14) = \overset{F}{\overbrace{x^2}} + \overset{O}{\overbrace{14x}} + \overset{I}{\overbrace{2x}} + \overset{L}{\overbrace{28}} = x^2 + 16x + 28$$

Our result checks, so $x^2 + 16x + 28 = (x + 2)(x + 14)$.

Note: Since multiplication is commutative, the order in which we list the factors in the product does not matter. For example, we can also write this answer as $x^2 + 16x + 28 = (x + 14)(x + 2)$.

b. Notice that $x^2 + 11x + 28$ has the same first and last terms as the trinomial in part a. The difference is the middle term. Looking at the pairs of factors of 28 listed in the solution to part a, the factors 4 and 7 have the required sum of 11. Use this information to finish factoring the trinomial on your own. Click here to check your answer, or watch this interactive video for the complete solution.

c. Try factoring this trinomial on your own. Click here to check your answer, or watch this interactive video for the complete solution.

You Try It   **Work through this You Try It problem.**

Work Exercise 1 in this textbook or in the *MyMathLab* ® **Study Plan.**

Recall that a **prime number** is a whole number greater than 1 whose only whole number factors are 1 and itself. For example, the first ten prime numbers are 2, 3, 5, 7, 11, 13, 17, 19, 23, and 29. As with numbers, there are also *prime polynomials*.

Definition   Prime Polynomial

A polynomial is a **prime polynomial** if its only factors over the integers are 1 and itself.

For example, the binomial $2x + 5$ is a prime polynomial because, over the integers, it cannot be written as the product of any two factors other than 1 and itself. The binomial $2x + 6$ is not prime because it can be factored into $2(x + 3)$.

When deciding if a polynomial is prime, consider only factors over the integers. For example, even though $2x + 5 = 2(x + 2.5)$, the binomial $2x + 5$ is prime. Because 2.5 is not an integer, $2(x + 2.5)$ is not factored over the integers.

Example 2 Recognizing a Prime Trinomial of the Form $x^2 + bx + c$

Factor $x^2 + 8x + 14$.

Solution Follow the three-step process.

Step 1. We need to find two integers whose product is 14 and whose sum is 8. We list the pairs of factors for 14:

Factors of 14	Sum of Factors
$1 \cdot 14 = 14$	$1 + 14 = 15$
$2 \cdot 7 = 14$	$2 + 7 = 9$

Step 2. Since neither of the two pairs of factors has a sum of 8, we cannot factor the trinomial into the form $(x + n_1)(x + n_2)$, and the trinomial is prime.

You Try It   Work through this You Try It problem.

Work Exercise 7 in this textbook or in the **MyMathLab**® Study Plan.

In Example 1, both coefficients b and c were positive integers, but the coefficients can also be negative. When this happens, we must consider the sign of the factors of c as we try to find the pair of factors whose sum is b.

Example 3   Factoring Trinomials of the Form $x^2 + bx + c$

 Factor.

a. $x^2 - 10x + 24$ **b.** $x^2 - 2x - 24$ **c.** $x^2 + 5x - 24$

Solution   Follow the three-step process.

a. Step 1. We need to find two integers whose product is 24 and whose sum is -10. The only way that the product can be positive and the sum can be negative is if both integers are negative. We begin by listing the pairs of negative factors for 24:

Negative Factors of 24	Sum of Factors
$(-1)(-24) = 24$	$(-1) + (-24) = -25$
$(-2)(-12) = 24$	$(-2) + (-12) = -14$
$(-3)(-8) = 24$	$(-3) + (-8) = -11$
$(-4)(-6) = 24$	$(-4) + (-6) = -10$

Step 2. Since -4 and -6 multiply to 24 and add to -10, we write

$$x^2 - 10x + 24 = (x - 4)(x - 6).$$

Step 3. Check the answer by multiplying. Click here to see the check.

b. We need to find two integers whose product is -24 and whose sum is -2. For the product to be negative, one of the integers must be positive and the other must be negative. For the sum to be negative, the integer with the larger absolute value must be negative. We list such pairs of factors for -24:

Factors of -24	Sum of Factors
$1(-24) = -24$	$1 + (-24) = -23$
$2(-12) = -24$	$2 + (-12) = -10$
$3(-8) = -24$	$3 + (-8) = -5$
$4(-6) = -24$	$4 + (-6) = -2$

Use this information to finish factoring the trinomial on your own. Click here to check your answer, or watch this interactive video for the complete solution.

c. Try factoring this trinomial on your own. Click here to check your answer, or watch this interactive video for the complete solution.

You Try It   Work through this You Try It problem.

Work Exercises 2–6 in this textbook or in the **MyMathLab**® Study Plan.

My interactive video summary

A polynomial is **factored completely** if it is written as the product of all prime polynomials. For example, we factored each trinomial in Example 3 as a product of prime binomials, so we factored each trinomial completely.

In Section 5.1, we learned how to factor out the greatest common factor from a polynomial. If a trinomial has a common factor, we factor it out first.

Example 4 Factoring Trinomials with a Common Factor

Factor completely.

a. $5y^2 - 25y + 20$ **b.** $3m^3 - 12m^2 - 96m$

Solution

a. The three terms have a GCF of 5. We first factor out the 5 and then factor the remaining trinomial.

$$\text{Begin with the original trinomial:} \quad 5y^2 - 25y + 20$$
$$\text{Write each term as a product of the GCF 5:} \quad = 5 \cdot y^2 - 5 \cdot 5y + 5 \cdot 4$$
$$\text{Factor out the 5:} \quad = 5(y^2 - 5y + 4)$$

The trinomial $y^2 - 5y + 4$ is of the form $x^2 + bx + c$, so we use the three-step process. We need to find two integers whose product is 4 and whose sum is -5. Since the product is positive and the sum is negative, both integers must be negative. The pairs of negative factors for 4 are

Negative Factors of 4	Sum of Factors
$(-1)(-4) = 4$	$(-1) + (-4) = -5$
$(-2)(-2) = 4$	$(-2) + (-2) = -4$

Since -1 and -4 multiply to 4 and add to -5, we have

$$5(y^2 - 5y + 4) = 5(y - 1)(y - 4).$$

Therefore, $5y^2 - 25y + 20 = 5(y - 1)(y - 4).$

Check the answer by multiplying. Click here to see the check.

My video summary

b. Try factoring this trinomial on your own. Remember to factor out the GCF first. Click here to check your answer, or watch this video for the complete solution.

You Try It   Work through this You Try It problem.

Work Exercises 13–16 in this textbook or in the *MyMathLab*® Study Plan.

A trinomial will sometimes have a leading coefficient of -1, such as $-x^2 - 8x + 9$. Since it is much easier to factor a trinomial when the leading coefficient is 1, we often begin by factoring out -1.

Example 5 Factoring a Trinomial with a Leading Coefficient of -1

My video summary

Factor $-x^2 - 8x + 9$.

Solution Since the leading coefficient is -1, we factor it out.

$$-x^2 - 8x + 9 = -(x^2 + 8x - 9)$$

The trinomial $x^2 + 8x - 9$ is of the form $x^2 + bx + c$. Try to finish factoring this trinomial on your own. Click here to check your answer, or watch this video for a complete solution.

You Try It   Work through this You Try It problem.

Work Exercises 17–18 in this textbook or in the *MyMathLab*® Study Plan.

We can factor a trinomial in two variables using the same approach as that for factoring a trinomial in one variable. For example, we factor a trinomial of the form $x^2 + bxy + cy^2$ just like we would factor $x^2 + bx + c$. The difference is that the second term of each binomial factor must contain the y variable. So, we find two integers, n_1 and n_2, whose product is the coefficient c and whose sum is the coefficient b. Then we write

$$x^2 + bxy + cy^2 = (x + n_1 y)(x + n_2 y).$$

Example 6 Factoring a Trinomial of the Form $x^2 + bxy + cy^2$

Factor $m^2 + 10mn - 56n^2$.

Solution We need to find two integers whose product is -56 and whose sum is 10. For the product to be negative, one of the integers must be positive and the other must be negative. For the sum to be positive, the integer with the larger absolute value must be positive. We list such pairs of factors for -56:

Factors of -56	Sum of Factors
$(-1)(56) = -56$	$-1 + 56 = 55$
$(-2)(28) = -56$	$-2 + 28 = 26$
$(-4)(14) = -56$	$-4 + 14 = 10$
$(-7)(8) = -56$	$-7 + 8 = 1$

Since -4 and 14 multiply to -56 and add to 10, we write

$$m^2 + 10mn - 56n^2 = (m - 4n)(m + 14n).$$

Check the answer by multiplying. Click here to see the check.

You Try It   Work through this You Try It problem.

Work Exercises 11–12 in this textbook or in the *MyMathLab*® Study Plan.

OBJECTIVE 2:   FACTOR TRINOMIALS OF THE FORM $ax^2 + bx + c$ USING TRIAL AND ERROR

If the leading coefficient of a trinomial is not 1, we must consider the coefficient as we look for the binomial factors. Work through this animation.

Once again, we can use FOIL to see how to factor this type of trinomial. For example, look at the product $(2x + 5)(3x + 2)$.

$$(2x + 5)(3x + 2) = \overset{\text{F}}{\overbrace{2x \cdot 3x}} + \overset{\text{O}}{\overbrace{2x \cdot 2}} + \overset{\text{I}}{\overbrace{5 \cdot 3x}} + \overset{\text{L}}{\overbrace{5 \cdot 2}} = 6x^2 + 19x + 10$$

The product results in a trinomial of the form $ax^2 + bx + c$ with $a \neq 1$. Study the terms of the trinomial and the terms of each binomial factor, and then answer the following questions:

- Where does the term $6x^2$ in the trinomial come from? Click here to check.
- Where does the term 10 in the trinomial come from? Click here to check.
- Where does the term $19x$ in the trinomial come from? Click here to check.

Reversing the multiplication, the trinomial $6x^2 + 19x + 10$ factors into the product $(2x + 5)(3x + 2)$. Using FOIL, the "first" terms in our factors must multiply to $6x^2$. The "last" terms must multiply to 10. And, the "outside" and "inside" products must add to $19x$.

$$
\begin{array}{ccccccc}
6x^2 & + & 19x & + & 10 & = & (2x + 5)(3x + 2) \\
\uparrow & & \uparrow & & \uparrow & & \\
2x \cdot 3x & & 2x \cdot 2 + 5 \cdot 3x & & 5 \cdot 2 & &
\end{array}
$$

If a trinomial of the form $ax^2 + bx + c$ can be factored, it will factor to the form $(m_1x + n_1)(m_2x + n_2)$. We can find the integers $m_1, m_2, n_1,$ and n_2 by using the following *trial-and-error strategy*.

Trial-and-Error Strategy for Factoring Trinomials of the Form $ax^2 + bx + c$

Step 1. Find all pairs of factors for the leading coefficient a.

Step 2. Find all pairs of factors for the constant term c.

Step 3. By trial and error, check different combinations of factors from step 1 and factors from step 2 in the form $(\Box x + \Box)(\Box x + \Box)$ until the correct middle term bx is found by adding the "outside" and "inside" products. If no such combination of factors exists, the trinomial is prime.

Step 4. Check your answer by multiplying out the factored form.

Example 7 Factoring Trinomials of the Form $ax^2 + bx + c$ Using Trial and Error

Factor $2x^2 + 13x + 15$.

Solution We follow the trial-and-error strategy. Since all of the coefficients are positive, we only need to check combinations of positive factors.

Step 1. The leading coefficient a is 2. The only pair of positive factors is $2 \cdot 1$.

Step 2. The constant term c is 15. The pairs of positive factors are: $1 \cdot 15$ and $3 \cdot 5$.

Step 3. Using the form $(\Box x + \Box)(\Box x + \Box)$, we try different combinations of the possible factors until we find one that gives the middle term $13x$.

Trial 1: $(2x + 15)(1x + 1)$. The "outside" product is $2x \cdot 1 = 2x$ and the "inside" product is $15 \cdot 1x = 15x$, giving a sum of $2x + 15x = 17x$. This is not the correct combination.

Trial 2: $(2x + 1)(1x + 15)$. The "outside" product is $2x \cdot 15 = 30x$ and the "inside" product is $1 \cdot 1x = x$, giving a sum of $30x + x = 31x$. This is not the correct combination.

Trial 3: $(2x + 5)(1x + 3)$. The "outside" product is $2x \cdot 3 = 6x$ and the "inside" product is $5 \cdot 1x = 5x$, giving a sum of $6x + 5x = 11x$. Again, this is not the correct combination.

Trial 4: $(2x + 3)(1x + 5)$. The "outside" product is $2x \cdot 5 = 10x$ and the "inside" product is $3 \cdot 1x = 3x$, giving a sum of $10x + 3x = 13x$. This is the correct combination.

Step 4. Check the answer by multiplying. Click here to see the check.

The answer checks, so $2x^2 + 13x + 15 = (2x + 3)(x + 5)$.

You Try It   Work through this You Try It problem.

Work Exercise 21 in this textbook or in the *MyMathLab*® Study Plan.

Now let's try factoring trinomials with negative coefficients.

Example 8   Factoring Trinomials of the Form $ax^2 + bx + c$ Using Trial and Error

 Factor.

a. $10x^2 + 11x - 6$ **b.** $8x^2 - 34x + 21$

Solution

a. We follow the trial-and-error strategy.

  Step 1. The leading coefficient a is 10. For convenience, we will only consider the positive factors of 10, which are $10 \cdot 1$ and $5 \cdot 2$.

  Step 2. The constant term c is -6. In this case, we must consider the negative sign. The pairs of factors are $1 \cdot (-6), -1 \cdot 6, 2 \cdot (-3)$, and $-2 \cdot 3$.

  Step 3. We can use the form $(\square x + \square)(\square x + \square)$ with different combinations of the factors from step 2 until we find one that gives a middle term of $11x$. Try to complete this problem on your own. Click here to check your answer, or watch this interactive video for the complete solution.

b. Try factoring this trinomial on your own. Click here to check your answer, or watch this interactive video for the complete solution.

You Try It   Work through this You Try It problem.

Work Exercises 22–24 in this textbook or in the *MyMathLab*® Study Plan.

We will now explore a prime trinomial in Example 9.

Example 9   Recognizing a Prime Trinomial of the Form $ax^2 + bx + c$

Factor $6x^2 - 2x + 5$.

Solution   We follow the trial-and-error strategy.

Step 1. The leading coefficient a is 6.
  The positive factors of 6 are $6 \cdot 1$ and $3 \cdot 2$.

Step 2. The constant term c is 5. Notice that the middle term has a negative coefficient, so we must consider the negative sign when choosing factors. Since the middle term comes from adding the outside and inside products of the binomial factors, we only need to consider the pairs of two negative factors for 5, which are $-1 \cdot (-5)$.

Step 3. Using the form $(\square x + \square)(\square x + \square)$, we try different combinations of the factors until we find one that gives the middle term $-2x$.

Trial 1: $(6x - 1)(1x - 5)$. The "outside" product is $6x \cdot (-5) = -30x$ and the "inside" product is $-1 \cdot 1x = -x$, giving a sum of $-30x + (-x) = -31x$. This is not the correct combination.

Trial 2: $(6x - 5)(1x - 1)$. The "outside" product is $6x \cdot (-1) = -6x$ and the "inside" product is $-5 \cdot 1x = -5x$, giving a sum of $-6x + (-5x) = -11x$. This is not the correct combination.

Trial 3: $(3x - 1)(2x - 5)$. The "outside" product is $3x \cdot (-5) = -15x$ and the "inside" product is $-1 \cdot 2x = -2x$, giving a sum of $-15x + (-2x) = -17x$. This is not the correct combination.

Trial 4: $(3x - 5)(2x - 1)$. The "outside" product is $3x \cdot (-1) = -3x$ and the "inside" product is $-5 \cdot 2x = -10x$, giving a sum of $-3x + (-10x) = -13x$. This is not the correct combination.

We have checked all the possible combinations of binomial factors, and none result in the middle term $-2x$. This means that $6x^2 - 2x + 5$ is a prime trinomial.

You Try It   **Work through this You Try It problem.**

Work Exercise 25 in this textbook or in the *MyMathLab*® Study Plan.

We factor a trinomial of the form $ax^2 + bxy + cy^2$ just like we would factor $ax^2 + bx + c$. The difference is that the second term of each binomial factor must contain the y variable. So, we use trial and error to check different combinations of factors for a and factors for c in the form $(\square x + \square y)(\square x + \square y)$ until the correct middle term bxy is found.

Example 10 Factoring a Trinomial of the Form $ax^2 + bxy + cy^2$

Factor $7p^2 - 11pq + 4q^2$.

Solution The leading coefficient is 7. The only pair of positive factors of 7 is $7 \cdot 1$.

The coefficient of the last term is 4. Notice that the middle term is negative. Since this negative term comes from adding the outside and inside products of the binomial factors, we only need to consider the pairs of negative factors of 4. These factors are $-1(-4)$ and $-2(-2)$.

Using the form $(\square p + \square q)(\square p + \square q)$, we try different combinations of the factors until we find one that gives the middle term $-11pq$.

Trial 1: $(7p - 1q)(1p - 4q)$. The "outside" product is $7p \cdot (-4q) = -28pq$ and the "inside" product is $-1q \cdot 1p = -pq$, giving a sum of $-28pq + (-pq) = -29pq$. This is not the correct combination.

Trial 2: $(7p - 4q)(1p - 1q)$. The "outside" product is $7p \cdot (-1q) = -7pq$ and the "inside" product is $-4q \cdot 1p = -4pq$, giving a sum of $-7pq + (-4pq) = -11pq$. This is the correct combination.

Since we have found the correct factored form, we do not need to check any more combinations. So, $7p^2 - 11pq + 4q^2 = (7p - 4q)(p - q)$.

 CAUTION When factoring a trinomial in two variables, make sure that the middle term contains both variables. For example, $7p^2 - 11p + 4q^2$ cannot be factored because the middle term does not contain the variable q.

You Try It   Work through this You Try It problem.

Work Exercise 26 in this textbook or in the **MyMathLab**® Study Plan.

OBJECTIVE 3:   FACTOR TRINOMIALS OF THE FORM $ax^2 + bx + c$ USING THE ac METHOD

The trial-and-error method for factoring trinomials can be time consuming. A second method, known as the ac **method**, can also be used when factoring trinomials of the form $ax^2 + bx + c$. We continue to factor over integers. Watch this animation to see how the ac method works.

The ac Method for Factoring Trinomials of the Form $ax^2 + bx + c$

Step 1. Multiply $a \cdot c$.

Step 2. Find two integers, n_1 and n_2, whose product is ac and whose sum is b. So, $n_1 \cdot n_2 = ac$ and $n_1 + n_2 = b$. If no such pair of integers exists, the trinomial is prime.

Step 3. Rewrite the middle term as the sum of two terms using the integers found in step 2. So, $ax^2 + bx + c = ax^2 + n_1x + n_2x + c$.

Step 4. Factor by grouping.

Step 5. Check your answer by multiplying out the factored form.

The ac method is also known as the **grouping method** or the **expansion method**. It is favored by some because its approach is more systematic, but it can still be time consuming. We suggest that you practice using both methods and then choose the one that you like best.

Example 11 Factoring Trinomials of the Form $ax^2 + bx + c$ Using the ac Method

Factor each trinomial using the ac method. If the trinomial is prime, state this as your answer.

a. $2x^2 + 9x - 18$ **b.** $6x^2 - 23x + 20$ **c.** $5x^2 + x + 6$

Solution

a. For $2x^2 + 9x - 18$, we have $a = 2$, $b = 9$, and $c = -18$. We follow the five-step process for the ac method.

  Step 1. $a \cdot c = 2(-18) = -36$.

  Step 2. We need to find two integers whose product is $ac = -36$ and whose sum is $b = 9$. For the product to be negative, one of the integers must be positive and the other must be negative. For the sum to be positive, the integer with the larger absolute value must be positive. We list such pairs of factors for -36:

Factors of -36	Sum of Factors
$-1 \cdot 36 = -36$	$-1 + 36 = 35$
$-2 \cdot 18 = -36$	$-2 + 18 = 16$
$-3 \cdot 12 = -36$	$-3 + 12 = 9$ ← This is the pair.
$-4 \cdot 9 = -36$	$-4 + 9 = 5$
$-6 \cdot 6 = -36$	$-6 + 6 = 0$

From the list above, -3 and 12 are the integers we need.

Step 3. $2x^2 + 9x - 18 = 2x^2 - 3x + 12x - 18.$

Step 4.

Begin with the new polynomial from step 3: $2x^2 - 3x + 12x - 18$

Group the first two terms and last two terms: $= (2x^2 - 3x) + (12x - 18)$

Factor out the GCF from each group: $= x(2x - 3) + 6(2x - 3)$

Factor out the common binomial factor: $= (2x - 3)(x + 6)$

Step 5. Check the answer by multiplying. Click here to see the check.

The answer checks, so $2x^2 + 9x - 18 = (2x - 3)(x + 6)$.

b. For $6x^2 - 23x + 20$, we have $a = 6$, $b = -23$, and $c = 20$. We follow the ac method.

Step 1. $a \cdot c = 6 \cdot 20 = 120$

Step 2. We need to find two integers whose product is $ac = 120$ and whose sum is $b = -23$. Since the product is positive and the sum is negative, both integers must be negative. We list the pairs of negative factors for 120:

Negative Factors of 120	Sum of Factors
$(-1)(-120) = 120$	$-1 + (-120) = -121$
$(-2)(-60) = 120$	$-2 + (-60) = -62$
$(-3)(-40) = 120$	$-3 + (-40) = -43$
$(-4)(-30) = 120$	$-4 + (-30) = -34$
$(-5)(-24) = 120$	$-5 + (-24) = -29$
$(-6)(-20) = 120$	$-6 + (-20) = -26$
$(-8)(-15) = 120$	$-8 + (-15) = -23$ ← This is the pair.
$(-10)(-12) = 120$	$-10 + (-12) = -22$

From the list above, -8 and -15 are the integers we need.

Try to complete this problem on your own. Click here to check your answer, or watch this interactive video for the complete solution.

c. Try factoring $5x^2 + x + 6$ on your own. Click here to check your answer, or watch this interactive video for the complete solution.

You Try It **Work through this You Try It problem.**

Work Exercises 27–30 in this textbook or in the MyMathLab® Study Plan.

In Example 10, we used trial and error to factor a trinomial of the form $ax^2 + bxy + cy^2$. We can also use the ac method to factor such trinomials, but we must be careful to include the second variable throughout the process.

Example 12 Factoring a Trinomial of the Form $ax^2 + bxy + cy^2$ Using the ac Method

Factor $6m^2 - 7mn - 10n^2$ using the ac method.

Solution For this trinomial, we have $a = 6$, $b = -7$, and $c = -10$.

Step 1. $a \cdot c = 6(-10) = -60$.

Step 2. We need to find two integers whose product is $ac = -60$ and whose sum is $b = -7$. These integers are 5 and -12. Do you see why?

Step 3. $6m^2 - 7mn - 10n^2 = 6m^2 + 5mn - 12mn - 10n^2$

Step 4. Begin with the polynomial
from step 3: $6m^2 + 5mn - 12mn - 10n^2$

Group the first two terms
and last two terms: $= (6m^2 + 5mn) + (-12mn - 10n^2)$

Factor out the GCF from
each group: $= m(6m + 5n) - 2n(6m + 5n)$

Factor out the common
binomial factor: $= (6m + 5n)(m - 2n)$

Step 5. Check the answer by multiplying. Click here to see the check.

The answer checks, so $6m^2 - 7mn - 10n^2 = (6m + 5n)(m - 2n)$.

You Try It   **Work through this You Try It problem.**

Work Exercise 31 in this textbook or in the *MyMathLab*® Study Plan.

When a trinomial has a common factor, we factor it out first.

Example 13 Factoring Trinomials with a Common Factor

My interactive video summary

Factor completely. $24t^5 - 52t^4 - 20t^3$

Solution The terms in this trinomial have a GCF of $4t^3$. We first factor out $4t^3$ and then factor the remaining trinomial.

Begin with the original trinomial: $24t^5 - 52t^4 - 20t^3$

Write each terms as a product of the GCF $4t^3$: $= 4t^3 \cdot 6t^2 - 4t^3 \cdot 13t - 4t^3 \cdot 5$

Factor out the $4t^3$: $= 4t^3(6t^2 - 13t - 5)$

Try to finish the problem on your own by factoring $6t^2 - 13t - 5$. You can use your preferred method, the ac method or trial and error. Click here to check your answer, or watch this interactive video for the complete solution.

You Try It   **Work through this You Try It problem.**

Work Exercises 33–36 in this textbook or in the *MyMathLab*® Study Plan.

Consider the trinomial $-2x^2 + 9x + 35$. Notice that the three terms do not have a common factor. However, when a trinomial has a negative number for a leading coefficient, we typically begin by factoring out a -1 like we did in Example 5. We do this because it is much easier to factor a trinomial when the leading coefficient is positive.

Example 14 Factoring Trinomials with a Common Factor

My interactive video summary

 Factor completely. $-2x^2 + 9x + 35$

Solution Since the leading coefficient is negative, we factor out a -1.

$$-2x^2 + 9x + 35 = -1(2x^2 - 9x - 35) = -(2x^2 - 9x - 35)$$

Try to finish the problem on your own by factoring $2x^2 - 9x - 35$. You can use either the *ac* method or trial and error. Click here to check your answer, or watch this interactive video for the complete solution.

You Try It   **Work through this You Try It problem.**

Work Exercises 37–39 in this textbook or in the *MyMathLab*® Study Plan.

OBJECTIVE 4:   FACTOR TRINOMIALS USING SUBSTITUTION

In the previous objective, we factored second-degree trinomials of the form $ax^2 + bx + c$. Another name for *second-degree trinomial* is **quadratic trinomial**.

Using **substitution**, we can sometimes factor more complicated trinomials. Consider the trinomial $x^{10} - 7x^5 + 12$. Notice that the degree of the leading term is twice the degree of the middle term.

$$\underset{\underset{x^{10}}{\uparrow}}{\text{Degree } 2\cdot 5} \quad - \quad \underset{\underset{7x^5}{\uparrow}}{\text{Degree 5}} \quad + \quad 12$$

Trinomials like this are called **quadratic in form** because the variable part of the first term can be written as the square of the variable factor in the middle term.

$$x^{10} - 7x^5 + 12 = \left(x^5\right)^2 - 7\left(x^5\right) + 12$$

If we substitute a new variable, say u, for the variable factor in the middle term, the result is a second-degree, or quadratic, trinomial.

Substitute u for x^5: $\left(x^5\right)^2 - 7\left(x^5\right) + 12 = u^2 - 7u + 12$

We then factor this trinomial, if possible. Click here to see how to factor $u^2 - 7u + 12$.

Factor: $u^2 - 7u + 12 = (u - 4)(u - 3)$

Finally, we write the factored form in terms of the original variable.

Substitute x^5 for u: $(u - 4)(u - 3) = (x^5 - 4)(x^5 - 3)$

So, $x^{10} - 7x^5 + 12 = (x^5 - 4)(x^5 - 3)$.

We can use the following steps to factor a trinomial that is quadratic in form:

Using Substitution to Factor Trinomials That Are Quadratic in Form

Step 1. Write the variable part of the leading term as the square of the variable factor of the middle term.

Step 2. Substitute u for the variable part from step 1.

Step 3. Factor the resulting trinomial, if possible.

Step 4. Substitute the variable factor from step 1 for u in the factored form. Simplify this result if necessary.

Step 5. Check your answer by multiplying out the factored form.

Example 15 Factoring Trinomials Using Substitution

Use substitution to factor each trinomial.

a. $x^4 - 11x^2 - 26$ **b.** $2(x + y)^2 + 13(x + y) + 21$

Solution

a.   Begin with the original trinomial: $x^4 - 11x^2 - 26$

Rewrite the first term in terms of x^2: $= (x^2)^2 - 11(x^2) - 26$

Substitute u for x^2: $= u^2 - 11u - 26$

Factor: $= (u - 13)(u + 2)$

Substitute x^2 for u: $= (x^2 - 13)(x^2 + 2)$

Click here to see the check.

 b. Substitute u for $x + y$ to result in the quadratic trinomial $2u^2 + 13u + 21$. Try to finish the factoring process on your own. Click here to check your answer, or watch this video for the complete solution.

My video summary

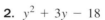

You Try It   Work through this You Try It problem.

Work Exercises 41–44 in this textbook or in the **MyMathLab** Study Plan.

5.2 Exercises

You Try It   In Exercises 1–12, factor each trinomial, or state that the trinomial is prime.

1. $x^2 + 15x + 50$ **2.** $y^2 + 3y - 18$

3. $w^2 - 18w + 56$ **4.** $z^2 - 4z - 32$

You Try It

5. $n^2 + 38n + 240$ **6.** $w^2 - 12w - 108$

7. $p^2 + 8p - 15$ **8.** $x^2 + 10x - 144$

You Try It **9.** $y^2 - 42y + 392$ **10.** $m^2 + 16m + 54$

 11. $x^2 - 20xy + 36y^2$ **12.** $p^2 - 18pq - 40q^2$

You Try It

You Try It

In Exercises 13–20, factor completely.

13. $8x^2 + 8x - 96$

14. $2y^4 + 34y^3 + 120y^2$

You Try It

15. $6x^3y^2 - 72x^2y^3 + 192xy^4$

16. $5m^3 - 15m^2 + 20m$

17. $-z^2 - 4z + 96$

18. $35 + 2w - w^2$

19. $-5x^4 + 90x^3 - 360x^2$

20. $25w^3x^2 - 75w^3x - 450w^3$

You Try It

In Exercises 21–26, use trial and error to factor each trinomial, or state that the trinomial is prime.

21. $6x^2 + 31x + 35$

22. $5y^2 - 3y - 8$

23. $8w^2 + 14w - 15$

24. $6p^2 - 19p + 3$

You Try It

25. $4z^2 - 23z - 18$

26. $12m^2 - 17mn + 6n^2$

You Try It

In Exercises 27–32, use the *ac* method to factor each trinomial, or state that the trinomial is prime.

27. $12m^2 - 47m + 40$

28. $9x^2 + 18x + 5$

You Try It

29. $8x^2 - 6x - 9$

30. $10y^2 - 23y + 12$

31. $12p^2 - 52pq - 9q^2$

32. $7m^2 + 32mn + 9n^2$

You Try It

In Exercises 33–40, factor completely.

You Try It

33. $24x^2 - 4x - 60$

34. $18y^5 - 93y^4 + 84y^3$

35. $36x^4y^2 + 60x^3y^3 - 75x^2y^4$

36. $90m^5n^3 + 306m^5n^2 + 108m^5n$

You Try It

37. $-20m^2 - 7m + 6$

38. $12 - x - 20x^2$

39. $-16wx^2 - 68wx - 42w$

40. $18x^3 + 108x^2 + 120x$

You Try It

In Exercises 41–44, use substitution to factor each trinomial.

41. $x^6 - x^3 - 20$

42. $6x^8 - 5x^4 - 25$

43. $5(x + y)^2 - 38(x + y) - 120$

44. $9(m - n)^2 + 78(m - n) - 27$

5.3 **Special-Case Factoring; A General Factoring Strategy**

THINGS TO KNOW

Before working through this section, be sure you are familiar with the following concepts:

VIDEO   ANIMATION   INTERACTIVE

You Try It

1. Multiply Two Binomials Using Special
Product Rules (Section 4.3, Objective 4)

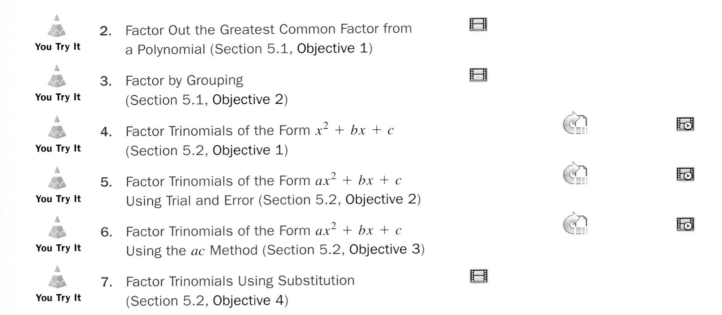

You Try It 2. Factor Out the Greatest Common Factor from a Polynomial (Section 5.1, Objective 1)

You Try It 3. Factor by Grouping (Section 5.1, Objective 2)

You Try It 4. Factor Trinomials of the Form $x^2 + bx + c$ (Section 5.2, Objective 1)

You Try It 5. Factor Trinomials of the Form $ax^2 + bx + c$ Using Trial and Error (Section 5.2, Objective 2)

You Try It 6. Factor Trinomials of the Form $ax^2 + bx + c$ Using the ac Method (Section 5.2, Objective 3)

You Try It 7. Factor Trinomials Using Substitution (Section 5.2, Objective 4)

OBJECTIVES

1 Factor the Difference of Two Squares

2 Factor Perfect Square Trinomials

3 Factor the Sum or Difference of Two Cubes

4 Factor Polynomials Completely

OBJECTIVE 1:   **FACTOR THE DIFFERENCE OF TWO SQUARES**

In this section, we emphasize the idea of *form*. For example, the standard form for a linear equation in two variables is $Ax + By = C$, where A and B are not both zero. The equations $y = 2x$ and $5 - 4y = -3x$ may look completely different and have different coefficients, but they are both linear equations in two variables because they can be written in the same standard form, $Ax + By = C$.

Equation	Standard Form	Coefficients
$y = 2x$	$2x - y = 0$	$A = 2, B = -1, C = 0$
$5 - 4y = -3x$	$3x - 4y = -5$	$A = 3, B = -4, C = -5$

The idea of *form* is particularly helpful with factoring. In Section 5.2, we learned how to factor trinomials with specific forms. In Section 4.3, we used special product rules to multiply two binomials. In that section, recognizing how two binomial factors fit a special form allowed us to quickly find the product. Since factoring is really just the reverse of multiplication, we can use the results of special binomial products to help us factor certain expressions.

Recall from Section 4.3 that the product of a binomial and its conjugate is the difference of the squares of the terms in the binomial. So, $(A + B)(A - B) = A^2 - B^2$. We use this result to form a rule for factoring the difference of two squares.

> **Factoring the Difference of Two Squares**
>
> If A and B are real numbers, variables, or algebraic expressions, then the difference of their squares can be factored into the product of the sum and difference of the two quantities.
>
> $$A^2 - B^2 = (A + B)(A - B)$$

Remember to focus on *form*. The quantities A and B could be numbers, variables, or complex algebraic expressions.

Example 1 Factoring the Difference of Two Squares

Factor each expression completely.

a. $y^2 - 25$　　　　b. $5x^2 - 45$　　　　c. $25x^2 - 36$

Solution

a. Start by rewriting the expression as the difference of two squares.

$$y^2 - 25 = y^2 - 5^2$$

Here we have $A = y$ (a variable) and $B = 5$ (a real number). Applying the difference of two squares rule, we get

$$y^2 - 25 = \overbrace{y^2}^{A=y} - \overbrace{5^2}^{B=5} = \underbrace{(y + 5)(y - 5)}_{(A + B)(A - B)}.$$

b. Start by factoring out the greatest common factor, 5.

$$5x^2 - 45 = 5 \cdot x^2 - 5 \cdot 9$$
$$= 5(x^2 - 9)$$

Next, rewrite the expression in parentheses as the difference of two squares.

$$5x^2 - 45 = 5(x^2 - 9) = 5(x^2 - 3^2)$$

Here we have $A = x$ (a variable) and $B = 3$ (a real number). Applying the difference of two squares rule, we get

$$5x^2 - 45 = 5(x^2 - 9) = 5(\overbrace{x^2}^{A=x} - \overbrace{3^2}^{B=3}) = 5\underbrace{(x + 3)(x - 3)}_{(A + B)(A - B)}.$$

c. Start by rewriting the expression as the difference of two squares.

$$25x^2 - 36 = (5x)^2 - 6^2$$

Here we have $A = 5x$ (an algebraic expression) and $B = 6$ (a real number). Try to complete this factorization on your own. Click here to check your answer.

You Try It　Work through this You Try It problem.

Work Exercises 1–3 in this textbook or in the _MyMathLab_®️ Study Plan.

In order to factor the difference of two squares, both quantities must be **perfect squares**. A number is a perfect square if it is the square of an integer. Click here to review some perfect square integers. A monomial with a perfect square coefficient and variables raised to even powers is also a perfect square (Example: $25x^2$). An algebraic expression is a perfect square if it is raised to an even power (Example: $(3z + 4)^2$).

The sum of two squares of the form $A^2 + B^2$ does not factor over the real numbers. However, an expression involving the sum of two squares will still factor if it contains a greatest common factor or if it can be written in a different factorable form. For example, $4x^2 + 36 = 4(x^2 + 9)$ where the GCF, 4, has been factored out.

My video summary

Example 2 Factoring the Difference of Two Squares

 Factor each expression completely.

a. $81m^2 - 49n^2$ **b.** $(x - 4)^2 - 25$

Solution Rewrite each expression as the difference of two perfect squares.

a. $81m^2 - 49n^2 = (9m)^2 - (7n)^2$ **b.** $(x - 4)^2 - 25 = (x - 4)^2 - 5^2$

Factor each difference of two squares, simplifying each factor if necessary. Click here to check your answers, or watch this video for a detailed solution.

You Try It   Work through this You Try It problem.

Work Exercises 4–8 in this textbook or in the *MyMathLab* ® Study Plan.

OBJECTIVE 2:   FACTOR PERFECT SQUARE TRINOMIALS

In Section 4.3, we had two special product rules for squaring the sum or difference of two terms. The results of these products were called perfect square trinomials. We can reverse these special product rules to factor perfect square trinomials.

> **Factoring Perfect Square Trinomials**
>
> If A and B are real numbers, variables, or algebraic expressions, then
> $$A^2 - 2AB + B^2 = (A - B)(A - B) = (A - B)^2 \text{ and}$$
> $$A^2 + 2AB + B^2 = (A + B)(A + B) = (A + B)^2.$$

Again, we focus on the form of the trinomial. To be a perfect square trinomial, the first and last terms must be perfect squares, and the middle term must be twice the product of the two quantities being squared. To factor a perfect square trinomial, we first identify the quantities being squared in the first and last terms and then apply the appropriate rule.

Example 3 Factoring Perfect Square Trinomials

Factor each expression completely.

a. $y^2 + 14y + 49$ **b.** $9x^2 - 30x + 25$

Solution

a. The first term is a perfect square, $y^2 = (y)^2$, and the last term is also a perfect square, $49 = 7^2$. Since $14y = 2(y)(7)$, we have a perfect square trinomial with a positive middle term.

$$A^2 + 2AB + B^2 = (A + B)^2$$
$$y^2 + 14y + 49 = (y)^2 + 2(y)(7) + 7^2 = (y + 7)^2$$

b. We write $9x^2 = 3^2x^2 = (3x)^2$ and $25 = 5^2$. Since $30x = 2(3x)(5)$, we have a perfect square trinomial with a negative middle term.

$$A^2 - 2AB + B^2 = (A - B)^2$$
$$9x^2 - 30x + 25 = (3x)^2 - 2(3x)(5) + 5^2 = (3x - 5)^2$$

You Try It   Work through this You Try It problem.

Work Exercises **9–12** in this textbook or in the *MyMathLab*® Study Plan.

Example 4 Factoring Perfect Square Trinomials

Factor each expression completely.

a. $16x^2 + 24xy + 9y^2$ **b.** $m^4 - 12m^2 + 36$

Solution

a. Write the first and last terms as perfect squares to determine A and B. Now complete the factorization on your own. Click here to check your answer, or watch this interactive video for a detailed solution.

b. Using rules for exponents, we can write $m^4 = (m^2)^2$. Now complete the factorization on your own. Click here to check your answer, or watch this interactive video for a detailed solution.

You Try It   Work through this You Try It problem.

Work Exercises **13–16** in this textbook or in the *MyMathLab*® Study Plan.

OBJECTIVE 3:   FACTOR THE SUM OR DIFFERENCE OF TWO CUBES

Like the difference of two squares, the difference of two cubes is the result of a special product rule. In fact, unlike the sum of two squares, the sum of two cubes can be factored as well. These new factor rules are given below.

Factoring the Sum and Difference of Two Cubes

If A and B are real numbers, variables, or algebraic expressions, then the sum or difference of their cubes can be factored as follows:

$$A^3 + B^3 = (A + B)(A^2 - AB + B^2)$$
$$A^3 - B^3 = (A - B)(A^2 + AB + B^2)$$

We can check these rules using two special products. Click here to see the steps.

In order to use these rules, both terms in the expression must be **perfect cubes**. A real number is a perfect cube if it is the cube of an integer. Click here to review some perfect cube integers. A monomial with a perfect cube coefficient and variables raised to powers that are multiples of 3 is also considered a perfect cube (Example: $125y^3$). An algebraic expression is a perfect cube if it is raised to a power that is a multiple of 3 (Example: $(4m - 5)^3$).

Notice that the form of the results can help us remember how to factor the sum or difference of two cubes. Click here to see the details.

To factor the sum or difference of two cubes, first we identify the quantities being cubed and then apply the appropriate rule.

Example 5 Factoring the Sum or Difference of Two Cubes

Factor each expression completely.

a. $x^3 + 125$ **b.** $27y^3 - 8$

Solution

a. The first term is a perfect cube, $x^3 = (x)^3$, and the last term is also a perfect cube, $125 = 5^3$. We have the sum of two cubes, so we apply the rule for $A^3 + B^3$.

$$A^3 + B^3 = (A + B)(A^2 - AB + B^2)$$
$$x^3 + 125 = x^3 + 5^3 = (x + 5)(x^2 - x(5) + 5^2)$$
$$= (x + 5)(x^2 - 5x + 25)$$

b. Since $27y^3 = (3y)^3$ and $8 = 2^3$, we have the difference of two cubes. Apply the rule for $A^3 - B^3$.

$$A^3 - B^3 = (A - B)(A^2 + AB + B^2)$$
$$27y^3 - 8 = (3y)^3 - 2^3 = (3y - 2)((3y)^2 + (3y)(2) + 2^2)$$
$$= (3y - 2)(9y^2 + 6y + 4)$$

You Try It   **Work through this You Try It problem.**

Work Exercises 17–20 in this textbook or in the *MyMathLab*® Study Plan.

Example 6 Factoring the Sum or Difference of Two Cubes

Factor each expression completely.

a. $64y^3 - 216z^3$ **b.** $8x^3y^2 + y^5$

Solution

a. First, factor out the GCF, 8. Write the remaining two terms as perfect cubes to determine the quantities $A = 2y$ and $B = 3z$. Try to complete the factorization on your own. Click here to see the rest of the solution.

b. First, factor out the GCF. Using rules for exponents, we write $y^5 = y^2 \cdot y^3$. Now complete this factorization on your own. Click here to check your answer, or watch this video for a detailed solution.

You Try It   Work through this You Try It problem.

Work Exercises 21–24 in this textbook or in the *MyMathLab*® Study Plan.

OBJECTIVE 4:   FACTOR POLYNOMIALS COMPLETELY

Now we know several techniques and rules for factoring polynomials. We can use this knowledge to create a general strategy for factoring polynomials completely. A polynomial is factored completely if all its polynomial factors, other than monomials, are prime.

General Strategy for Factoring Polynomials Completely

Step 1. If necessary, factor out the greatest common factor. If the leading coefficient is negative, factor out a common factor with a negative coefficient.

Step 2. Select a strategy based on the number of terms.

  a. If there are two terms, try to use one of the following special factor rules:

$$\text{Difference of two squares:} \quad A^2 - B^2 = (A + B)(A - B)$$
$$\text{Sum of two cubes:} \quad A^3 + B^3 = (A + B)\left(A^2 - AB + B^2\right)$$
$$\text{Difference of two cubes:} \quad A^3 - B^3 = (A - B)\left(A^2 + AB + B^2\right)$$

  b. If there are three terms, see if the trinomial is a perfect square trinomial. If so, factor using one of these special factor rules:

$$A^2 + 2AB + B^2 = (A + B)^2$$
$$A^2 - 2AB + B^2 = (A - B)^2$$

  If the trinomial is not a perfect square, try factoring by using trial and error or the *ac* method.

  c. If there are four or more terms, try factoring by grouping.

Step 3. Check to see if any factors can be factored further. Check each factor, other than monomial factors, to make sure they are prime.

Step 4. Check your answer by multiplying. Multiply out your result to see if it equals the original expression.

Example 7 Factoring a Polynomial Completely

Factor each expression completely.

a. $4x^3 - 20x^2 + 16x$ **b.** $-3m^2n + 6m$ **c.** $3y^3z^2 - 12y$

Solution

a. Follow the general factoring strategy.

  Step 1. Looking at each term, we can factor out a common factor of $4x$.

$$4x^3 - 20x^2 + 16x = 4x\left(x^2 - 5x + 4\right)$$

  Step 2. The trinomial factor has three terms. The first and last terms can be written as perfect squares, $x^2 = (x)^2$ and $4 = 2^2$. However, the middle term is not twice the product of the quantities being squared. Since $2(x)(2) = 4x \neq 5x$, this is not one of our special forms.

Since the leading coefficient of the trinomial factor is 1, we need two factors whose product is the constant, 4, and whose sum is the middle coefficient, -5. Since the constant is positive, the two factors have the same sign. And since the middle term is negative, the two factors must be negative.

Factor 1	Factor 2	Sum
-1	-4	-5
-2	-2	-4

The required factors are -1 and -4. We can factor the trinomial as $x^2 - 5x + 4 = (x - 1)(x - 4)$. So, $4x^3 - 20x^2 + 16x = 4x(x - 1)(x - 4)$.

Step 3. All the factors, other than the monomial factor, are prime, so our factorization is complete.

Step 4. Check by multiplying.

$$\begin{aligned}
\text{Factored expression:} \quad & 4x(x - 1)(x - 4) \\
\text{FOIL the binomial product:} \quad & = 4x(x \cdot x - 4 \cdot x - 1 \cdot x + 1 \cdot 4) \\
\text{Simplify:} \quad & = 4x(x^2 - 4x - x + 4) \\
& = 4x(x^2 - 5x + 4) \\
\text{Distribute } 4x: \quad & = 4x \cdot x^2 - 4x \cdot 5x + 4x \cdot 4 \\
\text{Simplify:} \quad & = 4x^3 - 20x^2 + 16x \checkmark
\end{aligned}$$

b. Follow the general factoring strategy.

Step 1. Looking at each term, we can factor out a common factor of $-3m$. Notice that the common factor has a negative coefficient because the leading term of our polynomial is negative.

$$\begin{aligned}
-3m^2n + 6m &= -3m \cdot mn - (-3m) \cdot 2 \\
&= -3m(mn - 2)
\end{aligned}$$

Step 2. Although there is a binomial factor, the two terms cannot be written as squares or cubes, so our special factor rules do not apply.

Step 3. All the factors, other than the monomial factor, are prime, so our factorization is complete.

$$-3m^2n + 6m = -3m(mn - 2)$$

Step 4. $\quad -3m(mn - 2) = -3m(mn) + 3m \cdot 2$
$$= -3m^2n + 6m \checkmark$$

My video summary

c. Follow the general factoring strategy.

Step 1. Looking at each term, we can factor out a common factor of $3y$.

$$\begin{aligned}
3y^3z^2 - 12y &= 3y \cdot y^2z^2 - 3y \cdot 4 \\
&= 3y(y^2z^2 - 4)
\end{aligned}$$

Step 2. There is a binomial factor. Examining the two terms, we can write perfect squares, $y^2z^2 = (yz)^2$ and $4 = 2^2$. So, we have the difference of two squares. Complete this factorization on your own. Click here to check your answer, or watch this video for a complete solution.

You Try It Work through this You Try It problem.

Work Exercises 25–29 in this textbook or in the *MyMathLab*® Study Plan.

Example 8 Factoring a Polynomial Completely

Factor each expression completely.

My interactive video summary **a.** $2x^3 - 5x^2 - 8x + 20$ **b.** $3a^2 - 10a - 8$ **c.** $3z^2 + z - 1$

Solution

    a. Use the general factoring strategy. Other than 1 or -1, there is no common factor. There are four terms, so we consider factoring by grouping. We try grouping the first two terms and the last two terms.

$$\underbrace{2x^3 - 5x^2} - \underbrace{8x + 20}$$

From the first two terms, we can factor out x^2. From the last two terms, we can factor out -4.

$$2x^3 - 5x^2 - 8x + 20 = \left(2x^3 - 5x^2\right) + (-8x + 20)$$
$$= x^2(2x - 5) - 4(2x - 5)$$

Try to complete this factorization on your own. Click here to see the full solution, or watch this interactive video.

   b.–c. Use the general factoring strategy. Other than 1 or -1, there is no common factor. There are three terms in both cases, but neither is a perfect square trinomial. Try factoring on your own using trial and error or the *ac* method. Click here to check your answers, or watch this interactive video for complete solutions.

 You Try It Work through this You Try It problem.

Work Exercises 30–32 in this textbook or in the *MyMathLab*® Study Plan.

Example 9 Factoring a Polynomial Completely

My interactive video summary Factor each expression completely.

a. $(m - 2)^2 - (n + 4)^2$ **b.** $-45p^2 + 150pq - 125q^2$

c. $27x^5y^5 + 64x^2y^5$ **d.** $8z^3x - 27y + 8z^3y - 27x$

e. $3x^2 - 7x - 6$ **f.** $4a^2 - 12ab + 9b^2 - c^2$

Solution Try to complete each factorization on your own. Click here to check your answers, or watch this interactive video to see fully worked solutions.

You Try It Work through this You Try It problem.

Work Exercises 33–44 in this textbook or in the *MyMathLab*® Study Plan.

5.3 Exercises

You Try It

In Exercises 1–8, factor each difference of two squares. Assume that variables represent whole numbers.

1. $x^2 - 4$

2. $10x^2 - 1000$

3. $4x^2 - 9$

4. $(m + 2)^2 - 81$

You Try It
5. $x^4 - 1$

6. $x^2y^2 - 16z^2$

7. $4x^2 - 49y^2$

8. $x^{10} - 25$

You Try It

In Exercises 9–16, factor the perfect square trinomials.

9. $x^2 + 4x + 4$

10. $x^2 - 30x + 225$

11. $9x^2 + 24x + 16$

You Try It
12. $4x^2 - 12x + 9$

13. $9x^2 - 12xy + 4y^2$

14. $x^4 - 18x^2 + 81$

15. $4a^2 + 20ab + 25b^2$

16. $4x^2y^2 + 20xy + 25$

You Try It

In Exercises 17–24, factor the sum or difference of two cubes.

17. $x^3 - 1$

18. $x^3 + 1$

19. $8z^3 + 125$

20. $125 - 216n^3$

21. $x^3 + 8y^3$

22. $27x^3 - y^3$

23. $x^6y^3 + z^3$

24. $2a^4b^2 - 16ab^5$

You Try It

In Exercises 25–45, factor completely. If a polynomial is prime, then state this as your answer.

25. $64x^3 + 8x^2 - 12x$

26. $4x^3y^2 + 2x^2y^3$

27. $x^2y - y^3$

28. $2x^3 + x^2y - 18x - 9y$

29. $6x^2 - 13x - 5$

30. $12y^2 - 7y + 3$

You Try It
31. $(x + 1)^2 - (x - 1)^2$

32. $x^4 - y^2$

33. $12x^2 - 16x + 48$

34. $x^2 - r^2 + 6x + 9$

35. $(x + y)^2 + 2(x + y) + 1$

36. $x^2 - y^2 - x + y$

You Try It
37. $(x + 1)^2 - (y - 1)^2$

38. $4x^2 - 12xy + 9y^2$

39. $x^3(t^3 - 8) + y^3(t^3 - 8)$

40. $6x^2 - 8xy + 3x - 4y$

41. $2x^5 - 32x$

42. $6b^2 - 27b - 105$

43. $2x^4 + x^2 - 3$

44. $7x^2 - 14x - 168$

45. $-9x^2 + 15x + 36$

5.4 Polynomial Equations and Models

THINGS TO KNOW

Before working through this section, be sure that you are familiar with the following concepts:

VIDEO   ANIMATION   INTERACTIVE

You Try It

1. Solve Linear Equations in One Variable (Section 1.1, Objective 2)

You Try It

2. Use Linear Equations to Solve Application Problems (Section 1.1, Objective 4)

You Try It

3. Express Equations of Functions Using Function Notation (Section 2.3, Objective 1)

You Try It

4. Factor Polynomials Completely (Section 5.3, Objective 4)

OBJECTIVES

1 Solve Polynomial Equations by Factoring

2 Find the Zeros of a Polynomial Function

3 Use Polynomial Equations and Models to Solve Application Problems

OBJECTIVE 1: SOLVE POLYNOMIAL EQUATIONS BY FACTORING

A **polynomial equation** results when we set two polynomials equal to each other. Some examples of polynomial equations are

$$2x - 7 = 4, \quad 3x^2 + 5x = x - 2, \quad \text{and} \quad 2x^3 + 7 = 3x^2 - x.$$

A polynomial equation is in **standard form** if one side equals zero and the other side is a simplified polynomial written in descending order.

Definition   Polynomial Equation

A **polynomial equation** in standard form is written as

$$P(x) = 0,$$

where $P(x)$ is a simplified polynomial in descending order.

The standard forms of the above polynomial equations are

$$2x - 11 = 0, \quad 3x^2 + 4x + 2 = 0, \quad \text{and} \quad 2x^3 - 3x^2 + x + 7 = 0.$$

The **degree of a polynomial equation** in standard form is the same as the highest degree of any of its terms. Notice that a polynomial of degree one, as in our first example, is a linear equation. We learned how to solve these types of equations in Section 1.1. To solve polynomial equations of degree 2 or higher, we can use the factoring techniques discussed in this chapter. To do so, we rely on the **zero product property**.

My video summary

Zero Product Property

If A and B are real numbers or algebraic expressions and $A \cdot B = 0$, then $A = 0$ or $B = 0$.

Recall that factors are quantities that are multiplied together to form a product. The zero product property tells us that if the product of two factors equals zero, then at least one of the factors must equal zero. This property extends to any number of factors. Watch this video to see how we can use the zero product property to solve a polynomial equation.

Example 1 Using the Zero Product Property to Solve a Polynomial Equation

Solve $(x + 3)(x - 7) = 0$.

Solution We have a product of two factors that is equal to zero. Using the zero product property, we set each factor equal to zero and solve the resulting equations.

$$x + 3 = 0 \quad \text{or} \quad x - 7 = 0$$
$$x = -3 \qquad\qquad x = 7$$

To check that these values are solutions, substitute them back into the original equation to see if a true statement results.

Check $x = -3$:
$$(x + 3)(x - 7) = 0$$
$$(-3 + 3)(-3 - 7) \overset{?}{=} 0$$
$$(0)(-10) \overset{?}{=} 0$$
$$0 = 0 \quad \text{True}$$

Check $x = 7$:
$$(x + 3)(x - 7) = 0$$
$$(7 + 3)(7 - 7) \overset{?}{=} 0$$
$$(10)(0) \overset{?}{=} 0$$
$$0 = 0 \quad \text{True}$$

Both values check, so the solution set is $\{-3, 7\}$.

You Try It   **Work through this You Try It problem.**

Work Exercises 1–5 in this textbook or in the *MyMathLab*® Study Plan.

CAUTION   Note that the zero product property only works when the product equals zero. For instance, if we have $A \cdot B = 8$, it is *not* true that one or both of the factors must be equal to 8. In fact, we can multiply $2 \cdot 4$ to get 8, and neither factor is equal to 8. Click here to see another example.

To solve a polynomial equation using the zero product property, one side of the equation must be zero and the other side must be written in factored form. We can combine our factoring techniques with the zero product property to help us solve polynomial equations.

Solving Polynomial Equations by Factoring

Step 1. Write the equation in standard form so that one side is zero and the other side is a simplified polynomial written in descending order.

Step 2. Factor the polynomial completely.

Step 3. Set each distinct factor with a variable equal to zero, and solve the resulting equations.

Step 4. Check each solution in the original equation.

CAUTION   When writing a solution set, we only include distinct solutions. This is why in step 3 we only set each distinct variable factor equal to zero.

In this section, we only focus on linear equations and polynomial equations of degree 2 or higher that can be factored over the set of integers.

Example 2 Solving Polynomial Equations by Factoring

Solve each equation by factoring.

a. $y^2 + 2y - 15 = 0$ **b.** $z^3 + z^2 = z + 1$

Solution

a. Step 1. The equation is already in standard form, so we move on to step 2.

Step 2. Using the general strategy for factoring, we see that the polynomial has three terms, but it is not written in one of the special forms. Since the leading coefficient is 1, we use trial and error. We need two numbers whose product is the constant -15 and whose sum is the middle coefficient 2. Since the product is negative, the two numbers have opposite signs. Since the sum is positive, the number with the larger absolute value is positive.

Factor 1	Factor 2	Sum
-1	15	14
-3	5	2

The required numbers are -3 and 5, which can be used to factor the polynomial.

$$y^2 + 2y - 15 = 0$$
$$(y - 3)(y + 5) = 0$$

Step 3. Set each factor equal to zero and solve the equations.

$$y - 3 = 0 \quad \text{or} \quad y + 5 = 0$$
$$y = 3 \qquad\qquad y = -5$$

Step 4. Check these values on your own to confirm that the solution set is $\{-5, 3\}$.

My video summary

 b. Write the equation in standard form by subtracting $z + 1$ from both sides.

Write the original equation:	$z^3 + z^2 = z + 1$
Subtract $z + 1$ from both sides:	$z^3 + z^2 - (z + 1) = 0$
Use the distributive property:	$z^3 + z^2 - z - 1 = 0$

Factor the left side, set each variable factor equal to zero, and solve the resulting equations on your own. Click here to check your answer, or watch this video for a detailed solution.

You Try It   Work through this You Try It problem.

Work Exercises 6–12 in this textbook or in the *MyMathLab*® Study Plan.

Example 3 Solving Polynomial Equations by Factoring

Solve each equation by factoring.

My interactive video summary 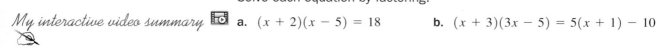 **a.** $(x + 2)(x - 5) = 18$ **b.** $(x + 3)(3x - 5) = 5(x + 1) - 10$

Solution We follow the four-step process for solving polynomial equations by factoring.

a. **Step 1.** Begin by simplifying both sides; then write the equation in standard form.

Begin with the original equation: $(x + 2)(x - 5) = 18$

Expand the left side: $x^2 - 5x + 2x - 10 = 18$

Simplify: $x^2 - 3x - 10 = 18$

Subtract 18 from both sides: $x^2 - 3x - 28 = 0$

Solve the equation on your own. Click here to check your answer, or watch this interactive video for a detailed solution.

b. Begin by simplifying both sides; then write the equation in standard form. Solve the equation on your own. Click here to check your answer, or watch this interactive video for a detailed solution.

You Try It   Work through this You Try It problem.

Work Exercises 13–18 in this textbook or in the *MyMathLab*® Study Plan.

Example 4 Solving Polynomial Equations by Factoring

Solve each equation by factoring.

My interactive video summary

a. $\dfrac{3}{2} k^2 = 10 - \dfrac{1}{4} k$ b. $0.02x^2 + 0.07x = 0.3$

Solution

a. Since we want to factor over integers only, first we clear the fractions by multiplying both sides of the equation by the LCD. In this case, the LCD is 4. Multiplying both sides of the equation by 4 gives us

$$4\left(\frac{3}{2} k^2\right) = 4\left(10 - \frac{1}{4} k\right)$$
$$6k^2 = 40 - k.$$

Continue solving the equation using the four-step process. Click here to check your answer, or watch this interactive video for a detailed solution.

b. Begin by clearing the decimals; then continue solving the equation using the four-step process. Click here to check your answer, or watch this interactive video for a detailed solution.

You Try It   Work through this You Try It problem.

Work Exercises 19–22 in this textbook or in the *MyMathLab*® Study Plan.

My interactive video summary

Example 5 Solving Polynomial Equations by Factoring

Solve the equations by factoring.

a. $4x^2 = 64x$ b. $4x - x^2 = 2x - 15$ c. $-h^3 + h^2 = -6h$

Solution

a. Although there is a common factor of x on both sides of the equation, we do not divide both sides of the equation by x. Click here to see why. Continue following the four-step process to solve the equation. Click here to check your answer, or watch this interactive video for a detailed solution.

b–c. In these two cases, the leading coefficient is negative. We can either multiply both sides of the equation by -1 to change all the signs, or write the polynomial on the other side of the equation. Try to solve these equations on your own. Click here to check your answers, or watch this interactive video for detailed solutions.

You Try It   **Work through this You Try It problem.**

Work Exercises 23–26 in this textbook or in the _MyMathLab_® Study Plan.

We have seen how to solve linear equations in Section 1.1 and have used factoring to solve polynomial equations of higher degrees. Graphs and tables can also be used to solve polynomial equations. In Chapter 8 we will look at different methods for solving polynomial equations of degree 2, called **quadratic equations**.

If $P(x)$ is prime, does this mean the polynomial equation $P(x) = 0$ has no solutions? Click here to find out.

OBJECTIVE 2:   FIND THE ZEROS OF A POLYNOMIAL FUNCTION

A **zero**, or **root**, of a function is a value for the independent variable that makes the value of the function equal to zero.

Definition   Zero of a Function

If c is a real number such that $f(c) = 0$, then c is called a **zero**, or root, of the function f.

When studying function notation in Section 2.3, we found that $y = f(x)$ is the value of the dependent variable y for a given value of the independent variable x. In Section 2.4, we found x-intercepts by letting $y = 0$, or $f(x) = 0$, and solving for x. Therefore, the zeros of a function are the same as the x-intercepts of the graph of the function.

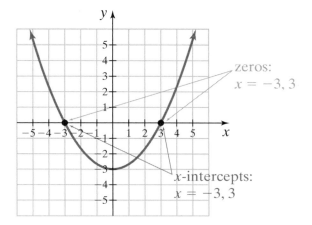

Figure 1

Example 6 Finding the Zeros of a Polynomial Function

Find the zeros for each polynomial function using its graph.

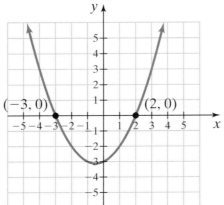

a.

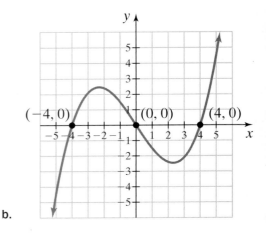

b.

Solution

a. The zeros of a polynomial function are the same as the *x*-intercepts of its graph. Using the graph, the *x*-intercepts are -3 and 2. Therefore, the zeros of the function are -3 and 2.

b. Since the *x*-intercepts are -4, 0, and 4, the zeros of the function are -4, 0, and 4.

You Try It Work through this You Try It problem.

Work Exercises 27–28 in this textbook or in the **MyMathLab** Study Plan.

When solving polynomial equations in standard form, we are actually looking for the zeros of the polynomial function $P(x)$.

Zeros of a Polynomial Function

Let $P(x)$ be a simplified polynomial function. If c is a real number such that $P(c) = 0$, then c is called a **zero**, or **root**, of the polynomial function. The zeros of a polynomial function are the solutions to the equation $P(x) = 0$.

Use this animation to see how to find the zeros of a polynomial function.

Example 7 Finding the Zeros of a Polynomial Function

Find the zeros for each polynomial function.

My interactive video summary **a.** $g(x) = 12x^3 - 8x^2 - 32x$ **b.** $h(x) = \dfrac{5}{6}x^2 - 3x - \dfrac{4}{3}$

c. $f(x) = 3x^2 - 24x + 48$ **d.** $p(x) = x^3 + 2x^2 - 9x - 18$

Solution

a. First, we set the function equal to zero and solve the resulting equation.

$$\text{Begin with } g(x): \quad g(x) = 12x^3 - 8x^2 - 32x$$
$$\text{Set } g(x) \text{ equal to } 0: \quad 0 = 12x^3 - 8x^2 - 32x$$
$$\text{Factor:} \quad 0 = 4x(x - 2)(3x + 4)$$

Click here to see the factoring details. Using the zero product property, set each factor equal to zero and solve the resulting equations.

$$4x = 0 \quad \text{or} \quad x - 2 = 0 \quad \text{or} \quad 3x + 4 = 0$$
$$x = 0 \qquad\qquad x = 2 \qquad\qquad 3x = -4$$
$$x = -\frac{4}{3}$$

The zeros are $-\dfrac{4}{3}$, 0, and 2. Click here to see the check.

b. First, we set the function equal to zero and solve the resulting equation.

$$\text{Begin with } h(x): \quad h(x) = \frac{5}{6}x^2 - 3x - \frac{4}{3}$$
$$\text{Set } h(x) \text{ equal to } 0: \quad 0 = \frac{5}{6}x^2 - 3x - \frac{4}{3}$$

Since the equation involves fractions, we clear the fractions by multiplying both sides of the equation by the LCD 6.

$$\text{Multiply by the LCD:} \quad 0 = 5x^2 - 18x - 8$$
$$\text{Factor:} \quad 0 = (x - 4)(5x + 2)$$

Click here to see the factoring details. Using the zero-product property, set each factor equal to zero and solve the resulting equations.

$$x - 4 = 0 \quad \text{or} \quad 5x + 2 = 0$$
$$x = 4 \qquad\qquad 5x = -2$$
$$x = -\frac{2}{5}$$

The zeros are $-\dfrac{2}{5}$ and 4. Click here to see the check.

c–d. Try to find these solutions on your own. Click here to check your answers, or watch this interactive video for detailed solutions.

You Try It   Work through this You Try It problem.

Work Exercises 29–36 in this textbook or in the *MyMathLab*® Study Plan.

Using factoring to find the zeros of a polynomial function should not be too surprising. From the factor theorem, we can say that $(x - c)$ is a factor of a polynomial function $P(x)$ if $P(c) = 0$. Therefore, if c is a zero of a polynomial function, then $x - c$ is a factor of the polynomial function. The following is a summary of the connection between zeros, x-intercepts, and factors.

> **Zeros, x-Intercepts, and Factors**
>
> Let $P(x)$ be a simplified polynomial function and c be a real number such that $P(c) = 0$. The following statements are all equivalent:
>
> **1.** c is a zero of the function.
>
> **2.** c is an x-intercept of the graph of the function.
>
> **3.** $x - c$ is a factor of the function.

OBJECTIVE 3: ## USE POLYNOMIAL EQUATIONS AND MODELS TO SOLVE APPLICATION PROBLEMS

Some real-world situations can be modeled by polynomial equations. To solve these types of applications, we follow the same problem-solving strategy used for linear equations.

Recall that we need to pay attention to the feasible domain when working with applications. Based on the problem's context, not every solution to an equation will be a solution to the problem. We will need to discard any solutions that do not make sense.

Example 8 Falling Object

The Burj Dubai, the world's tallest building at 2683 feet, has an observation deck on the 124th floor. An object is thrown upward with an initial velocity of 16 feet per second off the edge of the observation deck. The height of the object h, in feet, after t seconds is given by the function

$$h(t) = -16t^2 + 16t + 1440.$$

How long will it take for the object to hit the ground?

Solution

Step 1. Define the Problem.   An object is being thrown upward from an observation deck. Given the initial velocity and a function that describes the height of the object after some time, we must find the time it takes for the object to hit the ground.

Step 2. Assign Variables.   We are given that h is the height of the object in feet and t is the time in seconds after the object is thrown. Height is given as a function of time through the relation $h(t) = -16t^2 + 16t + 1440$.

Step 3. Translate into an Equation.   When the object hits the ground, it has a height of 0. Therefore, we need to solve the equation $h(t) = 0$. Substituting the expression for $h(t)$, we write $-16t^2 + 16t + 1440 = 0$.

Step 4. Solve the Equation.

$$\begin{aligned}
\text{Begin with the original equation:} \quad & -16t^2 + 16t + 1440 = 0 \\
\text{Factor out the GCF:} \quad & -16\left(t^2 - t - 90\right) = 0 \\
\text{Factor the trinomial:} \quad & -16(t - 10)(t + 9) = 0
\end{aligned}$$

Set each variable factor equal to zero and solve the resulting equations.

$$t - 10 = 0 \quad \text{or} \quad t + 9 = 0$$
$$t = 10 \qquad\qquad t = -9$$

Step 5. Check the Reasonableness of Your Answer. Since the time for the object to fall cannot be negative, we discard the negative solution. The only reasonable solution is $t = 10$ seconds.

Step 6. Answer the Question. The object will hit the ground 10 seconds after it is thrown.

You Try It **Work through this You Try It problem.**

Work Exercises 37–38 in this textbook or in the *MyMathLab*®️ Study Plan.

In Section 1.4 we explored common formulas for geometric figures such as triangles. **Right triangles** are triangles with a 90° angle, or **right angle**. The **hypotenuse** of a right triangle is the side opposite the right angle and is the longest of the three sides. The other two sides are called the **legs** of the triangle.

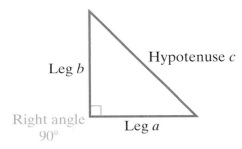

Figure 2

Pythagorean Theorem

For right triangles, the sum of the squares of the lengths of the legs of the triangle equals the square of the length of the hypotenuse.

$$a^2 + b^2 = c^2$$

Example 9 Zip Line Adventure

Zip line rides are popular activities at many vacation destinations. A zip line consists of a pulley mounted on a cable and set at an incline. The zip line uses gravity to propel a rider from one end to the other. For one such ride, the length of the zip line is 30 feet shorter than seven times the rise of the line. The run of the line is 30 feet longer than six times the rise of the line. See Figure 3. Find the length of the zip line.

My video summary

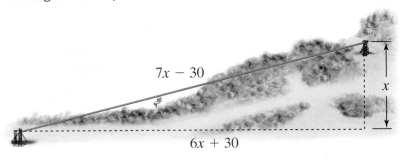

Figure 3

Solution Use the problem-solving strategy. Figure 3 shows us that we can use a right triangle to describe this situation. One leg is the rise of the line, and the other leg is the run. The length of the zip line is the hypotenuse. We can use the Pythagorean theorem to find the length of the line.

Begin with the Pythagorean theorem:

$$a^2 + b^2 = c^2$$

Substitute expressions for the lengths:

$$(6x + 30)^2 + (x)^2 = (7x - 30)^2$$

Expand using special products:

$$36x^2 + 360x + 900 + x^2 = 49x^2 - 420x + 900$$

Simplify the left side:

$$37x^2 + 360x + 900 = 49x^2 - 420x + 900$$

Finish solving this problem on your own. Be careful! In this problem, the value for the variable does not directly answer the question. The length of the zip line is given in terms of x, so once we know the value for x, we can find the length of the zip line. Click here to check your answer, or watch this video for a detailed solution.

You Try It   **Work through this You Try It problem.**

Work Exercises 39–40 in this textbook or in the *MyMathLab*® Study Plan.

Example 10 Seating Area

A club wants to remodel its rectangular stage area to make room for seating on three sides. The stage measures 10 meters by 20 meters. The width of the seating area is the same on all three sides, and the total combined area of the stage and seating area is 648 square meters. See Figure 4. Find the width of the seating area x.

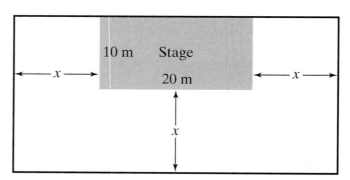

10 m   Stage

20 m

x

x x

Figure 4

Solution   Use the problem-solving strategy. We know that the width of the seating area is x. The combined width of the stage area and seating area is $10 + x$ because we are adding one seating width to the width of the stage. The combined length is $20 + 2x$ because we are adding *two* seating widths to the length of the stage. We can use the formula for the area of a rectangle to write an equation for the total combined area.

Begin with the rectangle area formula: $A = lw$

Substitute: $648 = (20 + 2x)(10 + x)$

Solve the equation to find the width of the seating area x. Click here to check your answer, or watch this video for a detailed solution.

You Try It   Work through this You Try It problem.

Work Exercises 41–42 in this textbook or in the *MyMathLab*® Study Plan.

5.4 Exercises

In Exercises 1–5, solve each equation using the zero product property.

You Try It

1. $(x + 1)(x - 2) = 0$ **2.** $w(w + 6) = 0$ **3.** $(2x + 1)(x + 3) = 0$

4. $5(2x - 1)(4x + 9) = 0$ **5.** $(2r - 3)(r + 1)(3r + 5) = 0$

In Exercises 6–26, use factoring techniques to solve each equation.

You Try It

6. $x^2 + 3x + 2 = 0$ **7.** $2q^2 - 5q - 3 = 0$ **8.** $x^3 + 3x^2 + 2x = 0$

9. $y^2 - 3y = 28$ **10.** $x^2 + 3 = 12$ **11.** $2w^2 + 14 = 11w$

12. $2x^2 + 5x + 3 = 1$ **13.** $(x + 1)(x - 2) = 4$ **14.** $6m^3 + 5m^2 - 6m = 6m^2 - m$

15. $(x - 1)(3x + 4) = -4$ **16.** $x(3x + 1) = 2(x + 5)$ **17.** $x(x^2 + 2) + 8 = 6(x^2 + x) - 4x^2$

18. $(2x + 1)(x - 4) = (x - 3)(x + 8)$ **19.** $\dfrac{x^2}{4} - \dfrac{5x}{6} - \dfrac{2}{3} = 0$

You Try It

20. $0.4x^2 + 0.3x = 1$ **21.** $0.06h^2 + 0.05h = 0.5$ **22.** $\dfrac{m^2}{10} = \dfrac{m}{5} + \dfrac{3}{2}$

23. $x^2 = 7x$ **24.** $25k^2 = 9$ **25.** $-5y^2 = 40y$

You Try It

26. $3x - 2x^2 = 3x^2 - 2$

In Exercises 27–28, find the zeros for the function in each graph.

You Try It

27.

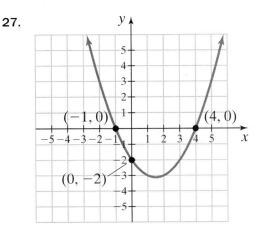

28.

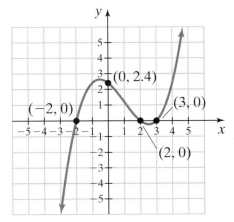

You Try It

 In Exercises 29–36, find the zeros of each polynomial function.

29. $p(x) = 9(x + 2)(x - 3)$ **30.** $p(x) = x(x - 1)(x - 2)$

31. $p(x) = x^2 - 9$ **32.** $p(x) = x^2 + 2x - 15$

 33. $p(x) = 3x^2 + 6x + 3$ **34.** $p(x) = 2x^3 + x^2 - 18x - 9$

35. $p(x) = x^4 - 13x^2 + 36$ **36.** $p(x) = \left(x^2 + 2x + 1\right)(x - 5)$

In Exercises 37–41, solve each application problem.

You Try It

37. Observation Deck The Shanghai World Financial Center has the world's highest outdoor observation deck at 1555 feet. An object is thrown upward with an initial velocity of 20 feet per second off the edge of the observation deck. The height h of the object in feet after t seconds is given by the function

$$h(t) = -16t^2 + 20t + 1555.$$

How long will it take for the object to be 691 feet above the ground?

38. Humans vs. Zombies During one semester at Truman State University, 300 students participated in a one-week Humans vs. Zombies game. The number of zombies, z, after d days of playing can be modeled by the function

$$z(d) = -d^2 + 60d - 75.$$

You Try It

How many days did it take for the number of zombies to reach 200?

 39. Signal Tower Support A 120-ft signal tower sits on the roof of a tall building. For support, a guy wire is attached to the top of the tower and anchored to the roof. The length of the wire is 30 feet more than twice the distance between the anchor and the base of the tower. How long is the wire?

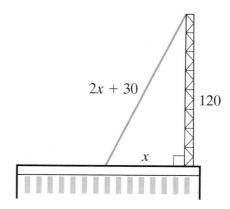

40. Traveling Cars Two cars leave Tombstone's Ace Hardware store at the same time. One car moves due east at 45 mph, whereas the other car moves due north at 60 mph. How long does it take before the cars are 300 miles apart?

You Try It

41. Bulletin Board   Elena wants to place a border of uniform width on the bulletin board in her classroom. If the bulletin board measures 48 inches by 72 inches and Elena wants to have a usable area of 2772 square inches, what is the widest possible border?

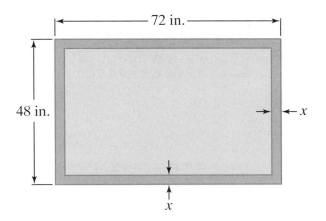

42. Perimeter of a Table Top   The area of a table top is 30 sq. ft., and twice the length is 1 ft. shorter than four times the width. Find the perimeter of the table top.

CHAPTER SIX

Rational Expressions, Equations, and Functions

CHAPTER SIX CONTENTS

6.1 Introduction to Rational Expressions and Functions

6.2 Multiplying and Dividing Rational Expressions

6.3 Adding and Subtracting Rational Expressions

6.4 Complex Rational Expressions

6.5 Rational Equations and Models

6.6 Variation

6.1 Introduction to Rational Expressions and Functions

THINGS TO KNOW

Before working through this section, be sure you are familiar with the following concepts:

VIDEO   ANIMATION   INTERACTIVE

You Try It

1. Find the Domain and Range of a Relation (Section 2.2, Objective 2)

You Try It

2. Evaluate Functions (Section 2.3, Objective 2)

You Try It

3. Factor Polynomials Completely (Section 5.3, Objective 4)

OBJECTIVES

1 Find the Domain of a Rational Function

2 Evaluate Rational Functions

3 Simplify Rational Expressions

OBJECTIVE 1: FIND THE DOMAIN OF A RATIONAL FUNCTION

Recall from Section R.1 that a number is a rational number if it can be written as the quotient $\frac{p}{q}$ of two integers p and q, as long as $q \neq 0$. A *rational expression* has a similar definition.

Definition   Rational Expression

A rational expression is an expression that can be written as the quotient $\frac{P}{Q}$ of two polynomials P and Q, as long as $Q \neq 0$.

Some examples of rational expressions are

$$\frac{5 - x}{3}, \quad \frac{x + 2}{x - 8}, \quad \frac{x^2 - x - 12}{x^2 - 3x - 4}, \quad \text{and} \quad \frac{x + 3y}{x - 3y}.$$

The first three expressions are *rational expressions in one variable*, x. The last expression is a *rational expression in two variables*, x and y.

In the definition of a rational expression, the statement $Q \neq 0$ means that the polynomial Q in the denominator cannot itself be zero and no value of the variable(s) can cause Q to equal zero. For example, if we evaluate $\frac{x + 2}{x - 8}$ for $x = 8$, the result is $\frac{8 + 2}{8 - 8} = \frac{10}{0}$, which is undefined. So, we restrict $\frac{x + 2}{x - 8}$ to values of x such that $x \neq 8$.

Identifying such restrictions is how we find the domain of a *rational function*.

Definition   Rational Function

A **rational function** is a function defined by a rational expression.

The function $f(x) = \frac{x + 2}{x - 8}$ is a rational function because it is defined by the rational expression $\frac{x + 2}{x - 8}$. The **domain of a rational function** is the set of all real numbers except for those values that cause the denominator to equal zero. From our previous discussion, we know that the domain of $f(x) = \frac{x + 2}{x - 8}$, using set-builder notation, is $\{x | x \neq 8\}$. Using interval notation, this domain is $(-\infty, 8) \cup (8, \infty)$. The set-builder notation is less messy. This is typical when stating the domain of rational functions, so primarily we will use set-builder notation in this section.

Finding the Domain of a Rational Function

To find the domain of a rational function, set the denominator equal to zero. Then solve the resulting equation and exclude the solutions from the domain. If the equation has no real solution, then the domain is the set of all real numbers.

Example 1 Finding the Domain of a Rational Function

Find the domain of $f(x) = \dfrac{2x + 9}{5x - 4}$.

Solution We follow the process for finding the domain of a rational function.

Set the polynomial in the denominator equal to 0: $\quad 5x - 4 = 0$

Add 4 to both sides: $\qquad\qquad 5x = 4$

Divide both sides by 5: $\qquad\qquad x = \dfrac{4}{5}$

We exclude the value $\dfrac{4}{5}$ from the domain, so the domain of f is $\left\{ x \middle| x \neq \dfrac{4}{5} \right\}$.

Click here to see the domain in interval notation.

You Try It Work through this You Try It problem.

Work Exercises 1–4 in this textbook or in the ***MyMathLab***® Study Plan.

Example 2 Finding the Domain of a Rational Function

 Find the domain of $g(x) = \dfrac{x^2 + 2x - 15}{x^2 + 5x - 24}$.

Solution We follow the same process.

Set the polynomial in the denominator equal to 0: $\quad x^2 + 5x - 24 = 0$

Finish finding the domain by solving this equation and excluding the solutions from the domain. Click here to check your answer, or watch this video for a complete solution.

You Try It Work through this You Try It problem.

Work Exercises 5–8 in this textbook or in the ***MyMathLab***® Study Plan.

Example 3 Finding the Domain of a Rational Function

 Find the domain of each rational function.

a. $R(x) = \dfrac{5x - 8}{7}$ $\qquad\qquad$ b. $h(x) = \dfrac{2x - 1}{x^2 + 4}$

Solution Try to find each domain on your own. Click here to check your answers, or watch this video for detailed solutions.

You Try It Work through this You Try It problem.

Work Exercises 9–10 in this textbook or in the ***MyMathLab***® Study Plan.

OBJECTIVE 2: EVALUATE RATIONAL FUNCTIONS

We evaluate rational functions just like we evaluated functions in Section 2.3. We substitute the given value for the variable and simplify.

My video summary

Example 4 Evaluating a Rational Function

 If $R(x) = \dfrac{5x^2 - 9}{7x + 3}$, evaluate each of the following.

a. $R(1)$ **b.** $R(-3)$

Solution

a. Substitute 1 for x in the function R: $\quad R(1) = \dfrac{5(1)^2 - 9}{7(1) + 3}$

$\qquad\qquad$ Simplify the exponent: $\qquad = \dfrac{5(1) - 9}{7(1) + 3}$

$\qquad\qquad$ Simplify the multiplication: $\qquad = \dfrac{5 - 9}{7 + 3}$

$\qquad\qquad\qquad$ Add and subtract: $\qquad = \dfrac{-4}{10}$

$\qquad\qquad$ Simplify the fraction: $\qquad = -\dfrac{2}{5}$

$\quad$ So, $R(1) = -\dfrac{2}{5}$. The value of R is $-\dfrac{2}{5}$ when x is 1.

b. Substitute -3 for x in the function R: $\quad R(-3) = \dfrac{5(-3)^2 - 9}{7(-3) + 3}$

$\quad$ Finish simplifying on your own. Click here to check your answer, or watch this video to see the complete solution.

You Try It Work through this You Try It problem.

Work Exercises 11–12 in this textbook or in the **MyMathLab**® Study Plan.

OBJECTIVE 3: SIMPLIFY RATIONAL EXPRESSIONS

A fraction is written in **lowest terms**, or **simplest form**, if the numerator and denominator have no common factors other than 1. For example, the fraction $\dfrac{5}{8}$ is in lowest terms because 5 and 8 have no common factors (except 1). However, the fraction $\dfrac{6}{10}$ is not in lowest terms because 6 and 10 have a common factor of 2.

Recall that multiplying or dividing both the numerator and denominator of a fraction by the same nonzero value results in an equivalent fraction. Therefore, we can simplify a fraction by dividing both the numerator and the denominator by their greatest common factor. So, $\dfrac{6}{10} = \dfrac{6 \div 2}{10 \div 2} = \dfrac{3}{5}$.

We can also use **cancellation** to simplify a fraction. We write both the numerator and denominator in terms of their prime factorizations. We then "cancel" each common factor by dividing both the numerator and denominator by the common factor. This process is shown by placing slash marks through each common factor and then replacing the factor with 1, the result of dividing the common factor by itself. Separately in the numerator and denominator, we multiply the 1's by any remaining factors to obtain in the simplified fraction.

$$\frac{6}{10} = \frac{2 \cdot 3}{2 \cdot 5} = \frac{\overset{1}{\cancel{2}} \cdot 3}{\underset{1}{\cancel{2}} \cdot 5} = \frac{1 \cdot 3}{1 \cdot 5} = \frac{3}{5}$$

We simplify rational expressions in the same way that we simplify fractions. We cancel common factors.

Cancellation Principle for Rational Expressions

If P, Q, and R are polynomials, then $\dfrac{P \cdot R}{Q \cdot R} = \dfrac{P \cdot \overset{1}{\cancel{R}}}{Q \cdot \underset{1}{\cancel{R}}} = \dfrac{P \cdot 1}{Q \cdot 1} = \dfrac{P}{Q}$ for $Q \neq 0$ and $R \neq 0$.

The cancellation principle leads to the following steps for simplifying rational expressions.

Simplifying Rational Expressions

Step 1. Factor the numerator and denominator completely.

Step 2. Cancel each common factor by dividing both the numerator and denominator by the common factor.

Step 3. Separately in the numerator and denominator, multiply any factors that are not cancelled to obtain the simplified rational expression. If all factors in the numerator cancel, the numerator will be 1.

In most cases, we will not multiply out the polynomial factors in a simplified rational expression. Instead, we will leave them in factored form.

Example 5 Simplifying a Rational Expression

Simplify $\dfrac{3x^2 - 12x}{5x - 20}$.

Solution We follow the three-step process.

$$\text{Begin with the original expression:} \quad \frac{3x^2 - 12x}{5x - 20}$$

$$\text{Factor the numerator and denominator:} \quad = \frac{3x(x - 4)}{5(x - 4)}$$

$$\text{Cancel the common factor } x - 4: \quad = \frac{3x\overset{1}{\cancel{(x - 4)}}}{5\underset{1}{\cancel{(x - 4)}}}$$

$$\text{Write the simplified rational expression:} \quad = \frac{3x}{5} \text{ for } x \neq 4$$

Do not forget about the restrictions for a rational expression. In Example 5, $x = 4$ causes the denominator of the original rational expression $\dfrac{3x^2 - 12x}{5x - 20}$ to equal 0. However, $x = 4$ does not cause the denominator of the simplified rational expression $\dfrac{3x}{5}$ to equal 0. So, the two expressions are not equal unless we include the restriction $x \neq 4$ for $\dfrac{3x}{5}$. For the rest of this section, we will not list such restrictions, but note that the restrictions are still needed for equality.

You Try It Work through this You Try It problem.

Work Exercises 13–14 in this textbook or in the *MyMathLab*® Study Plan.

Example 6 Simplifying a Rational Expression

Simplify $\dfrac{x^2 - 2xy - 15y^2}{x^2 + 5xy + 6y^2}$.

Solution We follow the three-step process.

Begin with the original expression: $\dfrac{x^2 - 2xy - 15y^2}{x^2 + 5xy + 6y^2}$

Factor the numerator and denominator: $= \dfrac{(x - 5y)(x + 3y)}{(x + 2y)(x + 3y)}$

Cancel the common factor $x + 3y$: $= \dfrac{(x - 5y)\cancel{(x + 3y)}^{1}}{(x + 2y)\cancel{(x + 3y)}_{1}}$

Write the simplified rational expression: $= \dfrac{x - 5y}{x + 2y}$

You Try It Work through this You Try It problem.

Work Exercises 15–16 in this textbook or in the *MyMathLab*® Study Plan.

CAUTION The cancellation principle for rational expressions applies only to common factors of the numerator and denominator. It is incorrect to cancel terms. In Example 6, $\dfrac{x - 5y}{x + 2y}$ is in simplest form because the numerator $x - 5y$ and the denominator $x + 2y$ have no common factor (except 1). Because the x- and y-variables are not factors of the numerator and denominator, they cannot be cancelled.

$$\frac{x - 5y}{x + 2y} \neq \frac{\cancel{x}^{1} - 5\cancel{y}}{\cancel{x}_{1} + 2\cancel{y}} \neq \frac{1 - 5}{1 + 2} \quad \text{Incorrect cancellation!}$$

$\dfrac{x - 5y}{x + 2y}$ is not equivalent to $\dfrac{1 - 5y}{1 + 2y}$ or $\dfrac{1 - 5}{1 + 2}$.

Note: When simplifying a rational expression, replacing each cancelled factor with a factor of 1 can lead to unnecessary clutter. For this reason, it is okay to leave these factors of 1 unwritten (just like we leave a leading coefficient of 1 unwritten). For example, we could write the simplified rational expression in Example 6 without using factors of 1 as follows:

$$\frac{x^2 - 2xy - 15y^2}{x^2 + 5xy + 6y^2} = \frac{(x - 5y)(x + 3y)}{(x + 2y)(x + 3y)} = \frac{(x - 5y)\cancel{(x + 3y)}}{(x + 2y)\cancel{(x + 3y)}} = \frac{x - 5y}{x + 2y}$$

However, if every factor in the numerator cancels, be sure to include a 1 in the numerator of the simplified rational expression.

Example 7 Simplifying a Rational Expression

 Simplify $\dfrac{2m^2 + m - 15}{2m^3 - 5m^2 - 18m + 45}$.

Solution We follow the three-step process.

Begin with the original expression: $\dfrac{2m^2 + m - 15}{2m^3 - 5m^2 - 18m + 45}$

Factor the numerator and denominator: $= \dfrac{(2m - 5)(m + 3)}{(2m - 5)(m + 3)(m - 3)}$

Finish simplifying this rational expression on your own. Click here to check your answer, or watch this video for a complete solution.

You Try It   Work through this You Try It problem.

Work Exercises 17–18 in this textbook or in the *MyMathLab*® Study Plan.

Example 8 Simplifying a Rational Expression

 Simplify $\dfrac{w^3 + 64}{w^2 - 4w + 16}$.

Solution Try to simplify this rational expression on your own. Click here to check your answer, or watch this video for a complete solution.

You Try It   Work through this You Try It problem.

Work Exercises 19–20 in this textbook or in the *MyMathLab*® Study Plan.

The commutative property of addition states that a sum is not affected by the order of the terms. So, $a + b = b + a$. Sometimes when factoring the numerator and denominator of a rational expression, the terms of common factors may be arranged in a different order. We can still cancel these common factors.

Example 9 Simplifying a Rational Expression

Simplify $\dfrac{w^2 - y^2}{2xy + 2xw}$.

Solution

Begin with the original expression: $\dfrac{w^2 - y^2}{2xy + 2xw}$

Factor the numerator and denominator: $= \dfrac{(w + y)(w - y)}{2x(y + w)}$

Cancel the common factors $w + y = y + w$: $= \dfrac{\cancel{(w + y)}(w - y)}{2x\cancel{(y + w)}}$

Write the simplified rational expression: $= \dfrac{w - y}{2x}$

(EBook Screens 6.1-1–6.1-22)

You Try It   Work through this You Try It problem.

Work Exercises 21–22 in this textbook or in the *MyMathLab*® Study Plan.

If the numerator and denominator of a rational expression have factors that are opposite polynomials, we can simplify by first factoring -1 from one of them.

Example 10 Simplifying a Rational Expression

Simplify $\dfrac{4x - 7}{7 - 4x}$.

Solution   We recognize that $4x - 7$ and $7 - 4x$ are opposite polynomials, so we factor -1 from $7 - 4x$:

$$7 - 4x = -1(-7 + 4x) = -1(4x - 7)$$

This gives

$$\frac{4x - 7}{7 - 4x} = \frac{4x - 7}{-1(4x - 7)} = \frac{\cancel{4x - 7}}{-1\cancel{(4x - 7)}} = \frac{1}{-1} = -1.$$

You Try It   Work through this You Try It problem.

Work Exercises 23–24 in this textbook or in the *MyMathLab*® Study Plan.

Example 10 shows us that opposite factors in the numerator and denominator of a rational expression will cancel, leaving a factor of -1.

Example 11 Simplifying a Rational Expression

My video summary

Simplify $\dfrac{2x^2 - x - 3}{9 - 4x^2}$.

Solution

Begin with the original expression: $\dfrac{2x^2 - x - 3}{9 - 4x^2}$

Factor the numerator and denominator: $= \dfrac{(2x - 3)(x + 1)}{(3 - 2x)(3 + 2x)}$

Notice that $2x - 3$ and $3 - 2x$ are opposite polynomials. Finish simplifying the rational expression on your own. Click here to check your answer, or watch this video for a complete solution.

You Try It   Work through this You Try It problem.

Work Exercises 25–26 in this textbook or in the *MyMathLab*® Study Plan.

Example 12 Simplifying Rational Expressions

My interactive video summary

Simplify each rational expression.

a. $\dfrac{x^2 + x - 12}{x^2 + 9x + 20}$ b. $\dfrac{x^3 + 1}{x + 1}$ c. $\dfrac{x^2 - x - 2}{2x - x^2}$

(EBook Screens 6.1-1–6.1-22)

Solution Try to simplify each rational expression on your own. Click here to check your answers, or watch this interactive video for a complete solution.

You Try It   Work through this You Try It problem.

Work Exercises 27–30 in this textbook or in the *MyMathLab*® Study Plan.

6.1 Exercises

In Exercises 1–10, find the domain of each rational function. Write your answer in set-builder notation.

You Try It

1. $f(x) = \dfrac{10}{x + 2}$

2. $g(x) = \dfrac{x - 4}{x - 6}$

You Try It

3. $h(x) = \dfrac{3x - 7}{5x + 2}$

4. $s(t) = \dfrac{3t - 7}{2t}$

5. $f(x) = \dfrac{x + 9}{(x + 2)(x - 5)}$

6. $g(x) = \dfrac{8x}{x^2 + 3x - 88}$

7. $R(x) = \dfrac{x + 4}{3x^2 + 10x - 8}$

8. $h(x) = \dfrac{3x - 1}{x^3 - 4x^2 - 12x}$

You Try It

9. $f(x) = \dfrac{7x - 2}{5}$

10. $R(x) = \dfrac{x + 4}{2x^2 + 1}$

You Try It

In Exercises 11–12, evaluate the given rational function as indicated. If the function value is undefined, state this as your answer.

11. If $f(x) = \dfrac{x^2 - 5x - 14}{5 - x}$, find $f(-2)$, $f(3)$, and $f(5)$.

12. If $g(t) = \dfrac{2t^3 - 7}{t^2 + 5}$, find $g(-1)$, $g(0)$, and $g(2)$.

In Exercises 13–30, simplify each rational expression. If the expression is already in simplest form, state this as your answer.

You Try It
13. $\dfrac{9a - 27}{5a - 15}$

14. $\dfrac{8x + 48}{9x^3 + 54x^2}$

You Try It
15. $\dfrac{3z^2 - 2z - 8}{3z^2 - 14z - 24}$

16. $\dfrac{x^2 + 3xy - 4y^2}{2x^2 + 5xy - 12y^2}$

You Try It

17. $\dfrac{w^3 + 6w^2 - 27w}{2w^4 - 18w^2}$

18. $\dfrac{4x^2 - 7x - 2}{8x^3 + 2x^2 + 4x + 1}$

You Try It **19.** $\dfrac{36x^2 - 42x + 49}{216x^3 + 343}$

20. $\dfrac{x^3 - 125}{3x - 15}$

21. $\dfrac{x^2 - 9}{9 + x}$

You Try It

23. $\dfrac{x - 13}{13 - x}$

You Try It

25. $\dfrac{x^2 - 4}{-2 - x}$

27. $\dfrac{q^2 - q - 42}{q^2 - 9q + 14}$

You Try It

29. $\dfrac{m^3 - 4m^2 + 6m - 24}{m^2 - 8m + 16}$

22. $\dfrac{5x + 6}{3 + 5x}$

24. $\dfrac{7k - 14}{28 - 14k}$

26. $\dfrac{y^2 - 8y - 20}{100 - y^2}$

28. $\dfrac{a^3 - 8}{a^2 - 4}$

30. $\dfrac{x^2 + x - 6}{x + 3}$

6.2 Multiplying and Dividing Rational Expressions

THINGS TO KNOW

Before working through this section, be sure you are familiar with the following concepts:

VIDEO   ANIMATION   INTERACTIVE

You Try It
1. Simplify Exponential Expressions Using the Quotient Rule (Section 4.1, Objective 2)

You Try It
2. Factor Polynomials Completely (Section 5.3, Objective 4)

You Try It
3. Simplify Rational Expressions (Section 6.1, Objective 3)

OBJECTIVES

1 Multiply Rational Expressions

2 Divide Rational Expressions

OBJECTIVE 1:   MULTIPLY RATIONAL EXPRESSIONS

Recall that to multiply rational numbers (fractions), we multiply straight across the numerators, multiply straight across the denominators, and simplify. If $\dfrac{p}{q}$ and $\dfrac{r}{s}$ are rational numbers, then

$$\frac{p}{q} \cdot \frac{r}{s} = \frac{p \cdot r}{q \cdot s} = \frac{pr}{qs}.$$

For example,

$$\frac{5}{6} \cdot \frac{3}{10} = \frac{5 \cdot 3}{6 \cdot 10} = \frac{15}{60} = \frac{1 \cdot 15}{4 \cdot 15} = \frac{1 \cdot \cancel{15}^1}{4 \cdot \cancel{15}_1} = \frac{1}{4}.$$

(EBook Screens 6.2-1–6.2-18)

It is often easier to simplify fractions if we write the numerator and the denominator as a product of prime factors and cancel any factors common to a numerator and a denominator.

$$\frac{5}{6}\cdot\frac{3}{10} = \frac{5}{2\cdot3}\cdot\frac{3}{2\cdot5} = \frac{\cancel{5}^{1}}{2\cdot3_1}\cdot\frac{\cancel{3}^{1}}{2\cdot\cancel{5}_1} = \frac{1}{2\cdot2} = \frac{1}{4}$$

We follow the same approach when multiplying rational expressions. We factor all numerators and denominators completely, cancel any common factors, multiply remaining factors in the numerators, and multiply remaining factors in the denominators. So, if $\frac{P}{Q}$ and $\frac{R}{S}$ are rational expressions, then

$$\frac{P}{Q}\cdot\frac{R}{S} = \frac{PR}{QS}.$$

We keep the result in factored form just as we did when simplifying rational expressions.

Multiplying Rational Expressions

Step 1. Factor each numerator and denominator completely into prime factors.

Step 2. Cancel by dividing the numerators and denominators by common factors.

Step 3. Multiply remaining factors in the numerators and multiply remaining factors in the denominators.

 Factoring is a critical step in the multiplication process. If necessary, review the factoring techniques discussed in Chapter 5.

Example 1 Multiplying Rational Expressions

Multiply $\dfrac{40x-10}{8x}\cdot\dfrac{7}{12x-3}$.

Solution We follow the three-step process for multiplying rational expressions.

Begin with the original expression: $\dfrac{40x-10}{8x}\cdot\dfrac{7}{12x-3}$

Factor numerators and denominators: $=\dfrac{10(4x-1)}{8x}\cdot\dfrac{7}{3(4x-1)}$

$=\dfrac{2\cdot5(4x-1)}{2\cdot2\cdot2x}\cdot\dfrac{7}{3(4x-1)}$

Cancel common factors: $=\dfrac{\cancel{2}\cdot5\cancel{(4x-1)}}{\cancel{2}\cdot2\cdot2x}\cdot\dfrac{7}{3\cancel{(4x-1)}}$

Multiply remaining factors: $=\dfrac{5\cdot7}{2\cdot2x\cdot3}$

$=\dfrac{35}{12x}$

 You Try It Work through this You Try It problem.

Work Exercises 1–4 in this textbook or in the *MyMathLab*® Study Plan.

Example 2 Multiplying Rational Expressions

My video summary

 Multiply $\dfrac{x^2 - 9}{x^2 + 2x - 35} \cdot \dfrac{3x - 15}{x + 3}$.

Solution Follow the three-step process for multiplying rational expressions.

Begin with the original expression: $\dfrac{x^2 - 9}{x^2 + 2x - 35} \cdot \dfrac{3x - 15}{x + 3}$

Factor numerators and denominators: $= \dfrac{(x - 3)(x + 3)}{(x + 7)(x - 5)} \cdot \dfrac{3(x - 5)}{(x + 3)}$

Continue by cancelling common factors, multiplying remaining factors in the numerator, and multiplying remaining factors in the denominator. Click here to check your answer, or watch this video for a complete solution.

You Try It Work through this You Try It problem.

Work Exercises 5–8 in this textbook or in the MyMathLab® Study Plan.

Example 3 Multiplying Rational Expressions

My interactive video summary

 Multiply $\dfrac{2x^2 + 3x - 2}{3x^2 - 2x - 1} \cdot \dfrac{3x^2 + 4x + 1}{2x^2 + x - 1}$.

Solution Work through this interactive video to complete the solution, or click here to check your answer.

You Try It Work through this You Try It problem.

Work Exercises 9–12 in this textbook or in the MyMathLab® Study Plan.

Common factors are not limited to constants or linear binomial factors. It is possible to have other common polynomial factors, as shown in the next example.

Example 4 Multiplying Rational Expressions

Multiply $\dfrac{3x^2 + 9x + 27}{x^2 - 3x - 10} \cdot \dfrac{x - 5}{x^3 - 27}$.

Solution Follow the same three-step process for multiplying rational expressions.

Begin with the original expression: $\dfrac{3x^2 + 9x + 27}{x^2 - 3x - 10} \cdot \dfrac{x - 5}{x^3 - 27}$

Factor numerators and denominators: $= \dfrac{3(x^2 + 3x + 9)}{(x - 5)(x + 2)} \cdot \dfrac{(x - 5)}{(x - 3)(x^2 + 3x + 9)}$

Notice that there is a common trinomial factor in a numerator and denominator that is prime. These factors cancel just like any other common factor.

Cancel common factors: $\dfrac{3\cancel{(x^2+3x+9)}}{\cancel{(x-5)}(x+2)} \cdot \dfrac{\cancel{(x-5)}}{(x-3)\cancel{(x^2+3x+9)}}$

Multiply remaining factors: $= \dfrac{3}{(x+2)(x-3)}$

You Try It   Work through this You Try It problem.

Work Exercises 13–14 in this textbook or in the *MyMathLab*® Study Plan.

Sometimes factors in the numerator and denominator are additive inverses. The quotient of additive inverses equals -1. This can be seen by factoring out -1 from either the numerator or denominator and cancelling the common factor. For example,

$$\frac{3x-5}{5-3x} = \frac{3x-5}{-1(-5+3x)} = \frac{3x-5}{-1(3x-5)} = \frac{\cancel{3x-5}}{-1\cancel{(3x-5)}} = \frac{1}{-1} = -1.$$

Usually, we want the factors to have positive leading coefficients. If a leading coefficient is negative, we can factor out a negative constant.

Example 5 Multiplying Rational Expressions

Multiply $\dfrac{3x^2+10x-8}{2x-3x^2} \cdot \dfrac{4x+1}{x+4}$.

Solution Multiply the expressions on your own. Remember to factor out a negative constant if the leading coefficient is negative. Click here to check your answer, or watch this video for a complete solution.

You Try It   Work through this You Try It problem.

Work Exercises 15–16 in this textbook or in the *MyMathLab*® Study Plan.

As we saw when factoring in Section 5.2, rational expressions may contain more than one variable.

Example 6 Multiplying Rational Expressions

Multiply $\dfrac{x^2+xy}{3x+y} \cdot \dfrac{3x^2+7xy+2y^2}{x^2-y^2}$.

Solution Remember to factor out the GCF first. For polynomials of the form $ax^2+bxy+cy^2$, remember that the second term in each binomial factor must contain the variable y.

Begin with the original expression: $\dfrac{x^2+xy}{3x+y} \cdot \dfrac{3x^2+7xy+2y^2}{x^2-y^2}$

Factor the numerators and denominators: $= \dfrac{x(x+y)}{3x+y} \cdot \dfrac{(3x+y)(x+2y)}{(x+y)(x-y)}$

Finish the multiplication on your own. Click here to check your answer, or watch this video for a complete solution.

You Try It   **Work through this You Try It problem.**

Work Exercises 17–18 in this textbook or in the *MyMathLab*® Study Plan.

OBJECTIVE 2:   DIVIDE RATIONAL EXPRESSIONS

To divide rational numbers, recall that if $\dfrac{p}{q}$ and $\dfrac{r}{s}$ are rational numbers such that q, r, and s are not zero, then

$$\frac{p}{q} \div \frac{r}{s} = \frac{p}{q} \cdot \frac{s}{r} = \frac{ps}{qr}.$$

The numbers $\dfrac{r}{s}$ and $\dfrac{s}{r}$ are reciprocals of each other.

We follow the same approach when dividing rational expressions. To find the quotient of two rational expressions, we find the product of the first rational expression and the reciprocal of the second rational expression. So, if $\dfrac{P}{Q}$ and $\dfrac{R}{S}$ are rational expressions, then

$$\frac{P}{Q} \div \frac{R}{S} = \frac{P}{Q} \cdot \frac{S}{R} = \frac{PS}{QR}.$$

Dividing Rational Expressions

Step 1. Change the division to multiplication and replace the divisor by its reciprocal.

Step 2. Multiply the expressions.

Example 7 Dividing Rational Expressions

Divide $\dfrac{4y^2 - 100}{5y^2} \div \dfrac{y + 5}{10}$.

Solution

Begin with the original expression: $\dfrac{4y^2 - 100}{5y^2} \div \dfrac{y + 5}{10}$

Change to multiplication by the reciprocal: $= \dfrac{4y^2 - 100}{5y^2} \cdot \dfrac{10}{y + 5}$

Factor: $= \dfrac{4\left(y^2 - 25\right)}{5y^2} \cdot \dfrac{10}{y + 5}$

$= \dfrac{2 \cdot 2(y - 5)(y + 5)}{5y^2} \cdot \dfrac{2 \cdot 5}{y + 5}$

Cancel common factors: $= \dfrac{2 \cdot 2(y - 5)\cancel{(y + 5)}}{\cancel{5}y^2} \cdot \dfrac{2 \cdot \cancel{5}}{\cancel{y + 5}}$

Multiply remaining factors: $= \dfrac{8(y - 5)}{y^2}$

You Try It   Work through this You Try It problem.

Work Exercises 19–21 in this textbook or in the **MyMathLab**® Study Plan.

Example 8 Dividing Rational Expressions

Divide $\dfrac{2x^2 + 21x + 40}{3x^2 + 23x - 8} \div \dfrac{4x^2 + 16x + 15}{x + 2}$.

Solution   Change the division to multiplication by the reciprocal:

Begin with the original expression: $\dfrac{2x^2 + 21x + 40}{3x^2 + 23x - 8} \div \dfrac{4x^2 + 16x + 15}{x + 2}$

Change to multiplication
by the reciprocal: $= \dfrac{2x^2 + 21x + 40}{3x^2 + 23x - 8} \cdot \dfrac{x + 2}{4x^2 + 16x + 15}$

Factor the numerators and denominators and then multiply. Click here to check your answer, or watch this video for a complete solution.

You Try It   Work through this You Try It problem.

Work Exercises 22–24 in this textbook or in the **MyMathLab**® Study Plan.

Example 9 Dividing Rational Expressions

Divide $\dfrac{x^3 - 8}{2x^2 - x - 6} \div \dfrac{x^2 + 2x + 4}{6x^2 + 11x + 3}$.

Solution   Work through this interactive video to complete the solution, or click here to check your answer.

You Try It   Work through this You Try It problem.

Work Exercises 25–26 in this textbook or in the **MyMathLab**® Study Plan.

Example 10 Dividing Rational Expressions

Divide $\dfrac{x^3 - 8y^3}{3x + y} \div \dfrac{4x - 8y}{6x^2 + 17xy + 5y^2}$.

Solution

Begin with the original
expression: $\dfrac{x^3 - 8y^3}{3x + y} \div \dfrac{4x - 8y}{6x^2 + 17xy + 5y^2}$

Change to multiplication: $= \dfrac{x^3 - 8y^3}{3x + y} \cdot \dfrac{6x^2 + 17xy + 5y^2}{4x - 8y}$

Factor: $= \dfrac{(x - 2y)(x^2 + 2xy + 4y^2)}{3x + y} \cdot \dfrac{(3x + y)(2x + 5y)}{4(x - 2y)}$

My video summary

My interactive video summary

My video summary

Finish the problem on your own. Click here to check your answer, or watch the video for a complete solution.

You Try It Work through this You Try It problem.

Work Exercises 27–28 in this textbook or in the *MyMathLab*® Study Plan.

If we need to multiply or divide more than two rational expressions, first we change any divisions to multiplication using the appropriate reciprocal. Then we follow the three-step process for multiplying rational expressions.

Example 11 Multiplying and Dividing Rational Expressions

My video summary Perform the indicated operations.

$$\frac{x^2 + x - 6}{x^2 + x - 42} \cdot \frac{x^2 + 12x + 35}{x^2 - x - 2} \div \frac{x + 7}{x^2 + 8x + 7}$$

Solution Try this problem on your own. Click here to check your answer. Remember to first change all divisions to multiplication using the reciprocal of the divisor. To see a complete solution, watch the video.

You Try It Work through this You Try It problem.

Work Exercises 29–34 in this textbook or in the *MyMathLab*® Study Plan.

6.2 Exercises

In Exercises 1–18, multiply the rational expressions.

You Try It

1. $\dfrac{x - 3}{x + 5} \cdot \dfrac{4x + 20}{9x - 27}$

2. $\dfrac{5x}{3x + 6} \cdot \dfrac{x + 2}{7}$

3. $\dfrac{2x + 2}{4x} \cdot \dfrac{5x - 35}{8x + 16}$

4. $\dfrac{3x + 9}{2x^2 - 6x} \cdot \dfrac{x^2 - 3x}{5x + 15}$

You Try It

5. $\dfrac{x - 6}{x^2 + 3x - 10} \cdot \dfrac{2x - 4}{x^2 - 36}$

6. $\dfrac{x^2 + 12x + 36}{x + 1} \cdot \dfrac{4x + 3}{x^2 + 2x - 24}$

7. $\dfrac{3x - 21}{x + 5} \cdot \dfrac{x^2 - 2x - 15}{4x^2 + 12x}$

8. $\dfrac{3x^2 + 12x}{x - 1} \cdot \dfrac{x^2 - 5x + 4}{x + 2}$

You Try It

9. $\dfrac{6x^2 - x - 1}{x^2 - 7x - 18} \cdot \dfrac{3x^2 + 10x + 8}{3x^2 + 10x + 3}$

10. $\dfrac{x^2 + 4x + 4}{3x^2 + 3x - 18} \cdot \dfrac{5x^2 - 45}{x^2 - 3x - 10}$

11. $\dfrac{4x^2 + 20x - 24}{6x^2 + 7x - 3} \cdot \dfrac{4x^2 + 20x + 21}{9x^2 + 51x - 18}$

12. $\dfrac{6x^2 - 33x - 120}{x^2 + 9x + 14} \cdot \dfrac{x^2 + 16x + 63}{30x^2 - 222x - 144}$

13. $\dfrac{x + 10}{2x^2 - 4x + 8} \cdot \dfrac{3x^3 + 24}{x - 5}$
You Try It

14. $\dfrac{20x^3 + 30x^2 - 10x}{x^2 - 2x - 8} \cdot \dfrac{x^2 - 5x + 4}{2x^2 + 3x - 1}$

You Try It 15. $\dfrac{12 - 2x - 2x^2}{x + 4} \cdot \dfrac{2x - 9}{2x^2 - 3x - 2}$

16. $\dfrac{7 - 28x}{2x - 7} \cdot \dfrac{x^2 + 7x + 12}{4x^2 + 11x - 3}$

17. $\dfrac{x^3 + y^3}{2y^2} \cdot \dfrac{x + 3y}{-x^2 + xy - y^2}$

18. $\dfrac{a^3 + a^2b + a + b}{2a^3 + 2a} \cdot \dfrac{18a^2}{6a^2 - 6b^2}$

You Try It

In Exercises 19–28, divide the rational expressions.

19. $\dfrac{a^2 - 16}{a + 3} \div \dfrac{2a + 8}{a - 3}$

20. $\dfrac{4x}{9} \div \dfrac{8x + 16}{9x + 18}$

You Try It

21. $\dfrac{2x}{5x + 30} \div \dfrac{x^2 + x}{12 - 6x}$

22. $\dfrac{4a^2}{3a^2 + 5a + 2} \div \dfrac{6a^2 - 14a}{2a^2 - 3a - 5}$

You Try It

23. $\dfrac{x^2 - 14x + 49}{x^2 - x - 42} \div \dfrac{x^2 - 49}{8}$

24. $\dfrac{3x - x^2}{x^3 - 27} \div \dfrac{x}{x^2 + 3x + 9}$

You Try It 25. $\dfrac{3y^2 - 3y - 60}{7y^2 - 37y + 10} \div \dfrac{12y^2 + 20y + 3}{2y^2 - y - 6}$

26. $\dfrac{16 + 6x - x^2}{3x^2 - 192} \div \dfrac{x^2 + 10x + 16}{x^2 + 16x + 64}$

27. $\dfrac{x^2 - y^2}{4x + 2y} \div \dfrac{x^2 - 3xy - 4y^2}{2x^2 - 7xy - 4y^2}$

28. $\dfrac{27x^3 - y^3}{4x} \div \dfrac{27x^2 + 9xy + 3y^2}{4x^2 + 16xy}$

You Try It

In Exercises 29–34, perform the indicated operations.

You Try It 29. $\dfrac{2xy}{x^3} \cdot \dfrac{xy^2}{4y} \div \dfrac{5y^3}{3x^2y}$

30. $\dfrac{8}{x} \div \dfrac{7xy}{x^4} \cdot \dfrac{5x^2}{x^7}$

31. $\dfrac{12x^2 + 8x + 1}{y^2 - 2y - 3} \cdot \dfrac{y^2 - 7y + 12}{6x^2 + 7x + 1} \div \dfrac{4x^2 - 4x - 3}{3x^2 - 2x - 5}$

32. $\dfrac{4a^2 - 64}{2a^2 - 4a} \div \dfrac{a^3 + 4a^2}{5a^2 - 10a} \cdot \dfrac{6a^3 + 4a^2}{4a^2 - 16a}$

33. $\dfrac{x + 2}{7x + 2} \cdot \dfrac{3x + 8}{x^2 + x - 2} \cdot \dfrac{7x^2 - 5x - 2}{5x + 10}$

34. $\dfrac{2x - 12}{4x} \div \dfrac{x^2 - 36}{10x} \div \dfrac{x + 1}{x + 6}$

6.3 Adding and Subtracting Rational Expressions

THINGS TO KNOW

Before working through this section, be sure you are familiar with the following concepts:

VIDEO   ANIMATION   INTERACTIVE

You Try It

1. Add Polynomials (Section 4.2, Objective 4)

You Try It

2. Subtract Polynomials (Section 4.2, Objective 5)

You Try It

3. Factor Polynomials Completely
   (Section 5.3, Objective 4)

You Try It

4. Simplify Rational Expressions
   (Section 6.1, Objective 3)

OBJECTIVES

1 Add and Subtract Rational Expressions with Common Denominators

2 Find the Least Common Denominator of Rational Expressions

3 Add and Subtract Rational Expressions with Unlike Denominators

---

OBJECTIVE 1:   ADD AND SUBTRACT RATIONAL EXPRESSIONS WITH COMMON DENOMINATORS

Recall that when adding or subtracting fractions, we need a common denominator. Then we add or subtract the numerators and keep the common denominator. To add or subtract rational numbers with common denominators, we follow this general approach:

$$\frac{p}{q} + \frac{r}{q} = \frac{p+r}{q} \quad \text{or} \quad \frac{p}{q} - \frac{r}{q} = \frac{p-r}{q}$$

For example,

$$\frac{3}{7} + \frac{12}{7} = \frac{3+12}{7} = \frac{15}{7} \quad \text{or} \quad \frac{10}{3} - \frac{8}{3} = \frac{10-8}{3} = \frac{2}{3}.$$

The same is true when working with rational expressions. Once we have a common denominator, we add or subtract the numerators and keep the common denominator.

Adding and Subtracting Rational Expressions with Common Denominators

If $\dfrac{P}{Q}$ and $\dfrac{R}{Q}$ are rational expressions, then

$$\frac{P}{Q} + \frac{R}{Q} = \frac{P+R}{Q} \quad \text{and} \quad \frac{P}{Q} - \frac{R}{Q} = \frac{P-R}{Q}.$$

Example 1 Adding and Subtracting Rational Expressions with Common Denominators

Add or subtract.

a. $\dfrac{3y}{5} + \dfrac{8y}{5}$

b. $\dfrac{7x}{5s^2} - \dfrac{3x}{5s^2}$

Solution

a. The rational expressions have common denominators, so we write the sum of the numerators over the common denominator and simplify.

Begin with the original expression: $\dfrac{3y}{5} + \dfrac{8y}{5}$

Add the numerators: $= \dfrac{3y + 8y}{5}$

Simplify: $= \dfrac{11y}{5}$

b. The rational expressions have a common denominator, so we write the difference of the numerators over the common denominator and simplify.

$$\text{Begin with the original expression:} \quad \frac{7x}{5s^2} - \frac{3x}{5s^2}$$

$$\text{Subtract the numerators:} \quad = \frac{7x - 3x}{5s^2}$$

$$\text{Simplify:} \quad = \frac{4x}{5s^2}$$

You Try It　Work through this You Try It problem.

Work Exercises 1–2 in this textbook or in the **MyMathLab** Study Plan.

CAUTION　When subtracting rational expressions, it is a good idea to use grouping symbols around each numerator that involves more than one term. This will help remind us to subtract each term in the second numerator.

Example 2　Adding and Subtracting Rational Expressions with Common Denominators

Add or subtract.

a. $\dfrac{2x}{x + 3} + \dfrac{4x - 1}{x + 3}$

b. $\dfrac{8x + 3}{x - 7} - \dfrac{2x + 6}{x - 7}$

Solution

a. The rational expressions have common denominators, so we write the sum of the numerators over the common denominator and simplify.

$$\text{Begin with the original expression:} \quad \frac{2x}{x + 3} + \frac{4x - 1}{x + 3}$$

$$\text{Add the numerators:} \quad = \frac{2x + 4x - 1}{x + 3}$$

Finish simplifying. Check the resulting rational expression to see if there are any common factors that can be cancelled. Click here to check your answer.

b. The rational expressions have common denominators, so we write the difference of the numerators over the common denominator and simplify.

$$\text{Begin with the original expression:} \quad \frac{8x + 3}{x - 7} - \frac{2x + 6}{x - 7}$$

$$\text{Subtract the numerators:} \quad = \frac{(8x + 3) - (2x + 6)}{x - 7}$$

$$\text{Use the distributive property:} \quad = \frac{8x + 3 - 2x - 6}{x - 7}$$

Finish simplifying. Check the resulting rational expression to see if there are any common factors that can be cancelled. Click here to check your answer.

You Try It   Work through this You Try It problem.

Work Exercises 3–4 in this textbook or in the *MyMathLab*® Study Plan.

Example 3   Adding and Subtracting Rational Expressions with Common Denominators

My interactive video summary

 Add or subtract.

a. $\dfrac{4}{x^2 + 2x - 8} + \dfrac{x}{x^2 + 2x - 8}$

b. $\dfrac{x}{x + 2} - \dfrac{x - 3}{x + 2}$

c. $\dfrac{x^2 - 2}{x - 5} - \dfrac{4x + 3}{x - 5}$

Solution   Try to perform the operations on your own. Remember to check your result for any common factors. Click here to check your answers, or watch this interactive video for complete solutions.

You Try It   Work through this You Try It problem.

Work Exercises 5–8 in this textbook or in the *MyMathLab*® Study Plan.

OBJECTIVE 2:   FIND THE LEAST COMMON DENOMINATOR OF RATIONAL EXPRESSIONS

To add or subtract rational expressions without a common denominator, first we rewrite each rational expression as an equivalent expression using a common denominator. Typically, we use the least common denominator (LCD). To find the LCD, we first factor each denominator. Then we include each unique factor the largest number of times that it occurs in any denominator. We can illustrate this idea by first considering rational numbers.

To add $\dfrac{3}{10} + \dfrac{4}{15} + \dfrac{7}{4}$, first factor each denominator into its prime factors

$$10 = 2 \cdot 5$$
$$15 = 3 \cdot 5$$
$$4 = 2 \cdot 2$$

Next, form the LCD by including each unique factor the largest number of times that it occurs in any denominator. The factors 3 and 5 both occur at most once in any denominator, whereas the factor 2 occurs a maximum of two times in a denominator. So, our LCD would be $\text{LCD} = 3 \cdot 5 \cdot 2 \cdot 2 = 3 \cdot 5 \cdot 2^2 = 60$.

We follow a similar process to find the LCD of rational expressions.

Finding the Least Common Denominator (LCD) of Rational Expressions

Step 1. Factor each denominator completely.

Step 2. List each unique factor from any denominator.

Step 3. The least common denominator is the product of the unique factors, each raised to a power equivalent to the largest number of times that the factor occurs in any denominator.

 Don't forget numerical factors when forming the LCD.

Example 4 Finding the Least Common Denominator of Rational Expressions

Find the LCD of the rational expressions.

a. $\dfrac{5}{8x^2y^2}, \dfrac{7}{2xy^3z}$

b. $\dfrac{x+4}{2-x}, \dfrac{x-7}{x^2-8x+12}$

c. $\dfrac{z^2}{7z-21}, \dfrac{4}{4z^2-12z}$

d. $\dfrac{y+2}{y^2+2y-3}, \dfrac{2y}{y^2+5y+6}$

Solution

a. Follow the three-step process.

Step 1. $8x^2y^2 = 2\cdot2\cdot2\cdot x\cdot x\cdot y\cdot y$

$2xy^3z = 2\cdot x\cdot y\cdot y\cdot y\cdot z$

Step 2. We have unique factors of 2, x, y, and z. Make sure to include each factor the largest number of times that it occurs in any denominator.

Step 3. The factor z occurs at most once, the factor x occurs at most two times, and the factors 2 and y occur at most three times.

$$LCD = 2\cdot2\cdot2\cdot x\cdot x\cdot y\cdot y\cdot y\cdot z = 8x^2y^3z$$

b. Follow the three-step process.

Step 1. $2 - x = -1\cdot(x-2)$

$x^2 - 8x + 12 = (x-6)(x-2)$

Step 2. We have unique factors of -1, $x-2$, and $x-6$. Notice that we factored out a -1 in the first denominator. We will do this when the leading coefficient is negative to avoid including opposite factors in the LCD. In this case, we include -1 and $x-2$ instead of $x-2$ and $2-x$.

Step 3. The factors -1, $x-2$, and $x-6$ occur at most once.

$$LCD = -1\cdot(x-2)\cdot(x-6) = -(x-2)(x-6)$$

c. Follow the three-step process.

Step 1. $7z - 21 = 7\cdot(z-3)$

$4z^2 - 12z = 2\cdot2\cdot z\cdot(z-3)$

Step 2. We have unique factors of $7, 2, z$, and $z-3$.

Determine the largest number of times that each unique factor occurs and form the LCD. Click here to check your answer.

My video summary

d. Follow the three-step process.

Step 1. $y^2 + 2y - 3 = (y + 3)(y - 1)$

$y^2 + 5y + 6 = (y + 3)(y + 2)$

 Determine the unique factors and the largest number of times that each occurs. Then form the LCD. Click here to check your answer, or watch this video for a complete solution.

You Try It   **Work through this You Try It problem.**

Work Exercises 9–12 in this textbook or in the *MyMathLab*® Study Plan.

Example 5 Finding the Least Common Denominator of Rational Expressions

My interactive video summary

 Find the LCD of the rational expressions.

a. $\dfrac{4x}{10x^2 - 7x - 12}, \dfrac{2x - 3}{5x^2 - 11x - 12}$

b. $\dfrac{10 - x}{6x^2 + 5x + 1}, \dfrac{-4}{9x^2 + 6x + 1}, \dfrac{x^2 - 7x}{10x^2 - x - 3}$

Solution Try to find the LCD on your own. Remember to check your result for any common factors. Click here to check your answers, or watch this interactive video for complete solutions.

You Try It   **Work through this You Try It problem.**

Work Exercises 13–14 in this textbook or in the *MyMathLab*® Study Plan.

OBJECTIVE 3: ADD AND SUBTRACT RATIONAL EXPRESSIONS WITH UNLIKE DENOMINATORS

To add or subtract rational expressions with unlike denominators, first we write each fraction as an equivalent expression using the LCD. The key to doing this lies with our previous work on cancelling common factors in Section 6.1. Recall that we said

$$\frac{\text{common factor}}{\text{common factor}} = 1.$$

We were able to cancel common factors that occurred in the numerator and denominator because it essentially meant we were multiplying by 1.

When writing equivalent fractions, we are actually working the other way. We are going to multiply the original expression by 1, but in such a way as to obtain the desired LCD. To do this, we first ask the question,

"What should we multiply by the original denominator to get the LCD?"

Whatever the answer is, multiply this by both the numerator and denominator.

For example, if we want to rewrite $\dfrac{4}{7}$ as an equivalent fraction with a denominator of 56, we ask, "What do we multiply by 7 to get 56?" Since the answer is 8, we multiply the numerator and denominator by 8 to get $\dfrac{4}{7} = \dfrac{4}{7} \cdot \dfrac{8}{8} = \dfrac{32}{56}$. Notice that $\dfrac{8}{8} = 1$, so we have not changed the value, just the way it looks.

To work with rational expressions, we ask the same question—*"What do we multiply by the original denominator to get the LCD?"* Then multiply both the numerator and denominator by your answer.

Adding and Subtracting Rational Expressions with Unlike Denominators

Step 1. Find the LCD for all expressions being added or subtracted.

Step 2. Write equivalent expressions for each term using the LCD as the denominator.

Step 3. Add/subtract the numerators, but keep the denominator the same (the LCD).

Step 4. Simplify if possible.

We focus on the LCD because it is the easiest common denominator to use. Any common denominator will allow us to add or subtract rational expressions.

Example 6 Adding and Subtracting Rational Expressions

Perform the indicated operations and simplify.

a. $\dfrac{5}{2x^2} + \dfrac{7}{3x}$
 b. $\dfrac{3x}{x-3} - \dfrac{x-2}{x+3}$

Solution

a. We follow the four-step process.

Step 1. $2x^2 = 2 \cdot x \cdot x$

$3x = 3 \cdot x$

We have unique factors of 2, 3, and x. Make sure to include each factor the largest number of times that it occurs in any denominator.

$$LCD = 2 \cdot 3 \cdot x \cdot x = 6x^2$$

Step 2. In the first expression, we multiply the denominator by 3 to get the LCD, so we obtain an equivalent expression by multiplying both the numerator and denominator by 3. In the second expression, we multiply the denominator by $2x$ to get the LCD, so we obtain an equivalent expression by multiplying both the numerator and denominator by $2x$.

The overall expression becomes

$$\frac{5}{2x^2} + \frac{7}{3x} = \frac{5}{2x^2} \cdot \frac{3}{3} + \frac{7}{3x} \cdot \frac{2x}{2x}$$

$$= \frac{15}{6x^2} + \frac{14x}{6x^2}.$$

Step 3. With a common denominator, we can now add:

$$\frac{15}{6x^2} + \frac{14x}{6x^2} = \frac{15 + 14x}{6x^2} \text{ or } \frac{14x + 15}{6x^2}$$

Step 4. The expression cannot be simplified, so $\dfrac{5}{2x^2} + \dfrac{7}{3x} = \dfrac{14x + 15}{6x^2}$.

My video summary

b. We follow the four-step process.

Step 1. The denominators cannot be factored further.

We have unique factors of $x - 3$ and $x + 3$. Make sure to include each factor the largest number of times that it occurs in any denominator.

$$\text{LCD} = (x - 3)(x + 3)$$

Step 2. In the first expression, we multiply the denominator by $x + 3$ to get the LCD and obtain an equivalent expression by multiplying both the numerator and denominator by $x + 3$. In the second expression, we multiply the denominator by $x - 3$ to get the LCD and obtain an equivalent expression by multiplying both the numerator and denominator by $x - 3$.

So, we can write

$$\frac{3x}{x - 3} - \frac{x - 2}{x + 3} = \frac{3x}{x - 3} \cdot \frac{x + 3}{x + 3} - \frac{x - 2}{x + 3} \cdot \frac{x - 3}{x - 3}$$

$$= \frac{3x(x + 3)}{(x - 3)(x + 3)} - \frac{(x - 2)(x - 3)}{(x - 3)(x + 3)}.$$

Carry out the subtraction on your own and simplify if necessary. Click here to check your answer, or watch this video for a complete solution.

You Try It Work through this You Try It problem.

Work Exercises 15–22 in this textbook or in the *MyMathLab*® Study Plan.

Example 7 Adding and Subtracting Rational Expressions

Perform the indicated operations and simplify.

a. $2 + \dfrac{4}{x - 5}$

b. $\dfrac{x^2 - 2}{x^2 + 6x + 8} - \dfrac{x - 3}{x + 4}$

Solution

a. Follow the four-step process.

Step 1. Note that the first term is a rational expression since we can write it as $2 = \dfrac{2}{1}$. The LCD is $x - 5$.

Step 2. In the first expression, we multiply the denominator by $x - 5$ to get the LCD and obtain an equivalent expression by multiplying both the numerator and denominator by $x - 5$.

So, we can write
$$2 + \frac{4}{x-5} = \frac{2}{1} + \frac{4}{x-5}$$
$$= \frac{2}{1} \cdot \frac{x-5}{x-5} + \frac{4}{x-5}$$
$$= \frac{2(x-5)}{x-5} + \frac{4}{x-5}.$$

Step 3. With a common denominator, we can now add:

Add: $\quad = \dfrac{2(x-5)+4}{x-5}$

Distribute: $\quad = \dfrac{2x-10+4}{x-5}$

Simplify: $\quad = \dfrac{2x-6}{x-5}$

$$= \frac{2(x-3)}{x-5}$$

Step 4. The expression cannot be simplified further, so
$$2 + \frac{4}{x-5} = \frac{2(x-3)}{x-5}.$$

My video summary

b. Follow the four-step process.

Step 1. Factor the denominators and determine the LCD. The second denominator is prime, but the first denominator can be factored as $x^2 + 6x + 8 = (x+4)(x+2)$.
The LCD is $(x+2)(x+4)$.

Step 2. In the first expression, we already have the LCD. In the second expression, we multiply the denominator by $x+2$ to get the LCD and obtain an equivalent expression by multiplying both the numerator and denominator by $x+2$.

Doing this gives
$$\frac{x^2-2}{x^2+6x+8} - \frac{x-3}{x+4} = \frac{x^2-2}{(x+4)(x+2)} - \frac{x-3}{x+4} \cdot \frac{x+2}{x+2}$$
$$= \frac{x^2-2}{(x+4)(x+2)} - \frac{(x-3)(x+2)}{(x+4)(x+2)}.$$

🎞 Carry out the subtraction on your own and simplify if necessary. Click here to check your answer, or watch this video for a complete solution.

You Try It Work through this **You Try It** problem.

Work Exercises **23–28** in this textbook or in the *MyMathLab*® Study Plan.

Example 8 Adding and Subtracting Rational Expressions with Unlike Denominators

Perform the indicated operations and simplify.

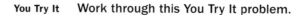

a. $\dfrac{3}{x-y} - \dfrac{x+5y}{x^2-y^2}$ $\qquad$ b. $\dfrac{x+4}{3x^2+20x+25} + \dfrac{x}{3x^2+16x+5}$

My interactive video summary

Solution Perform the operations on your own and simplify. Click here to check your answers, or watch this interactive video to see complete solutions.

You Try It Work through this You Try It problem.

Work Exercises 29–31 in this textbook or in the *MyMathLab*® Study Plan.

If we are adding or subtracting rational expressions whose denominators are opposites, we can find a common denominator by multiplying the numerator and denominator of either rational expression by -1 (but not both).

Example 9 Adding and Subtracting Rational Expressions with Unlike Denominators

Perform the indicated operation and simplify.

$$\frac{2y}{y-5} + \frac{y-1}{5-y}$$

Solution

$$\text{The denominators are opposites:} \quad \frac{2y}{y-5} + \frac{y-1}{5-y}$$

$$\text{Multiply the numerator and denominator of the second rational expression by } -1: \quad = \frac{2y}{y-5} + \frac{(-1)}{(-1)} \cdot \frac{y-1}{5-y}$$

$$\text{Distribute:} \quad = \frac{2y}{y-5} + \frac{-y+1}{-5+y}$$

$$\text{Rewrite the denominator in second expression:} \quad = \frac{2y}{y-5} + \frac{-y+1}{y-5}$$

$$\text{Add:} \quad = \frac{2y-y+1}{y-5}$$

$$\text{Simplify:} \quad = \frac{y+1}{y-5}$$

You Try It Work through this You Try It problem.

Work Exercises 32–33 in this textbook or in the *MyMathLab*® Study Plan.

Example 10 Adding and Subtracting Rational Expressions with Unlike Denominators

My video summary

Perform the indicated operations and simplify.

$$\frac{x+1}{x^2-6x+9} + \frac{3}{x-3} - \frac{6}{x^2-9}$$

Solution Following the four-step process, we start by finding the LCD for all the terms. $x^2 - 6x + 9 = (x-3)(x-3)$; $x - 3$ is prime; $x^2 - 9 = (x+3)(x-3)$

The LCD is $(x - 3)(x - 3)(x + 3)$.

Next, we rewrite each term as an equivalent expression using the LCD.

$$\frac{x + 1}{x^2 - 6x + 9} + \frac{3}{x - 3} - \frac{6}{x^2 - 9}$$

$$= \frac{x + 1}{(x - 3)(x - 3)} \cdot \frac{x + 3}{x + 3} + \frac{3}{x - 3} \cdot \frac{(x - 3)(x + 3)}{(x - 3)(x + 3)} - \frac{6}{(x + 3)(x - 3)} \cdot \frac{x - 3}{x - 3}$$

$$= \frac{(x + 1)(x + 3)}{(x - 3)(x - 3)(x + 3)} + \frac{3(x - 3)(x + 3)}{(x - 3)(x - 3)(x + 3)} - \frac{6(x - 3)}{(x - 3)(x - 3)(x + 3)}$$

Now we combine the three numerators and keep the common denominator.

$$= \frac{(x + 1)(x + 3) + 3(x - 3)(x + 3) - 6(x - 3)}{(x - 3)(x - 3)(x + 3)}$$

Finish simplifying the expression on your own. Click here to check your answer, or watch this video for a complete solution.

You Try It   Work through this You Try It problem.

Work Exercises 34–36 in this textbook or in the **MyMathLab**® Study Plan.

6.3 Exercises

In Exercises 1–8, add or subtract. Simplify if possible.

 You Try It   **1.** $\dfrac{7}{x} + \dfrac{9}{x}$

2. $\dfrac{5w}{3ab^2} - \dfrac{2w}{3ab^2}$

You Try It   **3.** $\dfrac{x + 3}{x - 4} + \dfrac{5x - 2}{x - 4}$

4. $\dfrac{4x - 5}{5x} - \dfrac{4x + 15}{5x}$

 You Try It   **5.** $\dfrac{2x}{x^2 - 3x - 28} - \dfrac{14}{x^2 - 3x - 28}$

6. $\dfrac{x^2}{3x + 1} - \dfrac{4}{3x + 1}$

7. $\dfrac{n^2}{n^2 - 64} + \dfrac{64 - 16n}{n^2 - 64}$

8. $\dfrac{x^2 - 8}{x - 3} + \dfrac{2x - 7}{x - 3}$

 In Exercises 9–14, find the LCD for the given rational expressions.

You Try It

9. $\dfrac{6}{5x^3y^2}, \dfrac{3}{4x^2y^5}$

10. $\dfrac{3x + 5}{x^2 - 9}, \dfrac{4x}{x^2 - 6x + 9}$

11. $\dfrac{n + 6}{n^2 - 5n - 14}, \dfrac{2n}{n^2 + 6n + 8}$

12. $\dfrac{7}{x^2 - 16}, \dfrac{x - 1}{3x}, \dfrac{2x}{20 - 5x}$

 You Try It

13. $\dfrac{2b}{6b^2 - 23b + 21}, \dfrac{3b - 1}{2b^2 + 7b - 15}$

14. $\dfrac{m^2 + 2m - 7}{10m^2 - 13m - 3}, \dfrac{m + 4}{25m^2 + 10m + 1}$

In Exercises 15–36, add or subtract. Simplify if possible.

15. $\dfrac{6}{49x} + \dfrac{5}{x^2}$

16. $\dfrac{7}{12x^2y} - \dfrac{13}{4xy}$

17. $\dfrac{8}{x-1} - \dfrac{6}{x+8}$

18. $\dfrac{x}{x+3} + \dfrac{2x-7}{x-3}$

19. $\dfrac{x-2}{x+4} - \dfrac{x+1}{x-4}$

20. $\dfrac{x}{x^2-36} + \dfrac{4}{x}$

21. $\dfrac{10}{a+b} + \dfrac{10}{a-b}$

22. $\dfrac{z+5}{z-9} - \dfrac{z-4}{z+6}$

23. $3 + \dfrac{x}{x-2}$

24. $\dfrac{x^2-18}{x^2-5x+6} - \dfrac{x-3}{x-2}$

25. $\dfrac{3x-4}{x+3} - \dfrac{6-24x}{x^2-9}$

26. $\dfrac{2(x+11)}{2x^2-x-10} + \dfrac{x+4}{x+2}$

27. $\dfrac{x-8}{x^2+12x+27} + \dfrac{x-9}{x^2-9}$

28. $\dfrac{6x}{x^2+4x-12} - \dfrac{x}{x^2-36}$

29. $\dfrac{9}{x-y} - \dfrac{x+3y}{x^2-y^2}$

30. $\dfrac{y-8}{2y^2+9y+7} - \dfrac{y+9}{2y^2-3y-35}$

31. $\dfrac{6x}{3x^2+4xy-4y^2} - \dfrac{3x}{x^2+xy-2y^2}$

32. $\dfrac{9}{x-3} + \dfrac{x}{3-x}$

33. $\dfrac{x-3}{4-x} - \dfrac{x}{x-4}$

34. $\dfrac{5}{y} - \dfrac{5}{y-3} + \dfrac{16}{(y-3)^2}$

35. $\dfrac{x-3}{x^2+12x+36} + \dfrac{1}{x+6} - \dfrac{2x+3}{2x^2+3x-54}$

36. $\left(\dfrac{1}{2} + \dfrac{3}{x}\right) - \left(\dfrac{1}{2} - \dfrac{1}{x}\right)$

6.4 Complex Rational Expressions

THINGS TO KNOW

Before working through this section, be sure you are familiar with the following concepts:

VIDEO   ANIMATION   INTERACTIVE

1. Use the Negative-Power Rule
(Section 4.1, Objective 4)

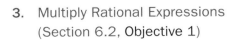
2. Simplify Rational Expressions
(Section 6.1, Objective 3)

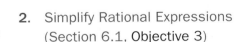
3. Multiply Rational Expressions
(Section 6.2, Objective 1)

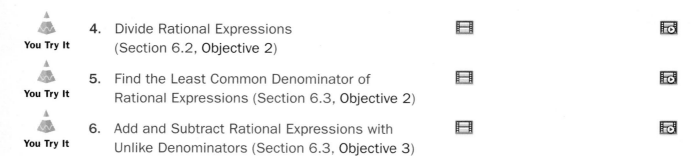

4. Divide Rational Expressions
   (Section 6.2, Objective 2)

You Try It

5. Find the Least Common Denominator of
   Rational Expressions (Section 6.3, Objective 2)

You Try It

6. Add and Subtract Rational Expressions with
   Unlike Denominators (Section 6.3, Objective 3)

You Try It

OBJECTIVES

1 Simplify Complex Rational Expressions by First Simplifying the Numerator and Denominator

2 Simplify Complex Rational Expressions by Multiplying by a Common Denominator

OBJECTIVE 1: **SIMPLIFY COMPLEX RATIONAL EXPRESSIONS BY FIRST SIMPLIFYING THE NUMERATOR AND DENOMINATOR**

Sometimes the numerator and/or denominator of a rational expression will contain one or more rational expressions. In this section, we learn to simplify such *complex rational expressions*.

Definition   Complex Rational Expression

A **complex rational expression**, or **complex fraction**, is a rational expression in which the numerator and/or denominator contain(s) rational expressions.

Some examples of complex rational expressions are

$$\frac{\dfrac{2}{9x}}{\dfrac{5}{6xy}}, \quad \frac{\dfrac{x+2}{x-4}}{\dfrac{x^2-4}{x+1}}, \quad \frac{\dfrac{1}{y}+\dfrac{y+1}{y-1}}{\dfrac{1}{y}-\dfrac{y-1}{y+1}}, \quad \frac{\dfrac{1}{x}+\dfrac{1}{y}}{z}, \quad \text{and} \quad \frac{z^{-2}+z^{-1}}{z^{-1}-z^{-2}}.$$

The rational expressions within the numerator and denominator are called **minor rational expressions** or **minor fractions**. The numerator and denominator of the complex rational expression are separated by the **main fraction bar**.

$$\text{Minor rational expressions} \begin{cases} \nearrow & \left.\dfrac{x+2}{x-4}\right\} \leftarrow \text{Numerator of the complex rational expression} \\ & \qquad\qquad \leftarrow \text{Main fraction bar} \\ \searrow & \left.\dfrac{x^2-4}{x+1}\right\} \leftarrow \text{Denominator of the complex rational expression} \end{cases}$$

To **simplify a complex rational expression**, we rewrite it in the form $\dfrac{P}{Q}, Q \neq 0$, where

P and Q are polynomials with no common factors. In this section, we show two methods that can be used to simplify complex rational expressions. We call these *Method I* and *Method II*. Method I results from recognizing that the main fraction bar is a division symbol. So, we divide the numerator by the denominator.

Example 1 Simplifying a Complex Rational Expression

Simplify $\dfrac{\dfrac{2}{9x}}{\dfrac{5}{6xy}}$.

Solution We divide the minor rational expression from the numerator by the minor rational expression from the denominator. To do this, we multiply the numerator by the reciprocal of the denominator.

$$\text{Begin with the original expression:} \quad \frac{\dfrac{2}{9x}}{\dfrac{5}{6xy}} = \frac{2}{9x} \div \frac{5}{6xy}$$

$$\text{Change to multiplication by the reciprocal:} \quad = \frac{2}{9x} \cdot \frac{6xy}{5}$$

$$\text{Factor:} \quad = \frac{2}{3 \cdot 3 \cdot x} \cdot \frac{2 \cdot 3 \cdot x \cdot y}{5}$$

$$\text{Cancel common factors:} \quad = \frac{2}{\cancel{3} \cdot 3 \cdot \cancel{x}} \cdot \frac{2 \cdot \cancel{3} \cdot \cancel{x} \cdot y}{5}$$

$$\text{Multiply the remaining factors:} \quad = \frac{4y}{15}$$

You Try It   Work through this You Try It problem.

Work Exercises 1–3 in this textbook or in the *MyMathLab*® Study Plan.

In addition to acting as a division symbol, the main fraction bar of a complex fraction also serves as a grouping symbol. If the numerator and denominator are not each written as single rational expressions, then they must be simplified as such before dividing.

Method I for Simplifying Complex Rational Expressions

Step 1. Simplify the expression in the numerator into a single rational expression.

Step 2. Simplify the expression in the denominator into a single rational expression.

Step 3. Divide the expression in the numerator by the expression in the denominator. To do this, multiply the minor rational expression in the numerator by the reciprocal of the minor rational expression in the denominator. Simplify if possible.

Example 2 Simplifying a Complex Rational Expression Using Method I

Use Method I to simplify each complex rational expression.

a. $\dfrac{\dfrac{1}{3} - \dfrac{1}{x}}{\dfrac{1}{9} - \dfrac{1}{x^2}}$

b. $\dfrac{4 - \dfrac{5}{x-1}}{\dfrac{6}{x-1} - 7}$

Solution

a. **Step 1.** The LCD for the minor fractions $\dfrac{1}{3}$ and $-\dfrac{1}{x}$ in the numerator is $3x$.

$$\frac{1}{3} - \frac{1}{x} = \frac{1}{3}\cdot\frac{x}{x} - \frac{1}{x}\cdot\frac{3}{3} = \frac{x}{3x} - \frac{3}{3x} = \frac{x-3}{3x}$$

Step 2. The LCD for the minor fractions $\dfrac{1}{9}$ and $-\dfrac{1}{x^2}$ in the denominator is $9x^2$.

$$\frac{1}{9} - \frac{1}{x^2} = \frac{1}{9}\cdot\frac{x^2}{x^2} - \frac{1}{x^2}\cdot\frac{9}{9} = \frac{x^2}{9x^2} - \frac{9}{9x^2} = \frac{x^2-9}{9x^2}$$

Step 3. Substitute the simplified rational expressions from steps 1 and 2 for the numerator and denominator of the complex fraction, and then divide.

$$\frac{\dfrac{1}{3} - \dfrac{1}{x}}{\dfrac{1}{9} - \dfrac{1}{x^2}} = \frac{\dfrac{x-3}{3x}}{\dfrac{x^2-9}{9x^2}} = \frac{x-3}{3x} \div \frac{x^2-9}{9x^2}$$

Change to multiplication by the reciprocal: $= \dfrac{x-3}{3x}\cdot\dfrac{9x^2}{x^2-9}$

Factor: $= \dfrac{x-3}{3x}\cdot\dfrac{3\cdot3\cdot x\cdot x}{(x+3)(x-3)}$

Cancel common factors: $= \dfrac{\cancel{x-3}}{3x}\cdot\dfrac{3\cdot3\cdot x\cdot x}{(x+3)\cancel{(x-3)}}$

Multiply remaining factors: $= \dfrac{3x}{x+3}$

My video summary

b. **Step 1.** The LCD for the numerator is $x - 1$.

$$4 - \frac{5}{x-1} = 4\cdot\frac{x-1}{x-1} - \frac{5}{x-1} = \frac{4(x-1)-5}{x-1} = \frac{4x-4-5}{x-1} = \frac{4x-9}{x-1}$$

Step 2. The LCD for the denominator is $x - 1$.

$$\frac{6}{x-1} - 7 = \frac{6}{x-1} - 7\cdot\frac{x-1}{x-1} = \frac{6-7(x-1)}{x-1} = \frac{6-7x+7}{x-1} = \frac{13-7x}{x-1}$$

Step 3. Try to finish simplifying the complex rational expression on your own. Click here to check your answer, or watch this video for the complete solution.

You Try It   Work through this You Try It problem.

Work Exercises 4–6 in this textbook or in the *MyMathLab*® Study Plan.

OBJECTIVE 2:   SIMPLIFY COMPLEX RATIONAL EXPRESSIONS BY
MULTIPLYING BY A COMMON DENOMINATOR

In Section 6.3, we saw that multiplying the numerator and denominator of a
rational expression by the same nonzero expression resulted in an equivalent
expression. For example, $\frac{3}{4} = \frac{3}{4} \cdot \frac{7}{7} = \frac{21}{28}$ because $\frac{7}{7} = 1$.

The second method for simplifying complex rational expressions is based on this
same concept. Let's take another look at the complex fraction from Example 1.

Example 3 Simplifying a Complex Rational Expression

Simplify $\dfrac{\frac{2}{9x}}{\frac{5}{6xy}}$.

Solution   Let's multiply the numerator and denominator of the complex rational
expression by $18xy$, the LCD of $\frac{2}{9x}$ and $\frac{5}{6xy}$.

$$\text{Begin with the original expression:} \quad \dfrac{\frac{2}{9x}}{\frac{5}{6xy}}$$

$$\text{Multiply the numerator and denominator by } 18xy: \quad = \dfrac{\frac{2}{9x}}{\frac{5}{6xy}} \cdot \dfrac{18xy}{18xy}$$

$$\text{Cancel:} \quad = \dfrac{\frac{2}{9\cancel{x}} \cdot \overset{2}{\cancel{18x}}y}{\frac{5}{6\cancel{xy}} \cdot \overset{3}{\cancel{18xy}}}$$

$$\text{Rewrite the remaining factors:} \quad = \dfrac{2 \cdot 2 \cdot y}{5 \cdot 3}$$

$$\text{Multiply:} \quad = \dfrac{4y}{15}$$

You Try It   Work through this You Try It problem.

Work Exercises 7–9 in this textbook or in the *MyMathLab*® Study Plan.

In Example 3, we multiplied the numerator and denominator of the complex rational
expression by the LCD of its minor rational expressions. Multiplying by the LCD is
the foundation of Method II for simplifying complex rational expressions.

Method II for Simplifying Complex Rational Expressions

Step 1. Determine the LCD of all the minor rational expressions within the complex rational expression.

Step 2. Multiply the numerator and denominator of the complex rational expression by the LCD from step 1.

Step 3. Simplify.

For comparison, let's revisit the complex rational expressions from Example 2.

Example 4 Simplifying a Complex Rational Expression by Using Method II

Use Method II to simplify each complex rational expression.

a. $\dfrac{\dfrac{1}{3} - \dfrac{1}{x}}{\dfrac{1}{9} - \dfrac{1}{x^2}}$
b. $\dfrac{4 - \dfrac{5}{x-1}}{\dfrac{6}{x-1} - 7}$

Solution

a. **Step 1.** The denominators of all the minor fractions are $3, x, 9,$ and x^2, so the LCD is $9x^2$.

Step 2. Multiply the numerator and denominator by $9x^2$.

$$\frac{\dfrac{1}{3} - \dfrac{1}{x}}{\dfrac{1}{9} - \dfrac{1}{x^2}} \cdot \frac{9x^2}{9x^2}$$

Step 3. Use the distributive property:

$$= \frac{\dfrac{1}{3} \cdot 9x^2 - \dfrac{1}{x} \cdot 9x^2}{\dfrac{1}{9} \cdot 9x^2 - \dfrac{1}{x^2} \cdot 9x^2}$$

Cancel:

$$= \frac{\dfrac{1}{\cancel{3}} \cdot \overset{3}{\cancel{9}}x^2 - \dfrac{1}{\cancel{x}} \cdot 9\overset{x}{\cancel{x^2}}}{\dfrac{1}{\cancel{9}} \cdot \cancel{9}x^2 - \dfrac{1}{\cancel{x^2}} \cdot 9\cancel{x^2}}$$

Simplify:

$$= \frac{3x^2 - 9x}{x^2 - 9}$$

Factor each polynomial:

$$= \frac{3x(x-3)}{(x+3)(x-3)}$$

Cancel common factors:

$$= \frac{3x\cancel{(x-3)}}{(x+3)\cancel{(x-3)}}$$

Simplify:

$$= \frac{3x}{x+3}$$

My video summary

b. **Step 1.** The LCD for all the minor fractions is $x - 1$.

Step 2. Multiply the numerator and denominator by $x - 1$.

$$\frac{4 - \dfrac{5}{x - 1}}{\dfrac{6}{x - 1} - 7} \cdot \frac{x - 1}{x - 1}$$

Step 3. Try to finish simplifying the complex rational expression on your own. Click here to check your answer, or watch this video for the complete solution.

You Try It   Work through this You Try It problem.

Work Exercises 10–12 in this textbook or in the *MyMathLab*® Study Plan.

As you have seen in Examples 1–4, Method I and Method II both give the same simplification. Have you developed a preference for one method over the other? Whether you have or not, we suggest that you practice using both methods for a while to discover when one method might be better suited than the other.

Example 5 Simplifying a Complex Rational Expression

My interactive video summary

 Simplify the complex rational expression using Method I or Method II.

$$\frac{-\dfrac{1}{x} - \dfrac{3}{x + 4}}{\dfrac{2}{x^2 + 4x} + \dfrac{2}{x}}$$

Solution Try simplifying on your own. Click here to check your answer, or watch this interactive video to see the complete solution.

You Try It   Work through this You Try It problem.

Work Exercises 13–24 in this textbook or in the *MyMathLab*® Study Plan.

Complex rational expressions can be written using negative exponents. For example, the expression $\dfrac{3^{-1} - x^{-1}}{3^{-2} - x^{-2}}$ is equivalent to the complex fraction $\dfrac{\dfrac{1}{3} - \dfrac{1}{x}}{\dfrac{1}{9} - \dfrac{1}{x^2}}$ from Example 2.

When a complex rational expression contains negative exponents, we can rewrite it as an equivalent expression with positive exponents by using the negative power rule. Then we simplify it by using Method I or Method II.

Example 6 Simplifying a Complex Rational Expression
Containing Negative Exponents

Simplify the complex rational expression.

$$\frac{1 - 9y^{-1} + 14y^{-2}}{1 + 3y^{-1} - 10y^{-2}}$$

 *My interactive video summary*

Solution Use the negative power rule to rewrite the expression with positive exponents.

$$\frac{1 - 9y^{-1} + 14y^{-2}}{1 + 3y^{-1} - 10y^{-2}} = \frac{1 - \dfrac{9}{y} + \dfrac{14}{y^2}}{1 + \dfrac{3}{y} - \dfrac{10}{y^2}}$$

Try to finish simplifying the complex rational expression on your own using Method I or Method II. Click here to check your answer, or watch this interactive video for the complete solution.

You Try It Work through this You Try It problem.

Work Exercises 25–26 in this textbook or in the *MyMathLab*® Study Plan.

6.4 Exercises

In Exercises 1–6, use Method I to simplify each complex rational expression.

You Try It

1. $\dfrac{\dfrac{14}{5x}}{\dfrac{7}{15x}}$

2. $\dfrac{\dfrac{x+7}{2}}{\dfrac{3x-1}{10}}$

3. $\dfrac{\dfrac{5x+5}{x^2-9}}{\dfrac{x^2-1}{x+3}}$

You Try It

4. $\dfrac{\dfrac{5}{6}+\dfrac{3}{4}}{\dfrac{8}{3}-\dfrac{5}{9}}$

5. $\dfrac{\dfrac{3}{x}+2}{\dfrac{9}{x^2}-4}$

6. $\dfrac{\dfrac{9}{x-4}+\dfrac{6}{x-5}}{\dfrac{5x-23}{x^2-9x+20}}$

In Exercises 7–12, use Method II to simplify each complex rational expression.

You Try It

7. $\dfrac{\dfrac{x+2}{15x}}{\dfrac{2x-1}{10x}}$

8. $\dfrac{\dfrac{8x}{x+y}}{\dfrac{2x^3}{y}}$

9. $\dfrac{\dfrac{3}{x^2-x-2}}{\dfrac{x+4}{x-2}}$

You Try It

10. $\dfrac{2-\dfrac{a}{b}}{\dfrac{a^2}{b^2}-4}$

11. $\dfrac{3-\dfrac{7}{x-2}}{\dfrac{8}{x-2}-5}$

12. $\dfrac{\dfrac{x+5}{x}-\dfrac{8}{x-1}}{\dfrac{x+1}{x}+\dfrac{x+1}{x-1}}$

In Exercises 13–26, simplify each complex rational expression using Method I or Method II.

13. $\dfrac{7+\dfrac{1}{x}}{7-\dfrac{1}{x}}$

14. $\dfrac{\dfrac{1}{x}+3}{\dfrac{1}{x^2}-9}$

15. $\dfrac{\dfrac{x+2}{x-6}-\dfrac{x+12}{x+5}}{x+82}$

16. $\dfrac{\dfrac{6}{x}+\dfrac{5}{y}}{\dfrac{5}{x}-\dfrac{6}{y}}$

17. $\dfrac{\dfrac{1}{16}-\dfrac{1}{x^2}}{\dfrac{1}{4}+\dfrac{1}{x}}$

18. $\dfrac{\dfrac{8}{x+4}-\dfrac{2}{x+7}}{\dfrac{x+8}{x+4}}$

You Try It

19. $\dfrac{\dfrac{2}{x+13}+\dfrac{1}{x-5}}{2-\dfrac{x+25}{x+13}}$

20. $\dfrac{\dfrac{x-5}{x+5}+\dfrac{x-5}{x-9}}{1+\dfrac{x+5}{x-9}}$

21. $\dfrac{1+\dfrac{4}{x}-\dfrac{5}{x^2}}{1-\dfrac{2}{x}-\dfrac{35}{x^2}}$

22. $\dfrac{\dfrac{x-4}{x^2-25}}{1+\dfrac{1}{x-5}}$

23. $\dfrac{\dfrac{x+3}{x-3}-\dfrac{x-3}{x+3}}{\dfrac{x-3}{x+3}+\dfrac{x+3}{x-3}}$

24. $\dfrac{\dfrac{3}{a^2}-\dfrac{1}{ab}-\dfrac{2}{b^2}}{\dfrac{2}{a^2}-\dfrac{5}{ab}+\dfrac{3}{b^2}}$

You Try It

25. $\dfrac{6x^{-1}+6y^{-1}}{xy^{-1}-x^{-1}y}$

26. $\dfrac{7x^{-1}-3y^{-1}}{49x^{-2}-9y^{-2}}$

6.5 Rational Equations and Models

THINGS TO KNOW

Before working through this section, be sure you are familiar with the following concepts:

| | VIDEO | ANIMATION | INTERACTIVE |

You Try It
1. Factor Polynomials Completely (Section 5.3, **Objective** 4)

You Try It
2. Solve Polynomial Equations by Factoring (Section 5.4, **Objective** 1)

You Try It
3. Find the Zeros of a Polynomial Function (Section 5.4, **Objective** 2)

You Try It
4. Use Polynomial Equations and Models to Solve Application Problems (Section 5.4, **Objective** 3)

You Try It
5. Find the Domain of a Rational Function (Section 6.1, **Objective** 1)

You Try It
6. Evaluate Rational Functions (Section 6.1, **Objective** 2)

You Try It
7. Add and Subtract Rational Expressions with Unlike Denominators (Section 6.3, **Objective** 3)

OBJECTIVES

1 Identify Rational Equations

2 Solve Rational Equations

3 Find the Zeros of a Rational Function

4 Use Rational Equations to Solve Application Problems

OBJECTIVE 1: IDENTIFY RATIONAL EQUATIONS

My interactive video summary

As we have seen before, we make a clear distinction between an equation and an expression. Watch this interactive video to practice distinguishing between the two. Recall that we *simplify* expressions and *solve* equations. We have previously defined linear equations and polynomial equations, and now define a **rational equation**.

Definition   Rational Equation

A rational equation results when we set two rational expressions equal to each other. Some examples of rational equations are

$$\frac{x+2}{x-7} + \frac{1}{x} = \frac{3x+4}{x-1} \quad \text{and} \quad x^{-2} + 3x^{-1} = 4.$$

Example 1 Identifying Rational Equations

Determine if each statement is a rational equation. If not, state why.

a. $\dfrac{3}{x-2} - \dfrac{x+2}{x+5} = \dfrac{5}{x+7}$

b. $\dfrac{3x^2 - 7x + 2}{x^3 - 3x + 1}$

c. $\dfrac{\sqrt{x-2}}{x+4} = \dfrac{x-7}{x+2}$

d. $3x^{-2} = \dfrac{x}{x-4}$

Solution

a. $\dfrac{3}{x-2} - \dfrac{x+2}{x+5}$ and $\dfrac{5}{x+7}$ are rational expressions. Therefore, this statement is a rational equation because we have two rational expressions set equal to each other.

b. This statement is not a rational equation because there is no equal sign. $\dfrac{3x^2 - 7x + 2}{x^3 - 3x + 1}$ is a rational *expression*.

c. $\dfrac{\sqrt{x-2}}{x+4}$ is not a rational expression because the numerator is not a polynomial (there is a variable under a radical). Therefore, this statement is not a rational equation.

d. $3x^{-2}$ and $\dfrac{x}{x-4}$ are rational expressions. Note that $3x^{-2} = 3 \cdot \dfrac{1}{x^2} = \dfrac{3}{x^2}$. Since we have two rational expressions set equal to each other, this statement is a rational equation.

You Try It   Work through this You Try It problem.

Work Exercises 1–4 in this textbook or in the *MyMathLab*® Study Plan.

OBJECTIVE 2:   SOLVE RATIONAL EQUATIONS

Before we begin solving rational equations in general, consider the following equation:

$$\frac{3}{2}x + \frac{2}{3} = -\frac{3}{4}$$

How would we go about solving this equation? One strategy would be to clear fractions by multiplying both sides of the equation by the LCD, then solving the resulting equation. Click here to see the solution.

The example above is a linear equation, but it can also be considered a rational equation because $\frac{3}{2}x + \frac{2}{3}$ and $-\frac{3}{4}$ are rational expressions. Often rational equations will contain a variable in at least one denominator, although this is not required. If any denominators contain variables, we must consider **restricted values**. Any value that makes the denominator of a rational expression equal to zero is a restricted value. Such values are not solutions of the rational equation and must be discarded if they appear as potential solutions.

The following steps can be used to solve rational equations.

Solving Rational Equations

Step 1. List all restricted values.

Step 2. Determine the LCD of all denominators in the equation.

Step 3. Multiply both sides of the equation by the LCD.

Step 4. Solve the resulting polynomial equation.

Step 5. Discard any restricted values and check the remaining solutions in the original equation.

Click here for an alternate but similar approach to solving rational equations.

Example 2 Solving Rational Equations

Solve $\dfrac{2}{x} - \dfrac{x-3}{2x} = 3$.

Solution We follow the five-step process:

Step 1. To find the restricted values, we look for values that make any denominator equal to zero.

$$\frac{2}{x} - \frac{x-3}{2x} = \frac{3}{1}$$

Examining the variable factors in the denominators, we see that the only restricted value is 0.

Step 2. There are unique factors of 2 and x in the denominators, each occurring at most once. The LCD is $2x$.

Step 3. Multiply both sides of the equation by the LCD.

$$\text{Multiply both sides by the LCD:}\quad 2x\left(\frac{2}{x} - \frac{x-3}{2x}\right) = 2x(3)$$

$$\text{Distribute:}\quad 2x\cdot\frac{2}{x} - 2x\cdot\frac{x-3}{2x} = 2x\cdot3$$

$$\text{Cancel common factors:}\quad 2\cancel{x}\cdot\frac{2}{\cancel{x}} - \cancel{2x}\cdot\frac{x-3}{\cancel{2x}} = 2x\cdot3$$

$$\text{Multiply:}\quad 4 - (x-3) = 6x$$

Step 4. Solve the resulting equation.

$$\text{Distribute:}\quad 4 - x + 3 = 6x$$

$$\text{Simplify:}\quad 7 - x = 6x$$

$$\text{Add } x \text{ to both sides:}\quad 7 = 7x$$

$$\text{Divide both sides by 7:}\quad 1 = x \quad \text{or} \quad x = 1$$

Step 5. The potential solution is not a restricted value, so we do not discard it. Check $x = 1$ in the original equation (the check is left to you). The solution set is $\{1\}$.

You Try It   Work through this **You Try It** problem.

Work Exercises 5–6 in this textbook or in the *MyMathLab*® Study Plan.

Example 3 Solving Rational Equations

Solve $\dfrac{4}{5} - \dfrac{3}{x-3} = \dfrac{1}{x}$.

Solution   We follow the five-step process:

Step 1. To find the restricted values, we look for values that make any denominator equal to zero. Set each denominator with a variable equal to zero and solve.

$$x - 3 = 0 \quad \text{or} \quad x = 0$$
$$x = 3$$

The restricted values are 0 and 3.

Step 2. There are unique factors of 5, x, and $x - 3$ in the denominators, each occurring at most once. The LCD is $5x(x - 3)$.

Step 3. Multiply both sides of the equation by the LCD.

$$\text{Multiply both sides by the LCD:}\quad 5x(x-3)\left(\frac{4}{5} - \frac{3}{x-3}\right) = 5x(x-3)\left(\frac{1}{x}\right)$$

$$\text{Distribute:}\quad 5x(x-3)\cdot\frac{4}{5} - 5x(x-3)\cdot\frac{3}{x-3} = 5x(x-3)\cdot\frac{1}{x}$$

$$\text{Cancel common factors:}\quad \cancel{5}x(x-3)\cdot\frac{4}{\cancel{5}} - 5x\cancel{(x-3)}\cdot\frac{3}{\cancel{x-3}} = 5\cancel{x}(x-3)\cdot\frac{1}{\cancel{x}}$$

$$\text{Multiply:}\quad 4x(x-3) - 15x = 5(x-3)$$

Finish solving the equation on your own. Click here to check your answer, or watch this video for a complete solution.

You Try It   Work through this You Try It problem.

Work Exercises 7–8 in this textbook or in the *MyMathLab* ® Study Plan.

Example 4 Solving Rational Equations

 Solve $\dfrac{m}{m+2} + \dfrac{5}{m-2} = \dfrac{20}{m^2-4}$.

Solution We follow the five-step process:

Step 1. To find the restricted values, look for values that make any denominator equal to zero. First, factor the denominators.

$$\frac{m}{m+2} + \frac{5}{m-2} = \frac{20}{(m-2)(m+2)}$$

Now set each variable factor equal to zero and solve.

$$m+2=0 \quad \text{or} \quad m-2=0$$
$$m=-2 \qquad\qquad m=2$$

The restricted values are -2 and 2.

Step 2. We have unique factors of $m-2$ and $m+2$ in the denominators, each occurring at most once. The LCD is $(m-2)(m+2)$.

Step 3. Multiply both sides of the equation by the LCD and cancel common factors.

$$(m-2)(m+2)\left(\frac{m}{m+2} + \frac{5}{m-2}\right) = (m-2)(m+2)\cdot\frac{20}{(m-2)(m+2)}$$

$$(m-2)(m+2)\cdot\frac{m}{m+2} + (m-2)(m+2)\cdot\frac{5}{m-2} = (m-2)(m+2)\cdot\frac{20}{(m-2)(m+2)}$$

$$(m-2)\cancel{(m+2)}\cdot\frac{m}{\cancel{m+2}} + \cancel{(m-2)}(m+2)\cdot\frac{5}{\cancel{m-2}} = \cancel{(m-2)}\cancel{(m+2)}\cdot\frac{20}{\cancel{(m-2)}\cancel{(m+2)}}$$

$$m(m-2) + 5(m+2) = 20$$

Finish solving the equation on your own, making sure to check for restricted values. Click here to check your answer, or watch this video for a complete solution.

You Try It   Work through this You Try It problem.

Work Exercises 9–10 in this textbook or in the *MyMathLab* ® Study Plan.

Example 5 Solving Rational Equations

Solve $\dfrac{2}{x-3} - \dfrac{4}{x^2-2x-3} = \dfrac{1}{x+1}$.

Solution Following the five-step process, we start by finding the restricted values. Factoring the denominators gives

$$\frac{2}{x - 3} - \frac{4}{(x - 3)(x + 1)} = \frac{1}{x + 1}.$$

Set each variable factor in the denominators equal to zero and solve.

$$x - 3 = 0 \quad \text{or} \quad x + 1 = 0$$
$$x = 3 \qquad\qquad x = -1$$

The restricted values are -1 and 3.

The LCD is $(x - 3)(x + 1)$. Multiplying both sides of the equation by the LCD gives the following polynomial equation (click here to see the steps):

$$2(x + 1) - 4 = x - 3$$

Solving the polynomial equation leads to $x = -1$. However, $x = -1$ is a restricted value, so it must be discarded. Since no other possible solution exists, the equation has no solution. The solution set is $\{\ \}$ or $\varnothing$.

You Try It Work through this You Try It problem.

Work Exercises 11–12 in this textbook or in the *MyMathLab*® Study Plan.

Example 6 Solving Rational Equations

Solve $\dfrac{3x^2 + 4x + 7}{x^2 - 9} = \dfrac{x^2 - 6x - 5}{x^2 - 9} + \dfrac{2x + 4}{x - 3}.$

My video summary

Solution Solve the equation on your own, remembering to watch for restricted values. Click here to check your answer, or watch this video for a complete solution.

You Try It Work through this You Try It problem.

Work Exercises 13–14 in this textbook or in the *MyMathLab*® Study Plan.

CAUTION

Note that encountering an identity does not mean that all real numbers are solutions to the equation. This shows the importance of writing the restricted values as our first step in solving rational equations.

A **ratio** is the quotient of two numbers or algebraic expressions, such as with slope $\left(\frac{\text{rise}}{\text{run}}\right)$. A rational number is the quotient of two integers, so rational numbers are ratios. Likewise, a rational expression is the quotient of two polynomials, so rational expressions are ratios. A **proportion** is a statement that two ratios are equal, such as $\dfrac{P}{Q} = \dfrac{R}{S}.$

When we have a rational equation that is a proportion, we can begin by **cross-multiplying** to get $P \cdot S = Q \cdot R$ and then solve the resulting polynomial equation.

Cross-Multiplying

$$\frac{P}{Q} \times \frac{R}{S}$$

$$P \cdot S = Q \cdot R$$

Note that any solutions that are restricted values in the original equation must be discarded. As we will see later in this section, we encounter proportions frequently in real-world problems involving work rates or distance traveled.

Example 7 Solving Rational Equations Involving Proportions

Solve $\dfrac{x + 2}{x - 2} = \dfrac{x}{x + 4}$.

Solution Because the equation is a proportion, we solve by cross-multiplying. First, we identify any restricted values by setting each denominator equal to zero and solving.

$$x - 2 = 0 \quad \text{or} \quad x + 4 = 0$$
$$x = 2 \qquad\qquad x = -4$$

We have two restricted values, -4 and 2.

Next, we cross-multiply and solve the resulting polynomial equation.

Begin with the original equation:
$$\frac{x + 2}{x - 2} = \frac{x}{x + 4}$$

Cross-multiply:
$$\frac{x + 2}{x - 2} \bowtie \frac{x}{x + 4}$$

$$(x + 2)(x + 4) = (x - 2)(x)$$

Multiply:
$$x^2 + 6x + 8 = x^2 - 2x$$

Subtract x^2 from both sides:
$$6x + 8 = -2x$$

Subtract $6x$ from both sides:
$$8 = -8x$$

Divide both sides by -8:
$$-1 = x$$

Since -1 is not a restricted value, the solution set is $\{-1\}$. The check is left to you.

You Try It   Work through this You Try It problem.

Work Exercises 15–16 in this textbook or in the *MyMathLab*® Study Plan.

OBJECTIVE 3:   FIND THE ZEROS OF A RATIONAL FUNCTION

In Section 5.4, we found the zeros of a polynomial function by setting the function equal to zero and solving the resulting equation. We follow the same procedure for finding the zeros of a rational function, but we must consider the restricted values first.

Let $R(x)$ be a rational function. If c is a real number in the domain of R such that $R(c) = 0$, then c is called a **zero**, or **root**, of the rational function. The zeros of a rational function are the solutions to the equation $R(x) = 0$.

Since a rational function is defined by a rational expression, solving $R(x) = 0$ is equivalent to solving $\dfrac{P(x)}{Q(x)} = 0$. The quotient will equal zero if the numerator is 0 and the denominator is not 0. Click here to see why.

This leads us to the following strategy for finding zeros of a rational function.

Finding the Zeros of a Rational Function

To find the **zeros**, or **roots**, of a rational function of the form $\dfrac{P(x)}{Q(x)}$, do the following:

Step 1. Find the restricted values by setting the denominator equal to zero and solving the resulting equation, $Q(x) = 0$.

Step 2. Set the numerator equal to zero and solve the resulting equation, $P(x) = 0$.

Step 3. Any solution from step 2 that is not a restricted value is a zero of the function.

Example 8 Finding the Zeros of a Rational Function

Find the zeros for each rational function.

a. $R(x) = \dfrac{3x^2 - 5x - 2}{x^2 - 4}$ b. $R(x) = \dfrac{x^2 - x - 30}{2x^2 - 15x + 18}$

c. $R(x) = \dfrac{3x^2 + 2x - 21}{6x^2 + 7x - 20}$

Solution

a. First, we find the restricted values of the function. Set the denominator equal to zero and solve the resulting equation.

$R(x) = \dfrac{3x^2 - 5x - 2}{x^2 - 4}$ $\longrightarrow$ Set denominator equal to 0: $\qquad x^2 - 4 = 0$

Factor: $\qquad (x + 2)(x - 2) = 0$

Set each factor equal to 0 and solve: $\quad x + 2 = 0 \quad$ or $\quad x - 2 = 0$

$$x = -2 \qquad\qquad x = 2$$

The restricted values are -2 and 2.

Next, we set the numerator equal to zero and solve.

$R(x) = \dfrac{3x^2 - 5x - 2}{x^2 - 4}$ $\longrightarrow$ Set numerator equal to 0: $\qquad 3x^2 - 5x - 2 = 0$

Factor: $\qquad (x - 2)(3x + 1) = 0$

Set each factor equal to 0 and solve: $\quad x - 2 = 0 \quad$ or $\quad 3x + 1 = 0$

$$x = 2 \qquad\qquad 3x = -1$$

$$x = -\dfrac{1}{3}$$

The solution $x = 2$ is a restricted value, so it must be discarded. The remaining solution, $x = -\dfrac{1}{3}$, is not a restricted value, so it is a zero of the function (the only one). What is the domain of the function? Click here to find out.

b. First, we find the restricted values of the function. Set the denominator equal to zero and solve the resulting equation.

Set denominator equal to 0: $2x^2 - 15x + 18 = 0$

Factor: $(x - 6)(2x - 3) = 0$

Set each factor equal to 0 and solve:

$$x - 6 = 0 \quad \text{or} \quad 2x - 3 = 0$$
$$x = 6 \quad \text{or} \qquad 2x = 3$$
$$x = \frac{3}{2}$$

The restricted values are 6 and $\frac{3}{2}$. Now set the numerator equal to zero and solve the resulting equation. Exclude any restricted values from these solutions to find the zeros of the function. Click here to check your answer, or here to see the rest of the solution.

My video summary

c. Find the restricted values, then set the numerator equal to zero and solve the resulting equation. Use these results to determine the zeros of the function. Click here to check your answer, or watch this video for a complete solution.

You Try It   Work through this You Try It problem.

Work Exercises 17–18 in this textbook or in the *MyMathLab*® Study Plan.

OBJECTIVE 4:   USE RATIONAL EQUATIONS TO SOLVE APPLICATION PROBLEMS

Rational equations appear frequently in applications. We follow our usual problem-solving strategy when solving application problems involving rational equations.

Example 9 Road Trip

Payton's car holds 13 gallons of gas and can travel 351 miles on a full tank. Currently she has 4 gallons of gas in her tank. Assuming the same gas mileage, how many gallons of gas will she need to purchase on a 540-mile road trip to Chicago?

Solution Follow the problem-solving strategy.

Step 1. We want to find the number of gallons of gas Payton needs to purchase for a 540-mile road trip. We know she can travel 351 miles on 13 gallons of gas, and we know she currently has 4 gallons of gas in the car.

Step 2. Let $x = $ the number of gallons purchased.

Step 3. Since we know her car's gas mileage and it remains constant, we can solve the problem using a rational equation in the form of a proportion.

$$\underbrace{\frac{\text{miles}}{\text{gallons}}}_{\text{full tank}} = \underbrace{\frac{\text{miles}}{\text{gallons}}}_{\text{road trip}}$$

$$\frac{351}{13} = \frac{540}{4 + x}$$

Step 4. Because we have a proportion, we cross-multiply and solve the resulting equation.

Cross multiply: $\dfrac{351}{13} \bcancel{\times} \dfrac{540}{4 + x}$

$$351(4 + x) = 13(540)$$

Distribute: $1404 + 351x = 7020$

Subtract 1404 from both sides: $351x = 5616$

Divide both sides by 351: $x = 16$

Step 5. 540 is about 1.5 times 351 and the gas mileage remains constant. We expect Payton will need about 1.5 tanks, or $1.5(13) = 19.5$ gallons for the trip. Her tank had 4 gallons already, so she would need about 15.5 more gallons. Our result is reasonable.

Step 6. Payton will need to purchase 16 gallons of gas to complete the road trip.

You Try It Work through this You Try It problem.

Work Exercises 19–21 in this textbook or in the *MyMathLab*® Study Plan.

Example 10 Boat Speed

 Emalie can travel 16 miles upriver in the same amount of time it takes her to travel 24 miles downriver. If the speed of the current is 4 mph, how fast can her boat travel in still water?

Solution Follow the problem-solving strategy.

Step 1. We want to find the speed of Emalie's boat in still water. We know she can travel 16 miles upriver in the same amount of time she can travel 24 miles downriver, and we know the speed of the current is 4 mph.

Step 2. Let r = the speed of Emalie's boat in still water. Then her speed upriver is $r - 4$ (because she goes against the current) and her speed downriver is $r + 4$ (because she goes with the current).

Step 3. This is a uniform motion problem, so we need the formula, $d = rt$.

Solving for time, we get $t = \dfrac{d}{r}$. We know the amount of time traveled is the same in either direction, so we have

$$\underbrace{\frac{\text{distance}}{\text{rate}}}_{\substack{\text{Time}\\\text{upriver}}} = \underbrace{\frac{\text{distance}}{\text{rate}}}_{\substack{\text{Time}\\\text{downriver}}}.$$

Substitute in the known distances and the expressions for the two rates (speeds). Try finishing the problem on your own. Click here to check your answer, or watch this video for a complete solution.

You Try It Work through this You Try It problem.

Work Exercises 22–23 in this textbook or in the *MyMathLab*® Study Plan.

My video summary

Example 11 Resistance in Parallel Circuits

In electronics, the total resistance R of a circuit containing two resistors in parallel is given by the formula $\dfrac{1}{R} = \dfrac{1}{R_1} + \dfrac{1}{R_2}$, where R_1 and R_2 are the two individual resistances. If the total resistance is 10 ohms and one resistor has twice the resistance of the other, find the resistance of each circuit.

Solution

Step 1. We want to find the individual resistances for a circuit containing two resistors in parallel. We know that the total resistance is 10 ohms and one resistor has twice the resistance as the other.

Step 2. We are given the resistance of one resistor in terms of the other. If we let $x =$ the resistance of the first resistor, then $2x =$ the resistance of the second.

Step 3. Using the given formula, we substitute in the total resistance and expressions for the individual resistances. This gives the equation

$$\frac{1}{10} = \frac{1}{x} + \frac{1}{2x}.$$

Step 4. The LCD is $10x$. Note that the only restricted value is $x = 0$.

Begin with the original equation: $\quad \dfrac{1}{10} = \dfrac{1}{x} + \dfrac{1}{2x}$

Multiply both sides by the LCD: $\quad 10x\left(\dfrac{1}{10}\right) = 10x\left(\dfrac{1}{x} + \dfrac{1}{2x}\right)$

Distribute: $\quad 10x \cdot \dfrac{1}{10} = 10x \cdot \dfrac{1}{x} + 10x \cdot \dfrac{1}{2x}$

Simplify: $\quad \cancel{10}x \cdot \dfrac{1}{\cancel{10}} = 10\cancel{x} \cdot \dfrac{1}{\cancel{x}} + \overset{5}{\cancel{10}}\cancel{x} \cdot \dfrac{1}{\cancel{2x}}$

$$x \cdot 1 = 10 \cdot 1 + 5 \cdot 1$$
$$x = 10 + 5$$
$$x = 15$$

Step 5. We can check our answer using our equation $\dfrac{1}{10} = \dfrac{1}{x} + \dfrac{1}{2x}$.

$$\frac{1}{15} + \frac{1}{2(15)} = \frac{1}{15} + \frac{1}{30} = \frac{2}{30} + \frac{1}{30} = \frac{3}{30} = \frac{1}{10}$$

The total resistance is 10 ohms, so this result is reasonable.

Step 6. The resistances are 15 ohms and $2(15) = 30$ ohms.

You Try It   **Work through this You Try It problem.**

Work Exercises 24–26 in this textbook or in the *MyMathLab*® Study Plan.

Example 12 Train Speed

Fatima rode an express train 223.6 miles from Boston to New York City and then rode a passenger train 218.4 miles from New York City to Washington, DC.

My video summary

If the express train travels 30 miles per hour faster than the passenger train and her total trip took 6.5 hours, what was the average speed of the express train?

Solution Follow the problem-solving strategy. Read through the problem and summarize the given information. If we let r = the average speed of the express train, then $r - 30$ is the average speed of the passenger train. We know the total time of Fatima's trip must be the sum of her time on the express train and her time on the passenger train. Thus, we can write

$$\underbrace{\text{time}}_{\text{Express}} + \underbrace{\text{time}}_{\text{Passenger}} = \underbrace{\text{time}}_{\text{Total}}.$$

Solving $d = r \cdot t$ for t gives $t = \dfrac{d}{r}$. This allows us to write

$$\underbrace{\frac{\text{distance}}{\text{rate}}}_{\text{Express time}} + \underbrace{\frac{\text{distance}}{\text{rate}}}_{\text{Passenger time}} = \text{total time}.$$

Substituting in the known distances, the total time, and expressions for the train speeds, we get

$$\underbrace{\frac{223.6}{r}}_{\substack{\text{Express} \\ \text{time}}} + \underbrace{\frac{218.4}{r-30}}_{\substack{\text{Passenger} \\ \text{time}}} = \underbrace{6.5}_{\text{Total time}}.$$

The LCD is $r(r - 30)$. Multiply both sides of the rational equation by the LCD to obtain a polynomial equation.

Original equation: $\dfrac{223.6}{r} + \dfrac{218.4}{r-30} = 6.5$

Multiply by the LCD: $r(r-30)\left(\dfrac{223.6}{r} + \dfrac{218.4}{r-30}\right) = r(r-30)(6.5)$

Distribute: $r(r-30) \cdot \dfrac{223.6}{r} + r(r-30) \cdot \dfrac{218.4}{r-30} = r(r-30)(6.5)$

Cancel common factors: $\cancel{r}(r-30) \cdot \dfrac{223.6}{\cancel{r}} + r\cancel{(r-30)} \cdot \dfrac{218.4}{\cancel{r-30}} = r(r-30)(6.5)$

Multiply: $223.6(r-30) + 218.4r = r(r-30)(6.5)$

Finish solving the equation on your own to answer the question. Click here to check your answer, or watch this video for a complete solution.

You Try It Work through this **You Try It** problem.

Work Exercises 27–28 in this textbook or in the *MyMathLab*® Study Plan.

For work problems involving multiple workers, such as people, copiers, pumps, etc., it is often helpful to consider **rate of work**.

> **Rate of Work**
>
> The **rate of work** is the number of jobs that can be completed in a given unit of time.
>
> If one job can be completed in t units of time, then the rate of work is given by $\dfrac{1}{t}$.

We can add rates but not times. For example, if it takes Avril 4 hours to paint a room and Anisa 2 hours to paint the same room, it would not take $4 + 2 = 6$ hours for them to paint the room together. It cannot take longer for two people to do a job than it would take either one alone. When dealing with two workers, we use the following formula:

$$\frac{1}{t_1} + \frac{1}{t_2} = \frac{1}{t}$$

Here, t_1 and t_2 are the individual times to complete one job, and t is the time to complete the job when working together.

Example 13 Emptying a Pool

My video summary

A small pump takes 8 more hours than a larger pump to empty a pool. Together the pumps can empty the pool in 3 hours. How long will it take the larger pump to empty the pool if it works alone?

Solution Follow the problem-solving strategy.

We want to find the time it will take the larger pump to empty the pool by itself. We know that the two pumps take 3 hours to empty the pool and the smaller pump takes 8 more hours than the larger pump to empty the pool by itself.

If we let $x =$ the time for the larger pump to empty the pool, then $x + 8 =$ the time for the smaller pump to empty the pool.

Since we are combining rates of work, we get

$$\underbrace{\frac{1}{t_1}}_{\text{Larger pump rate}} + \underbrace{\frac{1}{t_2}}_{\text{Smaller pump rate}} = \underbrace{\frac{1}{t}}_{\text{Combined rate}}.$$

Substituting in the given combined time and expressions for individual times, we get

$$\underbrace{\frac{1}{x}}_{\text{Larger pump rate}} + \underbrace{\frac{1}{x + 8}}_{\text{Smaller pump rate}} = \underbrace{\frac{1}{3}}_{\text{Combined rate}}.$$

The LCD is $3x(x + 8)$. Multiply both sides of the equation by the LCD and solve the resulting polynomial equation. Click here to check your answer, or watch this video for a complete solution.

You Try It Work through this You Try It problem.

Work Exercises 29–30 in this textbook or in the *MyMathLab*® Study Plan.

Example 14 Filling a Pond

A garden hose can fill a pond in 2 hours whereas an outlet pipe can drain the pond in 10 hours. If the outlet pipe is accidentally left open, how long would it take to fill the pond?

Solution We want to find the time required to fill the pond if the hose is running and the outlet pipe is open. Let t = the time required to fill the pond.

We are combining rates of work, but the rates are not working together. Rates that work towards completion of a task are positive actions and therefore are positive values. Rates that work against the completion of the task are negative actions and therefore are negative values.

$$\underbrace{\frac{1}{t_1}}_{\text{Garden hose rate}} + \underbrace{\frac{-1}{t_2}}_{\text{Outlet pipe rate}} = \underbrace{\frac{1}{t}}_{\text{Combined rate}}$$

Substituting in the given individual times, we get

$$\underbrace{\frac{1}{2}}_{\text{Garden hose rate}} + \underbrace{\frac{-1}{10}}_{\text{Outlet pipe rate}} = \underbrace{\frac{1}{t}}_{\text{Combined rate}}$$

or

$$\underbrace{\frac{1}{2}}_{\text{Garden hose rate}} - \underbrace{\frac{1}{10}}_{\text{Outlet pipe rate}} = \underbrace{\frac{1}{t}}_{\text{Combined rate}}.$$

Because the outlet pipe is working against the garden hose, we end up subtracting the rates. Note that this is similar to the situation when a boat is traveling upstream or downstream. When going downstream (with the current) we added the rates. However, when going upstream (against the current) we subtracted the rates because the current was working against the boat.

The LCD is $10t$. Multiply both sides of the equation by the LCD and solve the resulting equation for t.

Original equation:	$\dfrac{1}{2} - \dfrac{1}{10} = \dfrac{1}{t}$
Multiply by the LCD:	$10t\left(\dfrac{1}{2} - \dfrac{1}{10}\right) = 10t\left(\dfrac{1}{t}\right)$
Distribute:	$5t - t = 10$
Simplify:	$4t = 10$
Divide both sides by 4:	$t = \dfrac{10}{4} = 2.5$

It will take 2.5 hours to fill the pond with the outlet pipe open.

You Try It Work through this You Try It problem.

Work Exercises 31–32 in this textbook or in the *MyMathLab*® Study Plan.

Example 15 Prescription Drug Price

 The average retail price for prescription drugs in the United States for the years 2000–2007 can be approximated by the model

$$C(x) = \frac{140 + 23x - x^2}{2.88 + 0.116x - 0.004x^2}, \text{ where } x \text{ is the number of years after 2000.}$$

(*Source: Statistical Abstract, 2009*)

a. Use the model to approximate the average retail price for prescription drugs in 2005.

b. Use the model to predict the average retail price for prescription drugs in 2012.

c. Find and interpret the zeros of the rational function. Do the results make sense within the context of the problem?

Solution

a. The year 2005 corresponds to $x = 5$, so we evaluate $C(5)$.

$$C(5) = \frac{140 + 23(5) - (5)^2}{2.88 + 0.116(5) - 0.004(5)^2} = \frac{230}{3.36} \approx 68.45$$

Based on the model, the approximate average retail price for prescription drugs was $68.45 in 2005.

Finish working the problem on your own. Click here to check your answers, or watch this video for a complete solution.

You Try It Work through this You Try It problem.

Work Exercises 33–34 in this textbook or in the *MyMathLab*® Study Plan.

6.5 Exercises

In Exercises 1–4, determine if the statement is a rational equation. If not, state why.

1. $\dfrac{x - 3}{x^2 + \frac{2}{3}} = \dfrac{\sqrt{x^2 - 4x + 1}}{x - 5}$

2. $\dfrac{x^{-2} - 4}{x + 1} = \dfrac{x^3 + 2x}{3}$

3. $\dfrac{x^2 + 3x - 4}{2x + 5} = \dfrac{x - 2}{3x + 1}$

4. $\dfrac{x^3 - 4x^2 + 5x}{2x - \frac{2}{3}}$

In Exercises 5–16, solve the rational equation. If there is no solution, state this as your answer.

5. $\dfrac{4}{3x} - \dfrac{x + 1}{x} = \dfrac{2}{5}$

6. $\dfrac{x - 1}{x} = \dfrac{3}{4} - \dfrac{3}{2x}$

7. $\dfrac{3}{z} - \dfrac{9}{z - 5} = \dfrac{1}{4}$

8. $\dfrac{5}{3p} + \dfrac{2p}{p - 5} = -\dfrac{4}{3}$

9. $\dfrac{b}{b - 5} + \dfrac{2}{b + 5} = \dfrac{50}{b^2 - 25}$

10. $\dfrac{3}{3m + 2} - \dfrac{2}{m - 6} = \dfrac{1}{3m^2 - 16m - 12}$

You Try It

11. $\dfrac{1}{n+4} + \dfrac{3}{n+5} = \dfrac{1}{n^2+9n+20}$

12. $\dfrac{6}{y^2-25} = \dfrac{3}{y^2-5y}$

You Try It

13. $\dfrac{7x+2}{x^2-4} - \dfrac{3x+1}{x+2} = \dfrac{-3x^2+12x+4}{x^2-4}$

You Try It

14. $\dfrac{2w+3}{4w^2-9} + 4 = \dfrac{16w^2+2w-33}{4w^2-9}$

You Try It

15. $\dfrac{x+2}{x-5} = \dfrac{x-6}{x+1}$

16. $\dfrac{x}{3x+2} = \dfrac{x+5}{3x-1}$

You Try It

In Exercises 17 and 18, find the zeros of the rational function.

17. $R(x) = \dfrac{2x^2+5x-3}{x^2-9}$

18. $R(x) = \dfrac{6x^2-5x-6}{4x^2-4x-15}$

In Exercises 19–34, solve the application problems.

You Try It

19. Time to Complete an Order Olivia works at a craft store and can make 2 stone crafts in 45 minutes. If she has been working for 30 minutes on an order for 7 stone crafts, how much longer does she need to work to complete the order?

20. Commissions Nadia, a personal trainer, sells 4 training packages and earns $700 in commissions. How much would she make in commissions if she sold 13 training packages?

21. Similar Triangles For two similar triangles, such as those shown in the figure below, the ratios of the lengths of corresponding sides are equal. For example, $\dfrac{a}{A} = \dfrac{b}{B}$.

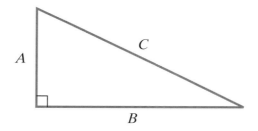
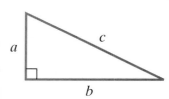

A right triangle with a hypotenuse measuring 10 inches and one leg measuring 4 inches has its hypotenuse increased to 12.5 inches. If the larger triangle is to be similar to the smaller triangle, by how much must the 4-inch leg be increased?

You Try It

22. Current Speed On a float trip, Kim traveled 30 miles downstream in the same amount of time that it would take her to row 12 miles upstream. If her speed in still water is 7 mph, find the speed of the current.

23. Plane Speed A plane can fly 500 miles against the wind in the same amount of time that it can fly 725 miles with the wind. If the wind speed is 45 mph, find the speed of the plane in still air.

You Try It

24. Resistors in Parallel If the total resistance of two circuits in parallel is 9 ohms and one circuit has three times the resistance of the other, find the resistance of each circuit. Use $\dfrac{1}{R} = \dfrac{1}{R_1} + \dfrac{1}{R_2}$, where R_1 and R_2 are the two individual resistances.

25. Focal Length The focal length f of a lens is given by the equation $\dfrac{1}{f} = \dfrac{1}{x_o} + \dfrac{1}{x_i}$, where x_o is the distance from the object to the lens, and x_i is the distance from the lens to the image of the object. If the focal length of a lens is 12 cm and an object is three times as far from the lens as its image, then how far is the object from the lens?

26. Resistance in a Circuit A circuit contains a 6-ohm resistor in parallel with a resistor of unknown resistance. The two parallel circuits are in series with a third resistor that has the same unknown resistance. If the total resistance is 6.4 ohms, solve the following equation for the unknown resistance.

$$\frac{6x}{6 + x} + x = 6.4$$

27. Walking Speed For exercise, Mae jogs 10 miles then walks an additional 2 miles to cool down. If her jogging speed is 4.5 miles per hour faster than her walking speed and her total exercise time is 3 hours, what is her walking speed?

28. Moving Sidewalk Choi can walk 108 m in 72 seconds. Standing on a moving sidewalk at Thailand's Suvarnabhumi International Airport, she can travel 108 m in 40 seconds. How long will it take her to travel 108 m if she walks on the moving sidewalk? Round your answer to the nearest tenth.

29. Staining a Deck It takes Shawn 2 more hours to stain a deck than Michelle. Together, it takes them 2.4 hours to complete the work. How long would it take Shawn to stain the deck by himself?

30. Lining Fields Together, Sari and Rilee can mark the lines on the soccer fields at a recreation center 18 minutes faster than if Rilee lines the fields on her own. If it takes Sari 56 minutes to line the fields by herself, how long would it take Rilee to do the job alone?

31. Filling a Hot Tub A garden hose can fill a hot tub in 180 minutes, whereas the drain on a hot tub can empty it in 300 minutes. If the drain is accidentally left open, how long will it take to fill the tub?

32. Emptying a Boat's Bilge A leak in the hull will fill a boat's bilge in 8 hours. The boat's bilge pump can empty a full bilge in 3 hours. How long will it take the pump to empty a full bilge if the boat is leaking?

33. Average Farm Size Based on data from the U.S. Department of Agriculture, the average number of acres per farm x years after 2000 can be approximated by the model

$$A(x) = \frac{2078x - 94{,}458}{13x - 2164}.$$

a. Use the model to estimate the average number of acres per farm in 2005.

b. Use the model to predict the average number of acres per farm in 2012.

c. Find and interpret the zero of the rational function. Does this result make sense within the context of the problem?

34. Newton's Method Newton's Method is a technique used to find solutions to an equation. For the equation $2x^2 + 3x - 4 = 0$, the function $z(x) = \dfrac{2x^2 + 4}{4x + 3}$ gives an estimate for a solution to the equation, where x is an initial guess. Each estimate is used as the next guess in the function until successive estimates are the same to a certain number of decimal places.

a. Using $x = 1$ as an initial guess for a solution to the equation, evaluate $z(1)$ to find a new estimate.

b. Use your exact result from part (a) to evaluate $z(x)$ and find a new estimate.

c. Find a third estimate using your result from part (b). Based on your results, estimate the solution to two decimal places.

6.5 Rational Equations and Models **291**

You Try It

You Try It

You Try It

You Try It

6.6 Variation

THINGS TO KNOW

Before working through this section, be sure you are familiar with the following concepts:

VIDEO   ANIMATION   INTERACTIVE

You Try It

1. Solve a Formula for a Given Variable (Section 1.4, **Objective 1**)

You Try It

2. Use Linear Models to Solve Application Problems; Direct Variation (Section 2.5, **Objective 6**)

OBJECTIVES

1 Solve Application Problems Involving Direct Variation

2 Solve Application Problems Involving Inverse Variation

3 Solve Application Problems Involving Combined Variation

OBJECTIVE 1: SOLVE APPLICATION PROBLEMS INVOLVING DIRECT VARIATION

In application problems, we are often concerned with exploring how one quantity varies with respect to other quantities. Variation equations allow us to show how one quantity changes with respect to one or more additional quantities. In this section we will discuss direct variation, inverse variation, and combined variation.

In Section 2.5, we introduced direct variation in the context of linear models. We can, however, have direct variation involving variables with powers other than 1. For example, if we say that y varies directly with the cube of x, then we have the equation $y = kx^3$, where k is the constant of variation. We now state our definition for direct variation more generally.

Direct Variation

For an equation of the form

$$y = kx,$$

we say that y **varies directly** with x, or y is **proportional to** x. The constant k is called the **constant of variation** or the **proportionality constant**.

For direct variation, the ratio of the two quantities is constant (the constant of variation). For example, consider $y = kx$ and $y = kx^3$.

$$\underbrace{y = kx}_{\substack{y \text{ varies directly} \\ \text{with } x}} \rightarrow \underbrace{\frac{y}{x}}_{\substack{\text{Ratio of the} \\ \text{quantities} \\ y \text{ and } x}} = \underbrace{k}_{\substack{\text{Constant of} \\ \text{variation}}} \qquad \underbrace{y = kx^3}_{\substack{y \text{ varies directly} \\ \text{with the cube} \\ \text{of } x}} \rightarrow \underbrace{\frac{y}{x^3}}_{\substack{\text{Ratio of the} \\ \text{quantities} \\ y \text{ and } x^3}} = \underbrace{k}_{\substack{\text{Constant of} \\ \text{variation}}}$$

Problems involving variation can generally be solved using the following guidelines.

> **Solving Variation Problems**
>
> **Step 1.** Translate the problem into an equation that models the situation.
>
> **Step 2.** Substitute given values for the variables into the equation and solve for the constant of variation, k.
>
> **Step 3.** Substitute the value for k into the equation to form the general model.
>
> **Step 4.** Use the general model to answer the question posed in the problem.

Example 1 Kinetic Energy

The kinetic energy of an object in motion varies directly with the square of its speed. If a van traveling at a speed of 30 meters per second has 945,000 joules of kinetic energy, how much kinetic energy does it have if it is traveling at a speed of 20 meters per second?

Solution Follow the guidelines for solving variation problems.

Step 1. We are told that the kinetic energy of an object in motion varies directly with the square of its speed. If we let $K =$ kinetic energy and $s =$ speed, we can translate the problem statement into the model

$$K = ks^2,$$

where k is the constant of variation.

Step 2. To determine the value of k, we use the fact that the kinetic energy is 945,000 joules when the velocity is 30 meters per second.

Substitute 945,000 for K and 30 for s: $945{,}000 = k(30)^2$

Simplify 30^2: $945{,}000 = 900k$

Divide both sides by 900: $1050 = k$

Step 3. The constant of variation is 1050, so the general model is $K = 1050s^2$.

Step 4. We want to determine the kinetic energy of the van if its speed is 20 meters per second. Substituting 20 for s, we find

$$K = 1050(20)^2 = 1050(400) = 420{,}000.$$

The van will have 420,000 joules of kinetic energy if it is traveling at a speed of 20 meters per second.

You Try It   **Work through this You Try It problem.**

Work Exercises 1–2 in this textbook or in the *MyMathLab*® Study Plan.

Example 2 Measuring Leanness

The Ponderal Index measure of leanness states that weight varies directly with the cube of height. If a "normal" person who is 1.2 m tall weighs 21.6 kg, how much will a "normal" person weigh if they are 1.8 m tall?

Solution Follow the guidelines for solving variation problems.

Step 1. We are told that weight varies directly with the cube of height. If we let $w =$ weight and $h =$ height, we can translate the problem statement into the model $w = kh^3$, where k is the constant of variation.

Step 2. To determine the value of k, we use the fact that a normal person who is 1.2 meters tall has a weight of 21.6 kg.

Substitute 1.2 for h and 21.6 for w: $21.6 = k(1.2)^3$

Simplify $(1.2)^3$: $21.6 = 1.728k$

Divide both sides by 1.728: $12.5 = k$

Step 3. The constant of variation is 12.5, so the general model is $w = 12.5h^3$.

Use the general model to determine the weight of a normal person who is 1.8 m tall. Click here to check your answer, or watch this video for a detailed solution.

You Try It   **Work through this You Try It problem.**

Work Exercises 3–4 in this textbook or in the *MyMathLab*® Study Plan.

OBJECTIVE 2:   SOLVE APPLICATION PROBLEMS INVOLVING INVERSE VARIATION

With our discussion of rational equations, we can now discuss **inverse variation**. Inverse variation means that one variable is a constant multiple of the reciprocal of another variable.

Inverse Variation

For equations of the form

$$y = \frac{k}{x} \quad \text{or} \quad y = k \cdot \frac{1}{x},$$

we say that y **varies inversely** with x, or y is **inversely proportional to** x. The constant k is called the constant of variation.

For inverse variation, the product of the two quantities is constant (the constant of variation). For example, consider $y = \dfrac{k}{x}$ and $y = \dfrac{k}{x^2}$.

$y = \dfrac{k}{x}$ $\rightarrow$ $\underbrace{xy}_{\substack{\text{Product of the}\\ \text{quantities}\\ y \text{ and } x}} = \underbrace{k}_{\substack{\text{Constant of}\\ \text{variation}}}$ $y = \dfrac{k}{x^2}$ $\rightarrow$ $\underbrace{x^2y}_{\substack{\text{Product of the}\\ \text{quantities}\\ y \text{ and } x^2}} = \underbrace{k}_{\substack{\text{Constant of}\\ \text{variation}}}$

y varies inversely with x y varies inversely with the square of x

Example 3 Density of an Object

For a given mass, the density of an object is inversely proportional to its volume. If 50 cubic centimeters of an object with a density of $28\dfrac{\text{g}}{\text{cm}^3}$ is compressed to 40 cubic centimeters, what would be its new density?

Solution Follow the guidelines for solving variation problems.

Step 1. We are told that the density of an object with a given mass varies inversely with its volume. If we let D = density and V = volume, we can translate the problem statement into the model

$$D = \frac{k}{V},$$

where k is the constant of variation.

Step 2. To determine the value of k, we use the fact that the density is $28\ \text{g/cm}^3$ when the volume is $50\ \text{cm}^3$.

Substitute 28 for D and 50 for V: $\quad 28 = \dfrac{k}{50}$

Multiply both sides by 50: $\quad (28)(50) = k$

Simplify: $\quad 1400 = k$

Step 3. The constant of variation is 1400, so the general model is $D = \dfrac{1400}{V}$.

Step 4. We want to determine the density of the object if the volume is compressed to $40\ \text{cm}^3$. Substituting 40 for V, we find

$$D = \frac{1400}{40} = 35\ \text{g/cm}^3.$$

The density of the compressed object would be $35\ \text{g/cm}^3$.

You Try It   **Work through this You Try It problem.**

Work Exercises 5–6 in this textbook or in the _MyMathLab_® Study Plan.

Example 4 Shutter Speed

My video summary

The shutter speed, S, of a camera varies inversely as the square of the aperture setting, f. If the shutter speed is 125 for an aperture of 5.6, what is the shutter speed if the aperture is 1.4?

Solution Follow the guidelines for solving variation problems.

Step 1. We are told that shutter speed varies inversely with the square of the aperture setting. Letting S = shutter speed and f = aperture setting, we can translate the problem statement into the model

$$S = \frac{k}{f^2},$$

where k is the constant of variation.

Step 2. To determine the value of k, we use the fact that an aperture setting of 5.6 corresponds to a shutter speed of 125.

Substitute 5.6 for f and 125 for S: $\quad 125 = \dfrac{k}{(5.6)^2}$

Simplify $(5.6)^2$: $\quad 125 = \dfrac{k}{31.36}$

Multiply both sides by 31.36: $\quad 3920 = k$

Step 3. The constant of variation is 3920, so the general model is $S = \dfrac{3920}{f^2}$.

Use the general model to determine the shutter speed for an aperture setting of 1.4. Click here to check your answer, or watch this video for a detailed solution.

You Try It   Work through this You Try It problem.

Work Exercises 7–8 in this textbook or in the *MyMathLab* Study Plan.

OBJECTIVE 3:   SOLVE APPLICATION PROBLEMS INVOLVING COMBINED VARIATION

When a variable is related to more than one other variable, we call this **combined variation**. Some examples of combined variation are

$$y = k\frac{x}{z} \qquad y = kxz \qquad y = k\frac{w^2x}{z}.$$

In the first example we say that y varies directly as x and inversely as z. In the second example we say that y varies directly as x and z. In the third example we say that y varies directly as x and the square of w, and inversely as z. In general, variables directly related to the dependent variable occur in the numerator while variables inversely related occur in the denominator.

$$y = k\frac{x}{z} \begin{matrix} \leftarrow \text{directly related} \\ \leftarrow \text{inversely related} \end{matrix} \qquad y = kxz \leftarrow \begin{matrix}\text{directly} \\ \text{related}\end{matrix} \qquad y = k\frac{w^2x}{z} \begin{matrix} \leftarrow \text{directly related} \\ \leftarrow \text{inversely related} \end{matrix}$$

We solve these problems the same way we solved other variation problems. First we find the constant of variation, then use it to help answer the question.

When a variable is directly proportional to the product of two or more other variables, such as $y = kxz$, this is often called **joint variation**. In this case we would say that y varies jointly as x and z.

Example 5 Volume of a Conical Tank

The number of gallons of a liquid that can be stored in a conical tank is directly proportional to the area of the base of the tank and its height (joint variation). A tank with a base area of 1200 square feet and a height of 15 feet holds 45,000 gallons of liquid. How tall must the tank be to hold 75,000 gallons of liquid if its base area is 1500 square feet?

Solution Follow the guidelines for solving variation problems.

Step 1. We are told that the volume is directly related to the area of the base of the tank and its height. If we let V = volume, B = base area, and h = height, we can translate the problem statement into the model

$$V = kBh \leftarrow \text{joint variation } (B \text{ and } h \text{ both direct}),$$

where k is the constant of variation.

Step 2. To determine the value of k, we use the fact that a tank with a base area of 1200 square feet and a height of 15 feet holds 45,000 gallons.

Substitute 1200 for B, 15 for h, and 45,000 for V: $45,000 = k(1200)(15)$

Simplify: $45,000 = 18,000k$

Divide both sides by 18,000: $2.5 = k$

Step 3. The constant of variation is 2.5, so the general model is $V = 2.5Bh$.

Step 4. We want to determine the height of a tank with a base area of 1500 square feet that holds 75,000 gallons. Substituting 75,000 for V and 1500 for B, we then solve for h.

Substitute 1500 for B and 75,000 for V: $75{,}000 = 2.5(1500)h$

Simplify: $75{,}000 = 3750k$

Divide both sides by 3750: $20 = h$

The tank would need to be 20 feet tall.

You Try It   Work through this You Try It problem.

Work Exercises 9–10 in this textbook or in the *MyMathLab*® Study Plan.

Example 6   Electrical Resistance

My video summary

The resistance of a wire varies directly with the length of the wire and inversely with the square of its radius. A wire with a length of 500 cm and a radius of 0.5 cm has a resistance of 15 ohms. Determine the resistance of an 800 cm piece of similar wire with a radius of 0.8 cm.

Solution   Follow the guidelines for solving variation problems.

Step 1. We are told that resistance varies directly with the length of the wire and inversely with the square of its radius. Letting R = resistance, L = length, and r = radius, we can translate the problem statement into the model

$$R = k\frac{L}{r^2} \begin{matrix} \leftarrow \text{length (direct)} \\ \leftarrow \text{square of the radius (inverse)} \end{matrix}$$

where k is the constant of variation.

Step 2. To determine the value of k, we use the fact that a wire of length 500 cm and radius 0.5 cm has a resistance of 15 ohms.

Substitute 500 for L, 0.5 for r, and 15 for R: $15 = k\dfrac{(500)}{(0.5)^2}$

Simplify: $15 = 2000k$

Divide both sides by 2000: $0.0075 = k$

Use the constant of variation to write a general model for this situation. Then use the model to determine the resistance of an 800 cm piece of similar wire with a radius of 0.8 cm. Click here to check your answer, or watch this video for a detailed solution.

You Try It   Work through this You Try It problem.

Work Exercises 11–12 in this textbook or in the *MyMathLab*® Study Plan.

Example 7   Burning Calories

For a fixed speed, the number of calories burned while jogging varies jointly with the weight of the jogger (in kg) and the time spent jogging (in minutes). If a 100-kg man jogs for 40 minutes and burns 490 calories, how many calories will a 130-kg man burn if he jogs for 60 minutes at the same speed?

Solution Letting C = calories burned, W = weight in kg, and T = time jogging in minutes, we have the model $C = kWT$, where k is the constant of variation. Solve the problem on your own. Click here to check your answer, or watch this video for a complete solution.

You Try It **Work through this You Try It problem.**

Work Exercises 13–14 in this textbook or in the *MyMathLab*® Study Plan.

6.6 Exercises

In Exercises 1–14, solve the variation problems.

You Try It

1. **Scuba Diving** The water pressure on a scuba diver is directly proportional to the depth of the diver. If the pressure on a diver is 13.5 psi when she is 30 feet below the surface, how far below the surface will she be when the pressure is 18 psi?

2. **Pendulum Length** The length of a simple pendulum varies directly with the square of its period. If a pendulum of length 2.25 meters has a period of 3 seconds, how long is a pendulum with a period of 8 seconds?

You Try It

3. **Water Flow Rate** For a fixed water flow rate, the amount of water that can be pumped through a pipe varies directly as the square of the diameter of the pipe. In one hour, a pipe with an 8-inch diameter can pump 400 gallons of water. Assuming the same water flow rate, how much water could be pumped through a pipe that is 12 inches in diameter?

4. **Falling Distance** The distance an object falls varies directly with the square of the time it spends falling. If a ball falls 19.6 meters after falling for 2 seconds, how far will it fall after 9 seconds?

You Try It

5. **Car Depreciation** The value of a car is inversely proportional to its age. If a car is worth $8100 when it is 4 years old, how old will it be when it is worth $3600?

6. **Electric Resistance** For a given voltage, the resistance of a circuit is inversely related to its current. If a circuit has a resistance of 5 ohms and a current of 12 amps, what is the resistance if the current is 20 amps?

You Try It

7. **Weight of an Object** The weight of an object within Earth's atmosphere varies inversely with the square of the distance of the object from Earth's center. If a low Earth orbit satellite weighs 100 kg on Earth's surface (6400 km), how much will it weigh in its orbit 800 km above Earth?

8. **Light Intensity** The intensity of a light varies inversely as the square of the distance from the light source. If the intensity from a light source 3 feet away is 8 lumens, what is the intensity at a distance of 2 feet?

You Try It

9. **Simple Interest** For a car loan using simple interest at a given rate, the amount of interest charged varies jointly with the loan amount and the time of the loan (in years). If a $15,000 car loan earns $2175 in simple interest over 5 years, how much interest will a $32,000 car loan earn over 6 years?

10. **Horsepower of a Pump** The horsepower of a water pump varies jointly with the weight of the water moved (in lb) and the length of the discharge head (in feet). A 3.1-horsepower pump can move 341 lb of water against a discharge head of 300 feet. What is the horsepower of a pump that moves 429 lb of water against a discharge head of 400 feet?

You Try It

11. **Gas Pressure** The pressure of a gas in a container varies directly with its temperature and inversely with its volume. At a temperature of 90 Kelvin and a volume of 10 cubic meters, the pressure is 32.4 kilograms per square meter. What is the pressure if the temperature is increased to 100 Kelvin and the volume is decreased to 8 cubic meters?

12. **Spindle Speed**   In milling operations, the spindle speed S (in revolutions per minute) is directly related to the cutting speed C (in feet per minute) and inversely related to the tool diameter D (in inches). A milling cut taken with a 2-inch high-speed drill and a cutting speed of 70 feet per minute has a spindle speed of 133.7 revolutions per minute. What is the spindle speed for a cut taken with a 4-inch high-speed drill and a cutting speed of 50 feet per minute?

You Try It

13. **Load on a Beam**   The load that can be supported by a rectangular beam varies jointly as the width of the beam and the square of its height, and inversely as the length of the beam. A beam 12 feet long, with a width of 6 inches and a height of 4 inches can support a maximum load of 900 pounds. If a similar board has a width of 8 inches and a height of 5 inches, how long must it be to support 1200 pounds?

14. **Windmill Power**   The power produced by a windmill varies directly with the square of its diameter and the cube of the wind speed. A fan with a diameter of 2.5 meters produces 270 watts of power if the wind speed is 3 m/s. How much power would a similar fan produce if it had a diameter of 4 meters and the wind speed was 4 m/s?

Radicals and Rational Exponents

CHAPTER SEVEN CONTENTS

7.1 **Radical Expressions**

7.2 **Radical Functions**

7.3 **Rational Exponents and Simplifying Radical Expressions**

7.4 **Operations with Radicals**

7.5 **Radical Equations and Models**

7.6 **Complex Numbers**

7.1 Radical Expressions

THINGS TO KNOW

Before working through this section, be sure you are familiar with the following concepts:

VIDEO   ANIMATION   INTERACTIVE

You Try It

1. Compute the Absolute Value of a Real Number (Section R.1, Objective 5)

You Try It

2. Simplify Numeric Expressions Containing Exponents and Radicals (Section R.2, Objective 2)

OBJECTIVES

1 Find Square Roots of Perfect Squares

2 Approximate Square Roots

3 Simplify Radical Expressions of the Form $\sqrt{a^2}$

4 Find Cube Roots

5 Find and Approximate nth Roots

OBJECTIVE 1:   FIND SQUARE ROOTS OF PERFECT SQUARES

Recall from Section R.2 that a **square root** of a non-negative real number a is a real number b that, when squared, results in a. So, b is a square root of a if $b^2 = a$.

Every positive real number has two square roots: one positive and one negative. For example, 2 and -2 are both square roots of 4 because

$$2^2 = 4 \quad \text{and} \quad (-2)^2 = 4.$$

2 is the *positive*, or *principal square root* of 4, and -2 is the *negative square root* of 4.

We use the **radical sign** $\sqrt{}$ to denote principal square roots. We place a negative sign in front of a radical sign, $-\sqrt{}$, to denote negative square roots. For example, $\sqrt{4}$ represents the principal square root of 4, whereas $-\sqrt{4}$ represents the negative square root of 4. So, we write $\sqrt{4} = 2$ and $-\sqrt{4} = -2$.

Definition   Principal and Negative Square Roots

A non-negative real number b is the **principal square root** of a non-negative real number a, denoted as $b = \sqrt{a}$, if $b^2 = a$.

A negative real number b is the **negative square root** of a non-negative real number a, denoted as $b = -\sqrt{a}$, if $b^2 = a$.

The expression $\sqrt{4}$ is called a *radical expression*. A **radical expression** is an expression that contains a radical sign. The expression beneath the radical sign is called the **radicand**.

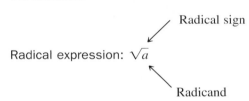

Radical sign

Radical expression: $\sqrt{a}$

Radicand

When the radicand is a perfect square, the square root will simplify to a rational number.

Example 1 Finding Square Roots

 Evaluate.

a. $\sqrt{64}$ **b.** $-\sqrt{169}$ **c.** $\sqrt{-100}$

d. $\sqrt{\dfrac{9}{25}}$ **e.** $\sqrt{0.81}$ **f.** $\sqrt{0}$

My video summary

Solution

a. We need to find the principal square root of 64. Since $8^2 = 64$, $\sqrt{64} = 8$.

b. We need to find the negative square root of 169.
Since $(-13)^2 = 169$, $-\sqrt{169} = -13$.

c. There is no real number that can be squared to result in -100 because the square of a real number will always be non-negative. So, $\sqrt{-100}$ is not a real number.

d.–f. Try to evaluate these square roots on your own. Click here or watch this video to check your work.

The square root of a negative number is not a real number.

You Try It Work through this You Try It problem.

Work Exercises 1–4 in this textbook or in the *MyMathLab®* Study Plan.

OBJECTIVE 2: APPROXIMATE SQUARE ROOTS

In Example 1, we saw that a square root simplifies to a rational number if the radicand is a perfect square. But what happens when the radicand is not a perfect square? In this case, the square root is an irrational number.

Consider $\sqrt{12}$. The radicand 12 is not a perfect square, so $\sqrt{12}$ is an irrational number. For such radical expressions, we can find decimal approximations using a calculator. Before doing this, consider what might be a reasonable result. Notice that 12 is between the perfect squares 9 and 16.

$$9 < 12 < 16$$

So, the principal square root of 12 should be between the principal square roots of 9 and 16.

$$\sqrt{9} < \sqrt{12} < \sqrt{16}, \text{ or}$$
$$3 < \sqrt{12} < 4$$

Figure 1 shows the TI-84 Plus calculator display for $\sqrt{12}$. Rounding to three decimal places, we have $\sqrt{12} \approx 3.464$. This approximation is between 3 and 4 as expected.

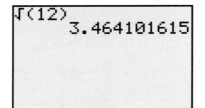

Figure 1
TI-84 Plus Display Approximating $\sqrt{12}$

Example 2 Approximating Square Roots

Use a calculator to approximate each square root. Round to three decimal places. Check that the answer is reasonable.

a. $\sqrt{5}$ b. $\sqrt{45}$ c. $\sqrt{103}$

Solution

a. The radicand 5 is between the perfect squares $2^2 = 4$ and $3^2 = 9$, so $\sqrt{4} < \sqrt{5} < \sqrt{9}$ or $2 < \sqrt{5} < 3$. The square root is between 2 and 3. From Figure 2, we get $\sqrt{5} \approx 2.236$, which is between 2 and 3.

b. 45 is between $6^2 = 36$ and $7^2 = 49$, so $\sqrt{36} < \sqrt{45} < \sqrt{49}$ or $6 < \sqrt{45} < 7$. Figure 2 shows us $\sqrt{45} \approx 6.708$, which is between 6 and 7.

c. 103 is between $10^2 = 100$ and $11^2 = 121$, so $\sqrt{100} < \sqrt{103} < \sqrt{121}$ or $10 < \sqrt{103} < 11$. In Figure 2, we see $\sqrt{103} \approx 10.149$, which is reasonable.

√(5)
        2.236067977
√(45)
        6.708203932
√(103)
        10.14889157

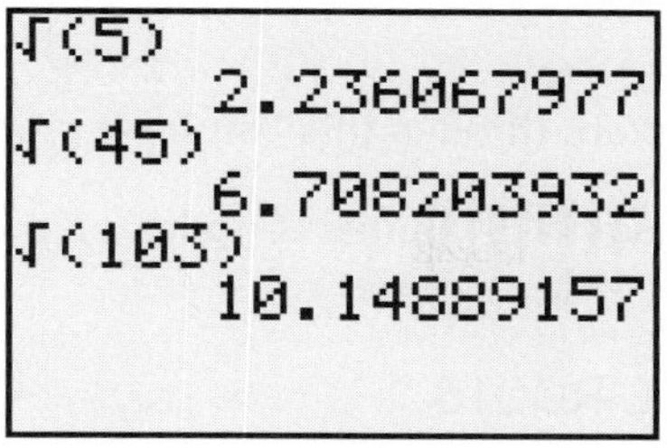

Figure 2
TI-84 Plus Display for Estimating $\sqrt{5}$, $\sqrt{45}$, and $\sqrt{103}$

You Try It   Work through this You Try It problem.

Work Exercises 5–6 in this textbook or in the *MyMathLab*® Study Plan.

OBJECTIVE 3:   SIMPLIFY RADICAL EXPRESSIONS OF THE FORM $\sqrt{a^2}$

Let's now consider square roots with variables in the radicand. A common misconception is to think that $\sqrt{a^2} = a$, but this is not necessarily true. To see why, substitute $a = -5$ in the expression $\sqrt{a^2}$ and simplify:

$$\text{Substitute } -5 \text{ for } a: \quad \sqrt{a^2} = \sqrt{(-5)^2}$$
$$\text{Simplify } (-5)^2: \qquad = \sqrt{25}$$
$$\text{Find the principal square root of } 25: \qquad = 5$$

The final result is not -5 but $|-5| = 5$. This illustrates the following square root property for simplifying radical expressions of the form $\sqrt{a^2}$.

Simplifying Radical Expressions of the Form $\sqrt{a^2}$

For any real number a,

$$\sqrt{a^2} = |a|.$$

When taking the square root of a base raised to the second power, the result will be the absolute value of the base.

Example 3 Simplifying Radical Expressions of the Form $\sqrt{a^2}$

My video summary

 Simplify.

a. $\sqrt{(-12)^2}$ b. $\sqrt{(2x-5)^2}$ c. $\sqrt{100x^2}$

d. $\sqrt{x^2 + 12x + 36}$ e. $\sqrt{9x^4}$ f. $\sqrt{y^6}$

Solution

a. $\sqrt{(-12)^2} = |-12| = 12$

b. $\sqrt{(2x-5)^2} = |2x-5|$. Since $2x - 5$ could be negative, the absolute value symbol is required to ensure a non-negative result.

c. Since $100x^2 = (10x)^2$, then $\sqrt{100x^2} = \sqrt{(10x)^2} = |10x|$ or $10|x|$.

d.–f. Try to simplify each square root on your own by first writing the radicand as a base raised to the second power. Include absolute value notation if necessary. Click here or watch this video to check your work.

You Try It   Work through this You Try It problem.

Work Exercises 7–14 in this textbook or in the **MyMathLab**® Study Plan.

OBJECTIVE 4:   FIND CUBE ROOTS

We can apply the process of finding square roots to other types of roots. For example, the **cube root** of a real number a is a real number b that, when cubed, results in a. So, b is the cube root of a if $b^3 = a$. For example, 2 is the cube root of 8 because $2^3 = 8$. To write the cube root of a, we use the notation $\sqrt[3]{a}$. The 3 in this radical expression indicates a cube root instead of a square root. Using this notation, we write $\sqrt[3]{8} = 2$.

Definition   **Cube Roots**

A real number b is the **cube root** of a real number a, denoted as $b = \sqrt[3]{a}$, if $b^3 = a$.

Recall that the cube of a negative number is a negative number. Unlike square roots, cube roots can have negative numbers in the radicand. For example, $\sqrt[3]{-64} = -4$ because $(-4)^3 = -64$. Every real number has one real cube root. If the radicand is positive, the cube root will be positive. If the radicand is negative, the cube root will be negative. Therefore, absolute value is not used when simplifying radical expressions of the form $\sqrt[3]{a^3}$.

Simplifying Radical Expressions of the Form $\sqrt[3]{a^3}$

For any real number a,

$$\sqrt[3]{a^3} = a.$$

Example 4   **Finding Cube Roots**

My video summary

📺 Simplify.

   a. $\sqrt[3]{125}$ **b.** $\sqrt[3]{-1000}$ **c.** $\sqrt[3]{x^{15}}$

   d. $\sqrt[3]{0.064}$ **e.** $\sqrt[3]{\dfrac{8}{27}}$ **f.** $\sqrt[3]{-64y^9}$

Solution

   a. $\sqrt[3]{125} = \sqrt[3]{5^3} = 5$
   b. $\sqrt[3]{-1000} = \sqrt[3]{(-10)^3} = -10$
   c. $\sqrt[3]{x^{15}} = \sqrt[3]{(x^5)^3} = x^5$

d.–f. Try to simplify each cube root on your own by first writing each radicand as a base raised to the third power. Click here or watch this video to check your work.

When simplifying expressions of the form $\sqrt[3]{a^3}$, do not use absolute value symbols. Doing so may lead to an incorrect result. For example, $\sqrt[3]{(-5)^3} = -5$, not $|-5| = 5$.

You Try It   **Work through this You Try It problem.**

Work Exercises 15–20 in this textbook or in the *MyMathLab* Study Plan.

OBJECTIVE 5:   FIND AND APPROXIMATE *n*th ROOTS

We know that 2 is a square root of 4 because $2^2 = 4$ and 2 is a cube root of 8 because $2^3 = 8$. In fact, 2 is a **4th root** of 16 because $2^4 = 16$, and 2 is a **5th root** of 32 because $2^5 = 32$. We denote each of these roots using radical signs as follows: $\sqrt{4} = 2$, $\sqrt[3]{8} = 2$, $\sqrt[4]{16} = 2$, and $\sqrt[5]{32} = 2$. These are called *n*th roots.

Definition   Principal *n*th Roots

If a and b are real numbers and n is an integer such that $n \geq 2$, then b is the **principal *n*th root** of a, denoted as $b = \sqrt[n]{a}$, if $b^n = a$.

Note: b must be non-negative when n is even.

In the notation $\sqrt[n]{a}$, n is called the **index** of the radical expression, and it indicates the type of root. For example, a cube root has an index of $n = 3$. If no index is shown, then it is understood to be $n = 2$ for a square root.

If n is an odd integer, then the radicand a can be any real number. However, if n is an even integer, then the radicand a must be non-negative. This leads to the following rule for simplifying radical expressions of the form $\sqrt[n]{a^n}$.

Simplifying Radical Expressions of the Form $\sqrt[n]{a^n}$

If n is an integer such that $n \geq 2$ and a is a real number, then

$$\sqrt[n]{a^n} = a \text{ if } n \text{ is odd.}$$
$$\sqrt[n]{a^n} = |a| \text{ if } n \text{ is even.}$$

Example 5 Finding *n*th Roots

My video summary

 Simplify.

a. $\sqrt[4]{81}$

b. $\sqrt[5]{-32}$

c. $\sqrt[6]{\dfrac{1}{64}}$

d. $\sqrt[5]{x^{15}}$

e. $\sqrt[6]{(x-7)^6}$

f. $\sqrt[4]{-1}$

Solution

a. $\sqrt[4]{81} = \sqrt[4]{3^4} = |3| = 3$

b. $\sqrt[5]{-32} = \sqrt[5]{(-2)^5} = -2$

c. $\sqrt[6]{\dfrac{1}{64}} = \sqrt[6]{\left(\dfrac{1}{2}\right)^6} = \left|\dfrac{1}{2}\right| = \dfrac{1}{2}$

d.–f. Try to simplify each *n*th root on your own by first writing each radicand as a base raised to the *n*th power. Click here or watch this video to check your work.

CAUTION When simplifying expressions of the form $\sqrt[n]{a^n}$, use absolute value symbols when n is even but not when n is odd.

You Try It Work through this You Try It problem.

Work Exercises 21–26 in this textbook or in the *MyMathLab*® Study Plan.

In Example 5 we saw that $\sqrt[4]{81} = 3$ because $81 = 3^4$. Since 81 is a perfect 4th power, its 4th root is a rational number. But what if the radicand of an nth root is not a perfect nth power? Such expressions represent irrational numbers. As with square roots, we can use a calculator to approximate these types of nth roots.

Example 6 Approximating *n*th Roots

Use a calculator to approximate each root. Round to three decimal places. Check that the answer is reasonable.

a. $\sqrt[3]{6}$ b. $\sqrt[4]{200}$ c. $\sqrt[5]{154}$

Solution

a. The radicand 6 is between the perfect cubes $1^3 = 1$ and $2^3 = 8$, so $\sqrt[3]{1} < \sqrt[3]{6} < \sqrt[3]{8}$ or $1 < \sqrt[3]{6} < 2$. From Figure 3, we get $\sqrt[3]{6} \approx 1.817$, which is between 1 and 2.

b. 200 is between $3^4 = 81$ and $4^4 = 256$, so $3 < \sqrt[4]{200} < 4$. In Figure 3, we see that $\sqrt[4]{200} \approx 3.761$, which is between 3 and 4.

c. 154 is between $2^5 = 32$ and $3^5 = 243$, so $2 < \sqrt[5]{154} < 3$. Figure 3 shows us that $\sqrt[5]{154} \approx 2.738$, which is reasonable.

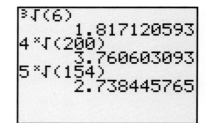

Figure 3 TI-84 Plus Display for Estimating $\sqrt[3]{6}$, $\sqrt[4]{200}$, and $\sqrt[5]{154}$

You Try It Work through this You Try It problem.

Work Exercises 27–30 in this textbook or in the *MyMathLab*® Study Plan.

7.1 Exercises

You Try It

In Exercises 1–4, evaluate each square root. If the answer is not a real number, state so.

  1. $\sqrt{100}$ **2.** $-\sqrt{49}$ **3.** $\sqrt{\dfrac{9}{121}}$ **4.** $\sqrt{0.09}$

In Exercises 5–6, use a calculator to approximate each square root. Round to three decimal places. Check that the answer is reasonable.

5. $\sqrt{55}$ **6.** $\sqrt{637}$

You Try It

You Try It

In Exercises 7–14, simplify. Include the absolute value symbol if necessary.

 7. $\sqrt{(-17)^2}$ **8.** $\sqrt{(8x)^2}$ **9.** $\sqrt{(6x - 5)^2}$

10. $\sqrt{16x^2}$ **11.** $\sqrt{81x^4}$ **12.** $\sqrt{w^{14}}$

13. $-\sqrt{121x^{10}}$ **14.** $\sqrt{x^2 + 16x + 64}$

You Try It

In Exercises 15–26, simplify. Include the absolute value symbol if necessary. If the answer is not a real number, state so.

 15. $\sqrt[3]{343}$ **16.** $\sqrt[3]{-125}$ **17.** $\sqrt[3]{-\dfrac{27}{64}}$

18. $\sqrt[3]{0.216}$ **19.** $\sqrt[3]{w^{21}}$ **20.** $\sqrt[3]{-8x^{15}}$

You Try It

 21. $\sqrt[5]{1024}$ **22.** $-\sqrt[4]{256}$ **23.** $\sqrt[4]{\dfrac{16}{625}}$

24. $\sqrt[5]{-243}$ **25.** $\sqrt[4]{x^{20}}$ **26.** $\sqrt[6]{(x - 4)^6}$

In Exercises 27–30, use a calculator to approximate each root. Round to three decimal places. Check that the answer is reasonable.

27. $\sqrt[3]{42}$ **28.** $\sqrt[4]{232}$ **29.** $\sqrt[5]{50}$ **30.** $\sqrt[6]{348}$

You Try It

7.2 Radical Functions

THINGS TO KNOW

Before working through this section, be sure you are familiar with the following concepts:

                                          VIDEO   ANIMATION   INTERACTIVE

You Try It
1. Find the Domain and Range of a Relation (Section 2.2, Objective 2)

You Try It
2. Evaluate Functions (Section 2.3, Objective 2)

You Try It
3. Graph Simple Functions by Plotting Points (Section 2.3, Objective 3)

You Try It
4. Find Square Roots of Perfect Squares (Section 7.1, Objective 1)

You Try It
5. Approximate Square Roots (Section 7.1, Objective 2)

You Try It 6. Simplify Radical Expressions of the Form $\sqrt{a^2}$
(Section 7.1, Objective 3)

You Try It 7. Find Cube Roots
(Section 7.1, Objective 4)

OBJECTIVES

1 Evaluate Radical Functions

2 Find the Domain of a Radical Function

3 Graph Functions That Contain Square Roots or Cube Roots

OBJECTIVE 1:   EVALUATE RADICAL FUNCTIONS

A **radical function** is a function that contains one or more radical expressions. We evaluate radical functions in the same way that we evaluated functions in Section 2.3. Substitute the given value for the variable and simplify using the order of operations.

Example 1 Evaluating Radical Functions

For the radical functions $f(x) = \sqrt{2x - 5}$, $g(x) = \sqrt[3]{5x + 9}$, and $h(x) = -3\sqrt[4]{x} + 2$, evaluate the following.

a. $f(15)$ **b.** $g(-2)$ **c.** $h(625)$

d. $g\left(-\dfrac{1}{5}\right)$ **e.** $f(0.5)$ **f.** $h(1)$

Solution

a. Recall that radical signs serve as grouping symbols as well as indicating the type of root. So, once we substitute the value for the variable, we simplify the radicand before finding the root.

Substitute 15 for x in the function f:	$f(15) = \sqrt{2(15) - 5}$
Multiply:	$= \sqrt{30 - 5}$
Subtract:	$= \sqrt{25}$
Evaluate the square root:	$= 5$ because $5^2 = 25$

b.
Substitute -2 for x in the function g:	$g(-2) = \sqrt[3]{5(-2) + 9}$
Multiply:	$= \sqrt[3]{-10 + 9}$
Add:	$= \sqrt[3]{-1}$
Evaluate the cube root:	$= -1$ because $(-1)^3 = -1$

c.
Substitute 625 for x in the function h:	$h(625) = -3\sqrt[4]{625} + 2$
Evaluate the 4th root:	$= -3(5) + 2$ because $5^4 = 625$
Multiply:	$= -15 + 2$
Add:	$= -13$

d.–f. Try to evaluate each radical function on your own. Click here or watch this interactive video to check your work.

You Try It **Work through this You Try It problem.**

Work Exercises 1–6 in this textbook or in the *MyMathLab*® Study Plan.

OBJECTIVE 2: FIND THE DOMAIN OF A RADICAL FUNCTION

We know that the radicand of a radical expression must be non-negative if the index of the radical expression is even. For example, $\sqrt[4]{-16}$ is not a real number. However, if the index of the radical expression is odd, then the radicand can be any real number. We use this information to find the domain of radical functions.

To find the domain of the function $f(x) = \sqrt{2x - 5}$ from Example 1, we recognize that the radicand of a square root must be non-negative. Setting the radicand greater than or equal to zero and solving will give the domain of this function:

$$\text{Set the radicand greater than or equal to } 0: \quad 2x - 5 \geq 0$$

$$\text{Subtract 5:} \qquad 2x \geq 5$$

$$\text{Divide by 2:} \qquad x \geq \frac{5}{2}$$

The domain of $f(x) = \sqrt{2x - 5}$ is $\left\{ x \mid x \geq \frac{5}{2} \right\}$ or, in interval notation, $\left[\frac{5}{2}, \infty \right)$.

To find the domain of $g(x) = \sqrt[3]{5x + 9}$, we recognize that the radicand of a cube root can be any real number. Since there are no restrictions for the radicand, the domain of $g(x) = \sqrt[3]{5x + 9}$ is the set of all real numbers.

Guideline to Finding the Domain of a Radical Function

If the index of a radical function is even, the radicand must be greater than or equal to zero.

If the index of a radical function is odd, the radicand can be any real number.

Example 2 Finding the Domain of a Radical Function

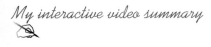

 Find the domain for each radical function.

a. $F(x) = \sqrt[4]{12 - 4x}$ **b.** $h(x) = \sqrt[5]{3x + 5}$ **c.** $G(x) = \sqrt[6]{5x + 7}$

Solution

a. The index is 4, which is even, so the radicand must be non-negative. We set the radicand greater than or equal to zero and solve:

$$\text{Set the radicand greater than or equal to } 0: \quad 12 - 4x \geq 0$$

$$\text{Subtract 12 from both sides of the inequality:} \qquad -4x \geq -12$$

$$\text{Divide both sides by } -4, \text{ reversing the inequality symbol:} \qquad x \leq 3$$

The domain is $\{ x \mid x \leq 3 \}$ or, in interval notation, $(-\infty, 3]$.

b.–c. Try to find the domain for each radical function on your own. Click here to check your answer, or watch this interactive video for a complete solution.

You Try It   **Work through this You Try It problem.**

Work Exercises 7–12 in this textbook or in the *MyMathLab*® Study Plan.

OBJECTIVE 3:   GRAPH FUNCTIONS THAT CONTAIN SQUARE ROOTS OR
CUBE ROOTS

In Section 2.3 we graphed simple functions by plotting points. We use the same strategy to graph the **square root function** $f(x) = \sqrt{x}$.

First we find several ordered pairs that belong to the function. The domain of the square root function is $\{x | x \geq 0\}$, or in interval notation $[0, \infty)$, so we evaluate the function for non-negative values (Table 1). Then we plot the points and connect them with a smooth curve (Figure 4).

x	$y = f(x) = \sqrt{x}$	(x, y)
0	$f(0) = \sqrt{0} = 0$	$(0, 0)$
1	$f(1) = \sqrt{1} = 1$	$(1, 1)$
2	$f(2) = \sqrt{2} \approx 1.414$	$(2, \sqrt{2})$
4	$f(4) = \sqrt{4} = 2$	$(4, 2)$
6	$f(6) = \sqrt{6} \approx 2.449$	$(6, \sqrt{6})$
9	$f(9) = \sqrt{9} = 3$	$(9, 3)$

Table 1

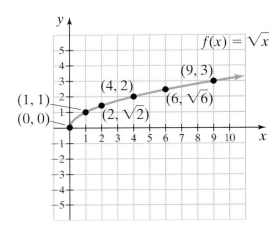

Figure 4

Other functions involving square roots will have a graph similar to the square root function, provided that the radicand is a linear expression.

Example 3 Graphing Functions That Contain Square Roots

My interactive video summary

Graph each function. Compare each graph to that of the square root function.

a. $F(x) = \sqrt{x + 1}$ **b.** $g(x) = \sqrt{x} + 1$ **c.** $h(x) = -\sqrt{x}$

Solution

a. The domain of $F(x) = \sqrt{x + 1}$ is $\{x | x \geq -1\}$, or using interval notation $[-1, \infty)$. (Do you see why?) We evaluate the function for values of x greater than or equal to -1.

To make the computations work out nicely, we choose values for x that will make the radicand a perfect square. Let $x = -1, 0, 3,$ and 8 (Table 2). Plotting the points and connecting them with a smooth curve gives the graph shown in Figure 5.

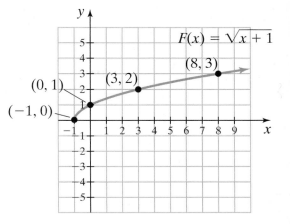

x	$y = F(x) = \sqrt{x+1}$	(x, y)
-1	$F(-1) = \sqrt{-1+1} = \sqrt{0} = 0$	$(-1, 0)$
0	$F(0) = \sqrt{0+1} = \sqrt{1} = 1$	$(0, 1)$
3	$F(3) = \sqrt{3+1} = \sqrt{4} = 2$	$(3, 2)$
8	$F(8) = \sqrt{8+1} = \sqrt{9} = 3$	$(8, 3)$

Table 2

Figure 5

The graph of $F(x) = \sqrt{x+1}$ looks like the graph of the square root function, but it is shifted one unit to the left. Click here to see the two graphs together on the same grid.

b. The domain of $g(x) = \sqrt{x} + 1$ is $\{x|x \geq 0\}$, or using interval notation $[0, \infty)$. We evaluate the function for non-negative values of x. For easier computations, we substitute perfect squares for x (Table 3).

x	$y = g(x) = \sqrt{x} + 1$	(x, y)
0	$g(0) = \sqrt{0} + 1 = 0 + 1 = 1$	$(0, 1)$
1	$g(1) = \sqrt{1} + 1 = 1 + 1 = 2$	$(1, 2)$
4	$g(4) = \sqrt{4} + 1 = 2 + 1 = 3$	$(4, 3)$
9	$g(9) = \sqrt{9} + 1 = 3 + 1 = 4$	$(9, 4)$

Table 3

Plot the points and connect them with a smooth curve. Next, compare this graph to the graph of the square root function. Click here to check your answer, or watch this interactive video for a complete solution.

c. Try to work this problem on your own. Click here to check your answer, or watch this interactive video for a complete solution.

You Try It   Work through this You Try It problem.

Work Exercises 13–15 in this textbook or in the *MyMathLab*® Study Plan.

Now let's graph the **cube root function** $f(x) = \sqrt[3]{x}$. The domain of the cube root function is the set of all real numbers $\mathbb{R}$, or using interval notation $(-\infty, \infty)$. We evaluate the function for both positive and negative values. For easier computations, we substitute perfect cubes for x (Table 4). Then we plot the points and connect them with a smooth curve (Figure 6).

x	$y = f(x) = \sqrt[3]{x}$	(x, y)
-8	$f(-8) = \sqrt[3]{-8} = -2$	$(-8, -2)$
-1	$f(-1) = \sqrt[3]{-1} = -1$	$(-1, -1)$
0	$f(0) = \sqrt[3]{0} = 0$	$(0, 0)$
1	$f(1) = \sqrt[3]{1} = 1$	$(1, 1)$
8	$f(8) = \sqrt[3]{8} = 2$	$(8, 2)$

Table 4

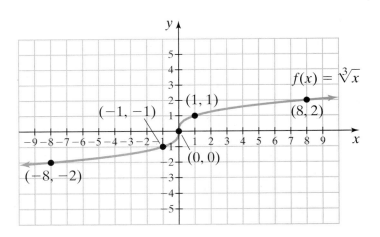

Figure 6

As with functions involving square roots, functions involving cube roots will have a graph similar to the cube root function if the radicand is a linear expression.

Example 4 Graphing Functions That Contain Cube Roots

Graph each function. Compare each graph to that of the cube root function.

a. $F(x) = \sqrt[3]{x - 2}$ **b.** $g(x) = \sqrt[3]{x} - 2$ **c.** $h(x) = -\sqrt[3]{x}$

My interactive video summary

Solution The domain of each function is the set of all real numbers $\mathbb{R}$ or $(-\infty, \infty)$, so we evaluate each function for positive and negative values. For easier computations, we choose values for x that make the radicand a perfect cube.

a. Let $x = -6, 1, 2, 3,$ and 10 (Table 5). Plotting the points and connecting them with a smooth curve gives the graph shown in Figure 7.

x	$y = F(x) = \sqrt[3]{x - 2}$	(x, y)
-6	$F(-6) = \sqrt[3]{-6 - 2} = \sqrt[3]{-8} = -2$	$(-6, -2)$
1	$F(1) = \sqrt[3]{1 - 2} = \sqrt[3]{-1} = -1$	$(1, -1)$
2	$F(2) = \sqrt[3]{2 - 2} = \sqrt[3]{0} = 0$	$(2, 0)$
3	$F(3) = \sqrt[3]{3 - 2} = \sqrt[3]{1} = 1$	$(3, 1)$
10	$F(10) = \sqrt[3]{10 - 2} = \sqrt[3]{8} = 2$	$(10, 2)$

Table 5

$$F(x) = \sqrt[3]{x} - 2$$

$(3, 1)$

$(1, -1)$

$(10, 2)$

$(2, 0)$

$(-6, -2)$

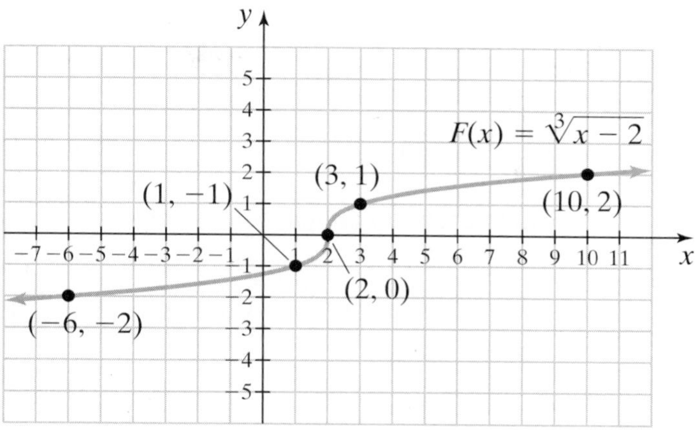

Figure 7

The graph of $F(x) = \sqrt[3]{x} - 2$ looks like the graph of the cube root function, but it is shifted two units to the right. Click here to view the two graphs together on the same grid.

b. For $g(x) = \sqrt[3]{x} - 2$, let $x = -8, -1, 0, 1$, and 8 (see Table 6).

x	$y = g(x) = \sqrt[3]{x} - 2$	(x, y)
-8	$F(-8) = \sqrt[3]{-8} - 2 = -2 - 2 = -4$	$(-8, -4)$
-1	$F(-1) = \sqrt[3]{-1} - 2 = -1 - 2 = -3$	$(-1, -3)$
0	$F(0) = \sqrt[3]{0} - 2 = 0 - 2 = -2$	$(0, -2)$
1	$F(1) = \sqrt[3]{1} - 2 = 1 - 2 = -1$	$(1, -1)$
8	$F(8) = \sqrt[3]{8} - 2 = 2 - 2 = 0$	$(8, 0)$

Table 6

Plot the points, connect them with a smooth curve, and compare the graph to the graph of the cube root function. Click here to check your answer, or watch this interactive video for a complete solution.

c. Try to work this problem on your own. Click here to check your answer, or watch this interactive video for a complete solution.

You Try It   Work through this You Try It problem.

Work Exercises 16–18 in this textbook or in the *MyMathLab* ® Study Plan.

7.2 **Exercises**

In Exercises 1–6, evaluate each radical function. If the function value is not a real number, state so.

You Try It

1. If $f(x) = \sqrt{x} - 7$, find $f(88)$.

2. If $F(x) = \sqrt{9x + 10}$, find $F\left(\dfrac{2}{3}\right)$.

3. If $g(x) = \sqrt[3]{x} + 5$, find $g(3)$.

4. If $G(x) = \sqrt[3]{4x} - 5$, find $G(-5.5)$.

5. If $h(x) = \sqrt[4]{2x} - 3$, find $h(8)$.

6. If $H(x) = 2\sqrt[4]{x}$, find $H(1)$.

You Try It

In Exercises 7–12, find the domain of each radical function.

7. $f(x) = \sqrt{x + 4}$

8. $F(x) = \sqrt{6 - 2x}$

9. $g(x) = \sqrt[3]{2x + 1}$

10. $G(x) = \sqrt[4]{2x + 5}$

11. $h(x) = \sqrt[4]{x} - 3$

12. $H(x) = \sqrt[5]{6x + 7}$

You Try It

In Exercises 13–18, graph each radical function. Compare the graph to that of the square root function or cube root function.

13. $f(x) = \sqrt{x - 2}$

14. $F(x) = \sqrt{x} - 3$

15. $g(x) = 2\sqrt{x}$

You Try It

16. $G(x) = \sqrt[3]{x + 1}$

17. $h(x) = \sqrt[3]{x} + 2$

18. $H(x) = -2\sqrt[3]{x}$

7.3 Rational Exponents and Simplifying Radical Expressions

THINGS TO KNOW

Before working through this section, be sure you are familiar with the following concepts:

	VIDEO	ANIMATION	INTERACTIVE

You Try It

1. Simplify Exponential Expressions Using the Product Rule (Section 4.1, Objective 1)

You Try It

2. Simplify Exponential Expressions Using the Quotient Rule (Section 4.1, Objective 2)

You Try It

3. Use the Negative-Power Rule (Section 4.1, Objective 4)

You Try It

4. Use the Power-to-Power Rule (Section 4.1, Objective 5)

You Try It

5. Use the Product-to-Power and Quotient-to-Power Rules (Section 4.1, Objective 6)

You Try It

6. Simplify Exponential Expressions Using a Combination of Rules (Section 4.1, Objective 7)

You Try It

7. Find Square Roots of Perfect Squares (Section 7.1, Objective 1)

You Try It

8. Simplify Radical Expressions of the Form $\sqrt{a^2}$ (Section 7.1, Objective 3)

You Try It

9. Find and Approximate nth Roots (Section 7.1, Objective 5)

OBJECTIVES

1 Use the Definition for Rational Exponents of the Form $a^{\frac{1}{n}}$

2 Use the Definition for Rational Exponents of the Form $a^{\frac{m}{n}}$

3 Simplify Exponential Expressions Involving Rational Exponents

4 Use Rational Exponents to Simplify Radical Expressions

5 Simplify Radical Expressions Using the Product Rule

6 Simplify Radical Expressions Using the Quotient Rule

OBJECTIVE 1:   USE THE DEFINITION FOR RATIONAL EXPONENTS OF THE FORM $a^{\frac{1}{n}}$

In Section 4.1 we defined an exponential expression as a constant or algebraic expression that is raised to a power. So far, we have only considered exponential expressions with integer powers. But what if the power is a rational number such as $\frac{1}{2}$? Suppose a and b are non-negative real numbers such that

$$b = a^{\frac{1}{2}}.$$

Squaring the expressions on both sides of the equal sign results in

$$(b)^2 = \left(a^{\frac{1}{2}}\right)^2.$$

Using the power-to-power rule for exponents, we multiply the powers in the expression on the right. This gives

$$b^2 = \left(a^{\frac{1}{2}}\right)^2 = a^{\frac{1}{2} \cdot 2} = a^1 = a.$$

We find that $b = a^{\frac{1}{2}}$ is a non-negative real number such that $b^2 = a$. However, this statement is similar to the definition for the principal square root of a: $b = \sqrt{a}$ if $b^2 = a$. So, we can conclude that

$$a^{\frac{1}{2}} = \sqrt{a}.$$

This idea can be used to define a **rational exponent of the form** $a^{\frac{1}{n}}$.

Definition   Rational Exponent of the Form $a^{\frac{1}{n}}$

If n is an integer such that $n \geq 2$ and if $\sqrt[n]{a}$ is a real number, then $a^{\frac{1}{n}} = \sqrt[n]{a}$.

The denominator n of the rational exponent is the index of the root. The base a of the exponential expression is the radicand of the root. If n is odd, then a can be any real number. If n is even, then a must be non-negative.

Example 1 Converting Exponential Expressions to Radical Expressions

⊞ Write each exponential expression as a radical expression. Simplify if possible.

a. $25^{\frac{1}{2}}$ **b.** $\left(-64x^3\right)^{\frac{1}{3}}$ **c.** $-100^{\frac{1}{2}}$ **d.** $(-81)^{\frac{1}{4}}$ **e.** $\left(7x^3y\right)^{\frac{1}{5}}$

My video summary

Solution

a. The index of the root is 2, the denominator of the rational exponent. So, the radical expression is a square root. The radicand is 25.

$$25^{\frac{1}{2}} = \sqrt{25} = 5$$

b. From the denominator of the rational exponent, the index is 3. So, we have a cube root. The radicand is $-64x^3$.

$$(-64x^3)^{\frac{1}{3}} = \sqrt[3]{-64x^3} = \sqrt[3]{(-4x)^3} = -4x$$

c. As in part (a), the radical expression is a square root. The negative sign is not part of the radicand because it is not part of the base. We can tell this because there are no grouping symbols around -100.

$$-100^{\frac{1}{2}} = -\sqrt{100} = -10$$

d.–e. Try to convert each expression on your own. Click here to check your answers, or watch this video for the complete solution.

You Try It   Work through this You Try It problem.

Work Exercises 1–6 in this textbook or in the *MyMathLab*® Study Plan.

Example 2 Converting Radical Expressions to Exponential Expressions

Write each radical expression as an exponential expression.

a. $\sqrt{5y}$

b. $\sqrt[3]{7x^2y}$

c. $\sqrt[4]{\dfrac{2m}{3n}}$

Solution

a. The radical expression is a square root, so the index is 2 and the denominator of the rational exponent is 2. The base of the exponential expression is the radicand $5y$.

$$\sqrt{5y} = (5y)^{\frac{1}{2}}$$

b. The index is 3 and the radicand is $7x^2y$, so the denominator of the rational exponent is 3 and the base of the exponential expression is $7x^2y$. Finish writing the exponential expression. Click here to check your answer, or watch this video for the complete solution.

c. Try to convert this expression on your own. Click here to check your answer, or watch this video for the complete solution.

You Try It   Work through this You Try It problem.

Work Exercises 13–15 in this textbook or in the *MyMathLab*® Study Plan.

OBJECTIVE 2:   USE THE DEFINITION FOR RATIONAL EXPONENTS OF THE FORM $a^{\frac{m}{n}}$

What if the numerator of the rational exponent is not 1? For example, consider the exponential expression $a^{\frac{2}{3}}$. From the power-to-power rule for exponents, we write

$$a^{\frac{2}{3}} = a^{\frac{1}{3} \cdot 2} = \left(a^{\frac{1}{3}}\right)^2 = \left(\sqrt[3]{a}\right)^2 \quad \text{or} \quad a^{\frac{2}{3}} = a^{2 \cdot \frac{1}{3}} = (a^2)^{\frac{1}{3}} = \sqrt[3]{a^2}.$$

(EBook Screens 7.3-1–7.3-32)

This idea can be used to define a rational exponent of the form $a^{\frac{m}{n}}$.

Definition   Rational Exponent of the Form $a^{\frac{m}{n}}$

If $\dfrac{m}{n}$ is a rational number in lowest terms, m and n are integers such that $n \geq 2$, and $\sqrt[n]{a}$ is a real number, then

$$a^{\frac{m}{n}} = \left(\sqrt[n]{a}\right)^m = \sqrt[n]{a^m}.$$

To simplify $a^{\frac{m}{n}}$ using the form $\left(\sqrt[n]{a}\right)^m$, we find the root first and the power second. Using the form $\sqrt[n]{a^m}$, we find the power first and the root second. Both forms may be used, but the form $\left(\sqrt[n]{a}\right)^m$ is usually easier because it involves smaller numbers.

Note: For the rest of this chapter, we assume that all variable factors of the radicand of a radical expression with an even index will be non-negative real numbers. Likewise, all variable factors of the base of an exponential expression containing a rational exponent with an even denominator will be non-negative real numbers. This assumption allows us to avoid using absolute value symbols when simplifying radical expressions.

My video summary

Example 3   Converting Exponential Expressions to Radical Expressions

Write each exponential expression as a radical expression. Simplify if possible.

a. $16^{\frac{3}{2}}$ **b.** $\left(\dfrac{y^3}{1000}\right)^{\frac{2}{3}}$ **c.** $-81^{\frac{3}{4}}$ **d.** $(-36)^{\frac{5}{2}}$ **e.** $(x^2 y)^{\frac{2}{5}}$

Solution

a. We use the form $a^{\frac{m}{n}} = \left(\sqrt[n]{a}\right)^m$. Since the denominator of the rational exponent is 2, the radical expression is a square root.

Begin with the original exponential expression: $16^{\frac{3}{2}}$
Use $a^{\frac{m}{n}} = \left(\sqrt[n]{a}\right)^m$ to rewrite as a radical expression: $= \left(\sqrt{16}\right)^3$
Simplify $\sqrt{16}$: $= (4)^3$
Simplify: $= 64$

b. $\left(\dfrac{y^3}{1000}\right)^{\frac{2}{3}} = \left(\sqrt[3]{\dfrac{y^3}{1000}}\right)^2 = \left(\sqrt[3]{\left(\dfrac{y}{10}\right)^3}\right)^2 = \left(\dfrac{y}{10}\right)^2 = \dfrac{y^2}{100}$

c. Note that the negative sign is not part of the base, so it goes in front of the radical expression.

$$-81^{\frac{3}{4}} = -\left(\sqrt[4]{81}\right)^3 = -(3)^3 = -27$$

d.–e. Try to convert each expression on your own. Click here to check your answers, or watch this video for the complete solution.

You Try It   Work through this You Try It problem.

Work Exercises 7–12 in this textbook or in the **MyMathLab**® **Study Plan.**

Example 4 Converting Radical Expressions to Exponential Expressions

Write each radical expression as an exponential expression.

a. $\sqrt[8]{x^5}$ b. $\left(\sqrt[5]{2ab^2}\right)^3$ c. $\sqrt[4]{(10x)^3}$

Solution

a. The index is 8, so the denominator of the rational exponent is 8. The radicand is a power of 5, so the numerator is 5.

$$\sqrt[8]{x^5} = x^{\frac{5}{8}}$$

b. The index is 5, so the denominator of the rational exponent is 5. The expression is raised to a power of 3, so the numerator is 3. Finish writing the exponential expression. Click here to check your answer, or watch this video for the complete solution.

c. Try to convert this expression on your own. Click here to check your answer, or watch this video for the complete solution.

You Try It   Work through this You Try It problem.

Work Exercises 16–18 in this textbook or in the **MyMathLab**® Study Plan.

If a rational exponent is negative, we can first use the negative-power rule from Section 4.1 to rewrite the expression with a positive exponent.

Example 5 Using the Negative-Power Rule with Negative Rational Exponents

Write each exponential expression with positive exponents. Simplify if possible.

a. $1000^{-\frac{1}{3}}$ b. $\dfrac{1}{81^{-\frac{1}{4}}}$ c. $125^{-\frac{2}{3}}$ d. $\dfrac{1}{8^{-\frac{4}{3}}}$ e. $(-25)^{-\frac{3}{2}}$

Solution

a.   Begin with the original exponential expression: $1000^{-\frac{1}{3}}$

Use the negative-power rule $a^{-n} = \dfrac{1}{a^n}$: $= \dfrac{1}{1000^{\frac{1}{3}}}$

Rewrite the rational exponent as a radical expression: $= \dfrac{1}{\sqrt[3]{1000}}$

Evaluate the root: $= \dfrac{1}{10}$

b.   Begin with the original exponential expression: $\dfrac{1}{81^{-\frac{1}{4}}}$

Use the negative-power rule $\dfrac{1}{a^{-n}} = a^n$: $= 81^{\frac{1}{4}}$

Rewrite the rational exponent as a radical expression: $= \sqrt[4]{81}$

Evaluate the root: $= 3$

c.–e. Try to rewrite and simplify each expression on your own. Click here to check your answers, or watch this video for the complete solution.

You Try It   Work through this You Try It problem.

Work Exercises 19–21 in this textbook or in the *MyMathLab*® Study Plan.

OBJECTIVE 3:   SIMPLIFY EXPONENTIAL EXPRESSIONS INVOLVING RATIONAL EXPONENTS

In Example 5 we used the negative-power rule for exponents to rewrite and simplify exponential expressions containing negative rational exponents. In Section 4.1 we used several other rules to simplify exponential expressions involving integer exponents. We can use these rules to simplify expressions involving rational exponents. For convenience, we repeat them here.

Rules for Exponents

Product Rule: $a^m \cdot a^n = a^{m+n}$

Quotient Rule: $\dfrac{a^m}{a^n} = a^{m-n} \quad (a \neq 0)$

Zero-Power Rule: $a^0 = 1 \quad (a \neq 0)$

Negative-Power Rule: $a^{-n} = \dfrac{1}{a^n}$ or $\dfrac{1}{a^{-n}} = a^n \quad (a \neq 0)$

Power-to-Power Rule: $(a^m)^n = a^{m \cdot n}$

Product-to-Power Rule: $(ab)^n = a^n b^n$

Quotient-to-Power Rule: $\left(\dfrac{a}{b}\right)^n = \dfrac{a^n}{b^n} \quad (b \neq 0)$

Recall that an exponential expression is **simplified** when

- No parentheses or grouping symbols are present.
- No zero or negative exponents are present.
- No powers are raised to powers.
- Each base occurs only once.

Example 6 Simplifying Expressions Involving Rational Exponents

My interactive video summary

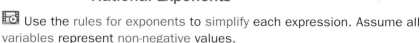 Use the rules for exponents to simplify each expression. Assume all variables represent non-negative values.

a. $x^{\frac{3}{8}} \cdot x^{\frac{1}{6}}$

b. $\dfrac{49^{\frac{7}{10}}}{49^{\frac{1}{5}}}$

c. $\left(64^{\frac{4}{9}}\right)^{\frac{3}{2}}$

d. $\left(32x^{\frac{5}{6}}y^{\frac{10}{9}}\right)^{\frac{3}{5}}$

e. $\left(\dfrac{125x^{\frac{5}{4}}}{y^{\frac{7}{8}}z^{\frac{9}{4}}}\right)^{\frac{4}{3}}, y \neq 0, z \neq 0$

f. $\left(4x^{\frac{1}{6}}y^{\frac{3}{4}}\right)^2\left(3x^{\frac{5}{9}}y^{-\frac{3}{2}}\right), y \neq 0$

Solution

a. Begin with the original expression: $x^{\frac{3}{8}} \cdot x^{\frac{1}{6}}$

Use the product rule for exponents: $= x^{\frac{3}{8} + \frac{1}{6}}$

Add: $\dfrac{3}{8} + \dfrac{1}{6} = \dfrac{9}{24} + \dfrac{4}{24} = \dfrac{13}{24}$: $= x^{\frac{13}{24}}$

b. Begin with the original expression: $\dfrac{49^{\frac{7}{10}}}{49^{\frac{1}{5}}}$

Use the quotient rule for exponents: $= 49^{\frac{7}{10} - \frac{1}{5}}$

Subtract: $\dfrac{7}{10} - \dfrac{1}{5} = \dfrac{7}{10} - \dfrac{2}{10} = \dfrac{5}{10} = \dfrac{1}{2}$: $= 49^{\frac{1}{2}}$

Use $a^{\frac{1}{n}} = \sqrt[n]{a}$ to rewrite as a radical expression: $= \sqrt{49}$

Simplify: $= 7$

c. Begin with the original expression: $\left(64^{\frac{4}{9}}\right)^{\frac{3}{2}}$

Use the power-to-power rule for exponents: $= 64^{\frac{4}{9} \cdot \frac{3}{2}}$

Multiply: $\dfrac{4}{9} \cdot \dfrac{3}{2} = \dfrac{\overset{2}{\cancel{4}}}{\underset{3}{\cancel{9}}} \cdot \dfrac{\overset{1}{\cancel{3}}}{\underset{1}{\cancel{2}}} = \dfrac{2}{3}$: $= 64^{\frac{2}{3}}$

Use $a^{\frac{m}{n}} = \left(\sqrt[n]{a}\right)^m$ to rewrite as a radical expression: $= \left(\sqrt[3]{64}\right)^2$

Simplify $\sqrt[3]{64}$: $= (4)^2$

Simplify: $= 16$

d.–f. Try to simplify each expression on your own. Click here to check your answers, or watch this interactive video for a complete solution.

You Try It   Work through this You Try It problem.

Work Exercises 22–28 in this textbook or in the *MyMathLab*® Study Plan.

OBJECTIVE 4:   USE RATIONAL EXPONENTS TO SIMPLIFY RADICAL EXPRESSIONS

Some radical expressions can be simplified by first writing them with rational exponents. We can use the following process.

Using Rational Exponents to Simplify Radical Expressions

Step 1. Convert each radical expression to an exponential expression with rational exponents.

Step 2. Simplify by writing fractions in lowest terms or using the rules of exponents, as necessary.

Step 3. Convert any remaining rational exponents back to a radical expression.

Example 7 Simplifying Radical Expressions

My interactive video summary

 Use rational exponents to simplify each radical expression. Assume all variables represent non-negative values.

a. $\sqrt[6]{y} \cdot \sqrt[3]{y}$

b. $\sqrt{\sqrt[3]{x}}$

c. $\sqrt[6]{x^4}$

d. $\sqrt[8]{25x^2y^6}$

e. $\sqrt[4]{49}$

f. $\dfrac{\sqrt[3]{x}}{\sqrt[4]{x}}, x \neq 0$

Solution

a. Rewrite each radical using rational exponents: $\sqrt[6]{y} \cdot \sqrt[3]{y} = y^{\frac{1}{6}} \cdot y^{\frac{1}{3}}$

Use the product rule for exponents: $= y^{\frac{1}{6} + \frac{1}{3}}$

Add: $\dfrac{1}{6} + \dfrac{1}{3} = \dfrac{1}{6} + \dfrac{2}{6} = \dfrac{3}{6} = \dfrac{1}{2}$: $= y^{\frac{1}{2}}$

Convert back to a radical expression: $= \sqrt{y}$

b. Rewrite the radicand $\sqrt[3]{x}$ using a rational exponent: $\sqrt{\sqrt[3]{x}} = \sqrt{x^{\frac{1}{3}}}$

Rewrite the square root using a rational exponent: $= \left(x^{\frac{1}{3}}\right)^{\frac{1}{2}}$

Use the power-to-power rule for exponents: $= x^{\frac{1}{3} \cdot \frac{1}{2}}$

Multiply: $= x^{\frac{1}{6}}$

Convert back to a radical expression: $= \sqrt[6]{x}$

c. Convert to a rational exponent: $\sqrt[6]{x^4} = x^{\frac{4}{6}}$

Write the rational exponent in lowest terms: $= x^{\frac{2}{3}}$

Convert back to a radical expression: $= \sqrt[3]{x^2}$

d.–f. Try to simplify each radical expression on your own. Click here to check your answers, or watch this interactive video for a complete solution.

You Try It Work through this You Try It problem.

Work Exercises 29–34 in this textbook or in the **MyMathLab** Study Plan.

OBJECTIVE 5:   SIMPLIFY RADICAL EXPRESSIONS USING THE PRODUCT RULE

We can develop a *product rule for radicals* that is similar to the product-to-power rule for exponents. Look at the radical expression $\sqrt[n]{ab}$.

Rewrite the radical as a rational exponent: $\sqrt[n]{ab} = (ab)^{\frac{1}{n}}$

Apply the product-to-power rule for exponents: $= a^{\frac{1}{n}}b^{\frac{1}{n}}$

Convert back to radical expressions: $= \sqrt[n]{a}\sqrt[n]{b}$

This **product rule for radicals** works in both directions: $\sqrt[n]{ab} = \sqrt[n]{a}\sqrt[n]{b}$ and $\sqrt[n]{a}\sqrt[n]{b} = \sqrt[n]{ab}$ So, we can use the rule to multiply radicals and to simplify radicals.

Product Rule for Radicals

If $\sqrt[n]{a}$ and $\sqrt[n]{b}$ are real numbers, then $\sqrt[n]{a}\sqrt[n]{b} = \sqrt[n]{ab}$.

CAUTION   The index on each radical must be the same in order to use the product rule for radicals.

Example 8 Using the Product Rule to Multiply Radicals

Multiply. Assume all variables represent non-negative values.

a. $\sqrt{2} \cdot \sqrt{5}$ b. $\sqrt{2} \cdot \sqrt{18}$ c. $\sqrt[5]{4x^2} \cdot \sqrt[5]{7y^3}$ d. $\sqrt[3]{2} \cdot \sqrt[3]{4}$

Solution In each case, we use the product rule for radicals and then simplify.

$$
\text{a.} \quad \underbrace{\sqrt{2} \cdot \sqrt{5}}_{\substack{\text{Original}\\\text{expression}}} = \underbrace{\sqrt{2 \cdot 5}}_{\substack{\text{Product rule}\\\text{for radicals}}} = \underbrace{\sqrt{10}}_{\substack{\text{Simplify the}\\\text{radicand}}}
$$

$$
\text{b.} \quad \underbrace{\sqrt{2} \cdot \sqrt{18}}_{\substack{\text{Original}\\\text{expression}}} = \underbrace{\sqrt{2 \cdot 18}}_{\substack{\text{Product rule}\\\text{for radicals}}} = \underbrace{\sqrt{36}}_{\substack{\text{Simplify the}\\\text{radicand}}} = \underbrace{6}_{\text{Evaluate}}
$$

$$
\text{c.} \quad \underbrace{\sqrt[5]{4x^2} \cdot \sqrt[5]{7y^3}}_{\substack{\text{Original}\\\text{expression}}} = \underbrace{\sqrt[5]{4x^2 \cdot 7y^3}}_{\substack{\text{Product rule}\\\text{for radicals}}} = \underbrace{\sqrt[5]{28x^2y^3}}_{\substack{\text{Simplify the}\\\text{radicand}}}
$$

$$
\text{d.} \quad \underbrace{\sqrt[3]{2} \cdot \sqrt[3]{4}}_{\substack{\text{Original}\\\text{expression}}} = \underbrace{\sqrt[3]{2 \cdot 4}}_{\substack{\text{Product rule}\\\text{for radicals}}} = \underbrace{\sqrt[3]{8}}_{\substack{\text{Simplify the}\\\text{radicand}}} = \underbrace{2}_{\text{Evaluate}}
$$

You Try It   Work through this You Try It problem.

Work Exercises 35–38 in this textbook or in the **MyMathLab**® Study Plan.

In Section 7.1 we simplified radical expressions of the form $\sqrt[n]{a^n}$. But when the radicand was not a perfect nth power, we approximated the irrational number. Now, we can use the product rule to simplify such radicals. A radical of the form $\sqrt[n]{a}$ is **simplified** if the radicand a has no factors that are perfect nth powers (other than 1 or −1). For example, $\sqrt{12}$ is not simplified because the perfect square 4 is a factor of the radicand 12. We simplify $\sqrt{12}$ as follows:

$$\text{Factor 12 into the product of 4 and 3:} \quad \sqrt{12} = \sqrt{4 \cdot 3}$$
$$\text{Use the product rule for radicals:} \quad = \sqrt{4} \cdot \sqrt{3}$$
$$\text{Evaluate } \sqrt{4}: \quad = 2\sqrt{3}$$

Since the radicand 3 has no perfect-square factors other than 1, this final result is simplified.

We can use the following process to simply radical expressions of the form $\sqrt[n]{a}$.

Using the Product Rule to Simplify Radical Expressions of the Form $\sqrt[n]{a}$

Step 1. Write the radicand as a product of two factors, one being the largest possible perfect nth power.

Step 2. Use the product rule for radicals to take the nth root of each factor.

Step 3. Simplify the nth root of the perfect nth power.

Example 9 Using the Product Rule to Simplify Radicals

Use the product rule to simplify. Assume all variables represent non-negative values.

a. $\sqrt{700}$ b. $\sqrt[3]{40}$ c. $\sqrt[4]{x^8 y^5}$ d. $\sqrt{50x^4 y^3}$

Solution We follow the three-step process.

a. The largest factor of 700 that is a perfect square is 100.

$$\begin{aligned} \text{Begin with the original expression:} \quad & \sqrt{700} \\ \text{Factor 700 into the product } 100 \cdot 7: \quad & = \sqrt{100 \cdot 7} \\ \text{Use the product rule for radicals:} \quad & = \sqrt{100} \cdot \sqrt{7} \\ \text{Evaluate } \sqrt{100}: \quad & = 10\sqrt{7} \end{aligned}$$

b. The largest factor of 40 that is a perfect cube is 8.

$$\begin{aligned} \text{Begin with the original expression:} \quad & \sqrt[3]{40} \\ \text{Factor 40 into the product } 8 \cdot 5: \quad & = \sqrt[3]{8 \cdot 5} \\ \text{Use the product rule for radicals:} \quad & = \sqrt[3]{8} \cdot \sqrt[3]{5} \\ \text{Evaluate } \sqrt[3]{8}: \quad & = 2\sqrt[3]{5} \end{aligned}$$

c. Note that x^8 is a perfect 4th power because $x^8 = (x^2)^4$. Also, $y^5 = y^4 \cdot y$. So, the largest factor of $x^8 y^5$ that is a perfect 4th power is $x^8 y^4$.

$$\begin{aligned} \text{Begin with the original expression:} \quad & \sqrt[4]{x^8 y^5} \\ \text{Factor } x^8 y^5 \text{ into the product } x^8 y^4 \cdot y: \quad & = \sqrt[4]{x^8 y^4 \cdot y} \\ \text{Use the product rule for radicals:} \quad & = \sqrt[4]{x^8 y^4} \cdot \sqrt[4]{y} \\ \text{Simplify } \sqrt[4]{x^8 y^4}: \quad & = x^2 y \sqrt[4]{y} \end{aligned}$$

My video summary

d. Try to simplify this radical expression on your own. Click here to check your answer, or watch this video for a complete solution.

You Try It   Work through this You Try It problem.

Work Exercises 39–44 in this textbook or in the *MyMathLab*® Study Plan.

Example 10 Using the Product Rule to Multiply and Simplify Radicals

Multiply and simplify. Assume all variables represent non-negative values.

a. $3\sqrt{10} \cdot 7\sqrt{2}$ b. $2\sqrt[3]{4} \cdot 5\sqrt[3]{6}$ c. $\sqrt[4]{18x^3} \cdot \sqrt[4]{45x^2}$

My interactive video summary

Solution First, use the product rule to multiply radicals. Then, simplify using the three-step process.

a.
$$\begin{aligned} \text{Begin with the original expression:} \quad & 3\sqrt{10} \cdot 7\sqrt{2} \\ \text{Rearrange factors to group radicals together:} \quad & = 3 \cdot 7 \cdot \sqrt{10} \cdot \sqrt{2} \\ \text{Use the product rule for radicals:} \quad & = 3 \cdot 7 \cdot \sqrt{10 \cdot 2} \\ \text{Multiply } 3 \cdot 7; \text{ multiply } 10 \cdot 2: \quad & = 21\sqrt{20} \\ \text{4 is the largest factor of 20 that is a perfect square:} \quad & = 21\sqrt{4 \cdot 5} \\ \text{Use the product rule for radicals:} \quad & = 21\sqrt{4} \cdot \sqrt{5} \\ \text{Simplify } \sqrt{4}: \quad & = 21 \cdot 2\sqrt{5} \\ \text{Multiply } 21 \cdot 2: \quad & = 42\sqrt{5} \end{aligned}$$

b.–c. Try to work these problems on your own. Click here to check your answers, or watch this interactive video for complete solutions.

You Try It   Work through this You Try It problem.

Work Exercises 45–50 in this textbook or in the *MyMathLab*® Study Plan.

CAUTION

Be careful not to confuse an exponent with the index of a radical, or vice versa. For example, consider the product of x^3 and $\sqrt{y}$, written as $x^3\sqrt{y}$. Now, consider the product of x and $\sqrt[3]{y}$, written as $x\sqrt[3]{y}$. Can you see how one expression might be confused with the other? When using such expressions, it is very important to write neatly. Consider using a multiplication symbol to make the expressions as clear as possible: $x^3 \cdot \sqrt{y}$ or $x \cdot \sqrt[3]{y}$.

OBJECTIVE 6:   SIMPLIFY RADICAL EXPRESSIONS USING THE QUOTIENT RULE

Next, we develop a *quotient rule for radicals*. Consider the radical expression $\sqrt[n]{\dfrac{a}{b}}$.

Rewrite the radical as a rational exponent: $\sqrt[n]{\dfrac{a}{b}} = \left(\dfrac{a}{b}\right)^{\frac{1}{n}}$

Apply the quotient-to-power rule for exponents: $= \dfrac{a^{\frac{1}{n}}}{b^{\frac{1}{n}}}$

Convert back to radical expressions: $= \dfrac{\sqrt[n]{a}}{\sqrt[n]{b}}$

We can use the **quotient rule for radicals** to simplify radical expressions involving fractions.

Quotient Rule for Radicals

If $\sqrt[n]{a}$ and $\sqrt[n]{b}$ are real numbers and $b \neq 0$, then $\sqrt[n]{\dfrac{a}{b}} = \dfrac{\sqrt[n]{a}}{\sqrt[n]{b}}$.

CAUTION

The index on each radical must be the same in order to use the quotient rule for radicals.

Like the product rule, the quotient rule works in both directions:

$$\sqrt[n]{\frac{a}{b}} = \frac{\sqrt[n]{a}}{\sqrt[n]{b}} \text{ and } \frac{\sqrt[n]{a}}{\sqrt[n]{b}} = \sqrt[n]{\frac{a}{b}}$$

We now add more requirements for a radical expression to be **simplified**.

Simplified Radical Expression

For a radical expression to be **simplified**, it must meet the following three conditions:

Condition 1. The radicand has no factor that is a perfect power of the index of the radical.

Condition 2. The radicand contains no fractions or negative exponents.

Condition 3. No denominator contains a radical.

As we saw in Example 9, we can use the product rule for radicals to help resolve issues with Condition 1. We use the quotient rule for radicals to resolve issues with Conditions 2 and 3.

Example 11 Using the Quotient Rule to Simplify Radicals

Use the quotient rule to simplify. Assume all variables represent non-negative values.

a. $\sqrt{\dfrac{25}{64}}$ b. $\sqrt[3]{\dfrac{16x^5}{27}}$ c. $\sqrt[4]{\dfrac{5x^4}{81}}$ d. $\sqrt{\dfrac{5x^7}{45x}}, \; x \neq 0$

Solution In each case, we use the quotient rule for radicals.

a.
$$\underbrace{\sqrt{\dfrac{25}{64}}}_{\substack{\text{Original}\\\text{expression}}} = \underbrace{\dfrac{\sqrt{25}}{\sqrt{64}}}_{\substack{\text{Quotient rule}\\\text{for radicals}}} = \underbrace{\dfrac{5}{8}}_{\substack{\text{Evaluate each}\\\text{square root}}}$$

b.
$$\underbrace{\sqrt[3]{\dfrac{16x^5}{27}}}_{\substack{\text{Original}\\\text{expression}}} = \underbrace{\dfrac{\sqrt[3]{16x^5}}{\sqrt[3]{27}}}_{\substack{\text{Quotient rule}\\\text{for radicals}}} = \underbrace{\dfrac{\sqrt[3]{8x^3 \cdot 2x^2}}{\sqrt[3]{27}}}_{\substack{\text{Write as product}\\\text{using perfect-cube}\\\text{factors}}} = \underbrace{\dfrac{\sqrt[3]{8x^3} \cdot \sqrt[3]{2x^2}}{\sqrt[3]{27}}}_{\substack{\text{Product rule}\\\text{for radicals}}} = \underbrace{\dfrac{2x\sqrt[3]{2x^2}}{3}}_{\substack{\text{Simplify the}\\\text{cube roots}}}$$

c.
$$\underbrace{\sqrt[4]{\dfrac{5x^4}{81}}}_{\substack{\text{Original}\\\text{expression}}} = \underbrace{\dfrac{\sqrt[4]{5x^4}}{\sqrt[4]{81}}}_{\substack{\text{Quotient rule}\\\text{for radicals}}} = \underbrace{\dfrac{\sqrt[4]{x^4} \cdot \sqrt[4]{5}}{\sqrt[4]{81}}}_{\substack{\text{Product rule}\\\text{for radicals}}} = \underbrace{\dfrac{x\sqrt[4]{5}}{3}}_{\substack{\text{Simplify the}\\\text{4th roots}}}$$

d. Here we simplify the fraction before using the quotient rule for radicals.

$$\underbrace{\sqrt{\dfrac{5x^7}{45x}}}_{\substack{\text{Original}\\\text{expression}}} = \underbrace{\sqrt{\dfrac{x^6}{9}}}_{\substack{\text{Simplify the}\\\text{fraction}}} = \underbrace{\dfrac{\sqrt{x^6}}{\sqrt{9}}}_{\substack{\text{Quotient rule}\\\text{for radicals}}} = \underbrace{\dfrac{x^3}{3}}_{\substack{\text{Simplify the}\\\text{square roots}}}$$

You Try It Work through this You Try It problem.

Work Exercises 51–56 in this textbook or in the **MyMathLab**® Study Plan.

In Example 11, we used the quotient rule for radicals to remove fractions from the radicand. In Example 12, we will use the quotient rule to remove radicals from a denominator.

Example 12 Using the Quotient Rule to Simplify Radicals

My interactive video summary

▣ Use the quotient rule to simplify. Assume all variables represent positive numbers.

a. $\dfrac{\sqrt{240x^3}}{\sqrt{15x}}$ b. $\dfrac{\sqrt[3]{-500z^2}}{\sqrt[3]{4z^{-1}}}$ c. $\dfrac{\sqrt{150m^9}}{\sqrt{3m}}$ d. $\dfrac{\sqrt{45x^5y^{-3}}}{\sqrt{20xy^{-1}}}$

Solution In each case, we use the quotient rule for radicals.

a. Begin with the original expression: $\dfrac{\sqrt{240x^3}}{\sqrt{15x}}$

Use the quotient rule for radicals: $= \sqrt{\dfrac{240x^3}{15x}}$

Simplify the radicand: $= \sqrt{16x^2}$

Simplify: $= 4x$

b. Begin with the original expression: $\dfrac{\sqrt[3]{-500z^2}}{\sqrt[3]{4z^{-1}}}$

Use the quotient rule for radicals: $= \sqrt[3]{\dfrac{-500z^2}{4z^{-1}}}$

Divide factors and subtract exponents in radicand: $= \sqrt[3]{\dfrac{-500}{4} \cdot z^{2-(-1)}}$

Simplify radicand: $= \sqrt[3]{-125z^3}$

Simplify the cube root: $= -5z$

c.–d. Try to simplify each radical expression on your own. Click here to check your answers, or watch this interactive video for complete solutions.

You Try It   Work through this You Try It problem.

Work Exercises 57–62 in this textbook or in the *MyMathLab*® Study Plan.

7.3 Exercises

In Exercises 1–12, write each exponential expression as a radical expression. Simplify if possible.

You Try It

1. $36^{\frac{1}{2}}$ **2.** $(-125)^{\frac{1}{3}}$ **3.** $-81^{\frac{1}{4}}$

4. $(27x^3)^{\frac{1}{3}}$ **5.** $(7xy)^{\frac{1}{5}}$ **6.** $(13xy^2)^{\frac{1}{3}}$

You Try It **7.** $100^{\frac{3}{2}}$ **8.** $-8^{\frac{4}{3}}$ **9.** $(-16)^{\frac{3}{4}}$

10. $\left(\dfrac{x^3}{64}\right)^{\frac{2}{3}}$ **11.** $\left(\dfrac{1}{8}\right)^{\frac{5}{3}}$ **12.** $\left(xy^2\right)^{37}$

You Try It

In Exercises 13–18, write each radical expression as an exponential expression.

13. $\sqrt{10}$ **14.** $\sqrt[3]{7m}$ **15.** $\sqrt[5]{\dfrac{2x}{y}}$

You Try It **16.** $\sqrt[7]{x^4}$ **17.** $\left(\sqrt[5]{3xy^3}\right)^4$ **18.** $\sqrt[9]{(2xy)^4}$

In Exercises 19–21, write each exponential expression with positive exponents. Simplify if possible.

You Try It **19.** $25^{-\frac{1}{2}}$ **20.** $\dfrac{1}{625^{-\frac{3}{4}}}$ **21.** $\left(\dfrac{27}{8}\right)^{-\frac{2}{3}}$

For Exercises 22–62, assume that all variables represent non-negative values.

 In Exercises 22–28, use the rules for exponents to simplify.
Write final answers in exponential form when necessary.

You Try It

 22. $x^{\frac{1}{3}} \cdot x^{\frac{1}{2}}$

23. $9^{\frac{3}{10}} \cdot 9^{\frac{1}{5}}$

24. $\dfrac{36^{\frac{3}{4}}}{36^{\frac{1}{4}}}$

25. $\left(x^{\frac{5}{3}}\right)^{\frac{9}{10}}$

26. $\left(9m^4 n^{-\frac{3}{2}}\right)^{\frac{1}{2}}$

27. $(32x^5 y^{-10})^{\frac{1}{5}}\left(xy^{-\frac{1}{2}}\right)$

28. $\left(\dfrac{16x^{\frac{2}{3}}}{81x^{\frac{5}{4}}y^{\frac{2}{3}}}\right)^{\frac{3}{4}}$

 In Exercises 29–34, use rational exponents to simplify each radical expression.
Write final answers in radical form when necessary.

You Try It

29. $\sqrt[5]{x^2} \cdot \sqrt[4]{x}$

30. $\sqrt[4]{\sqrt[3]{x^2}}$

31. $\sqrt[6]{x^3}$

32. $\dfrac{\sqrt{5}}{\sqrt[6]{5}}$

33. $\sqrt[8]{81}$

34. $\sqrt[10]{49x^6 y^8}$

 In Exercises 35–38, multiply.

You Try It **35.** $\sqrt[4]{5} \cdot \sqrt[4]{7}$

36. $\sqrt{3} \cdot \sqrt{12}$

37. $\sqrt[3]{5x^2} \cdot \sqrt[3]{25x}$

38. $\sqrt[4]{5x} \cdot \sqrt[4]{15y^3}$

 In Exercises 39–44, use the product rule to simplify.

You Try It

39. $\sqrt{72}$

40. $\sqrt[3]{56}$

41. $\sqrt{48x^3}$

42. $\sqrt[3]{-125x^6}$

43. $\sqrt{a^{15}}$

44. $\sqrt[4]{x^5 y^8 z^9}$

 In Exercises 45–50, multiply and simplify.

You Try It

45. $\sqrt{6} \cdot \sqrt{30}$

46. $4\sqrt{3x} \cdot 7\sqrt{6x}$

47. $\sqrt{20x^3 y} \cdot \sqrt{18xy^4}$

48. $\sqrt[3]{9x^2} \cdot \sqrt[3]{6x^2}$

49. $2x\sqrt[3]{5y^2} \cdot 3x\sqrt[3]{25y}$

50. $\sqrt[4]{8x^2} \cdot \sqrt[4]{6x^3}$

In Exercises 51–62, use the quotient rule to simplify.

51. $\sqrt{\dfrac{49}{81}}$

52. $\sqrt{\dfrac{12x^3}{25y^2}}$

53. $\sqrt[3]{\dfrac{11x^3}{8}}$

You Try It

 54. $\sqrt[3]{\dfrac{16x^9}{125y^6}}$

55. $\sqrt[4]{\dfrac{21x^5}{16}}$

56. $\sqrt{\dfrac{6x^5}{54x}}$

You Try It

 57. $\dfrac{\sqrt{45}}{\sqrt{5}}$

58. $\dfrac{\sqrt{96x^6 y^8}}{\sqrt{12x^2 y^5}}$

59. $\dfrac{\sqrt{120x^9}}{\sqrt{3x^{-1}}}$

60. $\dfrac{\sqrt[3]{5000}}{\sqrt[3]{5}}$

61. $\dfrac{\sqrt[3]{-72m}}{\sqrt[3]{3m^{-2}}}$

62. $\dfrac{\sqrt{3x^5 y^3}}{\sqrt{48xy^5}}$

7.4 Operations with Radicals

THINGS TO KNOW

Before working through this section, be sure you are familiar with the following concepts:

VIDEO   ANIMATION   INTERACTIVE

You Try It

1. Find Square Roots of Perfect Squares (Section 7.1, Objective 1)

You Try It

2. Find Cube Roots (Section 7.1, Objective 4)

You Try It

3. Find and Approximate *n*th Roots (Section 7.1, Objective 5)

You Try It

4. Simplify Radical Expressions Using the Product Rule (Section 7.3, Objective 5)

You Try It

5. Simplify Radical Expressions Using the Quotient Rule (Section 7.3, Objective 6)

OBJECTIVES

1 Add and Subtract Radical Expressions

2 Multiply Radical Expressions

3 Rationalizing Denominators of Radical Expressions

OBJECTIVE 1:   ADD AND SUBTRACT RADICAL EXPRESSIONS

As we learned in Chapter 4, when simplifying polynomial expressions we combine like terms. Recall that like terms have the same variables raised to the same corresponding exponents. For example, $3x^2y^3$ and $5x^2y^3$ are like terms. We can add or subtract these terms using the reverse of the distributive property.

$$3x^2y^3 + 5x^2y^3 = (3 + 5)x^2y^3 = 8x^2y^3$$
$$3x^2y^3 - 5x^2y^3 = (3 - 5)x^2y^3 = -2x^2y^3$$

$4xy^3$ and $2x^2y^2$ are not like terms, so we cannot add or subtract them. They have the same variables, but the variables in each term do not have the same corresponding exponents. This idea is true with radicals as well. We can only add and subtract **like radicals**.

> **Definition Like Radicals**
>
> Two radicals are **like** if they have the same index and the same radicand.

In Figure 8, $\sqrt{3}$ and $4\sqrt{3}$ are like radicals because they have the same index 2 and the same radicand 3. The terms $\sqrt[3]{5}$ and $\sqrt[3]{3}$ are **not** like radicals because the radicands are different. Likewise, $\sqrt{5}$ and $\sqrt[3]{5}$ are **not** like radicals because the indices are different.

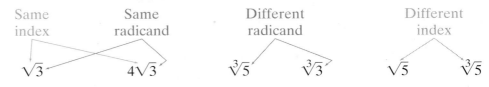

| Same index | Same radicand | Different radicand | Different index |

$\sqrt{3}$ $4\sqrt{3}$ $\sqrt[3]{5}$ $\sqrt[3]{3}$ $\sqrt{5}$ $\sqrt[3]{5}$

Like radicals   Not like radicals   Not like radicals

Figure 8

We add and subtract like radicals in the same way that we add and subtract like terms: reverse the distributive property to factor out the like radical and then simplify.

Example 1 Adding and Subtracting Radical Expressions

Add or subtract.

a. $\sqrt{11} + 6\sqrt{11}$ **b.** $7\sqrt{3} - 5\sqrt[4]{3}$ **c.** $\sqrt[3]{\dfrac{5}{8}} + 2\sqrt[3]{5}$

Solution

a. Begin with the original expression: $\sqrt{11} + 6\sqrt{11}$

$$\text{Identify like radicals:} = \sqrt{11} + 6\sqrt{11}$$
$$\text{Reverse distributive property:} = (1 + 6)\sqrt{11}$$
$$\text{Simplify:} = 7\sqrt{11}$$

b. The indices are different, so $\sqrt{3}$ and $\sqrt[4]{3}$ are not like radicals. We cannot simplify $7\sqrt{3} - 5\sqrt[4]{3}$ further since $\sqrt{3}$ and $\sqrt[4]{3}$ are not like radicals.

c. Begin with the original expression: $\sqrt[3]{\dfrac{5}{8}} + 2\sqrt[3]{5}$

$$\text{Quotient rule:} \quad \frac{\sqrt[3]{5}}{\sqrt[3]{8}} + 2\sqrt[3]{5}$$
$$\text{Simplify:} \quad \frac{\sqrt[3]{5}}{2} + 2\sqrt[3]{5}$$
$$\text{Identify like radicals:} = \frac{\sqrt[3]{5}}{2} + 2\sqrt[3]{5}$$
$$\text{Reverse distributive property:} = \left(\frac{1}{2} + 2\right)\sqrt[3]{5}$$
$$\text{Simplify:} = \frac{5}{2}\sqrt[3]{5} \quad \text{or} \quad \frac{5\sqrt[3]{5}}{2}$$

You Try It   Work through this You Try It problem.

Work Exercises 1–4 in this textbook or in the *MyMathLab*® Study Plan.

Example 2 Adding and Subtracting Radical Expressions

Add or subtract. Assume variables represent non-negative values.

a. $15\sqrt[3]{4x^2} - 9\sqrt[3]{4x^2}$

b. $2\sqrt{3x} + \sqrt{\dfrac{x}{4}}$

Solution

a. Begin with the original expression: $15\sqrt[3]{4x^2} - 9\sqrt[3]{4x^2}$

Identify like radicals: $= 15\sqrt[3]{4x^2} - 9\sqrt[3]{4x^2}$

Reverse distributive property: $= (15 - 9)\sqrt[3]{4x^2}$

Simplify: $= 6\sqrt[3]{4x^2}$

b. Begin with the original expression: $2\sqrt{3x} + \sqrt{\dfrac{x}{4}}$

Quotient rule: $= 2\sqrt{3x} + \dfrac{\sqrt{x}}{\sqrt{4}}$

Simplify: $= 2\sqrt{3x} + \dfrac{\sqrt{x}}{2}$

The radicands $3x$ and x are different, so $2\sqrt{3x}$ and $\dfrac{\sqrt{x}}{2}$ are not like radicals. We cannot simplify further.

You Try It   Work through this You Try It problem.

Work Exercises 5–8 in this textbook or in the **MyMathLab**®️ Study Plan.

CAUTION   Sometimes it is necessary to simplify radicals first before adding or subtracting.

Example 3 Adding and Subtracting Radical Expressions

Add or subtract.

a. $\sqrt{54} + 6\sqrt{72} - 3\sqrt{24}$

b. $\sqrt[3]{24} - \sqrt[3]{192} + 4\sqrt[3]{250}$

Solution

a. Begin with the original expression: $\sqrt{54} + 6\sqrt{72} - 3\sqrt{24}$

Factor: $= \sqrt{9\cdot 6} + 6\sqrt{36\cdot 2} - 3\sqrt{4\cdot 6}$

Product rule: $= \sqrt{9}\cdot\sqrt{6} + 6\sqrt{36}\cdot\sqrt{2} - 3\sqrt{4}\cdot\sqrt{6}$

Simplify radicals: $= 3\sqrt{6} + 6\cdot 6\sqrt{2} - 3\cdot 2\sqrt{6}$

Multiply factors: $= 3\sqrt{6} + 36\sqrt{2} - 6\sqrt{6}$

Collect like radicals: $= (3 - 6)\sqrt{6} + 36\sqrt{2}$

Simplify: $= -3\sqrt{6} + 36\sqrt{2}$

My video summary

b. Work the problem on your own. Click here to check your answer, or watch this video for a detailed solution.

You Try It   Work through this You Try It problem.

Work Exercises 9–12 in this textbook or in the **MyMathLab**®️ Study Plan.

Example 4 Adding and Subtracting Radical Expressions

Add or subtract. Assume variables represent non-negative values.

a. $\sqrt[3]{27m^5n^4} + 2mn\sqrt[3]{m^2n} - m\sqrt[3]{m^2n^4}$

b. $2a\sqrt{16ab^3} + 4\sqrt{9a^2b} - 5\sqrt{4a^3b^3}$

Solution

a.

$$\text{Original expression: } \sqrt[3]{27m^5n^4} + 2mn\sqrt[3]{m^2n} - m\sqrt[3]{m^2n^4}$$

$$\text{Factor: } = \sqrt[3]{27m^3n^3 \cdot m^2n} + 2mn\sqrt[3]{m^2n} - m\sqrt[3]{n^3 \cdot m^2n}$$

$$\text{Product rule: } = \sqrt[3]{27m^3n^3} \cdot \sqrt[3]{m^2n} + 2mn\sqrt[3]{m^2n} - m\sqrt[3]{n^3} \cdot \sqrt[3]{m^2n}$$

$$\text{Simplify radicals: } = 3mn\sqrt[3]{m^2n} + 2mn\sqrt[3]{m^2n} - mn\sqrt[3]{m^2n}$$

$$\text{Collect like radicals: } = (3 + 2 - 1)mn\sqrt[3]{m^2n}$$

$$\text{Simplify: } = 4mn\sqrt[3]{m^2n}$$

My video summary

b. Work the problem on your own. Click here to check your answer, or watch this video for a detailed solution.

You Try It   Work through this You Try It problem.

Work Exercises 13–16 in this textbook or in the *MyMathLab*® Study Plan.

Example 5 Adding and Subtracting Radical Expressions

Add or subtract. Assume variables represent non-negative values.

a. $\dfrac{\sqrt{45}}{6x} - \dfrac{4\sqrt{20}}{5x}$ **b.** $\dfrac{\sqrt[4]{a^5}}{3} + \dfrac{a\sqrt[4]{a}}{12}$ **c.** $\dfrac{3x^3\sqrt{24x^3y^3}}{2x\sqrt{3x^2y}} - \dfrac{x^2\sqrt{10xy^4}}{\sqrt{5y^2}}$

My interactive video summary

Solution Try to work the problems on your own. Remember to find a common denominator before adding or subtracting fractions. Click here to check your answer, or watch this interactive video for detailed solutions.

You Try It   Work through this You Try It problem.

Work Exercises 17–20 in this textbook or in the *MyMathLab*® Study Plan.

OBJECTIVE 2:   MULTIPLY RADICAL EXPRESSIONS

In Section 7.3 we saw how to multiply two radical expressions when each had only one term. Here we extend this idea to radical expressions with more than one term.

To multiply radical expressions, we follow the same approach as when multiplying polynomial expressions. We use the distributive property to multiply each term in the first expression by each term in the second. Then we simplify the resulting products and combine like terms and like radicals.

Example 6 Multiplying Radical Expressions

Multiply. Assume variables represent non-negative values.

a. $5\sqrt{2x}\left(3\sqrt{2x} - \sqrt{3}\right)$ **b.** $\sqrt[3]{2n^2}\left(\sqrt[3]{4n} + \sqrt[3]{5n}\right)$

Solution

a. Begin with the original expression: $5\sqrt{2x}\left(3\sqrt{2x} - \sqrt{3}\right)$

Distributive property: $= 5\sqrt{2x}\cdot 3\sqrt{2x} - 5\sqrt{2x}\cdot\sqrt{3}$

Rearrange factors: $= 5\cdot 3\cdot\sqrt{2x}\cdot\sqrt{2x} - 5\cdot\sqrt{2x}\cdot\sqrt{3}$

Multiply: $= 15\sqrt{4x^2} - 5\sqrt{6x}$

Simplify radical: $= 15\cdot 2x - 5\sqrt{6x}$

Simplify: $= 30x - 5\sqrt{6x}$

My video summary

b. Try this problem on your own. Click here to check your answer, or watch this video for a detailed solution.

You Try It Work through this You Try It problem.

Work Exercises 21–24 in this textbook or in the *MyMathLab* Study Plan.

Example 7 Multiplying Radical Expressions

Multiply. Assume variables represent non-negative values.

a. $\left(7\sqrt{2} - 2\sqrt{3}\right)\left(\sqrt{2} - 5\right)$ b. $\left(\sqrt{m} - 4\right)\left(3\sqrt{m} + 7\right)$

Solution

a. Use the FOIL method to multiply the radical expressions. The two first terms are $7\sqrt{2}$ and $\sqrt{2}$. The two outside terms are $7\sqrt{2}$ and -5. The two inside terms are $-2\sqrt{3}$ and $\sqrt{2}$. The two last terms are $-2\sqrt{3}$ and -5.

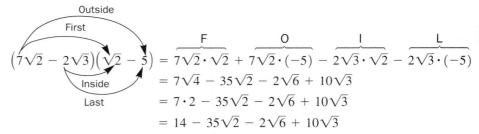

$$\left(7\sqrt{2} - 2\sqrt{3}\right)\left(\sqrt{2} - 5\right) = \overbrace{7\sqrt{2}\cdot\sqrt{2}}^{F} + \overbrace{7\sqrt{2}\cdot(-5)}^{O} - \overbrace{2\sqrt{3}\cdot\sqrt{2}}^{I} - \overbrace{2\sqrt{3}\cdot(-5)}^{L}$$

$$= 7\sqrt{4} - 35\sqrt{2} - 2\sqrt{6} + 10\sqrt{3}$$

$$= 7\cdot 2 - 35\sqrt{2} - 2\sqrt{6} + 10\sqrt{3}$$

$$= 14 - 35\sqrt{2} - 2\sqrt{6} + 10\sqrt{3}$$

My video summary

b. Try this problem on your own. Click here to check your answer, or watch this video for a detailed solution.

You Try It Work through this You Try It problem.

Work Exercises 25–30 in this textbook or in the *MyMathLab* Study Plan.

Example 8 Using Special Products to Multiply Radical Expressions

Multiply. Assume variables represent non-negative values.

a. $\left(\sqrt{y} + 3\right)\left(\sqrt{y} - 3\right)$ b. $\left(3\sqrt{x} - 2\right)^2$

Solution

a. $(\sqrt{y} + 3)(\sqrt{y} - 3)$ is a product of conjugates. We use the rule for $(A + B)(A - B)$ with $A = \sqrt{y}$ and $B = 3$.

Write the product of conjugates rule: $\qquad (A + B)(A - B) = A^2 - B^2$

Substitute $\sqrt{y}$ for A and 3 for B: $\quad (\sqrt{y} + 3)(\sqrt{y} - 3) = (\sqrt{y})^2 - (3)^2$

Simplify: $\qquad\qquad\qquad\qquad\qquad = y - 9$

My video summary

 b. This expression has the form of the square of a binomial difference $(A - B)^2$. Use the special product rule $(A - B)^2 = A^2 - 2AB + B^2$ with $A = 3\sqrt{x}$ and $B = 2$. Finish working this problem on your own. Click here to check your answer, or watch this video for a detailed solution.

You Try It **Work through this You Try It problem.**

Work Exercises 31–36 in this textbook or in the *MyMathLab*® Study Plan.

In Example 8, the product of conjugates involving square roots resulted in an expression without any radicals. This result is true in general and occurs because the product of conjugates equals the difference of two squares. We will make use of this result in the next objective.

OBJECTIVE 3: RATIONALIZING DENOMINATORS OF RADICAL EXPRESSIONS

In Section 7.3 we saw that a simplified radical expression has no radicals in the denominator. We used the quotient rule to eliminate radicals from the denominator of a radical expression. However, the quotient rule is not always enough. For example, simplifying $\sqrt{\dfrac{3}{32}}$ using the quotient rule and product rule yields the following:

$$\sqrt{\dfrac{3}{32}} = \underbrace{\dfrac{\sqrt{3}}{\sqrt{32}}}_{\substack{\text{Quotient} \\ \text{rule}}} = \underbrace{\dfrac{\sqrt{3}}{\sqrt{16} \cdot \sqrt{2}}}_{\text{Product rule}} = \dfrac{\sqrt{3}}{4\sqrt{2}}$$

We still have a radical in the denominator. Although there is nothing wrong with this, it is common practice to remove radicals from the denominator so that the denominator is a rational number. This process is called **rationalizing the denominator**.

Rationalizing a Denominator with One Term

To rationalize a denominator with a single radical of index n, multiply the numerator and denominator by a radical of index n so that the radicand in the denominator is a perfect nth power.

Multiplying the numerator and denominator by the same radical is equivalent to multiplying by 1 since $\dfrac{\sqrt{2}}{\sqrt{2}} = 1$, $\dfrac{\sqrt{11}}{\sqrt{11}} = 1$, $\dfrac{\sqrt{x}}{\sqrt{x}} = 1$, and so on. We are simply writing an equivalent expression without radicals in the denominator. This idea is like writing equivalent fractions with a least common denominator.

Example 9 Rationalizing Denominators with Square Roots

Rationalize the denominator.

a. $\dfrac{\sqrt{5}}{\sqrt{3}}$

b. $\sqrt{\dfrac{2}{5x}}$

Solution

a. Since the denominator contains a square root, we multiply the numerator and denominator by a square root so that the radicand in the denominator is a perfect square. $3 \cdot 3 = 3^2 = 9$ is a perfect square, so we multiply the numerator and denominator by $\sqrt{3}$.

$$\underbrace{\frac{\sqrt{5}}{\sqrt{3}}}_{\substack{\text{Original} \\ \text{expression}}} = \underbrace{\frac{\sqrt{5}}{\sqrt{3}} \cdot \frac{\sqrt{3}}{\sqrt{3}}}_{\substack{\text{Multiplying} \\ \text{by } 1}} = \underbrace{\frac{\sqrt{15}}{\sqrt{9}}}_{\substack{\text{Product} \\ \text{rule}}} = \underbrace{\frac{\sqrt{15}}{3}}_{\text{Simplify}}$$

b. Using the quotient rule, we write $\sqrt{\dfrac{2}{5x}} = \dfrac{\sqrt{2}}{\sqrt{5x}}$. Since the denominator contains a square root, we multiply the numerator and denominator by a square root so that the radicand in the denominator is a perfect square. $5x \cdot 5x = (5x)^2 = 25x^2$ is a perfect square, so we multiply the numerator and denominator by $\sqrt{5x}$.

$$\underbrace{\sqrt{\frac{2}{5x}}}_{\substack{\text{Original} \\ \text{expression}}} = \underbrace{\frac{\sqrt{2}}{\sqrt{5x}}}_{\substack{\text{Quotient} \\ \text{rule}}} = \underbrace{\frac{\sqrt{2}}{\sqrt{5x}} \cdot \frac{\sqrt{5x}}{\sqrt{5x}}}_{\substack{\text{Multiplying} \\ \text{by } 1}} = \underbrace{\frac{\sqrt{10x}}{\sqrt{25x^2}}}_{\substack{\text{Product} \\ \text{rule}}} = \underbrace{\frac{\sqrt{10x}}{5x}}_{\text{Simplify}}$$

You Try It Work through this You Try It problem.

Work Exercises 37–40 in this textbook or in the **MyMathLab**® Study Plan.

Why would we want to rationalize a denominator? Click here to find out.

Example 10 Rationalizing Denominators with Cube Roots or Fourth Roots

Rationalize the denominator.

a. $\sqrt[3]{\dfrac{11}{25x}}$

b. $\dfrac{\sqrt[4]{7x}}{\sqrt[4]{27y^2}}$

Solution

a. Using the quotient rule, we write $\sqrt[3]{\dfrac{11}{25x}} = \dfrac{\sqrt[3]{11}}{\sqrt[3]{25x}}$. For roots greater than 2, it is helpful to write the radicand of the denominator in exponential form.

$$\sqrt[3]{25x} = \sqrt[3]{5^2 x^1}$$

Since the denominator contains a cube root, we multiply the numerator and denominator by a cube root so that the radicand in the denominator is a perfect cube. To do this, we need the factors in the radicand to have exponents of 3 or multiples of 3. We have two 5's and one x, so we need one more 5 and two more x's to get $5^2x^1 \cdot 5^1x^2 = 5^3x^3 = (5x)^3$, which is a perfect cube. Therefore, we multiply the numerator and denominator by $\sqrt[3]{5^1x^2} = \sqrt[3]{5x^2}$.

$$\underbrace{\sqrt[3]{\frac{11}{25x}}}_{\substack{\text{Original} \\ \text{expression}}} = \underbrace{\frac{\sqrt[3]{11}}{\sqrt[3]{25x}}}_{\substack{\text{Quotient} \\ \text{rule}}} = \underbrace{\frac{\sqrt[3]{11}}{\sqrt[3]{25x}} \cdot \frac{\sqrt[3]{5x^2}}{\sqrt[3]{5x^2}}}_{\substack{\text{Multiplying} \\ \text{by 1}}} = \underbrace{\frac{\sqrt[3]{55x^2}}{\sqrt[3]{125x^3}}}_{\substack{\text{Product} \\ \text{rule}}} = \underbrace{\frac{\sqrt[3]{55x^2}}{5x}}_{\text{Simplify}}$$

My video summary

 b. Try this problem on your own. Click here to check your answer, or watch this video for a detailed solution.

You Try It   Work through this You Try It problem.

Work Exercises 41–42 in this textbook or in the *MyMathLab*® Study Plan.

CAUTION   Though not required, it is best to rationalize the denominator last when simplifying radical expressions since the radicand in the denominator will then involve simpler expressions.

Example 11 Rationalizing Denominators

 Simplify each expression first and then rationalize the denominator.

a. $\sqrt{\dfrac{3x}{50}}$ **b.** $\dfrac{\sqrt{18x}}{\sqrt{27xy}}$ **c.** $\sqrt[3]{\dfrac{-4x^5}{16y^5}}$

My video summary

Solution   Try these problems on your own. Click here to check your answers, or watch this video for detailed solutions.

You Try It   Work through this You Try It problem.

Work Exercises 43–44 in this textbook or in the *MyMathLab*® Study Plan.

The goal of rationalizing a denominator is to write an equivalent expression without radicals in the denominator. Earlier in this section, we noted that multiplying conjugates involving square roots results in an expression without any radicals. So, we can use the product of conjugates to rationalize the denominator of a radical expression whose denominator contains two terms involving one or more square roots.

Would we ever want to rationalize the numerator? Click here to find out.

Rationalizing a Denominator with Two Terms

To rationalize a denominator with two terms involving one or more square roots, multiply the numerator and denominator by the conjugate of the denominator.

Example 12 Rationalizing Denominators with Two Terms

Rationalize the denominator.

a. $\dfrac{2}{\sqrt{3} + 5}$ **b.** $\dfrac{7}{3\sqrt{x} - 4}$ **c.** $\dfrac{\sqrt{y} - 3}{\sqrt{y} + 2}$

(EBook Screens 7.4-1–7.4-25)

Solution

a. Since the denominator has two terms and involves a square root, we multiply the numerator and denominator by the conjugate of the denominator $\sqrt{3} - 5$.

Multiply numerator and denominator by $\sqrt{3} - 5$:
$$\frac{2}{\sqrt{3} + 5} = \frac{2}{\sqrt{3} + 5} \cdot \frac{\sqrt{3} - 5}{\sqrt{3} - 5}$$

Multiply numerators and multiply denominators:
$$= \frac{2(\sqrt{3} - 5)}{(\sqrt{3})^2 - (5)^2}$$

Simplify the denominator:
$$= \frac{2(\sqrt{3} - 5)}{3 - 25}$$

$$= \frac{2(\sqrt{3} - 5)}{-22}$$

Cancel common factor of 2:
$$= \frac{\overset{1}{\cancel{2}}(\sqrt{3} - 5)}{\underset{11}{\cancel{-22}}}$$

Simplify:
$$= -\frac{\sqrt{3} - 5}{11} \text{ or } \frac{5 - \sqrt{3}}{11}$$

My video summary

b. Since the denominator has two terms and involves a square root, we multiply the numerator and denominator by the conjugate of the denominator, $3\sqrt{x} + 4$.

$$\frac{7}{3\sqrt{x} - 4} = \underbrace{\frac{7}{3\sqrt{x} - 4} \cdot \frac{3\sqrt{x} + 4}{3\sqrt{x} + 4}}_{\substack{\text{Multiply numerator and} \\ \text{denominator by} \\ 3\sqrt{x} + 4}} = \underbrace{\frac{7(3\sqrt{x} + 4)}{(3\sqrt{x})^2 - (4)^2}}_{\substack{\text{Multiply numerators} \\ \text{and multiply} \\ \text{denominators}}} = \underbrace{\frac{7(3\sqrt{x} + 4)}{9x - 16}}_{\text{Simplify}} \text{ or } \underbrace{\frac{21\sqrt{x} + 28}{9x - 16}}_{\text{Distribute}}$$

We can leave the result in factored form, or we can distribute in the final step. Leaving the numerator factored until the end helps with canceling common factors, if they are present.

c. Work this problem on your own. Click here to check your answer, or watch this video for a detailed solution.

You Try It Work through this You Try It problem.

Work Exercises 45–50 in this textbook or in the **MyMathLab**® Study Plan.

7.4 Exercises

In Exercises 1–50, assume that all variables represent non-negative values and be sure to simplify your answer.

In Exercises 1–20, add or subtract, if possible.

1. $4\sqrt{7} + 9\sqrt{7}$

2. $\sqrt[3]{\frac{32}{27}} - \sqrt[3]{108}$

3. $3\sqrt{2} + 5\sqrt{6}$

4. $3\sqrt{7} - 4\sqrt[3]{6} + 2\sqrt{7} - 8\sqrt[3]{6}$

5. $\sqrt[3]{6x} - 7\sqrt[3]{6x}$

6. $\frac{4\sqrt{3x}}{9} + \frac{2\sqrt{3x}}{15} - \frac{7\sqrt{3x}}{6}$

7. $10\sqrt{5} - 3\sqrt{x} - 4\sqrt{5} + 9\sqrt{x}$

8. $6\sqrt{a} + 2\sqrt{b} - a\sqrt{3}$

336 **Chapter 7** Radicals and Rational Exponents

You Try It **9.** $\sqrt{45} - \sqrt{80}$

10. $3\sqrt[3]{32} + \sqrt[3]{500}$

 11. $5\sqrt{12} - \sqrt{75} + 3\sqrt{20}$

12. $6\sqrt[3]{54} + 2\sqrt{8} - 4\sqrt[3]{16}$

13. $2\sqrt[4]{81x^5} - 3\sqrt[4]{x^5}$
You Try It

14. $6\sqrt{20x^3} + 4x\sqrt{45x}$

15. $3a^2\sqrt{ab^3} + \sqrt{16a^5b^3} - 5b\sqrt{a^5b}$

16. $\sqrt[3]{81m^3n} - 4\sqrt[3]{3m^3n} + 3m\sqrt[3]{162n}$

17. $\dfrac{\sqrt{72}}{3y} + \dfrac{5\sqrt{18}}{4y}$

18. $\dfrac{\sqrt[3]{8a^4}}{4a} - \dfrac{\sqrt[3]{27a}}{7}$

You Try It

19. $\dfrac{\sqrt[4]{81a^7}}{3} - \dfrac{5a\sqrt[4]{16a^3}}{2}$

20. $\dfrac{5x^2\sqrt{128x^5y^3}}{\sqrt{64x^4y}} + \dfrac{2xy\sqrt{18x^3y}}{3\sqrt{y}}$

In Exercises 21–36, multiply.

 You Try It **21.** $\sqrt{5}(2x + \sqrt{3})$

22. $\sqrt{2x}(\sqrt{6x} - 5)$

23. $2\sqrt{3m}(m - 3\sqrt{5m})$

24. $\sqrt[3]{x}(\sqrt[3]{54x^2} - \sqrt[3]{x})$

25. $(8\sqrt{3} - 5)(\sqrt{3} - 1)$
You Try It

26. $(3 + \sqrt{5})(2 + \sqrt{7})$

27. $(\sqrt{x} + 4)(\sqrt{x} + 6)$

28. $(4 - 2\sqrt{3x})(5 + \sqrt{3x})$

29. $(3 - \sqrt{x})(4 - \sqrt{y})$
You Try It

30. $(\sqrt[3]{y} + 1)(\sqrt[3]{y} - 2)$

31. $(\sqrt{7z} + 3)(\sqrt{7z} - 3)$

32. $(\sqrt{5} + 3y)^2$

33. $(\sqrt{x} + 3)(3 - \sqrt{x})$

34. $(\sqrt{3a} - \sqrt{2b})(\sqrt{3a} + \sqrt{2b})$

35. $(2\sqrt{x} + 5)^2$

36. $(2\sqrt{m} - 3\sqrt{n})^2$

In Exercises 37–50, rationalize the denominator.

37. $\dfrac{4}{\sqrt{6}}$

38. $\dfrac{\sqrt{3}}{\sqrt{5}}$

You Try It 39. $\sqrt{\dfrac{9}{2x}}$

40. $\dfrac{5}{\sqrt{11x}}$

You Try It 41. $\sqrt[3]{\dfrac{9}{4x^2}}$

42. $\dfrac{\sqrt[4]{3b^5}}{\sqrt[4]{25a}}$

43. $\dfrac{8}{\sqrt{12x^3y^4}}$
You Try It

44. $\dfrac{\sqrt[3]{2x^2}}{\sqrt[3]{36y^5}}$

45. $\dfrac{5}{2 - \sqrt{3}}$

46. $\dfrac{-4x}{5 + \sqrt{6}}$

47. $\dfrac{-3}{\sqrt{x} + 1}$
You Try It

48. $\dfrac{\sqrt{a}}{3\sqrt{a} - \sqrt{b}}$

 49. $\dfrac{\sqrt{m} - 4}{\sqrt{m} - 7}$

50. $\dfrac{5\sqrt{3} + 2\sqrt{6}}{4\sqrt{6} - \sqrt{3}}$

7.5 Radical Equations and Models

THINGS TO KNOW

Before working through this section, be sure you are familiar with the following concepts:

| | | VIDEO | ANIMATION | INTERACTIVE |

You Try It

1. Solve Linear Equations in One Variable (Section 1.1, **Objective 2**)

You Try It

2. Solve a Formula for a Given Variable (Section 1.4, **Objective 1**)

You Try It

3. Use Formulas to Solve Application Problems (Section 1.4, **Objective 2**)

You Try It

4. Use the Power-to-Power Rule (Section 4.1, **Objective 5**)

You Try It

5. Solve Polynomial Equations by Factoring (Section 5.4, **Objective 1**)

You Try It

6. Multiply Radical Expressions (Section 7.4, **Objective 2**)

OBJECTIVES

1 Solve Equations Involving One Radical Expression

2 Solve Equations Involving Two Radical Expressions

3 Use Radical Equations and Models to Solve Application Problems

---

OBJECTIVE 1: SOLVE EQUATIONS INVOLVING ONE RADICAL EXPRESSION

Recall that when solving an equation in one variable, we find all values of the variable that make the equation true. All of these values together form the solution set of the equation. Previously, we have solved linear equations, polynomial equations, and rational equations. Now we learn how to solve *radical equations*.

> **Definition   Radical Equation**
>
> A **radical equation** is an equation that contains at least one radical expression with a variable in the radicand.

Some examples of radical equations are $\sqrt{2x + 1} = 3$, $\sqrt[3]{5x - 4} + 7 = 10$, $\sqrt[4]{19x - 2} = 2\sqrt[4]{x + 1}$, and $\sqrt{x + 9} - \sqrt{x - 6} = 3$.

The first two equations contain one radical expression, whereas the last two equations contain two radical expressions.

Not all equations that contain radical expressions are radical equations. If no radical expression contains a variable in the radicand, then the equation is not a radical equation. For example, $2x + 1 = \sqrt{3}$ is not a radical equation.

The key to solving a radical equation is to eliminate all of the radicals from the equation. To do this, we can use the following theorem:

Theorem   The Power Principle of Equality

If A and B represent algebraic expressions and n is a positive real number, then any solution to the equation $A = B$ is also a solution to the equation $A^n = B^n$.

Consider the equation $\sqrt{x} = 5$, which has one radical expression $\sqrt{x}$. The radical expression $\sqrt{x}$ is an **isolated radical expression** because it stands alone on one side of the equal sign. An isolated radical can be eliminated from an equation by raising both sides of the equation to the power of the index. In this case, we square both sides of the equation because the radical expression is a square root.

$$\text{Begin with the original equation:} \quad \sqrt{x} = 5$$
$$\text{Square both sides:} \quad \left(\sqrt{x}\right)^2 = (5)^2$$
$$\text{Simplify:} \quad x = 25$$

Checking this answer, we substitute 25 for x in the original equation:

$$\text{Substitute 25 for } x: \quad \sqrt{25} \overset{?}{=} 5$$
$$\text{Simplify:} \quad 5 = 5 \quad \text{True}$$

Since $x = 25$ satisfies the original equation, the solution set is $\{25\}$.

Sometimes when we raise both sides of an equation to an even power, the new equation will have solutions that are not solutions to the original equation. Such "solutions" are called **extraneous solutions**. They must be identified and excluded from the solution set. Click here for a more in-depth explanation of extraneous solutions.

We can use the following steps to solve equations involving one radical expression.

Solving Equations Involving One Radical Expression

Step 1. Isolate the radical expression. Use the properties of equality to get the radical expression by itself on one side of the equal sign.

Step 2. Eliminate the radical. Identify the index of the radical expression and raise both sides of the equation to the index power.

Step 3. Solve the resulting equation.

Step 4. Check each solution from step 3 in the original equation. Disregard any extraneous solutions.

Note: Extraneous solutions do not result from raising both sides of an equation to an odd power, but it is still good practice to check your answers anyway.

When solving radical equations, answers can fail to check for two reasons: An answer may be an extraneous solution, or an error may have been made while solving. When an answer is disregarded as an extraneous solution, it is important to make sure that no errors were made while solving.

Example 1 Solving an Equation Involving One Radical Expression

Solve $\sqrt{3x - 2} + 6 = 11$.

Solution We follow the four-step process.

Step 1. Isolate the radical expression by subtracting 6 from both sides.

Begin with the original equation: $\sqrt{3x - 2} + 6 = 11$

Subtract 6 from both sides: $\sqrt{3x - 2} + 6 - 6 = 11 - 6$

Simplify: $\sqrt{3x - 2} = 5$

Step 2. The radical expression is a square root, so we square both sides.

Square both sides: $\left(\sqrt{3x - 2}\right)^2 = (5)^2$

Simplify: $3x - 2 = 25$

Step 3. Solve the resulting linear equation.

Add 2 to both sides: $3x = 27$

Divide both sides by 3: $x = 9$

Step 4. Check $x = 9$ in the original equation.

Substitute 9 for x: $\sqrt{3(9) - 2} + 6 \overset{?}{=} 11$

Simplify beneath the radical: $\sqrt{25} + 6 \overset{?}{=} 11$

Evaluate the square root: $5 + 6 \overset{?}{=} 11$

Add: $11 = 11$ True

The answer checks in the original equation, so the solution set is $\{9\}$.

You Try It   Work through this You Try It problem.

Work Exercises 1–8 in this textbook or in the *MyMathLab*® Study Plan.

When solving a radical equation, it is important to isolate the radical before raising both sides to the index power. If this is not done, then the radical expression will not be eliminated. Click here to see an example.

Example 2 Solving an Equation Involving One Radical Expression

Solve $\sqrt{2x - 1} + 9 = 6$.

Solution

Step 1.   Begin with the original equation: $\sqrt{2x - 1} + 9 = 6$

Subtract 9 from both sides: $\sqrt{2x - 1} + 9 - 9 = 6 - 9$

Simplify: $\sqrt{2x - 1} = -3$

Step 2.   Square both sides: $\left(\sqrt{2x - 1}\right)^2 = (-3)^2$

Simplify: $2x - 1 = 9$

Step 3.   Add 1 to both sides: $2x = 10$

Divide both sides by 2: $x = 5$

Step 4. Check $x = 5$ in the original equation: $\sqrt{2(5) - 1} + 9 \overset{?}{=} 6$

Simplify beneath the radical: $\sqrt{9} + 9 \overset{?}{=} 6$

Evaluate the square root: $3 + 9 \overset{?}{=} 6$

Add: $12 = 6$ False

The answer $x = 5$ does not check in the original equation. It is an extraneous solution. Since this is the only possible solution, the equation has no real solution. The solution set is $\{\}$ or $\varnothing$.

Note: We might have noticed sooner that this equation would have no real solution. In step 1, when we isolated the radical, the resulting equation was $\sqrt{2x - 1} = -3$. This equation states that a principal square root equals a negative value, but this is impossible since principal square roots must be non-negative. So, this equation has no real solution.

You Try It   Work through this You Try It problem.

Work Exercises 9–10 in this textbook or in the *MyMathLab*®️ Study Plan.

Example 3 Solving an Equation Involving One Radical Expression

Solve $\sqrt{3x + 7} - x = 1$.

Solution   We follow the four-step process.

Step 1. Isolate the radical expression by adding x to both sides.

Begin with the original equation: $\sqrt{3x + 7} - x = 1$

Add x to both sides: $\sqrt{3x + 7} - x + x = 1 + x$

Simplify each side: $\sqrt{3x + 7} = x + 1$

Step 2. The radical expression is a square root, so we square both sides of the equation to eliminate the radical.

Square both sides: $\left(\sqrt{3x + 7}\right)^2 = (x + 1)^2$

Simplify each side: $3x + 7 = x^2 + 2x + 1$

Try to solve this polynomial equation on your own by using the four-step process from Section 5.4. Then finish solving the original equation by checking for extraneous solutions. Click here to check your answer, or watch this video for the complete solution.

You Try It   Work through this You Try It problem.

Work Exercises 11–16 in this textbook or in the *MyMathLab*®️ Study Plan.

So far, all of the radical equations solved have involved square roots. In Example 4, we look at an equation involving a cube root.

Example 4 Solving an Equation Involving One Radical Expression

 Solve $\sqrt[3]{3x^2 + 23x} + 10 = 12$.

Solution

Step 1. Begin with the original equation: $\qquad \sqrt[3]{3x^2 + 23x} + 10 = 12$

$\qquad$ Subtract 10 from both sides: $\quad \sqrt[3]{3x^2 + 23x} + 10 - 10 = 12 - 10$

$\qquad\qquad\qquad\qquad$ Simplify: $\qquad\qquad\qquad \sqrt[3]{3x^2 + 23x} = 2$

Step 2. The radical expression is a cube root, so we cube both sides of the equation to eliminate the radical.

$\qquad\qquad$ Cube both sides: $\quad \left(\sqrt[3]{3x^2 + 23x}\right)^3 = (2)^3$

$\qquad\qquad\qquad$ Simplify: $\qquad\quad 3x^2 + 23x = 8$

The result is a polynomial equation. Try to finish solving this equation on your own. Click here to check your answer, or watch this video for the complete solution.

You Try It Work through this You Try It problem.

Work Exercises 17–20 in this textbook or in the *MyMathLab*®️ Study Plan.

Recall from Section 7.2 that $a^{\frac{1}{n}} = \sqrt[n]{a}$. Sometimes equations containing one rational exponent can be solved using the same process for solving equations with one radical expression. We solve such an equation in Example 5.

Example 5 Solving an Equation Involving a Rational Exponent

 Solve $\left(x^2 - 9\right)^{\frac{1}{4}} + 3 = 5$.

Solution This equation is equivalent to $\sqrt[4]{x^2 - 9} + 3 = 5$. Try solving it on your own by using the four-step process. Click here to check your answer, or watch this video for a complete solution.

You Try It Work through this You Try It problem.

Work Exercises 21–24 in this textbook or in the *MyMathLab*®️ Study Plan.

OBJECTIVE 2: SOLVE EQUATIONS INVOLVING TWO RADICAL EXPRESSIONS

Solving an equation involving two radical expressions is similar to solving an equation involving one radical expression. However, it may be necessary to repeat the process for eliminating a radical in order to eliminate both radicals from the equation.

Solving Equations Involving Two Radical Expressions

Step 1. Isolate one of the radical expressions. Use the properties of equality to get a radical expression by itself on one side of the equal sign.

Step 2. Eliminate the radical from the isolated radical expression. Identify the index of the isolated radical expression and raise both sides of the equation to this index power.

Step 3. If all the radicals have been eliminated, then solve the resulting equation. Otherwise, repeat steps 1 and 2.

Step 4. Check each solution from step 3 in the original equation. Disregard any extraneous solutions.

Example 6 Solving an Equation Involving Two Radical Expressions

 Solve $\sqrt{x + 9} - \sqrt{x} = 1$.

Solution We follow the four-step process.

Step 1. We choose to isolate $\sqrt{x + 9}$.

Begin with the original equation: $\sqrt{x + 9} - \sqrt{x} = 1$

Add $\sqrt{x}$ to both sides: $\sqrt{x + 9} - \sqrt{x} + \sqrt{x} = 1 + \sqrt{x}$

Simplify: $\sqrt{x + 9} = 1 + \sqrt{x}$

Step 2. The isolated radical expression is a square root, so we square both sides.

Square both sides: $\left(\sqrt{x + 9}\right)^2 = \left(1 + \sqrt{x}\right)^2$

The right side has the form of the square of a binomial sum $(A + B)^2$, so we use the special product rule $(A + B)^2 = A^2 + 2AB + B^2$ with $A = 1$ and $B = \sqrt{x}$.

Use $(A + B)^2 = A^2 + 2AB + B^2$
on the right side: $\left(\sqrt{x + 9}\right)^2 = (1)^2 + 2(1)\left(\sqrt{x}\right) + \left(\sqrt{x}\right)^2$

Simplify: $x + 9 = 1 + 2\sqrt{x} + x$

Step 3. The equation still contains a radical expression, so we repeat steps 1 and 2. Try to finish solving this equation on your own. Remember to check for extraneous solutions. Click here to check your answer, or watch this video for the complete solution.

You Try It   Work through this You Try It problem.

Work Exercises 25–27 in this textbook or in the *MyMathLab*® Study Plan.

Example 7 Solving an Equation Involving Two Radical Expressions

 Solve $\sqrt{2x + 3} + \sqrt{x - 2} = 4$.

Solution Try solving this equation on your own by using the four-step process. Click here to check your answer, or watch this video for the complete solution.

(EBook Screens 7.5-1–7.5-26)

You Try It   Work through this You Try It problem.

Work Exercises 28–30 in this textbook or in the *MyMathLab*® Study Plan.

Example 8 Solving an Equation Involving Two Radical Expressions

Solve $\sqrt[3]{2x^2 - 9} + \sqrt[3]{3x - 11} = 0$.

Solution We follow the four-step process.

Step 1.   Begin with the original equation: $\sqrt[3]{2x^2 - 9} + \sqrt[3]{3x - 11} = 0$
Subtract $\sqrt[3]{3x - 11}$ from both sides: $\sqrt[3]{2x^2 - 9} = -\sqrt[3]{3x - 11}$

Step 2. The two radical expressions are cube roots, so we cube both sides.

$$\text{Cube both sides:} \quad \left(\sqrt[3]{2x^2 - 9}\right)^3 = \left(-\sqrt[3]{3x - 11}\right)^3$$
$$\text{Simplify:} \quad 2x^2 - 9 = -(3x - 11)$$
$$\text{Distribute the negative:} \quad 2x^2 - 9 = -3x + 11$$

Step 3. Both radicals are eliminated. Try to finish solving on your own.

Step 4. Click here to check your answer, or watch this video for the complete solution.

You Try It   Work through this You Try It problem.

Work Exercises 31–34 in this textbook or in the *MyMathLab*® Study Plan.

OBJECTIVE 3:   ## USE RADICAL EQUATIONS AND MODELS TO SOLVE APPLICATION PROBLEMS

Often real-world situations are modeled by formulas that involve radical expressions. For example, the formula $t = \dfrac{\sqrt{d}}{4}$ gives the time t, in seconds, that it takes a free-falling object to fall a distance of d feet.

In Section 1.4 we learned to solve a formula for a given variable. If the given variable is beneath a radical, then we must remove the radical by isolating the radical and raising both sides of the equation to the appropriate power.

Example 9 Solving a Formula for a Variable beneath a Radical

Solve each formula for the given variable.

a. Free-falling object: $t = \dfrac{\sqrt{d}}{4}$ for d.

b. Radius of a sphere: $r = \sqrt[3]{\dfrac{3V}{4\pi}}$ for V.

Solution

a. Begin with the original formula: $t = \dfrac{\sqrt{d}}{4}$

Multiply both sides by 4 to isolate the radical: $4(t) = 4\left(\dfrac{\sqrt{d}}{4}\right)$

Simplify: $4t = \sqrt{d}$

Square both sides: $(4t)^2 = \left(\sqrt{d}\right)^2$

Simplify: $16t^2 = d$

Solving for d, the formula for a free-falling object is $d = 16t^2$.

b. The variable V is already contained within the isolated radical expression. The radical expression is a cube root, so we eliminate the radical by cubing both sides. Try to finish solving the formula for V. Click here to check your answer, or watch this video for a complete solution.

You Try It   **Work through this You Try It problem.**

Work Exercises 35–40 in this textbook or in the *MyMathLab*® Study Plan.

We can use radical equations and models to solve a variety of application problems in many different disciplines. In Example 10, a radical equation is used to model the readability of written text.

Example 10 Assessing the Readability of Written Text

A SMOG grade for written text is a minimum reading grade level G that a reader must possess in order to fully understand the written text being graded. If w is the number of words that have three or more syllables in a sample of 30 sentences from a given text, then the SMOG grade for that text is given by the formula $G = \sqrt{w} + 3$. Use the SMOG grade formula to answer the following questions. (*Source:* readabilityformulas.com).

a. If a sample of 30 sentences contains 18 words with three or more syllables, then what is the SMOG grade for the text? If necessary, round to a whole number for the grade level.

b. If a text must have a tenth-grade reading level, then how many words with three or more syllables would be needed in the sample of 30 sentences?

Solution

a. There are 18 words with three or more syllables, so we substitute 18 for w in the given formula.

Begin with the original formula: $G = \sqrt{w} + 3$

Substitute 18 for w: $G = \sqrt{18} + 3$

Approximate the square root: $G \approx 4.24 + 3$

Simplify: $G \approx 7.24$

Rounding to the nearest whole number, we get $G = 7$. So, the SMOG grade for this text is a seventh-grade reading level.

b. The text must have a tenth-grade reading level, so we substitute 10 for G in the formula and solve for w.

Begin with the original formula: $G = \sqrt{w} + 3$

Substitute 10 for G: $10 = \sqrt{w} + 3$

Try to finish solving this problem on your own. Click here to check your answer, or watch this video for a complete solution.

You Try It **Work through this You Try It problem.**

Work Exercises 41–44 in this textbook or in the *MyMathLab*® Study Plan.

Example 11 Punting a Football

An important component of a good punt in football is **hang time**, which is the length of time that the punted ball remains in the air. If wind resistance is ignored, the relationship between the hang time t, in seconds, and the vertical height h, in feet, that the ball reaches can be modeled by the formula $t = \dfrac{\sqrt{h}}{2}$.

Use this formula to answer the following questions.

a. If the average hang time for an NFL punt is 4.6 seconds, then what is the vertical height for an average NFL punt? Round to the nearest foot.

b. Cowboys Stadium in Arlington, Texas, has a huge high-definition screen centered over most of the football field. The bottom of the screen is 90 feet above the field. What hang time would result in the ball hitting the screen? Round to the nearest hundredth of a second.

Solution

a. We substitute the hang time 4.6 for t in the formula and solve for h.

Begin with the original formula: $t = \dfrac{\sqrt{h}}{2}$

Substitute 4.6 for t: $4.6 = \dfrac{\sqrt{h}}{2}$

Multiply both sides by 2: $9.2 = \sqrt{h}$

Square both sides: $(9.2)^2 = (\sqrt{h})^2$

Simplify: $84.64 = h$

Rounding, the average NFL punt reaches a vertical height of about 85 feet.

b. We substitute the vertical height, 90 feet, for h and simplify.

Begin with the original formula: $t = \dfrac{\sqrt{h}}{2}$

Substitute 90 for h: $t = \dfrac{\sqrt{90}}{2}$

Approximate the square root: $t \approx \dfrac{9.4868}{2}$

Divide: $t \approx 4.7434$

Rounding, a hang time of about 4.74 seconds will result in a punt that hits the screen.

You Try It   Work through this You Try It problem.

Work Exercises 45–46 in this textbook or in the *MyMathLab*® Study Plan.

7.5 Exercises

In Exercises 1–34, solve each radical equation.

1. $\sqrt{x + 15} = 3$

2. $\sqrt{2x - 1} = 3$

3. $\sqrt{r} - 3 = -1$

4. $3\sqrt{m} + 5 = 7$

5. $\sqrt{z + 5} + 2 = 9$

6. $\sqrt{1 - x} + 1 = 4$

7. $\sqrt{4p + 1} - 3 = 2$

8. $5\sqrt{w - 9} - 2 = 8$

9. $\sqrt{q + 10} = -2$

10. $\sqrt{2x - 1} + 6 = 1$

11. $\sqrt{2m + 1} - m = -1$

12. $p - \sqrt{4p + 9} + 3 = 0$

13. $\sqrt{32 - 4x} - x = 0$

14. $\sqrt{7z + 2} + 3z = 2$

15. $2\sqrt{4y + 1} + 3y = 4y + 4$

16. $2q + 1 = \sqrt{q^3 + 17}$

17. $\sqrt[3]{2x - 1} = -2$

18. $\sqrt[3]{7 - 5w} + 1 = 4$

19. $\sqrt[3]{2n^2 + 15n} + 1 = 4$

20. $\sqrt[4]{x^2 - 6x} - 2 = 0$

21. $(x - 5)^{\frac{1}{2}} - 1 = 4$

22. $(2y + 3)^{\frac{1}{3}} + 7 = 9$

23. $(2w + 5)^{\frac{1}{4}} + 1 = 4$

24. $(p^2 - 19)^{\frac{1}{4}} + 2 = 5$

25. $\sqrt{x + 15} - \sqrt{x} = 3$

26. $\sqrt{x} - 1 = \sqrt{x - 5}$

27. $\sqrt{x - 64} + 8 = \sqrt{x}$

28. $\sqrt{6 - x} + \sqrt{5x + 6} = 6$

29. $\sqrt{3x + 1} - \sqrt{x + 4} = 1$

30. $\sqrt{6 + 5x} + \sqrt{3x + 4} = 2$

31. $\sqrt{7x - 4} = \sqrt{4x + 11}$

32. $\sqrt[3]{4y - 3} = \sqrt[3]{6y + 9}$

33. $\sqrt[3]{2x^2 + 5x} = \sqrt[3]{2x + 14}$

34. $\sqrt[4]{6w + 1} = \sqrt[4]{2w + 17}$

In Exercises 35–40, solve each formula for the given variable.

35. $T = \dfrac{\pi\sqrt{2L}}{4}$ for L

36. $d = k\sqrt[3]{E}$ for E

37. $r = \sqrt[3]{\dfrac{3V}{\pi h}}$ for h

38. $V = \sqrt{\dfrac{FR}{m}}$ for m

39. $A = P\sqrt{1 + r}$ for r

40. $T = \left(\dfrac{LH^2}{25}\right)^{\frac{1}{4}}$ for L

You Try It

In Exercises 41–46, use the given model to solve each application problem.

41. SMOG Grade   Recall from Example 10 that the formula $G = \sqrt{w} + 3$ gives the minimum reading grade level G needed to fully understand written text containing w words with three or more syllables in a sample of 30 sentences.

    a. If a sample of 30 sentences contains 51 words with three or more syllables, then what is the SMOG grade for the text? If necessary, round your answer to a whole number to find the grade level.

    b. If a text must have a ninth-grade reading level, then how many words with three or more syllables are needed in a sample of 30 sentences?

42. Measure of Leanness   The *ponderal index* is a measure of the "leanness" of a person. A person who is h inches tall and weighs w pounds has a ponderal index I given by $I = \dfrac{h}{\sqrt[3]{w}}$.

    a. Compute the ponderal index for a person who is 75 inches tall and weighs 190 pounds. Round to the nearest hundredth.

    b. What is a man's weight if he is 70 inches tall and has a ponderal index of 12.35? Round to the nearest whole number.

43. Body Surface Area   The Mosteller formula is used in the medical field to estimate a person's body surface area. The formula is $A = \dfrac{\sqrt{hw}}{60}$, where A is body surface area in square meters, h is height in centimeters, and w is weight in kilograms.

    a. Compute the body surface area of a person who is 178 cm tall and weighs 90.8 kg. Round to the nearest hundredth.

    b. If a woman is 165 cm tall and has body surface area of 1.72 m^2, how much does she weigh? Round to the nearest tenth.

44. Skid Marks   Under certain road conditions, the length of a skid mark S, in feet, is related to the velocity v, in miles per hour, by the formula $v = \sqrt{10S}$. Assuming these same road conditions, answer the following:

    a. Compute a car's velocity if it leaves a skid mark of 360 feet.

    b. If a car is traveling at 50 miles per hour when it skids, what will be the length of the skid mark?

You Try It

45. Hang Time   Recall from Example 11 that hang time t, in seconds, and the vertical height h, in feet, can be modeled by $t = \dfrac{\sqrt{h}}{2}$. This same formula can be used to model an athlete's hang time when jumping.

    a. If LeBron James has a vertical leap of 3.7 feet, what is his hang time? Round to the nearest hundredth.

    b. When Mark "Wild Thing" Wilson of the Harlem Globetrotters slam-dunked a regulation basketball on a 12-foot rim in front of an Indianapolis crowd, his hang time for the shot was approximately 1.16 seconds. What was the vertical distance of his jump? Round to the nearest tenth.

46. Distance to the Horizon   From a boat, the distance d, in miles, that a person can see to the horizon is modeled by the formula $d = \dfrac{3\sqrt{h}}{2}$, where h is the height, in feet, of eye level above the sea.

    a. From his ship, how far can a sailor see to the horizon if his eye level is 36 feet above the sea?

    b. How high is the eye level of a sailor who can see 12 miles to the horizon?

7.6 Complex Numbers

THINGS TO KNOW

Before working through this section, be sure you are familiar with the following concepts:

You Try It

1. Find Square Roots of Perfect Squares (Section 7.1, Objective 1)

You Try It

2. Simplifying Radical Expressions Using the Product Rule (Section 7.3, Objective 5)

OBJECTIVES

1 Simplify Powers of i

2 Add and Subtract Complex Numbers

3 Multiply Complex Numbers

4 Divide Complex Numbers

5 Simplify Radicals with Negative Radicands

OBJECTIVE 1:   SIMPLIFY POWERS OF i

So far we have learned about real number solutions to equations, but not every equation has real number solutions. For example, consider the equation $x^2 + 1 = 0$. We can check if a value is a solution by substituting the value for x and seeing if a true statement results. Is $x = -1$ a solution to this equation?

$$\text{Begin with the original equation:} \qquad x^2 + 1 = 0$$
$$\text{Substitute } -1 \text{ for } x: \quad (-1)^2 + 1 \overset{?}{=} 0$$
$$\text{Simplify:} \qquad 1 + 1 \overset{?}{=} 0$$
$$2 = 0 \quad \text{False}$$

Because $2 = 0$ is not a true statement, $x = -1$ is not a solution to the equation $x^2 + 1 = 0$. In fact, this equation has no real solution. Click here to see why.

To find the solution to equations such as $x^2 + 1 = 0$, we introduce a new number called the **imaginary unit i**.

> **Definition   Imaginary Unit i**
>
> The **imaginary unit i** is defined as
> $$i = \sqrt{-1}, \text{ where } i^2 = -1.$$

When working with the imaginary unit, we will encounter various powers of i. Let's consider some powers of i and look for patterns that will help us simplify them.

$$i^1 = \underbrace{i = \sqrt{-1}}_{\text{Defined}} \qquad\qquad i^2 = \underbrace{-1}_{\text{Defined}}$$

$$i^3 = \underbrace{i^2 \cdot i}_{\substack{\text{Product rule} \\ \text{for exponents}}} = \underbrace{(-1)}_{i^2 = -1} \cdot i = -i \qquad i^4 = \underbrace{i^2 \cdot i^2}_{\substack{\text{Product rule} \\ \text{for exponents}}} = \underbrace{(-1)}_{i^2 = -1} \cdot \underbrace{(-1)}_{i^2 = -1} = 1$$

$$i^5 = \underbrace{i^4 \cdot i}_{\substack{\text{Product rule} \\ \text{for exponents}}} = \underbrace{(1)}_{i^4 = 1} \cdot i = i \qquad i^6 = \underbrace{i^4 \cdot i^2}_{\substack{\text{Product rule} \\ \text{for exponents}}} = \underbrace{(1)}_{i^4 = 1} \cdot \underbrace{(-1)}_{i^2 = -1} = -1$$

$$i^7 = \underbrace{i^4 \cdot i^3}_{\substack{\text{Product rule} \\ \text{for exponents}}} = \underbrace{(1)}_{i^4 = 1} \cdot \underbrace{(-i)}_{i^3 = -i} = -i \qquad i^8 = \underbrace{i^4 \cdot i^4}_{\substack{\text{Product rule} \\ \text{for exponents}}} = \underbrace{(1)}_{i^4 = 1} \cdot \underbrace{(1)}_{i^4 = 1} = 1$$

$\vdots$

Notice that the powers of i follow the pattern $i, -1, -i, 1$. Based on this pattern, what is the value of i^0? Click here to find out. We can use the following procedure to simplify powers of i.

Simplifying i^n for $n > 4$

Step 1. Divide n by 4 and find the remainder r.

Step 2. Replace the exponent (power) on i by the remainder, $i^n = i^r$.

Step 3. Use the results $i^0 = 1$, $i^1 = i$, $i^2 = -1$, and $i^3 = -i$ to simplify if necessary.

Example 1 Simplifying Powers of i

Simplify.

 **a.** i^{17} **b.** i^{60} **c.** i^{39} **d.** $-i^{90}$ **e.** $i^{14} + i^{29}$

My video summary

Solution

a. Step 1. Divide the exponent by 4 and find the remainder:

$$\begin{array}{r} 4 \\ 4\overline{)17} \\ \underline{16} \\ 1 \end{array}$$
  ←— exponent, n
  ←— remainder, r

Step 2. Replace n by r: $i^{\overset{n}{17}} = i^{\overset{r}{1}}$

Step 3. Simplify if necessary: $i^{17} = i^1 = i$

b. Divide the exponent 60 by 4 and find the remainder.

$$\begin{array}{r} 15 \\ 4\overline{)60} \\ \underline{4} \\ 20 \\ \underline{20} \\ 0 \end{array}$$
  ←— exponent, n
  ←— remainder, r

Replace n by r and simplify if necessary.

$$\overset{n}{\overbrace{i^{60}}} = \overset{r}{\overbrace{i^{0}}} = 1$$

c.–e. Try to simplify these powers of i on your own. Click here to check your answers, or watch this video to see detailed solutions.

You Try It **Work through this You Try It problem.**

Work Exercises 1–5 in this textbook or in the *MyMathLab*® Study Plan.

With the imaginary unit, we can now expand our number system from the set of real numbers to the set of complex numbers.

Complex Numbers

The set of all numbers of the form

$$a + bi,$$

where a and b are real numbers and i is the imaginary unit, is called the set of **complex numbers**. The number a is called the **real part**, and the number b is called the **imaginary part**.

If $b = 0$, then the complex number is a purely real number. If $a = 0$, then the complex number is a purely imaginary number. Figure 10 illustrates the relationships between complex numbers.

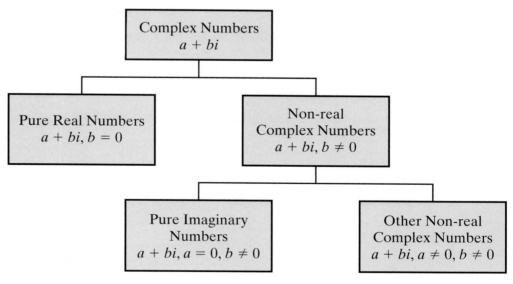

Figure 10 The Complex Number System

Figure 10 shows us that all real numbers are complex numbers, but not all complex numbers are real numbers. This distinction is important, particularly when solving equations. For example, $x^2 + 1 = 0$ has no *real* solutions, but it does have two *complex* solutions.

A complex number of the form $a + bi$ is written in **standard form**, and is the typical way to write complex numbers. Below are examples of complex numbers written in standard form.

	Standard form	Real part, $a = 7$	Imaginary part, $b = 0$	
Real Number:	7 =	7 +	0	i

	Standard form	Real part, $a = 0$	Imaginary part, $b = 3$	
Imaginary Number:	$3i$ =	0 +	3	i

	Standard form	Real part, $a = 4$	Imaginary part, $b = 9$	
Non-real Complex Number:	$4 + 9i$ =	4 +	9	i

	Standard form	Real part, $a = \frac{1}{3}$	Imaginary part, $b = \frac{2}{3}$	
	$\frac{1}{3} + \frac{2}{3}i$ =	$\frac{1}{3}$ +	$\frac{2}{3}$	i

OBJECTIVE 2:   ADD AND SUBTRACT COMPLEX NUMBERS

To add or subtract complex numbers, we combine the real parts and combine the imaginary parts.

Adding and Subtracting Complex Numbers

To add complex numbers, add the real parts and add the imaginary parts.

$$(a + bi) + (c + di) = (a + c) + (b + d)i$$

To subtract complex numbers, subtract the real parts and subtract the imaginary parts.

$$(a + bi) - (c + di) = (a - c) + (b - d)i$$

Example 2 Adding and Subtracting Complex Numbers

Perform the indicated operations.

 **a.** $(3 + 5i) + (2 - 7i)$　　　**b.** $(3 + 5i) - (2 - 7i)$

c. $(-3 - 4i) + (2 - i) - (3 + 7i)$

Solution

a.

Original expression:	$(3 + 5i) + (2 - 7i)$
Remove parentheses:	$= 3 + 5i + 2 - 7i$
Collect like terms:	$= 3 + 2 + 5i - 7i$
Combine real parts:	$= 5 + 5i - 7i$
Combine imaginary parts:	$= 5 - 2i$

b.

Original expression:	$(3 + 5i) - (2 - 7i)$
Change to add the opposite:	$= (3 + 5i) + (-2 + 7i)$
Remove parentheses:	$= 3 + 5i - 2 + 7i$

Finish working the problem on your own. Click here to check your answer, or watch this video to see the full solution.

c. Try to work the problem on your own. Click here to check your answer, or watch this video for a detailed solution.

You Try It   **Work through this You Try It problem.**

Work Exercises 6–13 in this textbook or in the *MyMathLab*⃝ᴿ Study Plan.

OBJECTIVE 3:   MULTIPLY COMPLEX NUMBERS

When multiplying complex numbers, we use the distributive property and the FOIL method just as when multiplying polynomials. Remember that $i^2 = -1$ when simplifying.

Example 3 Multiplying Complex Numbers

Multiply.

a. $-4i(3 - 8i)$ **b.** $(3 + 4i)(6 + 11i)$

Solution

a. Multiply using the distributive property.

$$
\begin{aligned}
\text{Original expression: } & -4i(3 - 8i) \\
\text{Distribute: } &= -4i \cdot 3 + (-4i)(-8i) \\
\text{Multiply: } &= -12i + 32i^2 \\
\text{Replace } i^2 \text{ with } -1\text{: } &= -12i + 32\underbrace{(-1)}_{i^2\,=\,-1} \\
\text{Simplify: } &= -12i - 32 \\
\text{Write in standard form: } &= -32 - 12i
\end{aligned}
$$

b. Multiply using the FOIL method.

$$
\begin{aligned}
(3 - 4i)(6 + 11i) &= \overset{\text{F}}{3 \cdot 6} + \overset{\text{O}}{3 \cdot 11i} - \overset{\text{I}}{4i \cdot 6} - \overset{\text{L}}{4i \cdot 11i} \\
&= 18 + \underbrace{33i - 24i}_{\text{Collect like terms}} - 44i^2 \\
&= 18 + 9i - \underbrace{44i^2}_{i^2\,=\,-1} \\
&= 18 + 9i - 44(-1) \\
&= 18 + \underbrace{9i + 44}_{\text{Collect like terms}} \\
&= 62 + 9i
\end{aligned}
$$

Example 4 Multiplying Complex Numbers

▣ Multiply $(4 - 3i)(7 + 5i)$.

Solution Work the problem on your own. Click here to check your answer, or watch this video for a detailed solution.

You Try It   **Work through this You Try It problem.**

Work Exercises 14–19 in this textbook or in the *MyMathLab*⃝ᴿ Study Plan.

Example 5 Special Products Involving Complex Numbers

Multiply.

a. $(4 + 2i)^2$ **b.** $\left(\sqrt{3} - 5i\right)^2$

Solution

a. We recognize this expression as the square of a binomial sum, where $A = 4$ and $B = 2i$. Use the special product rule to multiply.

$$\text{Write the square of a binomial sum rule:} \quad (A + B)^2 = A^2 + 2AB + B^2$$

$$\text{Substitute 4 for } A \text{ and } 2i \text{ for } B: \quad \underbrace{(4}_{A} + \underbrace{2i)^2}_{B} = \underbrace{(4)^2}_{A^2} + \underbrace{2(4)(2i)}_{2AB} + \underbrace{(2i)^2}_{B^2}$$

$$\text{Simplify:} \quad = 16 + 16i + 4i^2$$
$$\text{Replace } i^2 \text{ with } -1: \quad = 16 + 16i + 4(-1)$$
$$\text{Multiply:} \quad = 16 + 16i - 4$$
$$\text{Simplify:} \quad = 12 + 16i$$

We can also find this product using the FOIL method.

$$\text{Begin with the original expression:} \quad (4 + 2i)^2$$
$$\text{Write the expression as a product:} \quad = (4 + 2i)(4 + 2i)$$
$$\text{Multiply using the FOIL method:} \quad = \overset{F}{4 \cdot 4} + \overset{O}{4 \cdot 2i} + \overset{I}{2i \cdot 4} + \overset{L}{2i \cdot 2i}$$
$$\text{Simplify:} \quad = 16 + 8i + 8i + 4i^2$$
$$\text{Collect like terms:} \quad = 16 + 16i + 4i^2$$

At this point we would continue simplifying as before to find $(4 + 2i)^2 = 12 + 16i$.

My video summary

b. Try to work the problem on your own. Click here to check your answer, or watch this video for a detailed solution.

You Try It Work through this **You Try It** problem.

Work Exercises 20–21 in this textbook or in the **MyMathLab**® Study Plan.

In Section 4.3 we learned that a binomial sum and a binomial difference made from the same two terms are conjugates of each other. We also saw that the product of conjugates resulted in the difference of two squares. These results extend to our discussion of complex numbers as follows.

Complex Conjugates

The complex numbers $(a + bi)$ and $(a - bi)$ are called **complex conjugates** of each other. A complex conjugate is obtained by changing the sign of the imaginary part in a complex number. Also, $(a + bi)(a - bi) = a^2 + b^2$.

Notice that the product of complex conjugates is a *sum* of two squares rather than a difference and is always a real number. Click here to see why.

Example 6 Multiplying Complex Conjugates

Multiply $(-2 + 7i)(-2 - 7i)$.

Solution Since the two complex numbers are conjugates, we find the product using the result for the product of complex conjugates.

Identifying $a = -2$ and $b = 7$ we get

$$\overbrace{(a + bi)(a - bi)}\quad\quad\quad$$
$$(-2 + 7i)(-2 - 7i) = \overset{a^2}{(-2)^2} + \overset{b^2}{(7)^2} = 4 + 49 = 53.$$

We can also find the same result using the FOIL method. Click here to see how.

You Try It   **Work through this You Try It problem.**

Work Exercises 22–26 in this textbook or in the _MyMathLab_® Study Plan.

OBJECTIVE 4:   DIVIDE COMPLEX NUMBERS

When dividing complex numbers, the goal is to eliminate the imaginary part from the denominator and to express the quotient in standard form, $a + bi$. To do this, we multiply the numerator and denominator by the complex conjugate of the denominator.

Example 7 Dividing Complex Numbers

My video summary

Divide. Write the quotient in standard form.

$$\frac{1 - 3i}{5 - 2i}$$

Solution The denominator is $5 - 2i$, so its complex conjugate is $5 + 2i$. We multiply both the numerator and denominator by the complex conjugate and simplify to standard form.

Multiply numerator and denominator by $5 + 2i$:
$$\frac{1 - 3i}{5 - 2i} = \frac{1 - 3i}{5 - 2i} \cdot \frac{5 + 2i}{5 + 2i}$$

Multiply numerators and multiply denominators:
(remember that $(a + bi)(a - bi) = a^2 + b^2$)
$$= \frac{5 + 2i - 15i - 6i^2}{(5)^2 + (2)^2}$$

Simplify exponents:
(remember that $i^2 = -1$)
$$= \frac{5 + 2i - 15i - 6(-1)}{25 + 4}$$

Simplify the numerator and denominator:
$$= \frac{5 - 13i + 6}{29}$$

$$= \frac{11 - 13i}{29}$$

Write in standard form, $a + bi$:
$$= \frac{11}{29} - \frac{13}{29}i$$

You Try It   **Work through this You Try It problem.**

Work Exercises 27–30 in this textbook or in the _MyMathLab_® Study Plan.

Note: Remember that multiplying the numerator and denominator of an expression by the same quantity is the same as multiplying the expression by 1.

Example 8 Dividing Complex Numbers

 Divide. Write the quotient in standard form.

$$\frac{5 + 7i}{2i}$$

Solution First, we multiply the numerator and denominator by the complex conjugate of the denominator. The denominator is $2i = 0 + 2i$, so its complex conjugate is $0 - 2i = -2i$.

$$\frac{5 + 7i}{2i} = \overbrace{\frac{5 + 7i}{2i} \cdot \frac{-2i}{-2i}}^{\substack{\text{Multiply numerator} \\ \text{and denominator} \\ \text{by } -2i}} = \overbrace{\frac{-10i - 14i^2}{-4i^2}}^{\substack{\text{Multiply numerators} \\ \text{and multiply} \\ \text{denominators}}}$$

Finish simplifying on your own, and write the answer in standard form. Click here to check your answer, or watch this video for the complete solution.

You Try It   **Work through this You Try It problem.**

Work Exercise 31 in this textbook or in the *MyMathLab*®️ Study Plan.

In Example 8, we can get the same result if we multiply the numerator and denominator by $2i$. Click here to find out why.

OBJECTIVE 5:   SIMPLIFY RADICALS WITH NEGATIVE RADICANDS

In the next chapter, we will solve equations involving solutions with radicals having a negative radicand. Thus, we must first learn how to simplify a radical with a negative radicand such as $\sqrt{-49}$. By remembering that $\sqrt{-1} = i$, we can use the following rule to simplify this expression.

Square Root of a Negative Number

For any positive real number a,

$$\sqrt{-a} = \sqrt{-1} \cdot \sqrt{a} = i\sqrt{a}.$$

So, $\sqrt{-49} = \sqrt{-1} \cdot \sqrt{49} = i \cdot 7 = 7i$.

At this point, you might want to review how to simplify radicals using the product rule in Section 7.3.

Example 9 Simplifying a Square Root with a Negative Radicand

 Simplify.

a. $\sqrt{-81}$ b. $\sqrt{-48}$ c. $\sqrt{-108}$

My video summary

My video summary

Solution

a. $\sqrt{-81} = \underbrace{\sqrt{-1}}_{i} \cdot \sqrt{81} = i \cdot 9 = 9i$

b. $\sqrt{-48} = \underbrace{\sqrt{-1}}_{i} \cdot \sqrt{48} = i \cdot \sqrt{16} \cdot \sqrt{3}$
$= i \cdot 4\sqrt{3} = 4i\sqrt{3}$

c. Try this problem on your own. Click here to check your answer, or watch this video for a detailed solution.

You Try It **Work through this You Try It problem.**

Work Exercise 32 in this textbook or in the *MyMathLab*® Study Plan.

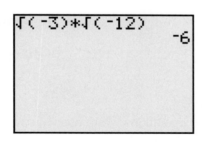

CAUTION When simplifying or performing operations involving radicals with a negative radicand and an even index, it is important to first write the numbers in terms of the imaginary unit i if possible.

The property $\sqrt{a} \cdot \sqrt{b} = \sqrt{ab}$ is only true when $a \geq 0$ and $b \geq 0$ so that $\sqrt{a}$ and $\sqrt{b}$ are real numbers. This property does not apply to non-real numbers. To find the correct answer if a or b are negative, we must first write each number in terms of the imaginary unit i.

$$\sqrt{-3} \cdot \sqrt{-12} = \sqrt{(-3)(-12)} = \sqrt{36} = 6 \quad \text{False}$$
$$\sqrt{-3} \cdot \sqrt{-12} = \underbrace{\sqrt{-1}}_{i} \cdot \sqrt{3} \cdot \underbrace{\sqrt{-1}}_{i} \cdot \sqrt{12} = i\sqrt{3} \cdot i\sqrt{12} = \underbrace{i^2}_{i^2 = -1}\sqrt{36} = -6 \quad \text{True}$$

We can use a graphing calculator to check the result. See Figure 11.

```
√(-3)∗√(-12)
                -6
```

Figure 11

Notice that in order to get the correct answer, we had to first write each number in terms of the imaginary unit i.

Example 10 Simplifying Expressions with Negative Radicands

Simplify.

a. $\sqrt{-8} + \sqrt{-18}$ **b.** $\sqrt{-8} \cdot \sqrt{-18}$

c. $\dfrac{6 + \sqrt{(6)^2 - 4(2)(5)}}{2}$ **d.** $\dfrac{4 - \sqrt{-12}}{4}$

Solution

a. $\sqrt{-8} + \sqrt{-18} = \underbrace{\sqrt{-1}}_{i} \cdot \sqrt{8} + \underbrace{\sqrt{-1}}_{i} \cdot \sqrt{18}$
$= i \cdot 2\sqrt{2} + i \cdot 3\sqrt{2}$
$= 2i\sqrt{2} + 3i\sqrt{2}$
$= 5i\sqrt{2}$

My video summary

b. $\sqrt{-8} \cdot \sqrt{-18} = \left(\underbrace{\sqrt{-1} \cdot \sqrt{8}}_{i}\right) \cdot \left(\underbrace{\sqrt{-1} \cdot \sqrt{18}}_{i}\right)$

$= (i \cdot 2\sqrt{2}) \cdot (i \cdot 3\sqrt{2})$

$= \underbrace{i^2}_{i^2 = -1} \cdot 6\sqrt{4}$

$= -1 \cdot 6 \cdot 2$

$= -12$

c.–d. Try these problems on your own. Click here to check your answer, or watch this video for detailed solutions.

You Try It   Work through this You Try It problem.

Work Exercises 33–40 in this textbook or in the **MyMathLab**® Study Plan.

7.6 Exercises

You Try It

In Exercises 1–5, write each power of i as i, -1, $-i$, or 1.

1. i^{41} **2.** i^{28} **3.** $-i^{19}$ **4.** $(-i)^7$ **5.** $i^{22} + i^{13}$

In Exercises 6–13, find the sum or difference. Write each answer in standard form, $a + bi$.

You Try It

6. $(3 - 2i) + (-7 + 9i)$ **7.** $(3 - 2i) - (-7 + 9i)$

8. $i - (1 + i)$ **9.** $5 + (2 - 3i)$

10. $(2 + 5i) - (2 - 5i)$ **11.** $(2 + 5i) + (2 - 5i)$

12. $[(-1 + 8i) - (3 - 4i)] + (9 - 4i)$ **13.** $(6 + 3i) - [(2 + 4i) + (5 - 2i)]$

In Exercises 14–21, perform the indicated operations. Write each answer in standard form.

You Try It **14.** $3i(7i)$ **15.** $2i(4 - 3i)$ **16.** $-i(1 - i)$

17. $(3 - 2i)(6 + i)$ **18.** $(-2 - i)(3 - 4i)$ **19.** $(5 + i)(2 + 3i)$

You Try It **20.** $(2 + 7i)^2$ **21.** $(6 - 2i)^2$

In Exercises 22–26, find the product of the complex number and its conjugate.

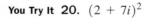
You Try It **22.** $5 - 2i$ **23.** $1 - i$ **24.** $\dfrac{1}{2} - 3i$ **25.** $\sqrt{5} + i$ **26.** $4i$

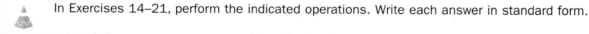

You Try It

In Exercises 27–31, write each quotient in standard form.

27. $\dfrac{2 - i}{3 + 4i}$

28. $\dfrac{1}{2 - i}$

29. $\dfrac{3i}{2 + 2i}$

You Try It

30. $\dfrac{5 + i}{5 - i}$

31. $\dfrac{2 - 3i}{5i}$

You Try It

In Exercises 32–40, write each expression in standard form.

32. $\sqrt{-320}$

33. $\sqrt{-36} - \sqrt{49}$

34. $\sqrt{-1} + 3 - \sqrt{-64}$

35. $\sqrt{-2} \cdot \sqrt{-18}$

36. $\left(\sqrt{-8}\right)^2$

37. $\left(i\sqrt{-4}\right)^2$

You Try It

38. $\dfrac{-4 - \sqrt{-20}}{2}$

39. $\dfrac{-3 - \sqrt{-81}}{6}$

40. $\dfrac{4 + \sqrt{-8}}{4}$

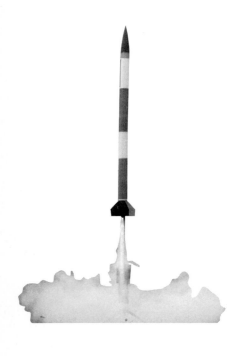

Quadratic Equations and Functions; Circles

CHAPTER EIGHT CONTENTS

8.1 **Solving Quadratic Equations**

8.2 **Quadratic Functions and Their Graphs**

8.3 **Applications and Modeling of Quadratic Functions**

8.4 **Circles**

8.5 **Polynomial and Rational Inequalities**

8.1 Solving Quadratic Equations

THINGS TO KNOW

Before working through this section, be sure you are familiar with the following concepts:

	VIDEO	ANIMATION	INTERACTIVE

You Try It

1. Factor Trinomials Using Substitution (Section 5.2, Objective 4) — VIDEO

You Try It

2. Solve Polynomial Equations by Factoring (Section 5.4, Objective 1) — INTERACTIVE

You Try It

3. Simplifying Radical Expression Using the Product Rule (Section 7.3, Objective 5) — VIDEO, INTERACTIVE

OBJECTIVES

1 Solve Quadratic Equations Using Factoring

2 Solve Quadratic Equations Using the Square Root Property

3 Solve Quadratic Equations by Completing the Square

4 Solve Quadratic Equations Using the Quadratic Formula

5 Use the Discriminant to Determine the Number and Type of Solutions to a Quadratic Equation

6 Solve Equations That Are Quadratic in Form

OBJECTIVE 1:   SOLVE QUADRATIC EQUATIONS USING FACTORING

In Section 1.1, we studied linear equations of the form $ax + b = c$, with $a \neq 0$. These equations were called first-degree polynomial equations. In this section we learn how to solve second-degree polynomial equations. These equations are called **quadratic equations**.

> **Definition   Quadratic Equation in One Variable**
>
> A **quadratic equation in one variable** is an equation that can be written in the form $ax^2 + bx + c = 0$, where a, b, and c are real numbers, and $a \neq 0$. Quadratic equations in this form are said to be in *standard form*.

As we have seen with polynomial equations, some quadratic equations can be solved quickly by factoring and by using the zero product property introduced in Section 5.4. Remember that the zero product property states that if two factors multiplied together are equal to zero, then at least one of the factors must be zero. We can use the following guidelines to solve certain quadratic equations by factoring.

> **Solving Quadratic Equations by Factoring**
>
> **Step 1.** Write the quadratic equation in the standard form $ax^2 + bx + c = 0$.
> **Step 2.** Factor the left-hand side.
> **Step 3.** Set each factor from step 2 equal to zero (zero product property), and solve the resulting linear equations.
> **Step 4.** Check each potential solution in the original equation and form the solution set.

Example 1 Solving a Quadratic Equation Using Factoring

 Solve $6x^2 - 17x = -12$.

Solution

$$\text{Original equation:} \qquad 6x^2 - 17x = -12$$
$$\text{Write in standard form:} \quad 6x^2 - 17x + 12 = 0$$

Factor the left-hand side. Then set each factor equal to zero and solve the resulting equations. The solution set is $\left\{ \dfrac{4}{3}, \dfrac{3}{2} \right\}$. Click here to see the check, or watch this video for the complete solution.

My video summary

You Try It   Work through this You Try It problem.

Work Exercises 1–6 in this textbook or in the *MyMathLab*® Study Plan.

OBJECTIVE 2:   SOLVE QUADRATIC EQUATIONS USING THE SQUARE ROOT PROPERTY

Consider the quadratic equation $x^2 = 25$. We could solve the equation using factoring and the zero product property to find the solution set $\{-5, 5\}$. These solutions are the positive and negative square roots of 25. Click here to see the details.

What about the equation $x^2 = 3$? We could subtract 3 from both sides to get $x^2 - 3 = 0$, but the expression on the left does not factor over the set of integers. However, if we write $3 = \left(\sqrt{3}\right)^2$, then we can express the left-hand side as the difference of two squares and factor.

Original equation:	$x^2 = 3$
Subtract 3 from both sides:	$x^2 - 3 = 0$
Write 3 as $\left(\sqrt{3}\right)^2$:	$x^2 - \left(\sqrt{3}\right)^2 = 0$
Factor the difference of two squares:	$\left(x - \sqrt{3}\right)\left(x + \sqrt{3}\right) = 0$
Use the zero product property:	$x - \sqrt{3} = 0$ or $x + \sqrt{3} = 0$
Solve each linear equation:	$x = \sqrt{3}$ or $x = -\sqrt{3}$

The solution set is $\left\{-\sqrt{3}, \sqrt{3}\right\}$. These are the positive and negative square roots of 3. This illustrates the **square root property**.

Square Root Property

If u is an algebraic expression and k is a real number, then $u^2 = k$ is equivalent to $u = -\sqrt{k}$ or $u = \sqrt{k}$. Equivalently, if $u^2 = k$ then $u = \pm\sqrt{k}$.

My video summary

Click here for an alternate explanation of the square root property using absolute value.

CAUTION

When solving equations of the form $u^2 = k$, we often simply say that we are "*taking the square root of both sides.*" However, because the square root of a number yields only one value, the principal root, many students forget to include the $\pm$. Remember that applying the square root property for $k \neq 0$ will result in *two* values—the positive and negative square roots of k.

To solve quadratic equations using the square root property, we use the following guidelines.

Solving Quadratic Equations Using the Square Root Property

Step 1. Write the equation in the form $u^2 = k$ to isolate the quantity being squared.

Step 2. Apply the square root property.

Step 3. Solve the resulting equations.

Step 4. Check the solutions in the original equation.

Example 2 Solving a Quadratic Equation Using the Square Root Property

Solve.

a. $x^2 - 16 = 0$ **b.** $2x^2 + 72 = 0$

c. $(x - 1)^2 = 9$ **d.** $2(x + 1)^2 - 17 = 23$

Solution

a. Isolate x^2, then apply the square root property.

$$\text{Original equation:} \quad x^2 - 16 = 0$$

$$\text{Add 16 to both sides:} \quad x^2 = 16 \leftarrow \text{square is isolated}$$

$$\text{Apply the square root property:} \quad x = \pm\sqrt{16}$$

$$\text{Simplify the radical:} \quad x = \pm 4$$

The solution set is $\{-4, 4\}$. The check is left to you.

b. Isolate x^2, then apply the square root property.

$$\text{Original equation:} \quad 2x^2 + 72 = 0$$

$$\text{Subtract 72 from both sides:} \quad 2x^2 = -72$$

$$\text{Divide both sides by 2:} \quad x^2 = -36 \leftarrow \text{square is isolated}$$

$$\text{Apply the square root property:} \quad x = \pm\sqrt{-36}$$

$$\text{Simplify the radical with a negative radicand:} \quad x = \pm\sqrt{-1} \cdot \sqrt{36}$$

$$x = \pm 6i$$

The solution set is $\{-6i, 6i\}$. The check is left to you.

My video summary
 c. In this case, we are squaring an algebraic expression. However, this will not change our process. We still isolate the square and apply the square root property.

$$\text{Original equation:} \quad (x - 1)^2 = 9 \leftarrow \text{square is isolated}$$

$$\text{Apply the square root property:} \quad x - 1 = \pm\sqrt{9}$$

Try to finish solving this equation on your own. Click here to check your answer, or watch this video for the complete solution.

d. Try to work this problem on your own. Remember to isolate the square first, then apply the square root property. Simplify radicals if possible. Click here to check your answer, or watch this video for the complete solution.

You Try It Work through this You Try It problem.

Work Exercises 7–12 in this textbook or in the **MyMathLab**® Study Plan.

In Example 2c, the quadratic equation $(x - 1)^2 = 9$ can be solved using the square root property because the left side of the equation is a perfect square and the right side is a constant. But what about quadratic equations such as $x^2 - 5x + 3 = 0$? Even if we get the square term by itself

$$x^2 = 5x - 3$$

we cannot apply the square root property because the right-hand side is not a constant. However, every quadratic equation can be written in the form $(x - h)^2 = k$ (as in Example 2c) by using a method known as **completing the square**.

OBJECTIVE 3:   SOLVE QUADRATIC EQUATIONS BY COMPLETING THE SQUARE

Consider the following perfect square trinomials:

$$x^2 + 2x + 1 = (x + 1)^2 \qquad x^2 - 6x + 9 = (x - 3)^2 \qquad x^2 - 7x + \frac{49}{4} = \left(x - \frac{7}{2}\right)^2$$

$$\left(\frac{1}{2}\cdot 2\right)^2 = 1 \qquad\qquad \left(\frac{1}{2}\cdot(-6)\right)^2 = 9 \qquad\qquad \left(\frac{1}{2}\cdot(-7)\right)^2 = \frac{49}{4}$$

In each case, notice the relationship between the coefficient of the linear term (x-term) and the constant term. The constant term of a perfect square trinomial is equal to the square of $\frac{1}{2}$ the linear coefficient.

To *complete the square* means to add an appropriate constant so that a binomial of the form $x^2 + bx$ becomes a perfect square trinomial. The appropriate constant is the square of half the linear coefficient, $\left(\frac{1}{2}\cdot b\right)^2$. For example, to complete the square given $x^2 + 10x$, we add $\left(\frac{1}{2}\cdot 10\right)^2 = 5^2 = 25$ so we can write

$$x^2 + 10x + 25 = (x + 5)^2.$$

Example 3 Completing the Square

What number must be added to make the binomial a perfect square trinomial?

a. $x^2 - 12x$ 　　　　　　　**b.** $x^2 + 5x$ 　　　　　　　**c.** $x^2 - \frac{3}{2}x$

Solution

a. The linear coefficient is -12, so we must add $\left(\frac{1}{2}(-12)\right)^2 = (-6)^2 = 36$ to complete the square. Thus, the expression $x^2 - 12x + 36$ is a perfect square trinomial and $x^2 - 12x + 36 = (x - 6)^2$.

b. The linear coefficient is 5, so we must add $\left(\frac{1}{2}\cdot 5\right)^2 = \left(\frac{5}{2}\right)^2 = \frac{25}{4}$ to complete the square: $x^2 + 5x + \frac{25}{4} = \left(x + \frac{5}{2}\right)^2$.

 *My video summary*

 c. Try to work this problem on your own. Click here to check your answer, or watch this video for the complete solution.

You Try It   Work through this You Try It problem.

Work Exercises 13–16 in this textbook or in the *MyMathLab*® Study Plan.

When writing the perfect square trinomial as a binomial squared, note that the first term of the binomial is x (the variable) and the second term is $\frac{1}{2}$ the linear coefficient from the trinomial. Consider the perfect square trinomials we saw earlier:

$$x^2 + 2x + 1 = (x + 1)^2 \qquad x^2 - 6x + 9 = (x - 3)^2 \qquad x^2 - 7x + \frac{49}{4} = \left(x - \frac{7}{2}\right)^2$$

$$\left(\frac{1}{2}\cdot 2\right) = 1 \qquad\qquad \left(\frac{1}{2}\cdot(-6)\right) = -3 \qquad\qquad \left(\frac{1}{2}\cdot(-7)\right) = -\frac{7}{2}$$

To solve a quadratic equation of the form $ax^2 + bx + c = 0$ by completing the square, where a, b, and c are real numbers, and $a \neq 0$, use the following guidelines:

Solving $ax^2 + bx + c = 0, a \neq 0$, by Completing the Square

Step 1. If $a \neq 1$, divide both sides of the equation by a.

Step 2. Move all constants to the right-hand side.

Step 3. Find $\dfrac{1}{2}$ times the coefficient of the x-term, square it, and add the result to both sides of the equation.

Step 4. The left-hand side is now a perfect square. Rewrite it as a binomial squared.

Step 5. Use the square root property and solve for x.

Example 4 Solving a Quadratic Equation by Completing the Square

Solve $x^2 - 8x + 2 = 0$ by completing the square.

Solution

Step 1. The leading coefficient is 1, so we proceed to step 2.

Step 2. Move all constants to the right-hand side.

$$x^2 - 8x = -2$$

Step 3. Multiply $\dfrac{1}{2}$ times the coefficient of the x-term, square the result, and add this to both sides of the equation.

$$\left(\frac{1}{2} \cdot (-8)\right)^2 = (-4)^2 = 16 \quad \rightarrow \quad x^2 - 8x + 16 = -2 + 16$$

$$x^2 - 8x + 16 = 14$$

Step 4. Rewrite the left-hand side as a perfect square: $(x - 4)^2 = 14$

Step 5. Use the square root property and solve for x.

Use the square root property: $x - 4 = \pm\sqrt{14}$

Add 4 to both sides: $x = 4 \pm \sqrt{14}$

The solution set is $\left\{4 - \sqrt{14}, 4 + \sqrt{14}\right\}$.

Example 5 Solving a Quadratic Equation by Completing the Square

Solve $2x^2 - 10x - 6 = 0$ by completing the square.

Solution

Step 1. Original equation: $2x^2 - 10x - 6 = 0$

Divide both sides by 2: $\dfrac{2x^2}{2} - \dfrac{10x}{2} - \dfrac{6}{2} = \dfrac{0}{2}$

Simplify: $x^2 - 5x - 3 = 0$

Step 2. Add 3 to both sides: $x^2 - 5x = 3$

Step 3. $\left(\dfrac{1}{2} \cdot (-5)\right)^2 = \left(-\dfrac{5}{2}\right)^2 = \dfrac{25}{4} \quad \rightarrow \quad x^2 - 5x + \dfrac{25}{4} = 3 + \dfrac{25}{4}$

$$x^2 - 5x + \dfrac{25}{4} = \dfrac{12}{4} + \dfrac{25}{4}$$

$$x^2 - 5x + \dfrac{25}{4} = \dfrac{37}{4}$$

Step 4. Write the left-hand side as a binomial squared: $\left(x - \dfrac{5}{2}\right)^2 = \dfrac{37}{4}$

Step 5. Use the square root property: $x - \dfrac{5}{2} = \pm\sqrt{\dfrac{37}{4}}$

Apply $\sqrt{\dfrac{a}{b}} = \dfrac{\sqrt{a}}{\sqrt{b}}$: $\quad x - \dfrac{5}{2} = \pm\dfrac{\sqrt{37}}{\sqrt{4}}$

Simplify: $\quad x - \dfrac{5}{2} = \pm\dfrac{\sqrt{37}}{2}$

Add $\dfrac{5}{2}$ to both sides and simplify: $\quad x = \dfrac{5}{2} \pm \dfrac{\sqrt{37}}{2} = \dfrac{5 \pm \sqrt{37}}{2}$

The solution set is $\left\{\dfrac{5 - \sqrt{37}}{2}, \dfrac{5 + \sqrt{37}}{2}\right\}$.

Example 6 Solving a Quadratic Equation by Completing the Square

 Solve $3x^2 - 18x + 19 = 0$ by completing the square.

Solution Try working this example on your own. Click here to check your answer, or watch this video for the complete solution.

You Try It   Work through this You Try It problem.

Work Exercises 17–21 in this textbook or in the **_MyMathLab_**® Study Plan.

OBJECTIVE 4:   SOLVE QUADRATIC EQUATIONS USING THE QUADRATIC FORMULA

We can solve any quadratic equation by completing the square. However, this process can be very time consuming. If we solve the general quadratic equation $ax^2 + bx + c = 0$ by completing the square, where a, b, and c are real numbers, and $a \neq 0$, we obtain a useful result known as the **quadratic formula.** The quadratic formula can be used to solve any quadratic equation and is often less time-consuming than completing the square. Work through this animation to see how to derive the quadratic formula by using completing the square to solve the general quadratic equation.

> **Quadratic Formula**
>
> The solutions to the quadratic equation $ax^2 + bx + c = 0$, $a \neq 0$, are given by the following formula:
>
> $$x = \frac{-b \pm \sqrt{b^2 - 4ac}}{2a}$$

Remember to write the quadratic equation in standard form before identifying the coefficients a, b, and c.

Example 7 Solving a Quadratic Equation Using the Quadratic Formula

Solve $2x^2 - 3x = 2$ using the quadratic formula.

Solution First, we write the equation in standard form.

$$2x^2 - 3x - 2 = 0$$

Next, we identify the coefficients.

$$2x^2 - 3x - 2 = 0$$

$$a = 2 \quad b = -3 \quad c = -2$$

Substitute 2 for a, -3 for b, and -2 for c in the quadratic formula.

Quadratic formula: $x = \dfrac{-b \pm \sqrt{b^2 - 4ac}}{2a}$

Substitute values for $a, b,$ and c: $= \dfrac{-(-3) \pm \sqrt{(-3)^2 - 4(2)(-2)}}{2(2)}$

Simplify: $= \dfrac{3 \pm \sqrt{9 + 16}}{4}$

$= \dfrac{3 \pm \sqrt{25}}{4}$

$= \dfrac{3 \pm 5}{4}$

There are two solutions to the equation, one from the $+$ sign and one from the $-$ sign.

$$x = \frac{3 + 5}{4} = \frac{8}{4} = 2 \quad \text{or} \quad x = \frac{3 - 5}{4} = \frac{-2}{4} = -\frac{1}{2}$$

The solution set is $\left\{ -\dfrac{1}{2}, 2 \right\}$.

My video summary

Example 8 Solving a Quadratic Equation Using the Quadratic Formula

◫ Solve $3x^2 + 2x - 2 = 0$ using the quadratic formula.

Solution The equation is in standard form, so we begin by identifying the coefficients.

$$3x^2 + 2x - 2 = 0$$

$$a = 3 \quad b = 2 \quad c = -2$$

Substitute 3 for a, 2 for b, and -2 for c in the quadratic formula.

Quadratic formula:
$$x = \frac{-b \pm \sqrt{b^2 - 4ac}}{2a}$$

Substitute values for a, b, and c:
$$= \frac{-(2) \pm \sqrt{(2)^2 - 4(3)(-2)}}{2(3)}$$

Simplify:
$$= \frac{-2 \pm \sqrt{4 + 24}}{6}$$

Finish simplifying to find the two solutions. Remember to simplify the radical first and then cancel any common factors. Click here to check your answer, or watch this video for the complete solution.

Example 9 Solving a Quadratic Equation Using the Quadratic Formula

Solve $4x^2 = x - 6$ using the quadratic formula.

Solution First, we write the equation in standard form.

$$4x^2 - x + 6 = 0$$

Next we identify the coefficients.

$$4x^2 - 1x + 6 = 0$$

$$a = 4 \quad b = -1 \quad c = 6$$

Substitute 4 for a, -1 for b, and 6 for c in the quadratic formula.

$$x = \frac{-b \pm \sqrt{b^2 - 4ac}}{2a} = \frac{-(-1) \pm \sqrt{(-1)^2 - 4(4)(6)}}{2(4)} = \frac{1 \pm \sqrt{1 - 96}}{8} = \frac{1 \pm \sqrt{-95}}{8}$$

The radicand is negative, so the solutions will be non-real complex numbers.

$$x = \frac{1 \pm \sqrt{-1} \cdot \sqrt{95}}{8} = \frac{1 \pm i\sqrt{95}}{8}$$

The solution set, in standard form $a + bi$, is $\left\{ \dfrac{1}{8} - \dfrac{\sqrt{95}}{8}i, \dfrac{1}{8} + \dfrac{\sqrt{95}}{8}i \right\}$.

Example 10 Solving a Quadratic Equation Using the Quadratic Formula

 Solve $14x^2 - 5x = 5x^2 + 7x - 4$ using the quadratic formula.

Solution Try to work this example on your own. Remember to write the equation in standard form first. Click here to check your answer, or watch this video for the complete solution.

You Try It Work through this You Try It problem.

Work Exercises 22–27 in this textbook or in the **MyMathLab**® Study Plan.

OBJECTIVE 5: USE THE DISCRIMINANT TO DETERMINE THE NUMBER AND TYPE OF SOLUTIONS TO A QUADRATIC EQUATION

In Example 9, the quadratic equation $4x^2 = x - 6$ had two non-real solutions. The solutions were non-real because the expression $b^2 - 4ac$ under the radical was a negative number. Given a quadratic equation of the form $ax^2 + bx + c = 0$, the expression $b^2 - 4ac$ is called the **discriminant**. Knowing the value of the discriminant can help us determine the number and type of solutions to a quadratic equation.

Discriminant

Given a quadratic equation $ax^2 + bx + c = 0$, $a \neq 0$, the expression $D = b^2 - 4ac$ is called the **discriminant**.

If $D > 0$, then the quadratic equation has two real solutions.
If $D < 0$, then the quadratic equation has two non-real solutions.
If $D = 0$, then the quadratic equation has exactly one real solution.

Example 11 Using the Discriminant

 Use the discriminant to determine the number and type of solutions to each of the following quadratic equations.

a. $3x^2 + 2x + 2 = 0$ **b.** $4x^2 + 1 = 4x$

Solution

a. The equation $3x^2 + 2x + 2 = 0$ is in standard form with $a = 3$, $b = 2$, and $c = 2$.

$$D = (2)^2 - 4(3)(2) = 4 - 24 = -20 < 0$$

Because the discriminant is less than zero, there are two non-real solutions.

b. The equation $4x^2 + 1 = 4x$ is not written in standard form. To write the equation in standard form, subtract $4x$ from both sides:

$$4x^2 - 4x + 1 = 0$$

Now the equation is in standard form with $a = 4$, $b = -4$, and $c = 1$.

$$D = (-4)^2 - 4(4)(1) = 16 - 16 = 0$$

Because the discriminant is zero, there is exactly one real solution.

(EBook Screens 8.1-1–8.1-28)

You Try It **Work through this You Try It problem.**

Work Exercises 28–31 in this textbook or in the **MyMathLab** Study Plan.

OBJECTIVE 6: SOLVE EQUATIONS THAT ARE QUADRATIC IN FORM

We have now seen several different techniques for solving quadratic equations. Click here to view a summary along with some advantages and disadvantages for each approach.

Sometimes equations that are not quadratic can be changed into a quadratic equation by using substitution. In Section 5.2 we used substitution to rewrite an algebraic expression into an easily factorable form. We use this same approach to rewrite some equations so that they look like quadratic equations. Equations of this type are said to be quadratic in form. These equations will have the form $au^2 + bu + c = 0$, $a \neq 0$, after an appropriate substitution.

Original Equation	Make an Appropriate Substitution		New Equation Is a Quadratic
$2x^4 - 11x^2 + 12 = 0$	Determine the proper substitution $\longrightarrow$	Let $u = x^2$, then $u^2 = x^4$.	New equation is a quadratic $\longrightarrow$ $2u^2 - 11u + 12 = 0$
$\left(\dfrac{1}{x-2}\right)^2 + \dfrac{2}{x-2} - 15 = 0$	Determine the proper substitution $\longrightarrow$	Let $u = \dfrac{1}{x-2}$, then $u^2 = \left(\dfrac{1}{x-2}\right)^2$.	New equation is a quadratic $\longrightarrow$ $u^2 + 2u - 15 = 0$
$x^{\frac{2}{3}} - 9x^{\frac{1}{3}} + 8 = 0$	Determine the proper substitution $\longrightarrow$	Let $u = x^{\frac{1}{3}}$, then $u^2 = \left(x^{\frac{1}{3}}\right)^2 = x^{\frac{2}{3}}$.	New equation is a quadratic $\longrightarrow$ $u^2 - 9u + 8 = 0$
$3x^{-2} - 5x^{-1} - 2 = 0$	Determine the proper substitution $\longrightarrow$	Let $u = x^{-1} = \dfrac{1}{x}$, then $u^2 = \left(x^{-1}\right)^2 = x^{-2}$.	New equation is a quadratic $\longrightarrow$ $3u^2 - 5u - 2 = 0$

Table 1

In Example 12, we solve each of the equations listed in Table 1.

Example 12 Solving Equations That Are Quadratic in Form

My interactive video summary

Solve each equation.

a. $2x^4 - 11x^2 + 12 = 0$ **b.** $\left(\dfrac{1}{x-2}\right)^2 + \dfrac{2}{x-2} - 15 = 0$

c. $x^{\frac{2}{3}} - 9x^{\frac{1}{3}} + 8 = 0$ **d.** $3x^{-2} - 5x^{-1} - 2 = 0$

Solution

a. To solve the equation $2x^4 - 11x^2 + 12 = 0$, we look for an appropriate substitution.

The middle term contains x^2, and we can rewrite the first term as $2(x^2)^2$ using rules for exponents. Therefore, an appropriate substitution would be $u = x^2$. Letting $u = x^2$, we get $u^2 = (x^2)^2 = x^4$. Substituting u for x^2 and u^2 for x^4, our equation becomes

$$2u^2 - 11u + 12 = 0,$$

which is a quadratic equation, in terms of u. We can solve this quadratic equation for u by using any of the approaches previously discussed. For this example, we will solve the quadratic equation by factoring.

$$\text{Quadratic equation:} \quad 2u^2 - 11u + 12 = 0$$
$$\text{Factor:} \quad (2u - 3)(u - 4) = 0$$

We now apply the zero product property and solve the resulting equations for u.

$$2u - 3 = 0 \quad \text{or} \quad u - 4 = 0$$
$$2u = 3 \qquad\qquad u = 4$$
$$u = \frac{3}{2}$$

Here we may be tempted to form a solution set. However, we have only solved for u. Since the original equation involved the variable x, we still need to solve for x. Because we said that $u = x^2$, we get

$$x^2 = \frac{3}{2} \quad \text{or} \quad x^2 = 4.$$

We solve these two equations for x by applying the square root property.

$$x^2 = \frac{3}{2} \qquad\qquad \text{or} \quad x^2 = 4$$
$$x = \pm\underbrace{\sqrt{\frac{3}{2}}}_{\substack{\text{Rationalize the} \\ \text{denominator}}} = \pm\frac{\sqrt{6}}{2} \qquad x = \pm\sqrt{4} = \pm 2$$

The solution set is $\left\{ \pm 2, \pm\dfrac{\sqrt{6}}{2} \right\}$.

b.–d. Try to solve these problems on your own. As a hint, look back at the suggested substitutions given earlier in Table 1. Click here to check your answers, or watch this interactive video to see complete solutions.

You Try It   Work through this You Try It problem.

Work Exercises 32–38 in this textbook or in the *MyMathLab*® Study Plan.

8.1 Exercises

You Try It   In Exercises 1–6, solve each equation by factoring.

1. $x^2 - 8x = 0$ **2.** $x^2 - x = 6$ **3.** $x^2 - 10x + 24 = 0$

4. $3x^2 + 8x - 3 = 0$ **5.** $8m^2 - 15 = 14m$ **6.** $28z^2 - 13z - 6 = 0$

You Try It   In Exercises 7–12, solve each equation using the square root property.

7. $x^2 - 64 = 0$ **8.** $x^2 + 64 = 0$ **9.** $3x^2 = 72$

10. $(x + 2)^2 - 9 = 0$ **11.** $(2x + 1)^2 + 4 = 0$ **12.** $3(x - 4)^2 + 2 = 8$

You Try It

In Exercises 13–16, decide what number must be added to each binomial to make a perfect square trinomial.

 13. $x^2 - 8x$ **14.** $x^2 + 10x$ **15.** $x^2 - 7x$ **16.** $x^2 + \dfrac{5}{3}x$

 You Try It

In Exercises 17–21, solve each quadratic equation by completing the square.

17. $x^2 - 8x - 2 = 0$ **18.** $x^2 + 7x + 14 = 0$ **19.** $2x^2 + 8 = -6x$

 20. $3x^2 = 7 - 24x$ **21.** $3x^2 + 5x + 12 = 0$

In Exercises 22–27, solve each quadratic equation using the quadratic formula.

You Try It **22.** $3x^2 + 8x - 3 = 0$ **23.** $x^2 - 8x - 2 = 0$ **24.** $4x^2 - x + 8 = 0$

You Try It **25.** $3x^2 = 1 + 4x$ **26.** $9x^2 - 6x = -1$ **27.** $5x^2 + 3x + 1 = 0$

In Exercises 28–31, use the discriminant to determine the number and type of the solutions to each quadratic equation. Do not solve the equations.

28. $x^2 + 3x + 1 = 0$ **29.** $4x^2 + 4x + 1 = 0$ **30.** $2x^2 + x = 5$ **31.** $3x^2 + \sqrt{12}x + 4 = 0$

 You Try It

In Exercises 32–38, solve the equation after making an appropriate substitution.

32. $x^4 - 6x^2 + 8 = 0$ **33.** $(13x - 1)^2 - 2(13x - 1) - 3 = 0$ **34.** $2x^{\frac{2}{3}} - 5x^{\frac{1}{3}} + 2 = 0$

You Try It **35.** $x^6 - x^3 = 6$ **36.** $\sqrt{x} - 3\sqrt[4]{x} - 4 = 0$ **37.** $3\left(\dfrac{1}{x-1}\right)^2 - \dfrac{5}{x-1} - 2 = 0$

38. $2x^{-2} - 3x^{-1} - 2 = 0$

8.2 Quadratic Functions and Their Graphs

THINGS TO KNOW

Before working through this section, be sure you are familiar with the following concepts:

VIDEO   ANIMATION   INTERACTIVE

 You Try It
1. Find x- and y-Intercepts (Section 2.1, Objective 5)

 You Try It
2. Find the Domain and Range of a Relation (Section 2.2, Objective 2)

 You Try It
3. Solve Quadratic Equations Using Factoring (Section 8.1, Objective 1)

 You Try It
4. Solve Quadratic Equations by Completing the Square (Section 8.1, Objective 3)

 You Try It
5. Solve Quadratic Equations Using the Quadratic Formula (Section 8.1, Objective 4)

OBJECTIVES

1 Identify the Characteristics of a Quadratic Function from Its Graph

2 Graph Quadratic Functions by Using Translations

3 Graph Quadratic Functions of the Form $f(x) = a(x - h)^2 + k$

4 Find the Vertex of a Quadratic Function by Completing the Square

5 Graph Quadratic Functions of the Form $f(x) = ax^2 + bx + c$ by Completing the Square

6 Find the Vertex of a Quadratic Function by Using the Vertex Formula

7 Graph Quadratic Functions of the Form $f(x) = ax^2 + bx + c$ by Using the Vertex Formula

OBJECTIVE 1: IDENTIFY THE CHARACTERISTICS OF A QUADRATIC FUNCTION FROM ITS GRAPH

In the previous section, we learned how to solve quadratic equations. In this section, we learn about *graphing quadratic functions*.

Definition Quadratic Function

A **quadratic function** is a second-degree polynomial function of the form $f(x) = ax^2 + bx + c$, where a, b, and c are real numbers and $a \neq 0$. Every quadratic function has a "u-shaped" graph called a **parabola.**

The function $f(x) = x^2$ is a quadratic function with $a = 1$, $b = 0$, and $c = 0$. Its graph is shown in Figure 1a. The function $g(x) = -x^2$ is a quadratic function with $a = -1$, $b = 0$, and $c = 0$. Its graph is shown in Figure 1b. Notice that both graphs are parabolas and have the characteristic "u-shape."

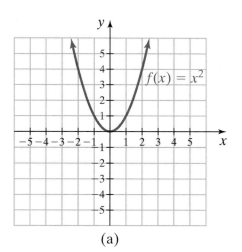
(a)

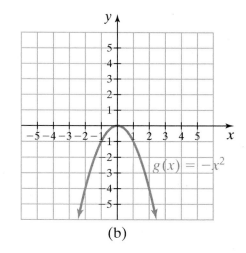
(b)

Figure 1
Graphs of $f(x) = x^2$ and $g(x) = -x^2$

A parabola either opens up (called **concave up**) or opens down (called **concave down**) depending on the leading coefficient, a. If $a > 0$, as in Figure 1a, the parabola will "open up." If $a < 0$, as in Figure 1b, the parabola will "open down."

The leading coefficient also affects the shape of the parabola. The function $f(x) = 2x^2$ is a quadratic function with $a = 2$, $b = 0$, and $c = 0$. Its graph is shown in Figure 2a. The function $g(x) = \frac{1}{2}x^2$ is a quadratic function with $a = \frac{1}{2}$, $b = 0$, and $c = 0$. Its graph is shown in Figure 2b. The graph of $y = x^2$ is shown in gray.

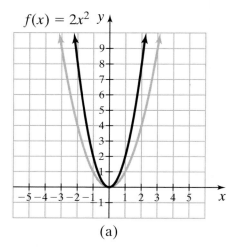
(a)

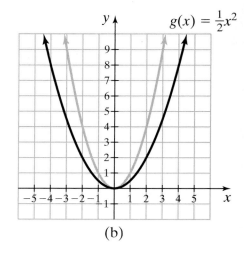
(b)

Figure 2

If $|a| > 1$, as in Figure 2a, the graph will be narrower than the graph of $y = x^2$. If $0 < |a| < 1$, as in Figure 2b, the graph will be wider than the graph of $y = x^2$.

Example 1 Determining the Shape of the Graph of a Quadratic Function

Without graphing, determine if the graph of each quadratic function opens up or down. Also determine if the graph will be wider or narrower than the graph of $y = x^2$.

a. $f(x) = 2x^2 - 3x + 5$ **b.** $g(x) = -\frac{2}{3}x^2 + 5x - 7$

Solution

a. Because the leading coefficient $a = 2$ is positive, the graph will open up. Since $|a| > 1$, the graph will be narrower than the graph of $y = x^2$.

b. Because the leading coefficient $a = -\frac{2}{3}$ is negative, the graph will open down. Since $0 < |a| < 1$, the graph will be wider than the graph of $y = x^2$.

You Try It   Work through this You Try It problem.

Work Exercises 1–4 in this textbook or in the *MyMathLab*® Study Plan.

Before we can sketch graphs of quadratic functions, we must be able to identify the five basic characteristics of a parabola: *vertex, axis of symmetry, y-intercept, x-intercepts,* and *domain and range.*

Every parabola has a **vertex**. If the parabola "opens up," the vertex (h, k) is the lowest point on the graph and the function will have a **minimum value**. The minimum value is the smallest possible value for the function and is given by the y-coordinate of the vertex, k. If the parabola "opens down," the vertex (h, k) is the highest point on the graph and the function will have a **maximum value**.

The maximum value is the largest possible value for the function and is given by the y-coordinate of the vertex, k. In either case, the y-coordinate of the vertex indicates the maximum or minimum value and the x-coordinate indicates *where* this occurs.

The **axis of symmetry** is an imaginary vertical line that passes through the vertex and divides the graph into two mirror images. Points on the graph that are the same horizontal distance from the axis of symmetry will have the same y-coordinate.

The domain of every quadratic function is all real numbers $(-\infty, \infty)$. The range is determined by k, the y-coordinate of the vertex, and whether the graph opens up or down. If the graph opens up, the range is all real numbers greater than or equal to k, or $[k, \infty)$. If the graph opens down, the range is all real numbers less than or equal to k, or $(-\infty, k]$.

The graph of every quadratic function $f(x)$ crosses the y-axis, so every parabola has a y-intercept. However, it may or may not have any x-intercepts. We find these intercepts in the usual way.

We summarize the five basic characteristics with the following:

1. Vertex
2. Axis of symmetry
3. y-intercept
4. x-intercept(s) or real zeros
5. Domain and range

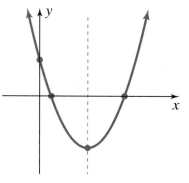

Work through this animation, clicking on each characteristic to get a detailed description.

Example 2   Finding the Characteristics of a Quadratic Function from Its Graph

Use the given graph of a quadratic function to find the following:

a. Vertex   **b.** Axis of symmetry   **c.** y-intercept

d. x-intercept(s)   **e.** Domain and range

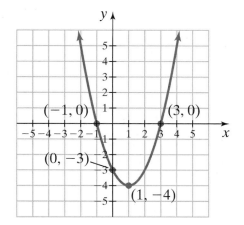

Solution

a. Because the graph "opens up," the vertex is the lowest point on the graph: $(h, k) = (1, -4)$.

b. The axis of symmetry is the vertical line that passes through the vertex. Since the x-coordinate of the vertex is $h = 1$, the equation of the axis of symmetry is $x = 1$.

c. The y-intercept is the y-coordinate of the point where the graph crosses the y-axis. The graph crosses the y-axis at the point $(0, -3)$, so the y-intercept is -3.

d. The x-intercepts are the x-coordinates of points where the graph crosses or touches the x-axis. The graph crosses the x-axis at the points $(-1, 0)$ and $(3, 0)$, so the x-intercepts are -1 and 3.

e. The domain of every quadratic function is all real numbers, or $(-\infty, \infty)$. To determine the range, first note that the graph has a minimum value because it opens up. The minimum value is the y-coordinate of the vertex, $k = -4$. So, the range is all real numbers greater than or equal to -4, or $[-4, \infty)$ in interval notation.

You Try It **Work through this You Try It problem.**

Work Exercises 5–6 in this textbook or in the *MyMathLab*® Study Plan.

OBJECTIVE 2: GRAPH QUADRATIC FUNCTIONS BY USING TRANSLATIONS

In mathematics, a **translation** is when every point on a graph is shifted the same distance in the same direction. We now examine translations of parabolas involving vertical or horizontal shifts.

In Section 2.3, we graphed simple functions by plotting points. We can use the same technique to graph the basic quadratic function $f(x) = x^2$. Table 2 gives some ordered pairs, and Figure 3 shows the resulting graph.

x	$y = x^2$	(x, y)
-3	$(-3)^2 = 9$	$(-3, 9)$
-2	$(-2)^2 = 4$	$(-2, 4)$
-1	$(-1)^2 = 1$	$(-1, 1)$
0	$(0)^2 = 0$	$(0, 0)$
1	$(1)^2 = 1$	$(1, 1)$
2	$(2)^2 = 4$	$(2, 4)$
3	$(3)^2 = 9$	$(3, 9)$

Table 2

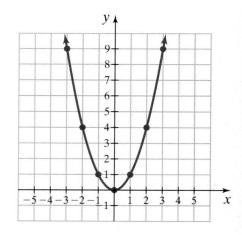

Figure 3
Graph of $f(x) = x^2$

Notice that the graph is "u-shaped" and opens up. The point $(0, 0)$, the lowest point on the graph, is the vertex of this parabola. The graph is symmetric about the vertical line $x = 0$. The domain is $(-\infty, \infty)$. From the graph we see that y can be any real number greater than or equal to 0, so the range is $[0, \infty)$. Similarly, we can use point plotting to graph other quadratic functions.

Example 3 Graphing a Quadratic Function with a Vertical Shift

Sketch the graph of $g(x) = x^2 + 2$. Compare the graph to the graph of $f(x) = x^2$.

Solution Table 3 shows that for every value of x, the y-coordinate of the function g is always 2 greater than the y-coordinate for f. The two functions are graphed in Figure 4. The graph of $g(x) = x^2 + 2$ is exactly the same as the graph of $f(x) = x^2$, except the graph of g is shifted *up* two units.

The point $(0, 2)$, now the lowest point on the graph, is the vertex of this parabola. The graph of g is symmetric about the vertical line $x = 0$. The domain is $(-\infty, \infty)$. From the graph, we see that y can be any real number greater than or equal to 2, so the range is $[2, \infty)$. Notice how the vertical shift affected the y-coordinate of the vertex and the range, but did not affect the x-coordinate of the vertex.

x	$f(x) = x^2$	$g(x) = x^2 + 2$
-2	$(-2)^2 = 4$	$(-2)^2 + 2 = 4 + 2 = 6$
-1	$(-1)^2 = 1$	$(-1)^2 + 2 = 1 + 2 = 3$
0	$(0)^2 = 0$	$(0)^2 + 2 = 0 + 2 = 2$
1	$(1)^2 = 1$	$(1)^2 + 2 = 1 + 2 = 3$
2	$(2)^2 = 4$	$(2)^2 + 2 = 4 + 2 = 6$

Table 3

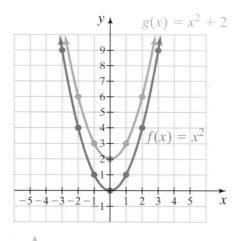

Figure 4
Graph of $g(x) = x^2 + 2$

You Try It Work through this You Try It problem.

Work Exercises 7–8 in this textbook or in the *MyMathLab*® Study Plan.

We see from Example 3 that if $k > 0$, the graph of $y = x^2 + k$ is the graph of $y = x^2$ shifted *up* k units. It follows that for $k < 0$, the graph of $y = x^2 + k$ is the graph of $y = x^2$ shifted *down* k units.

Vertical Shifts of Quadratic Functions

The graph of $y = x^2 + k$ is a parabola with the same shape as the graph of $y = x^2$, but it is shifted vertically $|k|$ units. The graph is shifted up if $k > 0$ and down if $k < 0$. The vertex of the parabola is $(0, k)$.

Example 4 Graphing a Quadratic Function with a Horizontal Shift

 Sketch the graph of $g(x) = (x - 2)^2$. Compare this graph to the graph of $f(x) = x^2$.

Solution Construct a table of values to find points on the graph. Use the points to sketch the graph. Click here for a sample table of values and to check your graph. Watch this video for a detailed solution. Confirm that the graph of $g(x) = (x - 2)^2$ is the same as the graph of $f(x) = x^2$ but shifted two units to the right.

You Try It Work through this You Try It problem.

Work Exercises 9–10 in this textbook or in the *MyMathLab*® Study Plan.

We see from Example 4 that if $h > 0$, the graph of $y = (x - h)^2$ is the graph of $y = x^2$ shifted *right* h units. It follows that for $h < 0$, the graph of $y = (x - h)^2$ is the graph of $y = x^2$ shifted *left* h units.

Horizontal Shifts of Quadratic Functions

The graph of $y = (x - h)^2$ is a parabola with the same shape as the graph of $y = x^2$, but it is shifted horizontally $|h|$ units. The graph is shifted right if $h > 0$ and left if $h < 0$. The vertex of the parabola is $(h, 0)$.

Example 5 Graphing a Quadratic Function with a Horizontal and Vertical Shift

Sketch the graph of $g(x) = (x + 1)^2 + 2$. Compare this graph to the graph of $f(x) = x^2$.

Solution Writing $g(x) = (x + 1)^2 + 2 = (x - (-1))^2 + 2$ we see that $h = -1$ and $k = 2$. Based on the results from Examples 3 and 4, the graph of $g(x) = (x + 1)^2 + 2$ will be the same as the graph of $f(x) = x^2$ but shifted left one unit (because $h = -1$) and up two units (because $k = 2$). See Figure 5. Click here for a sample table of values.

The point $(-1, 2)$, now the lowest point on the graph, is the vertex of this parabola. The graph of g is symmetric about the vertical line $x = -1$. The domain is $(-\infty, \infty)$. From the graph, we see that y can be any real number greater than or equal to 2, so the range is $[2, \infty)$.

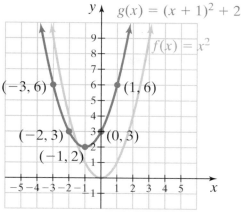

Figure 5

Graph of $g(x) = (x + 1)^2 + 2$

You Try It   Work through this You Try It problem.

Work Exercises 11–12 in this textbook or in the *MyMathLab*® Study Plan.

OBJECTIVE 3:   GRAPH QUADRATIC FUNCTIONS OF THE FORM $f(x) = a(x - h)^2 + k$

The quadratic function $f(x) = \dfrac{1}{2}(x - 3)^2 - 2$ has vertex $(3, -2)$ and is written in *standard form*.

Standard Form of a Quadratic Function

A quadratic function is in **standard form** if it is written as $f(x) = a(x - h)^2 + k$. The graph is a parabola with vertex (h, k). The parabola "opens up" if $a > 0$ or "opens down" if $a < 0$.

Since we can easily determine the coordinates of the vertex (h, k) when written this way, standard form is sometimes called **vertex form**.

Example 6 Graphing a Quadratic Function in the Form $f(x) = a(x - h)^2 + k$

Given the quadratic function $f(x) = \dfrac{1}{2}(x + 3)^2 - 2$, answer the following:

a. What are the coordinates of the vertex?

b. Does the graph "open up" or "open down"?

c. What is the equation of the axis of symmetry?

d. Find any x-intercepts.

e. Find the y-intercept.

f. Sketch the graph.

g. State the domain and range in interval notation.

Solution

a. First we change $x + 3$ to $x - (-3)$ to write the function in standard form. Then we find the vertex by determining the values for h and k.

$$f(x) = \frac{1}{2}(x + 3)^2 - 2 \quad \rightarrow \quad f(x) = \frac{1}{2}\overbrace{(x - (-3))}^{x + 3}{}^2 + (-2)$$

$$\boxed{a = \frac{1}{2}} \qquad \boxed{h = -3} \qquad \boxed{k = -2}$$

$$f(x) = a \quad (x - h)^2 + \quad k$$

Because $h = -3$ and $k = -2$, the vertex for this parabola is $(-3, -2)$.

b. Since the leading coefficient is $a = \frac{1}{2} > 0$, the parabola opens up.

c. The x-coordinate of the vertex is -3, so the equation of the axis of symmetry is $x = -3$.

d. We determine the x-intercepts (if any) by finding the real solutions to the equation $f(x) = 0$.

Write the original function: $\qquad\qquad f(x) = \frac{1}{2}(x + 3)^2 - 2$

Set $f(x)$ equal to 0: $\qquad \frac{1}{2}(x + 3)^2 - 2 = 0$

Add 2 to both sides: $\qquad \frac{1}{2}(x + 3)^2 = 2$

Multiply both sides by 2: $\qquad (x + 3)^2 = 4$

Square root property: $\qquad x + 3 = \pm 2$

Subtract 3 from both sides: $\qquad x = -3 \pm 2$

The x-intercepts are $-3 - 2 = -5$ and $-3 + 2 = -1$.

e. The y-intercept is found by evaluating $f(0)$.

$$f(0) = \frac{1}{2}(0 + 3)^2 - 2$$

$$= \frac{1}{2}(3)^2 - 2 = \frac{9}{2} - 2 = \frac{5}{2}$$

The y-intercept is $\frac{5}{2}$.

f. Before sketching a graph of the function, let's summarize what we know: The vertex is $(h, k) = (-3, -2)$; the graph opens up; the equation of the axis of symmetry is $x = -3$; there are two x-intercepts, -5 and -1; the y-intercept is $\frac{5}{2}$. From the vertex and intercepts, we can plot the points $(-3, -2)$, $(-5, 0)$, $(-1, 0)$, and $\left(0, \frac{5}{2}\right)$. The y-intercept lies three units to the right of the axis of symmetry. Because of the symmetry of the graph, we can move three units to left of the axis of symmetry to obtain another point on the graph: $\left(-6, \frac{5}{2}\right)$. Using this information, we can sketch the parabola, as shown in Figure 6.

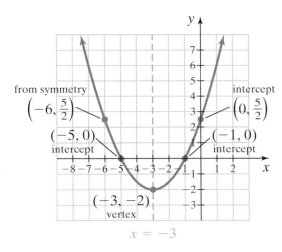

Figure 6

Graph of $f(x) = \dfrac{1}{2}(x + 3)^2 - 2$

g. The domain of every quadratic function is all real numbers $(-\infty, \infty)$. Because the parabola opens up, the graph has a minimum value at the vertex. The minimum value is the y-coordinate of the vertex, $k = -2$. So, the range is all real numbers greater than or equal to -2, or $[-2, \infty)$ in interval notation.

Example 7 Graphing a Quadratic Function in the Form $f(x) = a(x - h)^2 + k$

My video summary

 Given that the quadratic function $f(x) = -(x - 2)^2 - 4$ is in standard form, answer the following:

a. What are the coordinates of the vertex?

b. Does the graph "open up" or "open down"?

c. What is the equation of the axis of symmetry?

d. Find any x-intercepts.

e. Find the y-intercept.

f. Sketch the graph.

g. State the domain and range in interval notation.

Solution Try to answer the questions on your own. Click here to check your answers, or watch this video to see the complete solution.

You Try It   **Work through this You Try It problem.**

Work Exercises 13–18 in this textbook or in the *MyMathLab*® Study Plan.

OBJECTIVE 4:   FIND THE VERTEX OF A QUADRATIC FUNCTION BY COMPLETING THE SQUARE

In Section 8.1 we saw how to use completing the square to solve a quadratic equation. The process involved adding the same quantity to both sides of the equation. In this section, we use completing the square to write a quadratic function in standard form. Since we are dealing with a function instead of an equation, we take a slightly different approach. Rather than adding the same quantity to both sides, we will *add and subtract* the same quantity, which is the same as adding 0 to the function.

Writing $f(x) = ax^2 + bx + c$ in Standard Form by Completing the Square

Step 1. Group the variable terms together within parentheses.

Step 2. If $a \neq 1$, factor a out of the variable terms.

Step 3. Take half the coefficient of the x-term inside the parentheses, square it, and *add* it *inside* the parentheses. Multiply this value by a, then *subtract* from c.

Step 4. The expression inside the parentheses is now a perfect square. Rewrite it as a binomial squared and simplify the constant term outside of the parentheses.

Example 8 Writing a Quadratic Function in Standard Form

Write the function $f(x) = 2x^2 - 8x + 7$ in standard form and find the vertex.

Solution Watch this video, or use the following steps to write the function in standard form by completing the square.

Step 1. $f(x) = (2x^2 - 8x) + 7$

Step 2. $f(x) = 2(x^2 - 4x) + 7$

Factor out a, the coefficient of x^2, from the variable terms.

Step 3. $\left(\dfrac{1}{2}(-4)\right)^2 = (-2)^2 = 4$

Coefficient of x inside parentheses

Add inside parentheses

We are really adding $2(4) = 8$ to the function.

So we must subtract 8 as well.

$f(x) = 2(x^2 - 4x + 4) + 7 - 8$

 CAUTION Notice that although we added 4 inside the parentheses, we had to subtract 8 on the outside. Don't forget that a distributes to each term inside the parentheses so we must take it into account when determining what to subtract.

Step 4. $f(x) = 2(\underbrace{x^2 - 4x + 4}_{\text{Perfect square trinomial}}) + \underbrace{7 - 8}_{\text{Simplify}} \Rightarrow f(x) = 2(x - 2)^2 - 1$

Now we compare the function to the standard form $f(x) = a(x - h)^2 + k$, to determine the vertex. The vertex is $(h, k) = (2, -1)$.

You Try It Work through this You Try It problem.

Work Exercises 19–22 in this textbook or in the *MyMathLab*® Study Plan.

Example 9 Writing a Quadratic Function in Standard Form

Write the function $f(x) = -3x^2 - 24x$ in standard form and find the vertex.

Solution The constant term is $c = 0$ so we can write $f(x) = -3x^2 - 24x + 0$. Complete the square to write the function in standard form and determine the vertex. Click here to check your answer, or watch this video for the complete solution.

You Try It   Work through this You Try It problem.

Work Exercise 23 in this textbook or in the *MyMathLab*® Study Plan.

OBJECTIVE 5:   GRAPH QUADRATIC FUNCTIONS OF THE FORM
$f(x) = ax^2 + bx + c$ BY COMPLETING THE SQUARE

Once we have the function in standard form, we can sketch its graph.

Example 10   Graphing a Quadratic Function in the Form $f(x) = ax^2 + bx + c$

My video summary

⊞ Rewrite the quadratic function $f(x) = 2x^2 - 4x - 3$ in standard form, and then answer the following:

a. What are the coordinates of the vertex?

b. Does the graph "open up" or "open down"?

c. What is the equation of the axis of symmetry?

d. Find any x-intercepts.

e. Find the y-intercept.

f. Sketch the graph.

g. State the domain and range in interval notation.

Solution   Write the function in standard form by completing the square then try answering the questions as in Example 6. Click here to check your answers, or watch this video to see each step worked out in detail.

You Try It   Work through this You Try It problem.

Work Exercises 24–29 in this textbook or in the *MyMathLab*® Study Plan.

OBJECTIVE 6:   FIND THE VERTEX OF A QUADRATIC FUNCTION BY USING THE VERTEX FORMULA

My video summary

⊞ Just as we saw how the quadratic formula comes from solving a general quadratic equation by completing the square, we can use completing the square to establish a formula for the vertex. Work through the video to verify that the function $f(x) = ax^2 + bx + c$ is equivalent to $f(x) = a\left(x + \dfrac{b}{2a}\right)^2 + c - \dfrac{b^2}{4a}$, then compare this to the standard form $f(x) = a(x - h)^2 + k$.

$$f(x) = a\left(x - \left(-\frac{b}{2a}\right)\right)^2 + \underbrace{c - \frac{b^2}{4a}}$$

$$\downarrow \qquad\qquad \downarrow$$

$$f(x) = a(x - \quad h)^2 \quad + \quad k$$

We can see that the coordinates of the vertex must be $\left(-\dfrac{b}{2a}, c - \dfrac{b^2}{4a}\right)$.

It is not really necessary to memorize the formulas for both coordinates. Since we can find the value of the y-coordinate by evaluating the function at the x-coordinate, it is common to write the vertex as $\left(-\dfrac{b}{2a}, f\left(-\dfrac{b}{2a}\right)\right)$.

Formula for the Vertex of a Parabola

Given a quadratic function of the form $f(x) = ax^2 + bx + c$, $a \neq 0$, the vertex of the parabola is given by

$$(h, k) = \left(-\frac{b}{2a}, f\left(-\frac{b}{2a}\right)\right).$$

The axis of symmetry is the vertical line $x = -\dfrac{b}{2a}$.

Example 11 Using the Vertex Formula

My video summary

Use the vertex formula to find the vertex for each quadratic function.

a. $f(x) = 3x^2 - 12x - 4$ **b.** $f(x) = -\dfrac{1}{2}x^2 - 10x + 5$

Solution

a. We start by identifying the coefficients of the function: $a = 3$, $b = -12$, and $c = -4$

Using the vertex formula, the x-coordinate is $h = -\dfrac{b}{2a} = -\dfrac{(-12)}{2(3)} = \dfrac{12}{6} = 2$.

We find the y-coordinate by evaluating $k = f(h) = f(2)$.

$$f(2) = 3(2)^2 - 12(2) - 4$$
$$= 12 - 24 - 4 = -16$$

The vertex is $(h, k) = (2, -16)$.

b. Determine the vertex on your own. Click here to check your answer, or watch this video to see the complete solution.

You Try It Work through this You Try It problem.

Work Exercises 30–33 in this textbook or in the **MyMathLab**® Study Plan.

OBJECTIVE 7: GRAPH QUADRATIC FUNCTIONS OF THE FORM
$f(x) = ax^2 + bx + c$ BY USING THE VERTEX FORMULA

My video summary

Example 12 Graphing a Quadratic Function Using the Vertex Formula

Given the quadratic function $f(x) = -2x^2 - 4x + 5$, answer the following:

a. What are the coordinates of the vertex?

b. Does the graph "open up" or "open down"?

c. What is the equation of the axis of symmetry?

d. Find any x-intercepts.

e. Find the y-intercept.

f. Sketch the graph.

g. State the domain and range in interval notation.

Solution Find the vertex by using the vertex formula. Then try to answer the remaining questions as in Example 6. Click here to check your answers, or watch this video to see the complete solution.

You Try It   Work through this You Try It problem.

Work Exercises 34–39 in this textbook or in the MyMathLab® Study Plan.

8.2 Exercises

In Exercises 1–4, without graphing, determine if the graph of each quadratic function opens up or down. Also determine if the graph will be wider or narrower than the graph of $y = x^2$.

1. $f(x) = \dfrac{1}{4}x^2 - 3$

2. $f(x) = -2x^2 + 4x + 1$

3. $f(x) = -4 + 5x^2 - 2x$

4. $f(x) = -\dfrac{2}{3}x^2 + x + 7$

In Exercises 5–6, use the given graph of a quadratic function to find the following:

   a. Vertex   **b.** Axis of symmetry   **c.** y-intercept

   d. x-intercept(s)   **e.** Domain and range

5.

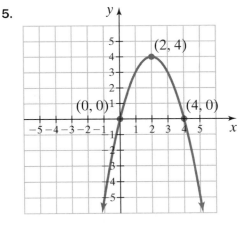

6.

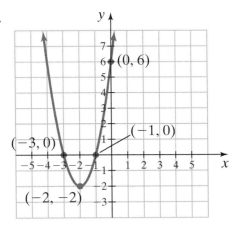

In Exercises 7–12, sketch the graph of each quadratic function by using translations (vertical or horizontal shifts). Compare each graph to the graph of $y = x^2$.

7. $f(x) = x^2 - 4$

8. $f(x) = x^2 + \dfrac{3}{2}$

9. $f(x) = (x + 3)^2$

10. $f(x) = (x - 4)^2$

11. $f(x) = (x - 1)^2 - 2$

12. $f(x) = (x + 2)^2 - 3$

In Exercises 13–18, use the given quadratic function to answer the following:

  a. What are the coordinates of the vertex?
  b. Does the graph "open up" or "open down"?
  c. What is the equation of the axis of symmetry?
  d. Find any x-intercepts.
  e. Find the y-intercept.
  f. Sketch the graph.
  g. State the domain and range in interval notation.

You Try It

13. $f(x) = (x - 2)^2 - 4$ **14.** $f(x) = -(x + 1)^2 - 9$

15. $f(x) = -2(x - 3)^2 + 2$ **16.** $f(x) = \frac{1}{2}(x + 2)^2 + 2$

17. $f(x) = -\frac{1}{4}(x - 4)^2 + 2$ **18.** $f(x) = 3\left(x + \frac{1}{3}\right)^2 - 4$

In Exercises 19–23, write the function in standard form and find the vertex.

You Try It   **19.** $f(x) = x^2 + 8x + 9$ **20.** $f(x) = 3x^2 - 6x + 2$

21. $f(x) = -2x^2 - 12x - 10$ **22.** $f(x) = \frac{1}{2}x^2 + 6x + 1$
You Try It

23. $f(x) = x^2 + 10x$

In Exercises 24–29, rewrite the quadratic function in standard form and then answer the following:

  a. Determine the vertex by completing the square.
  b. Does the graph "open up" or "open down"?
  c. What is the equation of the axis of symmetry?
  d. Find any x-intercepts.
  e. Find the y-intercept.
  f. Sketch the graph.
  g. State the domain and range in interval notation.

You Try It

24. $f(x) = x^2 + 6x - 7$ **25.** $f(x) = -x^2 - 3x + 4$

26. $f(x) = 4x^2 - 7x + 8$ **27.** $f(x) = -3x^2 - 6x + 1$

28. $f(x) = \frac{1}{4}x^2 - 2x + 1$ **29.** $f(x) = 2x^2 - 16x$

You Try It   In Exercises 30–33, use the vertex formula to find the vertex of the quadratic function.

30. $f(x) = x^2 + 5x - 3$ **31.** $f(x) = -x^2 + 2x - 5$

32. $f(x) = -4x^2 + 12x$ **33.** $f(x) = \frac{1}{2}x^2 + 7x + 4$

In Exercises 34–39, use the quadratic function to answer the following:

    a. Determine the vertex by using the vertex formula.   **b.** Does the graph "open up" or "open down"?

    c. What is the equation of the axis of symmetry?   **d.** Find any x-intercepts.

    e. Find the y-intercept.   **f.** Sketch the graph.

    g. State the domain and range in interval notation.

You Try It

 34. $f(x) = x^2 - 4x - 60$ **35.** $f(x) = 3x^2 + 6x - 4$ **36.** $f(x) = -x^2 + 2x - 6$

37. $f(x) = \dfrac{1}{2}x^2 + 6x + 1$ **38.** $f(x) = -3x^2 + 7x + 5$ **39.** $f(x) = x^2 - 8x$

8.3 Applications and Modeling of Quadratic Functions

THINGS TO KNOW

Before working through this section, be sure you are familiar with the following concepts:

                                                VIDEO   ANIMATION   INTERACTIVE

You Try It
  1. Use Linear Equations to Solve Application Problems (Section 1.1, **Objective 4**)

You Try It
  2. Use Formulas to Solve Application Problems (Section 1.4, **Objective 2**)

You Try It
  3. Use Linear Models to Solve Application Problems; Direct Variation (Section 2.5, **Objective 6**)

You Try It
  4. Use Polynomial Equations and Models to Solve Application Problems (Section 5.4, **Objective 3**)

You Try It
  5. Solve Quadratic Equations Using Factoring (Section 8.1, **Objective 1**)

You Try It
  6. Solve Quadratic Equations Using the Quadratic Formula (Section 8.1, **Objective 4**)

You Try It
  7. Graph Quadratic Functions of the Form $f(x) = ax^2 + bx + c$ Using the Vertex Formula (Section 8.2, **Objective 7**)

OBJECTIVES

1 Solve Applications Involving Unknown Numbers

2 Solve Applications Involving Projectile Motion

3 Solve Applications Involving Geometric Formulas

4 Solve Applications Involving Distance, Rate, and Time

5 Solve Applications Involving Work

6 Maximize Quadratic Functions to Solve Application Problems

7 Minimize Quadratic Functions to Solve Application Problems

OBJECTIVE 1: SOLVE APPLICATIONS INVOLVING UNKNOWN NUMBERS

In Section 1.1, we learned how to solve application problems involving linear equations. In this section, we follow the same six-step process to solve application problems involving quadratic equations.

Problem-Solving Strategy for Applications

Step 1. Define the Problem.   Read the problem carefully, or multiple times if necessary. Identify what you are trying to find and determine what information is available to help you find it.

Step 2. Assign Variables.   Choose a variable to assign to an unknown quantity in the problem. If other unknown quantities exist, express them in terms of the selected variable.

Step 3. Translate into an Equation.   Use the relationships among the known and unknown quantities to form an equation.

Step 4. Solve the Equation.   Determine the value of the variable and use the result to find any other unknown quantities in the problem.

Step 5. Check the Reasonableness of Your Answer.   Check to see if your answer makes sense within the context of the problem. If not, check your work for errors and try again.

Step 6. Answer the Question.   Write a clear statement that answers the question(s) posed.

Example 1 Find Two Numbers

The product of a number and 1 more than twice the number is 36.
Find the two numbers.

Solution

Step 1. We are looking for two numbers. We know that the product is 36 and the second number is 1 more than twice the first.

Step 2. Let x = the first number. Then $2x + 1$ = the second number.

Step 3. Because the product of the two numbers is 36, we get the equation $x(2x + 1) = 36$, which simplifies to the quadratic equation $2x^2 + x - 36 = 0$.

Step 4. Solve:

$$\text{Original equation:} \quad 2x^2 + x - 36 = 0$$

$$\text{Factor:} \quad (2x + 9)(x - 4) = 0$$

Apply the zero product property and solve for x:

$$2x + 9 = 0 \quad \text{or} \quad x - 4 = 0$$

$$2x = -9 \qquad\qquad x = 4$$

$$x = -\frac{9}{2}$$

Step 5. If $x = -\dfrac{9}{2}$, then the other number is $2\left(-\dfrac{9}{2}\right) + 1 = -8$. Since

$\left(-\dfrac{9}{2}\right)(-8) = 36$, this result is reasonable.

If $x = 4$, then the other number is $2(4) + 1 = 9$. Since $(4)(9) = 36$, this result is also reasonable.

Step 6. The two numbers are $-\dfrac{9}{2}$ and -8, or 4 and 9.

You Try It   **Work through this You Try It problem.**

Work Exercises 1–2 in this textbook or in the *MyMathLab*® Study Plan.

Example 2 Finding Consecutive Even Integers

Three consecutive *positive* even integers are such that the square of the third is 20 less than the sum of the squares of the first two. Find the positive integers.

Solution

Step 1. We are looking for three consecutive positive even integers. We know that the square of the third is 20 less than the sum of the squares of the first two.

Step 2. Let $x =$ the first positive even integer. Then $x + 2 =$ the second, and $x + 4 =$ the third.

Step 3. Because the square of the third is 20 less than the sum of the squares of the first two, we get the equation $\underbrace{(x + 4)^2}_{\substack{\text{Square of} \\ \text{the third}}} = \underbrace{x^2}_{\substack{\text{Square of} \\ \text{the first}}} + \underbrace{(x + 2)^2}_{\substack{\text{Square of} \\ \text{the second}}} \underbrace{- 20}_{\substack{\text{20 less} \\ \text{than the} \\ \text{sum}}}.$

Step 4. Solve:

$$
\begin{aligned}
\text{Original equation:} && (x + 4)^2 &= x^2 + (x + 2)^2 - 20 \\
\text{Expand:} && x^2 + 8x + 16 &= x^2 + x^2 + 4x + 4 - 20 \\
\text{Combine like terms:} && x^2 + 8x + 16 &= 2x^2 + 4x - 16 \\
\text{Write in standard form:} && 0 &= x^2 - 4x - 32 \\
\text{Factor:} && 0 &= (x - 8)(x + 4)
\end{aligned}
$$

Apply the zero product property and solve for x:

$$
\begin{aligned}
x - 8 &= 0 \quad \text{or} \quad x + 4 = 0 \\
x &= 8 \qquad\qquad\;\; x = -4
\end{aligned}
$$

Step 5. Since the consecutive even integers must be positive, we discard the negative solution. The only feasible solution remaining is $x = 8$ as the first positive even integer which would give 10 and 12 as the next two consecutive even integers.

Since $8^2 + 10^2 - 20 = 12^2$, this result is reasonable.

Step 6. The three consecutive positive even integers are 8, 10, and 12.

You Try It   Work through this You Try It problem.

Work Exercises 3–4 in this textbook or in the *MyMathLab*® Study Plan.

OBJECTIVE 2:   SOLVE APPLICATIONS INVOLVING PROJECTILE MOTION

An object launched, thrown, or shot vertically into the air with an initial velocity of v_0 meters per second (m/s) from an initial height of h_0 meters above the ground can be modeled by the equation $h = -4.9t^2 + v_0 t + h_0$. The variable h is the height above the ground (in meters) of the object (also known as a *projectile*) t seconds after its departure.

Example 3 Launch a Toy Rocket

A toy rocket is launched at an initial velocity of 14.7 m/s from a platform that sits 49 meters above the ground. The height h of the rocket above the ground at any time t seconds after launch is given by the equation $h = -4.9t^2 + 14.7t + 49$. When will the rocket hit the ground?

Solution

Step 1. We want to find the time when $h = 0$ (the rocket hits the ground) using the equation $h = -4.9t^2 + 14.7t + 49$.

Step 2. We know that h = height in meters and t = time in seconds.

Step 3. Using the equation $h = -4.9t^2 + 14.7t + 49$, set $h = 0$ and solve for t.

Step 4. Solve:

$$\text{Original equation:}\quad 0 = -4.9t^2 + 14.7t + 49$$
$$\text{Divide both sides by } -4.9:\quad 0 = t^2 - 3t - 10$$
$$\text{Factor:}\quad 0 = (t - 5)(t + 2)$$

Apply the zero product property and solve for t:

$$t - 5 = 0 \quad \text{or} \quad t + 2 = 0$$
$$t = 5 \qquad\qquad t = -2$$

Step 5. Because t represents the time (in seconds) after launch, its value cannot be negative. So, $t = -2$ seconds does not make sense. The only reasonable solution is $t = 5$ seconds.

Step 6. The rocket will hit the ground 5 seconds after launch.

You Try It   Work through this You Try It problem.

Work Exercises 5–6 in this textbook or in the *MyMathLab*® Study Plan.

OBJECTIVE 3:   SOLVE APPLICATIONS INVOLVING GEOMETRIC FORMULAS

Now let's explore application problems that involve geometric formulas and quadratic equations.

My interactive video summary

Example 4 Dimensions of a Rectangle

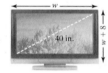 The length of a rectangle is 6 inches less than four times the width. Find the dimensions of the rectangle if the area of the rectangle is 54 square inches.

Solution Work through this interactive video to check that the dimensions of the rectangle are 12 inches by 4.5 inches.

You Try It Work through this You Try It problem.

Work Exercises 7–8 in this textbook or in the *MyMathLab*® Study Plan.

Example 5 Width of a High-Definition Television

Shayna bought a new 40-inch high-definition television. If the length of Shayna's television is 8 inches longer than the width, find the width of the television.

Solution

Step 1. We want to determine the width of the television. We know that the length is 8 inches more than the width, and we assume that the television is rectangular. The size of a television is the length of its *diagonal*, which in this case is 40 inches.

Step 2. Let w = width of the television. Then $w + 8$ = length of the television.

Step 3. We can create a quadratic equation using the Pythagorean theorem, $a^2 + b^2 = c^2$.

$$w^2 + (w + 8)^2 = 40^2$$

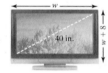

Step 4. Solve:

Original equation:	$w^2 + (w + 8)^2 = 40^2$
Square the binomial:	$w^2 + w^2 + 16w + 64 = 1600$
Combine like terms:	$2w^2 + 16w - 1536 = 0$
Divide both sides by 2:	$w^2 + 8w - 768 = 0$
Factor:	$(w - 24)(w + 32) = 0$

Apply the zero product property and solve for w:

$$w - 24 = 0 \quad \text{or} \quad w + 32 = 0$$
$$w = 24 \qquad\qquad w = -32$$

Step 5. Because w represents the width of the television, its value cannot be negative. Thus, $w = -32$ does not make sense. The only reasonable solution is $w = 24$ inches.

Step 6. The width of the television is 24 inches.

You Try It Work through this You Try It problem.

Work Exercises 9–10 in this textbook or in the *MyMathLab*® Study Plan.

OBJECTIVE 4:   SOLVE APPLICATIONS INVOLVING DISTANCE, RATE, AND TIME

In the next example, we use quadratic equations to solve an application problem involving distance, rate, and time.

Example 6 Speed of an Airplane

My video summary

 Kevin flew his new Cessna O-2A airplane from Jonesburg to Mountainview, a distance of 2560 miles. The average speed for the return trip was 64 mph faster than the average outbound speed. If the total flying time for the round trip was 18 hours, what was the plane's average speed on the outbound trip from Jonesburg to Mountainview?

Solution

Step 1. We are asked to find the average outbound speed of the plane from Jonesburg to Mountainview. We know that the distance traveled in each direction is 2560 miles and the total time of the trip is 18 hours. We also know that the speed on the return trip is 64 mph more than on the outbound trip. Because the problem involves distance, rate, and time, we will need the distance formula $d = r \cdot t$.

Step 2. Let r = speed of plane on the outbound trip. Then $r + 64$ = speed of plane on the return trip from Mountainview to Jonesburg.

Step 3. Since the total time of the trip was 18 hours, we write

$$\text{time}_{\text{outbound}} + \text{time}_{\text{return}} = \text{time}_{\text{total}}$$
$$\text{time}_{\text{outbound}} + \text{time}_{\text{return}} = 18$$

Because distance = rate · time, we also know that time = $\dfrac{\text{distance}}{\text{rate}}$.

Using the given distance and rate expressions, we get the following equation:

$$\frac{2560}{r} + \frac{2560}{r + 64} = 18$$

Solve this equation on your own. Be sure to check the reasonableness of your answer. Click here to check your answer, or watch this video for the complete solution.

You Try It   Work through this You Try It problem.

Work Exercises 11–12 in this textbook or in the *MyMathLab*® Study Plan.

OBJECTIVE 5:   SOLVE APPLICATIONS INVOLVING WORK

We can also use quadratic equations to solve application problems involving work.

Example 7 Monthly Sales Reports

My video summary

 Dawn can finish the monthly sales reports in 2 hours less time than it takes Adam. Working together, they were able to finish the sales reports in 8 hours. How long does it take each person to finish the monthly sales reports alone? (Round to the nearest minute.)

Solution

Step 1. We must find the time it takes for each person to finish the sales reports alone. We know that it takes Dawn 2 hours less than Adam to do the job alone, and it takes them 8 hours to do the job together. Because this problem involves combining rates of work, we use the equation

$$\frac{1}{t_1} + \frac{1}{t_2} = \frac{1}{t}.$$

Step 2. Let t_1 = the time it takes Dawn to complete the reports alone (in hours). Then $t_2 = t_1 + 2$ = the time it takes Adam to complete the reports alone. We know that it takes $t = 8$ hours to complete the job when they work together.

Step 3. Substituting the expressions for the individual times and 8 for the total time, we get the following equation:

$$\frac{1}{t_1} + \frac{1}{t_1 + 2} = \frac{1}{8}$$

Solve this equation for t_1. Confirm that it will take Dawn approximately 15 hours and 4 minutes to complete the reports on her own and it will take Adam 17 hours and 4 minutes to complete the same job alone.

Be sure to check that the sum of their rates is equal to $\frac{1}{8}$ (within rounding error). To see this problem worked out in detail, watch this video.

You Try It   **Work through this You Try It problem.**

Work Exercises 13–14 in this textbook or in the *MyMathLab*®️ Study Plan.

OBJECTIVE 6: MAXIMIZE QUADRATIC FUNCTIONS TO SOLVE APPLICATION PROBLEMS

Sometimes with application problems involving functions, we often need to find the *maximum* or *minimum* value of the function. For example, a builder with a fixed amount of fencing may wish to maximize the area enclosed, or an economist may want to minimize a cost function or maximize a profit function. Quadratic functions are relatively easy to maximize or minimize because we know a formula for finding the coordinates of the vertex. Recall that if $f(x) = ax^2 + bx + c$, $a \neq 0$, we know that the coordinates of the vertex are $\left(-\frac{b}{2a}, f\left(-\frac{b}{2a} \right) \right)$.

In addition, we know that if $a > 0$, the parabola opens *up* and the function has a **minimum** value at the vertex. If $a < 0$, the parabola opens *down* and the function has a **maximum** value at the vertex. See Figure 7.

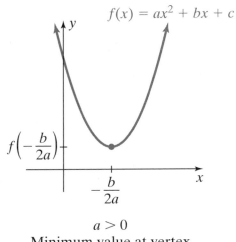

$$f(x) = ax^2 + bx + c$$

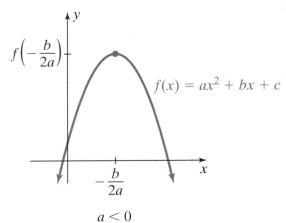

$a > 0$
Minimum value at vertex

$a < 0$
Maximum value at vertex

Figure 7

Projectile Motion

Example 8 Launching a Toy Rocket

A toy rocket is launched with an initial velocity of 44.1 meters per second from a platform located 1 meter above the ground. The height h of the object above the ground at any time t seconds after launch is given by the function $h(t) = -4.9t^2 + 44.1t + 1$. How long after launch did it take the rocket to reach its maximum height? What is the rocket's maximum height?

Solution Because the function is quadratic with $a = -4.9 < 0$, we know that the graph is a parabola that opens down. So, the function has a maximum value at the vertex. The t-coordinate of the vertex is $t = -\dfrac{b}{2a} = -\dfrac{44.1}{2(-4.9)} = \dfrac{-44.1}{-9.8} = 4.5$ seconds. Therefore, the rocket reaches its maximum height at 4.5 seconds after launch. The rocket's maximum height is

$$h(4.5) = -4.9(4.5)^2 + 44.1(4.5) + 1 = 100.225 \text{ meters.}$$

You Try It Work through this You Try It problem.

Work Exercises 15–17 in this textbook or in the *MyMathLab*® Study Plan.

Maximize Revenue and Profit

Revenue is the dollar amount received from selling x items at a price of p dollars per item. **Total revenue** is found by multiplying the number of units sold by the price per unit, $R = xp$. For example, if a child sells 50 cups of lemonade at a price of \$0.25 per cup, then the revenue generated is

$$R = \underbrace{(50)}_{x}\underbrace{(0.25)}_{p} = \$12.50.$$

A **demand equation** relates the quantity sold to the price per item, and we can write this equation as quantity in terms of price, such as $x = 2p - 6$, or as price in terms of quantity, such as $p = 0.5x + 3$. This allows us to write the revenue function in terms of a single variable, either x or p, as shown below:

$$R = xp \text{ and } x = 2p - 6 \text{ (or equivalently, } p = 0.5x + 3)$$
$$R(x) = x(0.5x + 3) = 0.5x^2 + 3x \leftarrow \text{revenue in terms of } x$$

or

$$R(p) = (2p - 6)p = 2p^2 - 6p \leftarrow \text{revenue in terms of } p$$

Example 9 Maximizing Shoe Revenue

 Records can be kept on the price of shoes and the number of pairs sold in order to gather enough data to reasonably model shopping trends for a particular type of shoe. Suppose the marketing and research department of a shoe company determined the price of a certain basketball shoe obeys the demand equation

$$p = -\frac{1}{50}x + 110.$$

a. According to the demand equation, how much should the shoes sell for if 500 pairs of shoes are sold? 1200 pairs of shoes?

b. What is the revenue if 500 pairs of shoes are sold? 1200 pairs of shoes?

c. How many pairs of shoes should be sold in order to maximize revenue? What is the maximum revenue?

d. What price should be charged in order to maximize revenue?

Solution

a. If 500 pairs of shoes are sold, the price should be

$$p(500) = -\frac{1}{50}(500) + 110 = -10 + 110 = \$100.$$

If 1200 pairs of shoes are sold, the price should be

$$p(1200) = -\frac{1}{50}(1200) + 110 = -24 + 110 = \$86.$$

b. Because $R = xp$, we can substitute the demand equation, $p = -\frac{1}{50}x + 110$, for p to get R as a function of x.

$$R(x) = x\underbrace{\left(-\frac{1}{50}x + 110\right)}_{p} \quad \text{or} \quad R(x) = -\frac{1}{50}x^2 + 110x$$

The revenue generated by selling 500 pairs of shoes is

$$R(500) = 500\left(-\frac{1}{50}(500) + 110\right)$$
$$= 500(-10 + 110)$$
$$= 500(100)$$
$$= \$50,000.$$

The revenue from selling 1200 pairs of shoes is

$$R(1200) = 1200\left(-\frac{1}{50}(1200) + 110\right)$$
$$= 1200(-24 + 110)$$
$$= 1200(86)$$
$$= \$103{,}200.$$

c. $R(x) = -\frac{1}{50}x^2 + 110x$ is a quadratic function with $a = -\frac{1}{50} < 0$ and $b = 110$. Because $a < 0$, the function has a maximum value at the vertex. Therefore, the value of x that produces the maximum revenue is

$$x = -\frac{b}{2a} = -\frac{110}{2\left(-\frac{1}{50}\right)} = \frac{-110}{\underbrace{-\frac{1}{25}}} = \frac{-110}{1}\cdot\left(-\frac{25}{1}\right) = 2750 \text{ pairs of shoes.}$$

Change to
multiplication
by the reciprocal

The maximum revenue is

$$R(2750) = 2750\left(-\frac{1}{50}(2750) + 110\right)$$
$$= 2750(55)$$
$$= \$151{,}250.$$

d. Using the demand equation, the price that should be charged to maximize revenue when selling 2750 pairs of shoes is $p(2750) = -\frac{1}{50}(2750) + 110 = \$55$.

You Try It Work through this You Try It problem.

Work Exercises 18–19 in this textbook or in the **MyMathLab** Study Plan.

Example 10 Maximizing Profit

My video summary

To sell x waterproof CD alarm clocks, WaterTime, LLC, has determined that the price in dollars must be $p = 250 - 2x$, which is the demand equation. Each clock costs \$2 to produce, with fixed costs of \$4000 per month, producing the cost function $C(x) = 2x + 4000$.

a. Express the revenue R as a function of x.
b. Express the profit P as a function of x.
c. Find the value of x that maximizes profit. What is the maximum profit?
d. What is the price of the alarm clock that will maximize profit?

Solution

a. The equation for revenue is $R = xp$. So, we substitute the demand equation $p = 250 - 2x$ for p to obtain the function:

$$R(x) = x\underbrace{(250 - 2x)}_{p} \quad \text{or} \quad R(x) = -2x^2 + 250x$$

b. Profit is equal to revenue minus cost.

$$P(x) = R(x) - C(x)$$
$$= -2x^2 + 250x - (2x + 4000)$$
$$= -2x^2 + 250x - 2x - 4000$$
$$= -2x^2 + 248x - 4000$$

c.–d. Try to finish the remaining parts on your own. Find the vertex to answer the questions in part (c), and use the demand equation to answer part (d). Click here to check your answers, or watch this video for a complete solution.

You Try It   Work through this You Try It problem.

Work Exercises 20–23 in this textbook or in the *MyMathLab*® Study Plan.

Maximize Area

Example 11 Maximizing Area

My video summary

Suppose you are asked to build a fence around a rectangular field that borders a river. You have 3000 feet of fencing available, but no fencing is required along the river. Find the dimensions of the field that maximizes the enclosed area. What is the maximum area?

Figure 8

Solution From Figure 8, let x = width of the field and y = length of the field. Since the field is rectangular, we find the area by multiplying length by width:

$$A = xy$$

Ideally, we want the area to depend on only one variable. So, we need to write x in terms of y, or write y in terms of x. The 3000 feet of fencing will be used for two widths and one length, giving us the equation

$$2x + y = 3000.$$

Solving this equation for y gives $y = -2x + 3000$. We can substitute $-2x + 3000$ for y to get the area as a function of x alone.

$$A(x) = x\underbrace{(-2x + 3000)}_{y} \quad \text{or} \quad A(x) = -2x^2 + 3000x$$

Our area function is quadratic with $a = -2 < 0$. Therefore, the function has a maximum value at its vertex. Try to finish working this problem on your own. Click here to check your answers, or watch this video for a complete solution.

You Try It   Work through this You Try It problem.

Work Exercises 24–25 in this textbook or in the *MyMathLab*® Study Plan.

OBJECTIVE 7:   MINIMIZE QUADRATIC FUNCTIONS TO SOLVE APPLICATION PROBLEMS

Let's now consider some situations in which we minimize a function.

Example 12 Minimizing Costs

A fabric manufacturer has daily production costs of $C = 7000 - 100x + 0.5x^2$, where C is the total cost (in dollars) and x is the number of units produced. How many units should be produced each day in order to minimize costs? What is the minimum daily cost?

Solution Writing the cost function in standard form, we have

$$C(x) = 0.5x^2 - 100x + 7000.$$

This is a quadratic function with $a = 0.5 > 0$, $b = -100$, and $c = 7000$. Because $a > 0$, the graph of the function is a parabola that *opens up*, and the function has a minimum value at the vertex. The x-coordinate of the vertex tells us the number of units to produce in order to minimize costs, and the y-coordinate tells us the minimum cost. To find these coordinates, we use the vertex formula

$$\left(-\frac{b}{2a}, f\left(-\frac{b}{2a}\right)\right).$$

$$x = -\frac{(-100)}{2(0.5)} \quad \text{and} \quad C(100) = 0.5(100)^2 - 100(100) + 7000$$
$$= \frac{100}{1} \qquad\qquad\qquad = 0.5(10{,}000) - 100(100) + 7000$$
$$= 100 \qquad\qquad\qquad = 5000 - 10{,}000 + 7000$$
$$\qquad\qquad\qquad\qquad = 2000$$

The manufacturer should produce 100 units each day to minimize costs. The minimum daily cost will be $2000.

You Try It   Work through this You Try It problem.

Work Exercises 26–27 in this textbook or in the *MyMathLab*® Study Plan.

Example 13 Exchange Rates

Between January 1, 2007, and February 1, 2009, the exchange rate between Canadian dollars and U.S. dollars can be approximated by $D(x) = 0.0017x^2 - 0.0397x + 1.2194$, where D is the number of Canadian dollars for one U.S. dollar x months after January 1, 2007. For the given time period, determine the month in which the exchange rate was a minimum. What was the minimum exchange rate? Round your answer to three decimal places. (*Source*: U.S. Federal Reserve, St. Louis, MO)

Solution $D(x) = 0.0017x^2 - 0.0397x + 1.2194$ is a quadratic function with $a = 0.0017 > 0$, $b = -0.0397$, and $c = 1.2194$. Because $a > 0$, the graph of the function is a parabola that *opens up*, and the function has a minimum value at the vertex. So, the value of x that produces the minimum exchange rate is

$$x = -\frac{b}{2a} = -\frac{(-0.0397)}{2(0.0017)} \approx 11.676.$$

How do we interpret this value for x? Since x represents the number of months after January 1, 2007, $x = 1$ indicates February 1, 2007, $x = 2$ means March 1, 2007, and so on. So, $x = 11$ represents December 1, 2007, and $x = 12$ indicates January 1, 2008. Since $11 < 11.676 < 12$, the minimum exchange rate occurred in December 2007 and was approximately

$$D(11.676) = 0.0017(11.676)^2 - 0.0397(11.676) + 1.2194$$
$$\approx \$0.988 \text{ Canadian dollars for one U.S. dollar.}$$

You Try It   Work through this You Try It problem.

Work Exercises 28–29 in this textbook or in the *MyMathLab*®️ Study Plan.

In economics, **long run average cost (LRAC)**, is the total long run cost divided by the number of units produced. Companies sometimes use the *LRAC* to determine if they should merge resources such as sales regions or franchises. If the *LRAC* is decreasing, then it would be best for the company to merge resources, provided that the merger does not cause the *LRAC* to increase. If the *LRAC* is increasing, a merger may not be beneficial since it would lead to higher costs. This idea is illustrated in the next example.

Example 14 Merging Resources

My video summary

 An account rep in one territory oversees $N = 20$ accounts and a second account rep in a nearby territory manages $N = 8$ accounts. The long run average cost function for their industry is $C = N^2 - 70N + 1400$.

a. Determine the long run average cost for $N = 20$ accounts and $N = 8$ accounts.
b. What number of accounts minimizes the long run average cost? What is the minimum long run average cost?
c. Should the two territories be merged into a single territory?

Solution Try to work this problem on your own. Click here to check your answer, or watch this video for a detailed solution.

You Try It   Work through this You Try It problem.

Work Exercises 30–31 in this textbook or in the *MyMathLab*®️ Study Plan.

8.3 Exercises

You Try It

1. **Finding a Number**   The product of some negative number and 5 less than three times that number is 12. Find the number.

2. **Finding a Number**   The square of a number plus the number is 132. What is the number?

You Try It

3. **Finding Integers**   Three consecutive odd integers are such that the square of the third integer is 15 more than the sum of the squares of the first two. Find the integers.

4. **Finding a Number**   The sum of the square of a number and the square of 7 more than a number is 169. What is the number?

You Try It

5. **Rocket Launch**   A toy rocket is launched from a platform 2.8 meters above the ground in such a way that its height, h (in meters), after t seconds is given by the equation $h = -4.9t^2 + 18.9t + 2.8$. How long will it take for the rocket to hit the ground?

6. **Projectile Motion**   Shawn threw a rock straight up from a cliff that was 24 feet above the water. If the height of the rock h, in feet, after t seconds is given by the equation $h = -16t^2 + 20t + 24$, how long will it take for the rock to hit the water?

You Try It

7. **Dimensions of a Rectangle**   The length of a rectangle is 1 cm less than three times the width. If the area of the rectangle is 30 cm^2, find the dimensions of the rectangle.

8. **Dimensions of a Rectangle**   The length of a rectangle is 1 inch less than twice the width. If the diagonal is 2 inches more than the length, find the dimensions of the rectangle.

You Try It

9. **Loading Ramp**   A loading ramp in a steel yard has a horizontal run that is 26 feet longer than its vertical rise. If the ramp is 30 feet long, what is the vertical rise? (round to two decimal places if necessary)

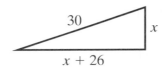

10. **Building a Walkway**   A 35- by 20-feet rectangular swimming pool is surrounded by a walkway of uniform width. If the total area of the walkway is 434 ft^2, how wide is the walkway?

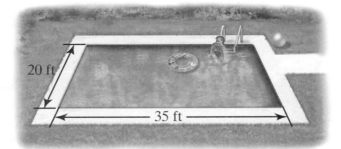

You Try It

11. **Speed of a Boat**   Logan rowed her boat upstream a distance of 9 miles and then rowed back to the starting point. The total time of the trip was 10 hours. If the rate of the current was 4 mph, find the average speed of the boat in still water.

12. **Speed of a Car** Imogene's car traveled 280 miles averaging a certain speed. If the car had gone 5 mph faster, the trip would have taken 1 hour less. Find the average speed.

You Try It

13. **Mowing Lawns** Twin brothers, Billy and Bobby, can mow their grandparent's lawn together in 56 minutes. Billy could mow the lawn by himself in 15 minutes less time than it would take Bobby. How long would it take Bobby to mow the lawn by himself?

14. **Working Together** Jeff and Kirk can build a 75-foot retaining wall together in 12 hours. Because Jeff has more experience, he could build the wall himself 4 hours quicker than Kirk. How long would it take Kirk (to the nearest minute) to build the wall by himself?

You Try It

15. **Maximum Height** A baseball player swings and hits a pop fly straight up in the air to the catcher. The height of the baseball in meters t seconds after it is hit is given by the quadratic function $h(t) = -4.9t^2 + 34.3t + 1$. How long does it take for the baseball to reach its maximum height? What is the baseball's maximum height?

16. **Maximum Height** An object is launched vertically in the air at 36.75 meters per second from a platform 10 meters above the ground. The height of the object above the ground (in meters) t seconds after launch is given by $h(t) = -4.9t^2 + 36.75t + 10$. How long does it take the object to reach its maximum height? What is the object's maximum height?

17. **Maximum Height of a Toy Rocket** A toy rocket is shot vertically into the air from a launching pad 5 feet above the ground with an initial velocity of 112 feet per second. The height h, in feet, of the rocket above the ground at t seconds after launch is given by the function $h(t) = -16t^2 + 112t + 5$. How long will it take the rocket to reach its maximum height? What is the maximum height?

18. **Maximum Revenue** The price p and the quantity x sold of a small flatscreen TV obeys the demand equation $p = -0.15x + 300$.

You Try It

    a. How much should be charged for the TV if there are 50 TVs in stock?

    b. What quantity x will maximize revenue? What is the maximum revenue?

    c. What price should be charged per TV in order to maximize revenue?

19. **Maximum Revenue** The dollar price for a barrel of oil sold at a certain oil refinery tends to obey the demand equation $p = -\frac{1}{10}x + 72$, where x is the number of barrels of oil on hand (in millions).

    a. How much should be charged per barrel of oil if there are 4 million barrels on hand?

    b. What quantity x will maximize revenue? What is the maximum revenue?

    c. What price should be charged per barrel in order to maximize revenue?

In Exercises 20 and 21, use the fact that profit is defined as revenue minus cost, or $P(x) = R(x) - C(x)$.

20. **Maximum Profit** Rite-Cut riding lawnmowers obey the demand equation $p = -\frac{1}{20}x + 1000$. The cost of producing x lawnmowers is given by the function $C(x) = 100x + 5000$.

You Try It

    a. Express the revenue R as a function of x.

    b. Express the profit P as a function of x.

    c. Find the value of x that maximizes profit. What is the maximum profit?

    d. What price should be charged per lawnmower to maximize profit?

21. **Maximum Profit** The CarryItAll minivan, a popular vehicle among soccer moms, obeys the demand equation $p = -\frac{1}{40}x + 8000$. The cost of producing x vans is given by the function $C(x) = 4000x + 20,000$.

    a. Express the revenue R as a function of x.

    b. Express the profit P as a function of x.

    c. Find the value of x that maximizes profit. What is the maximum profit?

    d. What price should be charged to maximize profit?

22. Maximum Profit   Silver Scooter, Inc., finds that it costs $200 to produce each motorized scooter and that the fixed costs are $1500 per day yielding the cost function $C(x) = 200x + 1500$. The price is given by $p = 600 - 5x$, where p is the price in dollars at which exactly x scooters will be sold. Find the quantity of scooters that Silver Scooter, Inc., should produce and the price it should charge to maximize profit. Find the maximum profit.

23. Maximum Profit   Amy, the owner of Amy's Pottery, can produce china pitchers at a cost of $5 each. She estimates her price function to be $p = 17 - 5x$, where p is the price at which exactly x pitchers will be sold per week. Find the number of pitchers that she should produce and the price that she should charge in order to maximize profit. Also find the maximum profit.

You Try It
24. Maximum Area   A farmer has 1800 feet of fencing available to enclose a rectangular area bordering a river. If no fencing is required along the river, find the dimensions of the fence that will maximize the area. What is the maximum area?

25. Maximum Area   Jim wants to build a rectangular parking lot along a busy street but has only 2500 feet of fencing available. If no fencing is required along the street, find the maximum area of the parking lot.

You Try It
26. Minimizing Cost   A manufacturer of denim jeans has daily production costs of $C = 0.2x^2 - 90x + 10{,}700$, where C is the total cost (in dollars) and x is the number of jeans produced. How many jeans should be produced each day in order to minimize costs? What is the minimum daily cost?

27. Minimizing Loss Function   Quality control analysts often use a *quadratic loss function* to determine costs associated with failing to meet product specifications. For a certain machine part, the loss function relating cost to diameter is $L = 4x^2 - 4x + 2.5$, where L is the total loss (in dollars) per part and x is the diameter of the machined part. Find the diameter that minimizes total loss per part. What is the minimum total loss?

You Try It
28. Minimum Exchange Rate   Between January 1, 1999, and January 1, 2009, the exchange rate between U.S. dollars and euros can be approximated by $D(x) = 0.003x^2 - 0.029x + 0.932$, where D is the number of U.S. dollars for one euro x years after January 1, 1999. For the given time period, find the year in which the exchange rate was at a minimum. What was the minimum exchange rate? Round your answer to three decimal places. (*Source:* U.S. Federal Reserve, St. Louis, MO)

29. Minimum Bankruptcies   Based on data from the Federal Judiciary, the number of business bankruptcy filings per year can be approximated by the model $B(x) = 3.9x^2 - 50.8x + 192.9$, where x is the number of years since 2000 and B is the number of business bankruptcy filings (in 1000s). During what year was the number of filings at a minimum? What was the approximate minimum number of filings to the nearest thousand?

You Try It

30. Merging Account Territories   An account manager in one territory oversees $N = 14$ accounts, and a second account manager in a nearby territory manages $N = 18$ accounts. The long run average cost function for their industry is $C = 484 - 44N + N^2$.

a. Find the long run average cost for $N = 14$ accounts and for $N = 18$ accounts.
b. How many accounts will minimize the long run average cost? What is the minimum long run average cost?
c. Should the two territories be merged into a single territory?

31. Bank Merger   One bank services $N = 150$ customers, and a second bank services $N = 80$ customers. The long run average cost function for their industry is $C = N^2 - 700N + 110{,}000$.

a. Find the long run average cost for $N = 150$ customers and for $N = 80$ customers.
b. What number of customers will minimize the long run average cost? What is the minimum long run average cost?
c. Should the two banks merge?

8.4 Circles

THINGS TO KNOW

Before working through this section, be sure you are familiar with the following concepts:

VIDEO   ANIMATION   INTERACTIVE

 You Try It
1. Plot Ordered Pairs in the Rectangular Coordinate System (Section 2.1, **Objective 1**)

 You Try It
2. Simplify Radical Expressions Using the Product Rule (Section 7.3, **Objective 5**)

 You Try It
3. Solve Quadratic Equations Using the Square Root Property (Section 8.1, **Objective 2**)

You Try It
4. Solve Quadratic Equations by Completing the Square (Section 8.1, **Objective 3**)

OBJECTIVES

1 Find the Distance between Two Points

2 Find the Midpoint of a Line Segment

3 Write the Standard Form of an Equation of a Circle

4 Sketch the Graph of a Circle Given in Standard Form

5 Write the General Form of a Circle in Standard Form and Sketching Its Graph

---

OBJECTIVE 1:   FIND THE DISTANCE BETWEEN TWO POINTS

Recall that the Pythagorean theorem states that the sum of the squares of the lengths of the two legs in a right triangle is equal to the square of the length of the hypotenuse, or $a^2 + b^2 = c^2$. See Figure 9.

My video summary

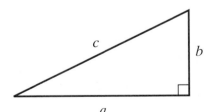

Figure 9
Pythagorean theorem
$a^2 + b^2 = c^2$

We can use the Pythagorean theorem to develop a formula for finding the distance between two points in a plane. Click here to see the steps, or watch this video for an explanation.

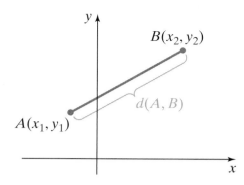

Figure 10

To find the distance $d(A, B)$ between the points $A(x_1, y_1)$ and $B(x_2, y_2)$ (see Figure 10), we use the **distance formula**.

Distance Formula

The distance $d(A, B)$ between two points (x_1, y_1) and (x_2, y_2) is given by
$$d(A, B) = \sqrt{(x_2 - x_1)^2 + (y_2 - y_1)^2}.$$

When using the distance formula, it does not matter which point is (x_1, y_1) and which is (x_2, y_2).

Example 1 Using the Distance Formula

Find the distance $d(A, B)$ between points $(-1, 5)$ and $(4, -5)$.

Solution Let $(x_1, y_1) = (-1, 5)$ and $(x_2, y_2) = (4, -5)$ and use the distance formula.

$$
\begin{aligned}
\text{Distance formula:} \quad d(A, B) &= \sqrt{(x_2 - x_1)^2 + (y_2 - y_1)^2} \\
\text{Substitute the coordinates:} \quad &= \sqrt{(4 - (-1))^2 + (-5 - 5)^2} \\
\text{Change to addition:} \quad &= \sqrt{(4 + 1)^2 + (-5 - 5)^2} \\
\text{Combine like terms:} \quad &= \sqrt{(5)^2 + (-10)^2} \\
\text{Square:} \quad &= \sqrt{25 + 100} \\
\text{Add:} \quad &= \sqrt{125} \\
\text{Simplify the radical:} \quad &= 5\sqrt{5}
\end{aligned}
$$

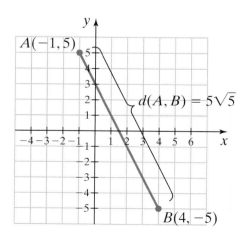

You Try It   **Work through this You Try It problem.**

Work Exercises 1–8 in this textbook or in the *MyMathLab*® Study Plan.

OBJECTIVE 2:   FIND THE MIDPOINT OF A LINE SEGMENT

The **midpoint** of a line segment is the point that lies exactly halfway between the endpoints of the line segment. The distance formula can be used to develop a formula for finding the midpoint. The coordinates of the midpoint are found by averaging the x-coordinates of the endpoints and averaging the y-coordinates of the endpoints.

Midpoint of a Line Segment

The **midpoint** of the line segment from $A(x_1, y_1)$ to $B(x_2, y_2)$ is the point with coordinates

$$\left(\frac{x_1 + x_2}{2}, \frac{y_1 + y_2}{2} \right).$$

The distance formula involves finding the difference between coordinates, but the midpoint formula involves finding the sum of the coordinates.

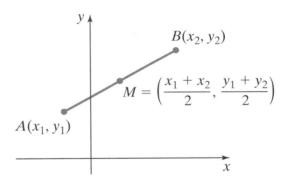

Figure 11

Because the midpoint lies halfway between the endpoints of a line segment (see Figure 11), the distance from the midpoint to either endpoint is half the distance between the endpoints.

$$d(A, M) = d(M, B) = \frac{d(A, B)}{2}$$

The distance between two points is a number, but the midpoint of a line segment is a point. When writing a midpoint, be sure to use an ordered pair.

Example 2 Finding the Midpoint of a Line Segment

Find the midpoint of the line segment with endpoints $(-3, 2)$ and $(4, 6)$.

Solution Let $(x_1, y_1) = (-3, 2)$ and $(x_2, y_2) = (4, 6)$ and use the midpoint formula, as shown on the following page.

$$\text{Midpoint formula:} \quad \left(\frac{x_1 + x_2}{2}, \frac{y_1 + y_2}{2}\right)$$

$$\text{Substitute the coordinates:} \quad = \left(\frac{-3 + 4}{2}, \frac{2 + 6}{2}\right)$$

$$\text{Simplify:} \quad = \left(\frac{1}{2}, \frac{8}{2}\right)$$

$$= \left(\frac{1}{2}, 4\right)$$

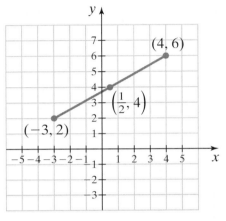

Figure 12

You Try It   Work through this You Try It problem.

Work Exercises 9–16 in this textbook or in the *MyMathLab*® Study Plan.

OBJECTIVE 3:   WRITE THE STANDARD FORM OF AN EQUATION OF A CIRCLE

A **circle** is the set of all points (x, y) in the coordinate plane that are a fixed distance r from a fixed point (h, k). The fixed distance r is called the **radius** of the circle, and the fixed point (h, k) is called the **center** of the circle. See Figure 13.

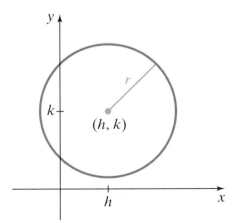

Figure 13
Circle with center (h, k)
and radius, r

The distance formula can be used to develop the standard form of the equation of a circle. Watch this animation to see how this is done.

Standard Form of the Equation of a Circle

The **standard form of the equation of a circle** with center (h, k) and radius r is

$$(x - h)^2 + (y - k)^2 = r^2.$$

Example 3 Writing the Standard Form of the Equation of a Circle

Write the standard form of the equation of a circle with center $(0, 0)$ and radius 3.

Solution The center is $(0, 0)$. Since the center is represented by (h, k) in our standard form equation, we know that $h = 0$ and $k = 0$. The radius is 3, so we have $r = 3$. Substitute these values into the equation.

$$\big(x - (0)\big)^2 + \big(y - (0)\big)^2 = 3^2$$
$$x^2 + y^2 = 9$$

The graph of the circle is shown in Figure 14.

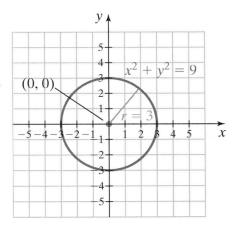

Figure 14

You Try It   Work through this You Try It problem.

Work Exercises 17–18 in this textbook or in the *MyMathLab*® Study Plan.

From Example 3, we see that the equation of a circle centered at the origin takes on a special form. The standard form of the equation of a circle centered at the origin with radius r is

$$x^2 + y^2 = r^2.$$

Example 4 Writing the Standard Form of the Equation of a Circle

My video summary

 Write the standard form of the equation of the circle with center $(-2, 3)$ and radius 6.

Solution The center is $(-2, 3)$. Since the center is (h, k) in our standard form equation, we know that $h = -2$ and $k = 3$. The radius is 6, so $r = 6$. Substitute these values into the equation $(x - h)^2 + (y - k)^2 = r^2$ and simplify. Click here to check your answer, or watch this video for the complete solution.

The graph of the circle is shown in Figure 15.

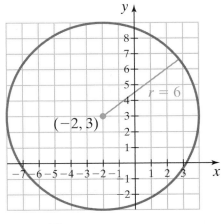

Figure 15

8.4   Circles   **407**

You Try It   Work through this You Try It problem.

Work Exercises 19–22 in this textbook or in the *MyMathLab*® Study Plan.

Example 5 Writing the Standard Form of the Equation of a Circle

My video summary

 Write the standard form of the equation of the circle with center $(0, -4)$ and radius $r = \sqrt{5}$.

Solution Determine h, k, and r. Then substitute these values into the standard form of the equation of a circle and simplify. Click here to check your answer, or watch this video for the complete solution.

You Try It   Work through this You Try It problem.

Work Exercises 23–24 in this textbook or in the *MyMathLab*® Study Plan.

OBJECTIVE 4:   SKETCH THE GRAPH OF A CIRCLE GIVEN IN STANDARD FORM

Example 6 Sketching the Graph of a Circle

Find the center and radius of the circle $x^2 + y^2 = 4$ and sketch its graph.

Solution We start by finding the center (h, k) and radius, r. The equation $x^2 + y^2 = 4$ has the form $(x - 0)^2 + (y - 0)^2 = 2^2$. Therefore, the circle is centered at the origin, $(h, k) = (0, 0)$, and has radius $r = 2$.

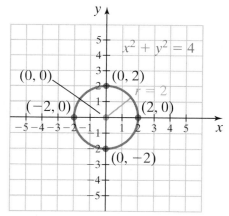

To sketch the graph of this circle, locate the center and then plot a few points on the circle. The easiest points to plot are located two units left and right from the center and two units up and down from the center.

<table>
<tr><td align="center">2 units
left</td><td align="center">2 units
right</td></tr>
<tr><td align="center">$(\underbrace{0 - 2}_{r}, 0) \to (-2, 0)$</td><td align="center">$(\underbrace{0 + 2}_{r}, 0) \to (2, 0)$</td></tr>
<tr><td align="center">2 units up</td><td align="center">2 units down</td></tr>
<tr><td align="center">$(0, \underbrace{0 + 2}_{r}) \to (0, 2)$</td><td align="center">$(0, \underbrace{0 - 2}_{r}) \to (0, -2)$</td></tr>
</table>

Figure 16
Graph of $x^2 + y^2 = 4$

Complete the graph by drawing the circle through these points, as shown in Figure 16.

You Try It   Work through this You Try It problem.

Work Exercises 25–26 in this textbook or in the *MyMathLab*® Study Plan.

Example 7 Sketching the Graph of a Circle

Find the center and radius of the circle $(x + 1)^2 + (y - 4)^2 = 25$ and sketch its graph.

Solution The equation $(x + 1)^2 + (y - 4)^2 = 25$ can be written as $(x - (-1))^2 + (y - 4)^2 = 5^2$. Comparing this equation to the standard form of the equation of a circle, we find that the circle has center $(h, k) = (-1, 4)$ and radius $r = 5$.

Locate the center, then plot a few points on the circle. Start with the points that are five units left and right of the center and five units up and down from the center.

$$\overbrace{(-1 - \underset{r}{\underline{5}}, 4)}^{\text{5 units left}} \rightarrow (-6, 4) \qquad \overbrace{(-1 + \underset{r}{\underline{5}}, 4)}^{\text{5 units right}} \rightarrow (4, 4)$$

$$\overbrace{(-1, 4 + \underset{r}{\underline{5}})}^{\text{5 units up}} \rightarrow (-1, 9) \qquad \overbrace{(-1, 4 - \underset{r}{\underline{5}})}^{\text{5 units down}} \rightarrow (-1, -1)$$

My video summary

Use these points to sketch the graph of the circle. Click here to check your answer.

You Try It   **Work through this You Try It problem.**

Work Exercises 27–30 in this textbook or in the *MyMathLab*®️ Study Plan.

Recall that each point on the graph of an equation is an ordered pair solution to the equation. We can improve our sketch by finding additional points. For example, we may want to find the intercepts (if any) and then plot those corresponding points.

Example 8 Sketching the Graph of a Circle

Find the center and the radius, and sketch the graph of the circle $(x - 1)^2 + (y + 2)^2 = 9$. Also find any intercepts.

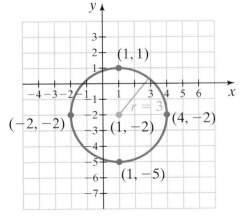

Solution The equation $(x - 1)^2 + (y + 2)^2 = 9$ can be written in standard form as $(x - 1)^2 + (y - (-2))^2 = 3^2$ with center $(h, k) = (1, -2)$ and radius $r = 3$.

To sketch the graph of this circle, locate the center, then plot a few points on the circle. The easiest points to plot are located three units left and right from the center, $(-2, -2)$ and $(4, -2)$ respectively, and three units up and down from the center, $(1, 1)$ and $(1, -5)$ respectively. After plotting these four points, complete the graph by drawing the circle through the points as shown in Figure 17.

Figure 17

Graph of $(x - 1)^2 + (y + 2)^2 = 9$

To find the x-intercepts, set $y = 0$ and solve the resulting quadratic equation for x. To find the y-intercepts, set $x = 0$ and solve the resulting quadratic equation for y. Click here to check your answer, or watch this video for the complete solution.

You Try It   Work through this You Try It problem.

Work Exercises 31–32 in this textbook or in the *MyMathLab* ® Study Plan.

OBJECTIVE 5:   WRITE THE GENERAL FORM OF A CIRCLE IN STANDARD FORM AND SKETCHING ITS GRAPH

Since a circle is completely defined by its center and radius, we only need these two pieces of information to write the equation of a circle or to sketch its graph. We saw in Examples 3–5 that given the center and radius, we can write the standard form of an equation of a circle. Then we used this standard form in Examples 6–8 to find the center and radius and then to sketch the graph of a circle.

Not all equations of circles are given in standard form. Consider the circle $(x + 3)^2 + (y - 1)^2 = 49$. This equation is in standard form, so we can quickly see that the center is $(h, k) = (-3, 1)$ and the radius is $r = \sqrt{49} = 7$. However, let's see what happens if we expand the equation.

$$
\begin{aligned}
\text{Original equation:} &\quad (x + 3)^2 + (y - 1)^2 = 49 \\
\text{Expand to remove parentheses:} &\quad x^2 + 6x + 9 + y^2 - 2y + 1 = 49 \\
\text{Collect like terms:} &\quad x^2 + 6x + y^2 - 2y + 10 = 49 \\
\text{Subtract 49 from both sides:} &\quad x^2 + 6x + y^2 - 2y - 39 = 0 \\
\text{Rearrange terms:} &\quad x^2 + y^2 + 6x - 2y - 39 = 0
\end{aligned}
$$

The equation $x^2 + y^2 + 6x - 2y - 39 = 0$ is written in **general form**. In this form, the center and radius are less obvious.

General Form of the Equation of a Circle

The **general form of the equation of a circle** is

$$Ax^2 + By^2 + Cx + Dy + E = 0,$$

where A, B, C, D, and E are real numbers and $A = B \neq 0$.

To find the center and radius of a circle from its equation in general form, we need to complete the square in both x and y to write the equation in standard form. Before working through Example 9, you might want to review completing the square from Section 8.1.

Example 9 Writing the General Form of a Circle in Standard Form and Sketching Its Graph

Write the equation $x^2 + y^2 + 10x + 9 = 0$ in standard form. Find the center, radius, and intercepts, then sketch the graph.

Solution The equation contains a linear term in x, but no linear term in y. So, we only need to complete the square in x. Start by rearranging terms to group the x terms together and move any constants to the right-hand side of the equation.

No linear y term so we already have a perfect square in y

Linear x term so we must complete the square in x

Original equation:	$x^2 + y^2 + 10x + 9 = 0$
Rearrange terms:	$x^2 + 10x + y^2 + 9 = 0$
Move constants to the right-hand side:	$x^2 + 10x + y^2 = -9$

To complete the square in x, we divide the coefficient of the linear term by 2, square the result, and add this to both sides of the equation. Since $\left(\frac{1}{2} \cdot 10\right)^2 = 5^2 = 25$, we need to add 25 to both sides, and then factor the left-hand side.

Complete the square in x: $\quad x^2 + 10x + 25 + y^2 = -9 + 25$

$$\left(\tfrac{1}{2} \cdot 10\right)^2 = 25$$

Factor the left-hand side: $\quad (x + 5)^2 + y^2 = 16$

Now the equation is written in standard form. Comparing to $(x - h)^2 + (y - k)^2 = r^2$, the center is $(h, k) = (-5, 0)$, and the radius is $r = \sqrt{16} = 4$.

To find the x-intercepts, we let $y = 0$ and solve for x.

Standard form:	$(x + 5)^2 + y^2 = 16$
Substitute 0 for y:	$(x + 5)^2 + (0)^2 = 16$
Simplify:	$(x + 5)^2 = 16$
Square root property:	$x + 5 = \pm\sqrt{16}$
Simplify:	$x + 5 = \pm 4$
Subtract 5 from both sides:	$x = -5 \pm 4$

The x-intercepts are $-5 - 4 = -9$ and $-5 + 4 = -1$. So, the points $(-9, 0)$ and $(-1, 0)$ are on the graph.

To find the y-intercepts, we let $x = 0$ and solve for y.

Original equation:	$x^2 + y^2 + 10x + 9 = 0$
Substitute 0 for x:	$(0)^2 + y^2 + 10(0) + 9 = 0$
Simplify:	$y^2 + 9 = 0$
Subtract 9 from both sides:	$y^2 = -9$

There is no real number whose square is -9, so the graph has no y-intercepts. The graph of the circle is shown in Figure 18.

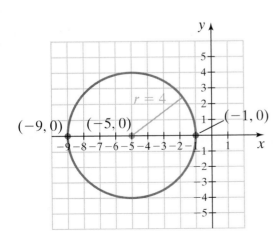

Figure 18
Graph of $x^2 + y^2 + 10x + 9 = 0$

You Try It   **Work through this You Try It problem.**

Work Exercises 33–34 in this textbook or in the *MyMathLab*® Study Plan.

Example 10   Writing the General Form of a Circle in Standard Form and Sketching Its Graph

Write the equation $x^2 + y^2 - 8x + 6y + 16 = 0$ in standard form; find the center, radius, and intercepts, and sketch the graph.

Solution   The equation contains linear terms in both x and y, so we will need to complete the square in both x and y. We rearrange the terms, complete the square, and move any constants to the right-hand side of the equation.

Rearrange the terms: $x^2 - 8x \qquad + y^2 + 6y \qquad = -16$

Complete the square in x $x^2 - 8x + 16 + y^2 + 6y + 9 = -16 + 16 + 9$
and y. Remember to add
16 and 9 to both sides:

$$\left(\tfrac{1}{2}\cdot(-8)\right)^2 = 16 \qquad \left(\tfrac{1}{2}\cdot 6\right)^2 = 9$$

Factor the left: $(x - 4)^2 + (y + 3)^2 = 9$

Now we have converted the general form of the circle into standard form with center $(h, k) = (4, -3)$ and radius $r = 3$. Watch this video to verify that there are no y-intercepts. The only x-intercept is $x = 4$. The graph of the circle is shown in Figure 19.

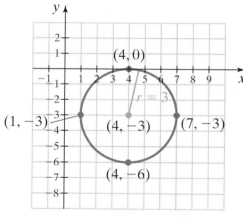

Figure 19
Graph of $x^2 + y^2 - 8x + 6y + 16 = 0$

You Try It   **Work through this You Try It problem.**

Work Exercises 35–40 in this textbook or in the *MyMathLab*® Study Plan.

Example 11   Writing the General Form of a Circle in Standard Form, Where $A \neq 1$ and $B \neq 1$

Write the equation $4x^2 + 4y^2 + 4x - 8y + 1 = 0$ in standard form. Find the center, radius, and intercepts. Then sketch the graph.

Solution Work through this animation to verify that this equation is equivalent to the following equation in standard form:

$$\left(x + \frac{1}{2}\right)^2 + (y - 1)^2 = 1; \text{ center } (h, k) = \left(-\frac{1}{2}, 1\right) \text{ and } r = \sqrt{1} = 1$$

Now work through the animation again to verify that the one x-intercept is found at $\left(-\frac{1}{2}, 0\right)$, and the y-intercepts are found at $\left(0, 1 - \frac{\sqrt{3}}{2}\right)$ and $\left(0, 1 + \frac{\sqrt{3}}{2}\right)$. The graph of the circle is shown in Figure 20.

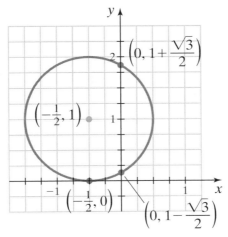

Figure 20
Graph of $4x^2 + 4y^2 + 4x - 8y + 1 = 0$

You Try It Work through this You Try It problem.

Work Exercises 41–44 in this textbook or in the *MyMathLab*® Study Plan.

 CAUTION Every equation of a circle can be written in general form. However, not every equation of the form $Ax^2 + By^2 + Cx + Dy + E = 0$ is the equation of a circle. For example, the equation $2x^2 + 2y^2 + 1 = 0$ has no graph, and the graph of $x^2 + y^2 + 2x - 4y + 5 = 0$ is a single point.

8.4 **Exercises**

In Exercises 1–8, find the distance $d(A, B)$ between points A and B.

You Try It

1. $A(2, 7)$; $B(5, 11)$

2. $A(1, 5)$; $B(-2, 1)$

3. $A(3, 5)$; $B(-2, -2)$

4. $A(2, -4)$; $B(-3, 6)$

5. $A\left(\frac{2}{3}, 5\right)$, $B\left(-1, \frac{1}{2}\right)$

6. $A\left(0, -\sqrt{2}\right)$; $B\left(\sqrt{3}, 0\right)$

7. $A(2, -3)$; $B(2, 5)$

8. $A(-1, 4)$; $B(5, 4)$

In Exercises 9–16, find the midpoint of the line segment with endpoints A and B.

9. $A(3, 7); B(5, 9)$

10. $A(2, -5); B(4, 1)$

You Try It **11.** $A(-1, 4); B(-2, -2)$

12. $A(-3, 0); B(0, 7)$

13. $A(0, 1); B\left(-3, \dfrac{1}{2}\right)$

14. $A\left(\sqrt{27}, 4\right); B\left(\sqrt{3}, 2\right)$

15. $A(1, 3); B(1, -5)$

16. $A(4, -2); B(-3, -2)$

In Exercises 17–24, write the standard form of the equation of each circle described.

You Try It **17.** Center $(0, 0)$, $r = 1$

18. Center $(0, 0)$, $r = \sqrt{3}$

19. Center $(-2, 3)$, $r = 4$
You Try It

20. Center $(0, 4)$, $r = 8$

21. Center $(1, -4)$, $r = \dfrac{3}{4}$

22. Center $\left(\dfrac{1}{5}, 3\right)$, $r = 2$

You Try It

23. Center $(3, 0)$, $r = \sqrt{2}$

24. Center $\left(-\dfrac{1}{4}, -\dfrac{1}{3}\right)$, $r = \dfrac{3}{4}$

In Exercises 25–30, find the center and radius of each circle and sketch its graph.

25. $x^2 + y^2 = 9$
You Try It

26. $(x + 3)^2 + y^2 = 4$

27. $(x - 1)^2 + (y + 5)^2 = 16$

28. $(x + 2)^2 + (y + 4)^2 = 36$

You Try It
29. $\left(x - \dfrac{1}{4}\right)^2 + \left(y + \dfrac{1}{2}\right)^2 = 4$

30. $(x + 4)^2 + (y - 5)^2 = 10$

In Exercises 31–32, find the center, radius, and intercepts of each circle and then sketch its graph.
You Try It

31. $(x - 4)^2 + (y + 7)^2 = 12$

32. $(x + 1)^2 + (y - 3)^2 = 20$

In Exercises 33–44, write the equation of each circle in standard form. Find the center, radius, and intercepts. Then sketch the graph.

You Try It **33.** $x^2 + y^2 + 6x + 5 = 0$

34. $x^2 + y^2 + 2y - 8 = 0$

35. $x^2 + y^2 + 2x - 4y + 1 = 0$
You Try It

36. $x^2 + y^2 - 10x + 6y + 18 = 0$

37. $x^2 + y^2 - 4x - 8y + 19 = 0$

38. $x^2 + y^2 - 6x - 12y - 5 = 0$

39. $x^2 + y^2 - 3x - y - \dfrac{1}{2} = 0$
You Try It

40. $x^2 + y^2 + \dfrac{2}{3}x - \dfrac{1}{2}y - \dfrac{7}{18} = 0$

 41. $2x^2 + 2y^2 - 4x + 8y + 2 = 0$

42. $16x^2 + 16y^2 - 16x - 8y - 11 = 0$

43. $144x^2 + 144y^2 - 72x - 96y - 551 = 0$

44. $36x^2 + 36y^2 + 12x + 72y - 35 = 0$

8.5 Polynomial and Rational Inequalities

THINGS TO KNOW

Before working through this section, be sure you are familiar with the following concepts:

You Try It

1. Factor by Grouping
   (Section 5.1, Objective 2)

You Try It

2. Factor Trinomials of the Form $x^2 + bx + c$
   (Section 5.2, Objective 1)

You Try It

3. Factor Trinomials of the form $ax^2 + bx + c$
   Using Trial and Error (Section 5.2, Objective 2)

You Try It

4. Factor Trinomials of the form $ax^2 + bx + c$
   Using the ac Method (Section 5.2, Objective 3)

You Try It

5. Solve Polynomial Equations by Factoring
   (Section 5.4, Objective 1)

OBJECTIVES

1 Solve Polynomial Inequalities

2 Solve Rational Inequalities

OBJECTIVE 1:   SOLVE POLYNOMIAL INEQUALITIES

In Section 1.2, we learned how to solve linear inequalities. In this objective, we learn how to solve **polynomial inequalities**.

> **Polynomial Inequality**
>
> A **polynomial inequality** is an inequality that can be written as
>
> $$P(x) < 0, P(x) > 0, P(x) \leq 0, \text{ or } P(x) \geq 0,$$
>
> where $P(x)$ is a polynomial function.

When solving polynomial inequalities, x-intercepts will play an important part. To understand why, consider the polynomial functions graphed in Figure 21.

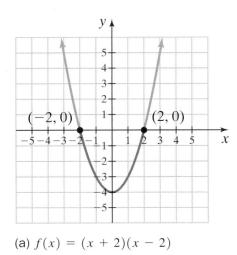

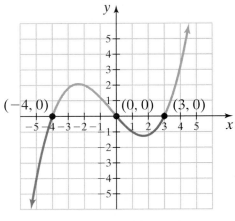

(a) $f(x) = (x + 2)(x - 2)$

(b) $g(x) = 0.1x(x + 4)(x - 3)$

Figure 21

From Figure 21, we see that for some x values, the value of the polynomial function will be positive; for others, it will be negative, and for still other x values, the resulting value of the polynomial function will be zero. The x values where the function value is zero are the x-intercepts, which are also **boundary points** that divide the real number line into intervals. Notice that in each interval, the function is always positive or always negative. This is true for all polynomials and is central to our approach to solving polynomial inequalities.

Example 1 Solving a Polynomial Inequality

Solve $x^3 - 3x^2 + 2x \geq 0$.

Solution Work through the following solution or watch this video.

First, factor the left-hand side to get $x(x - 1)(x - 2) \geq 0$. Second, find all real values of x that make the left-hand side *equal to* zero. These values are the boundary points. To find these boundary points, set the factored polynomial on the left equal to zero and solve for x.

Set the factored polynomial equal to 0: $x(x - 1)(x - 2) = 0$

Use the zero product property: $x = 0$ or $x - 1 = 0$ or $x - 2 = 0$

The boundary points are $x = 0$, $x = 1$, and $x = 2$.

Next, plot each boundary point on a number line. Because the expression $x(x - 1)(x - 2)$ is equal to zero at our three boundary points, we use a solid circle ● at each boundary point to indicate that the inequality $x(x - 1)(x - 2) \geq 0$ is satisfied at these points. (**Note:** If the inequality had been a strict inequality, such as $>$ or $<$, we would have used an open circle ○ to represent that the boundary points were *not* part of the solution. We will look at this in Example 2.) Notice that in Figure 22, we have naturally divided the number line into four intervals.

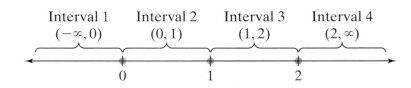

Figure 22

My video summary

The expression $x(x - 1)(x - 2)$ is equal to zero *only* at the three boundary points: 0, 1, and 2. So, in any of the four intervals shown in Figure 22, the expression $x(x - 1)(x - 2)$ must be either always *positive* or always *negative* throughout the entire interval. To check whether this expression is positive or negative on each interval, pick a number called a test value from each interval. The test value can be any point in the interval but not a boundary point. Possible test values are plotted in Figure 23.

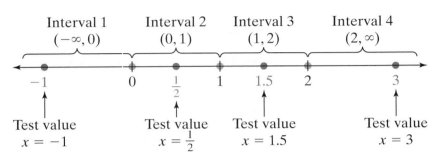

Figure 23

Substitute the test value into the expression $x(x - 1)(x - 2)$ and check to see if it yields a positive or negative value.

Interval	Test Value	Substitute Test Value into $x(x - 1)(x - 2)$	Comment
1. $(-\infty, 0)$	$x = -1$	$(-1)(-1 - 1)(-1 - 2) \Rightarrow (-)(-)(-) = -$	Expression is negative on $(-\infty, 0)$
2. $(0, 1)$	$x = \dfrac{1}{2}$	$\left(\dfrac{1}{2}\right)\left(\dfrac{1}{2} - 1\right)\left(\dfrac{1}{2} - 2\right) \Rightarrow (+)(-)(-) = +$	Expression is positive on $(0, 1)$
3. $(1, 2)$	$x = 1.5$	$(1.5)(1.5 - 1)(1.5 - 2) \Rightarrow (+)(+)(-) = -$	Expression is negative on $(1, 2)$
4. $(2, \infty)$	$x = 3$	$(3)(3 - 1)(3 - 2) \Rightarrow (+)(+)(+) = +$	Expression is positive on $(2, \infty)$

If the expression $x(x - 1)(x - 2)$ is positive on an interval, we place a "+" above the interval on the number line. If the expression $x(x - 1)(x - 2)$ is negative, we place a "−" above the interval. See Figure 24.

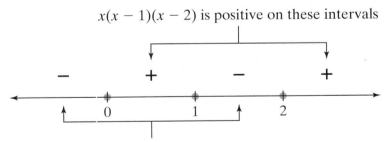

$x(x - 1)(x - 2)$ is negative on these intervals

Figure 24

In Figure 24, we see that $x(x - 1)(x - 2)$ is greater than zero on the intervals $(0, 1)$ and $(2, \infty)$ and is equal to zero at the boundary points $x = 0$, $x = 1$, and $x = 2$. So, the solution set for the inequality $x(x - 1)(x - 2) \geq 0$ is the interval $[0, 1] \cup [2, \infty)$.

You Try It   Work through this You Try It problem.

Work Exercises 1–6 in this textbook or in the *MyMathLab* Study Plan.

Example 1 illustrates the following approach for solving a factorable polynomial inequality.

Solving Polynomial Inequalities

Step 1. Move all terms to one side of the inequality, leaving zero on the other side.

Step 2. Factor the nonzero side of the inequality.

Step 3. Find all boundary points by setting the factored polynomial equal to zero.

Step 4. Plot the boundary points on a number line. If the inequality is $\leq$ or $\geq$, use a solid circle ●. If the inequality is $<$ or $>$, use an open circle ○.

Step 5. Now that the number line is divided into intervals, pick a test value from each interval.

Step 6. Substitute each test value into the polynomial and determine whether the expression is positive or negative on the corresponding interval.

Step 7. Find the intervals that satisfy the inequality.

Example 2 Solving a Polynomial Inequality

Solve $x^2 + 5x < 3 - x^2$.

Solution We use the steps for solving polynomial inequalities.

Step 1. Move all terms to one side of the inequality.

$$2x^2 + 5x - 3 < 0$$

Step 2. Factor.

$$(2x - 1)(x + 3) < 0$$

Step 3. Find boundary points by setting the factored polynomial equal to zero.

$$(2x - 1)(x + 3) < 0$$

Zero product property: $2x - 1 = 0$ or $x + 3 = 0$

$$2x = 1 \qquad\qquad x = -3$$

$$x = \frac{1}{2}$$

The boundary points are $x = -3$ and $x = \frac{1}{2}$.

Step 4. Plot the boundary points.

We use open circles to plot our boundary points because these points are *not* part of the solution. (**Note:** We are only looking for values that make the expression strictly less than zero.)

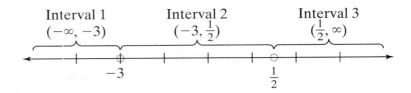

Step 5. We have three intervals. Pick a test value from each interval.

Interval 1: Test value $x = -4$

Interval 2: Test value $x = 0$

Interval 3: Test value $x = 1$

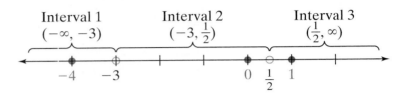

Step 6. Substitute each test value into the polynomial.

$x = -4$: $(2(-4) - 1)((-4) + 3) \Rightarrow (-)(-) = +,$

expression is positive on $(-\infty, -3)$.

$x = 0$: $(2(0) - 1)((0) + 3) \Rightarrow (-)(+) = -,$

expression is negative on $\left(-3, \dfrac{1}{2}\right)$.

$x = 1$: $(2(1) - 1)((1) + 3) \Rightarrow (+)(+) = +,$

expression is positive on $\left(\dfrac{1}{2}, \infty\right)$.

$(2x - 1)(x + 3)$ is negative on this interval

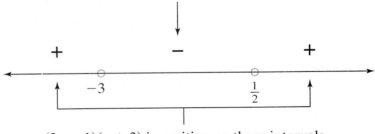

$(2x - 1)(x + 3)$ is positive on these intervals

Step 7. Find the intervals that satisfy the inequality.

Because we want x-values that result in a polynomial value less than zero (negative values), the solution set is the interval $\left(-3, \dfrac{1}{2}\right)$.

You Try It Work through this You Try It problem.

Work Exercises 7–10 in this textbook or in the *MyMathLab*® Study Plan.

OBJECTIVE 2:   SOLVE RATIONAL INEQUALITIES

We now extend our technique for solving polynomial inequalities to solve **rational inequalities**.

Rational Inequality

A **rational inequality** is an inequality that can be written as

$$R(x) < 0, R(x) > 0, R(x) \leq 0, \text{ or } R(x) \geq 0,$$

where $R(x) = \dfrac{P(x)}{Q(x)}$ is a rational function and $P(x)$ and $Q(x)$ are polynomial functions.

We solve rational inequalities in much the same way that we solve polynomial inequalities. The only differences are that we need the left-hand side to be a single rational expression and we find the boundary points by setting both the numerator and denominator equal to zero.

Solving Rational Inequalities

Step 1. Move all terms to one side of the inequality, leaving zero on the other side.

Step 2. Combine the terms into a single rational expression and factor the nonzero side of the inequality (both the numerator and denominator).

Step 3. Find all boundary points by setting the factored polynomials in the numerator and denominator equal to zero.

Step 4. Plot the boundary points on a number line. If the inequality is $\leq$ or $\geq$, use a solid circle ● for the boundary points from the numerator. If the inequality is $<$ or $>$, use an open circle ○. Boundary points from the denominator always have an open circle. If the same boundary point is obtained from both the numerator and denominator, then it is plotted with an open circle.

Step 5. Pick a test value from each interval.

Step 6. Substitute each test value into the rational expression and determine whether the expression is positive or negative on the corresponding interval.

Step 7. Find the intervals that satisfy the inequality.

Example 3 Solving a Rational Inequality

▣ Solve $\dfrac{x - 4}{x + 1} \geq 0$.

Solution   Using the steps for solving a rational inequality, work through the following solution, or watch this video.

Because the left-hand side of the inequality is already a single rational expression in completely factored form, we can skip steps 1 and 2 and start with step 3.

Step 3. Find the boundary points by setting the factored polynomials in the numerator and the denominator equal to zero.

Numerator: $x - 4 = 0$, so $x = 4$ is a boundary point.

Denominator: $x + 1 = 0$, so $x = -1$ is a boundary point.

Step 4. Plot the boundary points.

Since the inequality is a *greater than or equal to* inequality, we use a closed circle ● to represent the boundary point $x = 4$. Because we cannot divide by zero, we use an open circle ○ to represent the boundary point from the denominator, $x = -1$.

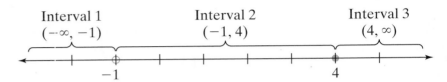

Step 5. Pick a test value from each interval.

Interval 1: Test value $x = -2$

Interval 2: Test value $x = 0$

Interval 3: Test value $x = 5$

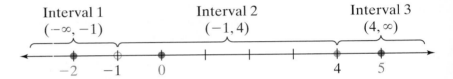

Step 6. Substitute each test value into the rational expression, and determine whether the expression is positive or negative on the corresponding interval.

$$x = -2: \frac{(-2 - 4)}{(-2 + 1))} \Rightarrow \frac{(-)}{(-)} = +,$$

expression is positive on the interval $(-\infty, -1)$.

$$x = 0: \frac{(-0 - 4)}{(0 + 1)} \Rightarrow \frac{(-)}{(+)} = -,$$

expression is negative on the interval $(-1, 4)$.

$$x = 5: \frac{(5 - 4)}{(5 + 1)} \Rightarrow \frac{(+)}{(+)} = +,$$

expression is positive on the interval $(4, \infty)$.

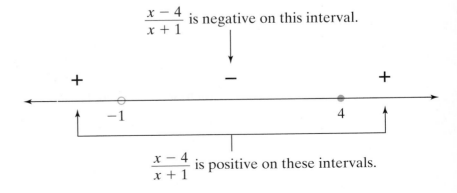

Step 7. Find the intervals that satisfy the inequality.

We look for x values that make the rational expression greater than or equal to zero. From the last step, we see that the solution set to the inequality is $(-\infty, -1) \cup [4, \infty)$.

Note that 4 is included as a solution but not -1 because $x = -1$ makes the denominator equal to zero.

You Try It   Work through this You Try It problem.

Work Exercises 11–14 in this textbook or in the *MyMathLab* ® Study Plan.

Example 4 Solving a Rational Inequality

My video summary

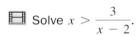 Solve $x > \dfrac{3}{x - 2}$.

Solution

Step 1. Subtract $\dfrac{3}{x - 2}$ from both sides of the inequality.

$$x - \frac{3}{x - 2} > 0$$

CAUTION You cannot multiply both sides of the inequality by $x - 2$ to clear the fraction because we do not know if $x - 2$ is positive or negative. So, we are not sure if we need to reverse the direction of the inequality.

All nonzero terms are on the left-hand side, so we combine the terms by finding a common denominator.

Rewrite the inequality: $\qquad x - \dfrac{3}{x - 2} > 0$

Rewrite using a common denominator of $x - 2$: $\qquad \dfrac{x(x - 2)}{x - 2} - \dfrac{3}{x - 2} > 0$

Combine terms: $\qquad \dfrac{x(x - 2) - 3}{x - 2} > 0$

Distribute: $\qquad \dfrac{x^2 - 2x - 3}{x - 2} > 0$

Step 2. Combine the terms into a single rational expression and factor the nonzero side of the inequality (both the numerator and denominator).

$$\frac{(x - 3)(x + 1)}{x - 2} > 0$$

Finish steps 3–7 on your own. Click here to check your answer, or watch this video to see the entire solution.

You Try It   Work through this You Try It problem.

Work Exercises 15–20 in this textbook or in the *MyMathLab* ® Study Plan.

Example 5 Solving a Rational Inequality

My video summary

 Solve $\dfrac{x + 1}{x - 2} > \dfrac{7x + 1}{x^2 + x - 6}$.

Solution Complete the seven-step process to solve the rational inequality. Click here to check your answer, or watch this video for the complete solution.

You Try It Work through this You Try It problem.

Work Exercises 21–22 in this textbook or in the *MyMathLab*® Study Plan.

8.5 Exercises

You Try It
In Exercises 1–10, solve each polynomial inequality. Express each solution using interval notation.

1. $(x - 1)(x + 3) \geq 0$ **2.** $x^2 + 4x - 21 < 0$

3. $x(3x + 2) \leq 0$ **4.** $2x^2 - 4x > 0$

5. $x^2 \leq 1$ **6.** $(x - 1)(x + 4)(x - 3) \geq 0$

7. $3x^2 + x < 3x + 1$ **8.** $2x^3 > 24x - 2x^2$

You Try It   **9.** $x^3 + x^2 - x \leq 1$ **10.** $x^3 \geq -2x^2 - x$

In Exercises 11–22, solve each rational inequality. Express each solution using interval notation.

You Try It

11. $\dfrac{x + 3}{x - 1} \leq 0$ **12.** $\dfrac{2 - x}{3x + 9} \geq 0$ **13.** $\dfrac{x}{x - 1} > 0$ **14.** $\dfrac{x^2 - 9}{x + 2} \leq 0$

You Try It
15. $\dfrac{4}{x + 1} \geq 2$ **16.** $\dfrac{x + 5}{2x - 3} > 1$ **17.** $\dfrac{x^2 - 8}{x + 4} \geq x$ **18.** $\dfrac{x}{2 - x} \leq \dfrac{1}{x}$

19. $\dfrac{x - 1}{x - 2} \geq \dfrac{x + 2}{x + 3}$ **20.** $\dfrac{x - 1}{x + 1} + \dfrac{x + 1}{x - 1} \leq \dfrac{x + 5}{x^2 - 1}$

21. $\dfrac{x + 1}{x + 4} < \dfrac{-11x - 17}{x^2 - x - 20}$ **22.** $\dfrac{x - 8}{x^2 + 5x + 4} < \dfrac{-21x}{(x + 1)(x - 3)(x + 4)}$

You Try It

CHAPTER NINE

Exponential and Logarithmic Functions and Equations

CHAPTER NINE CONTENTS

9.1 **Transformations of Functions**

9.2 **Composite and Inverse Functions**

9.3 **Exponential Functions**

9.4 **The Natural Exponential Function**

9.5 **Logarithmic Functions**

9.6 **Properties of Logarithms**

9.7 **Exponential and Logarithmic Equations**

9.8 **Applications of Exponential and Logarithmic Functions**

9.1 Transformations of Functions

THINGS TO KNOW

Before working through this section, be sure that you are familiar with the following concepts:

VIDEO   ANIMATION   INTERACTIVE

 You Try It
1. Graph Equations by Plotting Points (Section 2.1, Objective 4)

 You Try It
2. Find the Domain and Range of a Relation (Section 2.2, Objective 2)

 You Try It
3. Graph Simple Functions by Plotting Points (Section 2.3, Objective 3)

 You Try It
4. Graph Functions That Contain Square Roots or Cube Roots (Section 7.2, Objective 3)

 You Try It
5. Graph Quadratic Functions by Using Translations (Section 8.2, Objective 2)

424

OBJECTIVES

1 Use Vertical Shifts to Graph Functions

2 Use Horizontal Shifts to Graph Functions

3 Use Reflections to Graph Functions

4 Use Vertical Stretches and Compressions to Graph Functions

5 Use Horizontal Stretches and Compressions to Graph Functions

6 Use Combinations of Transformations to Graph Functions

In this section, we learn how to sketch the graphs of new functions using the graphs of known functions. Starting with the graph of a known function, we "transform" it into a new function by applying various transformations. Before we begin our discussion about transformations, it is critical that you know the graphs of some basic functions that have been explored earlier in this text. Take a moment to review these summaries of some basic functions. Click a function name to review its properties.

REVIEW OF THE BASIC FUNCTIONS

Click on any function to review its graph.

The identity function $f(x) = x$

The absolute value function $f(x) = |x|$

The square function $f(x) = x^2$

The cube function $f(x) = x^3$

The square root function $f(x) = \sqrt{x}$

The cube root function $f(x) = \sqrt[3]{x}$

OBJECTIVE 1: **USE VERTICAL SHIFTS TO GRAPH FUNCTIONS**

Example 1 Vertically Shifting a Function

Sketch the graphs of $f(x) = |x|$ and $g(x) = |x| + 2$.

My video summary

Solution Table 1 shows that for every value of x, the y-coordinate of the function g is always two greater than the y-coordinate of the function f. The two functions are sketched in Figure 1. The graph of $g(x) = |x| + 2$ is exactly the same as the graph of $f(x) = |x|$, except the graph of g is shifted *up* two units.

x	$f(x) = |x|$	$g(x) = |x| + 2$
-3	3	5
-2	2	4
-1	1	3
0	0	2
1	1	3
2	2	4
3	3	5

Table 1

(EBook Screens 9.1-1–9.1-39)

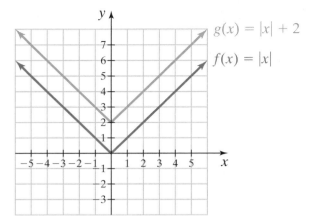

$g(x) = |x| + 2$

$f(x) = |x|$

Figure 1

We see from Example 1 that if c is a positive number, then $y = f(x) + c$ is the graph of f shifted *up* c units. It follows that for $c > 0$, the graph of $y = f(x) - c$ is the graph of f shifted *down* c units.

Vertical Shifts of Functions

If c is a positive real number,

The graph of $y = f(x) + c$ is obtained by shifting the graph of $y = f(x)$ vertically upward c units.

The graph of $y = f(x) - c$ is obtained by shifting the graph of $y = f(x)$ vertically downward c units.

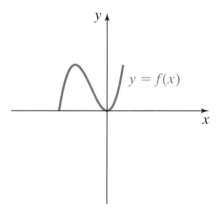

$y = f(x)$

Sketch $y = f(x) + c$.   Sketch $y = f(x) - c$.

Click to animate.   Click to animate.

You Try It   Work through this You Try It problem.

Work Exercises 1–7 in this textbook or in the _MyMathLab_® Study Plan.

OBJECTIVE 2: USE HORIZONTAL SHIFTS TO GRAPH FUNCTIONS

My video summary

To illustrate a horizontal shift, let $f(x) = x^2$ and $g(x) = (x + 2)^2$. Tables 2 and 3 show tables of values for f and g, respectively. The graphs of f and g are sketched in Figure 2. The graph of g is the graph of f shifted to the *left* two units.

x	$f(x) = x^2$
-2	4
-1	1
0	0
1	1
2	4

Table 2

x	$g(x) = (x + 2)^2$
-4	4
-3	1
-2	0
-1	1
0	4

Table 3

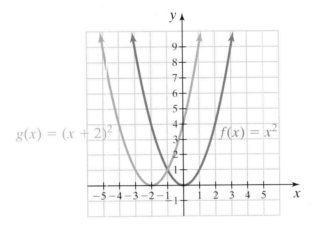

Figure 2

It follows that if c is a positive number, then $y = f(x + c)$ is the graph of f shifted to the *left* c units. For $c > 0$, the graph of $y = f(x - c)$ is the graph of f shifted to the *right* c units. At first glance, it appears that the rule for horizontal shifts is the opposite of what seems natural. Substituting $x + c$ for x causes the graph of $y = f(x)$ to be shifted to the left, whereas substituting $x - c$ for x causes the graph to shift to the right c units.

Horizontal Shifts of Functions

If c is a positive real number,

   The graph of $y = f(x + c)$ is obtained by shifting the graph of $y = f(x)$ horizontally to the left c units.

   The graph of $y = f(x - c)$ is obtained by shifting the graph of $y = f(x)$ horizontally to the right c units.

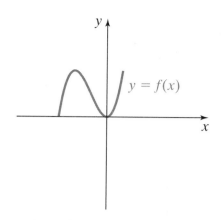

Sketch $y = f(x + c)$.   Sketch $y = f(x - c)$.

Click to animate.   Click to animate.

You Try It   **Work through this You Try It problem.**

Work Exercises 8–14 in this textbook or in the *MyMathLab*® Study Plan.

Example 2 Combining Horizontal and Vertical Shifts

 Use the graph of $y = x^3$ to sketch the graph of $g(x) = (x - 1)^3 + 2$.

Solution The graph of g is obtained by shifting the graph of the basic function $y = x^3$ first horizontally to the right one unit and then vertically upward two units. When doing a problem with multiple transformations, it is good practice to always perform the vertical transformation last. Click on the animate button to see how to sketch the graph of $g(x) = (x - 1)^3 + 2$.

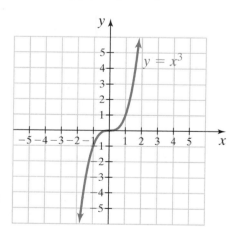

Click to animate.

You Try It   **Work through this You Try It problem.**

Work Exercises 15–21 in this textbook or in the *MyMathLab*® Study Plan.

OBJECTIVE 3:   USE REFLECTIONS TO GRAPH FUNCTIONS

My video summary

Given the graph of $y = f(x)$, what does the graph of $y = -f(x)$ look like?

Using a graphing utility with $y_1 = x^2$ and $y_2 = -x^2$, we can see that the graph of $y_2 = -x^2$ is the graph of $y_1 = x^2$ reflected about the **x-axis**.

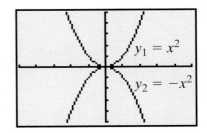

Reflections of Functions about the x-Axis

The graph of $y = -f(x)$ is obtained by reflecting the graph of $y = f(x)$ about the x-axis.

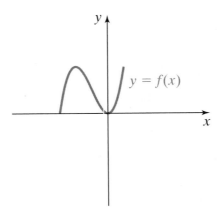

Sketch $y = -f(x)$.

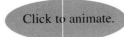

Functions can also be reflected about the y-axis. Given the graph of $y = f(x)$, the graph of $y = f(-x)$ will be the graph of $y = f(x)$ reflected about the y-axis. Using a graphing utility, we illustrate a y-axis reflection by letting $y_1 = \sqrt{x}$ and $y_2 = \sqrt{-x}$. You can see that the functions are mirror images of each other about the y-axis.

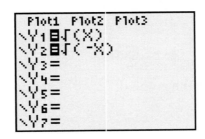

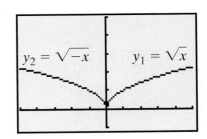

> **Reflections of Functions about the y-Axis**
>
> The graph of $y = f(-x)$ is obtained by reflecting the graph of $y = f(x)$ about the y-axis.

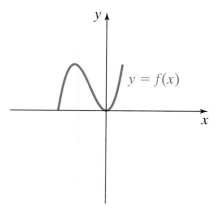

Sketch $y = f(-x)$.

Click to animate.

Example 3 Sketching Functions Using Reflections and Shifts

Use the graph of the basic function $y = \sqrt[3]{x}$ to sketch each graph.

a. $g(x) = -\sqrt[3]{x} - 2$ **b.** $h(x) = \sqrt[3]{1 - x}$.

Solution

a. Starting with the graph of $y = \sqrt[3]{x}$, we can obtain the graph of $g(x) = -\sqrt[3]{x} - 2$ by performing two transformations:

     1. Reflect about the x-axis.

     2. Vertically shift down two units.

See Figure 3.

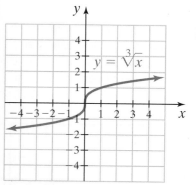

Reflect about the x-axis

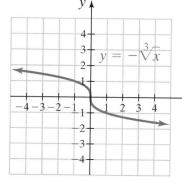

Vertical shift down two units

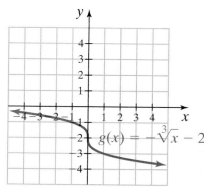

Start with the graph of the basic function $y = \sqrt[3]{x}$.

Sketch the graph of $y = -\sqrt[3]{x}$.

Sketch the graph of $g(x) = -\sqrt[3]{x} - 2$.

Figure 3

b. Starting with the graph of $y = \sqrt[3]{x}$, we can obtain the graph of $h(x) = \sqrt[3]{1 - x}$ by performing two transformations:

1. Horizontally shift left one unit.

2. Reflect about the y-axis.

See Figure 4.

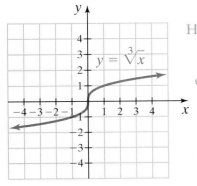
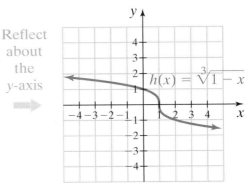

Start with the graph of the basic function $y = \sqrt[3]{x}$.

Sketch the graph of $y = \sqrt[3]{x + 1}$.

Replace x with $-x$, and sketch the graph of $h(x) = \sqrt[3]{-x + 1} = \sqrt[3]{1 - x}$.

Figure 4

You Try It   Work through this You Try It problem.

Work Exercises 22–33 in this textbook or in the **MyMathLab**® Study Plan.

OBJECTIVE 4:   USE VERTICAL STRETCHES AND COMPRESSIONS TO GRAPH FUNCTIONS

Example 4 Vertically Stretching and Compressing

Use the graph of $f(x) = x^2$ to sketch the graph of $g(x) = 2x^2$.

Solution Notice in Table 4 that for each value of x, the y-coordinate of g is two times as large as the corresponding y-coordinate of f. We can see in Figure 5 that the graph of $f(x) = x^2$ is vertically stretched by a factor of two to obtain the graph of $g(x) = 2x^2$. In other words, for each point (a, b) on the graph of f, the graph of g contains the point $(a, 2b)$.

x	$f(x) = x^2$	$g(x) = 2x^2$
-2	4	8
-1	1	2
0	0	0
1	1	2
2	4	8

Table 4

My video summary

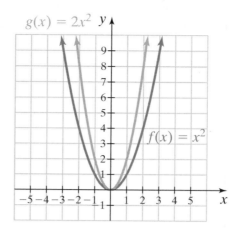

Figure 5

It follows from Example 4 that if $a > 1$, the graph of $y = af(x)$ is a **vertical stretch** of the graph of $y = f(x)$ and is obtained by multiplying each y-coordinate on the graph of f by a factor of a. If $0 < a < 1$, then the graph of $y = af(x)$ is a **vertical compression** of the graph of $y = f(x)$. Table 5 and Figure 6 show the relationship between the graphs of the functions $f(x) = x^2$ and $h(x) = \frac{1}{2}x^2$.

x	$f(x) = x^2$	$h(x) = \frac{1}{2}x^2$
-2	4	2
-1	1	$\frac{1}{2}$
0	0	0
1	1	$\frac{1}{2}$
2	4	2

Table 5

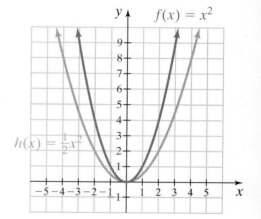

Figure 6

Vertical Stretches and Compressions of Functions

Suppose a is a positive real number:

The graph of $y = af(x)$ is obtained by multiplying each y-coordinate of $y = f(x)$ by a. If $a > 1$, the graph of $y = af(x)$ is a vertical stretch of the graph of $y = f(x)$. If $0 < a < 1$, the graph of $y = af(x)$ is a vertical compression of the graph of $y = f(x)$.

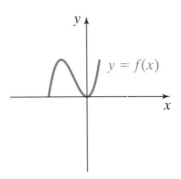

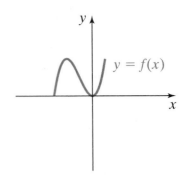

Sketch $y = af(x)$.
$(a > 1)$

Click to animate.

Sketch $y = af(x)$.
$(0 < a < 1)$

Click to animate.

You Try It Work through this You Try It problem.

Work Exercises 34–40 in this textbook or in the *MyMathLab*® Study Plan.

OBJECTIVE 5: USE HORIZONTAL STRETCHES AND COMPRESSIONS
TO GRAPH FUNCTIONS

My video summary

The final transformation to discuss is a horizontal stretch or compression. A function, $y = f(x)$, will be horizontally stretched or compressed when x is multiplied by a positive number, $a \neq 1$, to obtain the new function, $y = f(ax)$.

Horizontal Stretches and Compressions of Functions

If a is a positive real number,

For $a > 1$, the graph of $y = f(ax)$ is obtained by dividing each x-coordinate of $y = f(x)$ by a. The resultant graph is a horizontal compression.

For $0 < a < 1$, the graph of $y = f(ax)$ is obtained by dividing each x-coordinate of $y = f(x)$ by a. The resultant graph is a horizontal stretch.

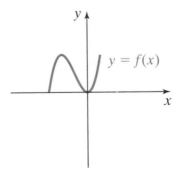

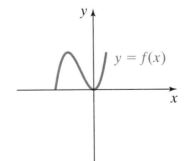

Sketch $y = f(ax)$.
$(a > 1)$

Click to animate.

Sketch $y = f(ax)$.
$(0 < a < 1)$

Click to animate.

Example 5 Horizontally Stretching and Compressing

Use the graph of $f(x) = \sqrt{x}$ to sketch the graphs of
$g(x) = \sqrt{4x}$ and $h(x) = \sqrt{\frac{1}{4}x}$.

Solution The graph of $f(x) = \sqrt{x}$ contains the ordered pairs $(0, 0), (1, 1), (4, 2)$.
To sketch the graph of $g(x) = \sqrt{4x}$, we must divide each previous x-coordinate
by 4. Therefore, the ordered pairs $(0, 0), \left(\frac{1}{4}, 1\right), (1, 2)$ must lie on the graph of g.
You can see that the graph of g is a horizontal compression of the graph of f. See
Figure 7.

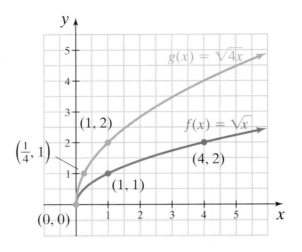

Figure 7
Graphs of $f(x) = \sqrt{x}$
and $g(x) = \sqrt{4x}$

Similarly, to sketch the graph of $h(x) = \sqrt{\frac{1}{4}x}$, we divide the x-coordinates of the
ordered pairs of f by $\frac{1}{4}$ to get the ordered pairs $(0, 0), (4, 1), (16, 2)$. You can see
that the graph of h is a horizontal stretch of the graph of f. See Figure 8.

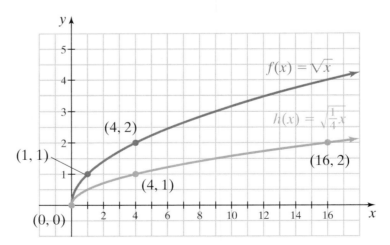

Figure 8
Graphs of $f(x) = \sqrt{x}$ and $g(x) = \sqrt{\frac{1}{4}x}$

You Try It   Work through this You Try It problem.

Work Exercises 41–45 in this textbook or in the *MyMathLab*® Study Plan.

OBJECTIVE 6: USE COMBINATIONS OF TRANSFORMATIONS TO GRAPH FUNCTIONS

You may encounter functions that combine many (if not all) of the transformations discussed in this section. When sketching a function that involves multiple transformations, it is important to follow a certain "order of operations." Following is the order in which each transformation is performed in this text:

1. Horizontal shifts
2. Horizontal stretches/compressions
3. Reflection about y-axis
4. Vertical stretches/compressions
5. Reflection about x-axis
6. Vertical shifts

Different ordering is possible for transformations 2 through 5, but you should always perform the horizontal shift first and the vertical shift last.

Example 6 Combining Transformations

Use transformations to sketch the graph of $f(x) = -2(x + 3)^2 - 1$.

Solution Watch the animation to see how to sketch the function $f(x) = -2(x + 3)^2 - 1$ as seen in Figure 9.

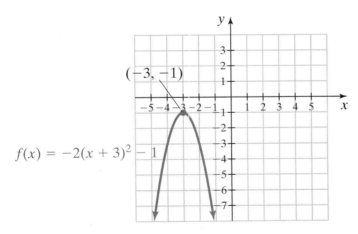

Figure 9
Graph of $f(x) = -2(x + 3)^2 - 1$

Example 7 Combining Transformations

My interactive video summary
 Use the graph of $y = f(x)$ to sketch each of the following functions.

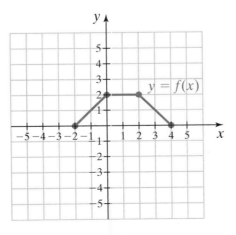

a. $y = -f(2x)$ b. $y = 2f(x - 3) - 1$ c. $y = -\dfrac{1}{2}f(2 - x) + 3$

Solution Watch the interactive video to see any one of the solutions worked out in detail.

a. The graph of $y = -f(2x)$ can be obtained from the graph of $y = f(x)$ using two transformations: (1) a horizontal compression and (2) a reflection about the x-axis. The resultant graph is shown in Figure 10.

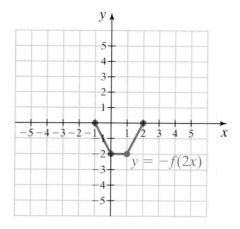

Figure 10
Graph of $y = -f(2x)$

b. The graph of $y = 2f(x - 3) - 1$ can be obtained from the graph of $y = f(x)$ using three transformations: (1) a horizontal shift to the right three units, (2) a vertical stretch by a factor of 2, and (3) a vertical shift down one unit. The resultant graph is shown in Figure 11.

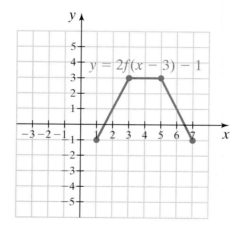

Figure 11
Graph of $y = 2f(x - 3) - 1$

c. The graph of $y = -\frac{1}{2}f(2 - x) + 3$ can be obtained from the graph of $y = f(x)$ using five transformations: (1) a horizontal shift to the left two units, (2) a reflection about the y-axis, (3) a vertical compression by a factor of $\frac{1}{2}$, (4) a reflection about the x-axis, and (5) a vertical shift up three units. The resultant graph is shown in Figure 12.

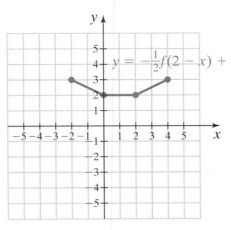

Figure 12

Graph of $y = -\dfrac{1}{2}f(2 - x) + 3$

You Try It   Work through this You Try It problem.

Work Exercises 46–55 in this textbook or in the *MyMathLab*® Study Plan

9.1 Exercises

In Exercises 1–5, use the graph of a known basic function and vertical shifting to sketch each function.

1. $f(x) = x^2 - 1$ **2.** $y = \sqrt{x} + 2$ **3.** $h(x) = \sqrt[3]{x} - 2$

You Try It

4. $y = |x| - 3$ **5.** $g(x) = x^3 + 1$

6. Use the graph of $y = f(x)$ to sketch the graph of $y = f(x) - 1$. Label at least three points on the new graph.

7. Use the graph of $y = f(x)$ to sketch the graph of $y = f(x) + 2$. Label at least three points on the new graph.

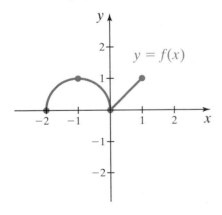

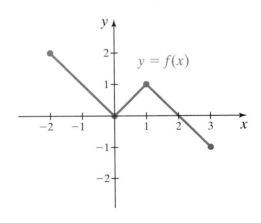

In Exercises 8–12, use the graph of a known basic function and horizontal shifts to sketch each function.

8. $f(x) = \sqrt[3]{x - 2}$ **9.** $y = \sqrt{x - 4}$ **10.** $h(x) = (x + 1)^3$

You Try It

11. $k(x) = |x - 1|$ **12.** $y = (x - 3)^2$

13. Use the graph of $y = f(x)$ to sketch the graph of $y = f(x - 2)$. Label at least three points on the new graph.

14. Use the graph of $y = f(x)$ to sketch the graph of $y = f(x + 2)$. Label at least three points on the new graph.

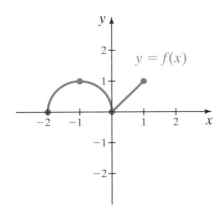

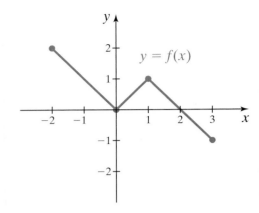

In Exercises 15–19, use the graph of a known basic function and a combination of horizontal and vertical shifts to sketch each function.

15. $y = (x + 1)^2 - 2$

16. $f(x) = (x - 3)^2 + 1$

17. $y = \sqrt{x + 3} + 2$

You Try It

18. $f(x) = |x + 2| + 2$

19. $y = \sqrt[3]{x + 1} - 1$

20. Use the graph of $y = f(x)$ to sketch the graph of $y = f(x - 2) - 1$. Label at least three points on the new graph.

21. Use the graph of $y = f(x)$ to sketch the graph of $y = f(x + 1) + 2$. Label at least three points on the new graph.

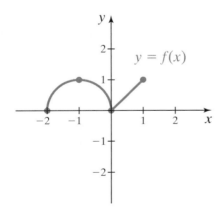

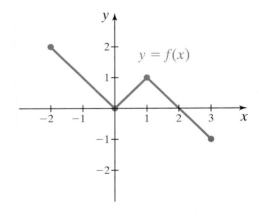

In Exercises 22–31, use the graph of a known basic function and a combination of horizontal shifts, reflections, and vertical shifts to sketch each function.

22. $g(x) = -x^2 - 2$

23. $h(x) = \sqrt{2 - x}$

24. $f(x) = |-1 - x|$

You Try It

25. $h(x) = \sqrt[3]{-x} - 2$

26. $g(x) = -x^3 + 1$

27. $h(x) = -\sqrt[3]{x} + 2$

28. $g(x) = (3 - x)^2$

29. $h(x) = -\sqrt{x} - 1$

30. $f(x) = -|x| + 1$

31. $g(x) = (1 - x)^3$

32. Use the graph of $y = f(x)$ to sketch the graph of $y = -f(x)$. Label at least three points on the new graph.

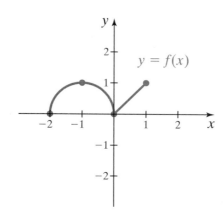

33. Use the graph of $y = f(x)$ to sketch the graph of $y = f(-x)$. Label at least three points on the new graph.

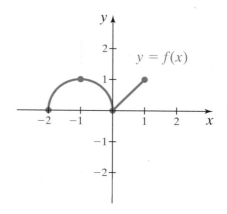

 In Exercises 34–38, use the graph of a known basic function and a vertical stretch or vertical compression to sketch each function.

You Try It

34. $f(x) = \dfrac{1}{4}|x|$

35. $g(x) = 2\sqrt{x}$

36. $f(x) = \dfrac{1}{3}x^3$

37. $f(x) = \dfrac{1}{2}\sqrt[3]{x}$

38. $g(x) = 3x^2$

39. Use the graph of $y = f(x)$ to sketch the graph of $y = 3f(x)$. Label at least three points on the new graph.

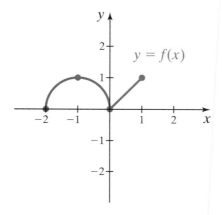

40. Use the graph of $y = f(x)$ to sketch the graph of $y = \frac{1}{2}f(x)$. Label at least three points on the new graph.

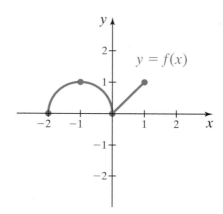

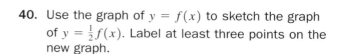 In Exercises 41–43, use the graph of a known basic function and a horizontal stretch or horizontal compression to sketch each function.

You Try It

41. $y = \left|\dfrac{1}{4}x\right|$

42. $f(x) = \sqrt{2x}$

43. $g(x) = \sqrt[3]{3x}$

44. Use the graph of $y = f(x)$ to sketch the graph of $y = f(2x)$. Label at least three points on the new graph.

45. Use the graph of $y = f(x)$ to sketch the graph of $y = f(\frac{1}{2}x)$. Label at least three points on the new graph.

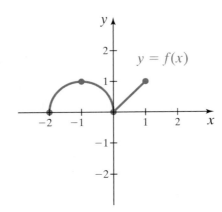

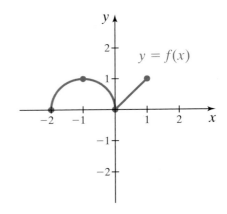

 In Exercises 46–50, use the graph of a known basic function and a combination of transformations to sketch each function.

You Try It

46. $f(x) = -(x - 2)^2 + 3$

47. $g(x) = \dfrac{1}{2}|x + 1| - 1$

48. $f(x) = 2\sqrt[3]{x} - 1$

49. $g(x) = -\dfrac{1}{2}(2 - x)^3 + 1$

50. $h(x) = 2\sqrt{4 - x} + 5$

 In Exercises 51–55, use the graph of $y = f(x)$ to sketch each function. Label at least three points on each graph.

51. $y = -f(-x) - 1$

52. $y = \dfrac{1}{2}f(2 - x)$

53. $y = -2f(x + 1) + 2$

54. $y = 3 - 3f(x + 3)$

55. $y = -f(1 - x) - 2$

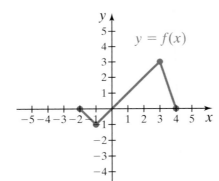

9.2 Composite and Inverse Functions

THINGS TO KNOW

Before working through this section, be sure that you are familiar with the following concepts:

VIDEO   ANIMATION   INTERACTIVE

You Try It

1. Find the Domain and Range of a Relation
   (Section 2.2, **Objective 2**)

You Try It

2. Evaluate Functions
   (Section 2.3, **Objective 2**)

You Try It

3. Graph Simple Functions by Plotting Points
   (Section 2.3, **Objective 3**)

OBJECTIVES

1 Form and Evaluate Composite Functions

2 Determine the Domain of Composite Functions

3 Determine If a Function Is One-to-One Using the Horizontal Line Test

4 Verify Inverse Functions

5 Sketch the Graphs of Inverse Functions

6 Find the Inverse of a One-to-One Function

OBJECTIVE 1:   FORM AND EVALUATE COMPOSITE FUNCTIONS

My video summary

Consider the functions $f(x) = x^2$ and $g(x) = 2x + 1$. How could we find $f(g(x))$? To find $f(g(x))$, we substitute $g(x)$ for x in the function f to get

$$f(g(x)) = f(2x + 1) = (2x + 1)^2.$$

Substitute $g(x)$ into f

The diagram in Figure 13 shows that given a number x, we first apply it to the function g to obtain $g(x)$. We then substitute $g(x)$ into f to get the result.

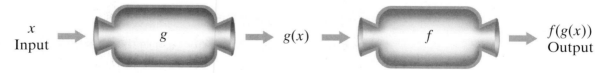

Figure 13 Composition of f and g

The function $f(g(x))$ is called a *composite function* because one function is "composed" of another function.

Definition   Composite Function

Given functions f and g, the **composite function**, $f \circ g$, (also called the **composition of f and g**) is defined by

$$(f \circ g)(x) = f(g(x)),$$

provided $g(x)$ is in the domain of f.

The composition of f and g does not equal the product of f and g:

$$(f \circ g)(x) \neq f(x)g(x).$$

Also, the composition of f and g does not necessarily equal the composition of g and f, although this equality does exist for certain pairs of functions.

Example 1 Forming and Evaluating Composite Functions

Let $f(x) = 4x + 1$, $g(x) = \dfrac{x}{x - 2}$ and $h(x) = \sqrt{x + 3}$.

a. Find the function $f \circ g$.

b. Find the function $g \circ h$.

c. Find the function $h \circ f \circ g$.

d. Evaluate $(f \circ g)(4)$, or state that it is undefined.

e. Evaluate $(g \circ h)(1)$, or state that it is undefined.

f. Evaluate $(h \circ f \circ g)(6)$, or state that it is undefined.

Solution Work through the interactive video to verify the following:

a. $(f \circ g)(x) = 4\left(\dfrac{x}{x - 2}\right) + 1 = \dfrac{5x - 2}{x - 2}$

b. $(g \circ h)(x) = \dfrac{\sqrt{x + 3}}{\sqrt{x + 3} - 2}$

c. $(h \circ f \circ g)(x) = \sqrt{\dfrac{5x - 2}{x - 2} + 3} = \sqrt{\dfrac{8x - 8}{x - 2}}$

d. $(f \circ g)(4) = 9$

e. $(g \circ h)(1)$ is undefined

f. $(h \circ f \circ g)(6) = \sqrt{10}$

You Try It   Work through this You Try It problem.

Work Exercises 1–16 in this textbook or in the *MyMathLab*® Study Plan.

Example 2 Evaluating Composite Functions Using a Graph

 *My interactive video summary*

 Use the graph to evaluate each expression:

a. $(f \circ g)(4)$

b. $(g \circ f)(-3)$

c. $(f \circ f)(-1)$

d. $(g \circ g)(4)$

e. $(f \circ g \circ f)(1)$

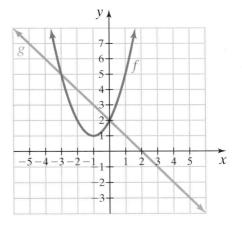

Solution Work through the interactive video to verify the following:

a. $(f \circ g)(4) = 2$ b. $(g \circ f)(-3) = -3$ c. $(f \circ f)(-1) = 5$

d. $(g \circ g)(4) = 4$ e. $(f \circ g \circ f)(1) = 5$

You Try It Work through this You Try It problem.

Work Exercises **17** and **18** in this textbook or in the *MyMathLab*® Study Plan.

OBJECTIVE 2: DETERMINE THE DOMAIN OF COMPOSITE FUNCTIONS

My video summary

Suppose f and g are functions. For a number x to be in the domain of $f \circ g$, x must be in the domain of g *and* $g(x)$ must be in domain of f. Follow these two steps to find the domain of $f \circ g$:

1. Find the domain of g.

2. Exclude from the domain of g all values of x for which $g(x)$ is not in the domain of f.

Example 3 Finding the Domain of a Composite Function

My interactive video summary

Let $f(x) = \dfrac{-10}{x - 4}$ and $g(x) = \sqrt{5 - x}$.

a. Find the domain of $f \circ g$.

b. Find the domain of $g \circ f$.

Solution

a. First, form the composite function $(f \circ g)(x) = \dfrac{-10}{\sqrt{5 - x} - 4}$.

To find the domain of $f \circ g$, we follow these two steps:

Step 1. Find the domain of g.

The domain of $g(x) = \sqrt{5 - x}$ is $(-\infty, 5]$. The domain of $f \circ g$ cannot contain any values of x that are not in this interval. In other words, the domain of $f \circ g$ is a subset of $(-\infty, 5]$.

Step 2. Exclude from the domain of g all values of x for which $g(x)$ is not in the domain of f.

All real numbers except 4 are in the domain of f. This implies that $g(x)$ cannot equal 4 because $g(x)$ equal to 4 would make the denominator of f, $x - 4$, equal to zero. Thus, we must exclude all values of x such that $g(x) = 4$.

$$\text{Substitute } \sqrt{5 - x} \text{ for } g(x): \qquad g(x) = 4$$
$$\text{Square both sides:} \quad \sqrt{5 - x} = 4$$
$$5 - x = 16$$
$$\text{Solve for } x: \qquad x = -11$$

We must *exclude* $x = -11$ from the domain of $f \circ g$. Therefore, the domain of $f \circ g$ is all values of x less than 5 such that $x \neq -11$, or the interval $(-\infty, -11) \cup (-11, 5]$. You may want to view the interactive video to view the solution in more detail.

b. You should carefully work through the interactive video to verify that the domain of $(g \circ f)(x) = \sqrt{5 + \dfrac{10}{x - 4}} = \sqrt{\dfrac{5x - 10}{x - 4}}$ is the interval $(-\infty, 2] \cup (4, \infty)$.

You Try It **Work through this You Try It problem.**

Work Exercises 19–24 in this textbook or in the *MyMathLab*®️ Study Plan.

OBJECTIVE 3: **DETERMINE IF A FUNCTION IS ONE-TO-ONE USING THE HORIZONTAL LINE TEST**

Later in this section, we examine a process for finding the inverse of a function, but keep the following in mind: We can find the inverse of many functions using this process, but that inverse will not always be a function. In this text, we are only interested in inverses that are functions, so we first develop a test to determine whether a function has an inverse *function*. When the word *inverse* is used throughout the remainder of this section, we assume that we are referring to the inverse that is a function.

My video summary

First, we must define the concept of **one-to-one** functions.

Definition **One-to-One Function**

A function f is **one-to-one** if for any values $a \neq b$ in the domain of f, $f(a) \neq f(b)$.

This definition suggests that a function is one-to-one if for any two *different* input values (domain values), the corresponding output values (range values) must be different. An alternate definition says that if two range values are the same, $f(u) = f(v)$, then the domain values must be the same; that is, $u = v$.

> **Alternate Definition of a One-to-One Function**
>
> A function f is **one-to-one** if for any two range values $f(u)$ and $f(v)$, $f(u) = f(v)$ implies that $u = v$.

The function sketched in Figure 14a is one-to-one because for any two distinct x-values in the domain $(a \neq b)$, the function values or range values are not equal $(f(a) \neq f(b))$. In Figure 14b, we see that the function $y = g(x)$ is *not* one-to-one because we can easily find two different domain values that correspond to the same range value.

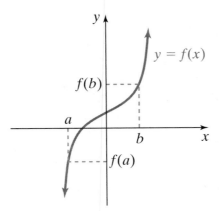

Figure 14a

An example of a one-to-one function

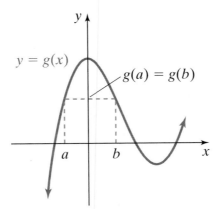

Figure 14b

An example of a function that is not one-to-one

My video summary

Notice in Figure 15 that the horizontal lines intersect the graph of $y = f(x)$ in at most one place, whereas we can find many horizontal lines that intersect the graph of $y = g(x)$ more than once. This gives us a visual example of how we can use horizontal lines to help us determine from the graph whether a function is one-to-one. Using horizontal lines to determine whether the graph of a function is one-to-one is known as the *horizontal line test*.

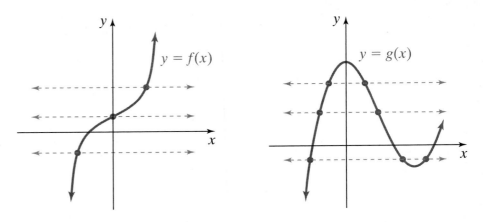

Figure 15 Drawing horizontal lines can help determine whether a graph represents a one-to-one function.

Horizontal Line Test

If every horizontal line intersects the graph of a function f at most once, then f is one-to-one.

Example 4 Determining Whether a Function Is One-to-One

 Determine whether each function is one-to-one.

a.

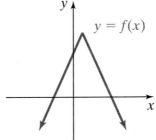

b.

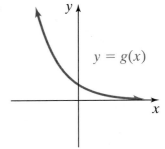

c. $f(x) = x^2 + 1, x \leq 0$

d. $f(x) = \begin{cases} 2x + 4 \text{ for } x \leq -1 \\ 2x - 6 \text{ for } x \geq 4 \end{cases}$

Solution The functions in parts b and c are one-to-one, whereas the functions in parts a and d are not one-to-one. Watch the animation to verify.

You Try It **Work through this You Try It problem.**

Work Exercises 25–39 in this textbook or in the *MyMathLab*® Study Plan.

OBJECTIVE 4: VERIFY INVERSE FUNCTIONS

My video summary

We are now ready to ask ourselves the question, "Why should we be concerned with one-to-one functions?"

Answer: Every one-to-one function has an inverse function!

To illustrate an inverse function, let's consider the function $F(C) = \frac{9}{5}C + 32$. This function is used to convert a given temperature in degrees Celsius into its equivalent temperature in degrees Fahrenheit. For example,

$$F(25) = \frac{9}{5}(25) + 32 = 45 + 32 = 77.$$

Thus, a temperature of 25°C corresponds to a temperature of 77°F. To convert a temperature of 77°F back into 25°C, we need a different function. The function that is used to accomplish this task is $C(F) = \frac{5}{9}(F - 32)$.

$$C(77) = \frac{5}{9}(77 - 32)$$

$$= \frac{5}{9}(45)$$

$$= 25$$

The function C is the *inverse* of the function F. In essence, these functions perform opposite actions (they "undo" each other). The first function converted 25°C into 77°F, whereas the second function converted 77°F back into 25°C.

Definition   Inverse Function

Let f be a one-to-one function with domain A and range B. Then f^{-1} is the **inverse function of f** with domain B and range A. Furthermore, if $f(a) = b$, then $f^{-1}(b) = a$.

My video summary

According to the definition of an inverse function, the domain of f is exactly the same as the range of f^{-1}, and the range of f is the same as the domain of f^{-1}. Figure 16 illustrates that if the function f assigns a number a to b, then the inverse function will assign b back to a. In other words, if the point (a, b) is an ordered pair on the graph of f, then the point (b, a) must be on the graph of f^{-1}.

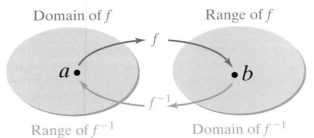

Figure 16

CAUTION   Do not confuse f^{-1} with $\frac{1}{f(x)}$. The negative 1 in f^{-1} is *not* an exponent!

As with our opening example using an inverse function to convert a Fahrenheit temperature back into a Celsius temperature, inverse functions "undo" each other. For example, it can be shown that if $f(x) = x^3$, then the inverse of f is $f^{-1}(x) = \sqrt[3]{x}$. Note that

$$f(2) = (2)^3 = 8 \quad \text{and} \quad f^{-1}(8) = \sqrt[3]{8} = 2.$$

The function f takes the number 2 to 8, whereas f^{-1} takes 8 back to 2. Observe what happens if we look at the composition of f and f^{-1} and the composition of f^{-1} and f at specified values:

x-value   Same as *x*-value

$$(f \circ f^{-1})(8) = f(f^{-1}(8)) = f(2) = 8$$

x-value   Same as *x*-value

$$(f^{-1} \circ f)(2) = f^{-1}(f(2)) = f^{-1}(8) = 2$$

Because of the "undoing" nature of inverse functions, we get the following **composition cancellation equations:**

Composition Cancellation Equations

$f(f^{-1}(x)) = x$ for all x in the domain of f^{-1}

and $f^{-1}(f(x)) = x$ for all x in the domain of f

These cancellation equations can be used to show whether two functions are inverses of each other. We can see from our example that if $f(x) = x^3$ and $f^{-1}(x) = \sqrt[3]{x}$, then

$$f(f^{-1}(x)) = f\left(\sqrt[3]{x}\right) = \left(\sqrt[3]{x}\right)^3 = x \text{ and } f^{-1}(f(x)) = f^{-1}\left(x^3\right) = \sqrt[3]{x^3} = x.$$

Example 5 Verify Inverse Functions

 Show that $f(x) = \dfrac{x}{2x + 3}$ and $g(x) = \dfrac{3x}{1 - 2x}$ are inverse functions using the composition cancellation equations.

Solution To show that f and g are inverses of each other, we must show that $(f \circ g)(x) = x$ and $(g \circ f)(x) = x$. Work through the interactive video to verify that both composition cancellation equations are satisfied.

You Try It Work through this You Try It problem.

Work Exercises 40–44 in this textbook or in the *MyMathLab*®️ Study Plan.

OBJECTIVE 5: SKETCH THE GRAPHS OF INVERSE FUNCTIONS

If f is a one-to-one function, then we know that it must have an inverse function, f^{-1}. Given the graph of a one-to-one function f, we can obtain the graph of f^{-1} by simply interchanging the coordinates of each ordered pair that lies on the graph of f. In other words, for any point (a, b) on the graph of f, the point (b, a) must lie on the graph of f^{-1}. Notice in Figure 17 that the points (a, b) and (b, a) are symmetric about the line $y = x$. Therefore, the graph of f^{-1} is a reflection of the graph of f about the line $y = x$. Figure 18 shows the graph of $f(x) = x^3$ and

$f^{-1}(x) = \sqrt[3]{x}$. You can see that if the functions have any points in common, they must lie along the line $y = x$.

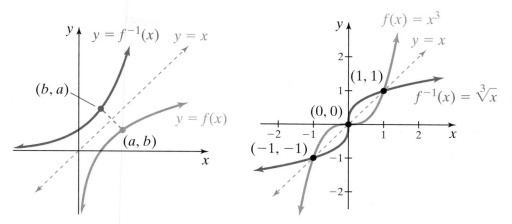

Figure 17 Graph of a one-to-one function and its inverse

Figure 18 Graph of $f(x) = x^3$ and $f^{-1}(x) = \sqrt[3]{x}$

Example 6 Sketch the Graph of a One-to-One Function and Its Inverse

 Sketch the graph of $f(x) = x^2 + 1, x \le 0$, and its inverse. Also state the domain and range of f and f^{-1}.

Solution The graphs of f and f^{-1} are sketched in Figure 19. Notice how the graph of f^{-1} is a reflection of the graph of f about the line $y = x$. Also notice that the domain of f is the same as the range of f^{-1}. Likewise, the domain of f^{-1} is equivalent to the range of f. View the animation to see exactly how to sketch f and f^{-1}.

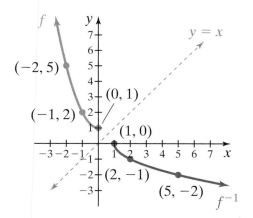

Domain of f: $(-\infty, 0]$ Domain of f^{-1}: $[1, \infty)$

Range of f: $[1, \infty)$ Range of f^{-1}: $(-\infty, 0]$

Figure 19 Graph of $f(x) = x^2 + 1, x \le 0$, and its inverse

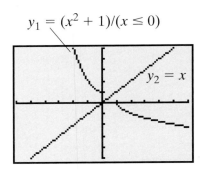

$$y_1 = (x^2 + 1)/(x \le 0)$$

$$y_2 = x$$

Using a TI-84Plus, we can sketch the functions from Example 6 by letting $y_1 = (x^2 + 1)/(x \le 0)$. We can draw the inverse function by typing the command **DrawInv** Y_1 in the calculator's main viewing window. The resultant graphs are shown here.

You Try It Work through this You Try It problem.

Work Exercises 45–50 in this textbook or in the *MyMathLab*® Study Plan.

OBJECTIVE 6: FIND THE INVERSE OF A ONE-TO-ONE FUNCTION

We are now ready to find the inverse of a one-to-one function algebraically. We know that if a point (x, y) is on the graph of a one-to-one function, then the point (y, x) is on the graph of its inverse function. We can use this information to develop a process for finding the inverse of a function algebraically simply by switching x and y in the original function to produce its inverse function.

We use as a motivating example the function $f(x) = x^2 + 1, x \le 0$, discussed in Example 6. To find the inverse of a one-to-one function, we follow the four-step process outlined here.

Step 1. Change $f(x)$ to y: $\quad y = x^2 + 1$

Step 2. Interchange x and y: $\quad x = y^2 + 1$

Step 3. $\qquad\qquad$ Solve for y: $\quad x - 1 = y^2$

$$\pm\sqrt{x - 1} = y$$

(Because the domain of f is $(-\infty, 0]$, the range of f^{-1} must be $(-\infty, 0]$. Therefore, we must use the negative square root or $y = -\sqrt{x - 1}$.)

Step 4. Change y to $f^{-1}(x)$: $\quad f^{-1}(x) = -\sqrt{x - 1}$

Thus, the inverse of $f(x) = x^2 + 1, x \le 0$, is $f^{-1}(x) = -\sqrt{x - 1}$.

Example 7 Find the Inverse of a Function

Find the inverse of the function $f(x) = \dfrac{2x}{1 - 5x}$, and state the domain and range of f and f^{-1}.

Solution Work through the animation, and follow the four-step process to verify that $f^{-1}(x) = \dfrac{x}{5x + 2}$. The domain of f is $\left(-\infty, \frac{1}{5}\right) \cup \left(\frac{1}{5}, \infty\right)$, whereas the domain of f^{-1} is $\left(-\infty, -\frac{2}{5}\right) \cup \left(-\frac{2}{5}, \infty\right)$. Because the range of f must be the domain of f^{-1} and the range of f^{-1} must be the domain of f, we get the following result:

Domain of f: $\left(-\infty, \dfrac{1}{5}\right) \cup \left(\dfrac{1}{5}, \infty\right)$ Domain of f^{-1}: $\left(-\infty, -\dfrac{2}{5}\right) \cup \left(-\dfrac{2}{5}, \infty\right)$

Range of f: $\left(-\infty, -\dfrac{2}{5}\right) \cup \left(-\dfrac{2}{5}, \infty\right)$ Range of f^{-1}: $\left(-\infty, \dfrac{1}{5}\right) \cup \left(\dfrac{1}{5}, \infty\right)$

You Try It   **Work through this You Try It problem.**

Work Exercises 51–59 in this textbook or in the **MyMathLab**® Study Plan.

Inverse Function Summary

1. f^{-1} exists if and only if the function f is one-to-one.
2. The domain of f is the same as the range of f^{-1}, and the range of f is the same as the domain of f^{-1}.
3. To verify that two one-to-one functions, f and g, are inverses of each other, we must use the composition cancellation equations to show that $f(g(x)) = g(f(x)) = x$.
4. The graph of f^{-1} is a reflection of the graph of f about the line $y = x$. That is, for any point (a, b) that lies on the graph of f, the point (b, a) must lie on the graph of f^{-1}.
5. To find the inverse of a one-to-one function, replace $f(x)$ with y, interchange the variables x and y, and solve for y. This is the function $f^{-1}(x)$.

9.2 Exercises

 In Exercises 1–8, let $f(x) = 3x + 1, g(x) = \dfrac{2}{x + 1}$, and $h(x) = \sqrt{x + 3}$.

You Try It

1. Find the function $f \circ g$.   **2.** Find the function $g \circ f$.

3. Find the function $f \circ h$.   **4.** Find the function $g \circ h$.

5. Find the function $h \circ f$.   **6.** Find the function $h \circ g$.

7. Find the function $f \circ f$.   **8.** Find the function $g \circ g$.

In Exercises 9–16, evaluate the following composite functions given that

$f(x) = 3x + 1, g(x) = \dfrac{2}{x + 1}$, and $h(x) = \sqrt{x + 3}$.

9. $(f \circ g)(0)$ **10.** $(f \circ h)(6)$ **11.** $(g \circ f)(1)$ **12.** $(g \circ h)(-2)$

13. $(h \circ f)(0)$ **14.** $(h \circ g)(3)$ **15.** $(f \circ f)(-1)$ **16.** $(g \circ g)(4)$

You Try It

In Exercises 17 and 18, use the graph to evaluate each expression.

 17. **a.** $(f \circ g)(1)$ **b.** $(g \circ f)(-1)$ **18.** **a.** $(f \circ g)(1)$ **b.** $(g \circ f)(-1)$

      c. $(g \circ g)(0)$ **d.** $(f \circ f)(1)$ **c.** $(g \circ g)(0)$ **d.** $(f \circ f)(1)$

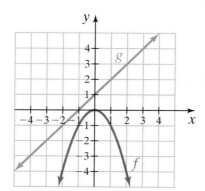 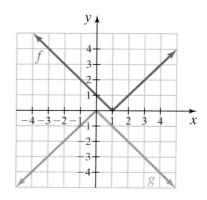

You Try It

 In Exercises 19–24, find the domain of $(f \circ g)(x)$ and $(g \circ f)(x)$.

19. $f(x) = x^2, g(x) = 2x - 1$ **20.** $f(x) = 3x - 5, g(x) = 2x^2 + 1$

21. $f(x) = x^2, g(x) = \sqrt{x}$ **22.** $f(x) = \dfrac{1}{x}, g(x) = x^2 - 4$

You Try It

23. $f(x) = \dfrac{3}{x + 1}, g(x) = \dfrac{x}{x - 2}$ **24.** $f(x) = \dfrac{2x}{x - 3}, g(x) = \dfrac{x + 1}{x - 1}$

In Exercises 25–39, determine whether each function is one-to-one.

25. $f(x) = 3x - 1$ **26.** $f(x) = 2x^2$ **27.** $f(x) = (x - 1)^2, x \geq 1$

28. $f(x) = (x - 1)^2, x \geq -1$ **29.** $f(x) = \dfrac{1}{x} - 2$ **30.** $f(x) = 4\sqrt{x}$

31. $f(x) = -2|x|$ **32.** $f(x) = 2$ **33.** $f(x) = (x + 1)^3 - 2$

34.

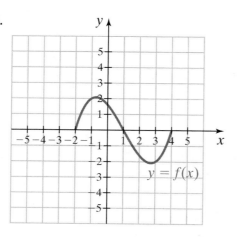

35.

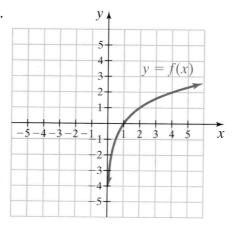

36.

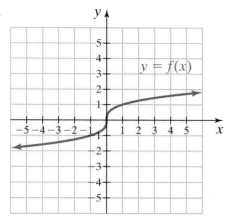

37.

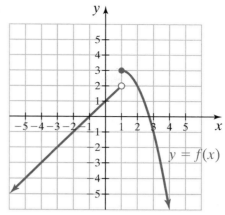

38.

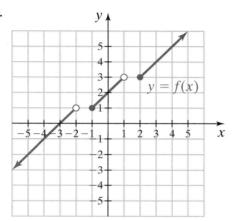

39.

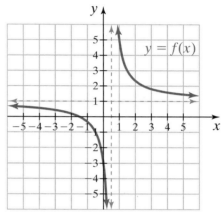

You Try It

In Exercises 40–44, use the composition cancellation equations to verify that f and g are inverse functions.

40. $f(x) = \dfrac{3}{2}x - 4$ and $g(x) = \dfrac{2x + 8}{3}$

41. $f(x) = (x - 1)^2, x \geq 1$, and $g(x) = \sqrt{x} + 1$

42. $f(x) = \dfrac{7}{x + 1}$ and $g(x) = \dfrac{7 - x}{x}$

43. $f(x) = \dfrac{x}{5 + 3x}$ and $g(x) = \dfrac{5x}{1 - 3x}$

44. $f(x) = 2\sqrt[3]{x - 1} + 3$ and $g(x) = \dfrac{(x - 3)^3}{8} + 1$

In Exercises 45–50, use the graph of f to sketch the graph of f^{-1}. Use the graphs to determine the domain and range of each function.

You Try It

45.

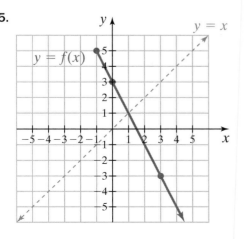

46.

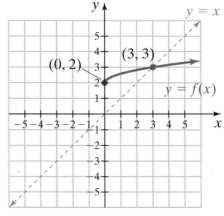

47.

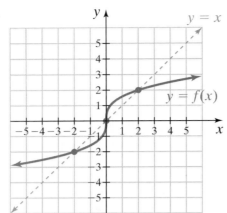

48.

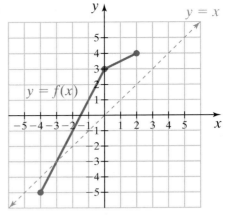

49.

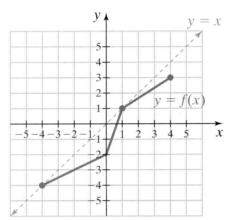

50.

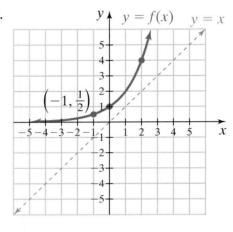

In Exercises 51–59, write an equation for the inverse function, and then state the domain and range of f and f^{-1}.

You Try It

51. $f(x) = \dfrac{1}{3}x - 5$

52. $f(x) = \dfrac{3x + 9}{7}$

53. $f(x) = \sqrt[3]{2x - 3}$

54. $f(x) = 1 - \sqrt[5]{x + 4}$

55. $f(x) = -x^2 - 2, x \geq 0$

56. $f(x) = (x + 3)^2 - 5, x \leq -3$

57. $f(x) = \dfrac{3}{x}$

58. $f(x) = \dfrac{1 - x}{2x}$

59. $f(x) = \dfrac{8x - 1}{7 - 5x}$

9.3 Exponential Functions

THINGS TO KNOW

Before working through this section, be sure that you are familiar with the following concepts:

VIDEO   ANIMATION   INTERACTIVE

You Try It

1. Use Combinations of Transformations to Graph Functions (Section 9.1, Objective 6)

You Try It

2. Determine If a Function Is One-to-One Using the Horizontal Line Test (Section 9.2, Objective 3)

OBJECTIVES

1 Use the Characteristics of Exponential Functions

2 Sketch the Graphs of Exponential Functions Using Transformations

3 Solve Exponential Equations by Relating the Bases

4 Solve Applications of Exponential Functions

OBJECTIVE 1: USE THE CHARACTERISTICS OF EXPONENTIAL FUNCTIONS

Many natural phenomena and real-life applications can be modeled using exponential functions. Before we define the exponential function, it is important to remember how to manipulate exponential expressions because this skill is necessary when solving certain equations involving exponents. In Section 7.3, expressions of the form b^r were evaluated for rational numbers r. For example,

$$3^2 = 9, \quad 4^{-2} = \frac{1}{4^2} = \frac{1}{16}, \quad \text{and} \quad 27^{-2/3} = \frac{1}{27^{2/3}} = \frac{1}{(\sqrt[3]{27})^2} = \frac{1}{(3)^2} = \frac{1}{9}.$$

In this section, we extend the meaning of b^r to include all **real** values of r by defining the exponential function $f(x) = b^x$.

Definition   **Exponential Function**

An **exponential function** is a function of the form $f(x) = b^x$, where x is any real number and $b > 0$ such that $b \neq 1$.

The constant, b, is called the base of the exponential function.

Notice in the definition that the base, b, must be positive and must not equal 1. If $b = 1$, then the function $f(x) = 1^x$ is equal to 1 for all x and is hence equivalent to the constant function $f(x) = 1$. If b were negative, then $f(x) = b^x$ would not be defined for all real values of x. For example, if $b = -4$, then $f\left(\frac{1}{2}\right) = (-4)^{1/2} = \sqrt{-4} = 2i$, which is *not* a positive real number.

My video summary

▣ Before we generalize the graph of $f(x) = b^x$, we create a table of values and sketch the graph of $y = b^x$ for $b = 2, 3, \frac{1}{2}$, and $\frac{1}{3}$. See Table 6.

Table 6

x	$y = 2^x$	$y = 3^x$	$y = \left(\frac{1}{2}\right)^x$	$y = \left(\frac{1}{3}\right)^x$
-2	$2^{-2} = \frac{1}{2^2} = \frac{1}{4}$	$3^{-2} = \frac{1}{3^2} = \frac{1}{9}$	$\left(\frac{1}{2}\right)^{-2} = 2^2 = 4$	$\left(\frac{1}{3}\right)^{-2} = 3^2 = 9$
-1	$2^{-1} = \frac{1}{2^1} = \frac{1}{2}$	$3^{-1} = \frac{1}{3^1} = \frac{1}{3}$	$\left(\frac{1}{2}\right)^{-1} = 2^1 = 2$	$\left(\frac{1}{3}\right)^{-1} = 3^1 = 3$
0	$2^0 = 1$	$3^0 = 1$	$\left(\frac{1}{2}\right)^0 = 1$	$\left(\frac{1}{3}\right)^0 = 1$
1	$2^1 = 2$	$3^1 = 3$	$\left(\frac{1}{2}\right)^1 = \frac{1}{2}$	$\left(\frac{1}{3}\right)^1 = \frac{1}{3}$
2	$2^2 = 4$	$3^2 = 9$	$\left(\frac{1}{2}\right)^2 = \frac{1}{4}$	$\left(\frac{1}{3}\right)^2 = \frac{1}{9}$

You can see from the graphs sketched in Figure 20 that all four functions intersect the y-axis at the point $(0, 1)$. This is true because $b^0 = 1$ for all nonzero values of b. For values of $b > 1$, the graph of $y = b^x$ increases rapidly as the values of x approach positive infinity ($b^x \to \infty$ as $x \to \infty$). In fact, the larger the base, the faster the graph will grow. Also, for $b > 1$, the graph of $y = b^x$ decreases quickly, approaching 0 as the values of x approach negative infinity ($b^x \to 0$ as $x \to -\infty$). Thus, the x-axis (the line $y = 0$) is a horizontal asymptote.

However, for $0 < b < 1$, the graph decreases quickly, approaching the horizontal asymptote $y = 0$ as the values of x approach positive infinity ($b^x \to 0$ as $x \to \infty$), whereas the graph increases rapidly as the values of x approach negative infinity ($b^x \to \infty$ as $x \to -\infty$). The preceding statements, along with some other characteristics of the graphs of exponential functions, are outlined on the following page.

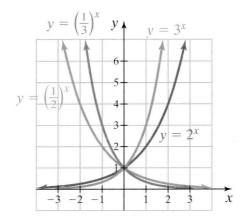

Figure 20
Graphs of $y = 2^x$, $y = 3^x$,
$y = \left(\frac{1}{2}\right)^x$, and $y = \left(\frac{1}{3}\right)^x$

Characteristics of Exponential Functions

For $b > 0, b \neq 1$, the exponential function with base b is defined by $f(x) = b^x$.

The domain of $f(x) = b^x$ is $(-\infty, \infty)$, and the range is $(0, \infty)$. The graph of $f(x) = b^x$ has one of the following two shapes:

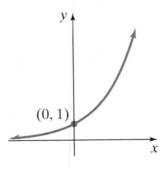

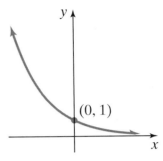

$f(x) = b^x, b > 1$ $f(x) = b^x, 0 < b < 1$

The graph intersects the y-axis at $(0, 1)$.

The graph intersects the y-axis at $(0, 1)$.

$b^x \to \infty$ as $x \to \infty$ $b^x \to 0$ as $x \to \infty$

$b^x \to 0$ as $x \to -\infty$ $b^x \to \infty$ as $x \to -\infty$

The line $y = 0$ is a horizontal asymptote.

The line $y = 0$ is a horizontal asymptote.

The function is one-to-one.   The function is one-to-one.

My video summary

Example 1 Sketching the Graph of an Exponential Function

Sketch the graph of $f(x) = \left(\frac{2}{3}\right)^x$.

Solution Because the base of the exponential function is $\frac{2}{3}$, which is between 0 and 1, the graph must approach the x-axis as the value of x approaches positive infinity. The graph intersects the y-axis at $(0, 1)$. We can find a few more points by choosing some negative and positive values of x:

$$f(-2) = \left(\frac{2}{3}\right)^{-2} = \left(\frac{3}{2}\right)^2 = \frac{9}{4}$$

$$f(-1) = \left(\frac{2}{3}\right)^{-1} = \left(\frac{3}{2}\right)^1 = \frac{3}{2}$$

$$f(1) = \left(\frac{2}{3}\right)^1 = \frac{2}{3}$$

$$f(2) = \left(\frac{2}{3}\right)^2 = \frac{4}{9}$$

We can complete the graph by connecting the points with a smooth curve. See Figure 21.

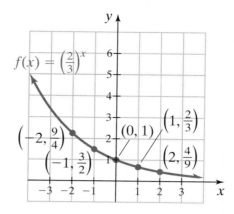

Figure 21
Graph of $f(x) = \left(\frac{2}{3}\right)^x$

You Try It   Work through this You Try It problem.

Work Exercises 1–5 in this textbook or in the *MyMathLab*® Study Plan.

Example 2 Determining an Exponential Function Given the Graph

Find the exponential function $f(x) = b^x$ whose graph is given as follows.

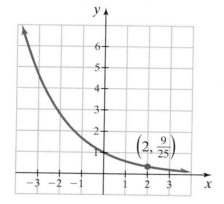

Solution From the point $\left(2, \frac{9}{25}\right)$, we see that $f(2) = \frac{9}{25}$. Thus,

$$\text{Write the exponential function } f(x) = b^x: \quad f(x) = b^x$$

$$\text{Evaluate } f(2): \quad f(2) = b^2$$

$$\text{The graph contains the point } \left(2, \frac{9}{25}\right): \quad f(2) = \frac{9}{25}$$

$$\text{Equate the two expressions for } f(2): \quad b^2 = \frac{9}{25}$$

Therefore, we are looking for a constant b such that $b^2 = \frac{9}{25}$. Using the square root property, we get

$$\sqrt{b^2} = \pm\sqrt{\frac{9}{25}}$$

$$b = \pm\frac{3}{5}.$$

By definition of an exponential function, $b > 0$; thus, $b = \frac{3}{5}$. Therefore, this is the graph of $f(x) = \left(\frac{3}{5}\right)^x$.

You Try It Work through this You Try It problem.

Work Exercises 6–12 in this textbook or in the *MyMathLab*® Study Plan.

OBJECTIVE 2: SKETCH THE GRAPHS OF EXPONENTIAL FUNCTIONS USING TRANSFORMATIONS

Often we can use the transformation techniques that are discussed in Section 9.1 to sketch the graph of exponential functions. For example, the graph of $f(x) = 3^x - 1$ can be obtained by vertically shifting the graph of $y = 3^x$ down one unit. You can see in Figure 22 that the y-intercept of $f(x) = 3^x - 1$ is $(0, 0)$ and the horizontal asymptote is the line $y = -1$.

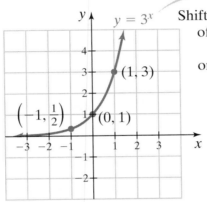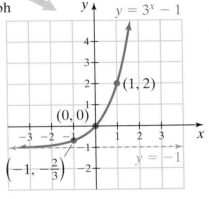

Figure 22
The graph of $f(x) = 3^x - 1$ can be obtained by vertically shifting the graph of $y = 3^x$ down one unit.

My video summary

Example 3 Using Transformations to Sketch an Exponential Function

Use transformations to sketch the graph of $f(x) = -2^{x+1} + 3$.

Solution Starting with the graph of $y = 2^x$, we can obtain the graph of $f(x) = -2^{x+1} + 3$ through a series of three transformations:

1. Horizontally shift the graph of $y = 2^x$ to the left one unit, producing the graph of $y = 2^{x+1}$.
2. Reflect the graph of $y = 2^{x+1}$ about the x-axis, producing the graph of $y = -2^{x+1}$.
3. Vertically shift the graph of $y = -2^{x+1}$ up three units, producing the graph of $f(x) = -2^{x+1} + 3$.

The graph of $f(x) = -2^{x+1} + 3$ is shown in Figure 23. Watch the video to see every step.

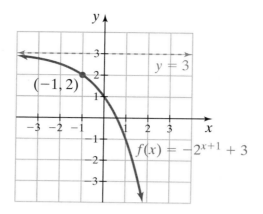

Figure 23
Graph of $f(x) = -2^{x+1} + 3$

Notice that the graph of $f(x) = -2^{x+1} + 3$ in Figure 23 has a y-intercept. We can find the y-intercept by evaluating $f(0) = -2^{0+1} + 3 = -2 + 3 = 1$. Also notice that the graph has an x-intercept. Can you find it? Recall that to find an x-intercept, we need to set $f(x) = 0$ and solve for x.

Write the original function:	$f(x) = -2^{x+1} + 3$
Substitute 0 for $f(x)$:	$0 = -2^{x+1} + 3$
Subtract 3 from both sides:	$-3 = -2^{x+1}$
Multiply both sides by -1:	$3 = 2^{x+1}$

We now have a problem! We are going to need a way to solve for a variable that appears in an exponent. (Stay tuned for Section 9.7.)

You Try It **Work through this You Try It problem.**

Work Exercises 13–20 in this textbook or in the *MyMathLab*® Study Plan.

OBJECTIVE 3: SOLVE EXPONENTIAL EQUATIONS BY RELATING THE BASES

One important property of all exponential functions is that they are one-to-one functions. You may want to review Section 9.2, which discusses one-to-one functions in detail. The function $f(x) = b^x$ is one-to-one because the graph of f passes the horizontal line test. In Section 9.2, the alternate definition of one-to-one is stated as follows:

> A function f is one-to-one if for any two range values
> $f(u)$ and $f(v)$, $f(u) = f(v)$ implies that $u = v$.

Using this definition and letting $f(x) = b^x$, we can say that if $b^u = b^v$, then $u = v$. In other words, if the bases of an exponential equation of the form $b^u = b^v$ are the same, then the exponents must be the same. Solving exponential equations with this property is known as the **method of relating the bases** for solving exponential equations.

Method of Relating the Bases

If an exponential equation can be written in the form $b^u = b^v$, then $u = v$.

Example 4 Using the Method of Relating the Bases to Solve Exponential Equations

Solve the following equations.

a. $8 = \dfrac{1}{16^x}$ **b.** $\dfrac{1}{27^x} = \left(\sqrt[4]{3}\right)^{x-2}$

Solution

a. Work through the animation to see how to obtain a solution of $x = -\dfrac{3}{4}$.

b. Work through the animation to see how to obtain a solution of $x = \dfrac{2}{13}$.

You Try It Work through this You Try It problem.

Work Exercises 21–30 in this textbook or in the *MyMathLab*® Study Plan.

OBJECTIVE 4: SOLVE APPLICATIONS OF EXPONENTIAL FUNCTIONS

Exponential functions are used to describe many real-life situations and natural phenomena. We now look at some examples.

Example 5 Learn to Hit a 3-Wood on a Golf Driving Range

Most golfers find that their golf skills improve dramatically at first and then level off rather quickly. For example, suppose that the distance (in yards) that a typical beginning golfer can hit a 3-wood after t weeks of practice on the driving range is given by the exponential function $d(t) = 225 - 100(2.7)^{-0.7t}$. This function has been developed after many years of gathering data on beginning golfers.

How far can a typical beginning golfer initially hit a 3-wood? How far can a typical beginning golfer hit a 3-wood after 1 week of practice on the driving range? After 5 weeks? After 9 weeks? Round to the nearest hundredth yard.

Solution Initially, when $t = 0$, $d(0) = 225 - 100(2.7)^0 = 225 - 100 = 125$ yards. Therefore, a typical beginning golfer can hit a 3-wood 125 yards.

After 1 week of practice on the driving range, a typical beginning golfer can hit a 3-wood $d(1) = 225 - 100(2.7)^{-0.7(1)} \approx 175.11$ yards. After 5 weeks of practice, $d(5) = 225 - 100(2.7)^{-0.7(5)} \approx 221.91$ yards. After 9 weeks of practice, $d(9) = 225 - 100(2.7)^{-0.7(9)} \approx 224.81$ yards.

Using a graphing utility, we can sketch the graph of $d(t) = 225 - 100(2.7)^{-0.7t}$. You can see from the graph in Figure 24 that the distance increases rather quickly and then tapers off toward a horizontal asymptote of 225 yards.

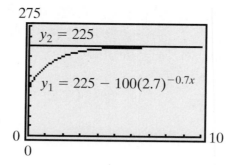

Figure 24
A TI-84 Plus was used to sketch the function $d(t) = 225 - 100(2.7)^{-0.7t}$ and the horizontal asymptote $y = 225$.

You Try It   Work through this You Try It problem.

Work Exercises 31–34 in this textbook or in the *MyMathLab* ® Study Plan.

Compound Interest

A real-life application of exponential functions is the concept of **compound interest**, or interest that is paid on *both principal and interest*. First, we take a look at how simple interest is accrued. If an investment of P dollars is invested at r percent annually (written as a decimal) using simple interest, then the interest earned after 1 year is Pr dollars. Adding this interest to the original investment yields a total amount, A, of

$$A = \underbrace{P}_{\substack{\text{Original} \\ \text{investment}}} + \underbrace{Pr}_{\substack{\text{Interest} \\ \text{earned}}} = P(1 + r).$$

If this amount is reinvested at the same interest rate, then the total amount after 2 years becomes

$$A = \underbrace{P(1 + r)}_{\substack{\text{Total investment} \\ \text{after 1 year}}} + \underbrace{P(1 + r)r}_{\substack{\text{Interest} \\ \text{earned}}} = P(1 + r)(1 + r) = P(1 + r)^2.$$

Reinvesting this amount for a third year gives an amount of $P(1 + r)^3$. Continuing this process for k years, we can see that the amount becomes $A = P(1 + r)^k$. This is an exponential function with base $1 + r$.

We can now modify this formula to obtain another formula that will model interest that is compounded periodically throughout the year(s). When interest is compounded periodically, then k no longer represents the number of years but rather the number of pay periods. If interest is paid n times per year for t years, then $k = nt$ pay periods. Thus, in the formula $A = P(1 + r)^k$, we substitute nt for k and get $A = P(1 + r)^{nt}$.

In the earlier simple interest model, the variable r was used to represent annual interest. In the periodically compounded interest model being developed here with n pay periods per year, the interest rate per pay period is no longer r but rather $\frac{r}{n}$. Thus, in the formula $A = P(1 + r)^{nt}$, we replace r with $\frac{r}{n}$ and get the periodic compound interest formula $A = P\left(1 + \frac{r}{n}\right)^{nt}$.

Periodic Compound Interest Formula

Periodic compound interest can be calculated using the formula

$$A = P\left(1 + \frac{r}{n}\right)^{nt},$$

where
A = Total amount after t years
P = Principal (original investment)
r = Interest rate per year
n = Number of times interest is compounded per year
t = Number of years

Example 6 Calculating Compound Interest

Which investment will yield the most money after 25 years?

Investment A: $12,000 invested at 3% compounded monthly
Investment B: $10,000 invested at 3.9% compounded quarterly

Solution Investment A: $P = 12,000, r = 0.03, n = 12, t = 25$:

$$A = 12,000\left(1 + \frac{0.03}{12}\right)^{12(25)} \approx \$25,380.23$$

Investment B: $P = 10,000, r = 0.039, n = 4, t = 25$:

$$A = 10,000\left(1 + \frac{0.039}{4}\right)^{4(25)} \approx \$26,386.77$$

Investment B will yield the most money after 25 years.

You Try It Work through this You Try It problem.

Work Exercises 35–37 in this textbook or in the *MyMathLab*® Study Plan.

Present Value

Sometimes investors want to know how much money to invest now to reach a certain investment goal in the future. This amount of money, P, is known as the present value of A dollars. To find a formula for present value, start with the formula for periodic compound interest and solve the formula for P:

Use the periodic compound interest formula: $\qquad A = P\left(1 + \frac{r}{n}\right)^{nt}$

Divide both sides by $\left(1 + \frac{r}{n}\right)^{nt}$: $\qquad \dfrac{A}{\left(1 + \frac{r}{n}\right)^{nt}} = \dfrac{P\left(1 + \frac{r}{n}\right)^{nt}}{\left(1 + \frac{r}{n}\right)^{nt}}$

Rewrite $\dfrac{1}{\left(1 + \frac{r}{n}\right)^{nt}}$ as $\left(1 + \frac{r}{n}\right)^{-nt}$: $\quad A\left(1 + \frac{r}{n}\right)^{-nt} = P$

The formula $P = A\left(1 + \frac{r}{n}\right)^{-nt}$ is known as the **present value formula.**

Present Value Formula

Present value can be calculated using the formula

$$P = A\left(1 + \frac{r}{n}\right)^{-nt},$$

where

P = Principal (original investment)
A = Total amount after t years
r = Interest rate per year
n = Number of times interest is compounded per year
t = Number of years

Example 7 Determining Present Value

Find the present value of $8000 if interest is paid at a rate of 5.6% compounded quarterly for 7 years. Round to the nearest cent.

Solution Using the present value formula $P = A\left(1 + \frac{r}{n}\right)^{-nt}$ with $A = \$8000$, $r = 0.056$, $n = 4$, and $t = 7$, we get

$$P = A\left(1 + \frac{r}{n}\right)^{-nt}$$

$$P = 8000\left(1 + \frac{0.056}{4}\right)^{-(4)(7)} \approx 5420.35.$$

Therefore, the present value of $8000 in 7 years at 5.6% compounded quarterly is $5420.35.

You Try It Work through this You Try It problem.

Work Exercises 38–40 in this textbook or in the *MyMathLab*® Study Plan.

9.3 Exercises

You Try It

In Exercises 1–5, sketch the graph of each exponential function. Label the y-intercept and at least two other points on the graph using both positive and negative values of x.

1. $f(x) = 4^x$

2. $f(x) = \left(\dfrac{1}{4}\right)^x$

3. $f(x) = \left(\dfrac{3}{2}\right)^x$

4. $f(x) = (.4)^x$

5. $f(x) = (2.7)^x$

In Exercises 6–12, determine the correct exponential function of the form $f(x) = b^x$ whose graph is given.

You Try It

6.

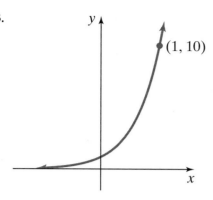

$(1, 10)$

7.

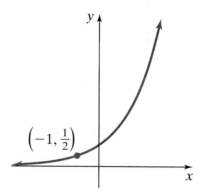

$\left(-1, \frac{1}{2}\right)$

8.

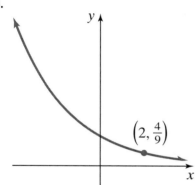

$\left(2, \frac{4}{9}\right)$

9.

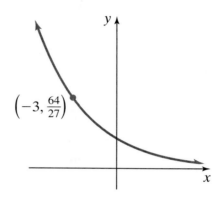

$\left(-3, \frac{64}{27}\right)$

10.

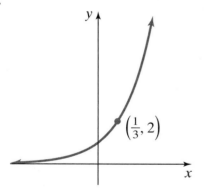

$\left(\frac{1}{3}, 2\right)$

11.

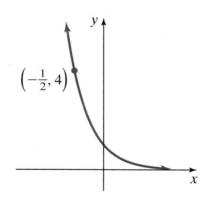

$\left(-\frac{1}{2}, 4\right)$

12.

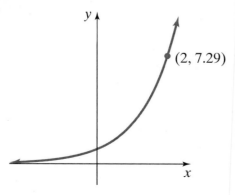

$(2, 7.29)$

In Exercises 13–20, use the graph of $y = 2^x$ or $y = 3^x$ and transformations to sketch each exponential function. Determine the domain and range. Also, determine the y-intercept, and find the equation of the horizontal asymptote.

You Try It

13. $f(x) = 2^{x-1}$

14. $f(x) = 3^x - 1$

15. $f(x) = -3^{x+2}$

16. $f(x) = -2^{x+1} - 1$

17. $f(x) = \left(\dfrac{1}{2}\right)^{x+1}$

18. $f(x) = \left(\dfrac{1}{3}\right)^x - 3$

19. $f(x) = 2^{-x} + 1$

20. $f(x) = 3^{1-x} - 2$

In Exercises 21–30, solve each exponential equation using the method of "relating the bases" by first rewriting the equation in the form $b^u = b^v$.

You Try It

21. $2^x = 16$

22. $3^{x-1} = \dfrac{1}{9}$

23. $\sqrt{5} = 25^x$

24. $\left(\sqrt[3]{3}\right)^x = 9$

25. $\dfrac{1}{\sqrt[5]{8}} = 2^x$

26. $\dfrac{9}{\sqrt[4]{3}} = \left(\dfrac{1}{27}\right)^x$

27. $(49)^x = \left(\dfrac{1}{7}\right)^{x-1}$

28. $\dfrac{125}{\sqrt[3]{5^x}} = \left(\dfrac{1}{25^x}\right)$

29. $\dfrac{3^{x^2}}{9^x} = 27$

30. $2^{x^3} = \dfrac{4^x}{2^{-x^2}}$

31. Typically, weekly sales will drop off rather quickly after the end of an advertising campaign. This drop in sales is known as *sales decay*. Suppose that the gross sales S, in hundreds of dollars, of a certain product is given by the exponential function $S(t) = 3000(1.5^{-0.3t})$, where t is the number of weeks after the end of the advertising campaign.

You Try It

Answer the following questions, rounding each answer to the nearest whole number:

  a. What was the level of sales immediately after the end of the advertising campaign when $t = 0$?
  b. What was the level of sales 1 week after the end of the advertising campaign?
  c. What was the level of sales 5 weeks after the end of the advertising campaign?

32. Most people who start a serious weight-lifting regimen initially notice a rapid increase in the maximum amount of weight that they can bench press. After a few weeks, this increase starts to level off. The following function models the maximum weight, w, that a particular person can bench press in pounds at the end of t weeks of working out.

$$w(t) = 250 - 120(2.7)^{-0.3t}$$

Answer the following questions, rounding each answer to the nearest whole number:

  a. What is the maximum weight that this person can bench press initially?
  b. What is the maximum weight that this person can bench press after 3 weeks of weight lifting?
  c. What is the maximum weight that this person can bench press after 7 weeks of weight lifting?

33. *Escherichia coli* bacteria reproduce by simple cell division, which is known as binary fission. Under ideal conditions, a population of E. coli bacteria can double every 20 minutes. This behavior can be modeled by the exponential function $N(t) = N_0(2^{0.05t})$, where t is in minutes and N_0 is the initial number of E. coli bacteria.

Answer the following questions, rounding each answer to the nearest bacteria:

  a. If the initial number of E. coli bacteria is five, how many bacteria will be present in 3 hours?
  b. If the initial number of E. coli bacteria is eight, how many bacteria will be present in 3 hours?
  c. If the initial number of E. coli bacteria is eight, how many bacteria will be present in 10 hours?

34. A wildlife-management research team noticed that a certain forest had no rabbits, so they decided to introduce a rabbit population into the forest for the first time. The rabbit population will be controlled by wolves and other predators. This rabbit population can be modeled by the function

$R(t) = \dfrac{960}{0.6 + 23.4(2.7)^{-0.045t}}$, where t is the number of weeks after the research team first introduced the rabbits into the forest.

Answer the following questions, rounding each answer to the nearest whole number:

a. How many rabbits did the wildlife-management research team bring into the forest?
b. How many rabbits can be expected after 10 weeks?
c. How many rabbits can be expected after the first year?
d. What is the expected rabbit population after 4 years? 5 years? What can the expected rabbit population approach as time goes on?

Use the periodic compound interest formula to solve Exercises 35–37.

You Try It

35. Suppose that $9000 is invested at 3.5% compounded quarterly. Find the total amount of this investment after 10 years. Round to the nearest cent.

36. Suppose that you have $5000 to invest. Which investment yields the greater return over a 10-year period: 7.35% compounded daily or 7.4% compounded quarterly?

37. Which investment yields the greatest return?

Investment A: $4000 invested for 5 years compounded semiannually (twice per year) at 8%

Investment B: $5000 invested for 4 years compounded quarterly at 4.5%

Use the present value formula to solve Exercises 38–40.

You Try It

38. Find the present value of $10,000 if interest is paid at a rate of 4.5% compounded semiannually for 12 years. Round to the nearest cent.

39. Find the present value of $1,000,000 if interest is paid at a rate of 9.5% compounded monthly for 8 years. Round to the nearest cent.

40. How much money would you have to invest at 10% compounded semiannually so that the total investment had a value of $2205 after 1 year? Round to the nearest cent.

9.4 **The Natural Exponential Function**

THINGS TO KNOW

Before working through this section, be sure you are familiar with the following concepts:

VIDEO   ANIMATION   INTERACTIVE

You Try It

1. Sketch the Graphs of Exponential Functions Using Transformations (Section 9.3, Objective 2)

You Try It

2. Solve Exponential Equations by Relating the Bases (Section 9.3, Objective 3)

You Try It

3. Solve Applications of Exponential Functions (Section 9.3, Objective 4)

OBJECTIVES

1 Use the Characteristics of the Natural Exponential Function

2 Sketch the Graphs of Natural Exponential Functions Using Transformations

3 Solve Natural Exponential Equations by Relating the Bases

4 Solve Applications of the Natural Exponential Function

OBJECTIVE 1:   USE THE CHARACTERISTICS OF THE NATURAL EXPONENTIAL FUNCTION

My video summary

We learned in the previous section that any positive number b, where $b \neq 1$, can be used as the base of an exponential function. However, there is one number that appears as the base in exponential applications more than any other number. This number is called the **natural base** and is symbolized using the letter e. The number e is an irrational number that is defined as the value of the expression $\left(1 + \frac{1}{n}\right)^n$ as n approaches infinity. Table 7 shows the values of the expression $\left(1 + \frac{1}{n}\right)^n$ for increasingly large values of n.

n	$\left(1 + \frac{1}{n}\right)^n$
1	2
2	2.25
10	2.5937424601
100	2.7048138294
1,000	2.7169239322
10,000	2.7181459268
100,000	2.7182682372
1,000,000	2.7182804693
10,000,000	2.7182816925
100,000,000	2.7182818149

Table 7

You can see from Table 7 that as the values of n get large, the value e (rounded to six decimal places) is 2.718282. The function $f(x) = e^x$ is called the **natural exponential function**. Because $2 < e < 3$, it follows that the graph of $f(x) = e^x$ lies between the graph of $y = 2^x$ and $y = 3^x$, as seen in Figure 25.

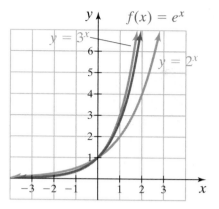

Figure 25

Graph of the natural exponential function $f(x) = e^x$

Characteristics of the Natural Exponential Function

The natural exponential function is the exponential function with base e and is defined as $f(x) = e^x$.

The domain of $f(x) = e^x$ is $(-\infty, \infty)$, and the range is $(0, \infty)$. The graph of $f(x) = e^x$ and some of its characteristics are stated here.

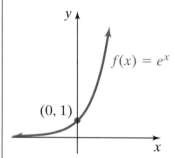

- The graph of $f(x) = e^x$ intersects the y-axis at $(0, 1)$.

- $e^x \to \infty$ as $x \to \infty$

- $e^x \to 0$ as $x \to -\infty$

- The line $y = 0$ is a horizontal asymptote.

- The function $f(x) = e^x$ is one-to-one.

It is important that you are able to use your calculator to evaluate various powers of e. Most calculators have an $\boxed{e^x}$ key. Find this special key on your calculator and evaluate the expressions in the following example.

Example 1 Evaluating the Natural Exponential Function

Evaluate each expression correctly to six decimal places.

a. e^2 **b.** $e^{-0.534}$ **c.** $1000e^{0.013}$

Solution Using the $\boxed{e^x}$ key on a calculator, we get

a. $e^2 \approx 7.389056$

b. $e^{-0.534} \approx 0.586255$

c. $1000e^{0.013} \approx 1013.084867$

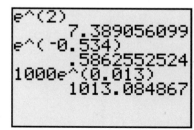

Screen Shot from TI-84 Plus

You Try It Work through this You Try It problem.

Work Exercises 1–5 in this textbook or in the _MyMathLab_® Study Plan.

OBJECTIVE 2: SKETCH THE GRAPHS OF NATURAL EXPONENTIAL FUNCTIONS USING TRANSFORMATIONS

Again, we can use the transformation techniques that are discussed in Section 9.1 to sketch variations of the natural exponential function.

Example 2 Using Transformations to Sketch Natural Exponential Functions

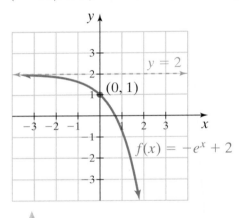 Use transformations to sketch the graph of $f(x) = -e^x + 2$. Determine the domain, range, and y-intercept, and find the equation of any asymptotes.

Solution We can sketch the graph of $f(x) = -e^x + 2$ through a series of the following two transformations.

Start with the graph of $y = e^x$.

1. Reflect the graph of $y = e^x$ about the x-axis, producing the graph of $y = -e^x$.
2. Vertically shift the graph of $y = -e^x$ up two units, producing the graph of $f(x) = -e^x + 2$.

The graph of $f(x) = -e^x + 2$ is shown in Figure 26. Watch the video to see each step. The domain of f is the interval $(-\infty, \infty)$. The range of f is the interval $(-\infty, 2)$. The y-intercept is 1, and the equation of the horizontal asymptote is $y = 2$.

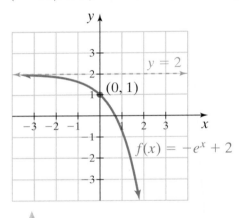

Figure 26
Graph of $f(x) = -e^x + 2$

You Try It Work through this You Try It problem.

Work Exercises 6–10 in this textbook or in the **MyMathLab**® Study Plan.

OBJECTIVE 3: SOLVE NATURAL EXPONENTIAL EQUATIONS BY RELATING THE BASES

Recall the method of relating the bases for solving exponential equations from Section 9.3. If we can write an exponential equation in the form of $b^u = b^v$, then $u = v$. This method for solving exponential equations certainly holds true for the natural base as illustrated in the following example.

Example 3 Using the Method of Relating the Bases to Solve Natural Exponential Equations

Use the method of relating the bases to solve each exponential equation:

a. $e^{3x-1} = \dfrac{1}{\sqrt{e}}$

b. $\dfrac{e^{x^2}}{e^{10}} = (e^x)^3$

My video summary

Solution Work through the interactive video to verify that the solutions are as follows:

a. $x = \dfrac{1}{6}$ **b.** $x = -2$ or $x = 5$

You Try It   Work through this You Try It problem.

Work Exercises 11–18 in this textbook or in the **MyMathLab**® Study Plan.

OBJECTIVE 4:   SOLVE APPLICATIONS OF THE NATURAL EXPONENTIAL FUNCTION

Continuous Compound Interest

Recall the periodic compound interest formula that is introduced in Section 9.3. Some banks use **continuous compounding**; that is, they compound the interest every fraction of a second every day! If we start with the formula for periodic compound interest, $A = P\left(1 + \frac{r}{n}\right)^{nt}$, and let n (the number of times the interest is compounded each year) approach infinity, we can derive the formula $A = Pe^{rt}$, which is the formula for continuous compound interest. Work through this animation to see exactly how this formula is derived.

Continuous Compound Interest Formula

Continuous compound interest can be calculated using the formula

$$A = Pe^{rt},$$

where

$A =$ Total amount after t years
$P =$ Principal
$r =$ Interest rate per year
$t =$ Number of years

Example 4   Calculating Continuous Compound Interest

How much money would be in an account after 5 years if an original investment of $6000 was compounded continuously at 4.5%? Compare this amount to the same investment that was compounded daily. Round to the nearest cent.

Solution First, the amount after 5 years compounded continuously is $A = Pe^{rt} = 6000e^{0.045(5)} \approx \$7513.94$. The same investment compounded daily yields an amount of $A = P\left(1 + \frac{r}{n}\right)^{nt} = 6000\left(1 + \frac{0.045}{365}\right)^{365(5)} \approx \$7513.83$. Continuous compound interest yields only $0.11 more interest after 5 years!

You Try It   Work through this You Try It problem.

Work Exercises 19–21 in this textbook orin the **MyMathLab**® Study Plan.

Present Value

Recall that the present value P is the amount of money to be invested now to obtain A dollars in the future. To find a formula for present value on money that is compounded continuously, we start with the formula for continuous compound interest and solve for P.

Write the continuous compound interest formula: $\quad A = Pe^{rt}$

Divide both sides by e^{rt}: $\quad \dfrac{A}{e^{rt}} = P\dfrac{e^{rt}}{e^{rt}}$

Rewrite $\dfrac{1}{e^{rt}}$ as e^{-rt}: $\quad Ae^{-rt} = P$

Present Value Formula

The present value of A dollars after t years of continuous compound interest, with interest rate r, is given by the formula

$$P = Ae^{-rt}.$$

Example 5 Calculating Present Value

Find the present value of $18,000 if interest is paid at a rate of 8% compounded continuously for 20 years. Round to the nearest cent.

Solution Using the present value formula $P = Ae^{-rt}$ with $A = \$18{,}000$, $r = 0.08$, and $t = 20$, we get

$$P = Ae^{-rt}$$
$$P = (18{,}000)e^{-(0.08)(20)} \approx \$3634.14.$$

You Try It Work through this You Try It problem.

Work Exercises 22–23 in this textbook or in the *MyMathLab*® Study Plan.

Exponential Growth Model

You have probably heard that some populations grow exponentially. Most populations grow at a rate proportional to the size of the population. In other words, the larger the population, the faster the population grows. With this in mind, it can be shown in a more advanced math course that the mathematical model that can describe population growth is given by the function $P(t) = P_0 e^{kt}$.

Exponential Growth

A model that describes the population, P, after a certain time, t, is

$$P(t) = P_0 e^{kt},$$

where $P_0 = P(0)$ is the initial population and $k > 0$ is a constant called the **relative growth rate**. (Note: k may be given as a percent.)

The graph of the exponential growth model is shown in Figure 27. Notice that the graph has a y-intercept of P_0.

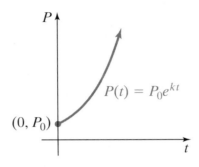

Figure 27
Graph of the exponential growth model
$P(t) = P_0 e^{kt}$

Example 6 Population Growth

My video summary

 The population of a small town follows the exponential growth model $P(t) = 900e^{0.015t}$, where t is the number of years after 1900.

Answer the following questions, rounding each answer to the nearest whole number:

a. What was the population of this town in 1900?

b. What was the population of this town in 1950?

c. Use this model to predict the population of this town in 2012.

Solution

a. The initial population was $P(0) = 900e^{0.015(0)} = 900$.

b. Because 1950 is 50 years after 1900, we must evaluate $P(50)$.
    $P(50) = 900e^{0.015(50)} \approx 1905$.

c. In the year 2012, we can predict that the population will be
    $P(112) = 900e^{0.015(112)} \approx 4829$.

Example 7 Determining the Initial Population

My video summary

 Twenty years ago, the State of Idaho Fish and Game Department introduced a new breed of wolf into a certain Idaho forest. The current wolf population in this forest is now estimated at 825, with a relative growth rate of 12%.

Answer the following questions, rounding each answer to the nearest whole number:

a. How many wolves did the Idaho Fish and Game Department initially introduce into this forest?

b. How many wolves can be expected after another 20 years?

Solution

a. The relative growth rate is 0.12, so we use the exponential growth model $P(t) = P_0 e^{0.12t}$. Because $P(20) = 825$, we get

$$\text{Substitute 20 for } t: \quad P(20) = P_0 e^{0.12(20)}$$
$$\text{Substitute 825 for } P(20): \quad 825 = P_0 e^{0.12(20)}$$
$$\text{Solve for } P_0: \quad P_0 = \frac{825}{e^{0.12(20)}} \approx 75.$$

Therefore, the Idaho Fish and Game Department initially introduced 75 wolves into the forest.

b. Because $P_0 = 75$, we can use the exponential growth model $P(t) = 75e^{0.12t}$. In another 20 years, the value of t will be 40. Thus, we must evaluate $P(40)$.

$$P(40) = 75e^{0.12(40)} \approx 9113$$

Therefore, we can expect the wolf population to be approximately 9113 in another 20 years.

You Try It **Work through this You Try It problem.**

Work Exercises 24–27 in this textbook or in the *MyMathLab*® Study Plan.

9.4 Exercises

You Try It

In Exercises 1–5, use a calculator to approximate each exponential expression to six decimal places.

1. e^3 **2.** $e^{-0.2}$ **3.** $e^{1/3}$ **4.** $100e^{-.123}$ **5.** $\sqrt{2}e^{\pi}$

You Try It

In Exercises 6–10, use transformations to sketch each exponential function. Determine the domain and range. Also, label the y-intercept, and find the equation of the horizontal asymptote.

6. $f(x) = e^{x-1}$ **7.** $f(x) = e^x - 1$ **8.** $f(x) = -e^{x+2}$

9. $f(x) = -e^{x+1} - 1$ **10.** $f(x) = e^{-x} - 2$

You Try It

In Exercises 11–18, solve each exponential equation using the method of relating the bases by first rewriting the equation in the form $e^u = e^v$.

11. $e^x = \dfrac{1}{e^2}$

12. $e^{5x+2} = \sqrt[3]{e}$

13. $\dfrac{1}{e^x} = \dfrac{\sqrt{e}}{e^{1-x}}$

14. $(e^{x^2})^2 = e^8$

15. $e^{x^2} = (e^x) \cdot e^{12}$

16. $e^{x^2} = \dfrac{e^3}{(e^x)^5}$

17. $\dfrac{e^{x^3}}{e^x} = \dfrac{e^{2x^2}}{e^2}$

18. $e^{x^3} = \dfrac{(e^{2x^2})^2 \cdot e^x}{e^4}$

You Try It

19. An original investment of $6000 earns 6.25% interest compounded continuously. What will the investment be worth in 2 years? in 20 years? Round to the nearest cent.

20. How much more will an investment of $10,000 earning 5.5% compounded continuously for 9 years earn compared to the same investment at 5.5% compounded quarterly for 9 years? Round to the nearest cent.

21. Suppose your great-great grandfather invested $500 earning 6.5% interest compounded continuously 100 years ago. How much would his investment be worth today? Round to the nearest cent.

You Try It

22. Find the present value of $16,000 if interest is paid at a rate of 4.5% compounded continuously for 10 years. Round to the nearest cent.

23. Which has the lower present value: (a) $20,000 if interest is paid at a rate of 5.18% compounded continuously for 2 years or (b) $25,000 if interest is paid at a rate of 3.8% compounded continuously for 30 months?

You Try It

24. The population of a rural city follows the exponential growth model $P(t) = 2000e^{0.035t}$, where t is the number of years after 1995.

    a. What was the population of this city in 1995?

    b. What is the relative growth rate as a percent?

    c. Use this model to approximate the population in 2030, rounding to the nearest whole number.

25. The relative growth rate of a certain bacteria colony is 25%. Suppose there are 10 bacteria initially.

    a. Find a function that describes the population of the bacteria after t hours.

    b. How many bacteria should we expect after 1 day? Round to the nearest whole number.

26. In 2006, the population of a certain American city was 18,221. If the relative growth rate has been 6% since 1986, what was the population of this city in 1986? Round to the nearest whole number.

27. In 1970, a wildlife resource management team introduced a certain rabbit species into a forest for the first time. In 2004, the rabbit population had grown to 7183. The relative growth rate for this rabbit species is 20%.

Answer the following questions, rounding each answer to the nearest whole number:

    a. How many rabbits did the wildlife resource management team introduce into the forest in 1970?

    b. How many rabbits can be expected in the year 2025?

9.5   **Logarithmic Functions**

THINGS TO KNOW

Before working through this section, be sure you are familiar with the following concepts:

        VIDEO   ANIMATION   INTERACTIVE

You Try It

1. Solve Polynomial Inequalities
(Section 8.5, **Objective** 1)

You Try It

2. Solve Rational Inequalities
(Section 8.5, **Objective** 2)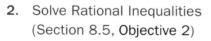

You Try It

3. Determine If a Function Is One-to-One
Using the Horizontal Line Test
(Section 9.2, **Objective** 3)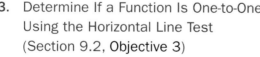

You Try It

4. Verify Inverse Functions
(Section 9.2, **Objective** 4)

You Try It

5. Sketch the Graphs of Inverse Functions
(Section 9.2, **Objective** 5)

You Try It

6. Find the Inverse of a One-to-One Function
(Section 9.2, **Objective** 6)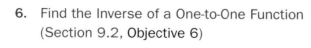

You Try It

7. Use the Characteristics of Exponential Functions
(Section 9.3, **Objective** 1)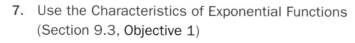

You Try It 8. Sketch the Graphs of Exponential Functions Using Transformations (Section 9.3, **Objective 2**)

You Try It 9. Solve Exponential Equations by Relating the Bases (Section 9.3, **Objective 3**)

OBJECTIVES

1 Use the Definition of a Logarithmic Function

2 Evaluate Logarithmic Expressions

3 Use the Properties of Logarithms

4 Use the Common and Natural Logarithms

5 Use the Characteristics of Logarithmic Functions

6 Sketch the Graphs of Logarithmic Functions Using Transformations

7 Find the Domain of Logarithmic Functions

OBJECTIVE 1:   **USE THE DEFINITION OF A LOGARITHMIC FUNCTION**

My video summary

Every exponential function of the form $f(x) = b^x$, where $b > 0$ and $b \neq 1$, is one-to-one and thus has an inverse function. (You may want to refer to Section 9.2 to review one-to-one functions and inverse functions.) Remember, given the graph of a one-to-one function f, the graph of the inverse function is a reflection about the line $y = x$. That is, for any point (a, b) that lies on the graph of f, the point (b, a) must lie on the graph of f^{-1}. In other words, the graph of f^{-1} can be obtained by simply switching the x and y coordinates of the ordered pairs of $f(x) = b^x$. Watch this video to see how to sketch the graph of $f(x) = b^x$ and its inverse.

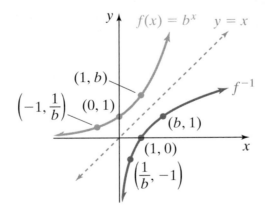

Figure 28
Graph of $f(x) = b^x, b > 1,$
and its inverse function

The graphs of f and f^{-1} are sketched in Figure 28, but what is the equation of the inverse of $f(x) = b^x$? To find the equation of f^{-1}, we follow the four-step process for finding inverse functions that is discussed in Section 9.2.

Step 1.   Change $f(x)$ to y : $y = b^x$

Step 2. Interchange x and y : $x = b^y$

Step 3.   Solve for y :   ??

Before we can solve for y, we must introduce the following definition:

Definition   Logarithmic Function

For $x > 0, b > 0$, and $b \neq 1$, the **logarithmic function** with base b is defined by

$$y = \log_b x \quad \text{if and only if} \quad x = b^y.$$

The equation $y = \log_b x$ is said to be in **logarithmic form**, whereas the equation $x = b^y$ is in **exponential form**. We can now continue to find the inverse of $f(x) = b^x$ by completing steps 3 and 4.

Step 3.   Solve for y: $x = b^y$ can be written as $y = \log_b x$

Step 4. Change y to $f^{-1}(x)$: $f^{-1}(x) = \log_b x$

In general, if $f(x) = b^x$ for $b > 0$ and $b \neq 1$, then the inverse function is $f^{-1}(x) = \log_b x$. For example, the inverse of $f(x) = 2^x$ is $f^{-1}(x) = \log_2 x$, which is read as "the log base 2 of x." We revisit the graphs of logarithmic functions later on in this section, but first it is very important to understand the definition of the logarithmic function and practice how to go back and forth writing exponential equations as logarithmic equations and vice versa.

My video summary

Example 1   Changing from Exponential Form to Logarithmic Form

Write each exponential equation as an equation involving a logarithm.

a. $2^3 = 8$ **b.** $5^{-2} = \dfrac{1}{25}$ **c.** $1.1^M = z$

Solution We use the fact that the equation $x = b^y$ is equivalent to the equation $y = \log_b x$.

a. $2^3 = 8$ is equivalent to $\log_2 8 = 3$.

b. $5^{-2} = \frac{1}{25}$ is equivalent to $\log_5 \frac{1}{25} = -2$.

c. $1.1^M = z$ is equivalent to $\log_{1.1} z = M$.

Note that the exponent of the original (exponential) equation ends up by itself on the right side of the second (logarithmic) equation. Therefore, a logarithmic expression can be thought of as describing the exponent of a certain exponential equation.

Watch the video to see this example worked out in more detail.

You Try It   Work through this You Try It problem.

Work Exercises 1–6 in this textbook or in the *MyMathLab*® Study Plan.

Example 2   Changing from Logarithmic Form to Exponential Form

My video summary

Write each logarithmic equation as an equation involving an exponent.

a. $\log_3 81 = 4$ **b.** $\log_4 16 = y$ **c.** $\log_{3/5} x = 2$

Solution We use the fact that the equation $y = \log_b x$ is equivalent to the equation $x = b^y$.

a. $\log_3 81 = 4$ is equivalent to $3^4 = 81$.

b. $\log_4 16 = y$ is equivalent to $4^y = 16$.

c. $\log_{3/5} x = 2$ is equivalent to $\left(\frac{3}{5}\right)^2 = x$.

Watch the video to see this example worked out in more detail.

You Try It **Work through this You Try It problem.**

Work Exercises 7–11 in this textbook or in the _MyMathLab_® Study Plan.

OBJECTIVE 2:   EVALUATE LOGARITHMIC EXPRESSIONS

Because a logarithmic expression represents the exponent of an exponential equation, it is possible to evaluate many logarithms by inspection or by creating the corresponding exponential equation. Remember that the expression $\log_b x$ is the exponent to which b must be raised in order get x. For example, suppose we are to evaluate the expression $\log_4 64$. To evaluate this expression, we must ask ourselves, "4 raised to what power is 64?" Because $4^3 = 64$, we conclude that $\log_4 64 = 3$. For some logarithmic expressions, it is often convenient to create an exponential equation and use the method of relating the bases for solving exponential equations. For more complex logarithmic expressions, additional techniques are required. These techniques are discussed in Sections 9.6 and 9.7.

Example 3 Evaluating Logarithmic Expressions

 Evaluate each logarithm:

a. $\log_5 25$ **b.** $\log_3 \dfrac{1}{27}$ **c.** $\log_{\sqrt{2}} \dfrac{1}{4}$

Solution

a. To evaluate $\log_5 25$, we must ask, "5 raised to what exponent is 25?" Because $5^2 = 25, \log_5 25 = 2$.

b. The expression $\log_3 \frac{1}{27}$ requires more analysis. In this case, we ask, "3 raised to what exponent is $\frac{1}{27}$?" Suppose we let y equal this exponent. Then $3^y = \frac{1}{27}$. To solve for y, we can use the method of relating the bases for solving exponential equations.

$$\text{Write the exponential equation:} \quad 3^y = \frac{1}{27}$$

$$\text{Rewrite 27 as } 3^3\text{:} \quad 3^y = \frac{1}{3^3}$$

$$\text{Use } \frac{1}{b^n} = b^{-n}\text{:} \quad 3^y = 3^{-3}$$

$$\text{Use the method of relating the bases:} \quad y = -3$$
$$(\text{If } b^u = b^v, \text{ then } u = v.)$$

Thus, $\log_3 \frac{1}{27} = -3$.

c. Watch the interactive video to verify that $\log_{\sqrt{2}} \dfrac{1}{4} = -4$. and to see each solution worked out in detail.

You Try It   **Work through this You Try It problem.**

Work Exercises 12–18 in this textbook or in the *MyMathLab*® Study Plan.

OBJECTIVE 3: USE THE PROPERTIES OF LOGARITHMS

Because $b^1 = b$ for any real number b, we can use the definition of the logarithmic function ($y = \log_b x$ if and only if $x = b^y$) to rewrite this expression as $\log_b b = 1$. Similarly, because $b^0 = 1$ for any real number b, we can rewrite this expression as $\log_b 1 = 0$. These two general properties are summarized as follows.

General Properties of Logarithms

For $b > 0$ and $b \neq 1$,

1. $\log_b b = 1$ and

2. $\log_b 1 = 0$.

In Section 9.2, we saw that a function f and its inverse function f^{-1} satisfy the following two composition cancellation equations:

$$f(f^{-1}(x)) = x \text{ for all } x \text{ in the domain of } f^{-1} \text{ and}$$
$$f^{-1}(f(x)) = x \text{ for all } x \text{ in the domain of } f.$$

If $f(x) = b^x$, then $f^{-1}(x) = \log_b x$. Applying the two composition cancellation equations we get

$$f(f^{-1}(x)) = b^{\log_b x} = x \text{ and}$$
$$f^{-1}(f(x)) = \log_b b^x = x.$$

Cancellation Properties of Exponentials and Logarithms

For $b > 0$ and $b \neq 1$,

1. $b^{\log_b x} = x$ and

2. $\log_b b^x = x$.

Example 4 Using the Properties of Logarithms

Use the properties of logarithms to evaluate each expression.

a. $\log_3 3^4$ **b.** $\log_{12} 12$ **c.** $7^{\log_7 13}$ **d.** $\log_8 1$

Solution

a. By the second cancellation property, $\log_3 3^4 = 4$.

b. Because $\log_b b = 1$ for all $b > 0$ and $b \neq 1$, $\log_{12} 12 = 1$.

c. By the first cancellation property, $7^{\log_7 13} = 13$.

d. Because $\log_b 1 = 0$ for all $b > 0$ and $b \neq 1$, $\log_8 1 = 0$.

You Try It   Work through this You Try It problem.

Work Exercises 19–26 in this textbook or in the **MyMathLab**® Study Plan.

OBJECTIVE 4:   USE THE COMMON AND NATURAL LOGARITHMS

There are two bases that are used more frequently than any other base. They are base 10 and base e. (Refer to Section 9.4 to review the natural base e.) Because our counting system is based on the number 10, the base 10 logarithm is known as the **common logarithm**. Instead of using the notation $\log_{10} x$ to denote the common logarithm, it is usually abbreviated without the subscript 10 as simply $\log x$. The base e logarithm is called the **natural logarithm** and is abbreviated as $\ln x$ instead of $\log_e x$. Most scientific calculators are equipped with a $\boxed{\log}$ key and a $\boxed{\ln}$ key. We can apply the definition of the logarithmic function for the base 10 and for the base e logarithm as follows.

Definition   **Common Logarithmic Function**

For $x > 0$, the **common logarithmic function** is defined by

$$y = \log x \quad \text{if and only if} \quad x = 10^y.$$

Definition   **Natural Logarithmic Function**

For $x > 0$, the **natural logarithmic function** is defined by

$$y = \ln x \quad \text{if and only if} \quad x = e^y.$$

Example 5 Changing from Exponential Form to Logarithmic Form

My video summary

Write each exponential equation as an equation involving a common logarithm or natural logarithm.

a. $e^0 = 1$ b. $10^{-2} = \dfrac{1}{100}$ c. $e^K = w$

Solution

a. $e^0 = 1$ is equivalent to $\ln 1 = 0$.
b. $10^{-2} = \frac{1}{100}$ is equivalent to $\log\left(\frac{1}{100}\right) = -2$.
c. $e^K = w$ is equivalent to $\ln w = K$.

Watch the video to see this example worked out in more detail.

You Try It   Work through this You Try It problem.

Work Exercises 27–31 in this textbook or in the **MyMathLab**® Study Plan.

Example 6 Changing from Logarithmic Form to Exponential Form

My video summary

Write each logarithmic equation as an equation involving an exponent.

a. $\log 10 = 1$ **b.** $\ln 20 = Z$ **c.** $\log(x - 1) = T$

Solution

a. $\log 10 = 1$ is equivalent to $10^1 = 10$.

b. $\ln 20 = Z$ is equivalent to $e^Z = 20$.

c. $\log(x - 1) = T$ is equivalent to $10^T = x - 1$.

Watch the video to see this example worked out in more detail.

You Try It   Work through this You Try It problem.

Work Exercises 32–35 in this textbook or in the *MyMathLab*® Study Plan.

Example 7 Evaluating Common and Natural Logarithmic Expressions

My video summary

Evaluate each expression without the use of a calculator.

a. $\log 100$ **b.** $\ln \sqrt{e}$ **c.** $e^{\ln 51}$ **d.** $\log 1$

Solution

a. $\log 100 = 2$ because $10^2 = 100$ by the definition of the logarithmic function or $\log 100 = \log 10^2 = 2$ by cancellation property (2).

b. $\ln \sqrt{e} = \ln e^{1/2} = \frac{1}{2}$ by cancellation property (2).

c. $e^{\ln 51} = 51$ by cancellation property (1).

d. $\log 1 = 0$ by general property (1).

Watch the video to see this example worked out in more detail.

You Try It   Work through this You Try It problem.

Work Exercises 36–43 in this textbook or in the *MyMathLab*® Study Plan.

OBJECTIVE 5:   USE THE CHARACTERISTICS OF LOGARITHMIC FUNCTIONS

To sketch the graph of a logarithmic function of the form $f(x) = \log_b x$, where $b > 0$ and $b \neq 1$, follow these three steps:

Step 1. Start with the graph of the exponential function $y = b^x$, labeling several ordered pairs.

Step 2. Because $f(x) = \log_b x$ is the inverse of $y = b^x$, we can find several points on the graph of $f(x) = \log_b x$ by reversing the coordinates of the ordered pairs of $y = b^x$.

Step 3. Plot the ordered pairs from step 2, and complete the graph of $f(x) = \log_b x$ by connecting the ordered pairs with a smooth curve. The graph of $f(x) = \log_b x$ is a reflection of the graph of $y = b^x$ about the line $y = x$.

My video summary
✎

Example 8 Sketching the Graph of a Logarithmic Function

Sketch the graph of $f(x) = \log_3 x$.

Solution

Step 1. The graph of $y = 3^x$ passes through the points $\left(-1, \frac{1}{3}\right)$, $(0, 1)$, and $(1, 3)$. See Figure 29.

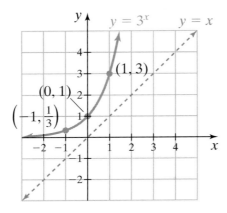

Figure 29
Graph of $y = 3^x$

Step 2. We reverse the three ordered pairs from step 1 to obtain the following three points: $\left(\frac{1}{3}, -1\right)$, $(1, 0)$, and $(3, 1)$.

Step 3. Plot the points $\left(\frac{1}{3}, -1\right)$, $(1, 0)$, and $(3, 1)$, and connect them with a smooth curve to obtain the graph of $f(x) = \log_3 x$. See Figure 30.

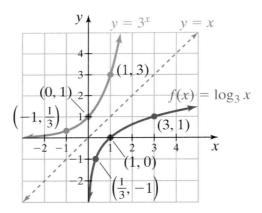

Figure 30
Graph of $y = 3^x$ and
$f(x) = \log_3 x$

Notice in Figure 30 that the y-axis is a vertical asymptote of the graph of $f(x) = \log_3 x$. Every logarithmic function of the form $y = \log_b x$, where $b > 0$ and $b \neq 1$ has a vertical asymptote at the y-axis. The graphs and the characteristics of logarithmic functions are outlined as follows.

Characteristics of Logarithmic Functions

For $b > 0, b \neq 1$, the logarithmic function with base b is defined by $y = \log_b x$. The domain of $f(x) = \log_b x$ is $(0, \infty)$, and the range is $(-\infty, \infty)$. The graph of $f(x) = \log_b x$ has one of the following two shapes.

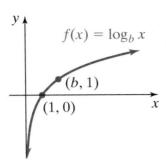

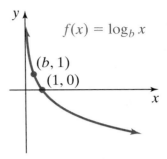

The graph intersects the x-axis at $(1, 0)$.

The graph contains the point $(b, 1)$.

The graph is increasing on the interval $(0, \infty)$.

The y-axis $(x = 0)$ is a vertical asymptote.

The function is one-to-one.

The graph intersects the x-axis at $(1, 0)$.

The graph contains the point $(b, 1)$.

The graph is decreasing on the interval $(0, \infty)$.

The y-axis $(x = 0)$ is a vertical asymptote.

The function is one-to-one.

You Try It   Work through this You Try It problem.

Work Exercises 44 and 45 in this textbook or in the **MyMathLab** Study Plan.

OBJECTIVE 6:   SKETCH THE GRAPHS OF LOGARITHMIC FUNCTIONS USING TRANSFORMATIONS

Often we can use the transformation techniques that are discussed in Section 9.1 to sketch the graph of logarithmic functions.

Example 9 Using Transformations to Sketch the Graph of a Logarithmic Function

My video summary

Sketch the graph of $f(x) = -\ln(x + 2) - 1$.

Solution Recall that the function $y = \ln x$ has a base of e, where $2 < e < 3$. This means that the graph of $y = \ln x$ is increasing on the interval $(0, \infty)$ and contains the points $(1, 0)$ and $(e, 1)$. Starting with the graph of $y = \ln x$, we can obtain the graph of $f(x) = -\ln(x + 2) - 1$ through the following series of transformations:

1. Shift the graph of $y = \ln x$ horizontally to the left two units to obtain the graph of $y = \ln(x + 2)$.

2. Reflect the graph of $y = \ln(x + 2)$ about the x-axis to obtain the graph of $y = -\ln(x + 2)$.

3. Shift the graph of $y = -\ln(x + 2)$ vertically down one unit to obtain the final graph of $f(x) = -\ln(x + 2) - 1$.

The graph of $f(x) = -\ln(x + 2) - 1$ is sketched in Figure 31. You can see from the graph that the domain of $f(x) = -\ln(x + 2) - 1$ is $(-2, \infty)$. The vertical asymptote is $x = -2$, and the x-intercept is $\left(\frac{1}{e} - 2, 0\right)$. Watch the video to see each step worked out in detail.

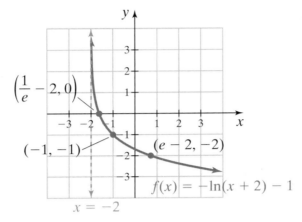

Figure 31
Graph of $f(x) = -\ln(x + 2) - 1$

 You Try It Work through this You Try It problem.

Work Exercises 46–51 in this textbook or in the **MyMathLab** Study Plan.

OBJECTIVE 7: FIND THE DOMAIN OF LOGARITHMIC FUNCTIONS

In Example 9, we sketched the function $f(x) = -\ln(x + 2) - 1$ and observed that the domain was $(-2, \infty)$. We do not have to sketch the graph of a logarithmic function to determine the domain. The domain of a logarithmic function consists of all values of x for which the argument of the logarithm is greater than zero. In other words, if $f(x) = \log_b[g(x)]$, then the domain of f can be found by solving the inequality $g(x) > 0$. For example, given the function $f(x) = -\ln(x + 2) - 1$ from Example 9, we can determine the domain by solving the linear inequality $x + 2 > 0$. Solving this inequality for x, we obtain $x > -2$. Thus, the domain of $f(x) = -\ln(x + 2) - 1$ is $(-2, \infty)$.

Example 10 is a bit more challenging because the argument of the logarithm is a rational expression.

Example 10 Finding the Domain of a Logarithmic Function with a Rational Argument

My interactive video summary 🎥 Find the domain of $f(x) = \log_5\left(\frac{2x - 1}{x + 3}\right)$.

Solution To find the domain of f, we must find all values of x for which the argument $\frac{2x - 1}{x + 3}$ is greater than zero. That is, you must solve the rational inequality $\left(\frac{2x - 1}{x + 3}\right) > 0$. See Section 8.5 if you need help remembering how to solve this inequality. By the techniques discussed in Section 8.5, we find that the solution to $\left(\frac{2x - 1}{x + 3}\right) > 0$ is $x < -3$ or $x > \frac{1}{2}$. Therefore, the domain of $f(x) = \log_5\left(\frac{2x - 1}{x + 3}\right)$ in set notation is $\left\{x | x < -3 \text{ or } x > \frac{1}{2}\right\}$. In interval notation, the domain is $(-\infty, -3) \cup \left(\frac{1}{2}, \infty\right)$. Watch the interactive video to see this problem worked out in detail.

You Try It   Work through this You Try It problem.

Work Exercises 52–58 in this textbook or in the *MyMathLab*® Study Plan.

9.5 Exercises

You Try It

In Exercises 1–6, write each exponential equation as an equation involving a logarithm.

1. $3^2 = 9$

2. $16^{1/2} = 4$

3. $2^{-3} = \dfrac{1}{8}$

4. $\sqrt{2}^{\pi} = W$

5. $\left(\dfrac{1}{3}\right)^t = 27$

6. $7^{5k} = L$

You Try It

In Exercises 7–11, write each logarithmic equation as an exponential equation.

7. $\log_5 1 = 0$

8. $\log_7 343 = 3$

9. $\log_{\sqrt{2}} 8 = 6$

10. $\log_4 K = L$

11. $\log_a (x - 1) = 3$

You Try It

In Exercises 12–18, evaluate each logarithm without the use of a calculator.

12. $\log_2 8$

13. $\log_6 \sqrt{6}$

14. $\log_3 \dfrac{1}{9}$

15. $\log_{\sqrt{5}} 25$

16. $\log_4 \left(\dfrac{1}{\sqrt[5]{64}}\right)$

17. $\log_{1/7} \sqrt[3]{7}$

18. $\log_{0.1} 100$

In Exercises 19–26, use the properties of logarithms to evaluate each expression without the use of a calculator.

19. $2^{\log_2 11}$

20. $\log_4 4$

21. $\log_9 1$

You Try It

22. $\log_7 7^{-3}$

23. $\log_a a, a > 1$

24. $5^{\log_5 M}, M > 0$

25. $\log_y 1, y > 0$

26. $\log_x x^{20}, x > 1$

You Try It

In Exercises 27–31, write each exponential equation as an equation involving a common logarithm or a natural logarithm.

27. $10^3 = 1000$

28. $e^{-1} = \dfrac{1}{e}$

29. $e^k = 2$

30. $10^e = M$

31. $e^{10} = Z$

You Try It

In Exercises 32–35, write each logarithmic equation as an exponential equation.

32. $\ln 1 = 0$

33. $\log (1,000,000) = 6$

34. $\log K = L$

35. $\ln Z = 4$

In Exercises 36–43, evaluate each expression without the use of a calculator, and then verify your answer using a calculator.

You Try It

36. $\log 10{,}000$

37. $\log\left(\dfrac{1}{1000}\right)$

38. $\ln 1$

39. $\ln\sqrt[3]{e^2}$

40. $10^{\log e}$

41. $e^{\ln 49}$

42. $\log 10^6$

43. $\ln e + \ln e^3$

In Exercises 44–51, sketch each logarithmic function. Label at least two points on the graph, and determine the domain and the equation of any vertical asymptotes.

You Try It

44. $h(x) = \log_4 x$

45. $g(x) = \log_{\frac{1}{3}} x$

46. $f(x) = \log_2(x) - 1$

47. $f(x) = \log_5(x - 1)$

48. $f(x) = -\ln(x)$

49. $y = \log_{1/2}(x + 1) + 2$

50. $y = \log_3(1 - x)$

51. $h(x) = -\dfrac{1}{2}\log_3(x + 3) + 1$

In Exercises 52–58, find the domain of each logarithmic function.

You Try It

52. $f(x) = \log(-x)$

53. $f(x) = \log_{1/4}(2x + 6)$

54. $f(x) = \ln(1 - 3x)$

55. $f(x) = \log_2(x^2 - 9)$

56. $f(x) = \log_7(x^2 - x - 20)$

57. $f(x) = \ln\left(\dfrac{x + 5}{x - 8}\right)$

58. $f(x) = \log\left(\dfrac{x^2 - x - 6}{x + 10}\right)$

9.6 Properties of Logarithms

THINGS TO KNOW

Before working through this section, be sure that you are familiar with the following concepts:

VIDEO   ANIMATION   INTERACTIVE

You Try It

1. Solve Exponential Equations by Relating the Bases (Section 9.3, Objective 3)

You Try It

2. Change from Exponential Form to Logarithmic Form (Section 9.5, Objective 1)

You Try It

3. Change from Logarithmic Form to Exponential Form (Section 9.5, Objective 1)

You Try It

4. Evaluate Logarithmic Expressions (Section 9.5, Objective 2)

You Try It **5.** Use the Common and Natural Logarithms
   (Section 9.5, Objective 4)

You Try It **6.** Find the Domain of Logarithmic Functions
   (Section 9.5, Objective 7)

OBJECTIVES

1 Use the Product Rule, Quotient Rule, and Power Rule for Logarithms

2 Expand and Condense Logarithmic Expressions

3 Solve Logarithmic Equations Using the Logarithm Property of Equality

4 Use the Change of Base Formula

OBJECTIVE 1: USE THE PRODUCT RULE, QUOTIENT RULE, AND POWER RULE FOR LOGARITHMS

In this section, we learn how to manipulate logarithmic expressions using properties of logarithms. Understanding how to use these properties will help us solve exponential and logarithmic equations that are encountered in the next section. Recall from Section 9.5 the general properties and cancellation properties of logarithms. We now look at three additional properties of logarithms.

My video summary

> **Properties of Logarithms**
>
> If $b > 0, b \neq 1$, u and v represent positive numbers and r is any real number, then
>
> $\log_b uv = \log_b u + \log_b v$ product rule for logarithms
>
> $\log_b \dfrac{u}{v} = \log_b u - \log_b v$ quotient rule for logarithms
>
> $\log_b u^r = r \log_b u$ power rule for logarithms

To prove the product rule and quotient rule for logarithms, we use properties of exponents and the method of relating the bases to solve exponential equations. The power rule for logarithms is a direct result of the product rule. Click on a video proof link above to see a proof for each of these properties.

My video summary

Example 1 Using the Product Rule

Use the product rule for logarithms to expand each expression. Assume $x > 0$.

a. $\ln (5x)$ **b.** $\log_2 (8x)$

Solution

a. Use the product rule for logarithms: $\ln (5x) = \ln 5 + \ln x$

b. Use the product rule for logarithms: $\log_2 (8x) = \log_2 8 + \log_2 x$

Use the definition of the logarithmic
function to rewrite $\log_2 8$ as 3
because $2^3 = 8$: $\log_2 (8x) = \quad 3 \quad + \log_2 x$

CAUTION $\log_b (u + v)$ is *not* equivalent to $\log_b u + \log_b v$.

You Try It Work through this You Try It problem.

My video summary

 Example 2 Using the Quotient Rule

Use the quotient rule for logarithms to expand each expression. Assume $x > 0$.

a. $\log_5 \left(\dfrac{12}{x} \right)$ b. $\ln \left(\dfrac{x}{e^5} \right)$

Solution

a. Use the quotient rule for logarithms: $\log_5 \left(\dfrac{12}{x} \right) = \log_5 12 - \log_5 x$

b. Use the quotient rule for logarithms: $\ln \left(\dfrac{x}{e^5} \right) = \ln x - \ln e^5$

     Use cancellation property (2) to
     rewrite $\ln e^5$ as 5: $= \ln x - 5$

CAUTION $\log_b (u - v)$ is *not* equivalent to $\log_b u - \log_b v$, and $\dfrac{\log_b u}{\log_b v}$ is *not* equivalent to $\log_b u - \log_b v$.

You Try It Work through this You Try It problem.

My video summary

 Example 3 Using the Power Rule

Use the power rule for logarithms to rewrite each expression. Assume $x > 0$.

a. $\log 6^3$ b. $\log_{1/2} \sqrt[4]{x}$

Solution

a. Use the power rule for logarithms: $\log 6^3 = 3 \log 6$

b.   Rewrite the fourth root of x
        using a rational exponent: $\log_{1/2} \sqrt[4]{x} = \log_{1/2} x^{1/4}$

     Use the power rule for logarithms: $= \dfrac{1}{4} \log_{1/2} x$

The process of using the power rule to simplify a logarithmic expression is often casually referred to as "bringing down the exponent."

CAUTION $(\log_b u)^r$ is *not* equivalent to $r \log_b u$.

You Try It Work through this You Try It problem.

Work Exercises 1–10 in this textbook or in the *MyMathLab*® Study Plan.

OBJECTIVE 2:   EXPAND AND CONDENSE LOGARITHMIC EXPRESSIONS

Sometimes it is necessary to combine several properties of logarithms to expand a logarithmic expression into the sum and/or difference of logarithms or to condense several logarithms into a single logarithm.

My interactive video summary   **Example 4 Expanding a Logarithmic Expression**

Use properties of logarithms to expand each logarithmic expression as much as possible.

a. $\log_7 \left(49 x^3 \sqrt[5]{y^2} \right)$ b. $\ln \left(\dfrac{(x^2 - 4)}{9 e^{x^3}} \right)$

Solution

a. Write the original expression: $\log_7 \left(49 x^3 \sqrt[5]{y^2} \right)$

Use the product rule: $= \log_7 49 + \log_7 x^3 \sqrt[5]{y^2}$

Use the product rule again: $= \log_7 49 + \log_7 x^3 + \log_7 \sqrt[5]{y^2}$

Rewrite $\sqrt[5]{y^2}$ using a rational exponent: $= \log_7 49 + \log_7 x^3 + \log_7 y^{2/5}$

Rewrite $\log_7 49$ as 2 and use the power rule: $= 2 + 3 \log_7 x + \dfrac{2}{5} \log_7 y$

b. Factor the expression in the numerator: $\ln \left(\dfrac{(x^2 - 4)}{9 e^{x^3}} \right) = \ln \left(\dfrac{(x - 2)(x + 2)}{9 e^{x^3}} \right)$

Use the quotient rule: $= \ln (x - 2)(x + 2) - \ln 9 e^{x^3}$

Use the product rule twice: $= \ln (x - 2) + \ln (x + 2) - \left[\ln 9 + \ln e^{x^3} \right]$

Use cancellation property (2) to rewrite $\ln e^{x^3}$ as x^3: $= \ln (x - 2) + \ln (x + 2) - \left[\ln 9 + x^3 \right]$

Simplify: $= \ln (x - 2) + \ln (x + 2) - \ln 9 - x^3$

Watch the interactive video to see this example worked out in detail.

You Try It   Work through this You Try It problem.

Work Exercises 11–20 in this textbook or in the *MyMathLab*® Study Plan.

My interactive video summary   **Example 5 Condensing a Logarithmic Expression**

Use properties of logarithms to rewrite each expression as a single logarithm.

a. $\dfrac{1}{2} \log (x - 1) - 3 \log z + \log 5$

b. $\dfrac{1}{3} (\log_3 x - 2 \log_3 y) + \log_3 10$

Solution

a. Write the original expression: $\dfrac{1}{2}\log(x-1) - 3\log z + \log 5$

Use the power rule twice: $= \log(x-1)^{1/2} - \log z^3 + \log 5$

Use the quotient rule: $= \log\dfrac{(x-1)^{1/2}}{z^3} + \log 5$

Use the product rule: $= \log\dfrac{5(x-1)^{1/2}}{z^3}$ or $\log\dfrac{5\sqrt{x-1}}{z^3}$

b. Write the original expression: $\dfrac{1}{3}(\log_3 x - 2\log_3 y) + \log_3 10$

Use the power rule: $= \dfrac{1}{3}(\log_3 x - \log_3 y^2) + \log_3 10$

Use the quotient rule: $= \dfrac{1}{3}\log_3\dfrac{x}{y^2} + \log_3 10$

Use the power rule: $= \log_3\left(\dfrac{x}{y^2}\right)^{1/3} + \log_3 10$

Use the product rule: $= \log_3\left[10\left(\dfrac{x}{y^2}\right)^{1/3}\right]$ or $\log_3\left[10\sqrt[3]{\dfrac{x}{y^2}}\right]$

Watch the interactive video to see this example worked out in detail.

You Try It   **Work through this You Try It problem.**

Work Exercises 21–30 in this textbook or in the *MyMathLab*® Study Plan.

OBJECTIVE 3:   SOLVE LOGARITHMIC EQUATIONS USING THE LOGARITHM PROPERTY OF EQUALITY

Remember that all logarithmic functions of the form $f(x) = \log_b x$ for $b > 0$ and $b \neq 1$ are one-to-one. In Section 9.2, the alternate definition of **one-to-one** stated that

> A function f is one-to-one if for any two range values $f(u)$ and $f(v)$, $f(u) = f(v)$ implies that $u = v$.

Using this definition and letting $f(x) = \log_b x$, we can say that if $\log_b u = \log_b v$, then $u = v$. In other words, if the bases of a logarithmic equation of the form $\log_b u = \log_b v$ are equal, then the arguments must be equal. This is known as the **logarithm property of equality**.

Logarithm Property of Equality

If a logarithmic equation can be written in the form $\log_b u = \log_b v$, then $u = v$. Furthermore, if $u = v$, then $\log_b u = \log_b v$.

The second statement of the logarithm property of equality says that if we start with the equation $u = v$, then we can rewrite the equation as $\log_b u = \log_b v$. This process is often casually referred to as "taking the log of both sides."

*My interactive video summary* **Example 6 Using the Logarithm Property of Equality to Solve Logarithmic Equations**

Solve the following equations.

a. $\log_7 (x - 1) = \log_7 12$

b. $2 \ln x = \ln 16$

Solution

a. Because the base of each logarithm is 7, we can use the logarithm property of equality to eliminate the logarithms.

$$\begin{aligned} \text{Write the original equation:} \quad & \log_7 (x - 1) = \log_7 12 \\ \text{If } \log_b u = \log_b v, \text{ then } u = v: \quad & (x - 1) = 12 \\ \text{Solve for } x: \quad & x = 13 \end{aligned}$$

b.
$$\begin{aligned} \text{Write the original expression:} \quad & 2 \ln x = \ln 16 \\ \text{Use the power rule:} \quad & \ln x^2 = \ln 16 \\ \text{If } \log_b u = \log_b v, \text{ then } u = v: \quad & x^2 = 16 \\ \text{Use the square root property:} \quad & x = \pm 4 \end{aligned}$$

The domain of $\ln x$ is $x > 0$; this implies that $x = -4$ is an extraneous solution, and hence, we must discard it. Therefore, this equation has only one solution, $x = 4$.

You Try It   Work through this You Try It problem.

Work Exercises 31–36 in this textbook or in the *MyMathLab*® Study Plan.

CAUTION Remember to check potential solutions in the *original* equation before forming the solution set.

OBJECTIVE 4:   USE THE CHANGE OF BASE FORMULA

Most scientific calculators are equipped with a $\boxed{\log}$ key and a $\boxed{\ln}$ key to evaluate common logarithms and natural logarithms. But how do we use a calculator to evaluate logarithmic expressions having a base other than 10 or e? The answer is to use the following **change of base formula**.

My video summary

Change of Base Formula

For any positive base $b \neq 1$ and for any positive real number u, then

$$\log_b u = \frac{\log_a u}{\log_a b},$$

where a is any positive number such that $a \neq 1$.

⊞ Click here to watch the video proof of the change of base formula.

The change of base formula allows us to change the base of a logarithmic expression into a ratio of two logarithms using any base we choose. For example, suppose we are given the logarithmic expression $\log_3 10$. We can use the change of base formula to write this logarithm as a logarithm involving any base we choose:

$$\log_3 10 = \frac{\log_7 10}{\log_7 3} \quad \text{or} \quad \log_3 10 = \frac{\log_2 10}{\log_2 3} \quad \text{or} \quad \log_3 10 = \frac{\log 10}{\log 3} \quad \text{or} \quad \log_3 10 = \frac{\ln 10}{\ln 3}$$

In each of the previous four cases, we introduced a new base (7, 2, 10, and e, respectively). However, if we want to use a calculator to get a numerical approximation of $\log_3 10$, then it really only makes sense to change $\log_3 10$ into an expression involving base 10 or base e because these are the only two bases most calculators can handle.

Note: $\log_3 10 = \dfrac{\log 10}{\log 3} \approx 2.0959 \quad \text{or} \quad \log_3 10 = \dfrac{\ln 10}{\ln 3} \approx 2.0959$

Example 7 Using the Change of Base Formula

Approximate the following expressions. Round each to four decimal places.

a. $\log_9 200$ **b.** $\log_{\sqrt{3}} \pi$

Solution

a. $\log_9 200 = \dfrac{\log 200}{\log 9} \approx 2.4114 \quad \text{or} \quad \log_9 200 = \dfrac{\ln 200}{\ln 9} \approx 2.4114$

b. $\log_{\sqrt{3}} \pi = \dfrac{\log \pi}{\log \sqrt{3}} \approx 2.0840 \quad \text{or} \quad \log_{\sqrt{3}} \pi = \dfrac{\ln \pi}{\ln \sqrt{3}} \approx 2.0840$

You Try It Work through this You Try It problem.

Work Exercises 37–40 in this textbook or in the **MyMathLab**® Study Plan.

Example 8 Using the Change of Base Formula and Properties of Logarithms

My video summary

Use the change of base formula and the properties of logarithms to rewrite as a single logarithm involving base 2.

$$\log_4 x + 3 \log_2 y$$

Solution

To use properties of logarithms, the base of each logarithmic expression must be the same. We use the change of base formula to rewrite $\log_4 x$ as a logarithmic expression involving base 2:

Use the change of base formula $\log_4 x = \dfrac{\log_2 x}{\log_2 4}$: $\log_4 x + 3 \log_2 y = \dfrac{\log_2 x}{\log_2 4} + 3 \log_2 y$

Rewrite $\log_2 4$ as 2 because $2^2 = 4$: $= \dfrac{\log_2 x}{2} + 3 \log_2 y$

Rewrite $\dfrac{\log_2 x}{2}$ as $\dfrac{1}{2} \log_2 x$: $= \dfrac{1}{2} \log_2 x + 3 \log_2 y$

Use the power rule: $= \log_2 x^{1/2} + \log_2 y^3$

Use the product rule: $= \log_2 x^{1/2} y^3 \quad \text{or} \quad \log_2 \sqrt{x}\, y^3$

Therefore, the expression $\log_4 x + 3\log_2 y$ is equivalent to $\log_2 \sqrt{x}y^3$. Note that we could have chosen to rewrite the original expression as a single logarithm involving base 4. Watch the video to see that the expression $\log_4 x + 3\log_2 y$ is also equivalent to $\log_4 xy^6$.

You Try It   Work through this You Try It problem.

Work Exercises 41–44 in this textbook or in the **MyMathLab**® Study Plan.

My video summary

Example 9 Using the Change of Base Formula to Solve Logarithmic Equations

Use the change of base formula and the properties of logarithms to solve the equation

$$2\log_3 x = \log_9 16.$$

Solution

Watch the video to see how the change of base formula and the power rule for logarithms can be used to solve this equation.

You Try It   Work through this You Try It problem.

Work Exercises 45–48 in this textbook or in the **MyMathLab**® Study Plan.

9.6 Exercises

In Exercises 1–10, use the product rule, quotient rule, or power rule to expand each logarithmic expression. Wherever possible, evaluate logarithmic expressions.

You Try It

1. $\log_4 (xy)$

2. $\log\left(\dfrac{9}{t}\right)$

3. $\log_5 y^3$

4. $\log_3 (27w)$

5. $\ln 5e^2$

6. $\log_9 \sqrt[4]{k}$

7. $\log 100P$

8. $\ln\left(\dfrac{e^5}{r}\right)$

9. $\log_{\sqrt{2}} 8x$

10. $\log_2\left(\dfrac{M}{32}\right)$

In Exercises 11–20, use the properties of logarithms to expand each logarithmic expression. Wherever possible, evaluate logarithmic expressions.

You Try It

11. $\log_7 x^2 y^3$

12. $\ln\dfrac{a^2 b^3}{c^4}$

13. $\log\dfrac{\sqrt{x}}{10y^3}$

14. $\log_3 9(x^2 - 25)$

15. $\log_2 \sqrt{4xy}$

16. $\log_5 \dfrac{\sqrt{5x^5}}{\sqrt[3]{25y^4}}$

17. $\ln\dfrac{\sqrt[5]{ez}}{\sqrt{x-1}}$

18. $\log_3 \sqrt[4]{\dfrac{x^3 y^5}{9}}$

19. $\ln\left[\dfrac{x+1}{(x^2-1)^3}\right]^{2/3}$

20. $\log\dfrac{(10x)^3 \sqrt{x-4}}{(x^2-16)^5}$

You Try It

In Exercises 21–30, use properties of logarithms to rewrite each expression as a single logarithm. Wherever possible, evaluate logarithmic expressions.

21. $\log_b A + \log_b C$

22. $\log_4 M - \log_4 N$

23. $2\log_8 x + \dfrac{1}{3}\log_8 y$

24. $\log 20 + \log 5$

25. $\log_2 80 - \log_2 5$

26. $\ln\sqrt{x} - \dfrac{1}{3}\ln x + \ln\sqrt[4]{x}$

27. $\log_5 (x - 2) + \log_5 (x + 2)$

28. $\log_9 (x^2 - 5x + 6) - \log_9 (x^2 - 4) + \log_9 (x + 2)$

29. $\log (x - 3) + 2\log (x + 3) - \log (x^3 + x^2 - 9x - 9)$

30. $\dfrac{1}{2}\left[\ln (x - 1)^2 - \ln (2x^2 - x - 1)^4\right] + 2\ln (2x + 1)$

You Try It

In Exercises 31–36, use the properties of logarithms and the logarithm property of equality to solve each logarithmic equation.

31. $\log_3 (2x + 1) = \log_3 11$

32. $\log_{11}\sqrt{x} = \log_{11} 6$

33. $2\log (x + 5) = \log 12 + \log 3$

34. $\ln 5 + \ln x = \ln 7 + \ln (3x - 2)$

35. $\log_7 (x + 6) - \log_7 (x + 2) = \log_7 x$

36. $\log_2 (x + 3) + \log_2 (x - 4) = \log_2 (x + 12)$

In Exercises 37–40, use the change of base formula and a calculator to approximate each logarithmic expression. Round your answers to four decimal places.

You Try It   **37.** $\log_4 51$ **38.** $\log_7 0.8$ **39.** $\log_{1/5} 72$ **40.** $\log_{\sqrt{7}} 100$

In Exercises 41–44, use the change of base formula and the properties of logarithms to rewrite each expression as a single logarithm in the indicated base.

You Try It

41. $\log_3 x + 4\log_9 w$, base 3

42. $\log_5 x + \log_{1/5} x^3$, base 5

43. $\log_{16} x^4 + \log_8 y^3 + \log_4 w^2$, base 2

44. $\log_{e^2} x^5 + \log_{e^3} x^6 + \log_{e^4} x^{12}$, base e

You Try It

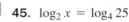

In Exercises 45–48, solve each logarithmic equation.

45. $\log_2 x = \log_4 25$

46. $\log_{1/3} x = \log_3 20$

47. $\log_5 x = \log_{\sqrt{5}} 6$

48. $2\ln x = \log_{e^3} 125$

9.7 Exponential and Logarithmic Equations

THINGS TO KNOW

Before working through this section, be sure that you are familiar with the following concepts:

| | VIDEO | ANIMATION | INTERACTIVE |

You Try It
1. Solve Exponential Equations by Relating the Bases (Section 9.3, **Objective 3**)

You Try It
2. Change from Exponential to Logarithmic Form (Section 9.5, **Objective 1**)

You Try It
3. Change from Logarithmic to Exponential Form (Section 9.5, **Objective 1**)

You Try It
4. Use the Cancellation Properties of Exponentials and Logarithms (Section 9.5, **Objective 3**)

You Try It
5. Expand and Condense Logarithmic Expressions (Section 9.6, **Objective 2**)

You Try It
6. Solve Logarithmic Equations Using the Logarithm Property of Equality (Section 9.6, **Objective 3**)

OBJECTIVES

1 Solve Exponential Equations

2 Solve Logarithmic Equations

In this section, we learn how to solve exponential and logarithmic equations. The techniques and strategies learned in this section help us solve applied problems that are discussed in Section 9.8. We start by developing a strategy to solve exponential equations.

OBJECTIVE 1: SOLVE EXPONENTIAL EQUATIONS

We have already solved exponential equations using the method of relating the bases. For example, we can solve the equation $4^{x+3} = \frac{1}{2}$ by converting the base on both sides of the equation to base 2. Click here to see how to solve this equation.

But suppose we are given an exponential equation in which the bases cannot be related, such as $2^{x+1} = 3$. Remember in Section 9.3, Example 3, we wanted to find the one x-intercept of the graph of $f(x) = -2^{x+1} + 3$ (Figure 32).

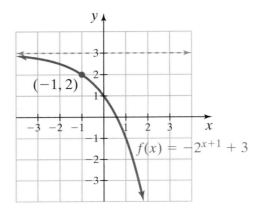

Figure 32
Graph of $f(x) = -2^{x+1} + 3$

To find the x-intercept of $f(x) = -2^{x+1} + 3$, we need to set $f(x) = 0$ and solve for x.

$$f(x) = -2^{x+1} + 3$$
$$0 = -2^{x+1} + 3$$
$$2^{x+1} = 3$$

My video summary

In Section 9.3, we could not solve this equation for x because we had not yet defined the logarithm. We can now use some properties of logarithms to solve this equation. Recall the following logarithmic properties.

If $u = v$, then $\log_b u = \log_b v$.	**logarithm property of equality**
$\log_b u^r = r \log_b u$	**power rule for logarithms**

We can use these two properties to solve the equation $2^{x+1} = 3$ and thus determine the x-intercept of $f(x) = -2^{x+1} + 3$. We solve the equation $2^{x+1} = 3$ in Example 1.

Example 1 Solving an Exponential Equation

Solve $2^{x+1} = 3$.

Solution

Write the original equation:	$2^{x+1} = 3$
Use the logarithm property of equality:	$\ln 2^{x+1} = \ln 3$
Use the power rule for logarithms:	$(x + 1)\ln 2 = \ln 3$
Use the distributive property:	$x \ln 2 + \ln 2 = \ln 3$
Subtract $\ln 2$ from both sides:	$x \ln 2 = \ln 3 - \ln 2$
Divide both sides by $\ln 2$:	$x = \dfrac{\ln 3 - \ln 2}{\ln 2}$

The solution to Example 1 verifies that the x-intercept of

$$f(x) = -2^{x+1} + 3 \text{ is } x = \frac{\ln 3 - \ln 2}{\ln 2} \approx 0.5850. \text{ See Figure 33.}$$

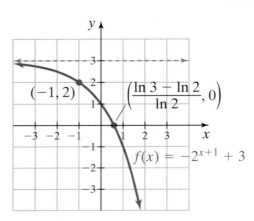

Figure 33 Graph of $f(x) = -2^{x+1} + 3$

When we cannot easily relate the bases of an exponential equation, as in Example 1, we use logarithms and their properties to solve them. The methods used to solve exponential equations are outlined as follows.

Solving Exponential Equations

- If the equation can be written in the form $b^u = b^v$, then solve the equation $u = v$.

- If the equation cannot easily be written in the form $b^u = b^v$,

  1. Use the logarithm property of equality to "take the log of both sides" (typically using base 10 or base e).
  2. Use the product rule of logarithms to "bring down" any exponents.
  3. Solve for the given variable.

Example 2 Solving Exponential Equations

 Solve each equation. For part b, round to four decimal places.

a. $3^{x-1} = \left(\dfrac{1}{27}\right)^{2x+1}$ **b.** $7^{x+3} = 4^{2-x}$

My interactive video summary

Solution

a. Watch the interactive video, or click here to see that the solution is

$$x = -\frac{2}{7}.$$

b. We cannot easily use the method of relating the bases because we cannot easily write both 7 and 4 using a common base. Therefore, we use logarithms to solve.

Write the original equation:	$7^{x+3} = 4^{2+x}$
Use the logarithm property of equality:	$\ln 7^{x+3} = \ln 4^{2-x}$
Use the power rule for logarithms:	$(x + 3)\ln 7 = (2 - x)\ln 4$
Use the distributive property:	$x \ln 7 + 3 \ln 7 = 2 \ln 4 - x \ln 4$
Add $x \ln 4$ to both sides, and subtract $3 \ln 7$ from both sides:	$x \ln 7 + x \ln 4 = 2 \ln 4 - 3 \ln 7$
Factor out an x from the left-hand side:	$x(\ln 7 + \ln 4) = 2 \ln 4 - 3 \ln 7$
Divide both sides by $\ln 7 + \ln 4$:	$x = \dfrac{2 \ln 4 - 3 \ln 7}{\ln 7 + \ln 4}$
Use the power rule for logarithms in the numerator, and use the product rule for logarithms in the denominator:	$= \dfrac{\ln 16 - \ln 343}{\ln 28}$
Use the quotient rule for logarithms to rewrite $\ln 16 - \ln 343$ as $\ln\left(\dfrac{16}{343}\right)$:	$= \dfrac{\ln\left(\dfrac{16}{343}\right)}{\ln 28}$
Use a calculator to round to four decimal places:	≈ -0.9199

You Try It   Work through this You Try It problem.

Work Exercises 1–10 in this textbook or in the *MyMathLab*® Study Plan.

Example 3 Solving Exponential Equations Involving the Natural Exponential Function

My interactive video summary

 Solve each equation. Round to four decimal places.

a. $25e^{x-5} = 17$ **b.** $e^{2x-1} \cdot e^{x+4} = 11$

Solution

a. Isolate the exponential term on the left by dividing both sides of the equation by 25.

Write the original equation: $25e^{x-5} = 17$

Divide both sides by 25: $e^{x-5} = \dfrac{17}{25}$

Use the logarithm property of equality: $\ln e^{x-5} = \ln \dfrac{17}{25}$

Use cancellation property (2) to rewrite $\ln e^{x-5}$ as $x - 5$: $x - 5 = \ln \dfrac{17}{25}$

Add 5 to both sides: $x = \ln \dfrac{17}{25} + 5$

Use a calculator to round to four decimal places: ≈ 4.6143

b. Write the original equation: $e^{2x-1} \cdot e^{x+4} = 11$

Use $b^m \cdot b^n = b^{m+n}$: $e^{(2x-1)+(x+4)} = 11$

Combine like terms in the exponent: $e^{3x+3} = 11$

Use the logarithm property of equality: $\ln e^{3x+3} = \ln 11$

Use cancellation property (2) to rewrite $\ln e^{3x+3}$ as $3x + 3$: $3x + 3 = \ln 11$

Subtract 3 from both sides: $3x = \ln 11 - 3$

Divide both sides by 3: $x = \dfrac{\ln 11 - 3}{3}$

Use a calculator to round to four decimal places: ≈ -0.2007

Watch the interactive video to see the solutions to this example worked out in detail.

You Try It **Work through this You Try It problem.**

Work Exercises 11–15 in this textbook or in the *MyMathLab*® Study Plan.

OBJECTIVE 2: SOLVE LOGARITHMIC EQUATIONS

We now turn our attention to solving logarithmic equations. In Section 9.6, we learned how to solve certain logarithmic equations by using the logarithm property of equality. That is, if we can write a logarithmic equation in the form $\log_b u = \log_b v$, then $u = v$. Before we look at an example, let's review three of the properties of logarithms.

Properties of Logarithms

If $b > 0, b \neq 1, u$ and v represent positive numbers and r is any real number, then

$$\log_b uv = \log_b u + \log_b v \quad \textbf{product rule for logarithms}$$

$$\log_b \frac{u}{v} = \log_b u - \log_b v \quad \textbf{quotient rule for logarithms}$$

$$\log_b u^r = r \log_b u \qquad\qquad \textbf{power rule for logarithms}$$

Example 4 Solving a Logarithmic Equation Using the Logarithm Property of Equality

 *My video summary*

 Solve $2 \log_5 (x - 1) = \log_5 64$.

Solution We can use the power rule for logarithms and the logarithmic property of equality to solve.

$$
\begin{aligned}
\text{Write the original equation:} \quad & 2 \log_5 (x - 1) = \log_5 64 \\
\text{Use the power rule:} \quad & \log_5 (x - 1)^2 = \log_5 64 \\
\text{Use the logarithm property of equality:} \quad & (x - 1)^2 = 64 \\
\text{Use the square root property:} \quad & x - 1 = \pm 8 \\
\text{Solve for } x: \quad & x = 1 \pm 8 \\
\text{Simplify:} \quad & x = 9 \quad \text{or} \quad x = -7
\end{aligned}
$$

Recall that the domain of a logarithmic function must contain only positive numbers; thus, $x - 1$ must be positive. Therefore, the solution of $x = -7$ must be discarded. The only solution is $x = 9$. You may want to review how to determine the domain of a logarithmic function, which is discussed in Section 9.5.

You Try It **Work through this You Try It problem.**

Work Exercises 16–19 in this textbook or in the *MyMathLab*® Study Plan.

CAUTION When solving logarithmic equations, it is important to always verify the solutions. Logarithmic equations often lead to extraneous solutions, as in Example 4.

When a logarithmic equation cannot be written in the form $\log_b u = \log_b v$, as in Example 4, we adhere to the steps outlined as follows:

Solving Logarithmic Equations

1. Determine the domain of the variable.

2. Use properties of logarithms to combine all logarithms, and write as a single logarithm, if needed.

3. Eliminate the logarithm by rewriting the equation in exponential form. Click here to review how to change from logarithmic form to exponential form, which is discussed in Section 9.5.

4. Solve for the given variable.

5. Check for any extraneous solutions. Verify that each solution is in the domain of the variable.

My video summary

Example 5 Solving a Logarithmic Equation

Solve $\log_4 (2x - 1) = 2$.

Solution The domain of the variable in this equation is the solution to the inequality $2x - 1 > 0$ or $x > \dfrac{1}{2}$. Thus, our solution must be greater than $\dfrac{1}{2}$. Because the equation involves a single logarithm, we can proceed to the third step.

Write the original equation:	$\log_4 (2x - 1) = 2$
Rewrite in exponential form:	$4^2 = 2x - 1$
Simplify:	$16 = 2x - 1$
Add 1 to both sides:	$17 = 2x$
Divide by 2:	$x = \dfrac{17}{2}$

Because the solution satisfies the inequality, there are no extraneous solutions. We can verify the solution by substituting $x = \dfrac{17}{2}$ into the original equation.

Check:	Write the original equation:	$\log_4 (2x - 1) = 2$
	Substitute $x = \dfrac{17}{2}$:	$\log_4 \left(2\left(\dfrac{17}{2}\right) - 1 \right) \overset{?}{=} 2$
	Simplify:	$\log_4 (17 - 1) \overset{?}{=} 2$
	This is a true statement because $4^2 = 16$:	$\log_4 (16) = 2$

You Try It   **Work through this You Try It problem.**

Work Exercises 20–24 in this textbook or in the *MyMathLab*® Study Plan.

Example 6 Solving a Logarithmic Equation

My interactive video summary

Solve $\log_2 (x + 10) + \log_2 (x + 6) = 5$.

Solution The domain of the variable in this equation is the solution to the compound inequality $x + 10 > 0$ and $x + 6 > 0$. The solution to this compound inequality is $x > -6$. (You may want to review compound inequalities from Section 1.3.)

Write the original equation:	$\log_2 (x + 10) + \log_2 (x + 6) = 5$
Use the product rule:	$\log_2 (x + 10)(x + 6) = 5$
Rewrite in exponential form:	$(x + 10)(x + 6) = 2^5$
Simplify:	$x^2 + 16x + 60 = 32$
Subtract 32 from both sides:	$x^2 + 16x + 28 = 0$
Factor:	$(x + 14)(x + 2) = 0$
Use the zero product property to solve:	$x = -14 \quad \text{or} \quad x = -2$

Because the domain of the variable is $x > -6$, we must *exclude* the solution $x = -14$. Therefore, the only solution to this logarithmic equation is $x = -2$. Work through the interactive video to see this solution worked out in detail.

You Try It   Work through this You Try It problem.

Work Exercises 25–30 in this textbook or in the *MyMathLab*® Study Plan.

Example 7 Solving a Logarithmic Equation

Solve $\ln (x - 4) - \ln (x - 5) = 2$. Round to four decimal places.

Solution The domain of the variable is the solution to the compound inequality $x - 4 > 0$ and $x - 5 > 0$. The solution to this compound inequality is $x > 5$. (You may want to review compound inequalities from Section 1.3.)

Write the original equation: $\ln (x - 4) - \ln (x - 5) = 2$

Use the quotient rule: $\ln \left(\dfrac{x - 4}{x - 5} \right) = 2$

Rewrite in exponential form: $e^2 = \dfrac{x - 4}{x - 5}$

Multiply both sides by $x - 5$: $e^2(x - 5) = x - 4$

Use the distributive property: $e^2 x - 5e^2 = x - 4$

Add $5e^2$ to both sides and subtract x from both sides: $e^2 x - x = 5e^2 - 4$

Factor out an x from the left-hand side: $x(e^2 - 1) = 5e^2 - 4$

Solve for x. Use a calculator to round to four decimal places: $x = \dfrac{5e^2 - 4}{e^2 - 1} \approx 5.1565$

We approximate the exact answer $x = \dfrac{5e^2 - 4}{e^2 - 1}$ in order to verify that the solution is in the domain of the variable. In this example, we see that 5.1565 is clearly greater than 5. In some cases, we may need to use the exact answer to verify a solution to a logarithmic equation. Click here to see this verification for Example 7.

You Try It   Work through this You Try It problem.

Work Exercises 31 and 32 in this textbook or in the *MyMathLab*® Study Plan.

9.7 Exercises

You Try It   In Exercises 1–15, solve each exponential equation. For irrational solutions, round to four decimal places.

1. $3^x = 5$

2. $2x^{x/3} = 19$

3. $4^{x^2 - 2x} = 64$

4. $3^{x+7} = -20$

5. $(1.52)^{-3x/7} = 11$

6. $\left(\dfrac{1}{5} \right)^{x-1} = 25^x$

7. $8^{4x-7} = 11^{5+x}$

8. $3(9)^{x-1} = (81)^{2x+1}$

9. $(3.14)^x = \pi^{1-2x}$

10. $7(2 - 10^{4x-2}) = 8$

You Try It

11. $e^x = 2$ **12.** $150e^{x-4} = 5$ **13.** $e^{x-3} \cdot e^{3x+7} = 24$

14. $2(e^{x-1})^2 \cdot e^{3-x} = 80$ **15.** $8e^{-x/3} \cdot e^x = 1$

In Exercises 16–30, solve each logarithmic equation.

You Try It

16. $\log_4 (x + 1) = \log_4 (6x - 5)$ **17.** $\log_3 (x^2 - 21) = \log_3 4x$

18. $2 \log_5 (3 - x) - \log_5 2 = \log_5 18$

19. $2 \ln x - \ln (2x - 3) = \ln 2x - \ln (x - 1)$

You Try It

20. $\log_2 (4x - 7) = 3$ **21.** $\log (1 - 5x) = 2$

22. $\log_3 (2x - 5) = -2$ **23.** $\log_x 3 = -1$

24. $\log_{x/2} 16 = 2$

You Try It

25. $\log_2 (x - 2) + \log_2 (x + 2) = 5$ **26.** $\log_7 (x + 9) + \log_7 (x + 15) = 1$

27. $\log_2 (3x + 1) - \log_2 (x - 2) = 2$ **28.** $\log_4 (x - 7) + \log_4 x = \dfrac{3}{2}$

29. $\ln 3 + \ln \left(x^2 + \dfrac{2x}{3} \right) = 0$

30. $\log_2 (x - 4) + \log_2 (x + 6) = 2 + \log_2 x$

In Exercises 31 and 32, solve each logarithmic equation. Round to four decimal places.

You Try It

31. $\ln x - \ln (x + 6) = 1$ **32.** $\ln (x + 3) - \ln (x - 2) = 4$

9.8 Applications of Exponential and Logarithmic Functions

THINGS TO KNOW

Before working through this section, be sure you are familiar with the following concepts:

VIDEO ANIMATION INTERACTIVE

You Try It
1. Solve Applications of Exponential Functions (Section 9.3, Objective 4)

You Try It
2. Solve Applications of the Natural Exponential Function (Section 9.4, Objective 4)

You Try It
3. Use the Properties of Logarithms (Section 9.5, Objective 3)

OBJECTIVES

1 Solve Compound Interest Applications

2 Solve Exponential Growth and Decay Applications

3 Solve Logistic Growth Applications

4 Use Newton's Law of Cooling

We have seen that exponential functions appear in a wide variety of settings, including biology, chemistry, physics, and business. In this section, we revisit some applications that are discussed previously in this chapter and then introduce several new applications. The difference between the applications presented earlier and the applications presented in this section is that we are now equipped with the tools necessary to solve for variables that appear as exponents. We start with applications involving **compound interest**. You may want to review periodic compound interest from Section 9.3 and continuous compound interest from Section 9.4 before proceeding.

OBJECTIVE 1:   SOLVE COMPOUND INTEREST APPLICATIONS

In Sections 9.3 and 9.4, the formula for compound interest and continuous compound interest are defined as follows.

Compound Interest Formulas

Periodic Compound Interest Formula

$$A = P\left(1 + \frac{r}{n}\right)^{nt}$$

Continuous Compound Interest Formula

$$A = Pe^{rt},$$

where

A = Total amount after t years
P = Principal (original investment)
r = Interest rate per year
n = Number of times interest is compounded per year
t = Number of years

My video summary

Example 1 Finding the Doubling Time

How long will it take (in years and months) for an investment to double if it earns 7.5% compounded monthly?

Solution We use the periodic compound interest formula $A = P\left(1 + \frac{r}{n}\right)^{nt}$ with $r = 0.075$ and $n = 12$ and solve for t. Notice that the principal is not given. As it turns out, any value of P will suffice. If the principal is P, then the amount needed to double the investment is $A = 2P$. We now have all of the information necessary to solve for t:

Use the periodic compound interest formula: $\qquad A = P\left(1 + \dfrac{r}{n}\right)^{nt}$

Substitute the appropriate values: $\qquad 2P = P\left(1 + \dfrac{0.075}{12}\right)^{12t}$

Divide both sides by P: $\qquad 2 = \left(1 + \dfrac{0.075}{12}\right)^{12t}$

Simplify within the parentheses: $\qquad 2 = (1.00625)^{12t}$

Use the logarithm property of equality: $\quad \ln 2 = \ln (1.00625)^{12t}$

Use the power rule: $\log_b u^r = r\log_b u$: $\quad \ln 2 = 12t \ln (1.00625)$

Divide both sides by $12 \ln (1.00625)$: $\qquad t = \dfrac{\ln 2}{12 \ln (1.00625)}$

Round to two decimal places: $\qquad t \approx 9.27$ years

Note that 0.27 years $= 0.27$ years $\times \frac{12\ \text{months}}{1\ \text{year}} = 3.24$ months. Because the interest is compounded at the end of each month, the investment will not double until 9 years and 4 months.

Example 2 Continuous Compound Interest

 Suppose an investment of \$5000 compounded continuously grew to an amount of \$5130.50 in 6 months. Find the interest rate, and then determine how long it will take for the investment to grow to \$6000. Round the interest rate to the nearest hundredth of a percent and the time to the nearest hundredth of a year.

Solution Because the investment is compounded continuously, we use the formula $A = Pe^{rt}$.

We are given that $P = 5000$, so $A = 5000e^{rt}$. In 6 months, or when $t = 0.5$ years, we know that $A = 5130.50$. Substituting these values into the compound interest formula will enable us to solve for r:

Substitute the appropriate values: $\qquad 5130.50 = 5000e^{r(0.5)}$

Divide by 5000: $\qquad \dfrac{5130.50}{5000} = e^{0.5r}$

Use the logarithm property of equality: $\qquad \ln\left(\dfrac{5130.50}{5000}\right) = \ln e^{0.5r}$

Use cancellation property (2) to rewrite $\ln e^{0.5r}$ as $0.5r$: $\qquad \ln\left(\dfrac{5130.50}{5000}\right) = 0.5r$

Divide by 0.5: $\quad r = \dfrac{\ln\left(\dfrac{5130.50}{5000}\right)}{0.5} \approx 0.05153$

Therefore, the interest rate is 5.15%. To find the time that it takes for the investment to grow to \$6000, we use the formula $A = Pe^{rt}$, with $A = 6000$, $P = 5000$, and $r = 0.0515$, and solve for t. Watch the video to verify that it will take approximately 3.54 years.

My video summary

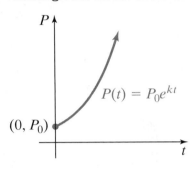

You Try It   Work through this **You Try It** problem.

Work Exercises 1–5 in this textbook or in the ***MyMathLab***® Study Plan.

OBJECTIVE 2:   SOLVE EXPONENTIAL GROWTH AND DECAY APPLICATIONS

In Section 9.4, the exponential growth model is introduced. This model is used when a population grows at a rate proportional to the size of its current population. This model is often called the uninhibited growth model. We review this exponential growth model and sketch the graph in Figure 34.

P

$$P(t) = P_0 e^{kt}$$

$(0, P_0)$

t

Figure 34
Graph of $P(t) = P_0 e^{kt}$ for $k > 0$

Exponential Growth

A model that describes the exponential uninhibited growth of a population, P, after a certain time, t, is

$$P(t) = P_0 e^{kt},$$

where $P_0 = P(0)$ is the initial population and $k > 0$ is a constant called the **relative growth rate**. (*Note:* k is sometimes given as a percent.)

Example 3   Population Growth

My video summary

The population of a small town grows at a rate proportional to its current size. In 1900, the population was 900. In 1920, the population had grown to 1600. What was the population of this town in 1950? Round to the nearest whole number.

Solution   Using the model $P(t) = P_0 e^{kt}$, we must first determine the constants P_0 and k. The initial population was 900 in 1900 so $P_0 = 900$. Therefore, $P(t) = 900 e^{kt}$. To find k, we use the fact that in 1920, or when $t = 20$, the population was 1600; thus,

Substitute $P(20) = 1600$: $P(20) = 900 e^{k(20)} = 1600$

$$900 e^{20k} = 1600$$

Divide by 900 and simplify: $e^{20k} = \dfrac{16}{9}$

Use the logarithm property of equality: $\ln e^{20k} = \ln \dfrac{16}{9}$

Use cancellation property (2) to rewrite
$\ln e^{20k}$ as $20k$: $20k = \ln \dfrac{16}{9}$

Divide by 20: $k = \dfrac{\ln\left(\frac{16}{9}\right)}{20}$

The function that models the population of this town at any time t is given by $P(t) = 900e^{\frac{\ln(16/9)}{20}t}$. To determine the population in 1950, or when $t = 50$, we evaluate $P(50)$:

$$P(50) = 900e^{\frac{\ln(16/9)}{20}(50)} \approx 3793$$

Exponential Decay

Some populations exhibit *negative exponential growth*. In other words, the population, quantity, or amount *decreases* over time. Such models are called **exponential decay** models. The only difference between an exponential growth model and an exponential decay model is that the constant, k, is less than zero. See Figure 35.

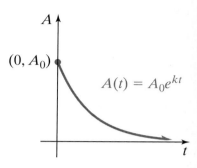

Figure 35

Graph of $A(t) = A_0 e^{kt}$ for $k < 0$

Exponential Decay

A model that describes the exponential decay of a population, quantity, or amount A, after a certain time, t, is

$$A(t) = A_0 e^{kt},$$

where $A_0 = A(0)$ is the initial quantity and $k < 0$ is a constant called the **relative decay constant**. (*Note:* k is sometimes given as a percent.)

 Half-Life

Every radioactive element has a half-life, which is the required time for a given quantity of that element to decay to half of its original mass. For example, the half-life of Cesium-137 is 30 years. Thus, it takes 30 years for any amount of Cesium-137 to decay to $\frac{1}{2}$ of its original mass. It takes an additional 30 years to decay to $\frac{1}{4}$ of its original mass and so on. See Figure 36, and view the animation that illustrates the half-life of Cesium-137.

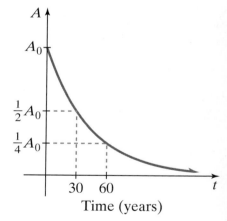

Figure 36

Half-life of Cesium-137

Example 4 Radioactive Decay

 Suppose that a meteorite is found containing 4% of its original Krypton-99. If the half-life of Krypton-99 is 80 years, how old is the meteorite? Round to the nearest year.

Solution We use the formula $A(t) = A_0 e^{kt}$, where A_0 is the original amount of Krypton-99. We first must find the constant k. To find k, we use the fact that the half-life of Krypton-99 is 80 years. Therefore, $A(80) = \frac{1}{2}A_0$. Because $A(80) = A_0 e^{k(80)}$, we can set $\frac{1}{2}A_0 = A_0 e^{k(80)}$ and solve for k.

$$\frac{1}{2}A_0 = A_0 e^{k(80)}$$

Divide both sides by A_0:
$$\frac{1}{2} = e^{80k}$$

Use the logarithm property of equality:
$$\ln\frac{1}{2} = \ln e^{80k}$$

Use cancellation property (2) to rewrite $\ln e^{80k}$ as $80k$:
$$\ln\frac{1}{2} = 80k$$

Divide both sides by 80:
$$\frac{\ln\frac{1}{2}}{80} = k$$

$\ln\frac{1}{2} = \ln 1 - \ln 2 = 0 - \ln 2 = -\ln 2$:
$$\frac{-\ln 2}{80} = k$$

Now that we know $k = \dfrac{-\ln 2}{80}$, our function becomes $A(t) = A_0 e^{\frac{-\ln 2}{80}t}$. To find out the age of the meteorite, we set $A(t) = 0.04A_0$ because the meteorite now contains 4% of the original amount of Krypton-99.

Substitute $0.04A_0$ for $A(t)$:
$$0.04A_0 = A_0 e^{\frac{-\ln 2}{80}t}$$

Divide both sides by A_0:
$$0.04 = e^{\frac{-\ln 2}{80}t}$$

Use the logarithm property of equality:
$$\ln 0.04 = \ln e^{\frac{-\ln 2}{80}t}$$

Use cancellation property (2) to rewrite $\ln e^{\frac{-\ln 2}{80}t}$ as $\dfrac{-\ln 2}{80}t$:
$$\ln 0.04 = \frac{-\ln 2}{80}t$$

Divide both sides by $\dfrac{-\ln 2}{80}$:
$$\frac{\ln 0.04}{\frac{-\ln 2}{80}} = t \approx 372 \text{ years}$$

The meteorite is about 372 years old.

You Try It   Work through this You Try It problem.

Work Exercises 6–12 in this textbook or in the *MyMathLab*® Study Plan.

OBJECTIVE 3:   SOLVE LOGISTIC GROWTH APPLICATIONS

The uninhibited exponential growth model $P(t) = P_0 e^{kt}$ for $k > 0$ is used when there are no outside limiting factors such as predators or disease that affect the population growth. When such outside factors exist, scientists often use a **logistic model** to describe the population growth. One such logistic model is described and sketched in Figure 37.

Logistic Growth

A model that describes the logistic growth of a population P at any time t is given by the function

$$P(t) = \frac{C}{1 + Be^{kt}},$$

where B, C, and k are constants with $C > 0$ and $k < 0$.

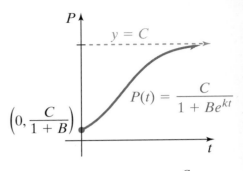

Figure 37 Graph of $P(t) = \dfrac{C}{1 + Be^{kt}}$

The number C is called the carrying capacity. In the logistic model, the population will approach the value of the carrying capacity over time but never exceed it. You can see in the graph sketched in Figure 37 that the graph of the logistic growth model approaches the horizontal asymptote $y = C$.

Example 5 Logistic Growth

 Ten goldfish were introduced into a small pond. Because of limited food, space, and oxygen, the carrying capacity of the pond is 400 goldfish. The goldfish population at any time t, in days, is modeled by the logistic growth function $F(t) = \dfrac{C}{1 + Be^{kt}}$. If 30 goldfish are in the pond after 20 days,

a. Find B.

b. Find k.

c. When will the pond contain 250 goldfish? Round to the nearest whole number.

Solution

a. The carrying capacity is 400; thus, $C = 400$. Also, initially (at $t = 0$) there were 10 goldfish, so $F(0) = 10$. Therefore,

Substitute $C = 400$ and $F(0) = 10$: $\quad 10 = \dfrac{400}{1 + Be^{k(0)}}$

Evaluate $e^0 = 1$: $\quad 10 = \dfrac{400}{1 + B}$

Multiply both sides by $1 + B$: $\quad 10 + 10B = 400$

Solve for B: $\quad B = 39$

b. Use the function $F(t) = \dfrac{400}{1 + 39e^{kt}}$ and the fact that $F(20) = 30$ (there are 30 goldfish after 20 days) to solve for k.

Substitute $F(20) = 30$: $\quad 30 = \dfrac{400}{1 + 39e^{k(20)}}$

Multiply both sides by $1 + 39e^{20k}$: $\quad 30(1 + 39e^{20k}) = 400$

Use the distributive property: $\quad 30 + 1170e^{20k} = 400$

Subtract 30 from both sides: $\quad 1170e^{20k} = 370$

Divide both sides by 1170: $\quad e^{20k} = \dfrac{370}{1170}$

Simplify: $\quad e^{20k} = \dfrac{37}{117}$

Use the logarithm property of equality: $\ln e^{20k} = \ln \dfrac{37}{117}$

Use cancellation property (2) to rewrite $\ln e^{20k}$ as $20k$: $20k = \ln \dfrac{370}{117}$

Divide both sides by 20: $k = \dfrac{\ln \frac{37}{117}}{20}$

c. Use the function $F(t) = \dfrac{400}{1 + 39e^{\frac{\ln(37/117)}{20}t}}$, and then find t when $F(t) = 250$.

By repeating the exact same process as in part (b), we find that it will take approximately 73 days until there are 250 goldfish in the pond. Watch the interactive video to verify the solution.

You Try It   **Work through this You Try It problem.**

Work Exercises 13–15 in this textbook or in the *MyMathLab*® Study Plan.

OBJECTIVE 4:   USE NEWTON'S LAW OF COOLING

Newton's law of cooling states that the temperature of an object changes at a rate proportional to the difference between its temperature and that of its surroundings. It can be shown in a more advanced course that the function describing Newton's law of cooling is given by the following.

Newton's Law of Cooling

The temperature T of an object at any time t is given by

$$T(t) = S + (T_0 - S)e^{kt},$$

where T_0 is the original temperature of the object, S is the constant temperature of the surroundings, and k is the cooling constant.

View the animation to see how this function behaves.

Example 6 Newton's Law of Cooling

My video summary

Suppose that the temperature of a cup of hot tea obeys Newton's law of cooling. If the tea has a temperature of 200°F when it is initially poured and 1 minute later has cooled to 189°F in a room that maintains a constant temperature of 69°F, determine when the tea reaches a temperature of 146°F. Round to the nearest minute.

Solution We start using the formula for Newton's law of cooling with $T_0 = 200$ and $S = 69$.

Use Newton's law of cooling formula: $T(t) = S + (T_0 - S)e^{kt}$

Substitute $T_0 = 200$ and $S = 69$: $T(t) = 69 + (200 - 69)e^{kt}$

Simplify: $T(t) = 69 + 131e^{kt}$

We now proceed to find k. The object cools to 189°F in 1 minute; thus, $T(1) = 189$. Therefore,

Substitute $T(1) = 189$:	$189 = 69 + 131e^{k(1)}$
Subtract 69 from both sides:	$120 = 131e^k$
Divide both sides by 131:	$\dfrac{120}{131} = e^k$
Use the logarithm property of equality:	$\ln\dfrac{120}{131} = \ln e^k$
Use cancellation property (2) to rewrite $\ln e^k$ as k:	$\ln\dfrac{120}{131} = k$

Now that we know the cooling constant $k = \ln\frac{120}{131}$, we can use the function $T(t) = 69 + 131e^{\ln\frac{120}{131}t}$ and determine the value of t when $T(t) = 146$.

Set $T(t) = 146$:	$146 = 69 + 131e^{\ln\frac{120}{131}t}$
Subtract 69 from both sides:	$77 = 131e^{\ln\frac{120}{131}t}$
Divide both sides by 131:	$\dfrac{77}{131} = e^{\ln\frac{120}{131}t}$
Use the logarithm property of equality:	$\ln\dfrac{77}{131} = \ln e^{\ln\frac{120}{131}t}$
Use cancellation property (2) to rewrite $\ln e^{\ln\frac{120}{131}t}$ as $\ln\dfrac{120}{131}t$:	$\ln\dfrac{77}{131} = \ln\dfrac{120}{131}t$
Divide both sides by $\ln\dfrac{120}{131}$:	$\dfrac{\ln\frac{77}{131}}{\ln\frac{120}{131}} = t$
Use a calculator to approximate the time rounded to the nearest minute:	$t \approx 6 \text{ minutes}$

So, it takes approximately 6 minutes for the tea to cool to 146°F.

The graph of $T(t) = 69 + 131e^{\ln\frac{120}{131}t}$, which describes the temperature of the tea t minutes after being poured, was created using a graphing utility. Note that the line $y = 69$, which represents the temperature of the surroundings, is a horizontal asymptote.

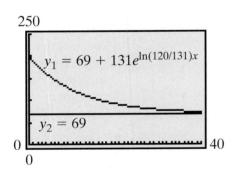

You Try It Work through this You Try It problem.

Work Exercises 16–18 in this textbook or in the *MyMathLab*® Study Plan.

9.8 Exercises

You Try It

1. Jimmy invests $15,000 in an account that pays 6.25% compounded quarterly. How long (in years and months) will it take for his investment to reach $20,000?

2. How long (in years and months) will it take for an investment to double at 9% compounded monthly?

3. How long will it take for an investment to triple if it is compounded continuously at 8%? Round to two decimal places.

4. What is the interest rate necessary for an investment to quadruple after 8 years of continuous compound interest? (Round to the nearest hundredth of a percent.)

You Try It

5. Marsha and Jan both invested money on March 1, 2005. Marsha invested $5000 at Bank A, where the interest was compounded quarterly. Jan invested $3000 at Bank B, where the interest was compounded continuously. On March 1, 2007, Marsha had a balance of $5468.12, whereas Jan had a balance of $3289.09. What was the interest rate at each bank? (Round to the nearest tenth of a percent.)

6. The population of Adamsville grew from 9000 to 15,000 in 6 years. Assuming uninhibited exponential growth, what is the expected population in an additional 4 years? Round to the nearest whole number.

7. During a research experiment, it was found that the number of bacteria in a culture grew at a rate proportional to its size. At 8:00 AM, there were 2000 bacteria present in the culture. At noon, the number of bacteria grew to 2400. How many bacteria will there be at midnight? Round to the nearest whole number.

8. A skull cleaning factory cleans animal skulls such as deer, buffalo, and other types of animal skulls using flesh-eating beetles to clean the skulls. The factory owner started with only 10 adult beetles. After 40 days, the beetle population grew to 30 adult beetles. Assuming uninhibited exponential growth, how long did it take before the beetle population reached 10,000 beetles? Round to the nearest whole number.

9. The population of a Midwest industrial town decreased from 210,000 to 205,000 in just 3 years. Assuming negative exponential growth and that this trend continues, what will the population be after an additional 3 years? Round to the nearest whole number.

10. A certain radioactive isotope is leaked into a small stream. Three hundred days after the leak, 2% of the original amount of the substance remained. Determine the half-life of this radioactive isotope. Round to the nearest whole number.

11. Radioactive Iodine-131 is a by-product of certain nuclear reactors. On April 26, 1986, one of the nuclear reactors in Chernobyl, Ukraine, a republic of the former Soviet Union, experienced a massive release of radioactive iodine. Fortunately, Iodine-131 has a very short half-life of 8 days. Estimate the percentage of the original amount of Iodine-131 released by the Chernobyl explosion on May 1, 1986, 5 days after the explosion. Round to two decimal places.

12. Superman is rendered powerless when exposed to 50 or more grams of kryptonite. A 500-year-old rock that originally contained 300 grams of kryptonite was recently stolen from a rock museum by Superman's enemies. The half-life of kryptonite is known to be 200 years.

    a. How many grams of kryptonite are still contained in the stolen rock? Round to two decimal places.

    b. For how many years can this rock be used by Superman's enemies to render him powerless? Round to the nearest whole number.

You Try It 13. The logistic growth model $H(t) = \dfrac{6000}{1 + 2e^{-0.65t}}$ represents the number of families that own a home in a certain small (but growing) Idaho city t years after 1980.

   a. What is the maximum number of families that will own a home in this city?

   b. How many families owned a home in 1980?

   c. In what year did 5920 families own a home?

14. The number of students that hear a rumor on a small college campus t days after the rumor starts is modeled by the logistic function $R(t) = \dfrac{3000}{1 + Be^{kt}}$. Determine the following if 8 students initially heard the rumor and 100 students heard the rumor after 1 day.

   a. What is the carrying capacity for the number of students who will hear the rumor?

   b. Find B.

   c. Find k.

   d. How long will it take for 2900 students to hear the rumor?

15. In 1999, 1500 runners entered the inaugural Run-for-Your-Life marathon in Joppetown, USA. In 2005, 21,500 runners entered the race. Because of the limited number of hotels, restaurants, and portable toilets in the area, the carrying capacity for the number of racers is 61,500. The number of racers at any time, t, in years, can be modeled by the logistic function $P(t) = \dfrac{C}{1 + Be^{kt}}$.

   a. What is the value of C?

   b. Find B.

   c. Find k.

   d. In what year should at least 49,500 runners be expected to run in the race? Round to the nearest year.

You Try It

16. Estabon poured himself a hot beverage that had a temperature of $198°F$ and then set it on the kitchen table to cool. The temperature of the kitchen was a constant $75°F$. If the drink cooled to $180°F$ in 2 minutes, how long will it take for the drink to cool to $100°F$?

17. Police arrive at a murder scene at 1:00 AM and immediately record the body's temperature, which was $92°F$. At 2:30 AM, after thoroughly inspecting and fingerprinting the area, they again took the temperature of the body, which had dropped to $85°F$. The temperature of the crime scene has remained at a constant $60°F$. Determine when the person was murdered. (Assume that the victim was healthy at the time of death. That is, assume that the temperature of the body at the time of death was $98.6°F$.)

18. Jodi poured herself a cold soda that had an initial temperature of $40°F$ and immediately went outside to sunbathe where the temperature was a steady $99°F$. After 5 minutes, the temperature of the soda was $47°F$. Jodi had to run back into the house to answer the phone. What is the expected temperature of the soda after an additional 10 minutes?

Conic Sections

APPENDIX A CONTENTS

Introduction to Conic Sections

A.1   **The Parabola**

A.2   **The Ellipse**

A.3   **The Hyperbola**

Introduction to Conic Sections

In this chapter, we focus on the geometric study of conic sections. Conic sections (or conics) are formed when a plane intersects a pair of right circular cones. The surface of the cones comprises the set of all line segments that intersect the outer edges of the circular bases of the cones and pass through a fixed point. The fixed point is called the **vertex** of the cone, and the line segments are called the **elements**. See Figure 1.

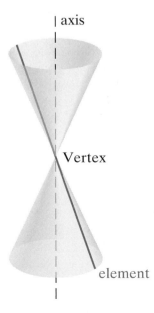

Figure 1

Pair of right circular cones

When a plane intersects a right circular cone, a conic section is formed. The four conic sections that can be formed are circles, ellipses, parabolas, and hyperbolas. Click on one of the following four animations to see how each conic section is formed.

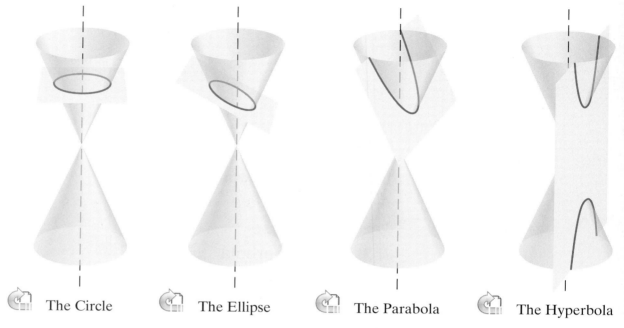

The Circle   The Ellipse   The Parabola   The Hyperbola

Because we studied circles in Section 8.4, they are not covered again in this chapter. It may, however, be useful to review circles before going on. Watch the video to review how to write the equation of a circle in standard form by completing the square. We also need to be able to complete the square to write each of the other conic sections in standard form. These conic sections are introduced in the following order:

Section A.1 The Parabola

Section A.2 The Ellipse

Section A.3 The Hyperbola

DEGENERATE CONIC SECTIONS

It is worth noting that when the circle, ellipse, parabola, or hyperbola is formed, the intersecting plane does not pass through the vertex of the cones. When a plane does intersect the vertex of the cones, a degenerate conic section is formed. The three degenerate conic sections are a point, a line, and a pair of intersecting lines. See Figure 2. We do not concern ourselves with degenerate conic sections in this chapter.

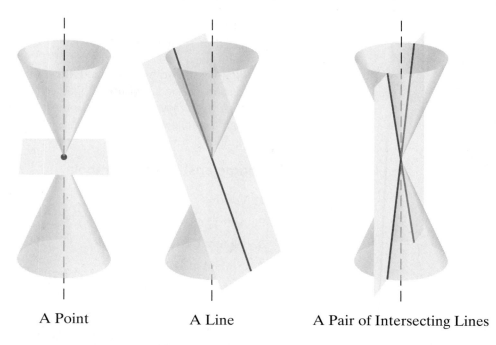

| A Point | A Line | A Pair of Intersecting Lines |

Figure 2 Degenerate conic sections

A.1 The Parabola

THINGS TO KNOW

Before working through this section, be sure you are familiar with the following concepts:

VIDEO   ANIMATION   INTERACTIVE

You Try It

1. Finding the Distance between Two Points (Section 8.4, **Objective** 1)

You Try It

2. Write the General Form of a Circle in Standard Form and Sketch Its Graph (Section 8.4, **Objective** 5)

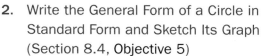
You Try It

3. Write the Equation of a Line from Given Information (Section 2.5, **Objective** 4)

OBJECTIVES

1 Work with the Equation of a Parabola with a Vertical Axis of Symmetry

2 Work with the Equation of a Parabola with a Horizontal Axis of Symmetry

3 Find the Equation of a Parabola Given Information about the Graph

4 Complete the Square to Find the Equation of a Parabola in Standard Form

5 Solve Applications Involving Parabolas

OBJECTIVE 1: **WORK WITH THE EQUATION OF A PARABOLA WITH A VERTICAL AXIS OF SYMMETRY**

In Section 8.2, we studied quadratic functions of the form $f(x) = ax^2 + bx + c$, $a \neq 0$. We learned that every quadratic function has a U-shaped graph called a *parabola*. You may want to review the different characteristics of a parabola. Work through the following animation, and click on each characteristic to get a detailed description.

Characteristics of a Parabola

1. Vertex
2. Axis of symmetry
3. y-Intercept
4. x-Intercept(s) or real zeros
5. Domain and range

My video summary

 In Section 8.2, we studied quadratic functions and parabolas from an algebraic point of view. We now look at parabolas from a geometric perspective. We see in the introduction to this chapter that when a plane is parallel to an element of the cone, the plane will intersect the cone in a parabola. (Click on the animation.)

The set of points that define the parabola formed by the intersection described previously is stated in the following geometric definition of the parabola.

The Parabola

> **Geometric Definition of the Parabola**
>
> A **parabola** is the set of all points in a plane equidistant from a fixed point F and a fixed line D. The fixed point is called the **focus**, and the fixed line is called the **directrix**.

Watch the video to see how to sketch the parabola seen in Figure 3 using the geometric definition of the parabola.

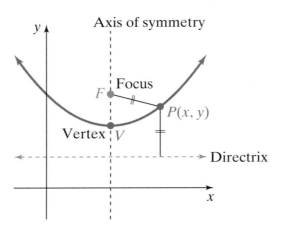

Figure 3

The distance from any point P to the focus is the same as the distance from point P to the directrix.

In Figure 3, we can see that for any point $P(x, y)$ that lies on the graph of the parabola, the distance from point P to the focus is exactly the same as the distance from point P to the directrix. Similarly, because the vertex, V, lies on the graph of the parabola, the distance from V to the focus must also be the same as the distance from V to the directrix. Therefore, if the distance from V to F is p units, then the distance from V to the directrix is also p units. If the coordinates of the vertex in Figure 3 are (h, k), then the coordinates of the focus must be $(h, k + p)$ and the equation of the directrix is $y = k - p$. We can use this information and the fact that the distance from $P(x, y)$ to the focus is equal to the distance from $P(x, y)$ to the directrix to derive the equation of a parabola.

Equation of a Parabola in Standard Form with a Vertical Axis of Symmetry

The equation of a parabola with a vertical axis of symmetry is $(x - h)^2 = 4p(y - k)$,

where

    the vertex is $V(h, k)$,

    $|p|$ = distance from the vertex to the focus = distance from the vertex to the directrix, the focus is $F(h, k + p)$, and the equation of the directrix is $y = k - p$.

The parabola opens *upward* if $p > 0$ or *downward* if $p < 0$.

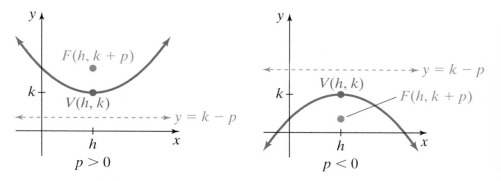

$p > 0$ $p < 0$

Example 1 Finding the Vertex, Focus, and Directrix of a Parabola and Sketching Its Graph

Find the vertex, focus, and directrix of the parabola $x^2 = 8y$ and sketch its graph.

Solution

Notice that we can rewrite the equation $x^2 = 8y$ as $(x - 0)^2 = 8(y - 0)$. We can now compare the equation $(x - 0)^2 = 8(y - 0)$ to the standard form equation $(x - h)^2 = 4p(y - k)$ to see that $h = 0$ and $k = 0$; hence, the vertex is at the origin $(0, 0)$. To find the focus and directrix, we need to find p.

$$4p = 8$$

Divide both sides by 4: $p = 2$

Because the value of p is positive, the parabola opens upward and the focus is located two units vertically *above* the vertex, whereas the directrix is the horizontal line located two units vertically *below* the vertex. The focus has coordinates $(0, 2)$, and the equation of the directrix is $y = -2$. The graph is shown in Figure 4.

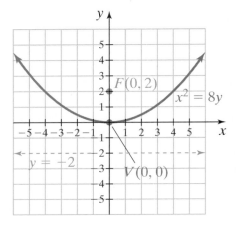

Figure 4

Example 2 Finding the Vertex, Focus, and Directrix of a Parabola and Sketching Its Graph

 Find the vertex, focus, and directrix of the parabola $-(x + 1)^2 = 4(y - 3)$ and sketch its graph.

Solution

Watch the video to verify that the vertex has coordinates $(-1, 3)$, the focus has coordinates $(-1, 2)$, and the equation of the directrix is $y = 4$. The graph of this parabola is shown in Figure 5.

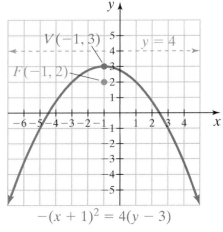

$-(x + 1)^2 = 4(y - 3)$ **Figure 5**

You Try It   Work through this You Try It problem.

Work Exercises 1–5 in this textbook or in the ***MyMathLab*** ® Study Plan.

OBJECTIVE 2:   WORK WITH THE EQUATION OF A PARABOLA WITH A HORIZONTAL AXIS OF SYMMETRY

In Examples 1 and 2, the graphs of both parabolas had vertical axes of symmetry. The graph of a parabola could also have a horizontal axis of symmetry and open "sideways." We derive the standard form of the parabola with a horizontal axis of symmetry in much the same way as we did with the parabola with a vertical axis of symmetry.

My video summary

Equation of a Parabola in Standard Form with a Horizontal Axis of Symmetry

The equation of a parabola with a horizontal axis of symmetry is
$(y - k)^2 4 = p(x - h),$
where

the vertex is $V(h, k)$,
$|p|$ = distance from the vertex to the focus = distance from the vertex to the directrix,
the focus is $F(h + p, k)$, and
the equation of the directrix is $x = h - p$.

The parabola opens *right* if $p > 0$ or *left* if $p < 0$.

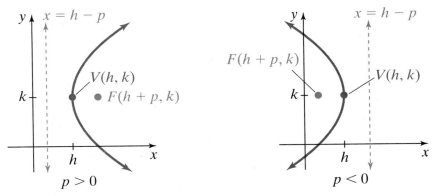

$p > 0$ $p < 0$

Example 3 Finding the Vertex, Focus, and Directrix of a Parabola and Sketching Its Graph

Find the vertex, focus, and directrix of the parabola $(y - 3)^2 = 8(x + 2)$ and sketch its graph.

Solution

Watch the video to verify that the vertex has coordinates $(-2, 3)$, the focus has coordinates $(0, 3)$, and the equation of the directrix is $x = -4$. The graph of this parabola is shown in Figure 6.

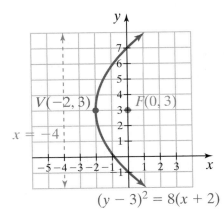

$(y - 3)^2 = 8(x + 2)$ **Figure 6**

You Try It   Work through this **You Try It** problem.

Work Exercises 6–10 in this textbook or in the *MyMathLab*® Study Plan.

My video summary

OBJECTIVE 3: FIND THE EQUATION OF A PARABOLA GIVEN INFORMATION ABOUT THE GRAPH

It is often necessary to determine the equation of a parabola given certain information. It is always useful to first determine whether the parabola has a vertical axis of symmetry or a horizontal axis of symmetry. Try to work through Examples 4 and 5. Then watch the solutions in the corresponding videos to determine whether you are correct.

Example 4 Finding the Equation of a Parabola

My video summary

Find the standard form of the equation of the parabola with focus $\left(-3, \frac{5}{2}\right)$ and directrix $y = \frac{11}{2}$.

Solution Watch the video to see that the equation of this parabola is $(x + 3)^2 = -6(y - 4)$. The graph is shown in Figure 7.

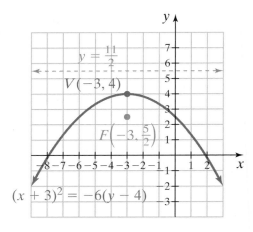

Figure 7

Example 5 Finding the Equation of a Parabola

My video summary

Find the standard form of the equation of the parabola with focus $(4, -2)$ and vertex $\left(\frac{13}{2}, -2\right)$.

Solution Watch the video to see that the equation of this parabola is $(y + 2)^2 = -10\left(x - \frac{13}{2}\right)$. The graph is shown in Figure 8.

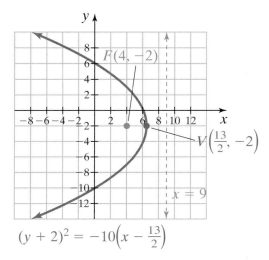

$$(y + 2)^2 = -10\left(x - \frac{13}{2}\right)$$

Figure 8

We can use a graphing utility to graph the parabola from Example 5 by solving the equation for y.

$$(y + 2)^2 = -10\left(x - \frac{13}{2}\right)$$

$$y + 2 = \pm\sqrt{-10\left(x - \frac{13}{2}\right)}$$

$$y = -2 \pm \sqrt{-10\left(x - \frac{13}{2}\right)}$$

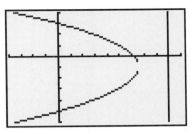

Figure 9

Using $y_1 = -2 + \sqrt{-10\left(x - \frac{13}{2}\right)}$, and $y_2 = -2 - \sqrt{-10\left(x - \frac{13}{2}\right)}$, we obtain the graph seen in Figure 9. *Note:* The directrix was created using the calculator's DRAW feature.

You Try It　Work through this You Try It problem.

Work Exercises 11–19 in this textbook or in the **MyMathLab**®　Study Plan.

OBJECTIVE 4:　COMPLETE THE SQUARE TO FIND THE EQUATION OF A PARABOLA IN STANDARD FORM

If the equation of a parabola is in the standard form of $(x - h)^2 = 4p(y - k)$ or $(y - k)^2 = 4p(x - h)$, it is not too difficult to determine the vertex, focus, and directrix and sketch its graph. However, the equation might not be given in standard form. If this is the case, we complete the square on the variable that is squared to rewrite the equation in standard form as in Example 6.

Example 6　Writing the Equation of a Parabola in Standard Form by Completing the Square

　Find the vertex, focus, and directrix and sketch the graph of the parabola $x^2 - 8x + 12y = -52$.

Solution　Because x is squared, we will complete the square on the variable x.

$$x^2 - 8x + 12y = -52$$
$$x^2 - 8x = -12y - 52$$
$$x^2 - 8x + 16 = -12y - 52 + 16$$
$$(x - 4)^2 = -12y - 36$$
$$(x - 4)^2 = -12(y + 3)$$

The equation is now in standard form with vertex $(4, -3)$ and $4p = -12$ so $p = -3$. The parabola must open down because the variable x is squared and $p < 0$. The focus is located three units below the vertex, whereas the directrix is three units above the vertex. Thus, the focus has coordinates $(4, -6)$, and the equation of the directrix is $y = 0$ or the x-axis. You can watch the video to see this solution worked out in detail. The graph is shown in Figure 10.

My video summary

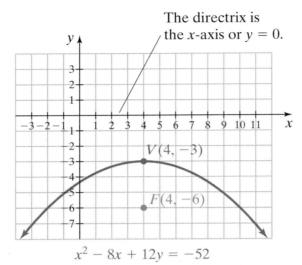

The directrix is the *x*-axis or $y = 0$.

$V(4, -3)$

$F(4, -6)$

$x^2 - 8x + 12y = -52$

Figure 10

You Try It   **Work through this You Try It problem.**

Work Exercises 20–23 in this textbook or in the *MyMathLab*® Study Plan.

OBJECTIVE 5:   SOLVE APPLICATIONS INVOLVING PARABOLAS

The Romans were one of the first civilizations to use the engineering properties of parabolic structures in their creation of arch bridges. The cables of many suspension bridges, such as the Golden Gate Bridge in San Francisco, span from tower to tower in the shape of a parabola.

Parabolic surfaces are used in the manufacture of many satellite dishes, search lights, car headlights, telescopes, lamps, heaters, and other objects. This is because parabolic surfaces have the property that incoming rays of light or radio waves traveling parallel to the axis of symmetry of a parabolic reflector or receiver will reflect off the parabolic surface and travel directly toward the antenna that is placed at the focus. See Figure 11. When a light source such as the headlight of a car is placed at the focus of a parabolic reflector, the light reflects off the surface outward, producing a narrow beam of light and thus maximizing the output of illumination.

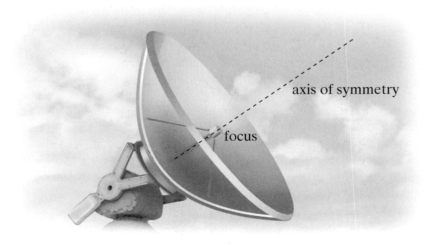

axis of symmetry

focus

Figure 11

Incoming rays reflect off the parabolic surface toward the antenna placed at the focus.

Example 7 Finding the Focus of a Parabolic Microphone

Parabolic microphones can be seen on the sidelines of professional sporting events so that television networks can capture audio sounds from the players on the field. If the surface of a parabolic microphone is 27 centimeters deep and has a diameter of 72 centimeters at the top, where should the microphone be placed relative to the vertex of the parabola?

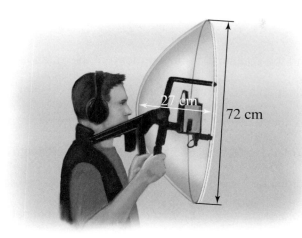

Solution We can draw a parabola with the vertex at the origin representing the center cross section of the parabolic microphone. The equation of this parabola in standard form is $x^2 = 4py$. Substitute the point $(36, 27)$ into the equation to get

$$x^2 = 4py$$
$$(36)^2 = 4p(27)$$
$$1296 = 108p$$
$$p = 12.$$

The microphone must be placed 12 centimeters from the vertex.

You Try It Work through this You Try It problem.

Work Exercises 24–28 in this textbook or in the *MyMathLab*® Study Plan.

A.1 Exercises

In Exercises 1–10, determine the vertex, focus, and directrix of the parabola and sketch its graph.

You Try It

1. $x^2 = 16y$

2. $x^2 = -8y$

3. $(x - 1)^2 = -12(y - 4)$

4. $(x + 3)^2 = 6(y - 1)$

5. $(x + 2)^2 = 5(y + 6)$

6. $y^2 = 4x$

You Try It

7. $y^2 = -8x$

8. $(y - 5)^2 = -4(x - 2)$

9. $(y + 3)^2 = 20(x - 4)$

10. $(y + 4)^2 = 9(x + 3)$

In Exercises 11–19, find the equation in standard form of the parabola described.

11. The focus has coordinates $(2, 0)$, and the equation of the directrix is $x = -2$.

You Try It **12.** The focus has coordinates $\left(0, -\frac{1}{2}\right)$, and the equation of the directrix is $y = \frac{1}{2}$.

13. The focus has coordinates $(3, -5)$, and the equation of the directrix is $y = -1$.

14. The focus has coordinates $(2, 4)$, and the equation of the directrix is $x = -4$.

15. The vertex has coordinates $\left(-\frac{11}{4}, -2\right)$, and the focus has coordinates $(-3, -2)$.

16. The vertex has coordinates $\left(4, -\frac{1}{4}\right)$, and the focus has coordinates $\left(4, \frac{1}{4}\right)$.

17. The vertex has coordinates $(-3, 4)$, and the equation of the directrix is $x = -7$.

18. Find the equations of the two parabolas in standard form that have a horizontal axis of symmetry and a focus at the point $(0, 4)$, and that pass through the origin.

19. Find the equations of the two parabolas in standard form that have a vertical axis of symmetry and a focus at the point $\left(2, \frac{3}{2}\right)$, and that pass through the origin.

In Exercises 20–23, determine the vertex, focus, and directrix of the parabola and sketch its graph.

20. $x^2 + 10x = 5y - 10$

You Try It **21.** $y^2 - 12y + 6x + 30 = 0$

22. $x^2 - 2x = -y + 1$

23. $y^2 + x + 6y = -10$

24. A parabolic eavesdropping device is used by a CIA agent to record terrorist conversations. The parabolic surface measures 120 centimeters in diameter at the top and is 90 centimeters deep at its center. How far from the vertex should the microphone be located?

You Try It

25. A parabolic space heater is 18 inches in diameter and 8 inches deep. How far from the vertex should the heat source be located to maximize the heating output?

26. A large NASA parabolic satellite dish is 22 feet across and has a receiver located at the focus 4 feet from its base. The satellite dish should be how deep?

27. A parabolic arch bridge spans 160 feet at the base and is 40 feet above the water at the center. Find the equation of the parabola if the vertex is placed at the point $(0, 40)$. Can a sailboat that is 35 feet tall fit under the bridge 30 feet from the center?

28. The cable between two 40-meter towers of a suspension bridge is in the shape of a parabola that just touches the bridge halfway between the towers. The two towers are 100 meters apart. Vertical cables are spaced every 10 meters along the bridge. What are the lengths of the vertical cables located 30 meters from the center of the bridge?

A.2 The Ellipse

THINGS TO KNOW

Before working through this section, be sure that you are familiar with the following concepts:

VIDEO   ANIMATION   INTERACTIVE

You Try It

1. Find the Distance between Two Points
(Section 8.4, Objective 1)

You Try It

2. Complete the Square to Find the Equation
of a Parabola in Standard Form
(Section A.1, Objective 4)

OBJECTIVES

1 Sketch the Graph of an Ellipse

2 Find the Equation of an Ellipse Given Information about the Graph

3 Complete the Square to Find the Equation of an Ellipse in Standard Form

4 Solve Applications Involving Ellipses

When a plane intersects a right circular cone at an angle between 0 and 90 degrees to the axis of the cone, the conic section formed is an ellipse. (Click on the animation.)

The set of points in the plane that defines an ellipse formed by the intersection of a plane and a cone described previously is stated in the following geometric definition.

Geometric Definition of the Ellipse

An **ellipse** is the set of all points in a plane, the sum of whose distances from two fixed points is a positive constant. The two fixed points, F_1 and F_2, are called the *foci*.

The Ellipse

The previous geometric definition implies that for any two points P and Q that lie on the graph of the ellipse, the sum of the distance between P and F_1 plus the distance between P and F_2 is equal to the sum of the distance between Q and F_1 plus the distance between Q and F_2. In symbols, we write $d(P, F_1) + d(P, F_2) = d(Q, F_1) + d(Q, F_2)$. See Figure 12.

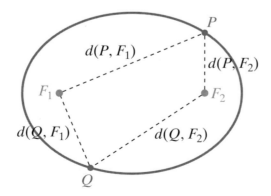

Figure 12

For any points P and Q on an ellipse,
$d(P, F_1) + d(P, F_2) = d(Q, F_1) + d(Q, F_2)$.

OBJECTIVE 1:   SKETCH THE GRAPH OF AN ELLIPSE

My video summary

An ellipse has two axes of symmetry. The longer axis, which is the line segment that connects the two vertices, is called the *major axis*. The foci are always located along the major axis. The shorter axis is called the *minor axis*. It is the line segment perpendicular to the major axis that passes through the center having endpoints that lie on the ellipse. Watch the video to see how to sketch the two ellipses seen in Figure 13. In Figure 13a, the ellipse has a horizontal major axis. The ellipse in Figure 13b has a vertical major axis.

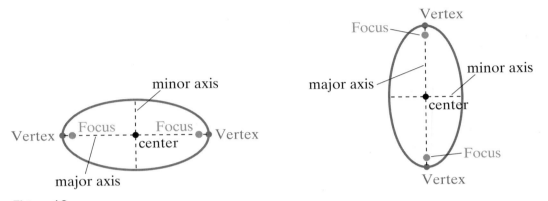

Figure 13

(a) Ellipse with horizontal major axis

(b) Ellipse with vertical major axis

Consider the ellipse with a horizontal major axis centered at (h, k), as shown in Figure 14, where $c > 0$ is the distance between the center and a focus, $a > 0$ is the distance between the center and one of the vertices, and $b > 0$ is the distance from the center to an endpoint of the minor axis.

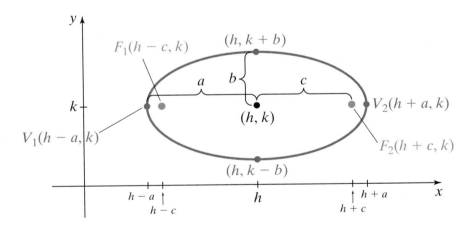

Figure 14

Before we can derive the equation of this ellipse, we must establish the following two facts:

Fact 1: The sum of the distances from any point on the ellipse to the two foci is $2a$.

Fact 2: $b^2 = a^2 - c^2$, or equivalently, $c^2 = a^2 - b^2$.

Once we have established these two facts, we can derive the equation of the ellipse. Click here to see this derivation. We now state the two standard equations of an ellipse.

Equation of an Ellipse in Standard Form with Center (h, k)

HORIZONTAL MAJOR AXIS

$$\frac{(x - h)^2}{a^2} + \frac{(y - k)^2}{b^2} = 1$$

- $a > b > 0$
- Foci: $F_1(h - c, k)$ and $F_2(h + c, k)$
- Vertices: $V_1(h - a, k)$ and $V_2(h + a, k)$
- Endpoints of minor axis: $(h, k - b)$ and $(h, k + b)$
- $c^2 = a^2 - b^2$

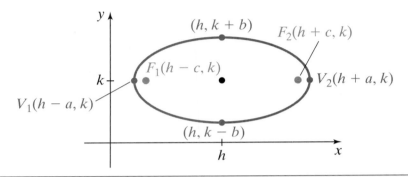

Equation of an Ellipse in Standard Form with Center (h, k)

VERTICAL MAJOR AXIS

$$\frac{(x - h)^2}{b^2} + \frac{(y - k)^2}{a^2} = 1$$

- $a > b > 0$
- Foci: $F_1(h, k - c)$ and $F_2(h, k + c)$
- Vertices: $V_1(h, k - a)$ and $V_2(h, k + a)$
- Endpoints of minor axis: $(h - b, k)$ and $(h + b, k)$
- $c^2 = a^2 - b^2$

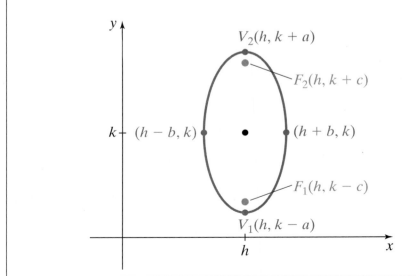

Notice that the two equations are identical except for the placement of a^2. If the equation of an ellipse is in standard form, we can quickly determine whether the ellipse has a horizontal major axis or a vertical major axis by looking at the denominator. If the larger denominator, a^2, appears under the x-term, then the ellipse has a horizontal major axis. If the larger denominator appears under the y-term, then the ellipse has a vertical major axis. If the denominators are equal ($a^2 = b^2$), then the ellipse is a circle!

If $h = 0$ and $k = 0$, then the ellipse is centered at the origin. Ellipses centered at the origin have the following equations.

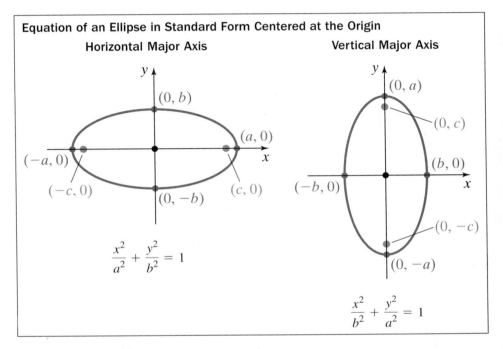

Equation of an Ellipse in Standard Form Centered at the Origin

Horizontal Major Axis

$$\frac{x^2}{a^2} + \frac{y^2}{b^2} = 1$$

Vertical Major Axis

$$\frac{x^2}{b^2} + \frac{y^2}{a^2} = 1$$

Example 1 Sketching the Graph of an Ellipse Centered at the Origin

My video summary

Sketch the graph of the ellipse $\dfrac{x^2}{25} + \dfrac{y^2}{4} = 1$, and label the center, foci, and vertices.

Solution Watch the video to see how to sketch the ellipse shown in Figure 15.

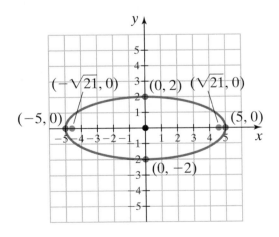

Figure 15

You Try It   Work through this You Try It problem.

Work Exercises 1–4 in this textbook or in the *MyMathLab*® Study Plan.

Example 2 Sketching the Graph of an Ellipse

 Sketch the graph of the ellipse $\dfrac{(x+2)^2}{20} + \dfrac{(y-3)^2}{36} = 1$, and label the center, foci, and vertices.

Solution   Note that the larger denominator appears under the *y*-term. This indicates that the ellipse has a vertical major axis. Watch the video to verify that the center of the ellipse is $(-2, 3)$. The foci have coordinates $(-2, -1)$ and $(-2, 7)$, the vertices have coordinates $(-2, -3)$ and $(-2, 9)$, and the coordinates of the minor axis are $(-2 - 2\sqrt{5}, 3)$ and $(-2 + 2\sqrt{5}, 3)$. The graph is shown in Figure 16.

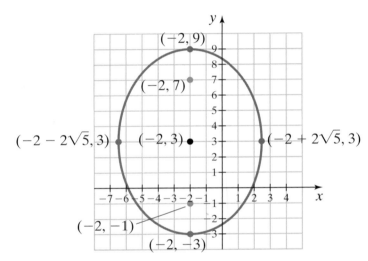

Figure 16

You Try It   Work through this You Try It problem.

Work Exercises 5–8 in this textbook or in the *MyMathLab*® Study Plan.

OBJECTIVE 2:   FIND THE EQUATION OF AN ELLIPSE GIVEN INFORMATION ABOUT THE GRAPH

It is often necessary to determine the equation of an ellipse given certain information. It is always useful to first determine whether the ellipse has a horizontal major axis or a vertical major axis. Try to work through Examples 3 and 4. Then watch the solutions to the corresponding videos to determine whether you are correct.

Example 3 Finding the Equation of an Ellipse

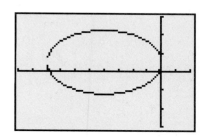 Find the standard form of the equation of the ellipse with foci at $(-6, 1)$ and $(-2, 1)$ such that the length of the major axis is eight units.

Solution Watch the video to verify that this is an ellipse centered at $(-4, 1)$ with a horizontal major axis such that $a = 4$ and $b = \sqrt{12}$. The equation in standard form is $\dfrac{(x + 4)^2}{4^2} + \dfrac{(y - 1)^2}{(\sqrt{12})^2} = 1$ or $\dfrac{(x + 4)^2}{16} + \dfrac{(y - 1)^2}{12} = 1$. The graph of this ellipse is shown in Figure 17.

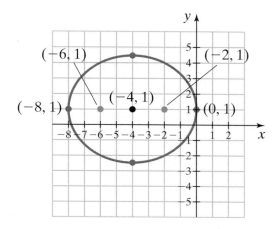

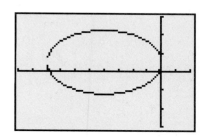

Figure 18

Figure 17

$$\dfrac{(x + 4)^2}{16} + \dfrac{(y - 1)^2}{12} = 1$$

We can use a graphing utility to graph the ellipse from Example 3 by solving the equation for y:

$$y = 1 \pm \sqrt{12\left(1 - \dfrac{(x + 4)^2}{16}\right)}$$

(Click here to see how to solve for y.)

Using $y_1 = 1 + \sqrt{12\left(1 - \dfrac{(x + 4)^2}{16}\right)}$ and $y_2 = 1 - \sqrt{12\left(1 - \dfrac{(x + 4)^2}{16}\right)}$,

we obtain the graph seen in Figure 18.

You Try It Work through this You Try It problem.

Work Exercises 9–18 in this textbook or in the *MyMathLab*® Study Plan.

Example 4 Finding the Equation of an Ellipse

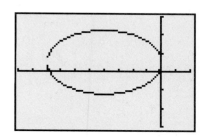 Determine the equation of the ellipse with foci located at $(0, 6)$ and $(0, -6)$ that passes through the point $(-5, 6)$.

Solution To find the equation, we can use Fact 1 and Fact 2 of ellipses. Watch the video to see how we can use these two facts to determine that the equation of this ellipse is $\dfrac{x^2}{45} + \dfrac{y^2}{81} = 1$.

You Try It Work through this You Try It problem.

Work Exercises 19 and 20 in this textbook or in the *MyMathLab*® Study Plan.

My video summary

OBJECTIVE 3: COMPLETE THE SQUARE TO FIND THE EQUATION OF AN ELLIPSE IN STANDARD FORM

If the equation of an ellipse is in the standard form of $\dfrac{(x-h)^2}{a^2} + \dfrac{(y-k)^2}{b^2} = 1$ or $\dfrac{(x-h)^2}{b^2} + \dfrac{(y-k)^2}{a^2} = 1$, it is not too difficult to determine the center and foci and sketch its graph. However, the equation might not be given in standard form. If this is the case, we complete the square on both variables to rewrite the equation in standard form as in Example 5.

Example 5 Writing the Equation of an Ellipse in Standard Form by Completing the Square

My video summary

Find the center and foci and sketch the ellipse
$36x^2 + 20y^2 + 144x - 120y - 396 = 0$.

Solution Rearrange the terms, leaving some room to complete the square and move any constants to the right-hand side:

Rearrange the terms: $36x^2 + 144x + 20y^2 - 120y = 396$

Then factor and complete the square.

Factor out 36 and 20: $36(x^2 + 4x\ \ ) + 20(y^2 - 6y\ \ ) = 396$

Complete the square on x and y. Remember to add $36 \cdot 4 = 144$ and $20 \cdot 9 = 180$ to the right side: $36(x^2 + 4x + 4) + 20(y^2 - 6y + 9) = 396 + 144 + 180$

Factor the left side, and simplify the right side:
$$36(x + 2)^2 + 20(y - 3)^2 = 720$$

Divide both sides by 720:
$$\frac{36(x + 2)^2}{720} + \frac{20(y - 3)^2}{720} = \frac{720}{720}$$

Simplify:
$$\frac{(x + 2)^2}{20} + \frac{(y - 3)^2}{36} = 1$$

The equation is now in standard form. Watch the video to see the solution to this example worked out in detail. Notice that this is the exact same ellipse that we sketched in Example 2. To see how to sketch this ellipse, refer to Example 2 or watch the last part of this video.

You Try It Work through this You Try It problem.

Work Exercises 21–24 in this textbook or in the *MyMathLab*® Study Plan.

OBJECTIVE 4: SOLVE APPLICATIONS INVOLVING ELLIPSES

Ellipses have many applications. The planets of our solar system travel around the Sun in an elliptical orbit with the Sun at one focus. Some comets, such as Halley's comet, also travel in elliptical orbits. See Figure 19. (Some comets are seen in our solar system only once because they travel in hyperbolic orbits.) We saw in Section A.1 that the parabola had a special reflecting property where incoming rays of light or radio waves traveling parallel to the axis of symmetry of a parabolic reflector or receiver will reflect off the parabolic surface and travel directly toward the antenna that is placed at the focus.

Figure 19

Planets travel around the Sun in an elliptical orbit.

Ellipses have a similar reflecting property. When light or sound waves originate from one focus of an ellipse, the waves will reflect off the surface of the ellipse and travel directly toward the other focus. See Figure 20. This reflecting property is used in a medical procedure called *sound wave lithotripsy* in which the patient is placed in an elliptical tank with the kidney stone placed at one focus. An ultrasound wave emitter is positioned at the other focus. The sound waves reflect off the walls of the tank directly to the kidney stone, thus obliterating the stone into fragments that are easily passed naturally through the patient's body. See Example 6.

Figure 20

Sound waves or light rays emitted from one focus of an ellipse reflect off the surface directly to the other focus.

Example 6 Position a Patient During Kidney Stone Treatment

A patient is placed in an elliptical tank that is 280 centimeters long and 250 centimeters wide to undergo sound wave lithotripsy treatment for kidney stones. Determine where the sound emitter and the stone should be positioned relative to the center of the ellipse.

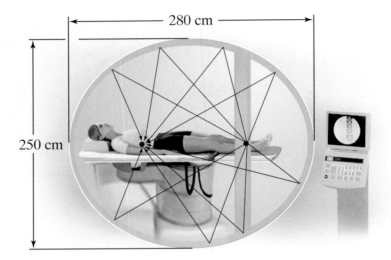

Solution The kidney stone and the sound emitter must be placed at the foci of the ellipse. We know that the major axis is 280 centimeters. Therefore, the vertices must be 140 centimeters from the center. Similarly, the endpoints of the minor axis are located 125 centimeters from the center.

To find c, we use the fact that $c^2 = a^2 - b^2$.

$$c^2 = 140^2 - 125^2$$
$$c^2 = 3975$$
$$c = \sqrt{3975} \approx 63.05$$

The stone and the sound emitter should be positioned approximately 63.05 centimeters from the center of the tank on the major axis.

You Try It **Work through this You Try It problem.**

Work Exercises 25–31 in this textbook or in the *MyMathLab*® Study Plan.

A.2 Exercises

You Try It

In Exercises 1–8, determine the center, foci, and vertices of the ellipse and sketch its graph.

1. $\dfrac{x^2}{16} + \dfrac{y^2}{4} = 1$

2. $\dfrac{x^2}{12} + \dfrac{y^2}{25} = 1$

3. $4x^2 + 9y^2 = 36$

4. $20x^2 + 5y^2 = 100$

You Try It

5. $\dfrac{(x-2)^2}{36} + \dfrac{(y-4)^2}{25} = 1$

6. $\dfrac{(x-1)^2}{9} + \dfrac{(y+4)^2}{49} = 1$

7. $9(x-5)^2 + 25(y+1)^2 = 225$

8. $(x+7)^2 + 9(y-2)^2 = 9$

In Exercises 9–18, determine the standard equation of each ellipse using the given graph or the stated information.

You Try It

9.

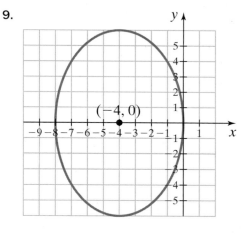

10.

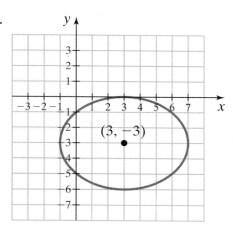

11. Foci at $(-5, 0)$ and $(5, 0)$; length of the major axis is fourteen units.

12. Foci at $(8, -1)$ and $(-2, -1)$; length of the major axis is twelve units.

13. Vertices at $(-6, 10)$ and $(-6, 0)$; length of the minor axis is eight units.

14. Center at $(4, 5)$; vertical minor axis with length sixteen units; $c = 6$.

15. Center at $(-1, 4)$; vertex at $(-1, 8)$; focus at $(-1, 7)$.

16. Vertices at $(2, -7)$ and $(2, 5)$; focus at $(2, 3)$.

17. Center at $(4, 1)$; focus at $(4, 8)$; ellipse passes through the point $(6, 1)$.

18. Center at $(1, 3)$; focus at $(1, 6)$; ellipse passes through the point $(2, 3)$.

In Exercises 19 and 20, determine the standard equation of each ellipse by first using Fact 1 to find a and then by using Fact 2 to find b.

You Try It

19. The foci have coordinates $(-2, 0)$ and $(2, 0)$. The ellipse contains the point $(2, 3)$.

20. The foci have coordinates $(0, -8)$ and $(0, 8)$. The ellipse contains the point $(-2, 6)$.

In Exercises 21–24, complete the square to write each equation in the form
$$\dfrac{(x-h)^2}{a^2} + \dfrac{(y-k)^2}{b^2} = 1 \text{ or } \dfrac{(x-h)^2}{b^2} + \dfrac{(y-k)^2}{a^2} = 1.$$
Determine the center, foci, and vertices of the ellipse and sketch its graph.

You Try It

21. $16x^2 + 20y^2 + 64x - 40y - 236 = 0$

22. $9y^2 + 16x^2 + 224x + 54y + 721 = 0$

23. $x^2 + 4x + 16y^2 - 32y + 4 = 0$

24. $50x^2 + y^2 + 20y = 0$

You Try It

25. An elliptical arch railroad tunnel 16 feet high at the center and 30 feet wide is cut through the side of a mountain. Find the equation of the ellipse if a vertex is represented by the point (0, 16).

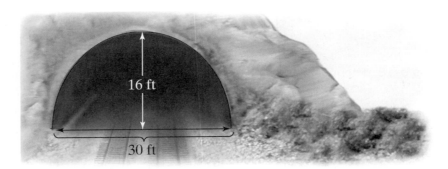

16 ft

30 ft

26. A rectangular playing field lies in the interior of an elliptical track that is 50 yards wide and 140 yards long. What is the width of the rectangular playing field if the width is located 10 yards from either vertex?

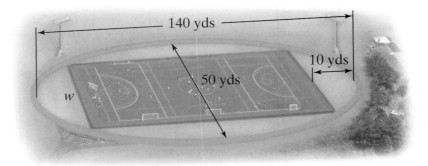

140 yds

50 yds

10 yds

w

27. A 78-inch by 36-inch door contains a decorative glass elliptical pattern. One vertex of the ellipse is located 9 inches from the top of the door. The other vertex is located 27 inches from the bottom of the door. The endpoints of the minor axis are located 9 inches from each side. Find the equation of the elliptical pattern. (Assume that the center of the elliptical pattern is at the origin.)

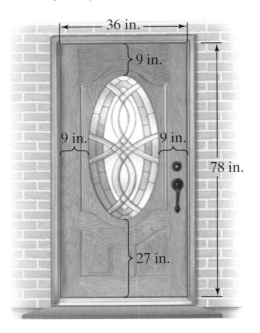

36 in.

9 in.

9 in. 9 in.

78 in.

27 in.

28. A patient is placed in an elliptical tank that is 180 centimeters long and 170 centimeters wide to undergo sound wave lithotripsy treatment for kidney stones. Determine where the sound emitter and the stone should be positioned relative to the center of the tank. (Round to the nearest hundredth of a centimeter.)

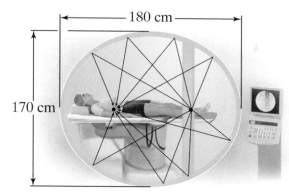

29. A window is constructed with the top half of an ellipse on top of a square. The square portion of the window has a 36-inch base. If the window is 50 inches tall at its highest point, find the height, h, of the window 14 inches from the center of the base. (Round the height to the nearest hundredth of an inch.)

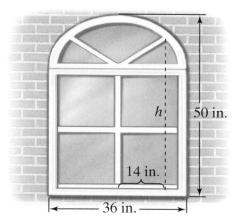

30. A government spy satellite is in an elliptical orbit around the Earth with the center of the Earth at a focus. If the satellite is 150 miles from the surface of the Earth at one vertex of the orbit and 500 miles from the surface of the Earth at the other vertex, find the equation of the elliptical orbit. (Assume that the Earth is a sphere with a diameter of 8000 miles.)

31. An elliptical arch bridge spans 160 feet. The elliptical arch has a maximum height of 40 feet. What is the height of the arch at a distance of 10 feet from the center?

A.3 The Hyperbola

THINGS TO KNOW

Before working through this section, be sure that you are familiar with the following concepts:

 You Try It **1.** Find the Distance between Two Points (Section 8.4, **Objective 1**)

 You Try It **2.** Write the Equation of a Line from Given Information (Section 2.5, **Objective 4**)

 You Try It **3.** Complete the Square to Find the Equation of an Ellipse in Standard Form (Section A.2, **Objective 3**)

OBJECTIVES

1 Sketch the Graph of a Hyperbola

2 Find the Equation of a Hyperbola in Standard Form

3 Complete the Square to Find the Equation of a Hyperbola in Standard Form

4 Solve Applications Involving Hyperbolas

When a plane intersects two right circular cones at the same time, the conic section formed is a hyperbola. (Click on the animation.)

The set of points in the plane that defines a hyperbola formed by the intersection of a plane and the cones described previously is stated in the following geometric definition.

Geometric Definition of the Hyperbola

A **hyperbola** is the set of all points in a plane, the difference of whose distances from two fixed points is a positive constant. The two fixed points, F_1 and F_2, are called the foci.

Notice that the previous geometric definition is very similar to the geometric definition of the ellipse. Recall that for a point to lie on the graph of an ellipse, the **sum** of the distances from the point to the two foci is constant. For a point to lie on the graph of a hyperbola, the **difference** between the distances from the point to the two foci is constant. Because subtraction is not commutative, we consider the absolute value of the difference in the distances between a point on the hyperbola and the foci to ensure that the constant is positive. Thus, for any two points P and Q that lie on the graph of a hyperbola, $|d(P, F_1) - d(P, F_2)| = |d(Q, F_1) - d(Q, F_2)|$. See Figure 21.

The Hyperbola

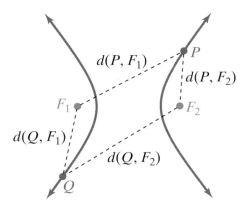

Figure 21
For any points P and Q that lie on the graph of a hyperbola, $|d(P, F_1) - d(P, F_2)| = |d(Q, F_1) - d(Q, F_2)|$.

OBJECTIVE 1: SKETCH THE GRAPH OF A HYPERBOLA

My video summary

 The graph of a hyperbola has two branches. These branches look somewhat like parabolas, but they are certainly not because the branches do not satisfy the geometric definition of the parabola. Every hyperbola has a center, two vertices, and two foci. The vertices are located at the endpoints of an invisible line segment called the **transverse axis**. The transverse axis is either parallel to the x-axis (horizontal transverse axis) or parallel to the y-axis (vertical transverse axis). The center of a hyperbola is located midway between the two vertices (or two foci). The hyperbola has another invisible line segment called the **conjugate axis** that passes through the center and lies perpendicular to the transverse axis. Each branch of the hyperbola approaches (but never intersects) a pair of lines called *asymptotes*. A **reference rectangle** is typically used as a guide to help sketch the asymptotes. The reference rectangle is a rectangle whose midpoints of each side are the vertices of the hyperbola or the endpoints of the conjugate axis. The asymptotes pass diagonally through opposite corners of the reference rectangle. Watch the video to see how to sketch the graphs of the two hyperbolas shown in Figure 22.

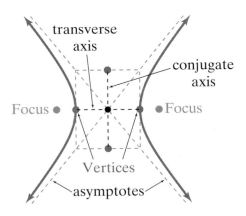

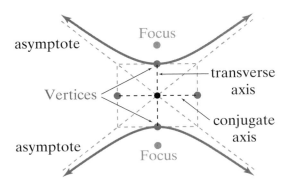

Figure 22 (a) Hyperbola with a horizontal transverse axis

(b) Hyperbola with a vertical transverse axis

To derive the equation of a hyperbola, consider a hyperbola with a horizontal transverse axis centered at (h, k). If the distance between the center and either vertex is $a > 0$, then the coordinates of the vertices are $V_1(h - a, k)$ and $V_2(h + a, k)$, and the length of the transverse axis is equal to $2a$. If $c > 0$ is the distance between the center and either foci, then the foci have coordinates $F_1(h - c, k)$ and $F_2(h + c, k)$. See Figure 23.

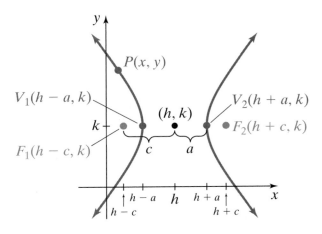

Figure 23

By the geometric definition of a hyperbola, we know that for any point $P(x, y)$ that lies on the hyperbola, the difference of the distances from P to the two foci is a constant. It can be shown that this constant is equal to $2a$, the length of the transverse axis. We now state this fact and denote it as Fact 1 for Hyperbolas.

Fact 1 for Hyperbolas

Given the foci of a hyperbola F_1 and F_2 and any point P that lies on the graph of the hyperbola, the difference of the distance between P and the foci is equal to $2a$. In other words, $|d(P, F_1) - d(P, F_2)| = 2a$. The constant $2a$ represents the length of the transverse axis.

Once we know that the constant stated in the geometric definition is equal to $2a$, we can derive the equation of a hyperbola. Click here to see this derivation. We now state the two standard equations of hyperbolas.

Equation of a Hyperbola in Standard Form with Center (h, k)

Horizontal Transverse Axis

$$\frac{(x - h)^2}{a^2} - \frac{(y - k)^2}{b^2} = 1$$

- Foci: $F_1(h - c, k)$ and $F_2(h + c, k)$
- Vertices: $V_1(h - a, k)$ and $V_2(h + a, k)$
- Endpoints of conjugate axis: $(h, k - b)$ and $(h, k + b)$
- $b^2 = c^2 - a^2$
- Asymptotes: $y - k = \pm\frac{b}{a}(x - h)$

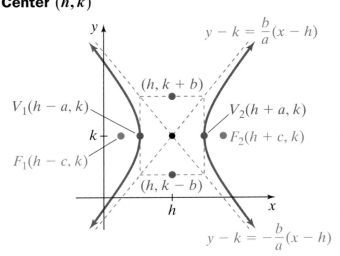

Equation of a Hyperbola in Standard Form with Center (h, k)

Vertical Transverse Axis

$$\frac{(y - k)^2}{a^2} - \frac{(x - h)^2}{b^2} = 1$$

- Foci: $F_1(h, k - c)$ and $F_2(h, k + c)$
- Vertices: $V_1(h, k - a)$ and $V_2(h, k + a)$
- Endpoints of conjugate axis: $(h - b, k)$ and $(h + b, k)$
- $b^2 = c^2 - a^2$
- Asymptotes: $y - k = \pm\frac{a}{b}(x - h)$

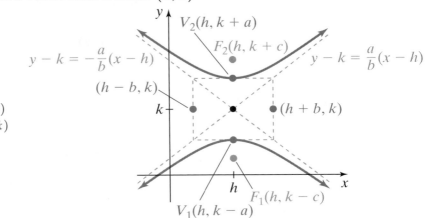

The two hyperbola equations are nearly identical except for two differences. The x-terms are positive and the y-terms are negative in the equation with a horizontal transverse axis. The signs are reversed in the equation with a vertical transverse axis. Also note that a^2 appears in the denominator of the positive squared term in each equation. We can derive the equations of the asymptotes by using the point-slope form of the equation of a line.

Note: If $h = 0$ and $k = 0$, then the hyperbola is centered at the origin. Hyperbolas centered at the origin have the following equations.

Standard Equations of a Hyperbola with the Center at the Origin

$$\frac{x^2}{a^2} - \frac{y^2}{b^2} = 1 \qquad\qquad \frac{y^2}{a^2} - \frac{x^2}{b^2} = 1$$

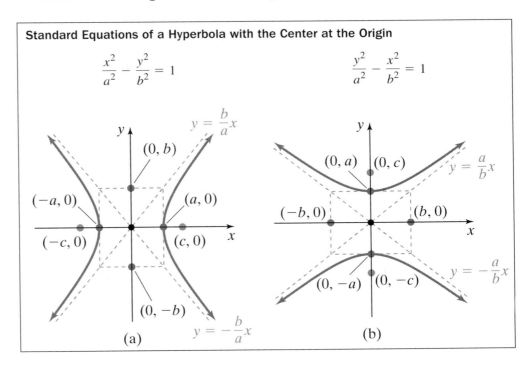

Example 1 Sketching the Graph of a Hyperbola in Standard Form

Sketch the following hyperbolas. Determine the center, transverse axis, vertices, and foci, and find the equations of the asymptotes.

a. $\dfrac{(y-4)^2}{36} - \dfrac{(x+5)^2}{9} = 1$ **b.** $25x^2 - 16y^2 = 400$

Solution Watch the interactive video to see how to sketch each hyperbola shown in Figure 24(a) and 24(b).

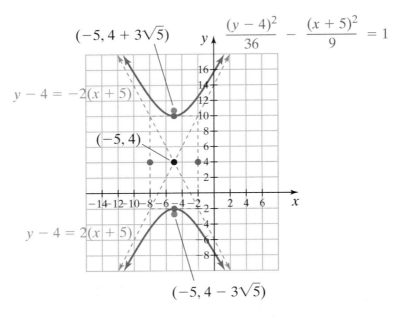

Figure 24(a)

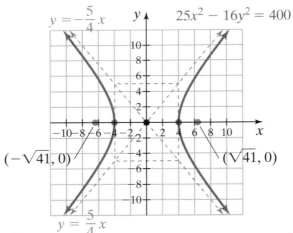

Figure 24(b)

You Try It Work through this You Try It problem.

Work Exercises 1–6 in this textbook or in the *MyMathLab*® Study Plan.

OBJECTIVE 2: FIND THE EQUATION OF A HYPERBOLA IN STANDARD FORM

My video summary

🎞 **Example 2 Finding the Equation of a Hyperbola**

Find the equation of the hyperbola with the center at $(-1, 0)$, a focus at $(-11, 0)$, and a vertex at $(5, 0)$.

Solution A focus and a vertex lie along the x-axis. This indicates that the hyperbola has a horizontal transverse axis. Because the center is at $(-1, 0)$, we know that the equation of the hyperbola is $\dfrac{(x+1)^2}{a^2} - \dfrac{y^2}{b^2} = 1$. Watch the video to verify that the equation in standard form is $\dfrac{(x+1)^2}{36} - \dfrac{y^2}{64} = 1$. The equations of the asymptotes are $y = -\dfrac{4}{3}(x+1)$ and $y = \dfrac{4}{3}(x+1)$. The graph of this hyperbola is shown in Figure 25.

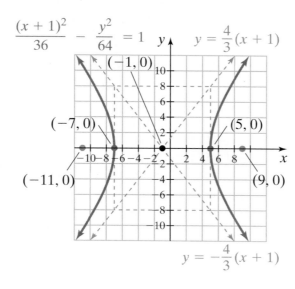

Figure 25

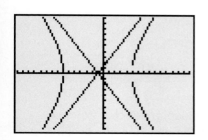

Figure 26
We can use a graphing utility to graph the hyperbola from Example 2 by solving the equation for y:

$$y = \pm \frac{4}{3}\sqrt{(x+1)^2 - 36}$$

(Click here to see how to solve for y.)

Using $y_1 = \dfrac{4}{3}\sqrt{(x+1)^2 - 36}$, $y_2 = -\dfrac{4}{3}\sqrt{(x+1)^2 - 36}$, $y_3 = -\dfrac{4}{3}(x+1)$, and $y_4 = -\dfrac{4}{3}(x+1)$, we obtain the graph seen in Figure 26.

You Try It　**Work through this You Try It problem.**

Work Exercises 7–14 in this textbook or in the *MyMathLab*®️ Study Plan.

OBJECTIVE 3:　COMPLETE THE SQUARE TO FIND THE EQUATION OF A HYPERBOLA IN STANDARD FORM

If the equation of a hyperbola is in the standard form of $\dfrac{(x-h)^2}{a^2} - \dfrac{(y-k)^2}{b^2} = 1$ or $\dfrac{(y-k)^2}{a^2} - \dfrac{(x-h)^2}{b^2} = 1$, then it is not too difficult to determine the center, vertices, foci, and asymptotes and sketch its graph. However, the equation might not be given in standard form. If this is the case, we may need to complete the square on both variables as in Example 3.

My video summary

Example 3 Writing the Equation of a Hyperbola in Standard Form by Completing the Square

Find the center, vertices, foci, and equations of asymptotes and sketch the hyperbola $12x^2 - 4y^2 - 72x - 16y + 140 = 0$.

Solution Rearrange the terms leaving some room to complete the square, and move any constants to the right-hand side:

Rearrange the terms: $\quad 12x^2 - 72x \quad\quad - 4y^2 - 16y \quad\quad = -140$

Then factor and complete the square.

Factor out 12 and -4: $\quad 12(x^2 - 6x \quad) - 4(y^2 + 4y \quad) = -140$

Complete the square on x and y. Remember to add $12 \cdot 9 = 108$ and $-4 \cdot 4 = -16$ to the right side: $\quad 12(x^2 - 6x + 9) - 4(y^2 + 4y + 4\ ) = -140 + 108 - 16$

Factor the left side, and simplify the right side: $\quad\quad\quad 12(x - 3)^2 - 4(y + 2)^2 = -48$

Divide both sides by -48: $\quad\quad\quad \dfrac{12(x - 3)^2}{-48} - \dfrac{4(y + 2)^2}{-48} = \dfrac{-48}{-48}$

Simplify: $\quad\quad\quad -\dfrac{(x - 3)^2}{4} + \dfrac{(y + 2)^2}{12} = 1$

Rewrite the equation: $\quad\quad\quad \dfrac{(y + 2)^2}{12} - \dfrac{(x - 3)^2}{4} = 1$

The equation is now in standard form. Watch the video to see this example worked out in detail. You should verify that this is the equation of a hyperbola with a vertical transverse axis with center $(3, -2)$. The vertices have coordinates $\left(3, -2 - 2\sqrt{3}\right)$ and $\left(3, -2 + 2\sqrt{3}\right)$. The foci have coordinates $(3, -6)$ and $(3, 2)$. The equations of the asymptotes are $y + 2 = -\sqrt{3}(x - 3)$ and $y + 2 = \sqrt{3}(x - 3)$.

You Try It   Work through this You Try It problem.

Work Exercises 15–20 in this textbook or in the *MyMathLab*® Study Plan.

OBJECTIVE 4:   SOLVE APPLICATIONS INVOLVING HYPERBOLAS

Hyperbolas have many applications. We saw in Section A.2 that the planets in our solar system and some comets, such as Halley's comet, travel through our solar system in elliptical orbits. However, some comets are seen only once in our solar system because they travel through the solar system on the path of a hyperbola with the Sun at a focus. On October 14, 1947, Chuck Yeager became the first person to break the sound barrier. As an airplane moves faster than the speed of sound, a cone-shaped shock wave is produced. The cone intersects the ground in the shape of a hyperbola. When two rocks are simultaneously tossed into a calm pool of water, ripples move outward in the form of concentric circles. These circles intersect in points that form a hyperbola. Hyperbolas can be used to locate ships by sending radio signals simultaneously from radio transmitters placed at some fixed distance apart. A device measures the difference in the time it takes the radio signals to reach the ship. The equation of a hyperbola can then be determined to describe the current path of the ship. If three transmitters are used, two hyperbolic equations can be determined. The precise location of the ship can be determined by finding the intersection of the two hyperbolas. This system of locating ships is known as long-range navigation or LORAN. See Example 4.

Example 4 Use a Hyperbola to Locate a Ship

One transmitting station is located 100 miles due east from another transmitting station. Each station simultaneously sends out a radio signal. The signal from the west tower is received by a ship $\frac{1600}{3}$ microseconds after the signal from the east tower. If the radio signal travels at 0.18 miles per microsecond, find the equation of the hyperbola on which the ship is presently located.

Solution Start by plotting the two foci of the hyperbola at points $F_1(-50, 0)$ and $F_2(50, 0)$. These two points represent the position of the two transmitting towers. Note that $c = 50$. Because the hyperbola is centered at the origin with a horizontal transverse axis, the equation must be of the form $\dfrac{x^2}{a^2} - \dfrac{y^2}{b^2} = 1$.

The difference in the distances from the two transmitters to the ship is $\left(\frac{1600}{3} \text{ microseconds}\right) \cdot \left(0.18 \, \frac{\text{miles}}{\text{microsecond}}\right) = 96$ miles. This distance represents the constant stated in the Fact 1 for Hyperbolas.

Therefore, $2a = 96$, so $a = 48$ or $a^2 = 2304$. To find b^2, we use the fact that $b^2 = c^2 - a^2$.

$$b^2 = c^2 - a^2$$
$$b^2 = 50^2 - 48^2$$
$$b^2 = 196$$

We now substitute the values of $a^2 = 2304$ and $b^2 = 196$ into the previous equation to obtain the equation $\dfrac{x^2}{2304} - \dfrac{y^2}{196} = 1$.

You Try It Work through this **You Try It** problem.

Work Exercises 21–24 in this textbook or in the _MyMathLab_® Study Plan.

A.3 Exercises

You Try It

In Exercises 1–6, determine the center, transverse axis, vertices, foci, and the equations of the asymptotes and sketch the hyperbola.

1. $\dfrac{x^2}{16} - \dfrac{y^2}{9} = 1$

2. $\dfrac{y^2}{9} - \dfrac{x^2}{16} = 1$

3. $\dfrac{(y-4)^2}{25} - \dfrac{(x-2)^2}{36} = 1$

4. $\dfrac{(x+1)^2}{9} - \dfrac{(y+3)^2}{49} = 1$

5. $20x^2 - 5y^2 = 100$

6. $20(x-1)^2 - 16(y-3)^2 = -320$

In Exercises 7–12, determine the standard equation of the hyperbola with the given characteristics and sketch the graph.

7. The center is at $(0,0)$, a focus is at $(5,0)$, and a vertex is at $(3,0)$.

You Try It

8. The center is at $(0,0)$, a focus is at $(0,10)$, and a vertex is at $(0,-6)$.

9. The center is at $(4,-4)$, a focus is at $(6,-4)$, and a vertex is at $(5,-4)$.

10. The center is at $(-6,-1)$, a focus is at $(-6,-9)$, and a vertex is at $(-6,-5)$.

11. The foci are at $(9,3)$ and $(9,9)$; the vertex is at $(9,8)$.

12. The vertices are at $(-2,-3)$ and $(10,-3)$; an asymptote has equation $y + 3 = \dfrac{7}{6}(x-4)$.

In Exercises 13 and 14, determine the equation of the hyperbola with the given characteristics and sketch the graph. *Hint:* $|d(P, F_1) - d(P, F_2)| = 2a$.

13. The hyperbola has foci with coordinates $F_1(0,-6)$ and $F_2(0,6)$ and passes through the point $P(8,10)$.

14. The hyperbola has foci with coordinates $F_1(-6,1)$ and $F_2(10,1)$ and passes through the point $P(10,13)$.

You Try It

In Exercises 15–20, complete the square to write each equation in the form $\dfrac{(x-h)^2}{a^2} - \dfrac{(y-k)^2}{b^2} = 1$ or $\dfrac{(y-k)^2}{a^2} - \dfrac{(x-h)^2}{b^2} = 1$. Determine the center, vertices, foci, endpoints of the conjugate axis, and the equations of the asymptotes of the hyperbola, and sketch its graph.

15. $x^2 - y^2 - 4x + 2y - 1 = 0$

16. $x^2 - y^2 + 8x - 6y + 9 = 0$

17. $y^2 - 9x^2 - 12y - 36x - 9 = 0$

18. $x^2 - 16y^2 + 10x + 64y - 55 = 0$

19. $25y^2 - 144x^2 + 1728x + 400y + 16 = 0$

20. $49y^2 - 576x^2 - 98y - 1152x - 28{,}751 = 0$

You Try It

21. A light on a wall produces a shadow in the shape of a hyperbola. If the distance between the two vertices of the hyperbola is 14 inches and the distance between the two foci is 16 inches, find the equation of the hyperbola.

22. This figure shows the hyperbolic orbit of a comet with the center of the Sun positioned at a focus, 100 million miles from the origin. (The units are in millions of miles.) The comet will be 50 million miles from the center of the Sun at its nearest point during the orbit. Find the equation of the hyperbola describing the comet's orbit.

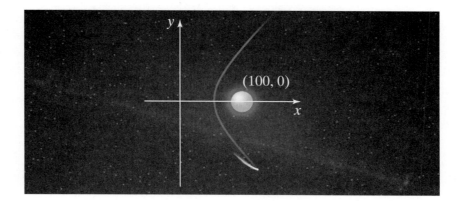

23. A nuclear power plant has a large cooling tower with sides curved in the shape of a hyperbola. The radius of the base of the tower is 60 meters. The radius of the top of the tower is 50 meters. The sides of the tower are 60 meters apart at the closest point located 90 meters above the ground.

   a. Find the equation of the hyperbola that describes the sides of the cooling tower. (Assume that the center is at the origin.)

   b. Determine the height of the tower. (Round your answer to the nearest meter.)

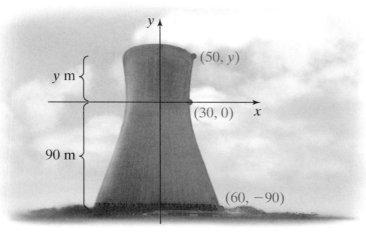

24. One transmitting station is located 80 miles north of another transmitting station. Each station simultaneously sends out a radio signal. The signal from the north station is received 200 microseconds after the signal from the south station. If the radio signal travels at 0.18 miles per microsecond, find the equation of the hyperbola on which the ship is presently located.

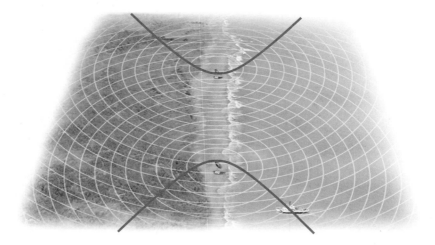

Sequences and Series

APPENDIX B CONTENTS

B.1 Introduction to Sequences and Series

B.2 Arithmetic Sequences and Series

B.3 Geometric Sequences and Series

B.4 The Binomial Theorem

B.1 Introduction to Sequences and Series

THINGS TO KNOW

Before working through this section, be sure that you are familiar with the following concept:

VIDEO   ANIMATION   INTERACTIVE

You Try It

1. Determine If Graphs Are Functions
(Section 2.2, Objective 4)

OBJECTIVES

1 Write the Terms of a Sequence

2 Write the Terms of a Recursive Sequence

3 Write the General Term for a Given Sequence

4 Compute Partial Sums of a Series

5 Determine the Sum of a Finite Series Written in Summation Notation

6 Write a Series Using Summation Notation

OBJECTIVE 1: WRITE THE TERMS OF A SEQUENCE

Consider the function $f(n) = 2n - 1$, where n is a natural number. The graph of this function consists of infinitely many ordered pairs of the form $(n, 2n - 1)$, where $n \geq 1$. Therefore, the ordered pairs that lie on the graph of this function are $(1, 1), (2, 3), (3, 5), (4, 7)$, and so on. A portion of the graph of this function can be seen in Figure 1.

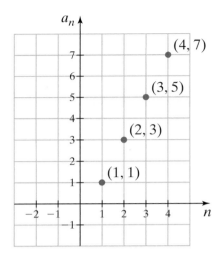

Figure 1

A portion of the graph of the function $f(n) = 2n - 1$, where n is a natural number.

Clearly, the graph seen in Figure 1 is a function because the graph passes the vertical line test. Any function whose domain is the set of natural numbers is called an **infinite sequence**. Instead of using the conventional function notation $f(n)$ to name a sequence, we will use a subscript notation, such as a_n (read as "a sub n"), to name our sequences. We now formally define a sequence.

Definition Sequence

A **finite sequence** is a function whose domain is the finite set $\{1, 2, 3, \ldots, n\}$, where n is a natural number.

An **infinite sequence** is a function whose domain is the set of all natural numbers.

The range values of a sequence are called the **terms** of the sequence.

We now rename the sequence $f(n) = 2n - 1$ as $a_n = 2n - 1$. The first four terms of this sequence are $a_1 = 1, a_2 = 3, a_3 = 5$, and $a_4 = 7$.

A graphing utility set to *sequence mode* can be used to sketch the graph of a sequence. The figure on the left shows a portion of the graph of the sequence $a_n = 2n - 1$.

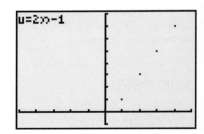

Before we start to find the terms of a sequence it is important to introduce factorial notation.

Definition   The Factorial of a Non-Negative Integer

The factorial of a non-negative integer n, denoted as $n!$, is the product of all positive integers less than or equal to n. Thus, $n! = n(n - 1) \cdot \cdots \cdot 3 \cdot 2 \cdot 1$.

Note: By definition, we say that zero factorial is equal to 1 or $0! = 1$.

Some examples of factorial notation are $5! = 5 \cdot 4 \cdot 3 \cdot 2 \cdot 1 = 120$ and $8! = 8 \cdot 7 \cdot 6 \cdot 5 \cdot 4 \cdot 3 \cdot 2 \cdot 1 = 40{,}320$. As you can see, the value of $n!$ gets large rather quickly. In fact, the value of 13! is over 6 billion, which is a rough estimate of the population of Earth! Example 1c illustrates how factorial notation can be used to define a sequence.

Example 1 Writing the Terms of a Sequence

Write the first four terms of each sequence whose nth term is given.

a. $a_n = 2n - 1$

b. $b_n = n^2 - 1$

c. $c_n = \dfrac{3^n}{(n-1)!}$

d. $d_n = (-1)^n 2^{n-1}$

Solution

a. To find the first four terms of the sequence, we evaluate $a_n = 2n - 1$ when n is 1, 2, 3, and 4.

$$a_1 = 2(1) - 1 = 2 - 1 = 1, \, a_2 = 2(2) - 1 = 4 - 1 = 3,$$
$$a_3 = 2(3) - 1 = 6 - 1 = 5, \quad \text{and} \quad a_4 = 2(4) - 1 = 8 - 1 = 7$$

Therefore, the first four terms of the sequence $a_n = 2n - 1$ are 1, 3, 5, and 7.

b. Work through the interactive video to verify that the first four terms of the sequence $b_n = n^2 - 1$ are 0, 3, 8, and 15.

c. To find the first four terms of the sequence $c_n = \frac{3^n}{(n-1)!}$, we evaluate c_n when n is 1, 2, 3, and 4.

$$c_1 = \frac{3^1}{(1-1)!} = \frac{3}{0!} = \frac{3}{1} = 3, \, c_2 = \frac{3^2}{(2-1)!} = \frac{9}{1!} = \frac{9}{1} = 9,$$

$$c_3 = \frac{3^3}{(3-1)!} = \frac{27}{2!} = \frac{27}{2}, \quad \text{and} \quad c_4 = \frac{3^4}{(4-1)!} = \frac{81}{3!} = \frac{81}{6} = \frac{27}{2}$$

Therefore, the first four terms of the sequence $c_n = \frac{3^n}{(n-1)!}$ are

$$3, 9, \tfrac{27}{2}, \text{ and } \tfrac{27}{2}.$$

d. Work through the interactive video to verify that the first four terms of the sequence $d_n = (-1)^n 2^{n-1}$ are $-1, 2, -4$, and 8.

Note: The sequence $d_n = (-1)^n 2^{n-1}$ is an example of an alternating sequence because the successive terms alternate in sign.

You Try It   Work through this You Try It problem.

Work Exercises 1–10 in this textbook or in the ***MyMathLab*** Study Plan.

OBJECTIVE 2:   WRITE THE TERMS OF A RECURSIVE SEQUENCE

Some sequences are defined recursively. A recursive sequence is a sequence in which each term is defined using one or more of its previous terms. Typically, the first term of a recursive sequence is given, followed by the formula for the nth term of the sequence. The following example illustrates two recursive sequences.

Example 2 Writing the Terms of a Recursive Sequence

Write the first four terms of each of the following recursive sequences.

a. $a_1 = -3, a_n = 5a_{n-1} - 1$ for $n \geq 2$

b. $b_1 = 2, b_n = \dfrac{(-1)^{n-1} n}{b_{n-1}}$ for $n \geq 2$

My interactive video summary

Solution

a. The first four terms of this recursive sequence are $-3, -16, -81,$ and -406. Work through this interactive video to verify.

b. The first four terms of this recursive sequence are $2, -1, -3,$ and $\frac{4}{3}$. Work through this interactive video to verify.

Arguably the most famous recursively defined sequence is the Fibonacci sequence named after the 13th-century Italian mathematician Leonardo of Pisa, also known as Fibonacci. The Fibonacci sequence is defined in Example 3.

Example 3 Writing the Terms of the Fibonacci Sequence

The Fibonacci sequence is defined recursively by $a_n = a_{n-1} + a_{n-2}$, where $a_1 = 1$ and $a_2 = 1$. Write the first eight terms of the Fibonacci sequence.

Solution We are given that $a_1 = 1$ and $a_2 = 1$. We use the recursive formula $a_n = a_{n-1} + a_{n-2}$ to find the next six terms starting with $n = 3$.

$$a_3 = a_2 + a_1 = 1 + 1 = 2$$
$$a_4 = a_3 + a_2 = 2 + 1 = 3$$
$$a_5 = a_4 + a_3 = 3 + 2 = 5$$
$$a_6 = a_5 + a_4 = 5 + 3 = 8$$
$$a_7 = a_6 + a_5 = 8 + 5 = 13$$
$$a_8 = a_7 + a_6 = 13 + 8 = 21$$

You can see in Example 3 that given the first two terms, each term of the Fibonacci sequence is the sum of the preceding two terms. We now write the first 12 terms of the Fibonacci sequence.

$$1, 1, 2, 3, 5, 8, 13, 21, 34, 55, 89, 144, \ldots$$

The numbers of this sequence are known as *Fibonacci numbers*. The Fibonacci sequence and Fibonacci numbers occur in many natural phenomena such as the spiral formation of seeds of various plants, the number of petals of a flower, and the formation of the branches of a tree. See Exercise 44.

You Try It Work through this **You Try It** problem.

Work Exercises 11–14 in this textbook or in the *MyMathLab*® Study Plan.

OBJECTIVE 3: WRITE THE GENERAL TERM FOR A GIVEN SEQUENCE

Sometimes the first several terms of a sequence are given without listing the nth term. When this occurs, we must try to determine a pattern and use deductive reasoning to establish a rule that describes the general term, or nth term, of the sequence. Example 4 illustrates two such sequences.

Example 4 Finding the General Term of a Sequence

Write a formula for the nth term of each infinite sequence, then use this formula to find the eighth term of the sequence.

a. $\dfrac{1}{1}, \dfrac{1}{2}, \dfrac{1}{3}, \dfrac{1}{4}, \dfrac{1}{5}, \ldots$

b. $-\dfrac{2}{1}, \dfrac{4}{2}, -\dfrac{8}{6}, \dfrac{16}{24}, -\dfrac{32}{120}, \ldots$

My video summary

Solution

a. The nth term of the sequence is $a_n = \frac{1}{n}$. Thus, the eighth term of this sequence is $a_8 = \frac{1}{8}$.

b. For this sequence, notice that the first term is negative and that terms alternate in sign. We can therefore represent the sign of each term as $(-1)^n$. Also, notice that the numerators are successive powers of 2. We now have the following pattern:

$$
\begin{array}{ccccc}
a_1 & a_2 & a_3 & a_4 & a_5 \\
\downarrow & \downarrow & \downarrow & \downarrow & \downarrow \\
-\dfrac{2}{1}, & \dfrac{4}{2}, & -\dfrac{8}{6}, & \dfrac{16}{24}, & -\dfrac{32}{120} \cdots \\
\downarrow & \downarrow & \downarrow & \downarrow & \downarrow \\
\dfrac{(-1)^1 2^1}{1}, & \dfrac{(-1)^2 2^2}{2}, & \dfrac{(-1)^3 2^3}{6}, & \dfrac{(-1)^4 2^4}{24}, & \dfrac{(-1)^5 2^5}{120}, \cdots
\end{array}
$$

Finally, if we factor each successive denominator we get $1 = 1$, $2 = 2 \cdot 1$, $6 = 3 \cdot 2 \cdot 1$, $24 = 4 \cdot 3 \cdot 2 \cdot 1$, and $120 = 5 \cdot 4 \cdot 3 \cdot 2 \cdot 1$. This suggests that the denominator of the nth term can be represented by $n!$

Therefore, the nth term of the sequence is $a_n = \dfrac{(-1)^n 2^n}{n!}$. The eighth term of this sequence is

$$
a_8 = \frac{(-1)^8 2^8}{8!} = \frac{256}{40{,}320} = \frac{2}{315}.
$$

If you would like to see this solution worked out in detail, watch this video.

You Try It   Work through this You Try It problem.

Work Exercises **15–22** in this textbook or in the ***MyMathLab***® Study Plan.

OBJECTIVE 4:   COMPUTE PARTIAL SUMS OF A SERIES

Suppose that we wanted to find the sum of the first four terms of the sequence $a_n = 2n - 1$. From Example 1 we saw that the first four terms of this sequence were $a_1 = 1, a_2 = 3, a_3 = 5$, and $a_4 = 7$. Therefore, the sum of the first four terms is $a_1 + a_2 + a_3 + a_4 = 1 + 3 + 5 + 7 = 16$. The expression $1 + 3 + 5 + 7$ is called a **series**.

Definition   Series

Let $a_1, a_2, a_3, \ldots$ be a sequence. The expression of the form $a_1 + a_2 + a_3 + \cdots + a_n$ is called a **finite series**.

The expression of the form $a_1 + a_2 + a_3 + \cdots + a_n + a_{n+1} + \cdots$ is called an **infinite series**.

The sum of the first n terms of a series is called the nth **partial sum** of the series and is denoted as S_n.

For the series $1 + 3 + 5 + 7 + 9 + \cdots + 2n - 1$, the first five partial sums are as follows:

$$S_1 = 1$$
$$S_2 = 1 + 3 = 4$$
$$S_3 = 1 + 3 + 5 = 9$$
$$S_4 = 1 + 3 + 5 + 7 = 16$$
$$S_5 = 1 + 3 + 5 + 7 + 9 = 25$$

It appears that $S_n = n^2$. In fact, it can be shown that for any positive integer n, the sum of the series $1 + 3 + 5 + 7 + 9 + \cdots + 2n - 1$ is equal to n^2.

Example 5 Computing Partial Sums of a Series

Given the general term of each sequence, find the indicated partial sum.

a. $a_n = \dfrac{1}{n}$, find S_3. **b.** $b_n = (-1)^n 2^{n-1}$, find S_5.

Solution

a. The first three terms are $a_1 = 1$, $a_2 = \frac{1}{2}$, and $a_3 = \frac{1}{3}$. Therefore, the partial sum, S_3, is $S_3 = 1 + \frac{1}{2} + \frac{1}{3} = \frac{11}{6}$.

b. The first five terms are $b_1 = -1$, $b_2 = 2$, $b_3 = -4$, $b_4 = 8$, and $b_5 = -16$. Therefore, the partial sum, S_5, is $S_5 = -1 + 2 + (-4) + 8 + (-16) = -11$.

You Try It **Work through this You Try It problem.**

Work Exercises 23–28 in this textbook or in the *MyMathLab*® Study Plan.

OBJECTIVE 5: **DETERMINE THE SUM OF A FINITE SERIES WRITTEN IN SUMMATION NOTATION**

My video summary

Writing out an entire finite series of the form $a_1 + a_2 + a_3 + \cdots + a_n$ can be quite tedious, especially if n is fairly large. Fortunately, there is a convenient way to express a finite series using a short-hand notation called *summation notation* (also called *sigma notation*). This notation involves the use of the uppercase Greek letter sigma, which is written as Σ.

Definition Summation Notation

If $a_1, a_2, a_3, \ldots$ is a sequence, then the finite series $a_1 + a_2 + a_3 + \cdots + a_n$

can be written in **summation notation** as $\displaystyle\sum_{i=1}^{n} a_i$. The infinite series $a_{n+1} + \cdots$

can be written as $\displaystyle\sum_{i=1}^{\infty} a_i$.

The variable i is called the **index of summation**. The number 1 is the **lower limit of summation** and n is the **upper limit of summation**.

The lower limit of summation, $i = 1$, below the sigma tells us which term to start with. The upper limit of summation, n, that appears above the sigma tells us which term of the sequence will be the last term to add. There is nothing special about the letter i to represent the index of summation. We will often use different letters such as j or k. Also, it is not necessary for the lower limit of summation to start at 1. In Examples 6b and 6c, the lower limits of summation are 2 and 0, respectively.

Example 6 Determining the Sum of a Series Written in Summation Notation

My interactive video summary

📹 Find the sum of each finite series.

a. $\displaystyle\sum_{i=1}^{5} i^2$ **b.** $\displaystyle\sum_{j=2}^{5} \frac{j-1}{j+1}$ **c.** $\displaystyle\sum_{k=0}^{6} \frac{1}{k!}$

(Round the sum to three decimal places.)

Solution

a. $\displaystyle\sum_{i=1}^{5} i^2 = 1^2 + 2^2 + 3^2 + 4^2 + 5^2 = 1 + 4 + 9 + 16 + 25 = 55$

b. $\displaystyle\sum_{j=2}^{5} \frac{j-1}{j+1} = \frac{2-1}{2+1} + \frac{3-1}{3+1} + \frac{4-1}{4+1} + \frac{5-1}{5+1}$

$$= \frac{1}{3} + \frac{2}{4} + \frac{3}{5} + \frac{4}{6} = \frac{21}{10}$$

c. $\displaystyle\sum_{k=0}^{6} \frac{1}{k!} = \frac{1}{0!} + \frac{1}{1!} + \frac{1}{2!} + \frac{1}{3!} + \frac{1}{4!} + \frac{1}{5!} + \frac{1}{6!}$

$$= 1 + 1 + \frac{1}{2} + \frac{1}{6} + \frac{1}{24} + \frac{1}{120} + \frac{1}{720} = \frac{1957}{720} \approx 2.718$$

You can work through this interactive video to see this solution worked out in detail.

Using a TI-84 Plus, we can calculate the sum obtained in Example 6c. Notice that this number is a good approximation of the number e. In fact, it can be shown that the exact value

of e is $e = \displaystyle\sum_{n=0}^{\infty} \frac{1}{n!}$.

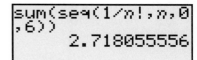

You Try It   Work through this You Try It problem.

Work Exercises 29–35 in this textbook or in the *MyMathLab*® Study Plan.

OBJECTIVE 6:   WRITE A SERIES USING SUMMATION NOTATION

Given the first several terms of a series, it is important to be able to rewrite the series using summation notation as in Example 7.

Example 7 Writing a Series Using Summation Notation

Rewrite each series using summation notation. Use 1 as the lower limit of summation.

a. $2 + 4 + 6 + 8 + 10 + 12$

b. $1 + 2 + 6 + 24 + 120 + 720 + \cdots + 3{,}628{,}800$

Solution

a. This series is the sum of six terms. Therefore, the lower limit of summation is 1 and the upper limit of summation is 6. Each term is a successive multiple of 2. So, one possible series is $\displaystyle\sum_{i=1}^{6} 2i$.

b. Notice that $1 = 1!, 2 = 2!, 6 = 3!, 24 = 4!, 120 = 5!, 720 = 6!$, and $3{,}628{,}800 = 10!$ Thus, a possible series is $\displaystyle\sum_{n=1}^{10} n!$.

You Try It   Work through this You Try It problem.

Work Exercises 36–44 in this textbook or in the *MyMathLab*® Study Plan.

B.1 Exercises

In Exercises 1–10, write the first four terms of each sequence.

You Try It

1. $a_n = 3n + 1$

2. $a_n = 4^n$

3. $a_n = \dfrac{4n}{n + 3}$

4. $a_n = (-4)^n$

5. $a_n = 5(n + 2)!$

6. $a_n = \dfrac{n^3}{(n + 1)!}$

7. $a_n = (-1)^n(5n)$

8. $a_n = \dfrac{3^n}{(-1)^{n+1} + 5}$

9. $a_n = \dfrac{(-1)^n}{(n + 5)(n + 6)}$

10. $a_n = \dfrac{(-1)^n(3)^{2n+1}}{(2n + 1)!}$

In Exercises 11–14, write the first four terms of each recursive sequence.

You Try It

11. $a_1 = 7, a_n = 3 + a_{n-1}$ for $n \geq 2$

12. $a_1 = -1, a_n = n - a_{n-1}$ for $n \geq 2$

13. $a_1 = 6, a_n = \dfrac{a_{n-1}}{n^2}$ for $n \geq 2$

14. $a_1 = -4, a_n = 1 - \dfrac{1}{a_{n-1}}$ for $n \geq 2$

In Exercises 15–22, write a formula for the general term, or nth term, for the given sequence. Then find the indicated term.

You Try It

15. $-1, 1, 3, 5, 7, \ldots; a_{11}$.

16. $\dfrac{1}{5}, \dfrac{2}{6}, \dfrac{3}{7}, \dfrac{4}{8}, \dfrac{5}{9}, \ldots; a_8$.

17. $1 \cdot 6, 2 \cdot 7, 3 \cdot 8, 4 \cdot 9, \ldots; a_7$.

18. $-2, 4, -8, 16, \ldots; a_7$.

19. $\dfrac{2}{5}, \dfrac{2}{25}, \dfrac{2}{125}, \dfrac{2}{625}, \ldots; a_6$.

20. $-6, 12, -24, 48, -96, \ldots; a_9$.

21. $-6, 24, -120, 720, \ldots; a_5.$

22. $\dfrac{3}{2}, \dfrac{9}{6}, \dfrac{27}{24}, \dfrac{81}{120}, \ldots; a_5.$

In Exercises 23–25, the first several terms of a sequence are given. Find the indicated partial sum.

23. $2, 4, 6, 8, 10, \ldots; S_4$

24. $3, -6, 9, -12, 15, -18, \ldots; S_9$

You Try It

25. $\dfrac{1}{2}, -\dfrac{1}{4}, \dfrac{1}{8}, -\dfrac{1}{16}, \ldots; S_5$

In Exercises 26–28, the general term of a sequence is given. Find the indicated partial sum.

26. $a_n = 3n + 8; S_6$

27. $a_n = (-1)^n \cdot (4n); S_6$

28. $a_1 = 4, a_n = a_{n-1} - 8$ for $n \geq 2; S_8$

In Exercises 29–35, find the sum of each series.

29. $\displaystyle\sum_{i=1}^{9} i$

30. $\displaystyle\sum_{i=1}^{6} (4i + 3)$

31. $\displaystyle\sum_{i=1}^{7} i(i + 2)$

32. $\displaystyle\sum_{i=1}^{21} 7$

You Try It

33. $\displaystyle\sum_{j=0}^{5} (j + 4)^2$

34. $\displaystyle\sum_{k=2}^{7} \dfrac{k!}{(k - 2)!}$

35. $\displaystyle\sum_{j=0}^{5} (j + 4)^2$

In Exercises 36–43, rewrite each series using summation notation. Use 1 as the lower limit of summation.

36. $1 + 2 + 3 + \cdots + 29$

37. $5 + 10 + 15 + \cdots + 50$

38. $1^2 + 2^2 + 3^2 + \cdots + 11^2$

39. $\dfrac{4}{5} + \dfrac{5}{6} + \dfrac{6}{7} + \cdots + \dfrac{12}{13}$

40. $2 + (-4) + 8 + (-16) + \cdots + (-256)$

41. $-\dfrac{1}{9} + \dfrac{1}{18} - \dfrac{1}{27} + \cdots + \dfrac{1}{54}$

42. $5 + \dfrac{5^2}{2} + \dfrac{5^3}{3} + \cdots + \dfrac{5^n}{n}$

43. $1 + 7 + \dfrac{7^2}{2!} + \dfrac{7^3}{3!} + \dfrac{7^4}{4!} + \cdots + \dfrac{7^n}{n!}$

44. The figure below shows the progression of the branching of a tree during each stage of development. Notice that the number of branches formed during a given stage is a Fibonacci number. Assuming that this branching pattern continues, how many branches will form during the 10th stage of development?

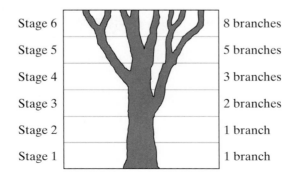

B.2 Arithmetic Sequences and Series

THINGS TO KNOW

Before working through this section, be sure that you are familiar with the following concepts:

VIDEO ANIMATION INTERACTIVE

You Try It

1. Solve Systems of Linear Equations in Two Variables by Substitution (Section 3.1, **Objective** 3)

You Try It

2. Solve Systems of Linear Equations in Two Variables by Elimination (Section 3.1, **Objective** 4)

You Try It

3. Determine the Sum of a Finite Series Written in Summation Notation (Section B.1, **Objective** 5)

OBJECTIVES

1 Determine If a Sequence Is Arithmetic

2 Find the General Term or a Specific Term of an Arithmetic Sequence

3 Compute the nth Partial Sum of an Arithmetic Series

4 Solve Applications of Arithmetic Sequences and Series

OBJECTIVE 1:   DETERMINE IF A SEQUENCE IS ARITHMETIC

In this section, we will work exclusively with a specific type of sequence known as an **arithmetic sequence**. A sequence is arithmetic if the difference in any two successive terms is constant. For example, the sequence

$$5, 9, 13, 17, \ldots$$

is arithmetic because the difference of any two successive terms is 4. The first term of this sequence is $a_1 = 5$ and the common difference is $d = 4$. Notice that we can rewrite the terms of this sequence as $5, 5 + 4, 5 + 2(4), 5 + 3(4), \ldots$

In general, given an arithmetic sequence with a first term of a_1 and a common difference of d, then the first n terms of the sequence are as follows:

$$a_1$$
$$a_2 = a_1 + d$$
$$a_3 = a_2 + d = \underbrace{(a_1 + d)}_{a_2} + d = a_1 + 2d$$
$$a_4 = a_3 + d = \underbrace{(a_1 + 2d)}_{a_3} + d = a_1 + 3d$$
$$\vdots$$
$$a_n = a_1 + (n - 1)d$$

Definition   Arithmetic Sequence

An **arithmetic sequence** is a sequence of the form $a_1, a_1 + d, a_1 + 2d,$ $a_1 + 3d, a_1 + 4d, \ldots,$ where a_1 is the first term of the sequence and d is the common difference. The general term, or nth term, of an arithmetic sequence has the form $a_n = a_1 + (n-1)d$.

Example 1   Determining If a Sequence Is Arithmetic

My interactive video summary

For each of the following sequences, determine if it is arithmetic. If the sequence is arithmetic, find the common difference.

a. $1, 4, 7, 10, 13, \ldots$

b. $b_n = n^2 - n$

c. $a_n = -2n + 7$

d. $a_1 = 14, a_n = 3 + a_{n-1}$

Solution   Watch this interactive video to verify that the sequences in parts (a), (c), and (d) are arithmetic. The sequence in part (b) is not arithmetic. Notice that the arithmetic sequence in part (d) is a recursive sequence.

It is worth noting that every arithmetic sequence is a linear function whose domain is the natural numbers. A portion of the graphs of the arithmetic sequences from Example 1a and Example 1c are seen in Figure 2. Notice that the ordered pairs of each sequence are collinear.

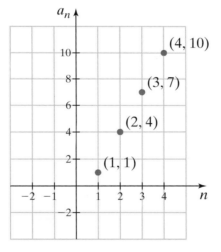

(a) A portion of the graph of the sequence $1, 4, 7, 10, 13, \ldots$

(b) A portion of the graph of the sequence $a_n = -2n + 7$

Figure 2   The graph of every arithmetic sequence is represented by a set of ordered pairs that lies on a straight line.

Note: When the common difference of an arithmetic sequence is positive, the terms of the sequence *increase* and the graph is represented by a set of ordered pairs that lie along a line with positive slope. When the common difference of an arithmetic sequence is negative, the terms of the sequence *decrease* and the graph is represented by a set of ordered pairs that lies along a line with negative slope.

You Try It   Work through this You Try It problem.

Work Exercises 1–6 in this textbook or in the *MyMathLab*® Study Plan.

OBJECTIVE 2: FIND THE GENERAL TERM OR A SPECIFIC TERM OF AN ARITHMETIC SEQUENCE

By the definition of an arithmetic sequence, the general term of an arithmetic sequence has the form $a_n = a_1 + (n - 1)d$. We can use this formula to find any term of an arithmetic sequence.

Example 2 Finding the General Term of an Arithmetic Sequence

My interactive video summary

 Find the general term of each arithmetic sequence, then find the indicated term of the sequence. (In part (c), only a portion of the graph is given. Assume that the domain of this sequence is all natural numbers.)

a. $11, 17, 23, 29, 35, \ldots ; a_{50}$

b. $2, 0, -2, -4, -6, \ldots ; a_{90}$

c. Find a_{31}.

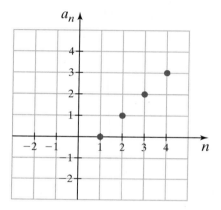

Solution

a. The first term of the sequence is $a_1 = 11$ and the common difference is $d = 6$. The general term is given by $a_n = 11 + (n - 1)(6) = 11 + 6n - 6 = 6n + 5$. Therefore, $a_{50} = 6(50) + 5 = 305$.

b. The first term of the sequence is $a_1 = 2$ and the common difference is $d = -2$. The general term is given by $a_n = 2 + (n - 1)(-2) = 2 - 2n + 2 = 4 - 2n$. Therefore, $a_{90} = 4 - 2(90) = -176$.

c. The first four terms of this sequence are $a_1 = 0, a_2 = 1, a_3 = 2$, and $a_4 = 3$. Thus, $a_1 = 0$ and the common difference is $d = 1$. The general term is given by $a_n = 0 + (n - 1)(1) = n - 1$. Thus, $a_{31} = 31 - 1 = 30$.

You may also watch this interactive video to see each of these solutions worked out in detail.

You Try It Work through this You Try It problem.

Work Exercises 7–12 in this textbook or in the *MyMathLab*® Study Plan.

Example 3 Finding a Specific Term of an Arithmetic Sequence

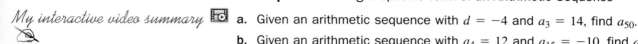

My interactive video summary

a. Given an arithmetic sequence with $d = -4$ and $a_3 = 14$, find a_{50}.

b. Given an arithmetic sequence with $a_4 = 12$ and $a_{15} = -10$, find a_{41}.

Solution

a. We are given that $d = -4$ and $a_3 = 14$. We can use this information to solve for a_1.

Use the formula for the general term of
an arithmetic sequence: $a_n = a_1 + (n - 1)d$
Substitute $n = 3$ and $d = -4$: $a_3 = a_1 + (3 - 1)(-4)$

Simplifying, we get $a_3 = a_1 - 8$. We can now substitute $a_3 = 14$ to solve for a_1.

Start with the formula for a_3: $a_3 = a_1 - 8$
Substitute $a_3 = 14$: $14 = a_1 - 8$
Add 8 to both sides: $22 = a_1$

Using the formula $a_n = a_1 + (n - 1)d$ with $a_1 = 22$ and $d = -4$, we can simplify to get $a_n = 26 - 4n$. Therefore, $a_{50} = 26 - 4(50) = -174$. Watch this interactive video to see every step of this solution.

b. Using the fact that $a_n = a_1 + (n - 1)d$, we get $a_4 = a_1 + (4 - 1)d = 12$ and $a_{15} = a_1 + (15 - 1)d = -10$. This gives us the following system of linear equations:

$$a_1 + 3d = 12$$
$$a_1 + 14d = -10$$

Using the method of substitution or the method of elimination that were discussed in Section 3.1 to solve this system, we get $a_1 = 18$ and $d = -2$. (Watch this interactive video to see how to solve this system using either method.) Using the formula $a_n = a_1 + (n - 1)d$ with $a_1 = 18$ and $d = -2$, we can find the general term.

Use the formula for the general term
of an arithmetic sequence: $a_n = a_1 + (n - 1)d$
Substitute $a_1 = 18$ and $d = -2$: $= 18 + (n - 1)(-2)$
Use the distributive property: $= 18 - 2n + 2$
Simplify: $= 20 - 2n$

The general term is $a_n = 20 - 2n$. Therefore, $a_{41} = 20 - 2(41) = -62$. Watch this interactive video to see this entire solution worked out in detail.

You Try It Work through this You Try It problem.

Work Exercises 13–18 in this textbook or in the **MyMathLab**® Study Plan.

OBJECTIVE 3: COMPUTE THE nth PARTIAL SUM OF AN ARITHMETIC SERIES

If $a_1, a_2, a_3, \ldots$ is an arithmetic sequence, then the expression $a_1 + a_2 + a_3 + \cdots + a_n + a_{n+1} + \cdots$ is called an **infinite arithmetic series** and can be written using summation notation as $\sum_{i=1}^{\infty} a_i$. Recall that the sum of the first n terms of a series is called the **nth partial sum** of the series and is given by $S_n = a_1 + a_2 + a_3 + \cdots + a_n$. We can also represent the nth partial sum using summation notation as $S_n = \sum_{i=1}^{n} a_i$. As you can see, the nth partial sum is simply the sum of a finite arithmetic series. Fortunately, there is a convenient formula for computing the nth partial sum of an arithmetic series.

Formula for the nth Partial Sum of an Arithmetic Series

The sum of the first n terms of an arithmetic series is called the **nth partial sum** of the series and is given by $S_n = \sum_{i=1}^{n} a_i = a_1 + a_2 + a_3 + \cdots + a_n$. This sum can be computed using the formula $S_n = \dfrac{n(a_1 + a_n)}{2}$.

My video summary

▣ Watch this video to see the derivation of this formula.

Note: The nth partial sum of an arithmetic series is simply the sum of a finite arithmetic series. An arithmetic series *must* be **finite** in order to compute the sum. This is not true for some other types of series. You will see how to find the sum of a special type of infinite series in Section B.3.

Example 4 Finding the Sum of an Arithmetic Series

▣ Find the sum of each arithmetic series.

a. $\displaystyle\sum_{i=1}^{20} (2i - 11)$

b. $-5 + (-1) + 3 + 7 + \cdots + 39$

My interactive video summary

Solution

a. We can use the formula $S_{20} = \dfrac{20(a_1 + a_{20})}{2}$ to compute the sum of the first 20 terms of this series.

$$\begin{array}{ll}
\text{Substitute } i = 1 \text{ in the} & \\
\text{formula } 2i - 11 \text{ to find } a_1: & a_1 = 2(1) - 11 = -9 \\
\text{Substitute } i = 20 \text{ in the} & \\
\text{formula } 2i - 11 \text{ to find } a_{20}: & a_{20} = 2(20) - 11 = 29
\end{array}$$

We now substitute $a_1 = -9$ and $a_{20} = 29$ into the formula $S_{20} = \dfrac{20(a_1 + a_{20})}{2}$.

$$\begin{array}{ll}
\text{Use the formula for 20th partial} & \\
\text{sum of an arithmetic series:} & S_{20} = \dfrac{20(a_1 + a_{20})}{2} \\
\\
\text{Substitute } a_1 = -9 \text{ and } a_{20} = 29: & = \dfrac{20(-9 + 29)}{2} \\
\\
\text{Simplify:} & = 200
\end{array}$$

Therefore, $\displaystyle\sum_{i=1}^{20} (2i - 11) = 200$. You may also watch this interactive video to see this solution worked out in detail.

b. Work through the interactive video to verify that the sum of this arithmetic series is 204.

You Try It   Work through this You Try It problem.

Work Exercises 19–27 in this textbook or in the *MyMathLab*® Study Plan.

OBJECTIVE 4:   SOLVE APPLICATIONS OF ARITHMETIC SEQUENCES AND SERIES

Example 5 Selling Newspaper Subscriptions

A local newspaper has hired teenagers to go door-to-door to try to solicit new subscribers. The teenagers receive $2 for selling the first subscription. For each additional subscription sold, the newspaper will pay the teenagers 10 cents more than what was paid for the previous subscription. How much will the teenagers get paid for selling the 100th subscription? How much money will the teenagers earn by selling 100 subscriptions?

Solution   The amount of money earned by selling one newspaper subscription can be represented by $a_1 = 2$. The money earned by selling the second subscription is $a_2 = 2.10$. The money earned by selling the third subscription is $a_3 = 2.20$. We see that the amount of money earned by selling n newspaper subscriptions is an arithmetic sequence with $a_1 = 2$ and $d = 0.10$. This sequence is defined by
$$a_n = 2 + (n - 1)(0.10) = 2 + (0.10)n - 0.10 = 0.10n + 1.90.$$

The cash earned by selling the 100th subscription is the 100th term of this sequence, or $a_{100} = 0.10(100) + 1.90 = 11.90$. Therefore, the teenagers are paid $11.90 for selling the 100th subscription.

To find the total amount earned by selling 100 subscriptions, we must find the sum of the series $\displaystyle\sum_{i=1}^{100} [(0.10)i + 1.90]$.

Using the formula $S_n = \dfrac{n(a_1 + a_n)}{2}$ with $n = 100$, $a_1 = 2$, and $a_{100} = 11.90$, we get
$$S_{100} = \frac{100(a_1 + a_{100})}{2} = \frac{100(2 + 11.90)}{2} = 50(13.90) = 695.$$ Thus, the teenagers will be paid $695 if they sell 100 subscriptions.

Example 6 Seats in a Theater

 A large multiplex movie house has many theaters. The smallest theater has only 12 rows. There are six seats in the first row. Each row has two seats more than the previous row. How many total seats are there in this theater?

Solution   Try solving this word problem on your own. When you are done, watch this video to see if you are correct, then work through the following "You Try It" problem.

You Try It   Work through this You Try It problem.

Work Exercises 28–35 in this textbook or in the *MyMathLab*® Study Plan.

B.2 Exercises

You Try It

In Exercises 1–6, determine if the sequence is arithmetic. If the sequence is arithmetic, find the common difference.

1. $8, 14, 20, 26, 32, \ldots$

2. $8, 11, 13, 16, 18, \ldots$

3. $a_n = \dfrac{3n + 1}{2}$

4. $a_n = n(n + 1)$

5. $a_1 = 8, a_n = 2 + a_{n-1}$

6. $a_1 = 5, a_n = 3a_{n-1} + 1$

You Try It

In Exercises 7–12, find the general term of each arithmetic sequence and then find the indicated term of the sequence. If the sequence is represented by a graph, assume that the domain of the sequence is all natural numbers.

7. $2, 7, 12, 17, \ldots; a_{10}$

8. $5, 1, -3, -7, \ldots; a_{31}$

9. $\dfrac{3}{2}, 3, \dfrac{9}{2}, 6, \dfrac{15}{2}, \ldots; a_{50}$

10. $5.0, 3.8, 2.6, 1.4, \ldots; a_{29}$

11. Find a_{17}.

12. Find a_{11}.

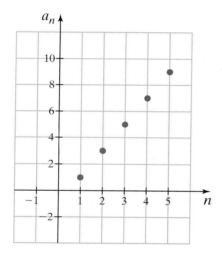

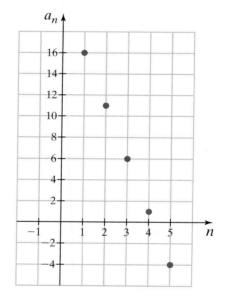

You Try It

13. Given an arithmetic sequence with $d = 3$ and $a_8 = 5$, find a_{30}.

14. Given an arithmetic sequence with $d = -5$ and $a_7 = 11$, find a_{22}.

15. Given an arithmetic sequence with $a_5 = 4$ and $a_{22} = 55$, find a_{36}.

16. Given an arithmetic sequence with $a_6 = 4$ and $a_{20} = -52$, find a_{33}.

17. Given an arithmetic sequence with $a_{16} = 30$ and $a_{30} = 65$, find a_9.

18. Given an arithmetic sequence with $a_8 = -6$ and $a_{19} = -\frac{45}{2}$, find a_{34}.

In Exercises 19–27, find the indicated sum.

You Try It

19. $\displaystyle\sum_{i=1}^{80} i$

20. $\displaystyle\sum_{j=1}^{10} (3j + 7)$

21. $7 + 10 + 13 + 16 + \cdots + 118$

22. $-14 + (-9) + (-4) + 1 + \cdots + 101$

23. $\displaystyle\sum_{i=3}^{14} (-7 - 9i)$

24. $1 + 11 + 21 + 31 + \cdots + a_{102}$

25. $6 + 14 + 22 + 30 + \cdots + (8n - 2)$

26. Find the sum of the first 100 odd integers.

You Try It

27. Find the sum of the first 100 even positive integers.

28. A large multiplex movie house has many theaters. The largest theater has 40 rows. There are 12 seats in the first row. Each row has four seats more than the previous row. How many total seats are there in this theater?

29. A stack of logs has 47 logs on the bottom layer. Each subsequent layer has nine fewer logs than the previous layer. If the top layer has two logs, how many total logs are there in the pile?

30. A middle school mathematics teacher accepts a teaching position that pays $31,000 per year. Each year, the expected raise is $1100. How much total money will this teacher earn teaching middle school mathematics over the first 12 years?

31. Suppose that you plan on taking a summer job selling magazine subscriptions. The magazine company will pay you $1 for selling the first subscription. For each additional subscription sold, the magazine company will pay you 15 cents more than what was paid for the previous subscription. How much will you earn by selling 200 magazine subscriptions?

32. Two companies have offered you a job. Alpha Company has offered you $35,000 per year with an annual raise of $2000. Beta Company has offered you a $46,000 annual salary with an annual raise of $800 per year. Which company will pay you more over the first 10 years?

33. A city fund-raiser raffle is raffling off 25 cash prizes. First prize is $5000. Each successive prize is $200 less than the preceding prize. What is the value of the 25th prize? What is the total amount of cash given out by this raffle?

34. Larry's Luxury Rental Car Company rents luxury cars for up to 18 days. The price is $300 for the first day, with the rental fee decreasing $7 for each additional day. How much will it cost to rent a luxury car for 18 days?

35. A ball thrown straight up in the air travels 48 inches in the first tenth of a second. In the next tenth of a second, the ball travels 44 inches. After each additional tenth of a second, the ball travels 4 inches less than it did during the preceding tenth of a second. How long will it take before the ball starts coming back down? What is the total distance that the ball has traveled when it has reached its maximum height?

B.3 Geometric Sequences and Series

THINGS TO KNOW

Before working through this section, be sure that you are familiar with the following concepts:

VIDEO   ANIMATION   INTERACTIVE

You Try It
1. Solve Applications of Exponential Functions (Section 9.3, **Objective 4**)

You Try It
2. Find the General Term or a Specific Term of an Arithmetic Sequence (Section B.2, **Objective 2**)

You Try It
3. Compute the nth Partial Sum of an Arithmetic Series (Section B.2, **Objective 3**)

OBJECTIVES

1 Write the Terms of a Geometric Sequence

2 Determine If a Sequence Is Geometric

3 Find the General Term or a Specific Term of a Geometric Sequence

4 Compute the nth Partial Sum of a Geometric Series

5 Determine If an Infinite Geometric Series Converges or Diverges

6 Solve Applications of Geometric Sequences and Series

OBJECTIVE 1:   WRITE THE TERMS OF A GEOMETRIC SEQUENCE

Suppose that you have agreed to work for Donald Trump on a particular job for 21 days. Mr. Trump gives you two choices of payment. You can be paid $100 for the first day and an additional $50 per day for each subsequent day. Or, you can choose to be paid 1 penny for the first day with your pay doubling each subsequent day. Which method of payment would you choose? (We will revisit this question later in this section. See Example 7.) Notice that each payment method can be represented by a sequence:

Payment Method 1: 100, 150, 200, 250, 300, . . .

Payment Method 2: 0.01, 0.02, 0.04, 0.08, 0.16, . . .

The first method of payment is an arithmetic sequence with $a_1 = 100$ and $d = 50$. The second method of payment is an example of a **geometric sequence**. Each term of this geometric sequence can be obtained by multiplying the previous term by 2. The number 2 in this case is called the **common ratio**. We can obtain this common ratio by dividing any term of the sequence (except the first term) by the previous term. That is, $r = \dfrac{a_2}{a_1} = \dfrac{a_3}{a_2} = \cdots = \dfrac{a_{n+1}}{a_n}$. The first term of the payment method 2 sequence is $a_1 = 0.01$ and the common ratio is $r = 2$.

Notice that we can rewrite the terms of this sequence as $0.01, (0.01)(2), (0.01)(2^2),$ $(0.01)(2^3), \ldots$. In general, given a geometric sequence with a first term of a_1 and a common ratio of r, then the first n terms of the sequence are

$$a_1$$
$$a_2 = a_1 r$$
$$a_3 = a_2 r = \underbrace{(a_1 r)}_{a^2} r = a_1 r^2$$
$$a_4 = a_3 r = \underbrace{(a_1 r^2)}_{a^3} r = a_1 r^3$$
$$\vdots$$
$$a_n = a_1 r^{n-1}.$$

Definition   Geometric Sequence

A **geometric sequence** is a sequence of the form $a_1, a_1 r, a_1 r^2, a_1 r^3, a_1 r^4, \ldots,$ where a_1 is the first term of the sequence and r is the common ratio such that $r = \dfrac{a_2}{a_1} = \dfrac{a_3}{a_2} = \cdots = \dfrac{a_{n+1}}{a_n}$ for all $n \geq 1$. The general term, or nth term, of a geometric sequence has the form $a_n = a_1 r^{n-1}$.

A portion of the two sequences representing the two payment methods are shown in Figure 3. The sequence representing payment method 1 (Figure 3a) is arithmetic. The ordered pairs of this sequence are collinear. The sequence representing payment method 2 (Figure 3b) is geometric. Notice that the ordered pairs of this sequence do not lie along a common line but, rather, lie on an exponential curve.

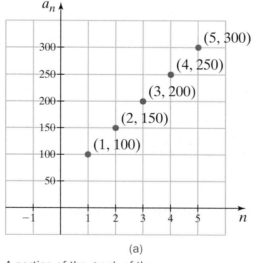

(a)

A portion of the graph of the sequence
$100, 150, 200, 250, 300, \ldots$

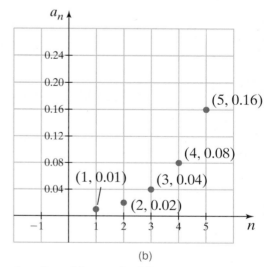

(b)

A portion of the graph of the sequence
$0.01, 0.02, 0.04, 0.08, 0.16, \ldots$

Figure 3 The graph of every arithmetic sequence is represented by a set of ordered pairs that lies on a straight line. The graph of a geometric sequence with $r > 0$ is represented by a set of ordered pairs that lies on an exponential curve.

If the first term and the common ratio of a geometric sequence are known, then we can determine the second term by multiplying the first term by the common ratio. The third term can be found by multiplying the second term by the common ratio. We continue this process to find the subsequent terms of the sequence.

Example 1 Writing a Geometric Sequence

My interactive video summary

a. Write the first five terms of the geometric sequence having a first term of 2 and a common ratio of 3.

b. Write the first five terms of the geometric sequence such that $a_1 = -4$ and $a_n = -5a_{n-1}$ for $n \geq 2$.

Solution

a. We are given that the first term is $a_1 = 2$ and the common ratio is $r = 3$. The second term is $a_2 = 2 \cdot 3 = 6$. The third term is $a_3 = 6 \cdot 3 = 18$. The fourth term is $a_4 = 18 \cdot 3 = 54$. The fifth term is $a_5 = 54 \cdot 3 = 162$. Therefore, the first five terms of this sequence are $2, 6, 18, 54$, and 162.

b. This sequence is defined recursively. The first term is $a_1 = -4$. To find a_2, substitute the value of 2 for n in the formula $a_n = -5a_{n-1}$ and simplify:

Given recursive formula:	$a_n = -5a_{n-1}$
Substitute:	$a_2 = -5a_{2-1}$
Simplify:	$= -5a_1$
Substitute $a_1 = -4$:	$= -5(-4)$
Multiply:	$= 20$

Therefore, $a_2 = 20$. We can follow this same procedure to find a_3, a_4, and a_5. You should verify that $a_3 = -100, a_4 = 500$, and $a_5 = -2500$ on your own, then watch this interactive video to see if you are correct.

You Try It   **Work through this You Try It problem.**

Work Exercises 1–5 in this textbook or in the *MyMathLab*® Study Plan.

OBJECTIVE 2:   DETERMINE IF A SEQUENCE IS GEOMETRIC

My interactive video summary

To determine if a given sequence is geometric, we must check to see if each term of the sequence can be obtained by multiplying the previous term by a common ratio r. That is, we must check to see if there exists a constant value r such that

$$r = \frac{a_{n+1}}{a_n} \text{ for any } n \geq 1.$$

Example 2 Determining If a Sequence Is Geometric

For each of the following sequences, determine if it is geometric. If the sequence is geometric, find the common ratio.

a. $2, 4, 6, 8, 10, \ldots$

b. $\dfrac{2}{3}, \dfrac{4}{9}, \dfrac{8}{27}, \dfrac{16}{81}, \dfrac{32}{243}, \ldots$

c. $12, -6, 3, -\dfrac{3}{2}, \dfrac{3}{4}, \ldots$

Solution

a. For this sequence, $\dfrac{a_2}{a_1} = \dfrac{4}{2} = 2$ and $\dfrac{a_3}{a_2} = \dfrac{6}{4} = \dfrac{3}{2}$. Since $\dfrac{a_2}{a_1} \neq \dfrac{a_3}{a_2}$, there does not exist a common ratio. Hence, this sequence is not geometric. (Note that this sequence is an arithmetic sequence.)

b. For this sequence, $\dfrac{a_2}{a_1} = \dfrac{\frac{4}{9}}{\frac{2}{3}} = \dfrac{4}{9} \cdot \dfrac{3}{2} = \dfrac{2}{3}$. Note that each term of this sequence (other than the first term) can be obtained by multiplying the previous term by $\dfrac{2}{3}$. Therefore, this sequence is geometric with a common ratio of $\dfrac{2}{3}$.

c. Try to determine if this sequence is geometric. Work through the interactive video to see if you are correct.

You Try It   Work through this You Try It problem.

Work Exercises 6–10 in this textbook or in the _MyMathLab_® Study Plan.

OBJECTIVE 3:   FIND THE GENERAL TERM OR A SPECIFIC TERM OF A GEOMETRIC SEQUENCE

By the definition of a geometric sequence, the general term, or nth term, of a geometric sequence has the form $a_n = a_1 r^{n-1}$. We can use this formula to find any term of a given geometric sequence.

Example 3 Finding the General Term of a Geometric Sequence

Find the general term of each geometric sequence.

a. $12, -6, 3, -\dfrac{3}{2}, \dfrac{3}{4}, \ldots$

b. $\dfrac{2}{3}, \dfrac{2}{9}, \dfrac{2}{27}, \dfrac{2}{81}, \dfrac{2}{243}, \ldots$

Solution

a. The first term of the sequence is $a_1 = 12$ and the common ratio is $r = \dfrac{a_2}{a_1} = \dfrac{-6}{12} = -\dfrac{1}{2}$. Therefore, $a_n = 12\left(-\dfrac{1}{2}\right)^{n-1}$.

b. The first term of the sequence is $a_1 = \dfrac{2}{3}$ and the common ratio is $r = \dfrac{a_2}{a_1} = \dfrac{\frac{2}{9}}{\frac{2}{3}} = \dfrac{1}{3}$. Therefore, $a_n = \left(\dfrac{2}{3}\right)\left(\dfrac{1}{3}\right)^{n-1}$.

Example 4 Finding a Specific Term of a Geometric Sequence

a. Find the seventh term of the geometric sequence whose first term is 2 and whose common ratio is -3.

b. Given a geometric sequence such that $a_6 = 16$ and $a_9 = 2$, find a_{13}.

My interactive video summary

Solution

a. We can use the formula $a_n = a_1 r^{n-1}$ with $a_1 = 2$ and $r = -3$ to find the general term. The general term is $a_n = 2(-3)^{n-1}$. Therefore, $a_7 = 2(-3)^{7-1} = 2(-3)^6 = 2(729) = 1458$.

b. Since $a_6 = 16$, we can substitute $n = 6$ into the formula $a_n = a_1 r^{n-1}$ to get $a_6 = a_1 r^5 = 16$. Similarly, we can substitute $n = 9$ into the formula $a_n = a_1 r^{n-1}$ to get $a_9 = a_1 r^8 = 2$. This gives the following two equations.

$$(1) \quad a_1 r^5 = 16$$
$$(2) \quad a_1 r^8 = 2$$

Divide both sides of equation (1) by r^5 to get $a_1 = \dfrac{16}{r^5}$. Now, substitute $a_1 = \dfrac{16}{r^5}$ into equation (2) and solve for a_1.

$$\text{Start with equation (2):} \qquad a_1 r^8 = 2$$

$$\text{Substitute } a_1 = \frac{16}{r^5}: \qquad \left(\frac{16}{r^5}\right) r^8 = 2$$

$$\frac{r^8}{r^5} = r^3: \qquad 16r^3 = 2$$

$$\text{Divide both sides by 16:} \qquad r^3 = \frac{1}{8}$$

$$\text{Take the cube root of both sides:} \qquad r = \frac{1}{2}$$

Now substitute $r = \dfrac{1}{2}$ into equation (1) to solve for a_1.

$$\text{Start with equation (1):} \qquad a_1 r^5 = 16$$

$$\text{Substitute } r = \frac{1}{2}: \qquad a_1 \left(\frac{1}{2}\right)^5 = 16$$

$$\left(\frac{1}{2}\right)^5 = \frac{1}{32}: \qquad a_1 \left(\frac{1}{32}\right) = 16$$

$$\text{Multiply both sides by 32:} \qquad a_1 = 512$$

We can now use the formula $a_n = a_1 r^{n-1}$ with $a_1 = 512$ and $r = \dfrac{1}{2}$ to find the general term. The general term is $a_n = 512\left(\dfrac{1}{2}\right)^{n-1}$. Therefore,

$$a_{13} = 512\left(\frac{1}{2}\right)^{13-1} = 512\left(\frac{1}{2}\right)^{12} = \frac{512}{2^{12}} = \frac{512}{4096} = \frac{1}{8}.$$

You may wish to work through this interactive video to see these solutions worked out in detail.

You Try It   Work through this You Try It problem.

Work Exercises 11–18 in this textbook or in the *MyMathLab*® Study Plan.

OBJECTIVE 4: COMPUTE THE nth PARTIAL SUM OF A GEOMETRIC SERIES

If $a_1, a_1r, a_1r^2, a_1r^3, \ldots$ is a geometric sequence, then the expression $a_1 + a_1r + a_1r^2 + a_1r^3 + \cdots + a_1r^{n-1} + \cdots$ is called an **infinite geometric series**

and can be written in summation notation as $\displaystyle\sum_{i=1}^{\infty} a_1 r^{i-1}$. Recall that the sum

of the first n terms of a series is called the **nth partial sum** of the series and is given by $S_n = a_1 + a_1r + a_1r^2 + a_1r^3 + \cdots + a_1r^{n-1}$. We can also represent the

nth partial sum using summation notation as $S_n = \displaystyle\sum_{i=1}^{n} a_1 r^{i-1}$. As you can see,

this nth partial sum is simply the sum of a finite geometric series. Fortunately, there is a convenient formula for computing the nth partial sum of a geometric series.

Formula for the nth Partial Sum of a Geometric Series

The sum of the first n terms of a geometric series is called the **nth partial sum** of the series and is given by

$$S_n = \sum_{i=1}^{n} a_1 r^{i-1} = a_1 + a_1r + a_1r^2 + a_1r^3 + \cdots + a_1r^{n-1}.$$

This sum can be computed using the formula $S_n = \dfrac{a_1(1 - r^n)}{1 - r}$ for $r \neq 1$.

 *My video summary*

🎞 Click here to see the derivation of this or watch this video.

Example 5 Computing the nth Partial Sum of a Geometric Series

 **a.** Find the sum of the series $\displaystyle\sum_{i=1}^{15} 5(-2)^{i-1}$.

b. Find the seventh partial sum of the geometric series $8 + 6 + \dfrac{9}{2} + \dfrac{27}{8} + \cdots$.

Solution

My interactive video summary

a. Using the formula $S_n = \dfrac{a_1(1 - r^n)}{1 - r}$ with $n = 15, a_1 = 5,$ and $r = -2$,

we get $S_{15} = \dfrac{5(1 - (-2)^{15})}{1 - (-2)} = \dfrac{5(1 - (-32{,}768))}{3} = \dfrac{5(32{,}769)}{3} =$

$\dfrac{163{,}845}{3} = 54{,}615$. You may also watch this interactive video to see this

solution worked out in detail.

b. Work through the interactive video to verify that the seventh partial sum of

this series is $S_7 = \dfrac{14{,}197}{512}$.

You Try It Work through this You Try It problem.

Work Exercises 19–23 in this textbook or in the *MyMathLab*® Study Plan.

OBJECTIVE 5: DETERMINE IF AN INFINITE GEOMETRIC SERIES CONVERGES OR DIVERGES

My video summary

Consider the infinite geometric series $\sum_{n=1}^{\infty} a_1 r^{n-1} = a_1 + a_1 r + a_1 r^2 + \cdots + a_1 r^{n-1} + \cdots$. Is it possible for a series of this form to have a finite sum? Is it possible to add infinitely many terms and get a finite sum? The answer is yes, it is possible, but it depends on the value of r. Before we determine the value(s) of r for which an infinite geometric series has a finite sum, let's look at an example. Consider the following geometric series

$$\frac{1}{2} + \frac{1}{3} + \frac{2}{9} + \frac{4}{27} + \frac{8}{81} + \cdots$$

Note that $a_1 = \frac{1}{2}$ and $r = \frac{2}{3}$. We can use the formula $S_n = \dfrac{a_1(1 - r^n)}{1 - r}$ to find the nth partial sum for any value of n of our choosing. The nth partial sums for $n = 5$, 10, 20, and 40 are given in Table 1 as well as the value of r^n.

n	$S_n = \dfrac{a_1(1 - r^n)}{1 - r} = \dfrac{\left(\frac{1}{2}\right)\left(1 - \left(\frac{2}{3}\right)^n\right)}{1 - \frac{2}{3}}$	$r^n = \left(\frac{2}{3}\right)^n$
5	1.3024691	0.1316872
10	1.4739877	0.0173415
20	1.4995489	0.0003007
40	1.4999999	0.0000001

Table 1

Looking at Table 1, it appears that as n increases, the value of S_n is getting closer to $1.5 = \dfrac{3}{2}$. Also notice that as n increases, the value of $r^n = \left(\dfrac{2}{3}\right)^n$ is getting closer to zero. In fact, for any value of r between -1 and 1, the value of r^n will always approach 0 as n approaches infinity. We say, "For values of r between -1 and 1, r^n approaches zero as n approaches infinity" and write: For $|r| < 1$, $r^n \to 0$ as $n \to \infty$. Thus, if $|r| < 1$, $S_n = \dfrac{a_1(1 - r^n)}{1 - r} \approx \dfrac{a_1(1 - 0)}{1 - r} = \dfrac{a_1}{1 - r}$ for large values of n. Therefore, given an infinite geometric series with $|r| < 1$, the sum of the series is given by $S = \dfrac{a_1}{1 - r}$.

Note: A formal proof of this formula requires calculus.

Formula for the Sum of an Infinite Geometric Series

Let $\sum_{n=1}^{\infty} a_1 r^{n-1} = a_1 + a_1 r + a_1 r^2 + a_1 r^3 + \cdots + a_1 r^{n-1} + \cdots$ be an infinite geometric series. If $|r| < 1$, then the sum of the series is given by $S = \dfrac{a_1}{1 - r}$.

Note that if $|r| < 1$, then the infinite geometric series has a finite sum and is said to **converge**. If $|r| \geq 1$, then the infinite geometric series does not have a finite sum and the series is said to **diverge**.

Example 6 Determining If an Infinite Geometric Series Converges or Diverges

My interactive video summary

 Determine whether each of the following series converges or diverges. If the series converges, find the sum.

a. $\displaystyle\sum_{n=1}^{\infty} \frac{1}{2}\left(\frac{2}{3}\right)^{n-1}$

b. $3 - \dfrac{6}{5} + \dfrac{12}{25} - \dfrac{24}{125} + \cdots$

c. $12 + 18 + 27 + \dfrac{81}{2} + \dfrac{243}{4} + \cdots$

Solution

a. This is an infinite geometric series with $|r| = \left|\dfrac{2}{3}\right| < 1$. Since $|r| < 1$, the infinite series must converge and, thus, must have a finite sum. The sum of the series is

$$S = \frac{a_1}{1-r} = \frac{\dfrac{1}{2}}{1 - \dfrac{2}{3}} = \frac{\dfrac{1}{2}}{\dfrac{1}{3}} = \frac{1}{2} \cdot \frac{3}{1} = \frac{3}{2}.$$

b. For this infinite geometric series, the common ratio is

$r = \dfrac{-\dfrac{6}{5}}{3} = -\dfrac{6}{5} \cdot \dfrac{1}{3} = -\dfrac{2}{5}$. Since $|r| = \left|-\dfrac{2}{5}\right| = \dfrac{2}{5} < 1$, the infinite series converges.

The sum is $S = \dfrac{a_1}{1-r} = \dfrac{3}{1 - \left(-\dfrac{2}{5}\right)} = \dfrac{3}{1 + \dfrac{2}{5}} = \dfrac{3}{\dfrac{7}{5}} = 3 \cdot \dfrac{5}{7} = \dfrac{15}{7}.$

c. This infinite series diverges. Do you know why? Work through the interactive video to see why this series diverges.

You Try It   Work through this You Try It problem.

Work Exercises 24–28 in this textbook or in the *MyMathLab*® Study Plan.

OBJECTIVE 6:   SOLVE APPLICATIONS OF GEOMETRIC SEQUENCES AND SERIES

In Example 7, we revisit the question that was presented at the beginning of this section.

Example 7 Choosing a Payment Method

Suppose that you have agreed to work for Donald Trump on a particular job for 21 days. Mr. Trump gives you two choices of payment. You can be paid $100 for the first day and an additional $50 per day for each subsequent day. Or, you can choose to be paid 1 penny for the first day with your pay doubling each subsequent day. Which method of payment yields the most income?

Solution Each payment method can be represented by a sequence.

$$\text{Payment method 1: } 100, 150, 200, 250, \ldots$$
$$\text{Payment method 2: } 0.01, 0.02, 0.04, 0.08, \ldots$$

To find out which method of payment will yield the greatest income, we must find the sum of the first 21 terms of each sequence.

Payment method 1 is an arithmetic sequence with $a_1 = 100$ and $d = 50$. Using the formula for the general term of an arithmetic sequence we get $a_n = 100 + (n-1)50 = 100 + 50n - 50 = 50n + 50$. Note that $a_{21} = 50(21) + 50 = 1100$. Using the formula for the nth partial sum of an arithmetic series with $n = 21$, we get $S_{21} = \dfrac{n(a_1 + a_{21})}{2} = \dfrac{21(100 + 1100)}{2} = \$12{,}600.$

Payment method 2 is a geometric sequence with $a_1 = 0.01$ and $r = 2$. Using the formula for the nth partial sum of a geometric series with $n = 21$, we get

$$S_{21} = \frac{a_1(1 - r^{21})}{1 - r} = \frac{(0.01)(1 - 2^{21})}{1 - 2} = \$20{,}971.51. \text{ Clearly, payment method 2}$$

is the better choice.

Example 8 Total Amount Given to a Local Charity

A local charity received \$8500 in charitable contributions during the month of January. Because of a struggling economy, it is projected that contributions will decline each month to 95% of the previous month's contributions. What are the expected contributions for the month of October? What is the total expected contributions that this charity can expect at the end of the year?

Solution The monthly contributions can be represented by a geometric sequence with $a_1 = 8500$ and $r = 0.95$. Thus, the contributions for the nth month is given by $a_n = 8500(0.95)^{n-1}$. The expected contributions for October, or when $n = 10$, are $a_{10} = 8500(0.95)^{10-1} \approx 5357.12$. Thus, the contributions for the month of October are expected to be about \$5357.12.

The total contributions for the year can be written as the following finite geometric series:

$$8500 + (8500)(0.95) + (8500)(0.95)^2 + \cdots + (8500)(0.95)^{11} = \sum_{i=1}^{12} 8500(0.95)^{i-1}$$

Using the formula for the nth partial sum of a geometric series with $n = 12$,

we get $S_{12} = \dfrac{a_1(1 - r^{12})}{1 - r} = \dfrac{8500(1 - 0.95^{12})}{1 - 0.95} \approx 78{,}138.79.$

Therefore, the charity can expect about \$78,138.79 in donations for the year.

You Try It   Work through this **You Try It** problem.

Work Exercises 29–33 in this textbook or in the **MyMathLab**® Study Plan.

Example 9 Expressing a Repeating Decimal as a Ratio of Two Integers

My interactive video summary

 Every repeating decimal number is a rational number and can therefore be represented by the quotient of two integers. Write each of the following repeating decimal numbers as the quotient of two integers.

a. $0.\overline{4}$

b. $0.2\overline{13}$

Solution

a. We can rewrite $0.\overline{4}$ as $0.44444\ldots = \dfrac{4}{10} + \dfrac{4}{100} + \dfrac{4}{1,000} + \dfrac{4}{10,000} +$

$\dfrac{4}{100,000} + \cdots$. This is an infinite geometric series with $a_1 = \dfrac{4}{10}$ and $r = \dfrac{1}{10}$.

Because $|r| = \left|\dfrac{1}{10}\right| < 1$, we know that the series converges.

Using the formula $S = \dfrac{a_1}{1 - r}$ we see that $0.\overline{4} = \dfrac{\dfrac{4}{10}}{1 - \dfrac{1}{10}} = \dfrac{\dfrac{4}{10}}{\dfrac{9}{10}} = \dfrac{4}{9}$.

b. Carefully work through the interactive video to see that $0.2\overline{13} = \dfrac{211}{990}$.

You Try It Work through this You Try It problem.

Work Exercises 34–35 in this textbook or in the *MyMathLab*® Study Plan.

Annuities

You Try It

In Section 9.3 we derived a formula for periodic compound interest. This formula is used to determine the future value of a *one-time* investment. (Click on the You Try It icon to see a periodic compound interest practice exercise.) Suppose that instead of investing one lump sum, you wish to invest equal amounts of money at steady intervals. An investment of equal amounts deposited at equal time intervals is called an **annuity**. If these equal deposits are made at the end of a compound period, the annuity is called an **ordinary annuity**.

For example, suppose that you want to invest $\$P$ at the end of each payment period at an annual rate r, in decimal form. Then the interest rate per payment period is $i = \dfrac{r}{\text{number of payment periods per year}}$. We now summarize the total amount of the ordinary annuity after the first k payment periods.

End of 1st payment period: $\underbrace{P}_{\text{1st payment}}$

End of 2nd payment period: $\underbrace{P}_{\text{1st payment}} + \underbrace{Pi}_{\substack{\text{Interest} \\ \text{earned on} \\ \text{1st payment}}} + \underbrace{P}_{\text{2nd payment}} = \underbrace{P(1 + i)}_{\substack{\text{Total amount} \\ \text{of 1st payment}}} + \underbrace{P}_{\text{2nd payment}}$

End of 3rd payment period:

$$\underbrace{P(1+i)}_{\substack{\text{Amount of}\\\text{1st payment}}} + \underbrace{Pi(1+i)}_{\substack{\text{Interest earned}\\\text{on amount of}\\\text{1st payment}}} + \underbrace{P}_{\text{2nd payment}} + \underbrace{Pi}_{\substack{\text{Interest}\\\text{earned on}\\\text{2nd payment}}} + \underbrace{P}_{\text{3rd payment}} = \underbrace{P(1+i)^2}_{\substack{\text{Total amount}\\\text{of 1st payment}}} + \underbrace{P(1+i)}_{\substack{\text{Total amount}\\\text{of 2nd payment}}} + \underbrace{P}_{\text{3rd payment}}$$

End of kth payment period: $\underbrace{P(1+i)^{k-1}}_{\substack{\text{Total amount}\\\text{of 1st payment}}} + \underbrace{P(1+i)^{k-2}}_{\substack{\text{Total amount}\\\text{of 2nd payment}}} + \cdots + \underbrace{P(1+i)}_{\substack{\text{Total amount}\\\text{of }(k-1)\text{st}\\\text{payment}}} + \underbrace{P}_{k\text{th payment}}$

The total amount of the ordinary annuity after k payment periods is
$A = P + P(1+i) + \cdots + P(1+i)^{k-2} + P(1+i)^{k-1}$.

This is a finite geometric series with $a_1 = P$ and a common ratio of $(1+i)$. Thus, the amount of the annuity after the kth payment is

$$A = \frac{P\left(1 - (1+i)^k\right)}{1 - (1+i)} = \frac{P\left(1 - (1+i)^k\right)}{-i} = \frac{P\left((1+i)^k - 1\right)}{i}.$$

Amount of an Ordinary Annuity after the kth Payment

The total amount of an ordinary annuity after the kth payment is given by the formula

$$A = \frac{P\left((1+i)^k - 1\right)}{i},$$

where

A = Total amount of annuity after k payments

P = Deposit amount at the end of each payment period

i = Interest rate per payment period

Example 10 Finding the Amount of an Ordinary Annuity

Chie and Ben decided to save for their newborn son Jack's college education. They decided to invest $200 every 3 months in an investment earning 8% interest compounded quarterly. How much is this investment worth after 18 years?

Solution This is an ordinary annuity with $P = \$200$ and $i = \dfrac{0.08}{4} = 0.02$. What

is k? See if you can determine k and use the formula $A = \dfrac{P((1+i)^k - 1)}{i}$ to

determine the total amount of this annuity. When you are done, watch this video to see if you are correct.

You Try It Work through this You Try It problem.

Work Exercises 36–38 in this textbook or in the *MyMathLab*® Study Plan.

My video summary

B.3 Exercises

In Exercises 1–5, write the first five terms of the geometric sequence with the given information.

1. The first term is 8 and the common ratio is 2.

2. The first term is 162 and the common ratio is $\frac{1}{3}$.

3. The first term is 25 and the common ratio is $-\frac{1}{5}$.

4. $a_n = 7a_{n-1}; a_1 = 3$

5. $a_n = -2a_{n-1}; a_1 = -4$

In Exercises 6–10, determine if the sequence is geometric. If the sequence is geometric, find the common ratio.

6. $4, 24, 144, 864, \ldots$

7. $-2, 2, -2, 2, \ldots$

8. $-3, 1, -1, -3, \ldots$

9. $2, -\dfrac{10}{3}, \dfrac{50}{9}, -\dfrac{250}{27}, \ldots$

10. $7.236, -5.7888, 4.63104, -3.704832, \ldots$

11. Determine the general term of the sequence $3, 6, 12, 24, \ldots$.

12. Determine the general term of the sequence $\frac{1}{2}, \frac{1}{8}, \frac{1}{32}, \frac{1}{128}, \ldots$.

13. Determine the general term of the sequence $\frac{1}{5}, -\frac{2}{15}, \frac{4}{45}, -\frac{8}{135}, \ldots$.

14. Find the seventh term of the geometric sequence whose first term is 5 and whose common ratio is 4.

15. Find the sixth term of the geometric sequence whose first term is 3804 and whose common ratio is $-\frac{1}{4}$.

16. Find the 11th term of the geometric sequence $\$5000, \$5050, \$5100.50, \ldots$.

17. Given a geometric sequence such that $a_4 = 108$ and $a_7 = 2916$, find a_{10}.

18. Given a geometric sequence such that $a_3 = 16$ and $a_8 = -\frac{1}{2}$, find a_{11}.

In Exercises 19–23, find the sum of each geometric series.

19. $\displaystyle\sum_{i=1}^{8} 7(-4)^{i-1}$

20. $\displaystyle\sum_{i=1}^{13} 2\left(\dfrac{3}{5}\right)^{i-1}$

21. $\displaystyle\sum_{i=1}^{10} 4(1.05)^{i-1}$

22. $2 + \dfrac{2}{3} + \dfrac{2}{9} + \dfrac{2}{27} + \cdots + \dfrac{2}{729}$

23. $1 - \dfrac{1}{2} + \dfrac{1}{4} - \dfrac{1}{8} + \cdots - \dfrac{1}{128}$

You Try It

In Exercises 24–28, determine if each infinite geometric series converges or diverges. If the series converges, find the sum.

You Try It

24. $-1 + \dfrac{1}{10} - \dfrac{1}{100} + \dfrac{1}{1000} - \cdots$

25. $343 + 49 + 7 + 1 + \cdots$

26. $\displaystyle\sum_{i=1}^{\infty} 7\left(\dfrac{1}{4}\right)^{i-1}$

27. $\displaystyle\sum_{i=1}^{\infty} \dfrac{1}{5}\left(3\right)^{i-1}$

28. $0.5 - 0.05 + 0.005 - 0.0005 + 0.00005 - \cdots$

29. Warren wanted to save money to purchase a new car. He started by saving $1 on the first of January. On the first of February, he saved $3. On the first of March he saved $9. So, on the first day of each month, he wanted to save three times as much as he did on the first day of the previous month. If Warren continues his savings pattern, how much will he need to save on the first day of September?

30. Suppose that you have accepted a job for 2 weeks that will pay $0.07 for the first day, $0.14 for the second day, $0.28 for the third day, and so on. What will your total earnings be after 2 weeks?

31. Mary has accepted a teaching job that pays $25,000 for the first year. According to the Teacher's Union, Mary will get guaranteed salary increases of 4 percent per year. If Mary plans to teach for 30 years, what will be her total salary earnings?

32. A child is given an initial push on a rope swing. On the first swing, the rope swings through an arc of 12 feet. On each successive swing, the length of the arc is 80% of the previous length. After 10 swings, what is the total length the rope will have swung? When the child stops swinging, what is the total length the rope will have swung?

33. Randy dropped a rubber ball from his apartment window from a height of 50 feet. The ball always bounces $\frac{3}{5}$ of the distance fallen. How far has the ball traveled once it is done bouncing?

You Try It

34. Rewrite the number $0.\overline{7}$ as the quotient of two integers.

35. Rewrite the number $0.3\overline{25}$ as the quotient of two integers.

36. Kip contributes $200 every month to his $401(k)$. What will the value of Kip's $401(k)$ be in 10 years if the yearly rate of return is assumed to be 12% compounded monthly?

You Try It

37. Mark and Lisa decide to invest $500 every 3 months in an education IRA to save for their son Beau's college education. What will the value of the IRA be after 10 years if the yearly assumed rate of return is 8% compounded quarterly?

38. Marv and Cindy decide to build a new home in 10 years. They will need $80,000 to purchase the lot of their dreams. How much should they save each month in an account that has an assumed yearly rate of return of 7% compounded monthly?

B.4 The Binomial Theorem

THINGS TO KNOW

Before working through this section, be sure that you are familiar with the following concepts:

VIDEO   ANIMATION   INTERACTIVE

You Try It

1. Multiply Two or More Polynomials
   (Section 4.3, Objective 5)

OBJECTIVES

1 Expand Binomials Raised to a Power Using Pascal's Triangle

2 Evaluate Binomial Coefficients

3 Expand Binomials Raised to a Power Using the Binomial Theorem

4 Find a Particular Term or a Particular Coefficient of a Binomial Expansion

OBJECTIVE 1: **EXPAND BINOMIALS RAISED TO A POWER USING PASCAL'S TRIANGLE**

My video summary

In this section, we will focus on expanding algebraic expressions of the form $(a + b)^n$, where n is an integer greater than or equal to zero. Because $(a + b)$ is a binomial, we call the expansion of $(a + b)^n$ a *binomial expansion*. Consider the expansion of $(a + b)^4$.

$$
\begin{aligned}
(a + b)^4 &= \underbrace{(a + b)(a + b)} \cdot \underbrace{(a + b)(a + b)} \\
&= (a^2 + 2ab + b^2)(a^2 + 2ab + b^2) \\
&= a^4 + 2a^3b + a^2b^2 + 2a^3b + 4a^2b^2 + 2ab^3 + a^2b^2 + 2ab^3 + b^4 \\
&= a^4 + 4a^3b + 6a^2b^2 + 4ab^3 + b^4
\end{aligned}
$$

Although the expansion of $(a + b)^4$ using the method above is not too complicated, it would not be desirable to use this method to expand $(a + b)^n$ for large values of n.

The goal in this section is to try to develop a method for expanding expressions of the form $(a + b)^n$ without actually performing all of the multiplication. We start by studying the expanded forms of $(a + b)^n$ for $n = 0, 1, 2, 3, 4,$ and 5.

$$
\begin{aligned}
n = 0 : (a + b)^0 &= & 1 \\
n = 1 : (a + b)^1 &= & 1a + 1b \\
n = 2 : (a + b)^2 &= & 1a^2 + 2ab + 1b^2 \\
n = 3 : (a + b)^3 &= & 1a^3 + 3a^2b + 3ab^2 + 1b^3 \\
n = 4 : (a + b)^4 &= & 1a^4 + 4a^3b + 6a^2b^2 + 4ab^3 + 1b^4 \\
n = 5 : (a + b)^5 &= 1a^5 + 5a^4b + 10a^3b^2 + 10a^2b^3 + 5ab^4 + 1b^5
\end{aligned}
$$

The coefficients of each expansion are highlighted in red. These coefficients are known as **binomial coefficients**. Before we determine a pattern for these coefficients, let's first observe the exponent pattern. Notice in each expansion of

$(a + b)^n$, there are always $n + 1$ terms. The sum of the exponents of each term is always equal to n. Also note that the first term is always a^n (or $a^n b^0$) and the last term is always b^n (or $a^0 b^n$). As we look at the terms of each expansion from left to right, the exponent of the first variable decreases by 1 and the exponent of the second variable increases by 1. Thus, the exponent pattern of the variables of each expansion is $a^n b^0, a^{n-1} b^1, a^{n-2} b^2, a^{n-3} b^3, \ldots, a^1 b^{n-1}, a^0 b^n$. The pattern for the binomial coefficients is less obvious. To see the pattern for the coefficients, we start by rewriting the six expansions of $(a + b)^n$ again, but this time we only write the coefficients. See Figure 4.

$n = 0$:   1

$n = 1$:   1   1

$n = 2$:   1   2   1

$n = 3$:   1   3   3   1

$n = 4$: 1   4   6   4   1

Figure 4

$n = 5$: 1   5   10   10   5   1

The coefficients of the expansions of $(a + b)^n$, also called Pascal's triangle

Notice that the coefficients in Figure 4 form a "triangle." This triangle is known as Pascal's triangle, named after the French mathematician, Blaise Pascal. The first and last number of each row of Pascal's triangle is 1. Every other number is equal to the sum of the two numbers directly above it. We can now write the next row of Pascal's triangle, which is the row corresponding to $n = 6$.

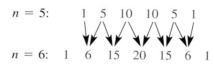

$$n = 5: \quad 1 \quad 5 \quad 10 \quad 10 \quad 5 \quad 1$$
$$n = 6: \quad 1 \quad 6 \quad 15 \quad 20 \quad 15 \quad 6 \quad 1$$

This new row of Pascal's triangle represents the coefficients of the expansion of $(a + b)^6$. Therefore,

$$(a + b)^6 = a^6 b^0 + 6a^5 b^1 + 15a^4 b^2 + 20a^3 b^3 + 15a^2 b^4 + 6a^1 b^5 + a^0 b^6$$
$$= a^6 + 6a^5 b + 15a^4 b^2 + 20a^3 b^3 + 15a^2 b^4 + 6ab^5 + b^6.$$

Notice the pattern of the exponents of each variable. The exponent of variable a starts with 6 and decreases by 1 for each successive term until it equals 0. The exponent of variable b starts with 0 and increases by 1 each term until it equals 6.

See if you can create Pascal's triangle for values of n up to 10. Click here to see if you are correct.

Example 1   Using Pascal's Triangle to Expand a Binomial Raised to a Power

My interactive video summary

Use Pascal's triangle to expand each binomial.

a. $(x + 2)^4$ **b.** $(x - 3)^5$ **c.** $(2x - 3y)^3$

Solution

a. We start by looking at the row of Pascal's triangle corresponding with $n = 4$. We see that this row is 1 4 6 4 1. Using these coefficients and the exponent pattern we get

$$(x + 2)^4 = 1\left(x^4 \cdot 2^0\right) + 4\left(x^3 \cdot 2^1\right) + 6\left(x^2 \cdot 2^2\right) + 4\left(x \cdot 2^3\right) + 1\left(x^0 \cdot 2^4\right)$$
$$= x^4 + 8x^3 + 24x^2 + 32x + 16.$$

b. The row of Pascal's triangle corresponding with $n = 5$ is 1 5 10 10 5 1. Using these coefficients and the exponent pattern we get

$$(x - 3)^5 = 1\left(x^5 \cdot (-3)^0\right) + 5\left(x^4 \cdot (-3)^1\right) + 10\left(x^3 \cdot (-3)^2\right)$$
$$+ 10\left(x^2 \cdot (-3)^3\right) + 5\left(x \cdot (-3)^4\right) + 1\left(x^0 \cdot (-3)^5\right)$$
$$= x^5 - 15x^4 + 90x^3 - 270x^2 + 405x - 243.$$

c. See if you can use Pascal's triangle to show that $(2x - 3y)^3 = 8x^3 - 36x^2y + 54xy^2 - 27y^3$. Work through the interactive video to see the solution.

Note: The terms of the expansion of the form $(a - b)^n$ will *always* alternate in sign with the sign of the first term being positive.

You Try It   **Work through this You Try It problem.**

Work Exercises 1–6 in this textbook or in the *MyMathLab*® Study Plan.

OBJECTIVE 2:   EVALUATE BINOMIAL COEFFICIENTS

Although Pascal's triangle is useful for determining the binomial coefficients of $(a + b)^n$ for fairly small values of n, it is not that useful for large values of n. For example, to find the binomial coefficients of the expansion of $(a + b)^{50}$ using Pascal's triangle, we would need to write the first 51 rows of the triangle to determine the coefficients. Fortunately, there is a convenient formula for the binomial coefficients. This formula requires the use of factorials. Recall, $n! = n \cdot (n - 1) \cdot (n - 2) \cdot \cdots \cdot 3 \cdot 2 \cdot 1$ and $0! = 1$. To establish a formula for the binomial coefficients, let's take another look at the expansion for $(a + b)^6$.

$$(a + b)^6 = 1a^6b^0 + 6a^5b^1 + 15a^4b^2 + 20a^3b^3 + 15a^2b^4 + 6a^1b^5 + 1a^0b^6$$

Table 2 shows the relationship between the variable parts of the expansion of the form $a^{n-r}b^r$ and the corresponding coefficients. Click on any of the coefficients in Table 2 to verify that the formula using factorial notation is true.

Variables	Coefficient	Variables	Coefficient
a^6b^0	$1 = \dfrac{6!}{0! \cdot 6!}$	a^2b^4	$15 = \dfrac{6!}{4! \cdot 2!}$
a^5b^1	$6 = \dfrac{6!}{1! \cdot 5!}$	a^1b^5	$6 = \dfrac{6!}{5! \cdot 1!}$
a^4b^2	$15 = \dfrac{6!}{2! \cdot 4!}$	a^0b^6	$1 = \dfrac{6!}{6! \cdot 0!}$
a^3b^3	$20 = \dfrac{6!}{3! \cdot 3!}$		

Table 2

You can see in Table 2 that for each pair of variables of the form $a^{n-r}b^r$, the corresponding binomial coefficient is of the form $\dfrac{n!}{r! \cdot (n - r)!}$. We will use the shorthand notation $\dbinom{n}{r}$, read as "n choose r," to denote a binomial coefficient.

Formula for a Binomial Coefficient

For non-negative integers n and r with $n \geq r$, the coefficient of the expansion of $(a + b)^n$ whose variable part is $a^{n-r}b^r$ is given by

$$\binom{n}{r} = \frac{n!}{r! \cdot (n - r)!}.$$

Example 2 Evaluating Binomial Coefficients

Evaluate each of the following binomial coefficients.

a. $\binom{5}{3}$ b. $\binom{4}{1}$ c. $\binom{12}{8}$

Solution

a. $\binom{5}{3} = \dfrac{5!}{3!(5 - 3)!} = \dfrac{5!}{3! \cdot 2!} = \dfrac{5 \cdot 4 \cdot 3!}{3! \cdot 2 \cdot 1} = \dfrac{20}{2} = 10$

b. $\binom{4}{1} = \dfrac{4!}{1!(4 - 1)!} = \dfrac{4!}{1! \cdot 3!} = \dfrac{4 \cdot 3!}{1 \cdot 3!} = \dfrac{4}{1} = 4$

c. $\binom{12}{8} = \dfrac{12!}{8!(12 - 8)!} = \dfrac{12!}{8! \cdot 4!} = \dfrac{12 \cdot 11 \cdot 10 \cdot 9 \cdot 8!}{8! \cdot 4 \cdot 3 \cdot 2 \cdot 1} = \dfrac{11,880}{24} = 495$

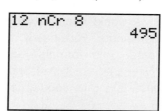

A calculator can be used to compute binomial coefficients. Typically, the key $\boxed{nCr}$ is used. The figure on the left shows the computation of $\binom{12}{8}$ using a graphing utility.

You Try It Work through this **You Try It** problem.

Work Exercises 7–10 in this textbook or in the *MyMathLab*® Study Plan.

OBJECTIVE 3: EXPAND BINOMIALS RAISED TO A POWER USING THE BINOMIAL THEOREM

Now that we know how to compute a binomial coefficient, we can state the Binomial Theorem.

Binomial Theorem

If n is a positive integer then,

$$(a + b)^n = \binom{n}{0}a^n + \binom{n}{1}a^{n-1}b + \binom{n}{2}a^{n-2}b^2 + \cdots + \binom{n}{n}b^n$$

$$= \sum_{i=0}^{n} \binom{n}{i}a^{n-i}b^i.$$

My interactive video summary

Example 3 Using the Binomial Theorem to Expand a Binomial Raised to a Power

 Use the Binomial Theorem to expand each binomial.

a. $(x - 1)^8$ **b.** $\left(\sqrt{x} + y^2\right)^5$

Solution

a. Work through the interactive video to verify that $(x - 1)^8 = x^8 - 8x^7 + 28x^6 - 56x^5 + 70x^4 - 56x^3 + 28x^2 - 8x + 1$ using the Binomial Theorem.

b. The expansion of $\left(\sqrt{x} + y^2\right)^5$ is as follows.

$$\left(\sqrt{x} + y^2\right)^5 = \binom{5}{0}(\sqrt{x})^5 + \binom{5}{1}(\sqrt{x})^4(y^2) + \binom{5}{2}(\sqrt{x})^3(y^2)^2 + \binom{5}{3}(\sqrt{x})^2(y^2)^3 + \binom{5}{4}(\sqrt{x})(y^2)^4 + \binom{5}{5}(y^2)^5$$

$$= 1 \cdot (\sqrt{x})^5 + 5 \cdot (\sqrt{x})^4(y^2) + 10 \cdot (\sqrt{x})^3(y^2)^2 + 10 \cdot (\sqrt{x})^2(y^2)^3 + 5 \cdot (\sqrt{x})(y^2)^4 + 1 \cdot (y^2)^5$$

$$= x^2\sqrt{x} + 5x^2y^2 + 10x\sqrt{x}y^4 + 10xy^6 + 5\sqrt{x}y^8 + y^{10}$$

You Try It Work through this You Try It problem.

Work Exercises 11–17 in this textbook or in the *MyMathLab*® Study Plan.

OBJECTIVE 4:   FIND A PARTICULAR TERM OR A PARTICULAR COEFFICIENT OF A BINOMIAL EXPANSION

We may want to find a particular term of a binomial expansion. Fortunately, we can use the Binomial Theorem to develop a formula for a particular term. We start by writing out the first several terms of $(a + b)^n$ using the Binomial Theorem.

$$(a + b)^n = \binom{n}{0}a^n + \binom{n}{1}a^{n-1}b + \binom{n}{2}a^{n-2}b^2 + \binom{n}{3}a^{n-3}b^3 + \cdots + \binom{n}{n}b^n$$

The first term is $\binom{n}{0}a^n$. The second term is $\binom{n}{1}a^{n-1}b$. The third term is $\binom{n}{2}a^{n-2}b^2$.

Following this pattern, we can see that the formula for the $(r + 1)$st term (for $r \geq 0$) is given by $\binom{n}{r}a^{n-r}b^r$.

Formula for the $(r + 1)$st Term of a Binomial Expansion

If n is a positive integer and if $r \geq 0$, then the $(r + 1)$st term of the expansion of $(a + b)^n$ is given by

$$\binom{n}{r}a^{n-r}b^r = \frac{n!}{r! \cdot (n - r)!}a^{n-r}b^r.$$

Example 4 Finding a Particular Term of a Binomial Expansion

My video summary

 Find the third term of the expansion of $(2x - 3)^{10}$.

Solution Since we want to find the third term of this expansion, we will use the formula for the $(r + 1)$st term, which is equal to $\binom{n}{r} a^{n-r} b^r$ for $r = 2$, $n = 10$, $a = 2x$, and $b = -3$. Therefore, the third term is

$$\binom{10}{2} (2x)^{10-2} (-3)^2 = 45(256x^8)(9) = 103{,}680x^8.$$

Watch this video to see every step of this solution.

Example 5 Finding a Particular Coefficient of a Binomial Expansion

My video summary

 Find the coefficient of x^7 in the expansion of $(x + 4)^{11}$.

Solution The formula for the $(r + 1)$st term of the expansion of $(x + 4)^{11}$ is given by the formula $\binom{11}{r} x^{11-r} 4^r$. The term containing x^7 occurs when $11 - r = 7$. Solving this equation for r we get $r = 4$. Therefore, the term involving x^7 is $\binom{11}{4} x^7 4^4$. Simplifying this expression we get $84{,}480x^7$. Thus, the coefficient of x^7 is 84,480. Watch this video to see this solution worked out in detail.

You Try It   Work through this You Try It problem.

Work Exercises 18–25 in this textbook or in the *MyMathLab*® Study Plan.

B.4 Exercises

In Exercises 1–6, use Pascal's triangle to expand each binomial.

You Try It

1. $(m + n)^6$ **2.** $(x + 5)^4$ **3.** $(x - y)^7$

4. $(x - 3)^5$ **5.** $(2x + 3y)^6$ **6.** $(3x^2 - 4y^3)^4$

In Exercises 7–10, evaluate each binomial coefficient.

You Try It

7. $\binom{7}{1}$ **8.** $\binom{10}{4}$ **9.** $\binom{7}{7}$ **10.** $\binom{23}{3}$

In Exercises 11–17, use the Binomial Theorem to expand each binomial.

You Try It

11. $(x + 2)^7$ **12.** $(x - 3)^6$ **13.** $(4x + 1)^5$

14. $(x + 3y)^4$ **15.** $(5x - 3y)^5$ **16.** $(x^4 + y^5)^6$

17. $\left(\sqrt{x} - \sqrt{2}\right)^4$

18. Find the sixth term of the expansion of $(x + 4)^9$.

19. Find the fifth term of the expansion of $(a - b)^8$.

20. Find the third term of the expansion of $(3c - d)^7$.

21. Find the seventh term of the expansion of $(3x + 2)^{10}$.

22. Find the coefficient of x^5 in the expansion of $(x - 3)^{11}$.

23. Find the coefficient of x^4 in the expansion of $(4x + 1)^{12}$.

24. Find the coefficient of x^0 in the expansion of $\left(x^2 + \frac{1}{x}\right)^{12}$.

25. Find the coefficient of x^{10} in the expansion of $\left(x - \frac{3}{\sqrt{x}}\right)^{19}$.

Basic Math Review — Fractions, Decimals, Proportions, Percents

APPENDIX C CONTENTS

C.1   **Fractions**

C.2   **Decimals**

C.3   **Proportions**

C.4   **Percents**

C.1   Fractions

Types of Fractions

Whole numbers are used to count whole units such as people, dogs, trees, and so on. To describe part of a whole unit, *fractions* are needed. For example, consider a pizza that is divided into twelve equally sized pieces. If you eat five of the pieces, then you have eaten five-twelfths, or $\frac{5}{12}$, of the pizza. The 12 is called the *denominator* and refers to the number of equal parts within the whole unit. The 5 is called the *numerator* and refers to the number of equal parts of interest, in this case 5.

Definition   Fraction

A **fraction** is a number of the form $\frac{a}{b}$, where a and b are integers and $b \neq 0$. a is called the **numerator**, and b is called the **denominator**.

Example 1 Using Fractions to Describe a Part of a Whole

Write a fraction to represent the shaded area of each figure.

a.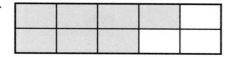
b.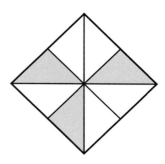

Solution

a. Seven of the 10 equal parts are shaded, so the fraction is $\dfrac{7}{10}$.

b. Three of the 8 equal parts are shaded, so the fraction is $\dfrac{3}{8}$.

A **proper fraction** is a fraction in which the numerator is smaller than the denominator. Proper fractions represent quantities less than 1. The fractions $\dfrac{7}{10}$ and $\dfrac{3}{8}$ from Example 1 are proper fractions.

An **improper fraction** is a fraction in which the numerator is larger than or equal to the denominator. Improper fractions represent quantities greater than or equal to 1. The fractions $\dfrac{5}{4}$ and $\dfrac{5}{5}$ are examples of improper fractions.

Example 2 Using Improper Fractions to Describe a Part of a Whole

Write an improper fraction to represent the shaded area of each figure.

a.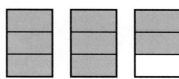
b.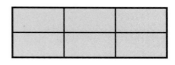

Solution

a. Each whole object is divided into 3 equal parts. A total of 8 parts are shaded, so the improper fraction is $\dfrac{8}{3}$.

b. The whole object is divided into 6 equal parts. All 6 parts are shaded, so the improper fraction is $\dfrac{6}{6}$.

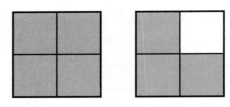

Figure 1

Mixed Numbers

Consider the shaded area in Figure 1, which represents the improper fraction $\frac{7}{4}$.

The shaded area represents one whole object and three-fourths of a second object. This represents the quantity *one and three-fourths*, which is known as a **mixed number**. A mixed number contains a whole-number part and a proper-fraction part. Symbolically, we denote "one and three-fourths" by $1\frac{3}{4}$, which means $1 + \frac{3}{4}$, but the + symbol is not written.

Since Figure 1 can be written as both $1\frac{3}{4}$ and $\frac{7}{4}$, then $1\frac{3}{4} = \frac{7}{4}$. In fact, mixed numbers can *always* be converted into improper fractions, and improper fractions can always be converted into mixed numbers.

Steps for Converting a Mixed Number into an Improper Fraction

Step 1. Multiply the whole number by the denominator of the fraction.

Step 2. Add the product from step 1 to the numerator of the original fraction.

Step 3. Write the sum from step 2 over the denominator of the original fraction.

Example 3 Converting Mixed Numbers into Improper Fractions

Convert $4\frac{5}{8}$ into an improper fraction.

Solution We follow the steps for converting a mixed number to an improper fraction.

$$4\frac{5}{8} = \frac{\overbrace{4 \cdot 8}^{\text{Step 1}} + \overbrace{5}^{\text{Step 2}}}{\underbrace{8}_{\text{Step 1}}} = \frac{32 + 5}{8} = \frac{37}{8}$$

To convert an improper fraction into a mixed number, we divide.

Steps for Converting an Improper Fraction into a Mixed Number (or Whole Number)

Step 1. Divide the numerator by the denominator.

Step 2. Use the quotient as the whole-number part of the mixed number. Place the remainder over the original denominator to form the fraction-part of the mixed number. The mixed number will take the form

$$\text{Quotient}\, \frac{\text{Remainder}}{\text{Original Denominator}}.$$

Note: If the denominator divides evenly into the numerator (with no remainder), then the improper fraction will be a whole number, not a mixed number.

Example 4 Converting Improper Fractions into Mixed Numbers

Convert $\dfrac{47}{5}$ into a mixed number.

Solution 47 divided by 5 is 9 with a remainder of 2. So, $\dfrac{47}{5} = 9\dfrac{2}{5}$.

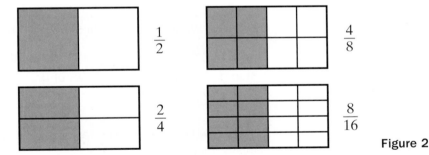

Equivalent Fractions

Definition   Equivalent Fraction

Fractions are called **equivalent fractions** if they represent the same part of a whole.

Consider the fractions shown in Figure 2. The shaded area is the same in all four figures, illustrating that the four fractions are equivalent. So, $\dfrac{1}{2} = \dfrac{2}{4} = \dfrac{4}{8} = \dfrac{8}{16}$.

$\dfrac{1}{2}$

$\dfrac{4}{8}$

$\dfrac{2}{4}$

$\dfrac{8}{16}$

Figure 2

Multiplying or dividing the numerator and denominator of a fraction by the same nonzero number results in an equivalent fraction.

Property of Equivalent Fractions

If a, b, and c are numbers, then

$$\frac{a}{b} = \frac{a \cdot c}{b \cdot c} \quad \text{and} \quad \frac{a}{b} = \frac{a \div c}{b \div c}$$

as long as b and c are not equal to zero.

Example 5 Finding Equivalent Fractions

Write each fraction as an equivalent fraction with the given denominator.

a. $\dfrac{7}{8}$; denominator 40

b. $\dfrac{18}{24}$; denominator 4

Solution

a. We multiply the current denominator 8 by 5 to obtain the new denominator 40, so we also multiply the numerator by 5.

$$\frac{7}{8} = \frac{7 \cdot 5}{8 \cdot 5} = \frac{35}{40}$$

b. To obtain the new denominator 4, we divide the current denominator 24 by 6. So, we also divide the numerator by 6.

$$\frac{18}{24} = \frac{18 \div 6}{24 \div 6} = \frac{3}{4}$$

Simplifying Fractions

A fraction can be written in infinitely many equivalent forms. Only one of these is the *simplest form*. For example, of all the equivalent fractions $\frac{1}{2} = \frac{2}{4} = \frac{4}{8} = \frac{8}{16}$ in Figure 2, the fraction $\frac{1}{2}$ is in simplest form.

Definition   Simplest Form

A fraction is in **simplest form**, or **lowest terms**, when the numerator and denominator have no common factors other than 1.

Steps for Writing a Fraction in Simplest Form

Step 1. Write the prime factorizations of the numerator and denominator.

Step 2. Divide out, or cancel, all of the common factors of the numerator and denominator.

Let's review finding the prime factorizations of a number, and then we will simplify fractions.

Example 6   Finding the Prime Factorization of a Number

Find the prime factorization of each number.

a. 80   **b.** 72

Solution

a.   **b.**

Example 7   Simplifying Fractions

Write each fraction in simplest form.

a. $\frac{8}{54}$ **b.** $\frac{40}{120}$

Solution

a. $\dfrac{8}{54} = \dfrac{2\cdot2\cdot2}{2\cdot3\cdot3\cdot3} = \dfrac{\overset{1}{\cancel{2}}\cdot2\cdot2}{\underset{1}{\cancel{2}}\cdot3\cdot3\cdot3} = \dfrac{1\cdot2\cdot2}{1\cdot3\cdot3\cdot3} = \dfrac{4}{27}$

b. $\dfrac{40}{120} = \dfrac{2\cdot2\cdot2\cdot5}{2\cdot2\cdot2\cdot3\cdot5} = \dfrac{\overset{1}{\cancel{2}}\cdot\overset{1}{\cancel{2}}\cdot\overset{1}{\cancel{2}}\cdot\overset{1}{\cancel{5}}}{\underset{1}{\cancel{2}}\cdot\underset{1}{\cancel{2}}\cdot\underset{1}{\cancel{2}}\cdot3\cdot\underset{1}{\cancel{5}}} = \dfrac{1\cdot1\cdot1\cdot1}{1\cdot1\cdot1\cdot3\cdot1} = \dfrac{1}{3}$

Multiplying Fractions

To multiply two fractions, we multiply the numerators, multiply the denominators, and simplify.

Multiplying Two Fractions

If $a, b, c,$ and d are numbers and b and d are not equal to zero, then

$$\frac{a}{b}\cdot\frac{c}{d} = \frac{a\cdot c}{b\cdot d}.$$

Example 8 Multiplying Fractions

Multiply and simplify.

a. $\dfrac{5}{7}\cdot\dfrac{8}{11}$
b. $\dfrac{4}{9}\cdot\dfrac{12}{20}$

Solution

a. $\dfrac{5}{7}\cdot\dfrac{8}{11} = \dfrac{\overbrace{5\cdot8}^{\text{Multiply numerators}}}{\underbrace{7\cdot11}_{\text{Multiply denominators}}} = \dfrac{40}{77}$

b. $\dfrac{4}{9}\cdot\dfrac{12}{20} = \dfrac{4\cdot12}{9\cdot20} = \dfrac{48}{180} = \dfrac{2\cdot2\cdot2\cdot2\cdot3}{2\cdot2\cdot3\cdot3\cdot5} = \dfrac{\overset{1}{\cancel{2}}\cdot\overset{1}{\cancel{2}}\cdot2\cdot2\cdot\overset{1}{\cancel{3}}}{\underset{1}{\cancel{2}}\cdot\underset{1}{\cancel{2}}\cdot\underset{1}{\cancel{3}}\cdot3\cdot5} = \dfrac{1\cdot1\cdot2\cdot2\cdot1}{1\cdot1\cdot1\cdot3\cdot5} = \dfrac{4}{15}$

Note: Example 8(b) can also be worked using cancellation first, which generally results in fewer shown steps. When multiplying fractions, if a number in a numerator has a factor in common with a number in a denominator, then this common factor can be cancelled from both numbers first before multiplying the fractions. Click here to see another way to solve Example 8(b).

Dividing Fractions

Before we divide fractions, let's review the definition of a *reciprocal* or *multiplicative inverse*.

Definition   Reciprocal or Multiplicative Inverse

Two numbers are **reciprocals** or **multiplicative inverses** if their product is 1.

For example, the reciprocal of $\dfrac{5}{8}$ is $\dfrac{8}{5}$ because $\dfrac{5}{8}\cdot\dfrac{8}{5} = 1$. Likewise, the reciprocal of 4 is $\dfrac{1}{4}$, and the reciprocal of $\dfrac{7}{10}$ is $\dfrac{10}{7}$ because $\dfrac{1}{4}\cdot4 = 1$ and $\dfrac{10}{7}\cdot\dfrac{7}{10} = 1,$

respectively. Recall that we can find the reciprocal of a fraction by switching its numerator and denominator.

To divide two fractions, multiply the first fraction by the reciprocal of the second fraction.

Dividing Two Fractions

If $a, b, c,$ and d are numbers and $b, c,$ and d are not equal to zero, then

$$\frac{a}{b} \div \frac{c}{d} = \frac{a}{b} \cdot \frac{d}{c} = \frac{a \cdot d}{b \cdot c}.$$

Example 9 Dividing Fractions

Divide and simplify.

a. $\dfrac{4}{9} \div \dfrac{12}{20}$

b. $\dfrac{2}{9} \div \dfrac{5}{12}$

Solution

a. $\dfrac{4}{9} \div \dfrac{12}{20} = \dfrac{4}{9} \cdot \dfrac{20}{12} = \dfrac{\overset{1}{\cancel{4}}}{9} \cdot \dfrac{20}{\underset{3}{\cancel{12}}} = \dfrac{1 \cdot 20}{9 \cdot 3} = \dfrac{20}{27}$

b. $\dfrac{2}{9} \div \dfrac{5}{12} = \dfrac{2}{9} \cdot \dfrac{12}{5} = \dfrac{2}{\underset{3}{\cancel{9}}} \cdot \dfrac{\overset{4}{\cancel{12}}}{5} = \dfrac{2 \cdot 4}{3 \cdot 5} = \dfrac{8}{15}$

Adding and Subtracting Fractions

Now let's add and subtract fractions. Consider $\dfrac{1}{2} + \dfrac{1}{3}$. Recall that fractions must be written with a common denominator before they can be added. This is illustrated in Figure 3.

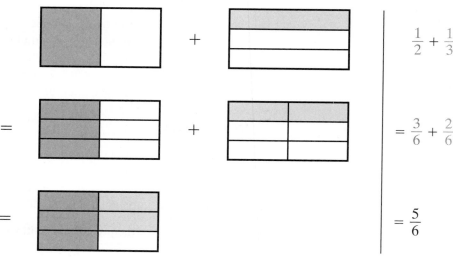

Figure 3

Typically we use the *least common denominator* of a group of fractions as a common denominator.

Definition   Least Common Denominator (LCD)

The **least common denominator (LCD)** of a group of fractions is the smallest number that is divisible by all the denominators in the group.

For example, the LCD of $\frac{1}{2}$ and $\frac{1}{3}$ is 6 because 6 is the smallest number that is divisible by both 2 and 3.

Sometimes the LCD is difficult to find, especially if the denominators are large numbers. The following steps will help:

Steps for Finding the LCD of a Group of Fractions

Step 1. Write out the prime factorization of each denominator in the group.

Step 2. Write down each distinct prime factor the greatest number of times it appears in any one factorization.

Step 3. Multiply the factors listed in step 2 to find the LCD.

Example 10 Finding a Least Common Denominator

Find the LCD of two fractions with denominators 90 and 120.

Solution   We use the steps for finding the LCD.

Step 1. We write out the prime factorization of each denominator:

$$90 = 2 \cdot 3 \cdot 3 \cdot 5$$
$$120 = 2 \cdot 2 \cdot 2 \cdot 3 \cdot 5$$

Step 2. There are three distinct prime factors: 2, 3, and 5. The greatest number of 2's in one factorization is three, so we include three 2's for the LCD. Likewise, the greatest number of 3's is two, and the greatest number of 5's is one, so we include two 3's and one 5 in the list for the LCD.

$$\text{LCD} = 2 \cdot 2 \cdot 2 \cdot 3 \cdot 3 \cdot 5$$

Step 3. Multiply. $\text{LCD} = 2 \cdot 2 \cdot 2 \cdot 3 \cdot 3 \cdot 5 = 360$

Now that we know how to find the LCD of two or more fractions, let's review how to add and subtract fractions.

Steps for Adding or Subtracting Fractions

Step 1. If necessary, find the LCD of the fractions.

Step 2. Write each fraction as an equivalent fraction with the LCD as its denominator.

Step 3. Add or subtract the numerators of the like fractions and write the result over the LCD.

Step 4. Simplify, if necessary.

Example 11 Adding and Subtracting Fractions

Add or subtract. Simplify if necessary.

a. $\dfrac{1}{6} + \dfrac{3}{4}$ **b.** $\dfrac{5}{8} - \dfrac{7}{12}$

Solution

a. The LCD is 12.

$$\text{Begin with the original problem:} \quad \frac{1}{6} + \frac{3}{4}$$

$$\text{Write each fraction with a denominator of LCD} = 12: \quad = \frac{2}{12} + \frac{9}{12}$$

$$\text{Add the numerators } 2 + 9 = 11; \text{ write the result over } 12: \quad = \frac{11}{12}$$

b. The LCD is 24.

$$\text{Begin with the original problem:} \quad \frac{5}{8} - \frac{7}{12}$$

$$\text{Write each fraction with a denominator of LCD} = 24: \quad = \frac{15}{24} - \frac{14}{24}$$

$$\text{Subtract the numerators } 15 - 14 = 1; \text{ write the result over } 24: \quad = \frac{1}{24}$$

Multiplying and Dividing Mixed Numbers

To multiply or divide two or more mixed numbers, we first convert each mixed number into an improper fraction.

Example 12 Multiplying and Dividing Mixed Numbers

Multiply or divide. Write the answer as a mixed number if possible.

a. $6\dfrac{1}{3} \cdot 4\dfrac{1}{2}$ **b.** $2\dfrac{3}{5} \div 3\dfrac{5}{7}$

Solution

a. Begin with the original problem: $6\dfrac{1}{3} \cdot 4\dfrac{1}{2}$

$$\text{Convert to improper fractions:} \quad = \frac{6 \cdot 3 + 1}{3} \cdot \frac{4 \cdot 2 + 1}{2} = \frac{19}{3} \cdot \frac{9}{2}$$

$$\text{Cancel and multiply:} \quad = \frac{19}{\overset{}{\underset{1}{\cancel{3}}}} \cdot \frac{\overset{3}{\cancel{9}}}{2} = \frac{19 \cdot 3}{1 \cdot 2} = \frac{57}{2}$$

$$\text{Convert back to mixed number:} \quad 28\frac{1}{2}$$

b. Begin with the original problem: $\quad 2\dfrac{3}{5} \div 3\dfrac{5}{7}$

Convert to improper fractions: $\quad = \dfrac{2 \cdot 5 + 3}{5} \div \dfrac{3 \cdot 7 + 5}{7} = \dfrac{13}{5} \div \dfrac{26}{7}$

Change to multiply by reciprocal: $\quad = \dfrac{13}{5} \cdot \dfrac{7}{26}$

Cancel and multiply: $\quad = \dfrac{\overset{1}{\cancel{13}}}{5} \cdot \dfrac{7}{\underset{2}{\cancel{26}}} = \dfrac{1 \cdot 7}{5 \cdot 2} = \dfrac{7}{10}$

Adding and Subtracting Mixed Numbers

To add two or more mixed numbers, first we can convert each mixed number into an improper fraction, just like we did to multiply and divide. However, because the equivalent improper fractions can involve very large numerators, it is usually easier to add the similar parts. This means that we can add the proper-fraction parts and add the whole-number parts.

Example 13 Adding Mixed Numbers

Add $1\dfrac{3}{8} + 4\dfrac{5}{12}$.

Solution It is usually easier to write the problem vertically rather than horizontally. We convert the proper-fraction parts to have common denominators. Then we add the proper-fraction parts and the whole-number parts. In this case, the LCD of the fraction parts is 24.

$$
\begin{aligned}
1\dfrac{3}{8} &= 1\dfrac{9}{24} \\
+4\dfrac{5}{12} &= +4\dfrac{10}{24} \\
\hline
&5\dfrac{19}{24}
\end{aligned}
$$

Whole-number parts ↓ · Proper-fraction parts

Sometimes when we add the proper-fraction parts of two mixed numbers, the result is an improper fraction. If this occurs, we must convert the improper-fraction part into a mixed number and "carry" the whole-number part of the result to the whole number parts of the original mixed numbers before finding the final sum.

This may sound confusing, so before we look at an example, consider the number $6\dfrac{8}{5}$.

This number is technically *not* a mixed number because the fraction part is not a proper fraction. However, we can convert this number into a mixed number as follows.

$$6\dfrac{8}{5} = 6 + \dfrac{8}{5} = 6 + 1\dfrac{3}{5} = 7\dfrac{3}{5}$$

Now we are ready to look at an example.

Example 14 Adding Mixed Numbers

Add $1\dfrac{5}{6} + 4\dfrac{7}{9}$.

Solution We write the problem vertically and convert the proper-fraction parts to have common denominators. Next, we add the fraction parts and the whole-number parts. In this case, the LCD of the fraction parts is 18.

$$
\begin{array}{rcl}
1\dfrac{5}{6} &=& 1\dfrac{15}{18}\\[2mm]
+4\dfrac{7}{9} &=& +4\dfrac{14}{18}\\[2mm]
\hline
&& 5\dfrac{29}{18}
\end{array}
$$

Since $5\dfrac{29}{18}$ is not simplified, we must convert it into a mixed number.

$$5\dfrac{29}{18} = 5 + \dfrac{29}{18} = 5 + 1\dfrac{11}{18} = 6\dfrac{11}{18}$$

So, $1\dfrac{5}{6} + 4\dfrac{7}{9} = 6\dfrac{11}{18}$.

Similar to adding, when subtracting two or more mixed numbers, we subtract the proper-fraction parts and subtract the whole-number parts. This is shown in the next example.

Example 15 Subtracting Mixed Numbers

Subtract $6\dfrac{3}{4} - 4\dfrac{2}{3}$.

Solution We write the problem vertically and convert the proper-fraction parts to have common denominators. Then we subtract the fraction parts and the whole-number parts. In this case, the LCD of the fraction parts is 12.

$$
\begin{array}{rcl}
6\dfrac{3}{4} &=& 6\dfrac{9}{12}\\[2mm]
-4\dfrac{2}{3} &=& -4\dfrac{8}{12}\\[2mm]
\hline
&& 2\dfrac{1}{12}
\end{array}
$$

So, $6\dfrac{3}{4} - 4\dfrac{2}{3} = 2\dfrac{1}{12}$.

In some cases, when we try to subtract the proper-fraction parts of two mixed numbers, the subtrahend may be larger than the minuend. If this occurs, we must borrow from the whole part.

Example 16 Subtracting Mixed Numbers

Add $7\dfrac{1}{5} - 4\dfrac{3}{8}$.

Solution First, we write the problem vertically and convert the proper-fraction parts to have common denominators. Then we subtract the proper-fraction parts and the whole-number parts. In this case, the LCD of the fraction parts is 40.

$$
\begin{aligned}
7\frac{1}{5} &= \quad 7\frac{8}{40} \\
-4\frac{3}{8} &= -4\frac{15}{40}
\end{aligned}
$$

When we try to subtract the fraction parts, we are faced with $\dfrac{8}{40} - \dfrac{15}{40}$, which we cannot do because 15 is larger than 8. So we must borrow.

$$7\frac{8}{40} = 7 + \frac{8}{40} = 6 + 1\frac{8}{40} = 6 + \frac{1\cdot 40 + 8}{40} = 6 + \frac{48}{40} = 6\frac{48}{40}.$$

We can subtract $\dfrac{48}{40} - \dfrac{15}{40}$, so the problem becomes

$$
\begin{aligned}
7\frac{1}{5} &= \quad 7\frac{8}{40} = \quad 6\frac{48}{40} \\
-4\frac{3}{8} &= -4\frac{15}{40} = -4\frac{15}{40} \\
& \qquad\qquad\qquad\qquad\quad 2\frac{33}{40}
\end{aligned}
$$

Therefore, $7\dfrac{1}{5} - 4\dfrac{3}{8} = 2\dfrac{33}{40}$.

C.2 Decimals

Decimal Notation

As with fractions, a **decimal number**, or simply a **decimal**, can be used to describe a part of a whole. The position of each digit in a decimal determines its place value. Figure 4 shows a **place-value chart**. In the chart, the digit 4 represents a value of four-tenths, or $\dfrac{4}{10}$. The digit 9 represents a value of nine-hundredths, or $\dfrac{9}{100}$.

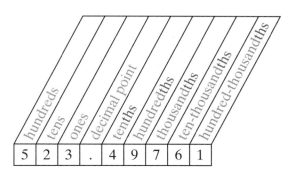

Figure 4

Place-Value Chart

Digits appearing in front of the decimal point form the **whole-number part** of the decimal number. Digits appearing behind the decimal point form the **decimal part** of the number. In the number 82.1356, for example, 82 is the whole number part and .1356 is the decimal part.

Whole-number part Decimal part

$$82 \, . \, 1356$$

Decimal point

Writing a Decimal Number in Words

We read or write decimals similar to the way we write whole numbers. For example, the whole number 1356 is written in words as *one thousand, three hundred fifty-six*. If these same four digits fall directly behind a decimal point, then we write the decimal number in the same way, but we follow it with the place value name of the digit in the last position. For the decimal number 0.1356, the last digit 6 has a "ten-thousandths" place value. So, 0.1356 is written as *one thousand three hundred fifty-six ten-thousandths.*

Writing a Decimal Number in Words

Step 1. Write the whole-number part in words.

Step 2. Write the word *and* for the decimal point.

Step 3. Write the decimal part as if it were a whole number and follow it with the place value name of the last digit.

The decimal number 82.1356 is written in words as *eighty-two and one thousand three hundred fifty-six ten-thousandths.*

82.1356

eighty-two and one thousand three hundred fifty-six ten-thousandths

Example 1 Writing Decimal Numbers in Words

Write each decimal number in words.

a. 5.23

b. 523.49761

Solution

a. The whole-number part is 5. The decimal part .23 has its last digit in the hundredths place. The number is *five and twenty-three hundredths*.

b. The whole-number part is 523. The last digit of the decimal part .49761 is in the hundred-thousandths place. The number is *five hundred twenty-three and forty-nine thousand, seven hundred sixty-one hundred-thousandths*.

Example 2 Writing Decimal Numbers in Words

Write each decimal number in words.

a. 0.407   **b.** 400.007

Solution

a. This number has no whole-number part. The last digit of the decimal part is in the thousandths place. The number is *four hundred seven thousandths*.

b. The whole-number part is 400. The last digit of the decimal part .007 is in the thousandths place. The number is *four hundred and seven thousandths*.

 There is a big difference between the values of 0.407 and 400.007 from Example 2, but when reading or writing these numbers, the only difference is the word *and*. Proper use of the word *and* is important!

Example 3 Converting from Word to Decimal Form

Write each number in decimal form.

a. *fifteen and seventy-eight hundredths*

b. *five hundred forty-one and two hundred seventeen thousandths*

Solution

a. The word *and* separates the whole-number part from the decimal part. The last digit of the decimal part is in the hundredths place. The whole-number part is 15 and the decimal part .78, so the decimal is 15.78.

Hundredths place

b. The last digit of the decimal part is in the thousandths place. The whole-number part is 541 and the decimal part is .217, so the decimal is 541.217.

Thousandths place

Example 4 Converting from Word to Decimal Form

Write each number in decimal form.

a. *three hundred and ninety-six thousandths*

b. *three hundred ninety-six thousandths*

Solution

a. The last digit of the decimal part is in the thousandths place. The word *and* separates the whole-number part 300 from the decimal part .096. The decimal is 300.096.

three hundred   and   ninety-six thousandths
3 0 0   ·   0 9 6

Thousandths place

b. The last digit of the decimal part is in the thousandths place. The word *and* does not appear in the written phrase, so there is no whole-number part. The decimal is 0.396.

Writing Decimals as Fractions or Mixed Numbers

If we know how to write decimals in words, then writing them as fractions or mixed numbers follows naturally. For example, we write the decimal 0.3 in words as *three tenths*, which is how we write the fraction $\frac{3}{10}$ in words. We write them the same way because they are equal in value: $0.3 = \frac{3}{10}$. Similarly, the decimal 5.371 and the mixed number $5\frac{371}{1000}$ are both written in words as *five and three hundred seventy-one thousandths* because $5.371 = 5\frac{371}{1000}$.

Example 5 Writing Decimals as Fractions or Mixed Numbers

Write each decimal as a fraction or mixed number. Be sure that your answer is written in simplest form.

a. 0.375 **b.** 7.8

Solution

a. In words, 0.375 is *three hundred seventy-five thousandths*. As a fraction, this is $\frac{375}{1000}$, which simplifies to $\frac{3}{8}$.

b. In words, 7.8 is *seven and eight tenths*. As a mixed number, this is $7\frac{8}{10}$, which simplifies to $7\frac{4}{5}$.

Rounding Decimals

We can approximate a decimal number by rounding it to a given place value.

Steps for Rounding a Decimal to the Right of the Decimal Point

Step 1. Find the digit to the right of the place being rounded to.

Step 2. If the digit from step 1 is greater than or equal to 5, then increase the digit in the given place value by 1 and drop all digits to its right. This is called **rounding up**.

Step 3. If the digit from step 1 is less than 5, then drop all digits to the right of the given place value. This is called **rounding down**.

Example 6 Rounding Decimals

a. Round 12.689 to the nearest tenth.

b. Round 7.256435 to the nearest thousandth.

Solution We follow the steps for rounding a decimal.

a. 12.689
       ↑ ↑
    Tenths Hundredths

The digit 6 is in the tenths place, where we want to round. The digit 8 is in the hundredths place, which is to the immediate right of the tenths place. Because 8 is larger than 5, we "round up" the tenths place from 6 to 7. So, 12.689 rounds to 12.7.

b. $7 . 2\, 5\, 6\, 4\, 3\, 5$

Thousandths   Ten-thousandths

The digit 6 is in the thousandths place, where we want to round. The digit 4 is in the ten-thousandths place, which is to the immediate right of the thousandths place. Because 4 is smaller than 5, we "round down" the thousandths place by keeping the 6 there and dropping all of the digits to its right. So, 7.256435 rounds to 7.256.

Adding and Subtracting Decimals

Addition and subtraction of decimals is similar to the addition and subtraction of whole numbers. We add or subtract the digits in corresponding place values, working from right to left and carrying or borrowing as necessary. To ensure that digits in corresponding place values are added or subtracted, we line up the decimal points vertically.

Steps for Adding or Subtracting Decimals

Step 1. Write the numbers vertically so that the decimal points line up.

Step 2. Add or subtract the corresponding place values, working from right to left. Carry or borrow as needed.

Step 3. Write the decimal point in the answer so that it lines up with the decimal points in the problem.

Note: If one number has more decimal places than the other, we can fill in the "missing" place values with zeros for placeholders.

Example 5   Adding and Subtracting Decimals

Add or subtract.

a. $26.81 + 5.467$ **b.** $26.81 - 5.467$

Solution We follow the steps for adding or subtracting decimals.

a.
$$
\begin{array}{r}
\overset{1\ 1}{} \quad \leftarrow \text{Carrying} \\
2\,6\,.\,8\,1\,\underset{}{0} \leftarrow \text{0 placeholder} \\
+\ \ 5\,.\,4\,6\,7 \\
\hline
3\,2\,.\,2\,7\,7 \\
\uparrow \\
\text{Align decimal points}
\end{array}
$$

b.
$$
\begin{array}{r}
\overset{7\ 10\ 10}{} \leftarrow \text{Borrowing} \\
2\,6\,.\,\cancel{8}\,\cancel{1}\,\cancel{0} \leftarrow \text{0 placeholder} \\
-\ \ \ 5\,.\,4\,6\,7 \\
\hline
2\,1\,.\,3\,4\,3 \\
\uparrow \\
\text{Align decimal points}
\end{array}
$$

Multiplying Decimals

We multiply decimals like we multiply whole numbers. The only difference is that we must write the decimal point in the proper place in the answer. Consider the following:

$$0.2 \cdot 0.14 = \frac{2}{10} \cdot \frac{14}{100} = \frac{2 \cdot 14}{10 \cdot 100} = \frac{28}{1000} = 0.028.$$

In the factors of the original problem, notice that there are a total of three digits to the right of decimal points. The final product also has three digits to the right of the decimal point.

$$0.2 \quad \cdot \quad 0.14 \quad = \quad 0.028$$

↑   ↑   ↑

1 decimal   2 decimal   3 decimal
place   places   places

The total number of digits to the right of the decimal points in the factors will equal the number of digits to the right of the decimal point in the product. We can use this pattern to simplify the multiplication process.

Steps for Multiplying Decimals

Step 1. Multiply the numbers as though no decimal points were present.

Step 2. Place the decimal point in the product so that the number of digits to its right is the same as the total number of digits to the right of the decimal points in the factors.

Example 6   Multiplying Decimals

Multiply.

a. $2.3 \cdot 1.7$ **b.** $1.75 \cdot 0.05$

Solution We follow the steps for multiplying decimals.

a. The factors have a total of two digits behind the decimal points, so the product will have two digits behind its decimal point.

$$
\begin{array}{r}
2.3 \leftarrow \text{1 decimal place} \\
\times\, 1.7 \leftarrow \text{1 decimal place} \\
\hline
1\,6\,1 \\
2\,3\,0 \\
\hline
3.9\,1 \leftarrow \text{2 decimal places}
\end{array}
$$

b. The factors have a total of three digits behind the decimal points, so the product will have three digits behind its decimal point.

$$
\begin{array}{r}
1.7\,5 \leftarrow \text{2 decimal places} \\
\times\quad 0.5 \leftarrow \text{1 decimal place} \\
\hline
0.8\,7\,5 \leftarrow \text{3 decimal places}
\end{array}
$$

Dividing Decimals

Dividing decimals is similar to dividing whole numbers. However, we must write the decimal point in the proper place as we did when multiplying decimals. Consider the following:

$$0.158 \div 0.2 = \frac{158}{1000} \div \frac{2}{10} = \frac{158}{1000} \cdot \frac{10}{2} = \frac{\overset{79}{\cancel{158}}}{\underset{100}{\cancel{1000}}} \cdot \frac{\overset{1}{\cancel{10}}}{\underset{1}{\cancel{2}}} = \frac{79 \cdot 1}{100 \cdot 1} = \frac{79}{100} = 0.79.$$

Steps for Dividing Decimals

Step 1. Move the decimal point in the divisor to the right until it is behind the last digit.

Step 2. Move the decimal point in the dividend to the right the same number of places as the decimal point was moved in step 1. Add zeros as placeholders if necessary.

Step 3. Divide. Place the decimal point in the quotient directly above the location of the moved decimal point in the dividend.

Note: The quotient will either be a terminating or repeating decimal.

Example 7 Dividing Decimals

Divide.

a. $4.325 \div 2.5$ **b.** $7.5 \div 0.33$

Solution We follow the steps for dividing decimals.

a. The decimal points in the divisor and dividend must be moved one place to the right.

$$
2.5\overline{)4.325} \quad \Rightarrow \quad
\begin{array}{r}
1.73 \\
25.\overline{)43.25} \\
\underline{25} \\
182 \\
\underline{175} \\
75 \\
\underline{75} \\
0
\end{array}
$$

So, $4.325 \div 2.5 = 1.73$.

b. The decimal points in the divisor and dividend must be moved two places to the right. We can place zeros behind the decimal point to continue dividing.

$$
0.33\overline{)7.50} \quad \Rightarrow \quad
\begin{array}{r}
22.7272.... \\
33.\overline{)750.0000} \\
\underline{66} \\
90 \\
\underline{66} \\
240 \\
\underline{231} \\
90 \\
\underline{66} \\
240 \\
\underline{231} \\
90 \\
\underline{66} \\
24
\end{array}
$$

The pattern "72" keeps repeating, so $7.5 \div 0.33 = 22.\overline{72}$.

Writing Fractions or Mixed Numbers as Decimals

To write a fraction or mixed number as a decimal, we follow this procedure:

Writing Fractions or Mixed Numbers as Decimals

To write a fraction as a decimal, divide the numerator by the denominator.

To write a mixed number as a decimal, divide the numerator of the proper-fraction part by the denominator, and add the whole-number part to the decimal result.

Note: The quotient will either be a terminating or repeating decimal.

Example 8 Writing Fractions or Mixed Numbers as Decimals

Write each fraction or mixed number as a decimal.

a. $\dfrac{5}{8}$ **b.** $\dfrac{3}{11}$ **c.** $13\dfrac{9}{25}$ **d.** $4\dfrac{2}{3}$

Solution For parts a and b, we divide the numerator by the denominator. Note that decimal points are understood to be behind the last digit of each whole number. We can place zeros behind the decimal point to continue dividing.

a.
$$\begin{array}{r} 0.625 \\ 8\overline{)5.000} \\ \underline{48} \\ 20 \\ \underline{16} \\ 40 \\ \underline{40} \\ 0 \end{array}$$

So, $\dfrac{5}{8} = 0.625$.

b.
$$\begin{array}{r} 0.2727... \\ 11\overline{)3.0000} \\ \underline{22} \\ 80 \\ \underline{77} \\ 30 \\ \underline{22} \\ 80 \\ \underline{77} \\ 3 \end{array}$$

The pattern "27" keeps repeating, so $\dfrac{3}{11} = 0.\overline{27}$.

For parts c and d, we divide the numerator of the proper-fraction part by the denominator and add the result to the whole-number part.

c.
$$\begin{array}{r} 0.36 \\ 25\overline{)9.00} \\ \underline{75} \\ 150 \\ \underline{150} \\ 0 \end{array}$$

So, $\dfrac{9}{25} = 0.36$. Adding the whole-number part 13 to this decimal result, we get

$13\dfrac{9}{25} = 13 + 0.36 = 13.36$.

d.
$$\begin{array}{r} 0.666... \\ 3\overline{)2.000} \\ \underline{18} \\ 20 \\ \underline{18} \\ 20 \\ \underline{18} \\ 2 \end{array}$$

The pattern "6" keeps repeating, so $\dfrac{2}{3} = 0.\overline{6}$. Adding the whole-number part 4 to this decimal result, we get $4\dfrac{2}{3} = 4 + 0.\overline{6} = 4.\overline{6}$.

C.3 Proportions

Ratios

If there are 15 men and 20 women in an algebra class, then we can say that the *ratio* of men to women is 15 to 20. A **ratio** is a comparison of two quantities. There are three common notations for expressing ratios. One uses the word *to* as we just saw, but we can also use *fraction notation* or *colon notation*. So, we can write the ratio of 15 to 20 as

$$15 \text{ to } 20 \qquad \text{or} \qquad \frac{15}{20} \qquad \text{or} \qquad 15:20$$

Fraction notation   Colon notation

The order of the quantities in a ratio is important. For example, the ratio of 15 to 20 is $15:20$ or $\frac{15}{20}$, *not* ~~20:15~~ and *not* ~~$\frac{20}{15}$~~.

Ratios behave like fractions, which is why we can use fraction notation. For example, the ratio of 15 to 20 is equivalent to the ratio of 3 to 4. We can see this using fraction notation and writing it in simplest form: $\dfrac{15}{20} = \dfrac{3 \cdot 5}{4 \cdot 5} = \dfrac{3 \cdot \cancel{5}}{4 \cdot \cancel{5}} = \dfrac{3}{4}$.

Simplifying a Ratio

To **simplify a ratio**, write the ratio in fraction notation and then in simplest form by cancelling all common factors and common units.

Example 1 Writing and Simplifying Ratios

a. A widescreen television has a length of 48 inches and a width of 27 inches. Write the ratio of length to width as a fraction in simplest form.

b. A recipe calls for $2\frac{1}{2}$ cups of milk and $\frac{3}{4}$ of a cup of sugar. Write the ratio of milk to sugar in simplest form.

Solution

a. $\dfrac{48 \text{ inches}}{27 \text{ inches}} = \dfrac{16 \cdot \cancel{3 \text{ inches}}}{9 \cdot \cancel{3 \text{ inches}}} = \dfrac{16}{9}$

b. $\dfrac{2\frac{1}{2} \text{ cups}}{\frac{3}{4} \text{ cups}} = \dfrac{2\frac{1}{2} \cancel{\text{cups}}}{\frac{3}{4} \cancel{\text{cups}}} = \dfrac{\frac{5}{2}}{\frac{3}{4}} = \dfrac{5}{2} \div \dfrac{3}{4} = \dfrac{5}{2} \cdot \dfrac{4}{3} = \dfrac{5}{\cancel{2}_1} \cdot \dfrac{\cancel{4}^2}{3} = \dfrac{10}{3}$

Rates

Rates are special kinds of ratios in which quantities of different types are compared. For example, consider a runner who ran 3 miles in 24 minutes. Writing this rate as a fraction, we have

$$\frac{3 \text{ miles}}{24 \text{ minutes}} \quad \text{which simplifies to} \quad \frac{1 \text{ mile}}{8 \text{ minutes}}.$$

Note that the units do not cancel.

A **unit rate** is a rate that has a denominator of 1 unit. A common unit rate is "miles per hour." For example, the speed of 65 miles per hour can be written as $\dfrac{65 \text{ miles}}{1 \text{ hour}}$.

Converting Rates to Unit Rates

To convert a rate to a unit rate, divide the quantity in the numerator by the quantity in the denominator. The units can be written as "<units of numerator> per <units of denominator>." For example, "miles per hour" are the units for the unit rate of $\dfrac{65 \text{ miles}}{1 \text{ hour}}$.

Example 2 Converting Rates to Unit Rates

a. A jet travels 900 miles in 2 hours. What is the unit rate in miles per hour?

b. A 50-ounce bottle of detergent costs $7.59. What is the unit price in $/oz?

Solution

a. $\dfrac{900 \text{ miles}}{2 \text{ hours}} = \dfrac{900 \div 2 \text{ miles}}{2 \div 2 \text{ hours}} = \dfrac{450 \text{ miles}}{1 \text{ hour}} = 450 \text{ miles per hour}$

b. $\dfrac{\$7.59}{50 \text{ ounces}} = \dfrac{\$7.59 \div 50}{50 \div 50 \text{ ounces}} = \dfrac{\$0.1518}{1 \text{ ounce}} = \$0.1518 \text{ per ounce}$

Proportions

Definition   Proportion

A **proportion** is a statement that two ratios are equal, such as $\dfrac{P}{Q} = \dfrac{R}{S}$.

The proportion $\dfrac{15}{20} = \dfrac{6}{8}$ is true because both ratios simplify to $\dfrac{3}{4}$. Notice that $8 \cdot 15 = 120$ and $20 \cdot 6 = 120$. The products $8 \cdot 15$ and $20 \cdot 6$ are called the **cross products** of this proportion. When a proportion is true, the cross products are equal. Consider this cross product visually as follows:

$$\dfrac{15}{20} \diagdown\!\!\!\!\!\diagup \dfrac{6}{8}$$

$$8 \cdot 15 = 20 \cdot 6$$

We can use **cross-multiplication** to find an unknown quantity in a proportion.

Example 3 Solving Proportions

Cross-multiply to solve each proportion.

a. $\dfrac{x}{11} = \dfrac{7.1}{2}$

b. $\dfrac{y}{3\frac{1}{2}} = \dfrac{5\frac{1}{5}}{2\frac{1}{3}}$

Solution

a. Begin with the original proportion: $\dfrac{x}{11} = \dfrac{7.1}{2}$

Cross-multiply: $2x = 11(7.1)$

Simplify: $2x = 78.1$

Divide both sides by 2: $\dfrac{2x}{2} = \dfrac{78.1}{2}$

Simplify: $x = 39.05$

b.

Begin with the original proportion: $\dfrac{y}{3\frac{1}{2}} = \dfrac{5\frac{1}{5}}{2\frac{1}{3}}$

Cross-multiply: $\left(2\frac{1}{3}\right)y = \left(3\frac{1}{2}\right)\left(5\frac{1}{5}\right)$

Change the mixed numbers to improper fractions: $\dfrac{7}{3}y = \left(\dfrac{7}{2}\right)\left(\dfrac{26}{5}\right)$

Simplify: $\dfrac{7}{3}y = \dfrac{91}{5}$

Multiply both sides by $\dfrac{3}{7}$: $\dfrac{3}{7}\left(\dfrac{7}{3}y\right) = \dfrac{3}{7}\left(\dfrac{91}{5}\right)$

Simplify: $y = \dfrac{39}{5} = 7\dfrac{4}{5}$

Writing proportions is a powerful tool in solving application problems. Given a ratio (or rate) of two quantities, a proportion can be used to determine an unknown quantity.

Example 4 Using Proportions to Solve Application Problems

A quality-control inspector examined a sample of 200 lightbulbs and found 18 of them to be defective. At this ratio, how many defective bulbs can the inspector expect in a shipment of 22,000 lightbulbs?

Solution Let x represent the number of defective bulbs expected in the shipment. Since we know the ratio of defective bulbs, we can write a proportion by setting the ratio of defective bulbs to total bulbs in the shipment equal to the ratio of defective bulbs to total bulbs in the sample.

$$\underset{\text{Shipment}}{\underbrace{\dfrac{x \text{ defective bulbs}}{22{,}000 \text{ total bulbs}}}} = \underset{\text{Sample}}{\underbrace{\dfrac{18 \text{ defective bulbs}}{200 \text{ total bulbs}}}}$$

Cross-multiply: $200x = 22{,}000(18)$

Simplify: $200x = 396{,}000$

Divide both sides by 200: $\dfrac{200x}{200} = \dfrac{396{,}000}{200}$

Simplify: $x = 1980$

The inspector can expect 1980 defective bulbs in a shipment of 22,000 lightbulbs.

Example 5 Using Proportions to Solve Application Problems

If 1 cup of skim milk contains 80 calories, how many calories are in an entire gallon of skim milk? Note that 1 gallon = 16 cups.

Solution Let x represent the number of calories in 1 gallon, or 16 cups, of milk. Since the rate of calories per cup is constant, we write a proportion.

$$\overset{\text{1 gallon}}{\underset{\downarrow}{}} \qquad \overset{\text{1 cup}}{\underset{\downarrow}{}}$$

$$\frac{x \text{ calories}}{16 \text{ cups}} = \frac{80 \text{ calories}}{1 \text{ cup}}$$

Cross-multiply: $\qquad 1x = 16(80)$

Simplify: $\qquad\qquad x = 1280$

There are 1280 calories in a gallon of skim milk.

Similar Triangles

Similar triangles have the same shape but not necessarily the same size. For similar triangles, corresponding angles are equal, and corresponding sides have lengths that are proportional. So, we can use proportions to find unknown lengths in similar triangles.

Example 6 Solving a Similar Triangle Problem

Find the unknown length n for the following similar triangles.

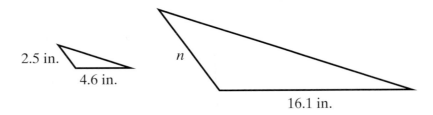

Solution Since the triangles are similar, their corresponding sides are proportional. Side n on the larger triangle corresponds to the 2.5-inch side on the smaller triangle. The 16.1-inch side on the larger triangle corresponds to the 4.6-inch side on the smaller triangle. So, we have the following proportion:

$$\frac{n \text{ inches}}{2.5 \text{ inches}} = \frac{16.1 \text{ inches}}{4.6 \text{ inches}}$$

Cross-multiply: $\qquad 4.6n = 2.5(16.1)$

Simplify: $\qquad\qquad 4.6n = 40.25$

Divide both sides by 4.6: $\qquad \dfrac{4.6n}{4.6} = \dfrac{40.25}{4.6}$

Simplify: $\qquad\qquad\qquad n = 8.75$

The unknown side length n is 8.75 inches long.

Example 7 Using Proportions to Solve Application Problems

A forest ranger wants to determine the height of a tree. She measures the tree's shadow as 84 feet long. Her own shadow at the same time is 7.5 feet. If she is 5.5 feet tall, how tall is the tree?

5.5 ft

84 ft

7.5 ft

Solution Since the shadows were measured at the same time, the ratio of the height-to- shadow length of the tree is proportional to the ratio of height-to-shadow length of the forest ranger. So, we have the following proportion:

$$\frac{x \text{ feet}}{84 \text{ feet}} = \frac{5.5 \text{ feet}}{7.5 \text{ feet}}$$

Cross-multiply: $7.5x = 84(5.5)$

Simplify: $7.5x = 462$

Divide both sides by 7.5: $\dfrac{7.5x}{7.5} = \dfrac{462}{7.5}$

Simplify: $x = 61.6$

The tree is 61.6 feet tall.

C.4 Percents

Converting Between Percents and Decimals

The word **percent** means "per hundred" or "out of 100." A percent is denoted by the **percent symbol** %. If you score an 86% on a test, what does it mean? It means that you receive 86 points out of every 100 points possible on the test.

We can write percents as ratios. For example, 86% can be written as the ratio 86:100 or $\dfrac{86}{100}$. Since $\dfrac{86}{100} = 0.86$, we can also say that 86% = 0.86. In fact, any percent can be written as a fraction or a decimal.

Since $100\% = \dfrac{100}{100} = 1$, multiplying or dividing by 100% is equivalent to multiplying or dividing by 1. So, a value does not change when it is multiplied or divided by 100%.

Writing a Percent as a Decimal

To write a percent as a decimal, divide the percent by 100% and write the answer in decimal form.

Note: This is the same as dropping the percent symbol (%) and moving the decimal point two places to the left.

Example 1 Writing Percents as Decimals

Write each percent as a decimal.

a. 72.4% **b.** 126% **c.** 1.5%

Solution

a. $72.4\% = \overbrace{\dfrac{72.4\%}{100\%}}^{\text{Divide by 100\%}} = \dfrac{72.4\%}{100\%} = \dfrac{72.4}{100} = 0.724$

b. $126\% = \overbrace{\dfrac{126\%}{100\%}}^{\text{Divide by 100\%}} = \dfrac{126\%}{100\%} = \dfrac{126}{100} = 1.26$

c. $1.5\% = \overbrace{\dfrac{1.5\%}{100\%}}^{\text{Divide by 100\%}} = \dfrac{1.5\%}{100\%} = \dfrac{1.5}{100} = 0.015$

Writing a Decimal as a Percent

To write a decimal as a percent, multiply the decimal by 100%.

Note: This is the same as moving the decimal point two places to the right and including the percent symbol (%).

Example 2 Writing Decimals as Percents

Write each decimal as a percent.

a. 0.891 **b.** 2.47 **c.** 0.0678

Solution

a. $0.891 = \overbrace{0.891(100\%)}^{\text{Multiply by 100\%}} = (0.891 \cdot 100)\% = 89.1\%$

b. $2.47 = \overbrace{2.47(100\%)}^{\text{Multiply by 100\%}} = (2.47 \cdot 100)\% = 247\%$

c. $0.0678 = \overbrace{0.0678(100\%)}^{\text{Multiply by 100\%}} = (0.0678 \cdot 100)\% = 6.78\%$

A decimal larger than 1 will yield a percent larger than 100%. Likewise, a percent larger than 100% will yield a decimal larger than 1.

Converting between Percents and Fractions

Writing a Percent as a Fraction or Mixed Number

To write a percent as a fraction or mixed number, divide the percent by 100% and write the answer as a simplified fraction or mixed number.

Note: This is the same as dropping the percent symbol (%), writing the number over 100, and simplifying the fraction.

Example 3 Writing Percents as Fractions or Mixed Numbers

Write each percent as a fraction or mixed number.

a. 154% **b.** 10.8% **c.** $15\dfrac{2}{3}\%$

Solution

a. $154\% = \dfrac{\overbrace{154\%}^{\text{Divide by 100\%}}}{100\%} = \dfrac{154\%}{100\%} = \dfrac{154}{100} = \dfrac{154 \div 2}{100 \div 2} = \dfrac{77}{50} = 1\dfrac{27}{50}$

b. $10.8\% = \dfrac{\overbrace{10.8\%}^{\text{Divide by 100\%}}}{100\%} = \dfrac{10.8\%}{100\%} = \dfrac{10.8}{100} = \dfrac{10.8 \cdot 10}{100 \cdot 10} = \dfrac{108}{1000} = \dfrac{108 \div 4}{1000 \div 4} = \dfrac{27}{250}$

c. $15\dfrac{2}{3}\% = \dfrac{\overbrace{15\frac{2}{3}\%}^{\text{Divide by 100\%}}}{100\%} = \dfrac{15\frac{2}{3}\%}{100\%} = \dfrac{\frac{47}{3}}{100} = \dfrac{47}{3} \cdot \dfrac{1}{100} = \dfrac{47}{300}$

Writing a Fraction as a Percent

To write a fraction or mixed number as a percent, multiply it by 100% and simplify.

Note: This is the same as multiplying by 100 and including the percent symbol (%).

Example 4 Writing Fractions or Mixed Numbers as Percents

Write each fraction or mixed number as a percent.

a. $\dfrac{3}{5}$ **b.** $5\dfrac{3}{8}$ **c.** $\dfrac{7}{9}$

Solution

a. $\dfrac{3}{5} = \overbrace{\dfrac{3}{5}(100\%)}^{\text{Multiply by 100\%}} = \left(\dfrac{3}{5} \cdot \dfrac{100}{1}\right)\% = \left(\dfrac{3}{\overset{}{\underset{1}{5}}} \cdot \dfrac{\overset{20}{\cancel{100}}}{1}\right)\% = 60\%$

b. $5\dfrac{3}{8} = \overbrace{5\dfrac{3}{8}(100\%)}^{\text{Multiply by 100\%}} = \left(\dfrac{43}{8} \cdot 100\right)\% = \left(\dfrac{43}{\underset{2}{8}} \cdot \overset{25}{\cancel{100}}\right)\% = \dfrac{1075}{2}\% = 537\dfrac{1}{2}\%$

c. $\frac{7}{9} = \overbrace{\frac{7}{9}(100\%)}^{\text{Multiply by 100\%}} = \left(\frac{7}{9} \cdot 100\right)\% = \left(\frac{700}{9}\right)\% = 77\frac{7}{9}\%$

In Example 4, we found exact percent representations of fractions. Often it is more practical to find approximations instead. For example, $77\frac{7}{9}\% = 77.7777\ldots\% \approx 77.78\%$.

Example 5 Writing Fractions as Approximate Percents

Write $\frac{1}{12}$ as a percent rounded to the nearest hundredth of a percent.

Solution

$$\frac{1}{12} = \overbrace{\frac{1}{12}(100\%)}^{\text{Multiply by 100\%}} = \left(\frac{1}{12} \cdot 100\right)\% = 8.3333\ldots\% \approx 8.33\%$$

Using Equations to Solve Percent Problems

Recall from Section R.3 that the word *of* indicates multiplication. For example, "three-fifths of forty is twenty-four" means

$$\frac{3}{5} \quad \text{of} \quad 40 \quad \text{is} \quad 24.$$
$$\frac{3}{5} \quad \cdot \quad 40 \quad = \quad 24$$

In Example 4(a), we found that $\frac{3}{5} = 60\%$. So, the statement "three-fifths of forty is twenty-four" is equivalent to the statement "sixty percent of forty is twenty-four."

$$60\% \quad \text{of} \quad 40 \quad \text{is} \quad 24.$$
$$60\% \quad \cdot \quad 40 \quad = \quad 24$$

When we solve percent problems, the word *of* means multiplication, the word *is* means equals, and the words *what* or *what number* refer to the unknown number. We can use equations that include percents, called **percent equations**, to solve percent problems.

Example 6 Translating a Statement into a Percent Equation

Translate each statement into a percent equation. Let x be the number.

a. 18 is 30% of what number?

b. 123% of 62 is what number?

c. 8.2 is what percent of 12.5?

Solution

a.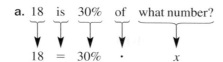

18   is   30%   of   what number?

$$18 \;=\; 30\% \;\cdot\; x$$

b. 123%   of   62   is   what number?

$$123\% \;\cdot\; 62 \;=\; x$$

c. 8.2   is   what percent   of   12.5?

$$8.2 \;=\; x \;\cdot\; 12.5$$

Each problem in Example 6 contains three quantities. We name each of the quantities in order to develop a **general equation for percents**. Let's review the statement "60% of 40 is 24." In this statement, 60% is the **percent**, 40 is the **base**, and 24 is the **amount**.

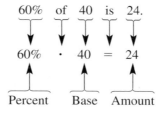

60%   of   40   is   24.

$$60\% \;\cdot\; 40 \;=\; 24$$

Percent   Base   Amount

General Equation for Percents

$$\text{Percent} \cdot \text{Base} = \text{Amount}$$

When we translate percent problems into equations, the percent can be identified by looking for the word *percent* or the % symbol. The base represents one whole and will usually follow the word *of*. The amount is the part compared to the whole and is normally isolated from the percent and the base on one side of the word is.

When using the general equation for percents, the percent must be written as a decimal or fraction in order to solve the problem.

Example 7 Using Equations to Solve Percent Problems

Use equations to solve each percent problem from Example 6.

a. 18 is 30% of what number?

b. 123% of 62 is what number?

c. 8.2 is what percent of 12.5?

Solution

a. Write the equation from Example 6: $18 = 30\% \cdot x$

          Convert 30% to a decimal: $18 = 0.30 \cdot x$

          Divide both sides by 0.30: $\dfrac{18}{0.30} = \dfrac{0.30 \cdot x}{0.30}$

                     Simplify: $60 = x$

So, 18 is 30% of 60.

b. Rewrite the equation from Example 6: $123\% \cdot 62 = x$

Convert 123% to a decimal: $1.23 \cdot 62 = x$

Multiply: $76.26 = x$

123% of 62 is 76.26.

c.   Rewrite the equation from Example 6: $8.2 = x \cdot 12.5$

Divide both sides of the equation by 12.5: $\dfrac{8.2}{12.5} = \dfrac{x \cdot 12.5}{12.5}$

Simplify: $0.656 = x$

Convert 0.656 to a percent: $65.6\% = x$

8.2 is 65.6% of 12.5.

Using Proportions to Solve Percent Problems

If the general equation for percents is true, we can say

$$\text{Percent} = \frac{\text{Amount}}{\text{Base}}.$$

If we then write the percent as a ratio over 100, we obtain the **general proportion for percents**.

General Proportion for Percents

$$\frac{\text{Amount}}{\text{Base}} = \frac{\text{Percent}}{100}$$

Note: When we write the percent over 100%, the % symbols cancel out. This is why the % symbols are left out of the general proportion for percents.

Example 8 Using Proportions to Solve Percent Problems

Use proportions to solve each percent problem.

a. 9% of 3250 is what number?

b. What percent of 47.32 is 106.47?

c. 74% of what number is 37?

Solution

a. In this case 9 is the percent, 3250 is the base, and we are looking for the amount. Let x represent the unknown number. The proportion is

$$\text{Amount} \rightarrow \frac{x}{3250} = \frac{9}{100} \leftarrow \text{Percent}$$
$$\text{Base} \rightarrow$$

Cross-multiply: $100x = 3250(9)$

Simplify: $100x = 29{,}250$

Divide both sides by 100: $\dfrac{100x}{100} = \dfrac{29{,}250}{100}$

Simplify: $x = 292.5$

9% of 3250 is 292.5.

b. Here we are looking for the percent. 47.32 is the base and 106.47 is the amount. Let x represent the percent. The proportion is

$$\text{Amount} \rightarrow \frac{106.47}{47.32} = \frac{x}{100} \leftarrow \text{Percent}$$
$$\text{Base} \rightarrow$$

Cross-multiply: $100(106.47) = 47.32x$

Simplify: $10{,}647 = 47.32x$

Divide both sides by 47.32: $\dfrac{10{,}647}{47.32} = \dfrac{47.32x}{47.32}$

Simplify: $225 = x$

225% of 47.32 is 106.47.

c. 74 is the percent, 37 is the amount, and we are looking for the base. The proportion is

$$\text{Amount} \rightarrow \frac{37}{x} = \frac{74}{100} \leftarrow \text{Percent}$$
$$\text{Base} \rightarrow$$

Cross-multiply: $100(37) = x(74)$

Simplify: $3700 = 74x$

Divide both sides by 74: $\dfrac{3700}{74} = \dfrac{74x}{74}$

Simplify: $50 = x$

74% of 50 is 37.

When using the general proportion for percents to find an unknown percent, be sure to label the answer with the % symbol and do not move the decimal point.

Percent Applications

Now we can apply what we have learned about percents to real settings. We can solve percent problems in two ways—using the general equation for percents or using the general proportion for percents. The key to solving percent applications is to identify the percent, the amount, and the base for the problem.

Example 9 Finding the Percent of Field Goals Made

A basketball team scored 45 of 80 attempted field goals during one game. What percent of the field goal attempts were made?

Solution We want to find the percent of field goals made. The base is the 80 attempted field goals, and the amount is the 45 field goals that were actually made. Both methods are illustrated:

Using the general equation for percents:

$$\underbrace{x}_{\text{Percent}} \cdot \underbrace{80}_{\text{Base}} = \underbrace{45}_{\text{Amount}}$$

Divide by 80: $x = \dfrac{45}{80}$

Simplify: $x = 0.5625$

Convert 0.5625 to a percent: $x = 56.25\%$

Using the general proportion for percents:

$$\text{Amount} \rightarrow \frac{45}{80} = \frac{x}{100} \leftarrow \text{Percent}$$
$$\text{Base} \rightarrow$$

Cross-multiply: $100(45) = 80x$

Simplify: $4500 = 80x$

Divide by 80: $\dfrac{4500}{80} = \dfrac{80x}{80}$

Simplify: $56.25 = x$

56.25% of the field goal attempts were made.

In Examples 9–14, both methods will be shown for solving the percent problem. However, when solving percent problems on your own, practice both methods and then choose the method that you prefer.

Example 10 Finding the Percent of Questions Answered Correctly

Deon took a history exam that consisted of 75 multiple-choice questions. He answered 9 of the questions incorrectly. What percent of the questions did Deon answer correctly?

Solution We want to find the percent of questions answered correctly. The base is 75 questions, and the amount is the $75 - 9 = 66$ questions answered correctly. Both methods are illustrated.

Using the general equation for percents:

Percent   Base   Amount

$$x \cdot 75 = 66$$

Divide by 75: $x = \dfrac{66}{75}$

Simplify: $x = 0.88$

Convert 0.88 to a percent: $x = 88\%$

Using the general proportion for percents:

Amount $\rightarrow \dfrac{66}{75} = \dfrac{x}{100} \leftarrow$ Percent
Base $\rightarrow$

Cross-multiply: $100(66) = 75x$

Simplify: $6600 = 75x$

Divide by 75: $\dfrac{6600}{75} = \dfrac{75x}{75}$

Simplify: $88 = x$

Deon answered 88% of the questions correctly.

Note: We could also solve Example 10 by first finding the percent of questions that Deon answered incorrectly. We could then subtract that percent from 100% to obtain the percent of questions answered correctly.

Example 11 Finding the Percent of a Fundraising Goal

At the beginning of a scholarship campaign, the college foundation set a goal to raise $25,000 in donations from businesses. By the end of the campaign, it had actually raised $32,800. What percent of the goal was actually raised?

Solution We need to find the percent of the goal that was actually raised. The base is the $25,000 goal, and the amount is the $32,800 that was actually raised. Both methods are illustrated.

Using the general equation for percents:

Percent   Base   Amount

$$x \cdot 25{,}000 = 32{,}800$$

Divide by 25,000: $x = \dfrac{32{,}800}{25{,}000}$

Simplify: $x = 1.312$

Convert 1.312 to a percent: $x = 131.2\%$

Using the general proportion for percents:

Amount $\rightarrow \dfrac{32{,}800}{25{,}000} = \dfrac{x}{100} \leftarrow$ Percent
Base $\rightarrow$

Cross-multiply: $100(32{,}800) = 25{,}000x$

Simplify: $3{,}280{,}000 = 25{,}000x$

Divide by 25,000: $\dfrac{3{,}280{,}000}{25{,}000} = \dfrac{25{,}000x}{25{,}000}$

Simplify: $131.2 = x$

The college foundation raised 131.2% of its goal.

Example 12 Finding the Number of Drivers Wearing Seatbelts

An observational study at a certain intersection found that of 650 drivers, 78% were wearing seatbelts. How many of the drivers were wearing seatbelts?

Solution We are looking for the amount of drivers who were wearing seatbelts. The percent is 78%, and the base is 650 drivers. Both methods are illustrated.

Using the general equation for percents:

Percent	Base	Amount

$$78\% \;\cdot\; 650 \;= x$$

Convert 78% to a decimal: $0.78 \cdot 650 = x$

Multiply: $507 = x$

Using the general proportion for percents:

Amount → $\dfrac{x}{650} = \dfrac{78}{100}$ ← Percent
Base →

Cross-multiply: $100x = 650(78)$

Simplify: $100x = 50{,}700$

Divide by 100: $\dfrac{100x}{100} = \dfrac{50{,}700}{100}$

Simplify: $x = 507$

Of the 650 drivers, 507 were wearing seatbelts.

Example 13 Calculating Withholdings and Take-Home Pay

Ben earns $682.73 per week at his job, but 32% of his pay is withheld for federal taxes, state taxes, social security, and so on.

a. How much money is withheld from Ben's weekly paycheck? Round to the nearest cent.

b. What is Ben's take home pay? Round to the nearest cent.

Solution

a. For this problem, we need to find the amount of money that is withheld from Ben's weekly paycheck. The percent is 32%, and the base is the $682.73 gross. Both methods are shown.

Using the general equation for percents:

Percent	Base	Amount

$$32\% \;\cdot\; 682.73 \;= x$$

Convert 32% to a decimal: $0.32 \cdot 682.73 = x$

Multiply: $218.4736 = x$

Round to hundredths: $218.47 \approx x$

Using the general proportion for percents:

Amount → $\dfrac{x}{682.73} = \dfrac{32}{100}$ ← Percent
Base →

Cross-multiply: $100x = 682.73(32)$

Simplify: $100x = 21{,}847.36$

Divide by 100: $\dfrac{100x}{100} = \dfrac{21{,}847.36}{100}$

Simplify: $x = 218.4736$

Round: $x \approx 218.47$

The amount withheld from Ben's weekly paycheck is $218.47.

b. Ben's take-home pay is $682.73 − $218.47 = $464.26.

Example 14 Calculating the Number of Calories in a Frozen Dinner

A frozen dinner advertises that 16% of its calories are from fat. If the dinner contains 65 calories from fat, what is the total number of calories in the dinner? Round to the nearest whole calorie.

Solution We are given the percent of the calories from fat (16%) and the amount of calories from fat (65 calories). We need to find the base, which is the total number of calories in the dinner. Both methods are shown.

Using the general equation for percents:

$$16\% \cdot x = 65$$

Convert 16% to a decimal: $0.16x = 65$

Divide by 0.16: $\dfrac{0.16x}{0.16} = \dfrac{65}{0.16}$

Simplify: $x = 406.25$

Round to a whole number: $x \approx 406$

Using the general proportion for percents:

Amount → $\dfrac{65}{x} = \dfrac{16}{100}$ ← Percent
Base →

Cross-multiply: $100(65) = x(16)$

Simplify: $6500 = 16x$

Divide by 16: $\dfrac{6500}{16} = \dfrac{16x}{16}$

Simplify: $406.25 = x$

Round: $406 \approx x$

The dinner contains about 406 calories total.

Percent of Increase and Percent of Decrease

Percents are often used to show how much a quantity has increased or decreased. Let's discuss **percent of increase** and **percent of decrease**.

Percent of Increase

Amount of increase = New amount − Original amount

$$\dfrac{\text{Percent of increase}}{100} = \dfrac{\text{Amount of increase}}{\text{Original amount}}$$

Example 15 Calculating the Percent of Increase

Over the last four years, Ethan's rent increased from $625 per month to $755 per month. What is the percent of increase?

Solution
Find the amount of increase: Amount of increase = $755 − $625 = $130.

Percent of increase → $\dfrac{x}{100} = \dfrac{130}{625}$ ← Amount of increase
← Original amount

Cross-multiply: $625x = 100(130)$

Simplify: $625x = 13,000$

Divide both sides by 625: $\dfrac{625x}{625} = \dfrac{13,000}{625}$

Simplify: $x = 20.8$

Ethan's rent has increased by 20.8%.

Example 16 Calculating the Amount of Increase

Last year, 16,528 students attended City Community College. This year, enrollment increased by 3.2%. How many students attend City Community College this year? Round to the nearest whole student.

Solution First, we find the amount of increase. Then, we add the amount of increase to last year's enrollment to find this year's enrollment.

Amount   Percent   Base

Amount of increase $= (3.2\%)(16{,}528) = (0.032)(16{,}528) = 528.896 \approx 529$
New enrollment $= 16{,}528 + 529 = 17{,}057$
This year's enrollment at City Community College is 17,057 students.

Percent of Decrease

$$\text{Amount of decrease} = \text{Original amount} - \text{New amount}$$

$$\frac{\text{Percent of decrease}}{100} = \frac{\text{Amount of decrease}}{\text{Original amount}}$$

Example 17 Calculating the Percent of Decrease

To attract more clients to her salon, Rachelle lowered the price of a haircut from $25 to $22.50. Find the percent of decrease for the price of a haircut at Rachelle's salon.

Solution
Find the amount of decrease: Amount of decrease $= \$25 - \$22.50 = \$2.50$.

Percent of increase $\rightarrow \dfrac{x}{100} = \dfrac{2.50}{25}$ ← Amount of increase
← Original amount

Cross-multiply: $25x = 100(2.50)$

Simplify: $25x = 250$

Divide both sides by 25: $\dfrac{25x}{25} = \dfrac{250}{25}$

Simplify: $x = 10$

The price of a haircut at Rachelle's salon decreased by 10%.

When calculating the percent of increase or percent of decrease, be sure to divide by the original amount, not the new amount.

Simple Interest

When we borrow money, the price that we pay for this privilege is called **interest**. Similarly, if we lend money to someone, we receive money for this generosity, which is also called **interest**. So, when we borrow money, we *pay* interest; when we loan or invest money, we *earn* interest. The amount of money that is borrowed or invested is called the **principal**. Usually, the amount of interest paid or earned is a percent of the principal over a given time period. The **interest rate** is the percent used to calculate the interest. Although there are different types of interest, we will look at only one type, called **simple interest**.

Simple Interest Formula

$$\text{Simple interest} = \text{Principal} \cdot \text{Rate} \cdot \text{Time}$$
$$I = P \cdot r \cdot t$$

When entering the interest rate r into the simple interest formula, it must be entered as a decimal, not a percent. For example, an interest rate of 7% would be entered into the formula as 0.07.

The interest rate r is given as a percent per unit of time. The time t must be entered in the same units. For example, if the interest rate is 7% per year, then the time must be in years. If the interest rate is 2% per month, then the time must be in months. Standard practice is to use annual (yearly) interest rates.

Example 18 Calculating Simple Interest

André borrowed $3000 for 9 months at a simple interest rate of 8.5% per year.

a. Find the interest on his loan.

b. Find the total amount that André had to pay back.

Solution

a. The interest rate is per year, but the time is given in months. So we must change one or the other. We choose to change 9 months to years:

$$t = 9 \text{ months} = \frac{9}{12} \text{ years} = \frac{3}{4} \text{ years.}$$

The principal is $P = \$3000$, and the interest rate is $r = 8.5\%$, or 0.085 per year. The interest is

$$I = Prt = 3000 \cdot (0.085) \cdot \frac{3}{4} = 191.25.$$

André paid $191.25 in interest.

b. André had to pay back the principal, plus the interest, so he paid back a total of $3000 + $191.25 = $3191.25.

Glossary

Absolute Value   The absolute value of a real number a, denoted as $|a|$, is the distance from 0 to a on the real number line.

Absolute Value Equation Property   If u is an algebraic expression and c is a real number such that $c > 0$ then $|u| = c$ is equivalent to $u = -c$ or $u = c$.

Absolute Value Inequality Property   If u is an algebraic expression and c is a real number such that $c > 0$ then (1) $|u| < c$ is equivalent to $-c < u < c$ and (2) $|u| > c$ is equivalent to $u < -c$ or $u > c$. Similar forms exist for the non-strict cases $|u| \leq c$ and $|u| \geq c$.

Additive Identity   The number 0 is called the additive identity because the sum of any number and 0 is that number.

Additive Inverses   Two numbers or expressions are additive inverses (also called *opposites*) if their sum is zero.

Algebraic Expression   A variable or a statement containing variables and/or constants that are added, subtracted, multiplied, and/or divided.

Angle   The geometric figure formed when two lines are extended from a fixed point.

Area   The measure of a surface.

Associative Property of Addition   The grouping of addition does not change the sum. $a + (b + c) = (a + b) + c$

Associative Property of Multiplication   The grouping of factors does not change the product. $a(bc) = (ab)c$

Axis of Symmetry   The axis of symmetry is an imaginary line that divides a graph into two mirror images.

Back-substitution   Taking known values of one or more variables and substituting them back into an equation involving the known variables and one unknown variable. The resulting equation is solved for the remaining variable.

Base   The quantity being raised to a power in an exponential expression.

Binomial   A simplified polynomial with two terms.

Boundary Line   A line that separates the ordered pair solutions to a linear inequality (in two variables) from the ordered pairs that are not solutions to the inequality.

Check Point   A point that satisifes the equation being graphed used to verify that the graph was constructed correctly. If the point lies on the graph, then it "checks." If the point does not lie on the graph, then it "does not check," which means an error has been made.

Circular Cylinder   A three-dimensional object that is made up of two equal circular ends and a curved side.

Coefficient   A number or symbol representing a number that is being multiplied by a variable, variable term, or other unknown quantity. Also called *numerical coefficient*.

Coinciding Lines   Two lines are coinciding if they have the same slope and the same y-intercept.

Combined Variation   When a variable is related to more than one other variable, we call this combined variation.

Common Factor   The expression w for expressions u and v if w divides both u and v evenly.

Common Denominator   Two fractions have a common denominator if their denominators are the same.

Common Logarithmic Function For $x > 0$, the common logarithmic function is defined by $y = \log x$ if and only if $x = 10^y$.

Commutative Property of Addition The property states that a sum is not affected by the order of the terms. $a + b = b + a$

Commutative Property of Multiplication The property states that the order of factors does not change the product. $ab = ba$

Complementary Angles Two angles are complementary if the sum of their measures is $90°$.

Complex Conjugate A complex conjugate is obtained by changing the sign of the imaginary part in a complex number.

Complex Fraction See *Complex Rational Expression*.

Complex Numbers The set of all numbers of the form $a + bi$, where a and b are real numbers and i is the imaginary unit (defined by $i^2 = -1$) is called the *set of complex numbers*.

Complex Rational Expression A rational expression in which the numerator and/or denominator contain(s) rational expressions. Also called *complex fraction*.

Composite Function The function $f(g(x))$ is called a *composite function* because one function f is applied to the result, $g(x)$, of another function.

Compound Inequality Two inequalities that are joined together using the words *and* or *or*.

Conditional Equation An equation that is true for some values of the variable(s) but not for others.

Conjugates Two expressions of the form $u + v$ and $u - v$ are conjugates of each other.

Consistent System A system of equations with at least one solution.

Constant A term that contains only a number; its value never changes. Also called *constant term*.

Constant of Variation (k) The (constant) ratio of two variables that vary directly. This is also called the *proportionality constant*.

Contradiction An equation for which no value of the variable can make the equation true. It has no solution, and its solution set is the empty $\{\}$ or null set $\varnothing$.

Coordinates The x- and y-coordinates of a point or ordered pair. The coordinates provide the position of the point in the coordinate plane.

Coordinate Axes The two perpendicular real number lines in a rectangular coordinate system.

Coordinate Plane The plane represented by the rectangular coordinate system. It is also called the *Cartesian plane* or the *xy-plane*.

Counting Number A counting number, or *natural number*, is an element of the set $N = \{1, 2, 3, 4, 5, \ldots\}$.

Cube An exponential expression that consists of a base with an exponent of 3.

Cube Root A real number b is the cube root of a real number a, denoted as $b = \sqrt[3]{a}$, if $b^3 = a$.

Decimal A representation of a real number using the base ten part of a whole. In decimal notation, the position of a digit determines its place value (which is given as a power of 10).

Decimal Point A period written between a whole number and a decimal fraction when writing a mixed number or proper fraction in decimal form (e.g., 4.391 or 0.391).

Denominator The expression below the fraction bar in a fraction or quotient.

Dependent Equations Two equations are dependent if they describe the same solution set. In two variables, the equations describe the same line. In three variables, the equations describe the same plane.

Dependent System A system of equations is dependent if an equation in the system can be obtained from one or more other equations in the system.

Dependent Variable A variable whose value depends on the value(s) of other variables. These are also called *output variables* because they represent the value that we get *out* after inputting values for the independent variables.

Digit A digit is any one of the ten numerals 0, 1, 2, 3, 4, 5, 6, 7, 8, 9 that is used to express real numbers.

Direct Variation For an equation of the form $y = kx$, we say that y varies directly with x, or y is proportional to x. The constant k is called the constant of variation or the proportionality constant.

Discriminant Given a quadratic equation of the form $ax^2 + bx + c = 0$, the expression $b^2 - 4ac$ is called the discriminant.

Dividend The quantity that you divide into in a division problem. This is also the numerator of a fraction.

Divisor The quantity that you divide by in a division problem. This is also the denominator of a fraction.

Domain   The set of all allowable values for the independent variable. These are the first coordinates in a set of ordered pairs that make up a relation and are also called *input values*.

Element   An individual number of a set or other object included in a set.

Elimination Method   A process used for solving systems of linear equations in two variables. It involves addition (or subtraction) to eliminate a variable. First, we multiply the equations by an appropriate constant so that the coefficients of the variable to be eliminated are opposites. Then the equations are added together to obtain a third equation with the desired variable eliminated.

Endpoint   A point where a graph does not continue in one direction.

Exponent   Also called *power*, it is a superscripted number that tells how many times a base expression is multiplied by itself.

Exponential Decay Model   A model that describes the exponential decay of a population, quantity, or amount after a certain time.

Exponential Function   An exponential function is a function of the form $f(x) = b^x$, where x is any real number and $b > 0$ such that $b \neq 1$. The constant, b, is called the *base* of the exponential function.

Extraneous Solutions   When both sides of an equation are modified, the new equation may have solutions that are not solutions to the original equation. Such solutions are called extraneous solutions.

Exterior Angle   An angle outside a figure.

Factor   One of two or more quantities whose product is a given quantity.

Factor Theorem   For a polynomial function $f(x)$, (1). If $f(c) = 0$, then is a $x - c$ is a factor of $f(x)$. (2) If $x - c$ is a factor of $f(x)$, then $f(c) = 0$.

Finite Set   A set with a definite, or limited, number of elements. It is (theoretically) possible to individually list all the elements of a finite set.

First-degree Equations   *See Linear Equation.*

FOIL   An acronym for multiplying binomials. It reminds us to multiply the two First terms, the two Outside terms, the two Inside terms, and the two Last terms.

Function   A special type of relation in which each value (or element) in the domain corresponds to exactly one value (or element) in the range.

Geometric Sequence   A geometric sequence is a sequence of the form $a_1, a_1r, a_1r^2, a_1r^3, \ldots$ where a_1 is the first term of the sequence and r is the common ratio such that $r = \frac{a_2}{a_1} = \frac{a_3}{a_2} = \ldots \frac{a_{n+1}}{a_n}$ for all $n \geq 1$.

Greatest Common Factor (GCF)   The expression with the largest coefficient and highest degree that divides each term of a polynomial evenly.

Grouping Symbols   Mathematical symbols used to group operations so they are treated as a single quantity. When simplifying, all operations within grouping symbols must be done first. Some examples of grouping symbols are parentheses ( ), brackets [ ], and braces { }. Operators such as absolute value | |; fraction bars, –; and radicals, $\sqrt{}$ are also treated as grouping symbols.

Half-plane   The set of ordered pair solutions to a linear inequality in two variables.

Horizontal Major Axis   An ellipse has a horizontal major axis if its major axis is parallel to the x-axis.

Hypotenuse   In a right triangle, the side opposite the right angle is called the hypotenuse of the triangle. It is the longest of the three sides.

Identity   An equation that is always true for all defined values of its variable.

Imaginary Unit i   The imaginary unit i is defined as $i = \sqrt{-1}$, where $i^2 = -1$.

Improper Fraction   An improper fraction is a fraction in which the absolute value of the numerator is greater than or equal to the absolute value of the denominator. The absolute value of an improper fraction is greater than or equal to 1.

Inconsistent System   A system of equations without a solution. If two lines in a system are parallel, then the system is inconsistent. When solving an inconsistent system, a contradiction (false statement) will result.

Independent Variable   A variable for which we can select values. These are also called *input variables* because they represent the set of values that we can arbitrarily *input* into a relation.

Index [of a radical expression]   In a radical expression $\sqrt[n]{a}$, n is called the index, and it indicates the type of root.

Inequality   A statement that two quantities are not equal. An inequality looks like an equation, but an inequality symbol is used instead of an equal sign.

Infinite Sequence   An infinite sequence is a function whose domain is the set of all natural numbers.

Infinite Set   A set with an endless, or unlimited, number of elements. It is not possible to individually list all the elements of an infinite set.

Integers   The set of integers is
$\{\ldots, -3, -2, -1, 0, 1, 2, 3, \ldots\}$.

Intercept   A point where a graph crosses or touches a coordinate axis.

Interior Angle   An angle inside a figure.

Intersection   For any two sets A and B, the intersection of A and B is given by $A \cap B$ and represents the elements that are common to both set A and set B.

Intersection Point   A point that two graphs share. These are points where graphs cross or touch each other.

Interval Notation   A shorthand way of expressing a set by using a pair of numbers to indicate an unbroken string of values.

Inverse Function   Let f be a one-to-one function with domain A and range B. Then the inverse function f^{-1} is the function with domain B and range A, such that if $f(a) = b$, then $f^{-1}(b) = a$.

Inverse Variation   For equations of the form $y = k/x$ or $y = k \cdot 1/x$, we say that y varies inversely with x or y is inversely proportional to x. The constant k is called the constant of variation.

Irrational Number   An irrational number is a real number that is not a rational number.

Leading Coefficient   The leading coefficient of a polynomial is the coefficient of the term with the highest degree.

Least Common Denominator (LCD)   For a set of fractions, the LCD is the least common multiple of the denominators of the fractions.

Legs   In a right triangle, the two sides that form the right angle are called the legs of the triangle.

Line   A graph of an equation that can be written in the form $Ax + By = C$ (A and B cannot both be zero). Lines are straight and continue forever in both directions.

Linear Coefficient   The linear coefficient is the coefficient of the linear term in a polynomial.

Linear Function   A function of the form $f(x) = ax + b$, where a and b are real numbers.

Linear Term   A variable term that consists of a constant times a variable raised to the first power.

Logarithmic Function   For $x > 0, b > 0$, and $b \neq 1$, the logarithmic function with base b is defined by $y = \log_b x$ if and only if $x = b^y$.

Long Division   A procedure to divide large or complex numbers in small steps. The procedure generally involves several cycles of division, multiplication, subtraction, and dropping down a digit.

Lowest Terms   A fraction is written in lowest terms, or *simplest form*, if the numerator and denominator have no common factors other than 1.

Major Axis   A line segment that passes through the foci of an ellipse and is the longer of the two axes of symmetry.

Member   Each object in a set is called an element or a member of the set.

Minor Axis   The minor axis of an ellipse is the shorter of the two axes of symmetry.

Minor Rational Expressions   The rational expressions within the numerator and denominator of a complex rational expression. Also called *minor fractions*.

Mixed Number   A mixed number is a number that is the sum of a whole number and a proper fraction. Mixed numbers contain a whole-number part and a proper-fraction part.

Monomial   A simplified term in which all variables are raised to non-negative integer powers and no variables appear in any denominator.

Multiplicative Identity Property   The number 1 is called the multiplicative identity because the product of any number and 1 is that number.

Natural Base   The natural base is the irrational number e, which is equal to $2.718281828\ldots$ The natural base appears very frequently in real-life applications.

Natural Logarithmic Function   For $x > 0$, the natural logarithmic function is defined by $y = \ln x$ if and only if $x = e^y$.

Natural Number   A natural number, or counting number, is an element of the set
$N = \{1, 2, 3, 4, 5, \ldots\}$.

Negative Power Rule   For any nonzero expression u and any integer $n \neq 0$, $u^{-n} = 1/u^n$.

Negative Square Root   A negative real number b is the negative square root of a non-negative real number a, denoted as $b = -\sqrt{a}$, if $b^2 = a$.

Non-negative   A real number is non-negative if it is 0 or larger. $\{x | x \geq 0\}$

Non-simplified Expression   An expression that can be written in a simpler form by combining like terms, removing common factors and so forth.

Non-strict Inequality   It contains one or both of the following inequality symbols: $\leq, \geq$. The possibility of equality is included in a non-strict inequality.

Null Set   A set that contains no elements. Also called *empty set*.

Numerator   The expression above the fraction bar in a fraction or quotient.

Numeric Expression   A numeric expression is a combination of real numbers and one or more operations.

One-to-One Function   A function f is one-to-one if for any values $a \neq b$ in the domain of f, $f(a) \neq f(b)$.

Opposite Factors   Two factors are opposite factors if they have opposite signs but the same absolute value.

Opposite Polynomials   Two polynomials are opposite polynomials if they add to zero. To find the opposite of a polynomial, change the sign of each term.

Opposite Reciprocals   Two numbers are said to be opposite reciprocals if their product equals -1. This occurs when the two numbers have opposite signs and their absolute values are reciprocals.

Order of Operations   A set of rules for evaluating expressions with more than one operation.

Ordered Pair (x, y) A pair of real number values that corresponds to a point in the coordinate plane.

Ordered Triple (x, y, z) A set of three values that gives the location of a point in three dimensions.

Origin   The point at which the x- and y-axes intersect in the rectangular coordinate system. The origin is represented by the ordered pair $(0, 0)$.

Parabola   A graph of an equation that can be written in the form $y = ax^2 + bx + c, a \neq 0$.

Parallel Lines   Two lines are parallel if they have the same slope and different y-intercepts.

Percent   Percent means "per hundred" or "out of 100."

Perimeter   The distance around a two-dimensional object.

Perpendicular Lines   Two lines are perpendicular if they meet at right angles $(90°)$. The product of their slopes will equal -1, so perpendicular lines have opposite-reciprocal slopes.

Place Value   The place value is the value of the position where a digit lies in a number.

Plane   An infinitely large two-dimensional flat surface with zero thickness. A plane is formed by the solution set of an equation of the form $Ax + By + Cz = 0$.

Point-Slope Form   Given the slope of a line m and a point (x_1, y_1) on the line, the point-slope form of the equation of the line is given by $y - y_1 = m(x - x_1)$.

Positive Factor of a Product   A positive number that is multiplied by other numbers, variables, or expressions to obtain the product.

Power   Also called *exponent*, it is a superscripted number that tells how many times a base number is multiplied by itself.

Power-to-Power Rule   When an exponential expression is raised to a power, we multiply the exponents.

Prime Number   A whole number greater than 1 whose only whole number factors are 1 and itself.

Principal nth Root   If a and b are real numbers and n is an integer such that $n \geq 2$, then b is the principal nth root of a, denoted as $b = \sqrt[n]{a}$, if $b^n = a$. Note: If n is even, then both a and b must be non-negative.

Principal Square Root   A non-negative real number b is the principal square root of a non-negative real number a, denoted as $b = \sqrt{a}$, if $b^2 = a$.

Product Rule for Radicals   If $\sqrt[n]{a}$ and $\sqrt[n]{b}$ are real numbers, then $\sqrt[n]{a}\sqrt[n]{b} = \sqrt[n]{ab}$.

Product-to-Power Rule   When raising a product to a power, we raise each factor of the base to the common exponent. $(ab)^n = a^n \times b^n$

Product Rule for Exponents   When multiplying exponential expressions with the same base, we add the exponents and keep the common base. $b^m \times b^n = b^{(m+n)}$

Proper Fraction   A proper fraction is a fraction in which the absolute value of the denominator is greater than the absolute value of the numerator. The absolute value of an improper fraction is less than 1.

Proper Subset   If every element of a set A is also an element of set B, but A and B are not equal sets, then set A is a proper subset or strict subset of set, denoted as $A \subset B$.

Proportion   A proportion is a statement that two ratios are equal, such as $\frac{P}{Q} = \frac{R}{S}$.

Pythagorean Theorem   For right triangles, the sum of the squares of the lengths of the legs of the triangle equals the square of the length of the hypotenuse. $a^2 + b^2 = c^2$

Quadrants   The four regions formed by the x- and y-axes in a rectangular coordinate system.

Quadratic Formula   The solution(s) to the general quadratic equation $ax^2 + bx + c = 0$ are given by the quadratic formula
$$x = \frac{-b \pm \sqrt{b^2 - 4ac}}{2a}.$$

Quotient   The result of division.

Quotient Rule for Exponents   When dividing exponential expressions with the same base, we subtract the denominator exponent from the numerator exponent and keep the common base. $b^n/b^m = b^{(n-m)}$

Quotient Rule for Radicals   If $\sqrt[n]{a}$ and $\sqrt[n]{b}$ are real numbers and $b \neq 0$, then $\dfrac{\sqrt[n]{a}}{\sqrt[n]{b}} = \sqrt[n]{\dfrac{a}{b}}$.

Quotient-to-Power Rule   When raising a quotient to a power, we raise both the numerator and denominator to the common exponent. $(a/b)^n = a^n/b^n$

Radical Equation   A radical equation is an equation that contains at least one radical expression with a variable in the radicand.

Radicand   The expression beneath the radical sign is called the radicand.

Range   The set of all values for the dependent variable. These are the second coordinates in the set of ordered pairs that define a relation and are also called *output values*.

Rate of Change   The ratio of the vertical change to the horizontal change when moving from one point on the graph to another. The slope of the line connecting the two points is equal to the rate of change.

Ratio   A ratio is a comparison of two quantities, usually in the form of a quotient.

Rational Expression   The ratio of two polynomials.

Rational Number   A number that can be written as the quotient p/q of two integers p and q, as long as $q \neq 0$. p/q is called the fraction form for the rational number. The decimal form of a rational number will either terminate or repeat.

Rationalizing a Denominator   The process of removing radicals so that the denominator contains only a rational number is called rationalizing the denominator.

Real Axis   A graph that represents the set of all real numbers. Also called the *number line*.

Real Number   A number that is either rational or irrational.

Reciprocal   The number by which you multiply another number so the product is 1. Also called the *multiplicative inverse*.

Remainder   The amount that is left over after division, when the divisor can no longer divide into the dividend wholly.

Remainder Theorem   If a polynomial $f(x)$ is divided by $x - c$, then the remainder is $f(c)$.

Repeating Decimal   A repeating decimal is a decimal in which a finite pattern of digits repeat without ending.

Restricted Value   Any value or expression that causes the denominator of a rational expression to equal zero must be restricted for that rational expression.

Rise   The vertical change between two points (x_1, y_1) and (x_2, y_2). We compute this by finding the difference of the y-coordinates: $y_2 - y_1$

Right Circular Cone   A right circular cone is a cone that has its vertex directly above the center of a circular base.

Right Triangle   A triangle with a $90°$ angle.

Root   If c is a real number such that $f(c) = 0$, then c is called a root of the function f. Also called a *zero* of the function.

Run   The horizontal change between two points (x_1, y_1) and (x_2, y_2). We compute this by finding the difference of the x-coordinates: $x_2 - x_1$

Scatter Plots   Graphs that display data as a set of points.

Scientific Notation   A number is written in scientific notation if it has the form $a \times 10^n$, where a is a real number such that $1 \leq |a| < 10$ and n is an integer.

Set   A collection of numbers or objects.

Simple Interest   It is found using the formula $I = Prt$, where I is the interest earned or paid, P is the principal (money invested or borrowed), r is the interest rate per time period (in decimal form), and t is the number of time periods that the money is invested or borrowed.

Simplest Form   A fraction is in simplest form, or lowest terms, when the numerator and denominator have no common factors other than 1.

Simplified Term   A term is a simplified term if it contains a single constant factor (possibly a simplified fraction) and if none of the variable factors can be combined using the rules for exponents.

Slope   A measure of the steepness of the line. For two points on the line, (x_1, y_1) and (x_2, y_2), the slope of the line is found using

$$m = (y_2 - y_1)/(x_2 - x_1).$$

Slope-intercept Form   A linear equation in two variables of the form $y = mx + b$ or $f(x) = mx + b$ is written in slope-intercept form, where m is the slope of the line and b is the y-intercept.

Solution   A value that, when substituted for the variable, makes the equation true.

Solution Set   The collection of *all* solutions to an equation or inequality.

Sphere   A sphere is a three-dimensional object such that all points on its surface are an equal distance r (called the *radius*) from a fixed center point.

Square   An exponential expression that consists of a base with an exponent of 2 is called the square of the base.

Square Root   If a is a non-negative real number and b is a real number such that $b^2 = a$, then b is a square root of a.

Standard Form of a Quadratic Function   A quadratic function is in standard form if it is written as $f(x) = a(x - h)^2 + k$. Also called *vertex form*.

Strict Inequality   It contains one or more of the following inequality symbols: $<, >, \neq$. There is no possibility of equality in a strict inequality.

Subset   If every element of set A is also an element of set B, then set A is a subset of set B, denoted as $A \subseteq B$.

Substitution   Substitution is a process in which a variable is substituted for an algebraic expression, or vice versa.

Sum   The result of addition.

Summation Notation   A convenient way to express the sum of a finite series of terms using a short-hand notation. Also called *sigma notation*.

Superset   If set A contains every element of set B, then set A is a superset of set B, denoted as $A \supseteq B$.

Supplementary Angle   Two angles are supplementary if the sum of their measures is $180°$.

Surface Area   The total area on the surface of a three-dimensional object.

Symmetry   A two-dimensional graph is symmetric if it has mirror images (reflections) on opposite sides of a dividing line.

Synthetic Division   A procedure for dividing two polynomials that uses only the coefficients of the terms. The procedure can only be used when the divisor is a first-degree binomial.

Term   A constant, a variable, or the product of a constant and one or more variables raised to powers.

Terminating Decimal   A terminating decimal is a decimal that ends, or has a finite number of digits after the decimal point.

Test Point   A point used to find the half-plane that contains the ordered pair solutions to a linear inequality in two variables.

Theorem   A mathematical rule that has been proven (or is to be proven) true.

Three-part Inequality   A statement in which three algebraic expressions are compared using two inequality symbols.

Translation   A translation is when every point on a graph is shifted the same distance in the same direction.

Trinomial   A simplified polynomial with three terms.

Undefined Fraction   When the numerator of a fraction is nonzero and the denominator is zero, the fraction is undefined.

Uniform Motion   A moving object traveling at a constant speed or rate.

Union   For any two sets A and B, the union of A and B is given by $A \cup B$ and represents the elements that are in set A or in set B.

Unique Factor   A factor that is distinct (different) from any other factor in a prime factorization.

Variable   A symbol (usually a letter) that represents a value that can change. For example, the variable x could represent 1 in one situation and 100 in another.

Vertex   The vertex of a parabola is the lowest point (if the parabola opens *up*) or the highest point (if the parabola opens *down*) of the graph of the parabola.

Vertical Line Test   If a vertical line intersects the graph of a relation at more than one point, then the relation is not a function. If every vertical line intersects the graph of a relation at no more than one point, then the relation is a function.

Vertical Major Axis   An ellipse has a vertical major axis if its major axis is parallel to the y-axis.

Whole Numbers   The set of whole numbers is $\{0, 1, 2, 3, \dots\}$.

x-Axis   The horizontal axis in the rectangular coordinate system.

x-Coordinate   The first number in an ordered pair. Also called the *abscissa*.

x-Intercept　The x-coordinate of a point where a graph crosses or touches the x-axis. (The y-coordinate is 0.)

y-Axis　The vertical axis in a rectangular coordinate system.

y-Coordinate　The second number in an ordered pair. Also called the *ordinate*.

y-Intercept　The y-coordinate of a point where a graph crosses or touches the y-axis. (The x-coordinate is 0.)

Zero-Power Rule　Any real number raised to the 0 power equals 1, except for 0^0, which is indeterminate.

CHAPTER R

R.1 Exercises

1. $\in$ **3.** $\notin$ **5.** $\in$ **7.** $\{3,4,5,6,7,8,9\}$ **9.** $\{21,22,23,...\}$ **11.** false **13.** true

15. natural number, whole number, integer, rational number, real number **17.** rational number, real number

19. **a.** 1 **b.** 0, 1 **c.** $-17, 0, 1$ **d.** $-17, -\dfrac{25}{29}, 0, 0.331, 1$ **e.** $-\sqrt{5}, \pi$

21. **23.** > **25.** > **27.** > **29.** true **31.** true **33.** 18 **35.** $\dfrac{3}{8}$

R.2 Exercises

1. 3 **3.** $-\dfrac{1}{18}$ **5.** 13 **7.** $\dfrac{4}{15}$ **9.** -22 **11.** $\dfrac{1}{8}$ **13.** 72 **15.** $\dfrac{1}{8}$ **17.** $\dfrac{1}{12}$ **19.** -9

21. -25 **23.** $-\dfrac{27}{20}$ **25.** 512 **27.** -64 **29.** 13 **31.** 6 **33.** 5 **35.** $c+ab$ **37.** $(4+a)+11$

39. $5x+40$ **41.** $-7m+7n-35$ **43.** Identity property of multiplication
45. Commutative property of addition **47.** -108 **49.** 9 **51.** 1223 **53.** 1 **55.** -35

R.3 Exercises

1. 225 **3.** 48 **5.** -7 **7.** 9 **9.** -21 **11.** -23 **13.** $-5z$ **15.** $13ab$ **17.** $-10a+30$

19. $4x^2-2xy-3y^2$ **21.** $9t-25$ **23.** $3x^2-2$ **25.** $x+10$ **27.** $5x+9$ **29.** $\dfrac{5}{8}x+12$

CHAPTER 1

1.1 Exercises

1. yes **3.** yes **5.** yes **7.** $x=13$ **9.** $a=8$ **11.** $m=-9$ **13.** $x=-3$ **15.** $t=10$

17. $x=-1$ **19.** $x=-18$ **21.** contradiction, $\varnothing$ **23.** Identity, $\mathbb{R}$ **25.** $n=-1$ **27.** $y=\dfrac{2}{29}$

29. $z=8$ **31.** $x=2.925$ **33.** Stocks: \$12,000, Bonds: \$8000 **35.** 6 hours **37.** \$159.95

1.2 Exercises

1. yes **3.** no **5.** no **7.** **9.** **11.** $(-\infty,1]$

13. $\left[\dfrac{3}{2},\dfrac{7}{3}\right]$ **15.** $\left(-\dfrac{3}{2},\infty\right)$, **17.** $\left(-\infty,-\dfrac{145}{46}\right]$,

19. $[-15,\infty)$ **21.** $(-\infty,7]$ **23.** $(-\infty,0)$ **25.** $(4,\infty)$ **27.** $(-\infty,\infty)$ **29.** $[-7,3)$
31. up to 1478 minutes **33.** 92 plush toys **35.** greater than \$131,450 and less than or equal to \$200,300

1.3 Exercises

1. $\{-2,0,6\}$ **3.** $\{-6,-4,-2,0,1,2,4,5,6,9,15\}$ **5.** $\{x\,|\,2\le x<9\}$ **7.** $(-8,3)$ **9.** $(1,4]$

11. $(-\infty,-3]\cup(4,\infty)$ **13.** $(-\infty,\infty)$ **15.** $(-\infty,-6)$ **17.** $\{-3,3\}$ **19.** $\left\{-\dfrac{5}{3},-\dfrac{2}{3}\right\}$ **21.** $\left\{-\dfrac{3}{7}\right\}$

23. $\left\{0,\dfrac{3}{2}\right\}$ **25.** $\left[-\dfrac{1}{3},\dfrac{1}{3}\right]$ **27.** $\left(-2,\dfrac{7}{2}\right)$ **29.** $\left\{\dfrac{7}{3}\right\}$ **31.** $(-\infty,-3)\cup(3,\infty)$ **33.** $\left(-\infty,\dfrac{9}{4}\right]\cup\left[\dfrac{15}{4},\infty\right)$

35. $\left(-\infty,\dfrac{4}{9}\right)\cup\left(\dfrac{4}{9},\infty\right)$ **37.** $\left[-1,\dfrac{5}{2}\right]$ **39.** $(-\infty,-8]\cup[-2,\infty)$ **41.** $(-\infty,\infty)$

1.4 Exercises

1. $w = \dfrac{A}{l}$ **3.** $h = \dfrac{3V}{\pi r^2}$ **5.** $r = \dfrac{E}{I} - R$ or $r = \dfrac{E - RI}{I}$ **7.** $h = \dfrac{A - 2\pi r^2}{2\pi r}$ **9.** $x = \sigma z + \mu$

11. 9 m by 18 m   **13.** 7 in.   **15.** $40°, 60°, 80°$ **17.** 14 quarters and 19 dimes

19. $8000 at 4% and $20,000 at 5%   **21.** 30 mph and 50 mph   **23.** 4 hours   **25.** 21.6%   **27.** $8\dfrac{1}{3}$ L

29. 28 ft to 68 ft

CHAPTER 2

2.1 Exercises

1. A: Quadrant II; B: y-axis; C: Quadrant I; D: Quadrant IV; E: x-axis; F: Quadrant III   **3. a.** no

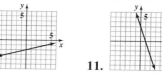

b. no   **c.** yes   **d.** yes   **5. a.** -6 **b.** 1   **7. a.** -4 **b.** $\dfrac{9}{2}$ **9.**   **11.**

13.   **15.**   **17.** x-intercepts: -3 and 5, y-intercept: 5

2.2 Exercises

1. independent: x, dependent: y **3.** Answers may vary. independent: x, dependent: y **5.** domain: $\{0, 1, 4\}$, range: $\{-2, -1, 0, 1, 2\}$ **7.** domain: $(-\infty, \infty)$, range: $[3, \infty)$ **9.** yes   **11.** yes   **13.** yes   **15.** no   **17.** yes   **19.** no   **21.** yes   **23.** no   **25. a.** independent: percent with access to clean water, dependent: percent undernourished   **b.** $\{(96, 3), (91, 7), (77, 15), (84, 12), (92, 8)\}$ **c.** domain: $\{96, 91, 77, 84, 92\}$, range: $\{3, 7, 15, 12, 8\}$ **d.** yes, it passes the vertical line test   **27. a.** yes, it passes the vertical line test **b.** independent: Year, dependent: Price per gallon   **c.** $2.60   **d.** 2015

2.3 Exercises

1. $f(x) = |2x - 5|$ **3.** $f(x) = -\sqrt{x} + 1$ **5.** $f(x) = -\dfrac{3}{4}x + 3$ **7.** $f(x) = 2x^2 - 5$ **9.** $f(x) = 2x^2 + 3$

11. 13   **13.** 1   **15.** 5   **17.** 54   **19.**   **21.**   **23.**

25. a. $10   **b.** about $0.09   **c.** about $46   **d.** domain: $[0, \infty)$, range: $[10, \infty)$ **e.** you must use over 311 text messages (approximately) to make it a better deal   **27. a.** March, June, and July **b.** February and December   **c.** January, April, May, August, September, October, and November **d.** September   **e.** $12,600   **f.** $8600   **g.** $4000   **h.** about $333.33   **29.** first graph

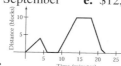

31.   **33. a.** $34,515, the amount of money owed after 0 monthly payments **b.** $20,709, the amount of money owed after 24 monthly payments

c. \$–6903, not possible, you can't owe a negative amount of money   **d.** \$13,806   **e.** 60 months

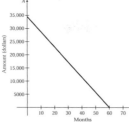

f. domain: $[0,60]$, range: $[0,34,515]$ **g.** 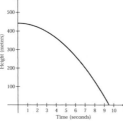   **35. a.** 442.225 m, the initial height of the object or the height of the building   **b.** 319.725 m, the height of the object after 5 seconds
c. –47.775 m, not possible, the object can't have a negative height   **d.** 0 m, when the object hits the ground

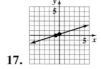

e. domain: $[0,9.5]$, range: $[0,442.225]$ **f.**

2.4 Exercises

 1.   **3.**   **5.**   **7.**   **9.**

 11.   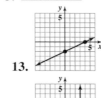 **13.**   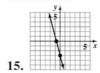 **15.**   **17.**   **19.**

 21.   **23.**

2.5 Exercises

1. $-\dfrac{8}{7}$ **3.** $\dfrac{1}{2}$ **5.** 0   **7.** slope: 4, y-intercept: -12 **9.** slope: $\dfrac{1}{3}$, y-intercept: 3,

11.   **13.** perpendicular   **15.** parallel   **17.** intersecting   **19.** $y = x - 2$, $x - y = 2$

21. $y = \dfrac{7}{9}x - \dfrac{8}{7}$, $49x - 63y = 72$ **23.** $y = 3x + 18$, $3x - y = -18$ **25.** $y = -\dfrac{4}{3}x + 12$, $4x + 3y = 36$

27. $y = -x + 3$, $x + y = 3$ **29.** $y = \dfrac{2}{15}x + \dfrac{16}{15}$, $2x - 15y = -16$ **31.** $y = -2$ **33.** $y = \dfrac{1}{4}x + \dfrac{7}{2}$

35. $y = -4x - 5$ **37. a.** $T(x) = 363.25x + 5132$ **b.** slope: tuition and fees increase \$363.25 per year, y-intercept: tuition and fees were \$5132 in 2004   **c.** \$9127.75   **39.** 7.2 inches

2.6 Exercises

1. a. no   **b.** no   **c.** yes   **3. a.** no   **b.** yes   **c.** no   **5.** 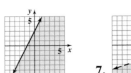   **7.**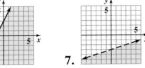

9.   **11.**   **13.**

CHAPTER 3

3.1 Exercises

1. a. yes   **b.** no   **3. a.** no   **b.** yes   **5.** $(-1,-2)$ **7.** $(-1,2)$ **9.** $(-3,2)$ **11.** $(-6,6)$

13. $(2,4)$ **15.** $(6,9)$ **17.** $(1,-1)$ **19. a.** $\{(x,y)\mid y=2x+6\}$ **b.** $(x,2x+6)$

21. inconsistent, $\varnothing$ **23.** $\left(-\dfrac{13}{16},\dfrac{13}{2}\right)$ **25.** inconsistent, $\varnothing$ **27.** Rodney: 26 points, Justin: 21 points

29. Egg McMuffin: 300 calories, hash browns: 150 calories   **31.** 40 in. by 110 in.

3.2 Exercises

1. yes: $(-1,1,-2)$, no: $(1,-1,2)$ **3.** $(-1,2,3)$ **5.** $\left(\dfrac{1}{4},-\dfrac{1}{2},\dfrac{3}{8}\right)$ **7.** $(-3,0,1)$ **9.** $(29,16,3)$

11. $(5,-2,-4)$ **13.** $\left(\dfrac{1}{2}y-\dfrac{1}{2}z-3,y,z\right)$ **15.** no solution   **17.** $(2,z-3,z)$

19. Hot dog: \$1.50, Hamburger: \$2.50, Chicken sandwich: \$3.75

21. Adult: 520, Children: 210, Senior citizen: 280   **23.** $y=-\dfrac{1}{80}x^2+\dfrac{37}{40}x-\dfrac{161}{16}$; 5.25 million

3.3 Exercises

1. Yessica: 9 mph, Zoe: 25 mph   **3.** 450 minutes   **5.** Plane: 550 mph, Wind: 50 mph   **7.** 75 yds by 120 yds

9. $22°$ and $68°$ **11.** 1.5 oz of 15% and 3 oz of 30%   **13.** $46\dfrac{2}{3}$ oz of Irish Breakfast tea and $53\dfrac{1}{3}$ oz of

buttered rum tea   **15.** \$950 at 3% and \$2050 at 4%   **17.** 39.4 ft, 58.8 ft, 68.8 ft

3.4 Exercises

1. a. yes   **b.** no   **c.** no   **3. a.** yes   **b.** no   **c.** yes   **5. a.** no   **b.** no   **c.** yes

7.   **9.**   **11.** 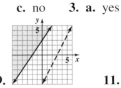   **13.** 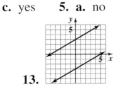   **15.**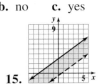

17.   **19.** 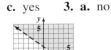   **21.**

CHAPTER 4

4.1 Exercises

1. $-3^5 = -243$ **3.** $20a^7b^9$ **5.** $5^4 = 625$ **7.** $-\dfrac{x^2y^4}{4}$ **9.** 1   **11.** 1   **13.** $\dfrac{1}{7^2} = \dfrac{1}{49}$ **15.** $\dfrac{n^7}{m^4}$

17. h^{32} **19.** y^{14} **21.** $256m^{12}$ **23.** $\dfrac{x^9}{8y^{12}}$ **25.** $\dfrac{a^6}{27b^{15}}$ **27.** $\dfrac{q^{36}}{p^{12}}$ **29.** $\dfrac{1}{a^2b^4c^3}$ **31.** $\dfrac{64}{m^{24}n^{18}}$

33. $-\dfrac{448a^4m^{21}}{b^6n^{15}}$ **35.** 6.3×10^6 **37.** 5.07×10^{-10} **39.** 0.000000315 **41.** $4,090,000,000$

43. 8×10^{11} **45.** 5×10^5 **47.** 4.26×10^2

4.2 Exercises

1. coefficient: 6, degree: 4   **3.** coefficient: $\dfrac{3}{8}$, degree: 1   **5.** coefficient: -1, degree: 2   **7.** 8   **9.** 6

11. $-\dfrac{2}{3}m^7 + m^5 - 9m^3 - 5m + 8$, degree: 7, leading coefficient: $-\dfrac{2}{3}$ **13.** $5x^2 - 9x + 4$, degree: 2, leading

coefficient: 5   **15.** 5   **17.** 124   **19.** $8x^3 - 3x^2 - 5x - 15$ **21.** $3w^3 + 4w^2 - 2w - 3$

23. $-\dfrac{2}{5}x^2 + \dfrac{17}{18}xy - \dfrac{5}{8}y^2$ **25.** $30p^2q + 11pq^2$ **27.** $2.3x^2 - 3.2x + 2.9$ **29.** $2x^3 - 14x^2 + 2x - 13$

31. $x^3 + 4x - 20$ **33.** $x^3 - 18x^2 - 4x - 4$

4.3 Exercises

1. $30x^7$ **3.** $24x^4y^5z^7$ **5.** $6.8a^5b^4 + 23.8a^4b^2$ **7.** $x^2 + 13x + 30$ **9.** $6x^2 + 19x + 10$

11. $3m^4 - 14m^2n^2 + 8n^4$ **13.** $x^2 - \dfrac{x}{6} - \dfrac{1}{6}$ **15.** $49x^2 - 81$ **17.** $x^2 - 16x + 64$ **19.** $49p^2 - 28pq + 4q^2$

21. $16z^2 + 40z + 25$ **23.** $x^3 + x^2 - 7x + 20$ **25.** $12a^3 - 17a^2b + 3ab^2 + 2b^3$ **27.** $10x^4 - 66x^3 - 112x^2$

29. $x^3 + 6x^2 + 11x + 6$ **31.** $x^3 + 12x^2 + 48x + 64$ **33.** $9x^2 + 12xy + 4y^2 - 25$ **35.** $2x^2 + 11x - 40$

4.4 Exercises

1. $5x^5 + 3x^3 - 2x^2 + 8$ **3.** $4x^2 + \dfrac{xy}{2} - y^2$ **5.** $x + 13$ **7.** $x^2 - x + \dfrac{6}{2x-3}$ **9.** $x^2 - 3x + 6 - \dfrac{34}{x+5}$

11. $3x^2 - 11x + 15 - \dfrac{3}{x^2-2}$ **13.** $9x^2 - 47 + \dfrac{2x+244}{x^2+5}$ **15.** $x + 4 + \dfrac{3}{x-5}$ **17.** $4x^2 - 8x + 21 - \dfrac{49}{x+2}$

19. $x^2 + 3x - 3$ **21.** $x^3 + 2x^2 + 9x + 18 - \dfrac{5}{x-2}$ **23.** $15x^2 + 20x - 35$ **25.** $3x - 1 + \dfrac{3}{x-2}$, $x \neq 2$ **27.** 8

29. -10 **31.** $(x-8)(x-3)$ **33.** $(x+1)$ is not a factor

CHAPTER 5

5.1 Exercises

1. $4(5x+2)$ **3.** $4xy(3x+1)$ **5.** $4x^2(x^2-5x+3)$ **7.** $4x^2y^2z^4(4yz^2 - 3x^2y^2 + 5xz)$ **9.** $-8x^2(x-6)$

11. $-2(2x^2 + 8x - 9)$ **13.** $(x+7)(11+3a)$ **15.** $(9y-7)(x^2+11)$ **17.** $(x+3)(y+5)$

19. $(3x+2)(3y+4)$ **21.** $(x+3)(5x+4y)$ **23.** $(x^2+4)(x+9)$ **25.** $(x^2+1)(x+8)$

5.2 Exercises

1. $(x+10)(x+5)$ **3.** $(w-14)(w-4)$ **5.** $(n+8)(n+30)$ **7.** prime   **9.** $(y-28)(y-14)$

11. $(x-18y)(x-2y)$ **13.** $8(x-3)(x+4)$ **15.** $6xy^2(x-8y)(x-4y)$ **17.** $-(z-8)(z+12)$

19. $-5x^2(x-12)(x-6)$ **21.** $(2x+7)(3x+5)$ **23.** $(2w+5)(4w-3)$ **25.** prime

27. $(3m-8)(4m-5)$ **29.** $(2x-3)(4x+3)$ **31.** $(2p-9q)(6p+q)$ **33.** $4(2x+3)(3x-5)$

35. $3x^2y^2(2x+5y)(6x-5y)$ **37.** $-(4m+3)(5m-2)$ **39.** $-2w(2x+7)(4x+3)$ **41.** $(x^3-5)(x^3+4)$

43. $(x+y-10)(5x+5y+12)$

5.3 Exercises

1. $(x-2)(x+2)$ **3.** $(2x-3)(2x+3)$ **5.** $(x-1)(x+1)(x^2+1)$ **7.** $(2x-7y)(2x+7y)$ **9.** $(x+2)^2$

11. $(3x+4)^2$ **13.** $(3x-2y)^2$ **15.** $(2a+5b)^2$ **17.** $(x-1)(x^2+x+1)$ **19.** $(2z+5)(4z^2-10z+25)$

21. $(x+2y)(x^2-2xy+4y^2)$ **23.** $(x^2y+z)(x^4y^2-x^2yz+z^2)$ **25.** $4x(2x+1)(8x-3)$

27. $y(x-y)(x+y)$ **29.** $(2x-5)(3x+1)$ **31.** $4x$ **33.** $4(3x^2-4x+12)$ **35.** $(x+y+1)^2$

37. $(x-y+2)(x+y)$ **39.** $(x+y)(x^2-xy+y^2)(t-2)(t^2+2t+4)$ **41.** $2x(x-2)(x+2)(x^2+4)$

43. $(x-1)(x+1)(2x^2+3)$ **45.** $-3(x-3)(3x+4)$

5.4 Exercises

1. $\{-1,2\}$ **3.** $\left\{-3,-\frac{1}{2}\right\}$ **5.** $\left\{-\frac{5}{3},-1,\frac{3}{2}\right\}$ **7.** $\left\{-\frac{1}{2},3\right\}$ **9.** $\{-4,7\}$ **11.** $\left\{2,\frac{7}{2}\right\}$ **13.** $\{-2,3\}$

15. $\left\{-\frac{1}{3},0\right\}$ **17.** $\{-2,2\}$ **19.** $\left\{-\frac{2}{3},4\right\}$ **21.** $\left\{-\frac{10}{3},\frac{5}{2}\right\}$ **23.** $\{0,7\}$ **25.** $\{-8,0\}$ **27.** $\{-1,4\}$

29. $\{-2,3\}$ **31.** $\{-3,3\}$ **33.** $\{-1\}$ **35.** $\{-3,-2,2,3\}$ **37.** 8 seconds   **39.** 130 feet   **41.** 3 inches

CHAPTER 6

6.1 Exercises

1. $\{x\mid x\neq-2\}$ **3.** $\left\{x\mid x\neq-\frac{2}{5}\right\}$ **5.** $\{x\mid x\neq-2,5\}$ **7.** $\left\{x\mid x\neq-4,\frac{2}{3}\right\}$ **9.** $\mathbb{R}$

11. $f(-2)=0$; $f(3)=-10$; $f(5)$ is undefined   **13.** $\frac{9}{5}$ **15.** $\frac{z-2}{z-6}$ **17.** $\frac{w+9}{2w(w+3)}$ **19.** $\frac{1}{6x+7}$

21. simplest form   **23.** -1 **25.** $2-x=-(x-2)$ **27.** $\frac{q+6}{q-2}$ **29.** $\frac{m^2+6}{m-4}$

6.2 Exercises

1. $\frac{4}{9}$ **3.** $\frac{5(x-7)(x+1)}{16x(x+2)}$ **5.** $\frac{2}{(x+5)(x+6)}$ **7.** $\frac{3(x-7)(x-5)}{4x(x+5)}$ **9.** $\frac{(2x-1)(3x+4)}{(x-9)(x+3)}$

11. $\frac{4(x-1)(2x+7)}{3(3x-1)^2}$ **13.** $\frac{3(x+10)(x+2)}{2(x-5)}$ **15.** $\frac{-2(x+3)(2x-9)}{(x+4)(2x+1)}$ **17.** $\frac{(x+y)(x+3y)}{-2y^2}$

19. $\dfrac{(a-4)(a-3)}{2(a+3)}$ **21.** $\dfrac{-12(x-2)}{5(x+1)(x+6)}$ **23.** $\dfrac{8}{(x+6)(x+7)}$ **25.** $\dfrac{3(y+4)(y-2)}{(7y-2)(6y+1)}$ **27.** $\dfrac{x-y}{2}$

29. $\dfrac{3x}{10}$ **31.** $\dfrac{(3x-5)(y-4)}{(y+1)(2x-3)}$ **33.** $\dfrac{3x+8}{5(x+2)}$

6.3 Exercises

1. $\dfrac{16}{x}$ **3.** $\dfrac{6x+1}{x-4}$ **5.** $\dfrac{2}{x+4}$ **7.** $\dfrac{n-8}{n+8}$ **9.** $20x^3y^5$ **11.** $(n-7)(n+2)(n+4)$

13. $(3b-7)(2b-3)(b+5)$ **15.** $\dfrac{6x+245}{49x^2}$ **17.** $\dfrac{2x+70}{(x-1)(x+8)}$ or $\dfrac{2(x+35)}{(x-1)(x+8)}$ **19.** $\dfrac{-11x+4}{(x+4)(x-4)}$

21. $\dfrac{20a}{(a+b)(a-b)}$ **23.** $\dfrac{4x-6}{x-2}$ or $\dfrac{2(2x-3)}{x-2}$ **25.** $\dfrac{3x+2}{x-3}$ **27.** $\dfrac{2x^2-11x-57}{(x+3)(x+9)(x-3)}$

29. $\dfrac{8x+6y}{x^2-y^2}$ or $\dfrac{2(4x+3y)}{(x+y)(x-y)}$ **31.** $\dfrac{-3x^2}{(3x-2y)(x+2y)(x-y)}$ **33.** $\dfrac{2x-3}{4-x}$ **35.** $\dfrac{2x^2-27x-45}{(x+6)^2(2x-9)}$

6.4 Exercises

1. 6   **3.** $\dfrac{5}{(x-3)(x-1)}$ **5.** $\dfrac{x}{3-2x}$ **7.** $\dfrac{2(x+2)}{3(2x-1)}$ **9.** $\dfrac{3}{(x+1)(x+4)}$ **11.** $\dfrac{3x-13}{18-5x}$ **13.** $\dfrac{7x+1}{7x-1}$

15. $\dfrac{1}{(x-6)(x+5)}$ **17.** $\dfrac{x-4}{4x}$ **19.** $\dfrac{3}{x-5}$ **21.** $\dfrac{x-1}{x-7}$ **23.** $\dfrac{6x}{x^2+9}$ **25.** $\dfrac{6}{x-y}$

6.5 Exercises

1. not a rational expression because the numerator is not a polynomial (there is a variable under a radical)   **3.** yes

5. $\left\{\dfrac{5}{21}\right\}$ **7.** $\{-15,-4\}$ **9.** $\{-12\}$ **11.** no solution   **13.** $\{x \mid x \neq -2, 2\}$ **15.** $\{2\}$ **17.** $x = \dfrac{1}{2}$

19. 127.5 min   **21.** 1 in.   **23.** 245 mph   **25.** 48 cm   **27.** 1.5 mph   **29.** 6 hours   **31.** 450 min

33. a. about 40 acres   **b.** about 34.6 acres   **c.** In 2045 there will be an average of 0 acres per farm. This does not make sense.

6.6 Exercises

1. 40 ft   **3.** 900 gallons   **5.** 9 years old   **7.** about 79 kg   **9.** $5568

11. 45 kilograms per square meter   **13.** 18.75 ft or less

CHAPTER 7

7.1 Exercises

1. 10   **3.** $\dfrac{3}{11}$ **5.** 7.416   **7.** 17   **9.** $|6x-5|$ **11.** $9x^2$ **13.** $-11|x^5|$ **15.** 7   **17.** $-\dfrac{3}{4}$

19. w^7 **21.** 4   **23.** $\dfrac{2}{5}$ **25.** $|x^5|$ **27.** 3.476   **29.** 2.187

7.2 Exercises

1. 9   **3.** 2   **5.** −1   **7.** $\{x \mid x \geq -4\}$ or $[-4, \infty)$ **9.** $\mathbb{R}$ or $(-\infty, \infty)$ **11.** $\{x \mid x \geq 0\}$ or $[0, \infty)$

13.   It looks like the graph of the square root function, but it is shifted two units to the right.

15. It looks like the graph of the square root function, but it is stretched by a factor of two.

17. It looks like the graph of the cube root function, but it is shifted two units up.

7.3 Exercises

1. $\sqrt{36}=6$ **3.** $-\sqrt[4]{81}=-3$ **5.** $\sqrt[5]{7xy}$ **7.** $\left(\sqrt{100}\right)^3=1000$ **9.** $\left(\sqrt[4]{-16}\right)^3$ is not a real number.

11. $\left(\sqrt[3]{\dfrac{1}{8}}\right)^5=\dfrac{1}{32}$ **13.** $10^{\frac{1}{2}}$ **15.** $\left(\dfrac{2x}{y}\right)^{\frac{1}{5}}$ **17.** $\left(3xy^3\right)^{\frac{4}{5}}$ **19.** $\dfrac{1}{25^{\frac{1}{2}}}=\dfrac{1}{5}$ **21.** $\left(\dfrac{8}{27}\right)^{\frac{2}{3}}=\dfrac{4}{9}$

23. $9^{\frac{1}{2}}=3$ **25.** $x^{\frac{3}{2}}$ **27.** $\dfrac{2x^2}{y^{\frac{5}{2}}}$ **29.** $\sqrt[20]{x^{13}}$ **31.** $\sqrt{x}$ **33.** $\sqrt{3}$ **35.** $\sqrt[4]{35}$ **37.** $5x$

39. $6\sqrt{2}$ **41.** $4x\sqrt{3x}$ **43.** $a^7\sqrt{a}$ **45.** $6\sqrt{5}$ **47.** $6x^2y^2\sqrt{10y}$ **49.** $30x^2y$ **51.** $\dfrac{7}{9}$

53. $\dfrac{x\sqrt[3]{11}}{2}$ **55.** $\dfrac{x\sqrt[4]{21x}}{2}$ **57.** 3 **59.** $2x^5\sqrt{10}$ **61.** $-2m\sqrt[3]{3}$

7.4 Exercises

1. $13\sqrt{7}$ **3.** $3\sqrt{2}+5\sqrt{6}$ **5.** $-6\sqrt[3]{6x}$ **7.** $6\sqrt{5}+6\sqrt{x}$ **9.** $-\sqrt{5}$ **11.** $5\sqrt{3}+6\sqrt{5}$ **13.** $3x\sqrt[4]{x}$

15. $2a^2b\sqrt{ab}$ **17.** $\dfrac{23\sqrt{2}}{4y}$ **19.** $-4a\sqrt[4]{a^3}$ **21.** $2x\sqrt{5}+\sqrt{15}$ **23.** $2m\sqrt{3m}-6m\sqrt{15}$ **25.** $29-13\sqrt{3}$

27. $x+10\sqrt{x}+24$ **29.** $12-3\sqrt{y}-4\sqrt{x}+\sqrt{xy}$ **31.** $7z-9$ **33.** $9-x$ **35.** $4x+20\sqrt{x}+25$

37. $\dfrac{2\sqrt{6}}{3}$ **39.** $\dfrac{3\sqrt{2x}}{2x}$ **41.** $\dfrac{\sqrt[3]{18x}}{2x}$ **43.** $\dfrac{4\sqrt{3x}}{3x^2y^2}$ **45.** $10+5\sqrt{3}$ **47.** $\dfrac{-3\sqrt{x}+3}{x-1}$

49. $\dfrac{m+3\sqrt{m}-28}{m-49}$

7.5 Exercises

1. $\{-6\}$ **3.** $\{4\}$ **5.** $\{44\}$ **7.** $\{6\}$ **9.** $\{\ \}$ or $\varnothing$ **11.** $\{4\}$ **13.** $\{4\}$ **15.** $\{2,6\}$

17. $\left\{-\dfrac{7}{2}\right\}$ **19.** $\left\{-9,\dfrac{3}{2}\right\}$ **21.** $\{30\}$ **23.** $\{38\}$ **25.** $\{1\}$ **27.** $\{64\}$ **29.** $\{5\}$ **31.** $\{5\}$

33. $\left\{-\dfrac{7}{2},2\right\}$ **35.** $L=\dfrac{8T^2}{\pi^2}$ **37.** $h=\dfrac{3V}{\pi r^3}$ **39.** $r=\sqrt{\dfrac{A^2}{P^2}}-1$ or $r=\dfrac{A^2-P^2}{P^2}$ **41. a.** 10th grade

b. 36 words   **43. a.** $2.12\ \text{m}^2$ **b.** 64.5 kg   **45. a.** 0.96 sec   **b.** 5.4 ft

7.6 Exercises

1. i **3.** i **5.** $-1+i$ **7.** $10-11i$ **9.** $7-3i$ **11.** 4 **13.** $-1+i$ **15.** $6+8i$ **17.** $20-9i$

19. $7+17i$ **21.** $32-24i$ **23.** 2 **25.** 6 **27.** $\dfrac{2}{25}-\dfrac{11}{25}i$ **29.** $\dfrac{3}{4}+\dfrac{3}{4}i$ **31.** $-\dfrac{3}{5}-\dfrac{2}{5}i$

33. $-7+6i$ **35.** -6 **37.** 4 **39.** $-\dfrac{1}{2}-\dfrac{3}{2}i$

CHAPTER 8

8.1 Exercises

1. $\{0,8\}$ **3.** $\{4,6\}$ **5.** $\left\{-\dfrac{3}{4},\dfrac{5}{2}\right\}$ **7.** $\{-8,8\}$ **9.** $\left\{-2\sqrt{6},2\sqrt{6}\right\}$ **11.** $\left\{-\dfrac{1}{2}-i,-\dfrac{1}{2}+i\right\}$ **13.** 16

15. $\dfrac{49}{4}$ **17.** $\left\{4-3\sqrt{2},4+3\sqrt{2}\right\}$ **19.** $\left\{-\dfrac{3}{2}-\dfrac{\sqrt{7}}{2}i,-\dfrac{3}{2}+\dfrac{\sqrt{7}}{2}i\right\}$ **21.** $\left\{-\dfrac{5}{6}-\dfrac{\sqrt{119}}{6}i,-\dfrac{5}{6}+\dfrac{\sqrt{119}}{6}i\right\}$

23. $\left\{4-3\sqrt{2},4+3\sqrt{2}\right\}$ **25.** $\left\{\dfrac{2-\sqrt{7}}{3},\dfrac{2+\sqrt{7}}{3}\right\}$ **27.** $\left\{-\dfrac{3}{10}-\dfrac{\sqrt{11}}{10}i,-\dfrac{3}{10}+\dfrac{\sqrt{11}}{10}i\right\}$

29. $D=0$, so there is exactly one real solution.   **31.** $D=-36<0$, so there are two non-real solutions.

33. $\left\{0,\dfrac{4}{13}\right\}$ **35.** $\left\{-\sqrt[3]{2},\sqrt[3]{3}\right\}$ **37.** $\left\{-2,\dfrac{3}{2}\right\}$

8.2 Exercises

1. up, wider   **3.** up, narrower   **5. a.** $(2,4)$ **b.** $x=2$ **c.** 0 **d.** 0 and 4   **e.** domain: $(-\infty,\infty)$ and

range: $(-\infty,4]$ **7.** It looks like the graph of $y=x^2$, but shifted down four units.

9. It looks like the graph of $y=x^2$, but shifted to the left three units.   **11.** It looks like the

graph of $y=x^2$, but shifted to the right one unit and down two units.   **13. a.** $(2,-4)$ **b.** up   **c.** $x=2$ **d.** 0

and 4   **e.** 0   **f.**   **g.** domain: $(-\infty,\infty)$ and range: $[-4,\infty)$ **15. a.** $(3,2)$ **b.** down   **c.**

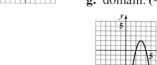

$x=3$ **d.** 2 and 4   **e.** -16 **f.**   **g.** domain: $(-\infty,\infty)$ and range: $(-\infty,2]$

17. a. $(4,2)$ **b.** down   **c.** $x=4$ **d.** $4-2\sqrt{2}\approx 1.1716$ and $4+2\sqrt{2}\approx 6.8284$ **e.** -2

f.   **g.** domain: $(-\infty,\infty)$ and range: $(-\infty,2]$ **19.** $f(x)=(x+4)^2-7$, $(-4,-7)$

21. $f(x)=-2(x+3)^2+8$, $(-3,8)$ **23.** $f(x)=(x+5)^2-25$, $(-5,-25)$ **25.** $f(x)=-\left(x+\dfrac{3}{2}\right)^2+\dfrac{25}{4}$

a. $\left(-\dfrac{3}{2},\dfrac{25}{4}\right)$ **b.** down   **c.** $x=-\dfrac{3}{2}$ **d.** -4 and 1 **e.** 4 **f.**

g. domain: $(-\infty,\infty)$ and range: $\left(-\infty,\dfrac{25}{4}\right]$ **27.** $f(x)=-3(x+1)^2+4$ **a.** $(-1,4)$ **b.** down   **c.** $x=-1$

d. $-1-\dfrac{2\sqrt{3}}{3}\approx-2.1547$ and $-1+\dfrac{2\sqrt{3}}{3}\approx0.1547$ **e.** 1 **f.**   **g.** domain: $(-\infty,\infty)$ and range: $(-\infty,4]$

29. $f(x)=2(x-4)^2-32$ **a.** $(4,-32)$ **b.** up   **c.** $x=4$ **d.** 0 and 8 **e.** 0

f.   **g.** domain: $(-\infty,\infty)$ and range: $[-32,\infty)$ **31.** $(1,-4)$ **33.** $\left(-7,-\dfrac{41}{2}\right)$

35. a. $(-1,-7)$ **b.** up   **c.** $x=-1$ **d.** $-1-\dfrac{\sqrt{21}}{3}\approx-2.5275$ and $-1+\dfrac{\sqrt{21}}{3}\approx0.5275$ **e.** -4

f.   **g.** domain: $(-\infty,\infty)$ and range: $[-7,\infty)$ **37. a.** $(-6,-17)$ **b.** up   **c.** $x=-6$

d. $-6-\sqrt{34}\approx-11.8310$ and $-6+\sqrt{34}\approx-0.1690$ **e.** 1 **f.**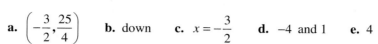
g. domain: $(-\infty,\infty)$ and range: $[-17,\infty)$ **39. a.** $(4,-16)$ **b.** up   **c.** $x=4$ **d.** 0 and 8 **e.** 0

f.   **g.** domain: $(-\infty,\infty)$ and range: $[-16,\infty)$

8.3 Exercises

1. $-\dfrac{4}{3}$ **3.** 1, 3, and 5 or 3, 5, and 7   **5.** 4 sec   **7.** $\dfrac{10}{3}$ cm by 9 cm   **9.** 3.76 ft   **11.** 5 mph

13. 120 min   **15.** 3.5 sec, 61.025 m   **17.** 3.5 sec, 201 ft   **19. a.** $71.60   **b.** 360, $12,960 million

c. $36   **21. a.** $R(x)=-\dfrac{1}{40}x^2+8000x$ **b.** $P(x)=-\dfrac{1}{40}x^2+4000x-20,000$ **c.** 80,000, $159,980,000

d. $6000   **23.** 1.2, $11, $7.20   **25.** 625 ft by 1250 ft, 781,250 ft^2 **27.** $\dfrac{1}{2}$, $1.50   **29.** 2006, 27,000

31. a. $27,500, $60,400   **b.** 350, –$12,500   **c.** yes

8.4 Exercises

1. 5 **3.** $\sqrt{74}$ **5.** $\dfrac{\sqrt{829}}{6}$ **7.** 8 **9.** $(4,8)$ **11.** $\left(-\dfrac{3}{2},1\right)$ **13.** $\left(-\dfrac{3}{2},\dfrac{3}{4}\right)$ **15.** $(1,-1)$

17. $x^2+y^2=1$ **19.** $(x+2)^2+(y-3)^2=16$ **21.** $(x-1)^2+(y+4)^2=\dfrac{9}{16}$ **23.** $(x-3)^2+y^2=2$

25. $(h,k)=(0,0)$, $r=3$, 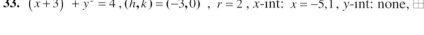 **27.** $(h,k)=(1,-5)$, $r=4$, **29.** $(h,k)=\left(\dfrac{1}{4},-\dfrac{1}{2}\right)$,

$r=2$, **31.** $(h,k)=(4,-7)$, $r=2\sqrt{3}$, x-int: none, y-int: none,

33. $(x+3)^2+y^2=4$, $(h,k)=(-3,0)$, $r=2$, x-int: $x=-5,1$, y-int: none,

35. $(x+1)^2+(y-2)^2=4$, $(h,k)=(-1,2)$, $r=2$, x-int: $x=-1$, y-int: $y=2\pm\sqrt{3}$,

37. $(x-2)^2+(y-4)^2=1$, $(h,k)=(2,4)$, $r=1$, x-int: none, y-int: none,

39. $\left(x-\dfrac{3}{2}\right)^2+\left(y-\dfrac{1}{2}\right)^2=3$, $(h,k)=\left(\dfrac{3}{2},\dfrac{1}{2}\right)$, $r=\sqrt{3}$, x-int: $x=\dfrac{3\pm\sqrt{11}}{2}$, y-int: $y=\dfrac{1\pm\sqrt{3}}{2}$,

41. $(x-1)^2+(y+2)^2=4$, $(h,k)=(1,-2)$, $r=2$, x-int: $x=1$, y-int: $y=-2\pm\sqrt{3}$,

43. $\left(x-\dfrac{1}{4}\right)^2+\left(y-\dfrac{1}{3}\right)^2=4$, $(h,k)=\left(\dfrac{1}{4},\dfrac{1}{3}\right)$, $r=2$, x-int: $x=\dfrac{3\pm4\sqrt{35}}{12}$, y-int: $y=\dfrac{4\pm9\sqrt{7}}{12}$,

8.5 Exercises

1. $(-\infty,-3]\cup[1,\infty)$ **3.** $\left[-\dfrac{2}{3},0\right]$ **5.** $[-1,1]$ **7.** $\left(-\dfrac{1}{3},1\right)$ **9.** $(-\infty,1]$ **11.** $[-3,1)$

13. $(-\infty,0)\cup(1,\infty)$ **15.** $(-1,1]$ **17.** $(-4,-2]$ **19.** $\left(-3,-\dfrac{1}{2}\right]\cup(2,\infty)$ **21.** $(-3,5)$

CHAPTER 9

9.1 Exercises

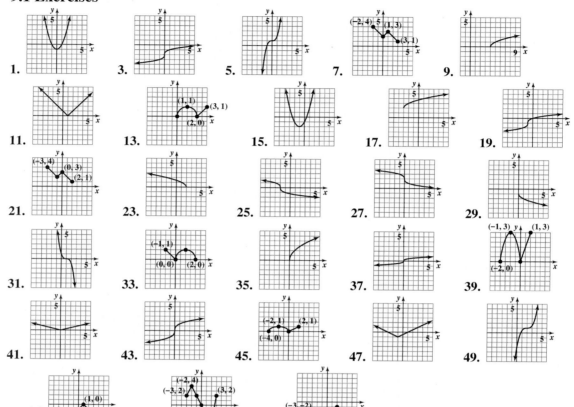

9.2 Exercises

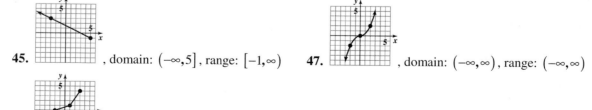

1. $(f \circ g)(x) = \dfrac{x+7}{x+1}$ **3.** $(f \circ h)(x) = 3\sqrt{x+3}+1$ **5.** $(h \circ f)(x) = \sqrt{3x+4}$ **7.** $(f \circ f)(x) = 9x+4$

9. 7   **11.** $\dfrac{2}{5}$ **13.** 2   **15.** -5 **17. a.** -4 **b.** 0   **c.** 2   **d.** -1 **19.** $(-\infty,\infty), (-\infty,\infty)$

21. $[0,\infty), (-\infty,\infty)$ **23.** $(-\infty,1)\cup(1,2)\cup(2,\infty), (-\infty,-1)\cup\left(-1,\dfrac{1}{2}\right)\cup\left(\dfrac{1}{2},\infty\right)$ **25.** yes   **27.** yes

29. yes   **31.** no   **33.** yes   **35.** yes   **37.** no   **39.** yes   **41.–43.** $(f \circ g)(x) = (g \circ f)(x) = x$

45.   , domain: $(-\infty,5]$, range: $[-1,\infty)$ **47.**   , domain: $(-\infty,\infty)$, range: $(-\infty,\infty)$

49.   , domain: $[-4,3]$, range: $[-4,4]$ **51.** $f^{-1}(x) = 3x+15$, domain f = range $f^{-1} = (-\infty,\infty)$,

range f = domain $f^{-1} = (-\infty,\infty)$ **53.** $f^{-1}(x) = \dfrac{x^3+3}{2}$, domain f = range $f^{-1} = (-\infty,\infty)$, range f = domain

$f^{-1} = (-\infty, \infty)$ **55.** $f^{-1}(x) = \sqrt{-x-2}$, domain f = range $f^{-1} = [0, \infty)$, range f = domain $f^{-1} = (-\infty, -2]$

57. $f^{-1}(x) = \dfrac{3}{x}$, domain f = range f^{-1} = range f = domain $f^{-1} = (-\infty, 0) \cup (0, \infty)$

59. $f^{-1}(x) = \dfrac{7x+1}{5x+8}$, domain f = range $f^{-1} = \left(-\infty, \dfrac{7}{5}\right) \cup \left(\dfrac{7}{5}, \infty\right)$, range f = domain $f^{-1} = \left(-\infty, -\dfrac{8}{5}\right) \cup \left(-\dfrac{8}{5}, \infty\right)$

9.3 Exercises

1.   **3.**   **5.**   **7.** $f(x) = 2^x$ **9.** $f(x) = \left(\dfrac{3}{4}\right)^x$

11. $f(x) = \left(\dfrac{1}{16}\right)^x$ **13.** , domain: $(-\infty, \infty)$, range: $(0, \infty)$, y-intercept: $\dfrac{1}{2}$, $y = 0$

15. , domain: $(-\infty, \infty)$, range: $(-\infty, 0)$, y-intercept: -9, $y = 0$ **17.** , domain: $(-\infty, \infty)$,

range: $(0, \infty)$, y-intercept: $\dfrac{1}{2}$, $y = 0$ **19.** , domain: $(-\infty, \infty)$, range: $(1, \infty)$, y-intercept: 2, $y = 1$

21. $\{4\}$ **23.** $\left\{\dfrac{1}{4}\right\}$ **25.** $\left\{-\dfrac{3}{5}\right\}$ **27.** $\left\{\dfrac{1}{3}\right\}$ **29.** $\{-1, 3\}$ **31. a.** \$300,000   **b.** \$265,640

c. \$163,299   **33. a.** 2560 bacteria   **b.** 4096 bacteria   **c.** 8,589,934,592 bacteria   **35.** \$12,752.18

37. Investment B   **39.** \$469,068.04

9.4 Exercises

1. 20.085537   **3.** 1.395612   **5.** 32.725881   **7.** , domain: $(-\infty, \infty)$, range: $(-1, \infty)$, $y = -1$

9. , domain: $(-\infty, \infty)$, range: $(-\infty, -1)$, $y = -1$ **11.** $\{-2\}$ **13.** $\left\{\dfrac{1}{4}\right\}$ **15.** $\{-3, 4\}$

17. $\{-1, 1, 2\}$ **19.** \$6798.89, \$20,942.06   **21.** \$332,570.82   **23.** (a) has the lower present value.

25. a. $P(t) = 10e^{0.25t}$ **b.** 4034 bacteria   **27. a.** 8 rabbits   **b.** 478,993 rabbits

9.5 Exercises

1. $\log_3 9 = 2$ **3.** $\log_2 \dfrac{1}{8} = -3$ **5.** $\log_{1/3} 27 = t$ **7.** $5^0 = 1$ **9.** $\sqrt{2}^6 = 8$ **11.** $a^3 = (x-1)$ **13.** $\dfrac{1}{2}$

15. 4   **17.** $-\dfrac{1}{3}$ **19.** 11   **21.** 0   **23.** 1   **25.** 0   **27.** $\log 1000 = 3$ **29.** $\ln 2 = k$ **31.** $\ln Z = 10$

33. $10^6 = 1,000,000$ **35.** $e^4 = Z$ **37.** -3 **39.** $\dfrac{2}{3}$ **41.** 49   **43.** 4   **45.** 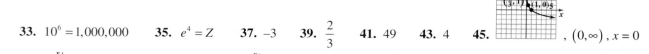, $(0,\infty)$, $x = 0$

47. 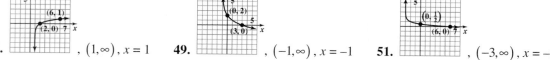, $(1,\infty)$, $x = 1$ **49.** , $(-1,\infty)$, $x = -1$ **51.** , $(-3,\infty)$, $x = -3$

53. $(-3,\infty)$ **55.** $(-\infty,-3)\cup(3,\infty)$ **57.** $(-\infty,-5)\cup(8,\infty)$

9.6 Exercises

1. $\log_4 x + \log_4 y$ **3.** $3\log_5 y$ **5.** $\ln 5 + 2$ **7.** $2 + \log P$ **9.** $6 + \log_{\sqrt{2}} x$ **11.** $2\log_7 x + 3\log_7 y$

13. $\dfrac{1}{2}\log x - 1 - 3\log y$ **15.** $1 + \dfrac{1}{2}\log_2 x + \dfrac{1}{2}\log_2 y$ **17.** $\dfrac{1}{5} + \dfrac{1}{5}\ln z - \dfrac{1}{2}\ln(x-1)$

19. $-\dfrac{4}{3}\ln(x+1) - 2\ln(x-1)$ **21.** $\log_b (AC)$ **23.** $\log_8 \left(x^2 \sqrt[3]{y}\right)$ **25.** 4   **27.** $\log_5 (x^2 - 4)$

29. $\log\left(\dfrac{x+3}{x+1}\right)$ **31.** $\{5\}$ **33.** $\{1\}$ **35.** $\{2\}$ **37.** 2.8362   **39.** -2.6572 **41.** $\log_3 (xw^2)$

43. $\log_2 (xyw)$ **45.** $\{5\}$ **47.** 36

9.7 Exercises

1. $\{1.4650\}$ **3.** $\{-1, 3\}$ **5.** $\{-13.3627\}$ **7.** $\{4.4841\}$ **9.** $\{0.3334\}$ **11.** $\{0.6931\}$

13. $\{-0.2055\}$ **15.** $\{-3.1192\}$ **17.** $\{7\}$ **19.** $\{2, 3\}$ **21.** $\left\{-\dfrac{99}{5}\right\}$ **23.** $\left\{\dfrac{1}{3}\right\}$ **25.** $\{6\}$ **27.** $\{9\}$

29. $\left\{-1, \dfrac{1}{3}\right\}$ **31.** $\{\ \}$ or $\varnothing$

9.8 Exercises

1. about 4 years 8 months   **3.** 13.73 years   **5.** Bank A: 4.5%, Bank B: 4.6%   **7.** 4147 bacteria
9. 200,119 people   **11.** 64.84%   **13. a.** 6000 families   **b.** 2000 families   **c.** 1988   **15. a.** 61,500

b. 40   **c.** $-\dfrac{\ln\left(\dfrac{43}{2}\right)}{6}$ **d.** 2009   **17.** about 11:52 PM

APPENDIX A

A.1 Exercises

1. vertex: (0, 0), focus: (0, 4), directrix: $y = -4$,

3. vertex: (1, 4), focus: (1, 1), directrix: $y = 7$,
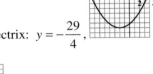

5. vertex: (−2, −6), focus: $\left(-2,-\dfrac{19}{4}\right)$, directrix: $y = -\dfrac{29}{4}$,

7. vertex: (0, 0), focus: (−2, 0), directrix: $x = 2$,

9. vertex: (4, −3), focus: (9, −3), directrix:

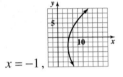

$x = -1$, **11.** $y^2 = 8x$ **13.** $(x-3)^2 = -8(y+3)$ **15.** $(y+2)^2 = -\left(x+\dfrac{11}{4}\right)$

17. $(y-4)^2 = 16(x+3)$ **19.** $(x-2)^2 = -2(y-2)$ and $(x-2)^2 = 8\left(y+\dfrac{1}{2}\right)$

21. vertex: (1, 6), focus: $\left(-\dfrac{1}{2},6\right)$, directrix: $x = \dfrac{5}{2}$,

23. vertex: (−1, −3), focus: $\left(-\dfrac{5}{4},-3\right)$,

directrix: $x = -\dfrac{3}{4}$, **25.** $\dfrac{81}{32}$ in. **27.** $-160(y-40) = x^2$, No

A.2 Exercises

1. center: (0, 0), foci: $\left(\sqrt{12},0\right)$ and $\left(-\sqrt{12},0\right)$, vertices (4, 0) and (−4, 0),

3. center: (0, 0), foci: $\left(\sqrt{5},0\right)$ and $\left(-\sqrt{5},0\right)$, vertices (3, 0) and (−3, 0),

5. center: (2, 4), foci: $\left(2+\sqrt{11},4\right)$ and $\left(2-\sqrt{11},4\right)$, vertices $(8,4)$ and $(-4,4)$,

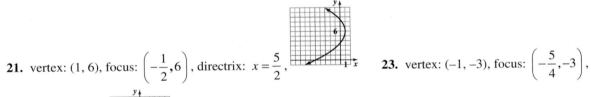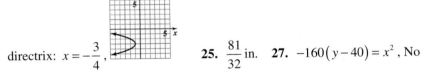

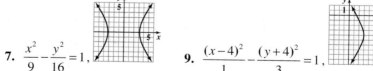

7. center: $(5, -1)$, foci: $(9,-1)$ and $(1,-1)$, vertices $(0,-1)$ and $(10,-1)$,

9. $\dfrac{(x+4)^2}{16} + \dfrac{y^2}{36} = 1$ **11.** $\dfrac{x^2}{49} + \dfrac{y^2}{24} = 1$ **13.** $\dfrac{(x+6)^2}{16} + \dfrac{(y-5)^2}{25} = 1$ **15.** $\dfrac{(x+1)^2}{7} + \dfrac{(y-4)^2}{16} = 1$

17. $\dfrac{(x-4)^2}{4} + \dfrac{(y-1)^2}{53} = 1$ **19.** $\dfrac{x^2}{16} + \dfrac{y^2}{12} = 1$ **21.** $\dfrac{(x+2)^2}{20} + \dfrac{(y-1)^2}{16} = 1$, center: $(-2, 1)$, foci: $(-4, 1)$ and $(0, 1)$,

vertices: $\left(-2-2\sqrt{5},1\right)$ and $\left(-2+2\sqrt{5},1\right)$,   **23.** $\dfrac{(x+2)^2}{16} + \dfrac{(y-1)^2}{1} = 1$, center: $(-2, 1)$,

foci: $\left(-2-\sqrt{15},1\right)$ and $\left(-2+\sqrt{15},1\right)$, vertices: $(-6,1)$ and $(2,1)$,   **25.** $\dfrac{x^2}{225} + \dfrac{y^2}{256} = 1$

27. about 25.8 yd   **29.** 44.80 in.   **31.** about 39.69 ft

A.3 Exercises

1. center: $(0, 0)$, transverse axis: $y = 0$, vertices: $(-4,0)$ and $(4,0)$, foci: $(-5,0)$ and $(5,0)$, asymptotes: $y = \pm\dfrac{3}{4}x$,

3. center: $(2, 4)$, transverse axis: $x = 2$, vertices: $(2,-1)$ and $(2,9)$,

foci: $\left(2,4-\sqrt{61}\right)$ and $\left(2,4+\sqrt{61}\right)$, asymptotes: $y-4 = \pm\dfrac{5}{6}(x-2)$,

5. center: $(0, 0)$, transverse axis: $y = 0$, vertices: $\left(-\sqrt{5},0\right)$ and $\left(\sqrt{5},0\right)$, foci: $(-5,0)$ and $(5,0)$, asymptotes:

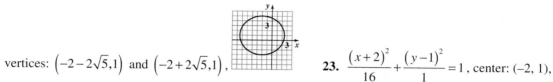

$y = \pm 2x$,   **7.** $\dfrac{x^2}{9} - \dfrac{y^2}{16} = 1$,   **9.** $\dfrac{(x-4)^2}{1} - \dfrac{(y+4)^2}{3} = 1$,

11. $\dfrac{(y-6)^2}{4} - \dfrac{(x-9)^2}{5} = 1$,   **13.** $\dfrac{y^2}{20} - \dfrac{x^2}{16} = 1$,

15. $\dfrac{(x-2)^2}{4} - \dfrac{(y-1)^2}{4} = 1$, center: (2, 1), vertices: (0, 1) and (4, 1), foci: $\left(2-2\sqrt{2},1\right)$ and $\left(2+2\sqrt{2},1\right)$, endpoints of

conjugate axis: (2, –1) and (2, 3), asymptotes: $y-1 = \pm(x-2)$, 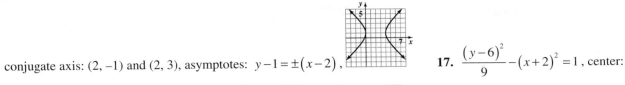   **17.** $\dfrac{(y-6)^2}{9} - (x+2)^2 = 1$, center:

(–2, 6), vertices: $(-2,3)$ and $(-2,9)$, foci: $\left(-2,6-\sqrt{10}\right)$ and $\left(-2,6+\sqrt{10}\right)$, endpoints of conjugate axis:

$(-3,6)$ and $(-1,6)$, asymptotes: $y-6 = \pm3(x+2)$, 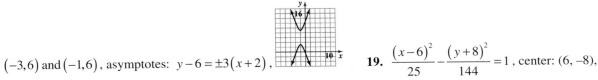   **19.** $\dfrac{(x-6)^2}{25} - \dfrac{(y+8)^2}{144} = 1$, center: (6, –8),

vertices: (1, –8) and (11, –8), foci: $(-7,-8)$ and $(19,-8)$, endpoints of conjugate axis: $(6,-20)$ and $(6,4)$,

asymptotes: $y+8 = \pm\dfrac{12}{5}(x-6)$, 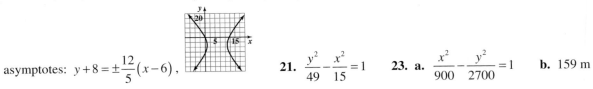   **21.** $\dfrac{y^2}{49} - \dfrac{x^2}{15} = 1$ **23. a.** $\dfrac{x^2}{900} - \dfrac{y^2}{2700} = 1$ **b.** 159 m

APPENDIX B

B.1 Exercises

1. 4, 7, 10, 13   **3.** 1, $\dfrac{8}{5}$, 2, $\dfrac{16}{7}$ **5.** 30, 120, 600, 3600   **7.** –5, 10, –15, 20   **9.** $-\dfrac{1}{42}, \dfrac{1}{56}, -\dfrac{1}{72}, \dfrac{1}{90}$

11. 7, 10, 13, 16   **13.** 6, $\dfrac{3}{2}$, $\dfrac{1}{6}$, $\dfrac{1}{96}$ **15.** $a_n = 2n-3$, $a_{11} = 19$ **17.** $a_n = n(n+5)$, $a_7 = 84$

19. $a_n = \dfrac{2}{5^n}$, $a_6 = \dfrac{2}{15,625}$ **21.** $a_1 = -6$, $a_n = (-1)^n(n+2)!$ for $n \ge 2$, $a_5 = -5040$ **23.** 20   **25.** $\dfrac{11}{32}$

27. 12   **29.** 45   **31.** 196   **33.** 271   **35.** 271   **37.** $\displaystyle\sum_{i=1}^{10} 5i$ **39.** $\displaystyle\sum_{i=1}^{9} \dfrac{i+3}{i+4}$ **41.** $\displaystyle\sum_{i=1}^{6} \dfrac{(-1)^i}{9i}$

43. $\displaystyle\sum_{i=1}^{n+1} \dfrac{7^{i-1}}{(i-1)!}$

B.2 Exercises

1. yes, 6   **3.** yes, $\dfrac{3}{2}$ **5.** yes, 2   **7.** $a_n = 5n-3$, $a_{10} = 47$ **9.** $a_n = \dfrac{3}{2}n$, $a_{50} = 75$

11. $a_n = 2n-1$, $a_{17} = 33$ **13.** 71   **15.** 97   **17.** $\dfrac{25}{2}$ **19.** 3240   **21.** 2375   **23.** –1002

25. no solution   **27.** 10,100   **29.** 147   **31.** $3185   **33.** $200, $65,000   **35.** 1.3 sec, 312 in.

B.3 Exercises

1. 8, 16, 32, 64, 128   **3.** 25, –5, 1, $-\dfrac{1}{5}$, $\dfrac{1}{25}$ **5.** –4, 8, –16, 32, –64   **7.** yes, –1   **9.** yes, $-\dfrac{5}{3}$

11. $a_n = 3 \cdot 2^{n-1}$ **13.** $a_n = \dfrac{1}{5}\left(-\dfrac{2}{3}\right)^{n-1}$ **15.** $-\dfrac{951}{256}$ **17.** 78,732   **19.** –91,749   **21.** ≈ 50.31

23. $\dfrac{85}{128}$ **25.** converges, $\dfrac{2401}{6}$ **27.** diverges   **29.** $6561   **31.** $\approx \$1,402,123.44$ **33.** 200 ft

35. $\dfrac{322}{990}$ **37.** $\approx \$30,200.99$

B.4 Exercises

1. $m^6 + 6m^5n + 15m^4n^2 + 20m^3n^3 + 15m^2n^4 + 6mn^5 + n^6$

3. $x^7 - 7x^6y + 21x^5y^2 - 35x^4y^3 + 35x^3y^4 - 21x^2y^5 + 7xy^6 - y^7$

5. $64x^6 + 576x^5y + 2160x^4y^2 + 4320x^3y^3 + 4860x^2y^4 + 2916xy^5 + 729y^6$ **7.** 7   **9.** 1

11. $x^7 + 14x^6 + 84x^5 + 280x^4 + 560x^3 + 672x^2 + 448x + 128$

13. $1024x^5 + 1280x^4 + 640x^3 + 160x^2 + 20x + 1$

15. $3125x^5 - 9375x^4y + 11,250x^3y^2 - 6750x^2y^3 + 2025xy^4 - 243y^5$ **17.** $x^2 - 4\sqrt{2}x^{\frac{3}{2}} + 12x - 8\sqrt{2}x^{\frac{1}{2}} + 4$

19. $70a^4b^4$ **21.** $1,088,640x^4$ **23.** 126,720   **25.** 19,779,228

Index

A

Abscissa, 49. *See also* Ordered pairs
Absolute values
  equations, 30–33
  inequalities, 33–35
  in order of operations, R-24–25
  overview, R-10
ac method, 212–215
Addition. *See also under specific concepts*
  in order of operations, R-23–25
  properties of, R-20–23
Addition method. *See* Elimination method
Addition property of inequality, 17
Additive identity, R-22
Additive inverses, R-13–14, R-22–23
Algebraic equations, defined, 2. *See also specific types*
Algebraic expressions
  defined, R-28, 2
  evaluating, R-28–29
  simplifying, R-29–31
  writing verbal descriptions as, R-31
Angles, 40, 141–142
Annuities, 575–576
Applications. *See* Problem-solving strategy for applications; *specific concepts*
Area
  of rectangles, 236–237, 390–391, 397–398
  of triangles, 38
Arithmetic sequences
  applications, 563–564
  determining if sequences are, 558–559
  finding general term of, 560
  finding specific term of, 561–562
Arithmetic series
  applications, 563–564
  computing nth partial sum of, 562–563
Associative property, R-20–21
Asymptotes, 456–457, 483, 539
Axes of symmetry, 374–375, 527

B

Back substitution, 126–127
Bases
  defined, 155
  formula for change of, 491–493

  method of relating the, 460–461, 470–471, 479
Binomial coefficients, 579–580, 581–582
Binomial differences, 180–181, 332–333
Binomial expansions
  finding particular coefficient of, 584
  finding particular term of, 583–584
  using Binomial Theorem, 582–583
  using Pascal's triangle, 579–581
Binomial sums, 180–181, 353–354
Binomial Theorem, defined, 582.
  See also Binomial expansions
Binomials. *See also* Polynomials
  dividing polynomials by, 190–192
  expansion. *see* Binomial expansions
  factor theorem, 194
  multiplying using FOIL method, 178–179
  multiplying using special product rules, 180–181
  remainder theorem, 193–194
Boundary points
  polynomial inequalities and, 416–419
  rational inequalities and, 420–423

C

Cancellation principle, 243–244, 245
Cancellation properties of exponentials and logarithms, 479–480
Carrying capacity, 507–508
Cartesian coordinate system, 48–50
Cartesian plane, 48
Celsius–Fahrenheit conversion, 448
Center of circles, 406, 410–414
Change of base formula, 491–493
Circles
  as conic sections, 514
  equation of, 406–408
  finding center and radius of, 410–414
  sketching graph of, 408–410
Coefficients
  binomial, 579–580, 581–582
  of binomial expansions, finding, 584
  leading. *see* Leading coefficients
  of monomials, finding, 169–170
Coinciding lines, 91–92, 108–109, 118
Combined variation, 296–298

Common factors, factoring trinomials with, 207, 214–215
Common logarithmic function, 480–481
Common ratio, 566–567
Commutative property, R-20, R-21
Complementary angles, 141
Completing the square
    finding equation of ellipses by, 532–533
    finding equation of hyperbolas by, 544–545
    finding equation of parabolas by, 521–522
    finding vertex of quadratic functions by, 381–382
    graphing quadratic functions by, 383
    solving quadratic equations by, 364–366
Complex conjugates, 354–355
Complex numbers
    adding and subtracting, 352
    dividing, 355–356
    multiplying, 353–355
    overview, 351–352
    simplifying powers of i, 349–352
Complex rational expressions
    simplifying using method I, 268–270
    simplifying using method II, 271–274
Composite functions
    determining domain of, 443–445
    forming and evaluating, 441–443
Composition cancellation equations, 450
Compound interest, 462–463, 471, 503–504
Compound linear inequalities, 28–30
Compressions
    horizontal, 436–437
    vertical, 434–436
Concentration. See Mixture problems
Conditional equations, 8
Conic sections, overview, 513–515. See also Circles; Ellipses; Hyperbolas; Parabolas
Conical tanks, volume of, 296–297
Conjugate axis, 539
Conjugates
    complex, 354–355
    defined, 180
    product of. see Product of conjugates
Consistent systems, 108–109, 126
Constant of variation
    in combined variation, 296
    in direct variation, 97–98, 292–293
    in inverse variation, 294
Constants, R-28
Continuous compound interest, 471, 503, 504
Contradictions
    determining if equations are, 8–9
    solving, 18–19
Coordinate axes, 48
Coordinate plane, 48

Coordinates. See also Ordered pairs
    defined, R-7
    finding unknown, 51–52
Costs, minimizing, 398–399
Counting numbers, denoting set of, R-2
Cross-multiplication, 606–607
Cube roots. See also Radical expressions; Radical functions
    defined, R-19, 304
    finding, 304
    graphing functions containing, 311–313
Cubes
    defined, R-19
    factoring sum/difference of two, 221–222
Cylinders, surface area of, 39–40

D
Decimals
    adding and subtracting, 601–602
    converting between percents and, 609–611
    converting between words and decimal form, 598–600
    dividing, 602–603
    in linear equations, 6–7
    multiplying, 601–603
    notation, 597–598
    rounding, 600–601
    in systems of linear equations, 116
    writing as fractions or mixed numbers, 599–600
    writing real numbers and fractions as, R-5
Degenerate conic sections, 514–515
Degrees
    of monomials, 169–170
    of polynomial equations, 227
    of polynomials, 170–171
Demand equations, 394–397
Density, 294–295
Dependent systems
    in three variables, 130–133
    in two variables, 108–109, 117–119
Dependent variables, identifying, 57–58
Depreciation, 70–71
Descartes, René, 48
Difference of two cubes, factoring, 221–222
Difference of two squares
    factoring, 218–220
    multiplying polynomials using, 180–181
    multiplying radical expressions using, 332–333
Direct variation, 97–98, 292–294
Directrix of parabolas, 516
Discriminants, 369
Distance between two points, 403–404
Distance traveled. See Uniform motion problems
Distributive property, R-21–22
Division, in order of operations, R-23–25

Domain
  of composite functions, 443–445
  of logarithmic functions, 484–485
  of quadratic functions, 374–376
  of radical functions, 309
  of rational functions, 241–242
  of relations, 58–60
Doubling time, 503–504

E

Electrical resistance, 297
Element symbol ($\in$), R-1–2
Elements
  of cones, 513
  of set, R-1–2, R-3
Elimination method
  for systems in three variables, 125–129
  for systems in two variables, 112–116
Ellipses
  applications, 533–535
  as conic sections, 514, 526
  equation in standard form, 528–529
  finding equation of, 530–533
  geometric defininition of, 526
  sketching graph of, 527–530
Ellipsis (. . .), for infinite sets, R-2
Empty sets, R-1–2, 26, 30, 32
Endpoints, 52
Equal sets, R-3–4
Equations
  defined, 2
  lines, 92–95
  for percents, 612–614
Equations in one variable, defined, 2.
  See also Linear equations in one variable
Equations in two variables
  defined, 48
  determining if ordered pair is solution to,
    50–51
  finding unknown coordinates in, 51–52
  finding x- and y-intercepts, 54–55
  graphing by plotting points, 52–54
Equivalent equations, 3–4
Equivalent fractions, 589–590
Equivalent functions, 67
Exchange rates, minimizing, 398–399
Exponential decay, 506–507
Exponential equations
  solving by relating the bases, 460–461
  solving using logarithm properties,
    495–498
Exponential expressions
  converting between radical and, 315–316,
    317–318
  defined, R-18, 155

in order of operations, R-23–25
  with rational exponents, simplifying, 319–320
  scientific notation, 164–167
  simplifying
    by multiplying the factors, R-18–19
    using combination of rules, 163–164
    using negative-power rule, 159–160
    using power-to-power rule, 160–161
    using product rule, 155–156
    using product-to-power rule, 161–162
    using quotient rule, 156–158
    using quotient-to-power rule, 162–163
    using zero-power rule, 158–159
Exponential form, 477, 478, 481
Exponential functions. *See also* Natural exponential
  functions
  applications, 461–464, 503–510
  characteristics, 455–459
  determining from graph, 458–459
  sketching graphs of, 457–460
  transformations, 459–460
Exponential growth model, 472–474, 505–506
Extraneous solutions, 339, 499

F

Factor theorem, 194
Factorials, 550–551
Factoring
  difference of two cubes, 221–222
  difference of two squares, 218–220
  GCF out from polynomials, 198–201
  perfect square trinomials, 220–221
  polynomials
    completely, general strategy, 223–225
    by grouping, 201–202
  solving polynomial equations by, 227–231
  solving quadratic equations by, 361
  sum of two cubes, 221–222
  trinomials. see *under* Trinomials
  zeros and, 233–234
Factorizations, prime, 590–591
Fahrenheit–Celsius conversion, 448
Falling objects, 72–73, 234–235,
  344–345
Feasible domain, 60
Fibonacci sequence, 552
Finite sequences, 550
Finite series, 553, 554–555
Finite sets, R-2
First-degree equations, 2
Foci
  of ellipses, 527
  of hyperbolas, 538
  of parabolas, 516, 522–524
FOIL method, 178–179

Formulas
    applications, 39–44
    solving for variable, 37–38
Fraction form, R-5
Fractions. *See also* Mixed numbers
    adding and subtracting, 592–594
    converting between percents and, 610–612
    dividing, 591–592
    dividing by, R-18
    equivalent, 589–590
    in linear equations, 6
    in linear inequalities, 18
    multiplying, 591
    order of operations and, R-24–25
    simplifying, 590–591
    in systems of linear equations, 116, 129–130
    types of, 586–588
    writing as decimals, 604
Free-falling objects, 344–345
Function notation, 66–68
Functions. *See also specific types*
    applications, 62–63, 72–74
    determining if graphs are, 61–62
    determining if relations are, 60–61
    evaluating, 68–69
    graphing by plotting points, 69–70
    interpreting graphs of, 70–72

G

General form for equation of circles, 410–414
General term
    of arithmetic sequences, 560
    of geometric sequences, 567, 569
    of sequences, in general, 552–553
Geocaching, 137–138
Geometric sequences
    applications, 573–576
    determining if sequences are, 568–569
    finding general term of, 569
    finding specific term of, 569–570
    graph of, 567
    overview, 566–568
    writing terms of, 568
Geometric series
    applications, 573–576
    computing nth partial sum of, 571
    infinite, 571, 572–573
Geometry problems
    using formulas for, 38–40
    using quadratic functions for, 390–391
    using systems of linear equations for, 140–142
Graphs and graphing
    circles, 408–410
    defined, R-7
    ellipses, 527–532

    equations by plotting points, 52–54
    exponential functions, 457–460
    of functions, interpreting, 70–72
    geometric sequences, 567
    horizontal shifts, 427–429
    horizontal stretches and compressions, 436–437
    hyperbolas, 539–542
    inverse functions, 450–452
    linear functions by plotting points, 79–80
    linear functions by using intercepts, 80–83
    linear inequalities, 102–104
    lines using slope and point, 89–91
    logarithmic functions, 482–484
    natural exponential functions, 470
    ordered pairs, 48–50
    parabolas, 520–521
    radical functions, 310–313
    real numbers, R-7–8
    scatter plots, 95–96
    simple functions by plotting points, 69–70
    solution set of inequalities, 14
    solving systems of linear equations by, 108–110
    systems of linear inequalities, 148–152
    vertical and horizontal lines, 83–84
    vertical line test, 61–62
Greatest common factor (GCF), 198–201
Grouping symbols, in order of operations, R-23–25

H

Half-life, 506–507
Hang time, 346
Horizontal asymptote, 456–457
Horizontal line test, 445–447
Horizontal lines
    graphing, 83–84
    slope of, 85–86, 87–88, 89
    writing equation of, 94
Horizontal shifts
    graphing functions using, 427–429
    of quadratic functions, 378–379
Horizontal stretches and compressions, 436–437
Hyperbolas
    applications, 545–546
    as conic sections, 514, 538
    equation in standard form, 540–541
    fact 1 for, 540, 546
    finding equation of, 542–545
    geometric defininition of, 538–539
    sketching graph of, 539–542
Hypotenuse problem, 235–236

I

Identities, determining if equations are, 8–9
Identity element of addition, R-22
Identity element of multiplication, R-22–23

Imaginary part, 351
Imaginary unit i, 349
Improper fractions, 587, 588–589
Inconsistent systems
    in three variables, 126, 130–131
    in two variables, 108–109, 116–117
Independent systems, 108–109
Independent variables, identifying, 57–58
Index of radical expressions, 305
Index of summation, 554–555
Infinite arithmetic series, 562
Infinite geometric series, 571, 572–573
Infinite sequences, 550
Infinite series, 553
Infinite sets, R-2
Input values. See Domain
Integers, R-4, R-6-7
Intercepts. See also x-intercepts; y-intercepts
    finding on graph, 54–55
    graphing linear functions by using, 80–83
    slope and, 88, 90–91
Interest, compound, 462–463, 471, 503–504
Intersecting lines, 91–92, 514–515
Intersection of two sets, 26, 27–28
Interval notation, 15–16
Inverse functions
    of one-to-one functions, 453–454
    sketching graphs of, 450–452
    summary, 453–454
    verifying, 448–450
Inverse property, R-22–23
Inverse variation, 294–296
Irrational numbers, R-5-7
Isolated variables, 3–4

J
Joint variation, 296

K
Kinetic energy, 293–294

L
Leading coefficients
    negative, 200, 207–208, 214–215
    of polynomials in one variable, 170–171
Leanness, 293–294
Least common denominator (LCD)
    of fractions, 593
    of rational expressions, 259–261
Leonardo of Pisa, 552
Like radicals, 328–329
Like terms, R-29–30
Line segments, midpoint of, 405–406
Linear equations, nonlinear equations vs., 2
Linear equations in one variable

applications, 9–11
contradictions vs. identities, 8–9
determining if value is solution to, 2–3
solving, 3–4, 7–8
    with decimals, 6–7
    with fractions, 6
    with non-simplified expressions, 5
    with variables on both sides, 5
Linear equations in two variables. See also Equations
    in two variables; Lines; Slope; Systems of linear
    equations in two variables
    applications, 95–98
    defined, 79
    graphing vertical and horizontal lines, 83–84
Linear equations in three variables, defined, 124
Linear functions
    graphing by plotting points, 79–80
    graphing by using intercepts, 80–83
Linear inequalities in one variable
    applications, 21–23
    compound, 28–30
    determining if value is solution to, 13
    solution sets
        graphing, 14
        interval notation for expressing, 15–16
    solving, 16–18
        with fractions, 18
        special cases, 18–19
        three-part, 19–20
Linear inequalities in two variables
    determining if ordered pair is solution to, 101–102
    graphing, 102–104
Lines. See also Horizontal lines; Vertical lines
    as degenerate conic sections, 514–515
    equation of, 92–95
    graphing using slope and point, 89–91
    relationship between two, 91–92
Logarithm property of equality, 490–491, 499
Logarithmic equations
    solving, general strategy, 498–501
    solving using logarithm property of equality,
        490–491, 499
Logarithmic expressions
    condensing, 489–490
    evaluating, 478–479, 481
    expanding, 488–489
    using change of base formula, 491–493
Logarithmic form, 477, 478, 481
Logarithmic functions
    applications, 503–510
    characteristics, 482–483
    common and natural, 480–481
    finding domain of, 484–485
    overview, 476–478
    transformations, 484

Logarithms
    common and natural, 480–481
    properties of, 479–480, 487–488
Logistic growth, 507–509
Long division of polynomials, 187–190
Long run average cost (LRAC), 399
Lower limit of summation, 554–555
Lowest terms, 590

M

Major axis of ellipses, 527
Maximum points, 52
Maximum value of quadratic functions, 374–375, 393–398
Members of set, R-1–2
Method of relating the bases, 460–461, 470–471, 479
Midpoint of line segments, 405–406
Minimum points, 52
Minimum value of quadratic functions, 374–375, 398–399
Minor axis of ellipses, 527
Minor rational expressions, 268
Mixed numbers
    adding and subtracting, 595–597
    converting between improper fractions and, 588–589
    converting between percents and, 610–612
    defined, 588
    multiplying and dividing, 594–595
    writing as decimals, 604
Mixture problems
    using formulas for, 42–43
    using systems of linear equations for, 142–144
Modeling. See Problem-solving strategy for applications; specific concepts
Monomials. See also Polynomials
    dividing polynomials by, 186–187
    finding coefficient and degree of, 169–170
    multiplying, 177
    multiplying polynomials by, 177–178
Multiplication
    in order of operations, R-23–25
    properties of, R-20–23
Multiplication property of inequality, 17
Multiplicative identity, R-22–23
Multiplicative inverses, 591, R-15–17, R-22–23

N

Natural base, 468
Natural exponential functions
    applications, 471–474
    characteristics, 468–469
    solving by relating the bases, 470–471
    solving equations involving, 498
    transformations, 470
Natural logarithmic function, 480–481
Natural numbers, R-2, R-4, R-6–7
Negative exponential growth, 506–507
Negative leading coefficients, 200, 207–208, 214–215
Negative radicands, 356–358
Negative real numbers, R-4, R-7
Negative square roots, 301. See also Square roots
Negative-power rule
    complex rational expressions, 273–274
    exponential expressions, 159–160
    rational exponents, 318–319
Newton's law of cooling, 509–510
Nonlinear equations, 2
Nonlinear inequalities, 5
Non-simplified expressions, equations with, 5
Non-strict inequalities, R-9, 13
Not-an-element symbol ($\notin$), R-1–2
Not-a-subset/superset symbols, R-3–4
Nth partial sums
    of arithmetic series, 562–563
    of geometric series, 571
    of series, in general, 553–554
Nth roots, 305–307. See also Radical expressions; Radical functions
Null (empty) sets, R-1–2, 26, 30, 32
Number line
    graphing solution set of inequalities on, 14, 15
    order of real numbers and, R-8–9
    plotting real numbers on, R-7–8
Numerical coefficients. See Coefficients

O

One-to-one functions
    defined, 445, 446, 460, 490
    determining if functions are, 445–447
    finding inverse of, 453–454
Opposites
    polynomials, 173, 246–248
    real numbers, R-13–14, R-22–23
    reciprocals, 91–92
Order of operations, R-23–25
Ordered pairs
    determining if solution to
        equations, 50–51
        inequalities, 101–102
        systems of linear equations, 107–108
        systems of linear inequalities, 147
    finding unknown coordinates in, 51–52
    plotting, 48–50
Ordered triples, 124–125

Ordered-pair notation, 118
Ordinary annuities, 575–576
Ordinate, 49. *See also* Ordered pairs
Origin (number line), R-7
Origin (rectangular coordinate system), 48–49
Output values. *See* Range

P

Parabolas
    applications, 522–524
    as conic sections, 514
    determining equation from graph, 520–521
    finding equation by completing the square,
        521–522
    geometric defininition of, 516
    with horizontal axis of symmetry, 519–520
    overview, 373–374, 374–375
    with vertical axis of symmetry, 516–518
Parallel lines
    in linear equations, 91–92, 94–95
    in systems of linear equations, 108–109, 117
    in systems of linear inequalities, 149–152
Parentheses
    in interval notation, 15–16
    in order of operations, R-23–25
Partial sums. *See* Nth partial sums
Pascal's triangle, 580–581
Percents
    applications, 615–618
    converting between decimals and, 609–611
    converting between fractions and, 610–612
    of increase and decrease, 618–620
    solving problems using equations, 612–614
    solving problems using proportions, 615
Perfect square trinomials
    factoring, 220–221
    forming, 180, 181
Perfect squares, 301–302
Perimeter of rectangles, 38, 39, 43–44, 140–141
Periodic compound interest, 462–463
Perpendicular lines, 91–92, 94–95
Pi (π), R-5–7
Place-value chart, 597
Planes, 125–126
Plotting. *See* Graphs and graphing; Points
Points
    as degenerate conic sections, 514–515
    distance between two, 403–404
    graphing equations in two variables by plotting,
        52–54
    graphing linear functions by plotting, 79–80
    graphing lines using slope and, 79–80
    graphing simple functions by plotting, 69–70
    writing equation of lines from two, 94

Point-slope form, 92–93
Polynomial equations
    applications, 234–237
    solving by factoring, 227–231
Polynomial functions
    adding and subtracting, 174–175
    dividing, 192–193
    evaluating for given value, 171–172
    factor theorem, 194
    finding zeros of, 231–234
    multiplying, 184
    remainder theorem, 193–194
Polynomial inequalities, 415–419
Polynomials. *See also* Binomials; Monomials;
    Trinomials
    adding, 172–173
    dividing by monomials, 186–187
    dividing using long division, 187–190
    dividing using synthetic division, 190–192
    factoring by grouping, 201–202
    factoring completely, general strategy,
        223–225
    factoring GCF out from, 198–201
    finding leading coefficient and degree of,
        170–171
    multiplying by monomials, 177–178
    multiplying two or more, 182–183
    prime, 205–207, 219–220
    subtracting, 173–174
Polynomials in one variable, 171
Ponderal Index measure of leanness, 293–294
Population growth, 472–474, 505–506
Positive numbers, on number line, R-7
Power principle of equality, 339
Power rule for logarithms, 487, 488
Powers of i, simplifying, 349–352
Power-to-power rule, 160–161
Prescription drug price model, 289
Present value, 463–464, 472
Prime factorizations, 590–591
Prime polynomials, 205–207, 219–220
Principal fourth roots, R-19
Principal nth roots, 305
Principal square roots, R-19, 301
Problem-solving strategy for applications
    of linear equations, 9
    of linear inequalities, 9
    of quadratic equations, 388
    of systems of linear equations, 119
Product of complex conjugates, 354–355
Product of conjugates
    multiplying binomials using, 180–181
    multiplying radical expressions using, 332–333
    rationalizing denominators using, 335–336

Product rule
    for exponents, 155–156
    for logarithms, 487–488
    for radicals, 321–324
Products of functions, 184
Product-to-power rule, 161–162
Profit, 22–23, 40–41, 76, 396–397
Projectile motion, 390, 394
Proper fractions, 587
Proper subsets, R-3–4
Proper supersets, R-3–4
Properties
    of equality, 3–4
    of inequalities, 17
Proportionality constant. See Constant of variation
Proportions
    applications, 607–609
    defined, 606, 280
    solving, 606–607
    solving percent problems using, 615
    solving rational equations that are, 280–281
Pythagorean theorem, 235–236, 390–391, 403

Q

Quadrants, 48–50
Quadratic equations in one variable
    determining number and type of solutions to, 369
    solving
        by completing the square, 364–366
        equations that are quadratic in form, 370–371
        using factoring, 361
        using quadratic formula, 366–369
        using square root property, 362–363
Quadratic formula, 366–369
Quadratic functions
    applications
        distance, rate, time, 391–392
        geometric formulas, 390–391
        maximizing, 393–398
        minimizing, 398–399
        projectile motion, 390, 394
        unknown numbers, finding, 388–389
        work, 392–393
    characteristics, 373–376
    finding vertex of, 381–382, 383–384
    graphing form $f(x) = a(x − h)^2 + k$, 379–381
    graphing form $f(x) = ax^2 + bx + c$, 383, 384–385
    graphing using translations, 376–379
    writing in standard form, 381–382
"Quadratic in form" equations, solving, 370–371
"Quadratic in form" trinomials, factoring, 215–216
Quotient rule for exponents, 156–158

Quotient rule for logarithms, 487, 488
Quotient rule for radicals, 324–326
Quotients of functions, 192–193
Quotient-to-power rule, 162–163

R

Radical equations
    applications, 344–346
    with one radical expression, 338–342
    with two radical expressions, 342–344
Radical expressions
    adding and subtracting, 328–331
    converting between exponential and, 315–316, 317–318
    defined, R-19, 301
    finding and approximating nth roots, 305–307
    finding cube roots, 304
    irrational numbers and, R-5–6, R-19
    multiplying, 331–333
    with negative radicands, 356–358
    in order of operations, R-23–25
    rationalizing denominators, 333–336
    simplifying, R-19–20
        using product rule, 321–324
        using quotient rule, 324–326
        using rational exponents, 320–321
    square roots with a^2 as radicand, 303
Radical functions
    evaluating, 308–309
    finding domain of, 309
    graphing, 310–313
Radical sign, 301
Radius
    of circles, 406, 410–414
    of spheres, 344–345
Range
    of quadratic functions, 374–376
    of relations, 58–60
Rates, 605–606
Rates of work, 287–288, 392–393
Rational equations
    applications
        combined variation, 296–298
        direct variation, 292–294
        inverse variation, 294–296
        others, 283–289
    identifying, 276
    solving, 277–281
Rational exponents
    with 1 as numerator, 315–316
    in exponential expressions, 319–320
    with numerator other than 1, 316–319
    in radical expressions, 320–321

Rational expressions. See also Complex rational
    expressions
        adding and subtracting
            with common denominators, 257–259
            with unlike denominators, 261–266
        defined, 241
        dividing, 253–255
        finding LCD of, 259–261
        multiplying, 249–252
        simplifying, 243–248
Rational functions
        evaluating, 242–243
        finding domain of, 241–242
        finding zeros of, 281–283
Rational inequalities, 420–423
Rational numbers, R-4–5, R-6–7
Ratios, 280, 605
Readability formula, 345–346
Real axis. See Number line
Real numbers
        absolute value of, R-10
        adding, R-12–13
        classifying, R-4–7
        complex numbers and, 351–352
        defined, R-6
        dividing, R-17–18
        exponential expressions, R-18–19
        multiplying, R-15–16
        opposites, R-13–14
        order of operations, R-23–25
        ordering, R-8–10
        plotting on number line, R-7–8
        properties of, R-20–23
        radical expressions, R-19–20
        reciprocals, R-15–17
        subtracting, R-14–15
Real part, 351
Reciprocals, R-15–16, R-22, 591
Rectangles
        area of, 236–237, 390–391, 397–398
        perimeter of, 38, 39, 43–44, 140–141
Rectangular coordinate system, 48–50
Recursive sequences, 551–552
Reference rectangle, 539
Reflecting property, 533–535
Reflections, graphing functions using, 429–433
Relating the bases, 460–461, 470–471, 479
Relations
        applications, 62–63
        determining if functions, 60–61
        finding domain and range of, 58–60
Relative decay constant, 506
Remainder theorem, 193–194

Repeating decimals, R-5
Resistance, 284–286, 297
Restricted values, 277
Revenue, maximizing, 394–396
Right triangles, 235–236
Roots (zeros). See under Zero(s)
Roots, nth, 305–307. See also Cube roots;
    Square roots
Roster method, R-2, R-3

S

Scale, R-7
Scatter plots, 95–96
Scientific notation, 164–167
Second-degree trinomials, 215–216
Sequences. See also Arithmetic sequences;
    Geometric sequences
        finding general term of, 552–553
        recursive, 551–552
        writing terms of, 549–552
Series. See also Arithmetic series;
    Geometric series
        computing partial sums of, 553–554
        determining sum of finite, 554–555
        writing using summation notation, 555–556
Set-builder notation, R-2, 14, 118, 132
Sets
        classifying real numbers, R-4–7
        identifying, R-1–4
        intersection of, 26, 27–28
        union of, 27
Shifts. See Horizontal shifts; Vertical shifts
Shutter speed, 295–296
Similar triangles, 608–609
Simple interest
        overview, 619–620
        using formulas for, 41
        using linear equations for, 98
        using systems of linear equations for, 143–144
Simplest form, 590
Simplified polynomials, 170
Simplified terms, 169
Slope
        defined, 85–86
        finding, 85–89
        graphing using point and, 89–91
        relationship between two lines and, 91–92
        writing equation of lines from given, 92–95
Slope-intercept form, 88–89
SMOG formula, 345–346
Solution sets
        defined, 3, 14
        of inequalities, 14–16

Solutions, defined, 3
Sound, 545–546
Special product rules for binomials
    multiplying binomials using, 180–181
    multiplying complex numbers using, 353–354
    multiplying radical expressions using, 332–333
    multiplying trinomials using, 183
Spheres, radius of, 344–345
Square brackets, in interval notation, 15–16
Square root property, 362
Square roots. *See also* Radical expressions;
  Radical functions
    with a^2 as radicand, 303
    approximating, 302–303
    graphing functions containing, 310–311
    of perfect squares, 301–302
    principal, R-19, 301
Squares, defined, R-19
Standard form
    for complex numbers, 351–352
    for equation of circles, 406–408
    for equation of ellipses, 528–529
    for equation of parabolas, 540–541
    for exponential expressions, 164–166
    for linear equations in two variables, 218
    for polynomial equations, 227
    for polynomials in one variable, 171
    for quadratic equations, 361
    for quadratic functions, 379–381
Straight-line depreciation, 70–71
Stretches
    horizontal, 436–437
    vertical, 434–436
Strict inequalities, R-9, 13
Strict subsets, R-3–4
Strict supersets, R-3–4
Subsets, R-3–4
Substitution method
    for equations quadratic in form, 370–371
    factoring trinomials using, 215–216
    solving systems of linear equations by, 111–112
Subtraction, in order of operations, R-23–25
Sum of two cubes, factoring, 221–222
Summation notation, 554–556
Supersets, R-3–4
Supplementary angles, 141–142
Surface area of cylinders, 39–40
Synthetic division of polynomials, 190–192
Systems of linear equations
    for solving geometry problems, 140–142
    for solving mixture problems, 142–144
    for solving uniform motion problems, 137–140
Systems of linear equations in two variables
    applications, 119–121. *see also* Systems of linear
      equations

determining if ordered pair is solution to, 107–108
    solving
        by elimination, 112–116
        by graphing, 108–110
        inconsistent and dependent systems, 116–119
        by substitution, 111–112
Systems of linear equations in three variables
    applications, 133–134
    determining if ordered triple is solution to,
      124–125
    solving
        with dependent linear equations, 132–133
        with fractions, 129–130
        in general, by elimination, 125–129
        with missing terms, 129–130
        with no solution, 130–131
Systems of linear inequalities in two variables
    determining if ordered pair is solution to, 147
    graphing, 148–152

T

Taxable income, 23
Temperature
    Celsius–Fahrenheit conversion, 448
    Newton's law of cooling, 509–510
Terminating decimals, R-5
Terms
    defined, R-29, 169
    of polynomials, 170
    of recursive sequences, 551–552
    of sequences. see Sequences
Three-part inequalities, 19–20, 23
Total resistance, 284–286
Transformations of functions
    combining, 438–439
    graphing exponential functions using, 459–460
    graphing logarithmic functions using, 484
    graphing natural exponential functions
      using, 470
    horizontal shifts, 427–429
    horizontal stretches and compressions,
      436–437
    reflections, 429–433
    vertical shifts, 425–427, 429
    vertical stretches and compressions, 434–436
Translations, 376–379
Transverse axis, 539
Trial and error, factoring trinomials using,
  208–212
Triangles
    angles of, 40
    area of, 38
    Pythagorean theorem, 235–236, 390–391, 403
    similar, 608–609
Trichotomy property, 33

Trinomials. *See also* Polynomials
  defined, 180
  determining if prime, 205–206, 210–211
  factoring $ax^2 + bx + c$ forms
    using ac method, 212–215
    using trial and error, 208–211
  factoring $ax^2 + bxy + cy^2$ forms
    using ac method, 213–214
    using trial and error, 211–212
  factoring using substitution, 215–216
  factoring $x^2 + bx + c$ forms, 203–208
  factoring $x^2 + bxy + cy^2$ forms, 207–208
  multiplying using special product rules, 183

U

Uniform motion problems
  using formulas for, 41–42
  using quadratic functions for, 391–392
  using rational equations for, 284–285,
    285–286
  using systems of linear equations for, 137–140
Union of two sets, 27
Unit distance, R-7, R-10
Unit rates, 605–606
Unknown numbers, using quadratic functions to find,
  388–389
Upper limit of summation, 554–555

V

Variables
  defined, R-28
  in formulas, solving for, 37–38
Vertex
  of cones, 513
  of parabolas, 374–376
  of quadratic functions, 381–384
Vertex form, 379
Vertex formula, 383–384
Vertical asymptote, 483
Vertical line test, 61–62
Vertical lines
  graphing, 83–84
  writing equation of, 94

Vertical shifts
  graphing functions using, 425–427, 429
  of quadratic functions, 377–378, 378–379
Vertical stretches and compressions, 434–436
Volume of conical tanks, 296–297

W

Whole numbers, R-4, R-6–7
Work, rates of, 287–288, 392–393

X

x-axis, 48–49, 49–50
x-coordinate, 49. *See also* Ordered pairs
x-intercepts. *See also* Intercepts
  of parabolas, 374–376
  polynomial inequalities and, 415–416
  zeros and, 231–232, 233–234
xy-plane, 48

Y

y-axis, 48–49
y-coordinate, 49. *See also* Ordered pairs
y-intercepts. *See also* Intercepts
  finding from equation, 88–89
  graphing lines using slope and, 90–91
  of parabolas, 374–376
  relationship between two lines and, 91–92
  writing equation of lines from given, 93

Z

Zero product property, 227–228, 361
Zero-power rule, 158–159
Zero(s)
  absolute value equations and, 31
  absolute value inequalities and, 34
  as additive identity, R-22
  of functions, 231
  on number line, R-7
  opposite of, R-13–14
  of polynomial functions, 231–234
  properties of equality and, 4
  of rational functions, 281–283